Design and Analysis of Experiments

A complete list of the titles in this series appears at the end of this volume.

Design and Analysis of Experiments

Volume 2
Advanced Experimental Design

KLAUS HINKELMANN

Virginia Polytechnic Institute and State University
Department of Statistics
Blacksburg, VA

OSCAR KEMPTHORNE

Iowa State University
Department of Statistics
Ames, IA

WILEY-
INTERSCIENCE

A JOHN WILEY & SONS, INC., PUBLICATION

Published by John Wiley & Sons, Inc., Hoboken, New Jersey.
Published simultaneously in Canada.

Library of Congress Cataloging-in-Publication Data is available.

ISBN 0-471-55177-5

Printed in the United States of America.

10 9 8 7 6 5 4 3 2 1

Contents

Preface

The project of revising Kempthorne's 1952 book *Design and Analysis of Experiments* started many years ago. Our desire was to not only make minor changes to what had become a very successful book but to update it and incorporate new developments in the field of experimental design. Our involvement in teaching this topic to graduate students led us soon to the decision to separate the book into two volumes, one for instruction at the MS level and one for instruction and reference at the more advanced level.

Volume 1 (Hinkelmann and Kempthorne, 1994) appeared as an *Introduction to Experimental Design*. It lays the philosophical foundation and discusses the principles of experimental design, going back to the ground-breaking work of the founders of this field, R. A. Fisher and Frank Yates. At the basis of this development lies the randomization theory as advocated by Fisher and the further development of these ideas by Kempthorne in the form of derived linear models. All the basic error control designs, such as completely randomized design, block designs, Latin square type designs, split-plot designs, and their associated analyses are discussed in this context. In doing so we draw a clear distinction among the three components of an experimental design: the error control design, the treatment design, and the sampling design.

Volume 2 builds upon these foundations and provides more details about certain aspects of error control and treatment designs and the connections between them. Much of the effort is concentrated on the construction of incomplete block designs for various types of treatment structures, including "ordinary" treatments, control and test treatments, and factorial treatments. This involves, by necessity, a certain amount of combinatorics and leads, almost automatically, to the notions of balancedness, partial balancedness, orthogonality, and uniformity. These, of course, are also generally desirable properties of experimental designs and aspects of their analysis.

In our discussion of ideas and methods we always emphasize the historical developments of and reasons for the introduction of certain designs. The development of designs was often dictated by computational aspects of the ensuing analysis, and this, in turn, led to the properties mentioned above. Even though in the age of powerful computers and the wide availability of statistical computer software these concerns no longer play the dominant feature, we remind the reader that such properties have general statistical appeal and often serve as starting points for new developments. Moreover, we caution the reader that not all software can be trusted all the time when it comes to the analysis of data from completely unstructured designs, apart from the fact that the interpretation of the results may become difficult and ambiguous.

The development and introduction of new experimental designs in the last 50 years or so has been quite staggering, brought about, in large part, by an ever-widening field of applications and also by the mathematical beauty and challenge that some of these designs present. Whereas many designs had their origin in agricultural field experiments, it is true now that these designs as well as modifications, extensions, and new developments were initiated by applications in almost all types of experimental research, including industrial and clinical research. It is for this reason that books have been written with special applications in mind. We, on the other hand, have tried to keep the discussion in this book as general as possible, so that the reader can get the general picture and then apply the results in whatever area of application is desired.

Because of the overwhelming amount of material available in the literature, we had to make selections of what to include in this book and what to omit. Many special designs or designs for special cases (parameters) have been presented in the literature. We have concentrated, generally speaking, on the more general developments and results, providing and discussing methods of constructing rather large classes of designs. Here we have built upon the topics discussed in Kempthorne's 1952 book and supplemented the material with more recent topics of theoretical and applications oriented interests. Overall, we have selected the material and chosen the depth of discussion of the various topics in order to achieve our objective for this book, namely to serve as a textbook at the advanced graduate level and as a reference book for workers in the field of experimental design. The reader should have a solid foundation in and appreciation of the principles and fundamental notions of experimental design as discussed, for example, in Volume 1. We realize that the material presented here is more than can be covered in a one-semester course. Therefore, the instructor will have to make choices of the topics to be discussed.

In Chapters 1 through 6 we discuss incomplete block and row–column designs at various degrees of specificity. In Chapter 1 we lay the general foundation for the notion and analysis of incomplete block designs. This chapter is essential because its concepts permeate through almost every chapter of the book, in particular the ideas of intra- and interblock analyses. Chapters 2 through 5 are devoted to balanced and partially balanced incomplete block designs, their special features and methods of construction. In Chapter 6 we present some other types of incomplete block designs, such as α-designs and control-test treatment comparison designs. Further, we discuss various forms of row–column designs as examples of the use of additional blocking factors.

In Chapters 7 through 13 we give a general discussion of the most fundamental and important ideas of factorial designs, beginning with factors at two levels (Chapters 7 through 9), continuing with the case of factors with three levels (Chapter 10) through the general case of symmetrical and asymmetrical factorial designs (Chapters 11 and 12), and concluding with the important concept of fractional factorial designs (Chapter 13). In these chapters we return often to the notion of incomplete block designs as we discuss various systems of confounding of interaction effects with block effects.

Additional topics involving factorial designs are taken up in Chapters 14 through 17. In Chapter 14 we discuss the important concept of main effect plans and their construction. This notion is then extended to supersaturated designs (Chapter 15) and incorporated in the ideas of search designs (Chapter 16) and robust-design or Taguchi experiments (Chapter 17). We continue with an extensive chapter about lattice designs (Chapter 18), where the notions of factorial and incomplete block designs are combined in a unique way. We conclude the book with a chapter on crossover designs (Chapter 19) as an example where the ideas of optimal incomplete row–column designs are complemented by the notion of carryover effects.

In making a selection of topics for teaching purposes the instructor should keep in mind that we consider Chapters 1, 7, 8, 10, and 13 to be essential for the understanding of much of the material in the book. This material should then be supplemented by selected parts from the remaining chapters, thus providing the student with a good understanding of the methods of constructing various types of designs, the properties of the designs, and the analyses of experiments based on these designs. The reader will notice that some topics are discussed in more depth and detail than others. This is due to our desire to give the student a solid foundation in what we consider to be fundamental concepts.

In today's computer-oriented environment there exist a number of software programs that help in the construction and analysis of designs. We have chosen to use the *Statistical Analysis System* (SAS) for these purposes and have provided throughout the book examples of input statements and output using various procedures in SAS, both for constructing designs as well as analyzing data from experiments based on these designs. For the latter, we consider, throughout, various forms of the analysis of variance to be among the most important and informative tools.

As we have mentioned earlier, Volume 2 is based on the concepts developed and described in Volume 1. Nevertheless, Volume 2 is essentially self-contained. We make occasional references to certain sections in Volume 1 in the form (I.xx.yy) simply to remind the reader about certain notions. We emphasize again that the entire development is framed within the context of randomization theory and its approximation by normal theory inference. It is with this fact in mind that we discuss some methods and ideas that are based on normal theory.

There exist a number of books discussing the same types of topics that we exposit in this book, some dealing with only certain types of designs, but perhaps present more details than we do. For some details we refer to these books in the text. A quite general, but less detailed discussion of various aspects of experimental design is provided by Cox and Reid (2000).

Even though we have given careful attention to the selection of material for this book, we would be remiss if we did not mention that certain areas are completely missing. For example, the reader will not find any discussion of Bayesian experimental design. This is, in part, due to our philosophical attitude toward the Bayesian inferential approach (see Kempthorne, 1984; Hinkelmann, 2001). To explain, we strongly believe that design of experiment is a Bayesian experimentation process, *not* a Bayesian inference process, but one in which the experimenter approaches the experiment with some beliefs, to which he accommodates the design. It is interesting to speculate whether precise mathematical formulation of informal Bayesian thinking will be of aid in design. Another area that is missing is that of sequential design. Here again, we strongly believe and encourage the view that most experimentation is sequential in an operational sense. Results from one, perhaps exploratory, experiment will often lead to further, perhaps confirmatory, experimentation. This may be done informally or more formally in the context of sequential probability ratio tests, which we do not discuss explicitly. Thus, the selection and emphases are to a certain extent subjective and reflect our own interests as we have taught over the years parts of the material to our graduate students.

As mentioned above, the writing of this book has extended over many years. This has advantages and disadvantages. My (K.H.) greatest regret, however, is that the book was not completed before the death of my co-author, teacher, and mentor, Oscar Kempthorne. I only hope that the final product would have met with his approval.

This book could not have been completed without the help from others. First, we would like to thank our students at Virginia Tech, Iowa State University, and the University of Dortmund for their input and criticism after being exposed to some of the material. K.H. would like to thank the Departments of Statistics at Iowa State University and the University of Dortmund for inviting him to spend research leaves there and providing him with support and a congenial atmosphere to work. We are grateful to Michele Marini and Ayca Ozol-Godfrey for providing critical help with some computer work. Finally, we will never be able to fully express our gratitude to Linda Breeding for her excellent expert word-processing skills and her enormous patience in typing the manuscript, making changes after changes to satisfy our and the publisher's needs. It was a monumental task and she did as well as anybody possibly could.

KLAUS HINKELMANN

Blacksburg, VA
May 2004

CHAPTER 1

General Incomplete Block Design

1.1 INTRODUCTION AND EXAMPLES

One of the basic principles in experimental design is that of reduction of experimental error. We have seen (see Chapters I.9 and I.10) that this can be achieved quite often through the device of blocking. This leads to designs such as randomized complete block designs (Section I.9.2) or Latin square type designs (Chapter I.10). A further reduction can sometimes be achieved by using blocks that contain fewer experimental units than there are treatments.

The problem we shall be discussing then in this and the following chapters is the comparison of a number of treatments using blocks the size of which is less than the number of treatments. Designs of this type are called *incomplete block designs* (see Section I.9.8). They can arise in various ways of which we shall give a few examples.

In the case of field plot experiments, the size of the plot is usually, though by no means always, fairly well determined by experimental and agronomic techniques, and the experimenter usually aims toward a block size of less than 12 plots. If this arbitrary rule is accepted, and we wish to compare 100 varieties or crosses of inbred lines, which is not an uncommon situation in agronomy, we will not be able to accommodate all the varieties in one block. Instead, we might use, for example 10 blocks of 10 plots with different arrangements for each replicate (see Chapter 18).

Quite often a block and consequently its size are determined entirely on biological or physical grounds, as, for example, a litter of mice, a pair of twins, an individual, or a car. In the case of a litter of mice it is reasonable to assume that animals from the same litter are more alike than animals from different litters. The litter size is, of course, restricted and so is, therefore, the block size. Moreover, if one were to use female mice only for a certain investigation, the block size would be even more restricted, say to four or five animals. Hence,

Design and Analysis of Experiments. Volume 2: Advanced Experimental Design
By Klaus Hinkelmann and Oscar Kempthorne
ISBN 0-471-55177-5

comparing more than this number of treatments would require some type of incomplete block design.

Suppose we wish to compare seven treatments, $T_1, T_2, T_3, T_4, T_5, T_6, T_7$, say, using female mice, and suppose we have several litters with four females. We then could use the following incomplete block design, which, as will be explained later, is a balanced incomplete block design:

	Animal			
Litter	**1**	**2**	**3**	**4**
1	T_1	T_4	T_7	T_6
2	T_3	T_6	T_5	T_7
3	T_7	T_1	T_2	T_5
4	T_1	T_2	T_3	T_6
5	T_2	T_7	T_3	T_4
6	T_5	T_3	T_4	T_1
7	T_2	T_4	T_5	T_6

Notice that with this arrangement every treatment is replicated four times, and every pair of treatments occurs together twice in the same block; for example, T_1 and T_2 occur together in blocks 3 and 4.

Many sociological and psychological studies have been done on twins because they are "alike" in many respects. If they constitute a block, then the block size is obviously two. A number of incomplete block designs are available for this type of situation, for example, Kempthorne (1953) and Zoellner and Kempthorne (1954).

Blocks of size two arise also in some medical studies, when a patient is considered to be a block and his eyes or ears or legs are the experimental units.

With regard to a car being a block, this may occur if we wish to compare brands of tires, using the wheels as the experimental units. In this case one may also wish to take the effect of position of the wheels into account. This then leads to an incomplete design with two-way elimination of heterogeneity (see Chapters 6 and I.10).

These few examples should give the reader some idea why and how the need for incomplete block designs arises quite naturally in different types of research. For a given situation it will then be necessary to select the appropriate design from the catalogue of available designs. We shall discuss these different types of designs in more detail in the following chapters along with the appropriate analysis.

Before doing so, however, it seems appropriate to trace the early history and development of incomplete block designs. This development has been a remarkable achievement, and the reader will undoubtedly realize throughout the next chapters that the concept of incomplete block designs is fundamental to the understanding of experimental design as it is known today.

The origins of incomplete block designs go back to Yates (1936a) who introduced the concept of balanced incomplete block designs and their analysis utilizing both intra- and interblock information (Yates, 1940a). Other incomplete block designs were also proposed by Yates (1936b, 1937a, 1940b), who referred to these designs as quasi-factorial or lattice designs. Further contributions in the early history of incomplete block designs were made by Bose (1939, 1942) and Fisher (1940) concerning the structure and construction of balanced incomplete block designs. The notion of balanced incomplete block design was generalized to that of partially balanced incomplete block designs by Bose and Nair (1939), which encompass some of the lattice designs introduced earlier by Yates. Further extensions of the balanced incomplete block designs and lattice designs were made by Youden (1940) and Harshbarger (1947), respectively, by introducing balanced incomplete block designs for eliminating heterogeneity in two directions (generalizing the concept of the Latin square design) and rectangular lattices some of which are more general designs than partially balanced incomplete block designs. After this there has been a very rapid development in this area of experimental design, and we shall comment on many results more specifically in the following chapters.

1.2 GENERAL REMARKS ON THE ANALYSIS OF INCOMPLETE BLOCK DESIGNS

The analysis of incomplete block designs is different from the analysis of complete block designs in that comparisons among treatment effects and comparisons among block effects are no longer orthogonal to each other (see Section I.7.3). This is referred to usually by simply saying that treatments and blocks are not orthogonal. This nonorthogonality leads to an analysis analogous to that of the two-way classification with unequal subclass numbers. However, this is only partly true and applies only to the analysis that has come to be known as the *intrablock analysis*.

The name of the analysis is derived from the fact that contrasts in the treatment effects are estimated as linear combinations of comparisons of observations in the same block. In this way the block effects are eliminated and the estimates are functions of treatment effects and error (intrablock error) only. Coupled with the theory of least squares and the Gauss–Markov theorem (see I.4.16.2), this procedure will give rise to the best linear unbiased intrablock estimators for treatment comparisons. Historically, this has been the method first used for analyzing incomplete block designs (Yates, 1936a). We shall derive the intrablock analysis in Section 1.3.

Based upon considerations of efficiency, Yates (1939) argued that the intrablock analysis ignores part of the information about treatment comparisons, namely that information contained in the comparison of block totals. This analysis has been called *recovery of interblock information* or *interblock analysis*.

Yates (1939, 1940a) showed for certain types of lattice designs and for the balanced incomplete block design how these two types of analyses can be combined to yield more efficient estimators of treatment comparisons. Nair (1944) extended these results to partially balanced incomplete block designs, and Rao (1947a) gave the analysis for any incomplete block design showing the similarity between the intrablock analysis and the combined intra- and interblock analysis.

The intrablock analysis, as it is usually presented, is best understood by assuming that the block effects in the underlying linear model are fixed effects. But for the recovery of interblock information the block effects are then considered to be random effects. This leads sometimes to confusion with regard to the assumptions in the combined analysis, although it should be clear from the previous remark that then the block effects have to be considered random effects for *both* the intra- and interblock analysis. To emphasize it again, we can talk about intrablock analysis under the assumption of either fixed or random block effects. In the first case ordinary least squares (OLS) will lead to best linear unbiased estimators for treatment contrasts. This will, at least theoretically, not be true in the second case, which is the reason for considering the interblock information in the first place and using the Aitken equation (see I.4.16.2), which is also referred to as *generalized (weighted) least squares*.

We shall now derive the intrablock analysis (Section 1.3), the interblock analysis (Section 1.7), and the combined analysis (Section 1.8) for the general incomplete block design. Special cases will then be considered in the following chapters.

1.3 THE INTRABLOCK ANALYSIS

1.3.1 Notation and Model

Suppose we have t treatments replicated $r_1, r_2, \ldots, r_t$ times, respectively, and b blocks with $k_1, k_2, \ldots, k_b$ units, respectively. We then have

$$\sum_{i=1}^{t} r_i = \sum_{j=1}^{b} k_j = n$$

where n is the total number of observations.

Following the derivation of a linear model for observations from a randomized complete block design (RCBD), using the assumption of additivity in the broad sense (see Sections I.9.2.2 and I.9.2.6), an appropriate linear model for observations from an incomplete block design is

$$y_{ij\ell} = \mu + \tau_i + \beta_j + e_{ij\ell} \tag{1.1}$$

$(i = 1, 2, \ldots, t;\ j = 1, 2, \ldots, b;\ \ell = 0, 1, \ldots, n_{ij})$, where τ_i is the effect of the ith treatment, β_j the effect of the jth block, and $e_{ij\ell}$ the error associated with the observation $y_{ij\ell}$. As usual, the $e_{ij\ell}$ contain both experimental and observational (sampling) error, that is, using notation established in Volume 1,

$$e_{ij\ell} = \epsilon_{ij\ell} + \eta_{ij\ell}$$

with $\epsilon_{ij\ell}$ representing experimental error and $\eta_{ij\ell}$ representing observational error. Also, based on previous derivations (see I.6.3.4), we can treat the $e_{ij\ell}$ as i.i.d. random variables with mean zero and variance $\sigma_e^2 = \sigma_\epsilon^2 + \sigma_\eta^2$. Note that because n_{ij}, the elements of the incidence matrix $\boldsymbol{N}$, may be zero, not all treatments occur in each block which is, of course, the definition of an incomplete block design.

Model (1.1) can also be written in matrix notation as

$$\boldsymbol{y} = \mu\mathfrak{I} + \boldsymbol{X}_\tau\boldsymbol{\tau} + \boldsymbol{X}_\beta\boldsymbol{\beta} + \boldsymbol{e} \tag{1.2}$$

where $\mathfrak{I}$ is a column vector consisting of n unity elements, $\boldsymbol{X}_\beta$ is the observation-block incidence matrix

$$\boldsymbol{X}_\beta = \begin{bmatrix} \mathfrak{I}_{k_1} & & & \\ & \mathfrak{I}_{k_2} & & \\ & & \ddots & \\ & & & \mathfrak{I}_{k_b} \end{bmatrix}$$

with $\mathfrak{I}_{k_j}$ denoting a column vector of k_j unity elements $(j = 1, 2, \ldots, b)$ and

$$\boldsymbol{X}_\tau = (\boldsymbol{x}_1, \boldsymbol{x}_2, \ldots, \boldsymbol{x}_t)$$

is the observation-treatment incidence matrix, where $\boldsymbol{x}_i$ is a column vector with r_i unity elements and $(n - r_i)$ zero elements such that $\boldsymbol{x}_i'\boldsymbol{x}_i = r_i$ and $\boldsymbol{x}_i'\boldsymbol{x}_{i'} = 0$ for $i \neq i' (i, i' = 1, 2, \ldots, t)$.

1.3.2 Normal and Reduced Normal Equations

The normal equations (NE) for μ, τ_i, and β_j are then

$$n\widehat{\mu} + \sum_{i=1}^{t} r_i\widehat{\tau}_i + \sum_{j=1}^{b} k_j\widehat{\beta}_j = G$$

$$r_i\widehat{\mu} + r_i\widehat{\tau}_i + \sum_{j=1}^{b} n_{ij}\widehat{\beta}_j = T_i \qquad (i = 1, 2, \ldots, t) \tag{1.3}$$

$$k_j\widehat{\mu} + \sum_{i=1}^{t} n_{ij}\widehat{\tau}_i + k_j\widehat{\beta}_j = B_j \qquad (j = 1, 2, \ldots, b)$$

where

$$T_i = \sum_{j\ell} y_{ij\ell} = i\text{th treatment total}$$

$$B_j = \sum_{i\ell} y_{ij\ell} = j\text{th block total}$$

$$G = \sum_i T_i = \sum_j B_j = \text{overall total}$$

Equations (1.3) can be written in matrix notation as

$$\begin{pmatrix} \mathfrak{I}'_n\mathfrak{I}_n & \mathfrak{I}'_n\boldsymbol{X}_\tau & \mathfrak{I}'_n\boldsymbol{X}_\beta \\ \boldsymbol{X}'_\tau\mathfrak{I}_n & \boldsymbol{X}'_\tau\boldsymbol{X}_\tau & \boldsymbol{X}'_\tau\boldsymbol{X}_\beta \\ \boldsymbol{X}'_\beta\mathfrak{I}_n & \boldsymbol{X}'_\beta\boldsymbol{X}_\tau & \boldsymbol{X}'_\beta\boldsymbol{X}_\beta \end{pmatrix} \begin{pmatrix} \widehat{\mu} \\ \widehat{\boldsymbol{\tau}} \\ \widehat{\boldsymbol{\beta}} \end{pmatrix} = \begin{pmatrix} \mathfrak{I}'_n\boldsymbol{y} \\ \boldsymbol{X}'_\tau\boldsymbol{y} \\ \boldsymbol{X}'_\beta\boldsymbol{y} \end{pmatrix} \tag{1.4}$$

which, using the properties of $\mathfrak{I}, \mathbf{X}_\tau, \mathbf{X}_\beta$, can be written as

$$\begin{bmatrix} \mathfrak{I}'_n\mathfrak{I}_n & \mathfrak{I}'_\tau\boldsymbol{R} & \mathfrak{I}'_b\boldsymbol{K} \\ \boldsymbol{R}\mathfrak{I}_t & \boldsymbol{R} & \boldsymbol{N} \\ \boldsymbol{K}\mathfrak{I}_b & \boldsymbol{N}' & \boldsymbol{K} \end{bmatrix} \cdot \begin{bmatrix} \widehat{\mu} \\ \widehat{\boldsymbol{\tau}} \\ \widehat{\boldsymbol{\beta}} \end{bmatrix} = \begin{bmatrix} G \\ \boldsymbol{T} \\ \boldsymbol{B} \end{bmatrix} \tag{1.5}$$

where

$\boldsymbol{R} = \text{diag}\,(r_i)$ $\quad t \times t$

$\boldsymbol{K} = \text{diag}\,(k_j)$ $\quad b \times b$

$\boldsymbol{N} = (n_{ij})$ $\quad t \times b$ (the incidence matrix)

$\boldsymbol{T}' = (T_1, T_2, \ldots, T_t)$

$\boldsymbol{B}' = (B_1, B_2, \ldots, B_b)$

$\boldsymbol{\tau}' = (\tau_1, \tau_2, \ldots, \tau_t)$

$\boldsymbol{\beta}' = (\beta_1, \beta_2, \ldots, \beta_b)$

and the $\mathfrak{I}$'s are column vectors of unity elements with dimensions indicated by the subscripts. From the third set of equations in (1.5) we obtain

$$\widehat{\mu}\mathfrak{I}_b + \widehat{\boldsymbol{\beta}} = \boldsymbol{K}^{-1}(\boldsymbol{B} - \boldsymbol{N}'\widehat{\boldsymbol{\tau}}) \tag{1.6}$$

Substituting (1.6) into the second set of (1.5), which can also be expressed as $\boldsymbol{N}\mathfrak{J}_b\widehat{\mu} + \boldsymbol{N}\widehat{\boldsymbol{\beta}} + \boldsymbol{R}\widehat{\boldsymbol{\tau}} = \boldsymbol{T}$ (since $\boldsymbol{N}\mathfrak{J}_b = \boldsymbol{R}\mathfrak{J}_t$), leads to the *reduced normal equations* (RNE) (see Section I.4.7.1) for $\boldsymbol{\tau}$

$$(\boldsymbol{R} - \boldsymbol{N}\boldsymbol{K}^{-1}\boldsymbol{N}')\widehat{\boldsymbol{\tau}} = \boldsymbol{T} - \boldsymbol{N}\boldsymbol{K}^{-1}\boldsymbol{B} \tag{1.7}$$

Standard notation for (1.7) is

$$\boldsymbol{C}\widehat{\boldsymbol{\tau}} = \boldsymbol{Q} \tag{1.8}$$

where

$$\boldsymbol{C} = \boldsymbol{R} - \boldsymbol{N}\boldsymbol{K}^{-1}\boldsymbol{N}' \tag{1.9}$$

and

$$\boldsymbol{Q} = \boldsymbol{T} - \boldsymbol{N}\boldsymbol{K}^{-1}\boldsymbol{B} \tag{1.10}$$

the (i, i') element of $\boldsymbol{C}$ being

$$c_{ii'} = \delta_{ii'}r_i - \sum_{j=1}^{b} \frac{n_{ij}n_{i'j}}{k_j}$$

with $\delta_{ii'} = 1$ for $i = i'$ and $= 0$ otherwise, and the ith element of $\boldsymbol{Q}$ being

$$Q_i = T_i - \sum_{j=1}^{b} \frac{n_{ij}B_j}{k_j}$$

And Q_i is called the ith *adjusted treatment total*, the adjustment being due to the fact that the treatments do not occur the same number of times in the blocks.

1.3.3 The *C* Matrix and Estimable Functions

We note that the matrix $\boldsymbol{C}$ of (1.9) is determined entirely by the specific design, that is, by the incidence matrix $\boldsymbol{N}$. It is, therefore, referred to as *the* $\boldsymbol{C}$ matrix (sometimes also as the *information matrix*) of that design. The $\boldsymbol{C}$ matrix is symmetric, and the elements in any row or any column of $\boldsymbol{C}$ add to zero, that is, $\boldsymbol{C}\mathfrak{J} = \boldsymbol{0}$, which implies that $r(\boldsymbol{C}) = \text{rank}(\boldsymbol{C}) \leq t - 1$. Therefore, $\boldsymbol{C}$ does not have an inverse and hence (1.8) cannot be solved uniquely. Instead we write a solution to (1.8) as

$$\widehat{\boldsymbol{\tau}} = \boldsymbol{C}^{-}\boldsymbol{Q} \tag{1.11}$$

where $\boldsymbol{C}^{-}$ is a generalized inverse for $\boldsymbol{C}$ (see Section 1.3.4).

If we write $\boldsymbol{C} = (\boldsymbol{c}_1, \boldsymbol{c}_2, \ldots, \boldsymbol{c}_t)$, where $\boldsymbol{c}_i$ is the ith column of $\boldsymbol{C}$, then the set of linear functions

$$\{\boldsymbol{c}_i'\boldsymbol{\tau}, i = 1, 2, \ldots, t\}$$

which span the totality of estimable functions of the treatment effects, has dimensionality $r(\boldsymbol{C})$. Let $\boldsymbol{c}'\boldsymbol{\tau}$ be an estimable function and $\boldsymbol{c}'\widehat{\boldsymbol{\tau}}$ its estimator, with $\widehat{\boldsymbol{\tau}}$ from (1.11). Then

$$\begin{aligned} E(\boldsymbol{c}'\widehat{\boldsymbol{\tau}}) &= E\left(\boldsymbol{c}'\boldsymbol{C}^-\boldsymbol{Q}\right) \\ &= \boldsymbol{c}'\boldsymbol{C}^- E(\boldsymbol{Q}) \\ &= \boldsymbol{c}'\boldsymbol{C}^-\boldsymbol{C}\boldsymbol{\tau} \end{aligned}$$

For $\boldsymbol{c}'\widehat{\boldsymbol{\tau}}$ to be an unbiased estimator for $\boldsymbol{c}'\boldsymbol{\tau}$ for any $\boldsymbol{\tau}$, we then must have

$$\boldsymbol{c}'\boldsymbol{C}^-\boldsymbol{C} = \boldsymbol{c}' \tag{1.12}$$

Since $\boldsymbol{C}\mathfrak{I} = 0$, it follows from (1.12) that $\boldsymbol{c}'\mathfrak{I} = 0$. Hence, only treatment contrasts are estimable. If $r(\boldsymbol{C}) = t - 1$, then all treatment contrasts are estimable. In particular, all differences $\tau_i - \tau_{i'} (i \neq i')$ are estimable, there being $t - 1$ linearly independent estimable functions of this type. Then the design is called a *connected design* (see also Section I.4.13.3).

1.3.4 Solving the Reduced Normal Equations

In what follows we shall assume that the design is connected; that is, $r(\boldsymbol{C}) = t - 1$. This means that $\boldsymbol{C}$ has $t - 1$ nonzero (positive) eigenvalues and one zero eigenvalue. From

$$\boldsymbol{C}\begin{bmatrix}1\\1\\\vdots\\1\end{bmatrix} = \boldsymbol{0} = 0\begin{bmatrix}1\\1\\\vdots\\1\end{bmatrix}$$

it follows then that $(1, 1, \ldots, 1)'$ is an eigenvector corresponding to the zero eigenvalue. If we denote the nonzero eigenvalues of $\boldsymbol{C}$ by $d_1, d_2, \ldots, d_{t-1}$ and the corresponding eigenvectors by $\boldsymbol{\xi}_1, \boldsymbol{\xi}_2, \ldots, \boldsymbol{\xi}_{t-1}$ with $\boldsymbol{\xi}_i'\boldsymbol{\xi}_i = 1 \quad (i = 1, 2, \ldots, t-1)$ and $\boldsymbol{\xi}_i'\boldsymbol{\xi}_{i'} = 0 (i \neq i')$, then we can write $\boldsymbol{C}$ in its spectral decomposition as

$$\boldsymbol{C} = \sum_{i=1}^{t-1} d_i \boldsymbol{\xi}_i \boldsymbol{\xi}_i' \tag{1.13}$$

or with $d_t = 0$ and $\boldsymbol{\xi}'_t = 1/\sqrt{t}(1, 1, \ldots, 1)$, alternatively as

$$\boldsymbol{C} = \sum_{i=1}^{t} d_i \boldsymbol{\xi}_i \boldsymbol{\xi}'_i \tag{1.14}$$

We note that $\boldsymbol{\xi}'_t \boldsymbol{\xi}_t = 1$ and $\boldsymbol{\xi}'_i \boldsymbol{\xi}_t = 0$ for $i = 1, 2, \ldots, t-1$.

We now return to (1.8) and consider a solution to these equations of the form given by (1.11). Although there are many methods of finding generalized inverses, we shall consider here one particular method, which is most useful in connection with incomplete block designs, especially balanced and partially balanced incomplete block designs (see following chapters). This method is based on the following theorem, which is essentially due to Shah (1959).

Theorem 1.1 Let $\boldsymbol{C}$ be a $t \times t$ matrix as given by (1.9) with $r(\boldsymbol{C}) = t - 1$. Then $\widetilde{\boldsymbol{C}} = \boldsymbol{C} + a\boldsymbol{\Im}\boldsymbol{\Im}'$, where $a \neq 0$ is a real number, admits an inverse $\widetilde{\boldsymbol{C}}^{-1}$, and $\widetilde{\boldsymbol{C}}^{-1}$ is a generalized inverse for $\boldsymbol{C}$.

Proof

(a) We can rewrite $\widetilde{C}$ as

$$\widetilde{\boldsymbol{C}} = \boldsymbol{C} + a\boldsymbol{\Im}\boldsymbol{\Im}' = \boldsymbol{C} + a \begin{bmatrix} 1 \\ 1 \\ \vdots \\ 1 \end{bmatrix} (1, 1, \ldots, 1) = \boldsymbol{C} + at\, \boldsymbol{\xi}_t \boldsymbol{\xi}'_t$$

and because of (1.13)

$$\widetilde{\boldsymbol{C}} = \sum_{i=1}^{t-1} d_i \boldsymbol{\xi}_i \boldsymbol{\xi}'_i + at\, \boldsymbol{\xi}_t \boldsymbol{\xi}'_t \tag{1.15}$$

Clearly, $\widetilde{\boldsymbol{C}}$ has nonzero roots $d_1, d_2, \ldots, d_{t-1}, d_t = at$ and hence is nonsingular. Then

$$\widetilde{\boldsymbol{C}}^{-1} = \sum_{i=1}^{t-1} \frac{1}{d_i} \boldsymbol{\xi}_i \boldsymbol{\xi}'_i + \frac{1}{at} \boldsymbol{\xi}_t \boldsymbol{\xi}'_t \tag{1.16}$$

(b) To show that $\widetilde{\boldsymbol{C}}^{-1} = \boldsymbol{C}^{-}$ we consider $\boldsymbol{C}\widetilde{\boldsymbol{C}}^{-1}\boldsymbol{C}$. From (1.13), (1.15), and (1.16) we have

$$\widetilde{\boldsymbol{C}}\widetilde{\boldsymbol{C}}^{-1} = \boldsymbol{I} = \sum_{i=1}^{t} \boldsymbol{\xi}_i \boldsymbol{\xi}'_i$$

and

$$C\widetilde{C}^{-1} = \sum_{i=1}^{t-1} \xi_i \xi_i' = I - \xi_t \xi_t' = I - \frac{1}{t}\mathfrak{I}\mathfrak{I}' \tag{1.17}$$

$$C\widetilde{C}^{-1}C = C$$

which implies

$$\widetilde{C}^{-1} = C^{-}$$

We remark here already that determining C^{-} for the designs in the following chapters will be based on (1.17) rather than on (1.14).

Substituting $\widetilde{C}^{-1}$ into (1.13) then yields a solution of the RNE (1.8); that is,

$$\widehat{\tau} = \widetilde{C}^{-1} Q \tag{1.18}$$

We note that because of (1.8) and (1.16)

$$\begin{aligned} E(\widehat{\tau}) &= E\left(\widetilde{C}^{-1}Q\right) \\ &= \widetilde{C}^{-1}E(Q) \\ &= \widetilde{C}^{-1}E(C\widehat{\tau}) \\ &= \widetilde{C}^{-1}C\tau \\ &= \left(I - \frac{1}{t}\mathfrak{I}\mathfrak{I}'\right)\tau \\ &= \begin{bmatrix} \tau_1 - \overline{\tau} \\ \tau_2 - \overline{\tau} \\ \vdots \\ \tau_t - \overline{\tau} \end{bmatrix} \end{aligned}$$

with $\overline{\tau} = 1/t \sum_i \tau_i$; that is, $E(\widehat{\tau})$ is the same as if we had obtained a generalized inverse of C by imposing the condition $\sum_i \widehat{\tau}_i = 0$. □

1.3.5 Estimable Functions of Treatment Effects

We know from the Gauss–Markov theorem (see Section I.4.16.2) that for any linear estimable function of the treatment effects, say $c'\tau$,

$$E(c'\widehat{\tau}) = c'\tau \tag{1.19}$$

is independent of the solution to the NE (see Section I.4.4.4). We have further

$$\operatorname{var}(c'\widehat{\tau}) = c'\,\widetilde{C}^{-1}\,c\,\sigma_e^2 \tag{1.20}$$

with a corresponding result (but same numerical value) for any other solution obtained, using an available software package (see Section 1.14). We shall elaborate on this point briefly.

Let us rewrite model (1.2) as

$$\begin{aligned} \boldsymbol{y} &= \mu \mathfrak{J} + \boldsymbol{X}_\beta \boldsymbol{\beta} + \boldsymbol{X}_\tau \boldsymbol{\tau} + \boldsymbol{e} \\ &= (\mathfrak{J}\, \boldsymbol{X}_\beta\, \boldsymbol{X}_\tau) \begin{pmatrix} \mu \\ \boldsymbol{\beta} \\ \boldsymbol{\tau} \end{pmatrix} + \boldsymbol{e} \\ &\equiv \boldsymbol{X\Theta} + \boldsymbol{e} \end{aligned} \tag{1.21}$$

with

$$\boldsymbol{X} = (\mathfrak{J} : \boldsymbol{X}_\beta : \boldsymbol{X}_\tau) \tag{1.22}$$

and

$$\boldsymbol{\Theta}' = (\mu, \boldsymbol{\beta}', \boldsymbol{\tau}')$$

The NE for model (1.21) are

$$\boldsymbol{X}'\boldsymbol{X}\boldsymbol{\Theta}^* = \boldsymbol{X}'\boldsymbol{y} \tag{1.23}$$

A solution to (1.23) is given by, say,

$$\boldsymbol{\Theta}^* = (\boldsymbol{X}'\boldsymbol{X})^- \boldsymbol{X}'\boldsymbol{y}$$

for some $(\boldsymbol{X}'\boldsymbol{X})^-$. Now $(\boldsymbol{X}'\boldsymbol{X})^-$ is a $(1 + b + t) \times (1 + b + t)$ matrix that we can partition conformably, using the form of $\boldsymbol{X}$ as given in (1.22), as

$$(\boldsymbol{X}'\boldsymbol{X})^- = \begin{pmatrix} \boldsymbol{A}_{\mu\mu} & \boldsymbol{A}_{\mu\beta} & \boldsymbol{A}_{\mu\tau} \\ \boldsymbol{A}'_{\mu\beta} & \boldsymbol{A}_{\beta\beta} & \boldsymbol{A}_{\beta\tau} \\ \boldsymbol{A}'_{\mu\tau} & \boldsymbol{A}'_{\beta\tau} & \boldsymbol{A}_{\tau\tau} \end{pmatrix} \tag{1.24}$$

Here, $\boldsymbol{A}_{\tau\tau}$ is a $t \times t$ matrix that serves as the variance–covariance matrix for obtaining

$$\text{var}(\boldsymbol{c}'\boldsymbol{\tau}^*) = \boldsymbol{c}'\, \boldsymbol{A}_{\tau\tau}\, \boldsymbol{c}\, \sigma_e^2 \tag{1.25}$$

For any estimable function $\boldsymbol{c}'\boldsymbol{\tau}$ we have $\boldsymbol{c}'\widehat{\boldsymbol{\tau}} = \boldsymbol{c}'\boldsymbol{\tau}^*$ and also the numerical values for (1.20) and (1.25) are the same. If we denote the (i, i') element of $\widehat{\boldsymbol{C}}^{-1}$ by $c^{ii'}$ and the corresponding element of $\boldsymbol{A}_{\tau\tau}$ in (1.24) by $a^{ii'}$, then we have, for example, for $\boldsymbol{c}'\boldsymbol{\tau} = \tau_i - \tau_{i'}$

$$\text{var}\,(\widehat{\tau}_i - \widehat{\tau}_{i'}) = \left(c^{ii} - 2c^{ii'} + c^{i'i'}\right)\sigma_e^2 = \left(a^{ii} - 2a^{ii'} + a^{i'i'}\right)\sigma_e^2 \tag{1.26}$$

For a numerical example and illustration of computational aspects, see Section 1.13.

1.3.6 Analyses of Variance

It follows from general principles (see Section I.4.7.1) that the two forms of analysis of variance are as given in Tables 1.1 and 1.2. We shall henceforth refer to the analysis of variance in Table 1.1 as the treatment-after-block ANOVA or T | B-ANOVA as it is associated with the ordered model

$$y = \mu \mathfrak{I} + X_\beta \boldsymbol{\beta} + X_\tau \boldsymbol{\tau} + e$$

whereas the analysis of variance in Table 1.2 is associated with the ordered model

$$y = \mu \mathfrak{I} + X_\tau \boldsymbol{\tau} + X_\beta \boldsymbol{\beta} + e$$

and hence shall be referred to as the block-after-treatment ANOVA or B | T-ANOVA. To indicate precisely the sources of variation and the associated sums of squares, we use the notation developed in Section I.4.7.2 for the general case as it applies to the special case of the linear model for the incomplete block

Table 1.1 T|B-ANOVA for Incomplete Block Design

Source	d.f.[a]	SS	$E(MS)$
$X_\beta \mid \mathfrak{I}$	$b-1$	$\sum_{j=1}^{b} \frac{B_j^2}{k_j} - \frac{G^2}{n}$	
$X_\tau \mid \mathfrak{I}, X_\beta$	$t-1$	$\sum_{i=1}^{t} \widehat{\tau}_i Q_i$	$\sigma_e^2 + \frac{\boldsymbol{\tau}' C \boldsymbol{\tau}}{t-1}$
$I \mid \mathfrak{I}, X_\beta, X_\tau$	$n-b-t+1$	Difference	σ_e^2
Total	$n-1$	$\sum_{ij\ell} y_{ij\ell}^2 - \frac{G^2}{n}$	

[a] d.f. = degrees of freedom.

Table 1.2 B|T-ANOVA for Incomplete Block Design

Source	d.f.	SS
$X_\tau \mid \mathfrak{I}$	$t-1$	$\sum_{i=1}^{t} \frac{T_i^2}{r_i} - \frac{G^2}{n}$
$X_\beta \mid \mathfrak{I}, X_\tau$	$b-1$	Difference
$I \mid \mathfrak{I}, X_\beta, X_\tau$	$n-b-t+1$	From Table 1.1
Total	$n-1$	$\sum_{ij\ell} y_{ij\ell}^2 - \frac{G^2}{n}$

design, thereby avoiding the commonly used but not always clearly understood terms *blocks ignoring treatments* for $(\boldsymbol{X}_\beta \mid \mathfrak{I})$, *treatments eliminating blocks* for $(\boldsymbol{X}_\tau \mid \mathfrak{I}, \boldsymbol{X}_\beta)$, and *blocks eliminating treatments* for $(\boldsymbol{X}_\beta \mid \mathfrak{I}, \boldsymbol{X}_\tau)$.

The T | B-ANOVA follows naturally from the development of the RNE for the treatment effects. It is the appropriate ANOVA for the intrablock analysis as it allows to test the hypothesis

$$H_0: \tau_1 = \tau_2 = \cdots = \tau_t$$

by means of the (approximate) F test (see I.9.2.5)

$$F = \frac{\text{SS}(\boldsymbol{X}_\tau \mid \mathfrak{I}, \boldsymbol{X}_\beta)/(t-1)}{\text{SS}(\boldsymbol{I} \mid \mathfrak{I}, \boldsymbol{X}_\beta, \boldsymbol{X}_\tau)/(n-b-t+1)} \tag{1.27}$$

Also MS(Error)= $\text{SS}(\boldsymbol{I} \mid \mathfrak{I}, \boldsymbol{X}_\beta, \boldsymbol{X}_\tau)/(n-b-t+1)$ is an estimator for σ_e^2 to be used for estimating $\text{var}(\boldsymbol{c}'\widehat{\boldsymbol{\tau}})$ of (1.20).

The usefulness of the B | T-ANOVA in Table 1.2 will become apparent when we discuss specific aspects of the combined intra- and interblock analysis in Section 1.10. At this point we just mention that $\text{SS}(\boldsymbol{X}_\beta \mid, \mathfrak{I}, \boldsymbol{X}_\tau)$ could have been obtained from the RNE for block effects. Computationally, however, it is more convenient to use the fact that $\text{SS}(\boldsymbol{I} \mid \mathfrak{I}, \boldsymbol{X}_\beta, \boldsymbol{X}_\tau) = \text{SS}(\boldsymbol{I} \mid \mathfrak{I}, \boldsymbol{X}_\tau, \boldsymbol{X}_\beta)$ and then obtain $\text{SS}(\boldsymbol{X}_\beta \mid \mathfrak{I}, \boldsymbol{X}_\tau)$ by subtraction.

Details of computational procedures using SAS PROC GLM and SAS PROC Mixed (*SAS*1999–2000) will be described in Section 1.14.

1.4 INCOMPLETE DESIGNS WITH VARIABLE BLOCK SIZE

In the previous section we discussed the intrablock analysis of the general incomplete block design; that is, a design with possibly variable block size and possibly variable number of replications. Although most designed experiments use blocks of equal size, k say, there exist, however, experimental situations where blocks of unequal size arise quite naturally. We shall distinguish between two reasons why this can happen and why caution may have to be exercised before the analysis as outlined in the previous section can be used:

1. As pointed out by Pearce (1964, p. 699):

    > With much biological material there are natural units that can be used as blocks and they contain plots to a number not under the control of the experimenter. Thus, the number of animals in a litter or the number of blossoms in a truss probably vary only within close limits.

2. Although an experiment may have been set up using a proper design, that is, a design with equal block size, missing plots due to accidents during the course of investigation will leave one for purpose of analysis with a design of variable block size.

In both cases there are two alternatives to handle the situation. In case 1 one may wish to reduce all blocks to a constant size, thereby reducing the number of experimental units available. If experimental units are at a premium, this may not be the most desirable course of action. The other alternative is to use the natural blocks and then use the analysis as given in the previous section. Before doing so we mention that its validity will depend on one very important assumption, and that is the constancy of the variance σ_e^2 for all blocks. In general, the size of σ_e^2 will depend on the size of the blocks: The larger the blocks, the larger σ_e^2 will be since it is in part a measure of the variability of the experimental units within blocks (see I.9.2.4). In fact, this is the reason for reducing the block size since it may also reduce the experimental error. Experience shows that such a reduction in σ_e^2 is not appreciable for only modest reduction in block size. It is therefore quite reasonable to assume that σ_e^2 is constant for blocks of different size if the number of experimental units varies only slightly.

In case 2 one possibility is to estimate the missing values and then use the analysis for the proper design. Such a procedure, however, would only be approximate. The exact analysis then would require the analysis with variable block size as in case 1. Obviously, the assumption of constancy of experimental error is satisfied here if is was satisfied for the original proper design.

1.5 DISCONNECTED INCOMPLETE BLOCK DESIGNS

In deriving the intrablock analysis of an incomplete block design in Section 1.3.4 we have made the assumption that the $\boldsymbol{C}$ matrix of (1.9) has maximal rank $t - 1$, that is, the corresponding design is a connected design. Although connectedness is a desirable property of a design and although most designs have this property, we shall encounter designs (see Chapter 8) that are constructed on purpose as disconnected designs. We shall therefore comment briefly on this class of designs.

Following Bose (1947a) a treatment and a block are said to be associated if the treatment is contained in that block. Two treatments are said to be connected if it is possible to pass from one to the other by means of a chain consisting alternately of treatments and blocks such that any two adjacent members of the chain are associated. If this holds true for any two treatments, then the design is said to be connected, otherwise it is said to be *disconnected* (see Section I.4.13.3 for a more formal definition and Srivastava and Anderson, 1970). Whether a design is connected or disconnected can be checked easily by applying the definition given above to the incidence matrix $\boldsymbol{N}$: If one can connect two nonzero elements of $\boldsymbol{N}$ by means of vertical and horizontal lines such that the vertices are at nonzero elements, then the two treatments are connected. In order to check whether a design is connected, it is sufficient to check whether a given treatment is connected to all the other $t - 1$ treatments. If a design is disconnected, it follows then that (possibly after suitable relabeling of the treatments) the matrix $\boldsymbol{N}\boldsymbol{N}'$ and hence $\boldsymbol{C}$ consist of disjoint block diagonal matrices such that the treatments associated with one of these submatrices are connected with each other.

Suppose $\boldsymbol{C}$ has m submatrices, that is,

$$\boldsymbol{C} = \begin{bmatrix} \boldsymbol{C}_1 & & & \\ & \boldsymbol{C}_2 & & \\ & & \ddots & \\ & & & \boldsymbol{C}_m \end{bmatrix}$$

where $\boldsymbol{C}_\nu$ is $t_\nu \times t_\nu$ $\left(\sum_{\nu=1}^m t_\nu = t\right)$. It then follows that rank $(\boldsymbol{C}_\nu) = t_\nu - 1(\nu = 1, 2, \ldots, m)$ and hence $\text{rank}(\boldsymbol{C}) = t - m$. The RNE is still of the form (1.8) with a solution given by (1.11), where in $\boldsymbol{C}^- = \widetilde{\boldsymbol{C}}^{-1}$ we now have, modifying Theorem 1.1,

$$\widetilde{\boldsymbol{C}} = \boldsymbol{C} + \begin{bmatrix} a_1\mathfrak{I}\mathfrak{I}' & & & \\ & a_2\mathfrak{I}\mathfrak{I}' & & \\ & & \ddots & \\ & & & a_m\mathfrak{I}\mathfrak{I}' \end{bmatrix}$$

Table 1.3 T|B-ANOVA for Disconnected Incomplete Block Design

Source	d.f.	SS
$\boldsymbol{X}_\beta \mid \mathfrak{I}$	$b - 1$	$\sum_j \frac{B_j^2}{k_j} - \frac{G^2}{n}$
$\boldsymbol{X}_\tau \mid \mathfrak{I}, \boldsymbol{X}_\beta$	$t - m$	$\sum_i \widehat{\tau}_i Q_i$
$\boldsymbol{I} \mid \mathfrak{I}, \boldsymbol{X}_\beta, \boldsymbol{X}_\tau$	$n - b - t + m$	Difference
Total	$n - 1$	$\sum_{ij\ell} y_{ij\ell}^2 - \frac{G^2}{n}$

Table 1.4 B|T-ANOVA for Disconnected Incomplete Block Design

Source	d.f.	SS
$\boldsymbol{X}_\tau \mid \mathfrak{I}$	$t - 1$	$\sum_i \frac{T_i^2}{r_i} - \frac{G^2}{n}$
$\boldsymbol{X}_\beta \mid \mathfrak{I}, \boldsymbol{X}_\tau$	$b - m$	Difference
$\boldsymbol{I} \mid \mathfrak{I}, \boldsymbol{X}_\tau, \boldsymbol{X}_\beta$	$n - t - b + m$	From Table 1.3
Total	$n - 1$	$\sum_{ij\ell} y_{ij\ell}^2 - \frac{G^2}{n}$

with $a_\nu (\nu = 1, 2, \ldots, m)$ arbitrary constants ($\neq 0$) and the $\mathfrak{J}\mathfrak{J}'$ matrices are of appropriate dimensions. Following the development in Section 1.3.6, this then leads to the ANOVA tables as given in Tables 1.3 and 1.4.

1.6 RANDOMIZATION ANALYSIS

So far we have derived the analysis of data from incomplete block designs using a Gauss–Markov linear model as specified in (1.1). We have justified the appropriate use of such an infinite population theory model in our earlier discussions of error control designs (see, e.g., Sections I.6.3 and I.9.2) as a substitute for a derived, that is, finite, population theory model that takes aspects of randomization into account. In this section we shall describe in mathematical terms the randomization procedure for an incomplete block design, derive an appropriate linear model, and apply it to the analysis of variance. This will show again, as we have argued in Section I.9.2 for the RCBD, that treatment effects and block effects cannot be considered symmetrically for purposes of statistical inference.

1.6.1 Derived Linear Model

Following Folks and Kempthorne (1960) we shall confine ourselves to proper (i.e., all $k_j = k$), equireplicate (i.e., all $r_i = r$) designs. The general situation is then as follows: We are given a set of b blocks, each of constant size k; a master plan specifies b sets of k treatments; these sets are assigned at random to the blocks; in each block the treatments are assigned at random to the experimental units (EU). This randomization procedure is described more formally by the following design random variables:

$$\alpha_j^u = \begin{cases} 1 & \text{if the } u\text{th set is assigned to the } j\text{th block} \\ 0 & \text{otherwise} \end{cases} \tag{1.28}$$

and

$$\delta_{j\ell}^{uv} = \begin{cases} 1 & \text{if the } uv \text{ treatment is assigned to the} \\ & \ell\text{th unit of the } j\text{th block} \\ 0 & \text{otherwise} \end{cases} \tag{1.29}$$

The uv treatment is one of the t treatments that, for a given design, has been assigned to the uth set.

Assuming additivity in the strict sense (see Section I.6.3), the conceptual response of the uv treatment assigned to the ℓth EU in the jth block can be written as

$$T_{j\ell uv} = U_{j\ell} + T_{uv} \tag{1.30}$$

where $U_{j\ell}$ is the contribution from the ℓth EU in the jth block and T_{uv} is the contribution from treatment uv. We then write further

$$T_{j\ell uv} = \overline{U}_{..} + (\overline{U}_{j.} - \overline{U}_{..}) + (U_{j\ell} - \overline{U}_{j.}) + \overline{T}_{..} + (T_{uv} - \overline{T}_{..})$$

$$= \mu + b_j + \tau_{uv} + u_{j\ell} \tag{1.31}$$

where

$\mu = \overline{U}_{..} + \overline{T}_{..}$ is the overall mean

$b_j = \overline{U}_{j.} - \overline{U}_{..}$ is the effect of the jth block $(j = 1, 2, \ldots, b)$

$\tau_{uv} = t_{uv} - \overline{T}_{..}$ is the effect of the uv treatment $(u = 1, 2, \ldots, b;\ v = 1, 2, \ldots, k)$

$u_{j\ell} = U_{j\ell} - \overline{U}_{j.}$ is the unit error $(\ell = 1, 2, \ldots, k)$

with $\sum_j b_j = 0 = \sum_{uv} \tau_{uv} = \sum_\ell u_{j\ell}$. We then express the observed response for the uv treatment, y_{uv}, as

$$y_{uv} = \sum_j \sum_\ell \alpha_j^u \, \delta_{j\ell}^{uv} \, T_{j\ell uv}$$

$$= \mu + \tau_{uv} + \sum_j \alpha_j^u b_j + \sum_j \sum_\ell \alpha_j^u \, \delta_{j\ell}^{uv} \, u_{j\ell}$$

$$= \mu + \tau_{uv} + \beta_u + \omega_{uv} \tag{1.32}$$

where

$$\beta_u = \sum_j \alpha_j^u \, b_j \tag{1.33}$$

is a random variable with

$$E(\beta_u) = 0 \qquad E(\beta_u^2) = \frac{1}{b} \sum_j b_j^2 \qquad E(\beta_u \beta_{u'}) = \frac{1}{b(b-1)} \sum_j b_j^2 \qquad (u \neq u')$$

Also,

$$\omega_{uv} = \sum_j \sum_\ell \alpha_j^u \, \delta_{j\ell}^{uv} \, u_{j\ell} \tag{1.34}$$

is a random variable with

$$E(\omega_{uv}) = 0 \qquad E(\omega_{uv}^2) = \frac{1}{bk}\sum_{j\ell} u_{j\ell}^2$$

$$E(\omega_{uv}\omega_{uv'}) = -\frac{1}{bk(k-1)}\sum_j \sum_\ell u_{j\ell}^2 \qquad (v \neq v')$$

$$E(\omega_{uv}\omega_{u'v'}) = 0 \qquad (u \neq u')$$

In deriving the properties of the random variables β_u and ω_{uv} we have used, of course, the familiar distributional properties of the design random variables α_j^u and $\delta_{j\ell}^{uv}$, such as

$$P(\alpha_j^u = 1) = \frac{1}{b}$$

$$P(\alpha_j^u = 1 \mid \alpha_{j'}^u = 1) = 0 \qquad (j \neq j')$$

$$P(\alpha_j^u = 1 \mid \alpha_{j'}^{u'} = 1) = \frac{1}{b(b-1)} \qquad (u \neq u', j \neq j')$$

$$P(\delta_{j\ell}^{uv} = 1) = \frac{1}{k}$$

$$P(\delta_{j\ell}^{uv} = 1) \mid (\delta_{j\ell'}^{uv} = 1) = 0 \qquad (\ell \neq \ell')$$

$$P(\delta_{j\ell}^{uv} = 1) \mid (\delta_{j\ell'}^{uv'} = 1) = \frac{1}{k(k-1)} \qquad (\ell \neq \ell', v \neq v')$$

$$P(\delta_{j\ell}^{uv} = 1) \mid (\delta_{j'\ell'}^{u'v'} = 1) = \frac{1}{k^2} \qquad (j \neq j', u \neq u')$$

and so on.

1.6.2 Randomization Analysis of ANOVA Tables

Using model (1.32) and its distributional properties as induced by the design random variables α_j^u and $\delta_{j\ell}^{uv}$, we shall now derive expected values of the sums of squares in the analyses of variance as given in Tables 1.1 and 1.2:

1. $$E(\text{SS Total}) = E\sum_{uv}(y_{uv} - \overline{y}_{..})^2$$

$$= E\sum_{uv}(\tau_{uv} + \beta_u + \omega_{uv})^2$$

$$= \sum_{uv}\tau_{uv}^2 + k\sum_j b_j^2 + \sum_{j\ell} u_{j\ell}^2$$

2. $$E[\mathrm{SS}(\boldsymbol{X}_\beta \mid \mathfrak{I})] = E\sum_{uv}(\overline{y}_{u.} - \overline{y}_{..})^2$$
$$= k\,E\sum_u\left(\overline{\tau}_{u.} + \beta_u + \frac{1}{k}\sum_v \omega_{uv}\right)^2$$
$$= k\sum_u \overline{\tau}^2_{u.} + k\sum_j b_j^2$$

3. $$E[\mathrm{SS}\,(\boldsymbol{I} \mid \mathfrak{I}, \boldsymbol{X}_\beta, \boldsymbol{X}_\tau)] = \frac{n-t-b+1}{b(k-1)}\sum_{j\ell} u^2_{j\ell}$$

since the incomplete block designs considered are unbiased.

4. $E[\mathrm{SS}(\boldsymbol{X}_\tau \mid \mathfrak{I}, \boldsymbol{X}_\beta)]$ can be obtained by subtraction.

5. To obtain $E[\mathrm{SS}(\boldsymbol{X}_\tau \mid \mathfrak{I})]$ let

$$\gamma^w_{uv} = \begin{cases} 1 & \text{if the } w\text{th treatment corresponds to the} \\ & uv \text{ index} \qquad (w = 1, 2, \ldots, t) \\ 0 & \text{otherwise} \end{cases}$$

$$\gamma^w_u = \begin{cases} 1 & \text{if the } w\text{th treatment occurs in the } u\text{th block} \\ 0 & \text{otherwise} \end{cases}$$

where

$$\sum_v \gamma^w_{uv} = \gamma^w_u$$

and

$$\sum_u \gamma^w_u = r$$

Then

$$E[\mathrm{SS}(\boldsymbol{X}_\tau \mid \mathfrak{I})] = E\left[\frac{1}{r}\sum_w\left(\sum_{uv}\gamma^w_{uv}y_{uv} - \frac{1}{t}\sum_w\sum_{uv}\gamma^w_{uv}y_{uv}\right)^2\right]$$

$$= E\left[\frac{1}{r}\sum_w\left(r\tau_w + \sum_u \gamma^w_u\,\beta_u + \sum_{uv}\gamma^w_{uv}\,\omega_{uv}\right)^2\right]$$

$$= r\sum_w \tau^2_w + \frac{1}{r}\sum_w E\left(\sum_u \gamma^w_u\,\beta_u\right)^2 + \frac{1}{r}\sum_w E\left(\sum_{uv}\gamma^w_{uv}\,\omega_{uv}\right)^2$$

$$= r \sum_w \tau_w^2 + \frac{1}{r} \sum_w E \left(\sum_u \gamma_u^w \beta_u^2 + \sum_{\substack{uu' \\ u \neq u'}} \gamma_u^w \gamma_{u'}^w \beta_u \beta_{u'} \right)$$

$$+ \frac{1}{r} \sum_w E \left(\sum_{uv} \gamma_{uv}^w \omega_{uv}^2 + \sum_u \sum_{\substack{vv' \\ u \neq v'}} \gamma_{uv}^w \gamma_{uv'}^w \omega_{uv} \omega_{uv'} \right.$$

$$\left. + \sum_{\substack{uu' \\ u \neq u'}} \sum_{vv'} \gamma_{uv}^w \gamma_{u'v'}^w \omega_{uv} \omega_{u'v'} \right)$$

Now

$$E\left[\sum_u \gamma_u^w \beta_u^2\right] = r \frac{1}{b} \sum_j b_j^2$$

$$E\left[\sum_{\substack{uu' \\ u \neq u'}} \gamma_u^w \gamma_{u'}^w \beta_u \beta_{u'}\right] = - \sum_{\substack{uu' \\ u \neq u'}} \gamma_u^w \gamma_{u'}^w \frac{1}{b(b-1)} \sum_j b_j^2$$

$$= - \sum_u \gamma_u^w (r - \gamma_u^w) \frac{1}{b(b-1)} \sum_j b_j^2$$

$$= - \frac{r(r-1)}{b(b-1)} \sum_j b_j^2$$

$$E\left[\sum_{uv} \gamma_{uv}^w \omega_{uv}^2\right] = r \frac{1}{bk} \sum_{j\ell} u_{u\ell}^2$$

$$E\left[\sum_u \sum_{\substack{vv' \\ v \neq v'}} \gamma_{uv}^w \gamma_{uv'}^w \omega_{uv}^w \omega_{uv'}\right] = 0 \quad \text{since } \gamma_{uv}^w \gamma_{uv'}^w = 0$$

$$E\left[\sum_{\substack{uu' \\ u \neq u'}} \gamma_{uv}^w \gamma_{u'v'}^w \omega_{uv} \omega_{u'v'}\right] = 0$$

and hence

$$E[(X_\tau \mid \mathfrak{I}] = r \sum_w \tau_w^2 + \frac{t(b-r)}{b(b-1)} \sum_j b_j^2 + \frac{t}{bk} \sum_{j\ell} u_{j\ell}^2$$

Thus, we have for the mean squares (MS) from Tables 1.1 and 1.2 the expected values under randomization theory as given in Tables 1.5 and 1.6, respectively.

Table 1.5 E(MS) for T|B-ANOVA

Source	E(MS)
$\boldsymbol{X}_\beta \mid \mathfrak{I}$	$\frac{k}{b-1}\sum_u \overline{\tau}_{u.}^2 + \frac{k}{b-1}\sum_j b_j^2$
$\boldsymbol{X}_\tau \mid \mathfrak{I}, \boldsymbol{X}_\beta$	$\frac{1}{b(k-1)}\sum_{j\ell} u_{j\ell}^2 + \left(\sum_{uv} \tau_{uv}^2 - k\sum_u \overline{\tau}_{u.}^2\right) \Big/ (t-1)$
$\boldsymbol{I} \mid \mathfrak{I}, \boldsymbol{X}_\beta, \boldsymbol{X}_\tau$	$\frac{1}{b(k-1)}\sum_{j\ell} u_{j\ell}^2$

Table 1.6 E(MS) for B|T-ANOVA

Source	E(MS)
$\boldsymbol{X}_\tau \mid \mathfrak{I}$	$\frac{t}{bk(t-1)}\sum_{j\ell} u_{j\ell}^2 + \frac{t(b-r)}{b(b-1)(t-1)}\sum_j b_j^2 + \frac{r}{t-1}\sum_w \tau_w^2$
$\boldsymbol{X}_\beta \mid \mathfrak{I}, \boldsymbol{X}_\tau$	$\frac{t-k}{b(b-1)k(k-1)}\sum_{j\ell} u_{j\ell}^2 + \frac{bk-t}{(b-1)^2}\sum_j b_j^2$
$\boldsymbol{I} \mid \mathfrak{I}, \boldsymbol{X}_\beta, \boldsymbol{X}_\tau$	$\frac{1}{b(k-1)}\sum_{j\ell} u_{j\ell}^2$

If we define

$$\frac{1}{b(k-1)}\sum_{j\ell} u_{j\ell}^2 = \sigma_u^2$$

and

$$\frac{1}{b-1}\sum_j b_j^2 = \sigma_\beta^2$$

we can then express the expected values for the three important mean squares in ANOVA Tables 1.5 and 1.6 as

$$E\left[\text{MS}(\boldsymbol{X}_\tau \mid \mathfrak{I}, \boldsymbol{X}_\beta)\right] = \sigma_u^2 + \frac{\sum_{uv} \tau_{uv}^2 - k\sum_u \overline{\tau}_{u.}^2}{t-1} \tag{1.35}$$

$$E\left[\text{MS}(\boldsymbol{X}_\beta \mid \mathfrak{I}, \boldsymbol{X}_\tau)\right] = \frac{t-k}{(b-1)k}\sigma_u^2 + \frac{n-t}{b-1}\sigma_\beta^2 \tag{1.36}$$

$$E\left[\text{MS}(\boldsymbol{I} \mid \mathfrak{I}, \boldsymbol{X}_\beta, \boldsymbol{X}_\tau)\right] = \sigma_u^2 \tag{1.37}$$

We make the following observations:

1. The quadratic form in the τ_{uv} in (1.35) is just a different way of writing $\boldsymbol{\tau}'\mathbf{C}\boldsymbol{\tau}$ in Table 1.1. Both expressions indicate that the quadratic form depends on the particular design chosen, and both equal zero when all the treatment effects are the same.
2. It follows from (1.35) and (1.37) that, based on the equality of the $E(\text{MS})$ under $H_0 : \tau_1 = \tau_2 = \cdots = \tau_t$, the ratio

$$\text{MS}(\boldsymbol{X}_\tau \mid \mathfrak{I}, \boldsymbol{X}_\beta)/\text{MS}(\boldsymbol{I} \mid \mathfrak{I}, \boldsymbol{X}_\beta, \boldsymbol{X}_\tau) \tag{1.38}$$

   provides a test criterion for testing the above hypothesis. In fact, Ogawa (1974) has shown that the asymptotic randomization distribution of (1.38) is an F distribution with $t-1$ and $n-t-b+1$ d.f. We interpret this again to mean that the usual F test is an approximation to the randomization test based on (1.38).
3. Considering (1.36) and (1.37), there does not exist an exact test for testing the equality of block effects. This is in agreement with our discussion in Section I.9.2 concerning the asymmetry of treatment and block effects.
4. For $k = t$ and $r = b$, that is, for the RCBD, the results of Tables 1.5 and 1.6 agree with those in Table 9.1 of Section I.9.2.
5. With treatment-unit additivity in the broad sense (see Section I.6.3.3) the expressions in (1.35), (1.36), and (1.37) are changed by adding $\sigma_\nu^2 + \sigma_\eta^2$ to the right-hand sides (recall that $\sigma_u^2 + \sigma_\nu^2 + \sigma_\eta^2 \equiv \sigma_\epsilon^2 + \sigma_\eta^2 \equiv \sigma_e^2$). Remarks (2) and (3) above remain unchanged.
6. For the recovery of interblock information (to be discussed in Section 1.7), we need to estimate σ_β^2 (or a function of σ_β^2). Clearly, under the assumption of additivity in the broad sense, this cannot be done considering that

$$E\left[\text{MS}\left(\boldsymbol{X}_\beta \mid \mathfrak{I}, \boldsymbol{X}_\tau\right)\right] = \sigma_\nu^2 + \sigma_\eta^2 + \frac{t-k}{(b-1)k}\,\sigma_u^2 + \frac{n-t}{b-1}\,\sigma_\beta^2$$

   It is for this reason only that we shall resort to the approximation

$$E\left[\text{MS}\left(\boldsymbol{X}_\beta \mid \mathfrak{I}, \boldsymbol{X}_\tau\right)\right] \approx \sigma_\nu^2 + \sigma_\eta^2 + \sigma_u^2 + \frac{n-t}{b-1}\,\sigma_\beta^2 = \sigma_e^2 + \frac{n-t}{b-1}\,\sigma_\beta^2 \tag{1.39}$$

   which is the expected value based on an infinite population theory model [see (1.49) and (1.50)].

For a different approach to randomization analysis, see Calinski and Kageyama (2000).

1.7 INTERBLOCK INFORMATION IN AN INCOMPLETE BLOCK DESIGN

1.7.1 Introduction and Rationale

As mentioned earlier, Yates (1939, 1940a) has argued that for incomplete block designs comparisons among block totals (or averages) contain some information about treatment comparisons, and he referred to this as *recovery of interblock information*. The basic idea is as follows.

Consider, for purposes of illustration, the following two blocks and their observations from some design:

$$\text{Block 1:} \quad y_{51}, \quad y_{31}, \quad y_{11}$$

$$\text{Block 2:} \quad y_{22}, \quad y_{42}, \quad y_{32}$$

Let

$$B_1 = y_{51} + y_{31} + y_{11}$$

and

$$B_2 = y_{22} + y_{42} + y_{32}$$

represent the block totals. Using model (1.1) we can write

$$\begin{aligned} B_1 - B_2 = {} & (\tau_5 + \tau_2 + \tau_1) - (\tau_2 + \tau_4 + \tau_3) \\ & + 3\beta_1 - 3\beta_2 + (e_{51} + e_{31} + e_{11}) \\ & - (e_{22} + e_{42} + e_{32}) \end{aligned}$$

Assuming now that the β_j are random effects with mean zero, we find

$$E(B_1 - B_2) = \tau_1 + \tau_5 - \tau_2 - \tau_4$$

It is in this sense that block comparisons contain information about treatment comparisons. We shall now formalize this procedure.

1.7.2 Interblock Normal Equations

Consider the model equation (1.1)

$$\boldsymbol{y} = \mu \mathfrak{I} + \boldsymbol{X}_\tau \boldsymbol{\tau} + \boldsymbol{X}_\beta \boldsymbol{\beta} + \boldsymbol{e}$$

where $\boldsymbol{\beta}$ is now assumed to be a random vector with $E(\boldsymbol{\beta}) = \boldsymbol{0}$ and $\text{var}(\boldsymbol{\beta}) = \sigma_\beta^2 \boldsymbol{I}$. As pointed out above the interblock analysis is based on block totals rather than on individual observations, that is, we now consider

$$\boldsymbol{X}'_\beta \boldsymbol{y} = \begin{pmatrix} k_1 \\ k_2 \\ \vdots \\ k_b \end{pmatrix} \mu + \boldsymbol{N}'\boldsymbol{\tau} + \boldsymbol{K}\boldsymbol{\beta} + \boldsymbol{X}'_\beta \boldsymbol{e} \tag{1.40}$$

We then have

$$E(\boldsymbol{X}'_\beta \boldsymbol{y}) = \begin{pmatrix} k_1 \\ k_2 \\ \vdots \\ k_b \end{pmatrix} \mu + \boldsymbol{N}'\boldsymbol{\tau} \tag{1.41}$$

and the variance–covariance matrix under what we might call a double error structure with both $\boldsymbol{\beta}$ and $\boldsymbol{e}$ being random vectors

$$\begin{aligned} \operatorname{var}(\boldsymbol{X}'_\beta \boldsymbol{y}) &= \boldsymbol{K}^2 \sigma_\beta^2 + \boldsymbol{K}\sigma_e^2 \\ &= \boldsymbol{K}\left(\boldsymbol{I} + \frac{\sigma_\beta^2}{\sigma_e^2}\boldsymbol{K}\right)\sigma_e^2 \\ &= \boldsymbol{L}\sigma_e^2 \quad \text{(say)} \end{aligned}$$

and

$$\boldsymbol{L} = \operatorname{diag}\{\ell_j\} = \operatorname{diag}\left\{k_j\left(1 + \frac{\sigma_\beta^2}{\sigma_e^2}k_j\right)\right\} = \operatorname{diag}\left\{\left(k_j \frac{w}{w'_j}\right)\right\} = \operatorname{diag}\{k_j \rho_j\}$$

with

$$w = \frac{1}{\sigma_e^2} \tag{1.42}$$

and

$$w'_j = \frac{1}{\sigma_e^2 + k_j \sigma_\beta^2} \tag{1.43}$$

and

$$\rho_j = \frac{w}{w'_j} = \frac{\sigma_e^2 + k_j \sigma_\beta^2}{\sigma_e^2} \tag{1.44}$$

The quantities w and w'_j of (1.42) and (1.43) are referred to as *intrablock* and *interblock weights*, respectively, as w is the reciprocal of the intrablock variance,

σ_e^2, and w_j' is the reciprocal of the interblock variance, that is, $\text{var}(B_j)$ on a per observation basis, or $\text{var}(B_j/k_j) = \sigma_e^2 + k_j\sigma_\beta^2$. We then use as our "observation" vector

$$z = \boldsymbol{L}^{-1/2}\boldsymbol{X}_\beta'\boldsymbol{y} \tag{1.45}$$

which has

$$\text{var}(z) = \boldsymbol{I}\sigma_e^2$$

and hence satisfies the Gauss–Markov conditions.

If we write (1.41) as

$$E(\boldsymbol{X}_\beta'\boldsymbol{y}) = (\boldsymbol{k}\ \boldsymbol{N}')\begin{pmatrix}\mu \\ \boldsymbol{\tau}\end{pmatrix}$$

with $\boldsymbol{k} = (k_1, k_2, \ldots, k_b)'$, we have from (1.45) that

$$E(z) = \boldsymbol{L}^{-1/2}(\boldsymbol{k}\ \boldsymbol{N}')\begin{pmatrix}\mu \\ \boldsymbol{\tau}\end{pmatrix}$$

The resulting NE, which we shall refer to as the *interblock* NE, is then given by

$$\begin{pmatrix}\boldsymbol{k}' \\ \boldsymbol{N}\end{pmatrix}\boldsymbol{L}^{-1}(\boldsymbol{k}\ \boldsymbol{N}')\begin{pmatrix}\mu^* \\ \boldsymbol{\tau}^*\end{pmatrix} = \begin{pmatrix}\boldsymbol{k}' \\ \boldsymbol{N}\end{pmatrix}\boldsymbol{L}^{-1}\boldsymbol{X}_\beta'\boldsymbol{y} \tag{1.46}$$

or explicitly as

$$\begin{bmatrix} \sum_j \frac{k_j^2}{\ell_j} & \sum_j n_{1j}\frac{k_j}{\ell_j} & \cdots & \sum_j n_{tj}\frac{k_j}{\ell_j} \\ \sum_j n_{1j}\frac{k_j}{\ell_j} & \sum_j \frac{n_{1j}^2}{\ell_j} & \cdots & \sum_j \frac{n_{1j}n_{tj}}{\ell_j} \\ \vdots & \vdots & & \vdots \\ \sum_j n_{tj}\frac{k_j}{\ell_j} & \sum_j \frac{n_{1j}n_{tj}}{\ell_j} & \cdots & \sum_j \frac{n_{tj}^2}{\ell_j} \end{bmatrix} \begin{bmatrix}\mu^* \\ \tau_1^* \\ \vdots \\ \tau_t^*\end{bmatrix} = \begin{bmatrix} \sum_j \frac{k_j}{\ell_j}B_j \\ \sum_j \frac{n_{1j}}{\ell_j}B_j \\ \vdots \\ \sum_j \frac{n_{tj}}{\ell_j}B_j \end{bmatrix} \tag{1.47}$$

It can be seen easily that the rank of the coefficient matrix in (1.47) is t. To solve the interblock NE, we take $\mu^* = 0$ and hence reduce the set to the following t

equations in $\tau_1^*, \tau_2^*, \ldots, \tau_t^*$ where we have used the fact that $\ell_j = k_j\rho_j$:

$$\begin{bmatrix} \sum_j n_{1j}^2 \frac{\rho_j^{-1}}{k_j} & \sum_j n_{1j}n_{2j}\frac{\rho_j^{-1}}{k_j} & \cdots & \sum_j n_{1j}n_{tj}\frac{\rho_j^{-1}}{k_j} \\ \sum n_{2j}n_{1j}\frac{\rho_j^{-1}}{k_j} & \sum n_{2j}^2\frac{\rho_j^{-1}}{k_j} & \cdots & \sum n_{2j}n_{tj}\frac{\rho_j^{-1}}{k_j} \\ \vdots & \vdots & & \vdots \\ \sum n_{tj}n_{1j}\frac{\rho_j^{-1}}{k_j} & \sum n_{tj}n_{2j}\frac{\rho_j^{-1}}{k_j} & \cdots & \sum n_{tj}^2\frac{\rho_j^{-1}}{k_j} \end{bmatrix} \begin{bmatrix} \tau_1^* \\ \tau_2^* \\ \vdots \\ \tau_t^* \end{bmatrix}$$

$$= \begin{bmatrix} \sum_j \frac{\rho_j^{-1}}{k_j} n_{1j}B_j \\ \sum_j \frac{\rho_j^{-1}}{k_j} n_{2j}B_j \\ \vdots \\ \sum_j \frac{\rho_j^{-1}}{k_j} n_{tj}B_j \end{bmatrix} \tag{1.48}$$

The solution to the equations (1.48) is referred to as the *interblock information* about the treatment effects, with

$$E(\tau_i^*) = \mu + \tau_i + \text{const.} \cdot \sum_{i'=1}^{t} (\mu + \tau_{i'})$$

Hence

$$E\left(\sum_i c_i \tau_i^*\right) = \sum_i c_i \tau_i \quad \text{for} \quad \sum c_i = 0$$

We note here that typically (see Kempthorne, 1952) the interblock analysis is derived not in terms of the "observations" z [as given in (1.45)] but rather in terms of the block totals $\boldsymbol{X}'_\beta \boldsymbol{y}$. The resulting NE are then simply obtained by using $\boldsymbol{L} = \boldsymbol{I}$ in (1.46) and subsequent equations. The reason why we prefer our description is the fact that then the intra- and interblock information can be combined additively to obtain the so-called combined analysis (see Section 1.8) rather than in the form of a weighted average (see Kempthorne, 1952).

1.7.3 Nonavailability of Interblock Information

We conclude this section with the following obvious remarks:

1. For the special case $k_j = t$ for all $j, r_i = b$ for all i and all $n_{ij} = 1$, we have, of course, the RCBD. Then the elements in the coefficient matrix of (1.48) are all identical, and so are the right-hand sides. Consequently, (1.48) reduces to a single equation

$$b \sum_i \tau_i^* = \sum_j B_j$$

and no contrasts among the τ_i are estimable. Expressed differently, any contrast among block totals estimates zero, that is, no interblock information is available.

2. For a design with $b < t$ (and such incomplete block designs exist as we shall see in Chapter 4; see also the example in Section 1.7.1), the rank of the coefficient matrix of (1.48), $\boldsymbol{N}\boldsymbol{L}^{-1}\boldsymbol{N}'$, is less than t. Hence not all $\mu + \tau_i$ are estimable using block totals, which means that interblock information is not available for all treatment contrasts.

1.8 COMBINED INTRA- AND INTERBLOCK ANALYSIS

1.8.1 Combining Intra- and Interblock Information

The two types of information about treatment effects that we derived in Sections 1.3.2 and 1.7.2 can be combined to yield the "best" information about estimable functions of treatment effects. All we need to do is to add the coefficient matrices from the intrablock RNE (1.8) and the interblock NE (1.48) and do the same for the corresponding right-hand sides. This will lead to a system of equations in $\tau_1^{**}, \tau_2^{**}, \ldots, \tau_t^{**}$, say, and the solution to these equations will lead to the combined intra- and interblock estimators for treatment contrasts.

In the following section we shall derive the equations mentioned above more directly using the method of generalized least squares, that is, by using the Aitken equations described in Section I.4.16.2

1.8.2 Linear Model

In order to exhibit the double error structure that characterizes the underlying assumptions for the combined analysis, we rewrite model (1.1) as

$$y_{j\ell} = \mu + \tau_{j\ell} + \beta_j + e_{j\ell} \tag{1.49}$$

where $j = 1, 2, \ldots, b$; $\ell = 1, 2, \ldots, k_j$; $\tau_{j\ell}$ denotes the effect of the treatment applied to the ℓth experimental unit in the jth block, the β_j are assumed to be i.i.d. random variables with $E(\beta_j) = 0$, $\text{var}(\beta_j) = \sigma_\beta^2$, and the $e_{j\ell}$ are i.i.d.

random variables with $E(e_{j\ell}) = 0$, $\text{var}(e_{j\ell}) = \sigma_e^2$. We then have

$$E(y_{j\ell}) = \mu + \tau_{j\ell} \tag{1.50}$$

and

$$\text{cov}\left(y_{j\ell}, y_{j'\ell'}\right) = \begin{cases} \sigma_\beta^2 + \sigma_e^2 & \text{for } j = j', \ell = \ell' \\ \sigma_\beta^2 & \text{for } j = j', \ell \neq \ell' \\ 0 & \text{otherwise} \end{cases} \tag{1.51}$$

To use matrix notation it is useful to arrange the observations according to blocks, that is, write the observation vector as

$$\boldsymbol{y} = \left(y_{11}, y_{12}, \ldots, y_{1k_1}, y_{21}, y_{22}, \ldots, y_{b1}, y_{b2}, \ldots, y_{bk_b}\right)'$$

Letting

$$\boldsymbol{X} = (\mathfrak{I} \quad \boldsymbol{X}_\tau)$$

we rewrite (1.50) as

$$E(\boldsymbol{y}) = \boldsymbol{X}\begin{pmatrix} \mu \\ \boldsymbol{\tau} \end{pmatrix} \tag{1.52}$$

and the variance–covariance (1.51) as

$$\text{var}(\boldsymbol{y}) = \begin{bmatrix} \boldsymbol{V}_1 & & & \\ & \boldsymbol{V}_2 & & \boldsymbol{0} \\ & & \ddots & \\ \boldsymbol{0} & & & \boldsymbol{V}_b \end{bmatrix} \sigma_e^2 \equiv \boldsymbol{V}\sigma_e^2 \tag{1.53}$$

where $\boldsymbol{V}_j$ is given by

$$\boldsymbol{V}_j = \boldsymbol{I}_{k_j} + \frac{\sigma_\beta^2}{\sigma_e^2} \mathfrak{I}_{k_j} \mathfrak{I}'_{k_j} \tag{1.54}$$

1.8.3 Normal Equations

Applying now the principles of least squares to the model (1.52) with covariance structure (1.53) yields the Aitken equations (see Section I.4.16):

$$(\boldsymbol{X}'\boldsymbol{V}^{-1}\boldsymbol{X})\begin{pmatrix} \widehat{\widehat{\mu}} \\ \widehat{\widehat{\tau}} \end{pmatrix} = \boldsymbol{X}'\boldsymbol{V}^{-1}\boldsymbol{y} \tag{1.55}$$

where

$$\boldsymbol{V}^{-1} = \text{diag}\left(\boldsymbol{V}_1^{-1}, \boldsymbol{V}_2^{-1}, \ldots, \boldsymbol{V}_b^{-1}\right)$$

and

$$\boldsymbol{V}_j^{-1} = \boldsymbol{I}_{k_j} - \frac{\sigma_\beta^2}{\sigma_e^2 + k_j \sigma_\beta^2} \; \mathfrak{I}_{k_j} \mathfrak{I}'_{k_j} \tag{1.56}$$

With

$$\frac{1}{\sigma_e^2} = w \qquad \frac{1}{\sigma_e^2 + k_j \sigma_\beta^2} = w'_j$$

and

$$\frac{w'_j}{w} = \rho_j^{-1}$$

Eq. (1.56) can be written as

$$\boldsymbol{V}_j^{-1} = \boldsymbol{I}_{k_j} - \frac{1 - \rho_j^{-1}}{k_j} \; \mathfrak{I}_{k_j} \mathfrak{I}'_{k_j}$$

and hence

$$\boldsymbol{V}^{-1} = \boldsymbol{I}_n - \operatorname{diag}\left(\frac{1 - \rho_1^{-1}}{k_1} \; \mathfrak{I}_{k_1} \mathfrak{I}'_{k_1}, \frac{1 - \rho_2^{-1}}{k_2} \; \mathfrak{I}_{k_2} \mathfrak{I}'_{k_2}, \ldots, \frac{1 - \rho_b^{-1}}{k_b} \; \mathfrak{I}_{k_b} \mathfrak{I}'_{k_b} \right) \tag{1.57}$$

Further, if we let $\left(1 - p_j^{-1}\right)/k_j = \delta_j (j = 1, 2, \ldots, b)$, we have

$$\boldsymbol{X}'\boldsymbol{V}^{-1} = \boldsymbol{X}' - \begin{bmatrix} \left(1 - \rho_1^{-1}\right) \mathfrak{I}'_{k_1} & \left(1 - \rho_2^{-1}\right) \mathfrak{I}'_{k_2} & \cdots & \left(1 - \rho_b^{-1}\right) \mathfrak{I}'_{k_b} & \\ \delta_1 n_{11} \mathfrak{I}'_{k_1} & \delta_2 n_{12} \mathfrak{I}'_{k_2} & \cdots & \delta_b n_{1b} \mathfrak{I}'_{k_b} & \\ \cdots & \cdots & \cdots & \cdots & \cdots \\ \delta_1 n_{t1} \mathfrak{I}'_{k_1} & \delta_2 n_{t2} \mathfrak{I}'_{k_2} & \cdots & \delta_b n_{tb} \mathfrak{I}'_{k_b} & \end{bmatrix}$$

and

$$\boldsymbol{X}'\boldsymbol{V}^{-1}\boldsymbol{X} = \begin{bmatrix} \sum k_j \rho_j^{-1} & \sum n_{1j} \rho_j^{-1} & \sum n_{2j} \rho_j^{-1} & \cdots & \sum n_{tj} \rho_j^{-1} \\ \sum n_{1j} \rho_j^{-1} & r_1 - \sum \delta_j n_{1j}^2 & -\sum \delta_j n_{1j} n_{2j} & \cdots & -\sum \delta_j n_{1j} n_{tj} \\ \cdots & \cdots & \cdots & \cdots & \cdots \\ \sum n_{tj} \rho_j^{-1} & -\sum \delta_j n_{1j} n_{tj} & -\sum \delta_j n_{2j} n_{tj} & \cdots & r_t - \sum \delta_j n_{tj}^2 \end{bmatrix} \tag{1.58}$$

$$
X'V^{-1}y = \begin{bmatrix} \sum \rho_j^{-1} B_j \\ T_1 - \sum \delta_j n_{1j} B_j \\ \vdots \\ T_t - \sum \delta_j N_{tj} B_j \end{bmatrix} \tag{1.59}
$$

By inspection one can verify that in the coefficient matrix (1.58) the elements in rows 2 to $t+1$ add up to the elements in row 1, which shows that (1.55) is not of full rank; in fact, it is of rank t. The easiest way to solve these equations then is to impose the condition $\widehat{\widehat{\mu}} = 0$. This means that we eliminate the first row and first column from (1.58) and the first element in (1.59) and solve the resulting system of t equations in the t unknowns $\widehat{\widehat{\tau}}_1, \widehat{\widehat{\tau}}_2, \ldots, \widehat{\widehat{\tau}}_t$. If we define $S = \text{diag}(\rho_j)$, then this system of equations resulting from (1.58) and (1.59) can be written as

$$
\left[R - NK^{-1}\left(I - S^{-1}\right)\right]\widehat{\widehat{\tau}} = T - NK^{-1}\left(I - S^{-1}\right)B \tag{1.60}
$$

which we write for short as

$$
A\widehat{\widehat{\tau}} = P \tag{1.61}
$$

with A and P as described above. The solution then is

$$
\widehat{\widehat{\tau}} = A^{-1}P \tag{1.62}
$$

or, in terms of a generalized inverse for the original set of NE (1.55)

$$
\begin{bmatrix} \widehat{\widehat{\mu}} \\ \widehat{\widehat{\tau}} \end{bmatrix} = \begin{bmatrix} 0 & \mathbf{0}' \\ \mathbf{0} & A^{-1} \end{bmatrix} X'V^{-1}y \tag{1.63}
$$

with

$$
E\left(\widehat{\widehat{\tau}}_i\right) = \mu + \tau_i \qquad (i = 1, 2, \ldots, t)
$$

If we denote the (i, i') element of A^{-1} by $a^{ii'}$, then

$$
\text{var}\left(\widehat{\widehat{\tau_i - \tau_{i'}}}\right) = \text{var}\left(\widehat{\widehat{\tau}}_i - \widehat{\widehat{\tau}}_{i'}\right) = \left(a^{ii} + a^{i'i'} - 2a^{ii'}\right)\sigma_e^2 \tag{1.64}
$$

More generally, the treatment contrast $\mathbf{c}'\boldsymbol{\tau}$ is estimated by $\mathbf{c}'\widehat{\widehat{\boldsymbol{\tau}}}$ with variance

$$
\text{var}\left(c'\widehat{\widehat{\tau}}\right) = c'A^{-1}c\sigma_e^2 \tag{1.65}
$$

Expression (1.65) looks deceptively simple, but the reader should keep in mind that the elements of $\boldsymbol{A}^{-1}$ depend on σ_β^2 and σ_e^2. We shall return to estimating (1.65) in Section 1.10.

Finally, we note that the equations (1.60) show a striking similarity to the intrablock NE (1.7), except that the system (1.60) is of full rank and the elements of its coefficient matrix depend on the unknown parameters σ_β^2 and σ_e^2.

1.8.4 Some Special Cases

As a special case of the above derivations we mention explicitly the equireplicate, proper design, that is, the design with all $r_i = r$ and all $k_j = k$. We then define

$$w' = \frac{1}{\sigma_e^2 + k\sigma_\beta^2} \tag{1.66}$$

and

$$\rho = \frac{w}{w'} \tag{1.67}$$

and write (1.60) as

$$\left[r\boldsymbol{I} - \frac{1}{k}\left(1 - \rho^{-1}\right)\boldsymbol{N}\boldsymbol{N}'\right]\widehat{\widehat{\boldsymbol{\tau}}} = \boldsymbol{T} - \frac{1}{k}\left(1 - \rho^{-1}\right)\boldsymbol{N}\boldsymbol{B} \tag{1.68}$$

We shall comment briefly on the set of equations (1.68) for two special cases:

1. If $\rho^{-1} = 0$, that is, $\sigma_\beta^2 = \infty$, then (1.68) reduces to the NE (1.7) for the intrablock analysis. This means, of course, that in this case no interblock information is available. This is entirely plausible and suggests further that for "large" σ_β^2 the interblock information is very weak and perhaps not worthwhile considering.
2. If $\rho^{-1} = 1$, that is, $\sigma_\beta^2 = 0$, the solution to (1.68) is the same as that obtained for the completely randomized design (CRD) with the restriction $\hat{\mu} = 0$. This, of course, is a formal statement and should not imply that in this case the observations should be analyzed as if a CRD had been used.

1.9 RELATIONSHIPS AMONG INTRABLOCK, INTERBLOCK, AND COMBINED ESTIMATION

On several occasions we have pointed out that there exist certain relationships among the different types of analysis for incomplete block designs. It is worthwhile to exposit this in a little detail.

1.9.1 General Case

For the full model

$$y = \mathfrak{I}_n\mu + X_\tau\tau + X_\beta\beta + e$$

with the double error structure we have

$$E(y) = \mathfrak{I}\mu + X_\tau\tau$$

and

$$\begin{aligned}\text{var}(y) &= X_\beta X'_\beta\sigma_\beta^2 + I\sigma_e^2\\ &= (I + \gamma X_\beta X'_\beta)\sigma_e^2\\ &\equiv V\sigma_e^2\end{aligned}$$

as in (1.53) with $\gamma = \sigma_\beta^2/\sigma_e^2$. Then, as explained in Section 1.8, the estimators of estimable functions are obtained from the Aitken equations by minimizing

$$(y - \mathfrak{I}\mu - X_\tau\tau)'V^{-1}(y - \mathfrak{I}\mu - X_\tau\tau) \tag{1.69}$$

with respect to μ and τ. To simplify the algebra, let $\Theta = \mathfrak{I}\mu + X_\tau\tau$ and then write $\psi = y - \Theta$. Expression (1.69) can now be written simply as $\psi'V^{-1}\psi$. We then write

$$\psi = P\psi + (I - P)\psi$$

where, again for brevity, we write

$$\begin{aligned}P = P_{x_\beta} &= X_\beta(X'_\beta X_\beta)^{-1}X'_\beta\\ &= X_\beta K^{-1}X'_\beta\end{aligned} \tag{1.70}$$

Then

$$\psi'V^{-1}\psi = [\psi'P + \psi'(I - P)]V^{-1}[P\psi + (I - P)\psi] \tag{1.71}$$

In order to expand the right-hand side of (1.71) we make use of the following results. From

$$(I - P)V = (I - P)(I + \gamma X_\beta X'_\beta) = (I - P)$$

[using (1.70)] it follows that

$$(I - P)V^{-1} = (I - P)$$

and hence

$$(\boldsymbol{I} - \boldsymbol{P})\boldsymbol{V}^{-1}\boldsymbol{P} = (\boldsymbol{I} - \boldsymbol{P})\boldsymbol{P} = \boldsymbol{0}$$

Thus, (1.71) reduces to

$$\boldsymbol{\psi}'\boldsymbol{V}^{-1}\boldsymbol{\psi} = \boldsymbol{\psi}'\boldsymbol{P}\boldsymbol{V}^{-1}\boldsymbol{P}\boldsymbol{\psi} + \boldsymbol{\psi}'(\boldsymbol{I} - \boldsymbol{P})\boldsymbol{\psi} \tag{1.72}$$

To handle the term $\boldsymbol{\psi}'\boldsymbol{P}\boldsymbol{V}^{-1}\boldsymbol{P}\boldsymbol{\psi}$ we note that $\boldsymbol{V}^{-1}$ in (1.57) can be rewritten as

$$\boldsymbol{V}^{-1} = \boldsymbol{I} - \gamma \boldsymbol{X}_\beta(\boldsymbol{I} + \gamma \boldsymbol{X}'_\beta \boldsymbol{X}_\beta)^{-1}\boldsymbol{X}'_\beta$$

so that

$$\begin{aligned}
\boldsymbol{P}\boldsymbol{V}^{-1}\boldsymbol{P} &= \boldsymbol{P} - \gamma \boldsymbol{X}_\beta(\boldsymbol{I} + \gamma \boldsymbol{X}'_\beta \boldsymbol{X}_\beta)^{-1}\boldsymbol{X}'_\beta \\
&= \boldsymbol{X}_\beta \left[\boldsymbol{K}^{-1} - \gamma(\boldsymbol{I} + \gamma \boldsymbol{K})^{-1}\right] \boldsymbol{X}'_\beta \\
&= \boldsymbol{X}_\beta \operatorname{diag}\left[\frac{1}{k_j(1 + \gamma k_j)}\right] \boldsymbol{X}'_\beta \\
&= \boldsymbol{X}_\beta(\boldsymbol{K} + \gamma \boldsymbol{K}^2)^{-1}\boldsymbol{X}'_\beta
\end{aligned}$$

Hence

$$\boldsymbol{\psi}'\boldsymbol{P}\boldsymbol{V}^{-1}\boldsymbol{P}\boldsymbol{\psi} = \left(\boldsymbol{X}'_\beta\boldsymbol{\psi}\right)' \left(\boldsymbol{K} + \gamma \boldsymbol{K}^2\right)^{-1} \left(\boldsymbol{X}'_\beta\boldsymbol{\psi}\right) \tag{1.73}$$

Then we note that $\boldsymbol{X}'_\beta\boldsymbol{\psi}$ is the vector of block totals of the vector $\boldsymbol{\psi}$. Since $\boldsymbol{\psi} = \boldsymbol{y} - \boldsymbol{\Theta}$, we have

$$\begin{aligned}
\operatorname{var}\left(\boldsymbol{X}'_\beta\boldsymbol{\psi}\right) &= \operatorname{var}\left(\boldsymbol{X}'_\beta\boldsymbol{y}\right) \\
&= \left(\boldsymbol{X}'_\beta\boldsymbol{X}_\beta\right)^2 \sigma_\beta^2 + \boldsymbol{X}'_\beta\boldsymbol{X}_\beta\sigma_e^2 \\
&= \sigma_e^2\left(\boldsymbol{K} + \gamma \boldsymbol{K}^2\right)
\end{aligned}$$

Hence $\boldsymbol{\psi}'\boldsymbol{P}\boldsymbol{V}^{-1}\boldsymbol{P}\boldsymbol{\psi}$ is equal to

$$Q_2 \equiv [\boldsymbol{B} - E(\boldsymbol{B})]'\left(\boldsymbol{K} + \gamma \boldsymbol{K}^2\right)^{-1} [\boldsymbol{B} - E(\boldsymbol{B})] \tag{1.74}$$

with $\boldsymbol{B} = \boldsymbol{X}'_\beta \boldsymbol{y}$ being the vector of block totals.

The second expression in (1.72) is actually the quadratic form that needs to be minimized to obtain the RNE for $\boldsymbol{\Theta}$ (see Section I.4.7.1), and this is equivalent to minimizing

$$Q_1 \equiv (\boldsymbol{y} - \boldsymbol{X}_\tau \boldsymbol{\tau})'(\boldsymbol{I} - \boldsymbol{P})(\boldsymbol{y} - \boldsymbol{X}_\tau \boldsymbol{\tau}) \tag{1.75}$$

We thus summarize: To obtain the intrablock estimator, we minimize Q_1; to obtain the interblock estimator, we minimize Q_2; and to obtain the combined estimator, we minimize $Q_1 + Q_2$.

1.9.2 Case of Proper, Equireplicate Designs

We now consider the case with $\boldsymbol{R} = r\boldsymbol{I}$ and $\boldsymbol{K} = k\boldsymbol{I}$. The intrablock NE are [see (1.7)]

$$\left(r\boldsymbol{I} - \frac{1}{k}\boldsymbol{N}\boldsymbol{N}'\right)\hat{\boldsymbol{\tau}} = \boldsymbol{Q}$$

For the interblock observational equations

$$\boldsymbol{B} = k\mathfrak{I}\mu + \boldsymbol{N}'\boldsymbol{\tau} + \text{error}$$

or, absorbing μ into $\boldsymbol{\tau}$, that is, replacing $\mathfrak{I}\mu + \boldsymbol{\tau}$ by $\boldsymbol{\tau}$,

$$\boldsymbol{B} = \boldsymbol{N}'\boldsymbol{\tau} + \text{error}$$

we have the interblock NE [see (1.47)]

$$\boldsymbol{N}\boldsymbol{N}'\boldsymbol{\tau}^* = \boldsymbol{N}\boldsymbol{B}$$

As we have pointed out earlier (see Section 1.8) and as is obvious from (1.72), the combined NE are

$$\left[\left(r\boldsymbol{I} - \frac{1}{k}\boldsymbol{N}\boldsymbol{N}'\right) + \frac{1}{k(1+\gamma k)}\boldsymbol{N}\boldsymbol{N}'\right]\hat{\hat{\boldsymbol{\tau}}} = \boldsymbol{Q} + \frac{1}{k(1+\gamma k)}\boldsymbol{N}\boldsymbol{B} \tag{1.76}$$

The form of (1.76) shows again two things:

1. The intrablock and interblock NE are related.
2. The matrix $\boldsymbol{N}\boldsymbol{N}'$ determines the nature of the estimators, both intrablock and interblock.

The properties of $\boldsymbol{NN}'$ can be exploited to make some statements about estimable functions of treatment effects. Being real symmetric, $\boldsymbol{NN}'$ is orthogonally diagonalizable. We know that

$$\left(r\boldsymbol{I} - \frac{1}{k}\boldsymbol{NN}'\right) \mathfrak{J} = \mathbf{0}$$

or

$$\boldsymbol{NN}'\mathfrak{J} = rk\mathfrak{J}$$

so that one root of $\boldsymbol{NN}'$ is $rk \equiv \delta_t$, say, with associated eigenvector $(1/\sqrt{t})\ \mathfrak{J} \equiv \boldsymbol{\xi}_t$, say. Suppose $\boldsymbol{\xi}_1, \boldsymbol{\xi}_2, \ldots, \boldsymbol{\xi}_{t-1}$ complete the full set of orthonormal eigenvectors with associated eigenvalues $\delta_1, \delta_2, \ldots, \delta_{t-1}$. Then with

$$\boldsymbol{O} = (\boldsymbol{\xi}_1, \boldsymbol{\xi}_2, \ldots, \boldsymbol{\xi}_t)$$

and

$$\boldsymbol{NN}'\boldsymbol{\xi}_i = \delta_i \boldsymbol{\xi}_i$$

the NE are equivalent to

$$\boldsymbol{O}'\left(r\boldsymbol{I} - \frac{1}{k}\boldsymbol{NN}'\right)\boldsymbol{OO}'\,\widehat{\boldsymbol{\tau}} = \boldsymbol{O}'\boldsymbol{Q} \tag{1.77}$$

and

$$\boldsymbol{O}'\boldsymbol{NN}'\boldsymbol{OO}'\boldsymbol{\tau}^* = \boldsymbol{O}'\boldsymbol{NB} \tag{1.78}$$

respectively. If we write

$$\boldsymbol{\nu} = \begin{bmatrix} \nu_1 \\ \nu_2 \\ \vdots \\ \nu_t \end{bmatrix} = \boldsymbol{O}'\boldsymbol{\tau}$$

then Eq. (1.77) and (1.78) reduce to

$$\left(r - \frac{\delta_i}{k}\right)\widehat{\nu_i} = a_i$$

and

$$\delta_i \nu_i^* = b_i$$

respectively, where

$$\boldsymbol{a} = (a_1, a_2, \ldots, a_t)' = \boldsymbol{O}'\boldsymbol{Q}$$

and

$$\boldsymbol{b} = (b_1, b_2, \ldots, b_t)' = \boldsymbol{O}'\boldsymbol{N}\boldsymbol{B}$$

We see then that we have both intrablock and interblock estimators of ν_i if δ_i is not equal to 0 or to rk. For the component ν_t, only the interblock estimator exists. If other roots are equal to rk, then the intrablock estimators for the corresponding treatment parameters do not exist. Similarly, if other roots are equal to zero, then the interblock estimators for the corresponding treatment parameters do not exist. The treatment parameters $\nu_1, \nu_2, \ldots, \nu_{t-1}$ are necessarily treatment contrasts. If the design is connected, then no $\delta_i (i = 1, 2, \ldots, t-1)$ will equal rk.

The combined NE (1.76) are now transformed to

$$\left[\left(r - \frac{\delta_i}{k}\right) + \frac{1}{k(1+\gamma k)}\,\delta_i\right]\widehat{\widehat{\nu}}_i = a_i + \frac{b_i}{k(1+\gamma k)} \qquad (i = 1, 2, \ldots, t) \quad (1.79)$$

We know that

$$\operatorname{var}(\boldsymbol{Q}) = \left(r\boldsymbol{I} - \frac{1}{k}\boldsymbol{N}\boldsymbol{N}'\right)\sigma_e^2$$

and

$$\begin{aligned}\operatorname{var}(\boldsymbol{N}\boldsymbol{B}) &= \boldsymbol{N}\ \operatorname{var}(\boldsymbol{B})\boldsymbol{N}' \\ &= \boldsymbol{N}\ \frac{k}{\sigma_e^2}(1+\gamma k)\boldsymbol{I}\boldsymbol{N}'\end{aligned}$$

Hence

$$\operatorname{var}(\boldsymbol{a}) = \sigma_e^2\ \operatorname{diag}\left(r - \frac{\delta_i}{k}\right)$$

and

$$\operatorname{var}(\boldsymbol{b}) = \sigma_e^2\ k(1+\gamma k)\operatorname{diag}(\delta_i)$$

So we see from (1.79) that combined estimation of the parameter vector $\boldsymbol{\nu}$ consists of combining intrablock and interblock estimators of components of $\boldsymbol{\nu}$, weighting inversely as their variances.

1.10 ESTIMATION OF WEIGHTS FOR THE COMBINED ANALYSIS

The estimator for the treatment effects as given by (1.61) depends on the weights w and w'_j as can be seen from (1.58) and (1.59). If these weights were known, or alternatively as is apparent from (1.68), if the ratios of the interblock variance and the intrablock variance,

$$\rho_j = \frac{w}{w'_j} = \frac{\sigma_e^2 + k_j\sigma_\beta^2}{\sigma_e^2}$$

were known, then the solution (1.63) to the NE (1.55) would lead to best linear unbiased estimators for estimable functions of treatment effects. Usually, however, these parameters are not known and have to be estimated from the data. If the estimators are used instead of the unknown parameters, then the solutions to the normal equations (1.55) lose some of their good properties. It is for this reason that the properties of the combined estimator have to be examined critically, in particular with regard to their dependence on the type of estimator for the ρ_j's, and with regard to the question of how the combined estimator compares with the intrablock estimator. Before we discuss these questions in some more detail, we shall outline the "classical" procedure for estimating the ρ_j. Since this method was proposed first by Yates (1940a) we shall refer to it as the *Yates procedure* or to the estimators as the *Yates estimators*.

1.10.1 Yates Procedure

One way to estimate w and w'_j is to first estimate σ_e^2 and σ_β^2 and then use these estimators to estimate w and w'_j. If the estimators are denoted by $\widehat{\sigma}_e^2$ and $\widehat{\sigma}_\beta^2$, respectively, then we estimate w and w'_j as

$$\widehat{w} = \frac{1}{\widehat{\sigma}_e^2} \qquad \widehat{w}'_j = \frac{1}{\widehat{\sigma}_e^2 + k_j\widehat{\sigma}_\beta^2} \qquad (j = 1, 2, \ldots, b) \tag{1.80}$$

Obviously, from Table 1.1

$$\widehat{\sigma}_e^2 = \text{MS}(\boldsymbol{I}|\mathfrak{I}, \boldsymbol{X}_\beta, \boldsymbol{X}_{\boldsymbol{\tau}}) \tag{1.81}$$

To estimate σ_β^2 we turn to Table 1.2. Under model (1.50) with covariance structure (1.51) we find [see also (1.39)]

$$E[\text{SS}(\boldsymbol{X}_\beta|\mathfrak{I}, \boldsymbol{X}_\tau)] = (b-1)\sigma_e^2 + \left(n - \sum_{ij}\frac{1}{r_i}n_{ij}^2\right)\sigma_\beta^2 \tag{1.82}$$

Hence it follows from (1.81) and (1.82) that

$$\widehat{\sigma}_\beta^2 = \frac{b-1}{n - \sum \frac{1}{r_i}n_{ij}^2}\left[\text{SS}(\boldsymbol{X}_\beta|\mathfrak{I}, \boldsymbol{X}_\tau) - \text{SS}(\boldsymbol{I}|\mathfrak{I}, \boldsymbol{X}_\beta, \boldsymbol{X}_\tau)\right] \tag{1.83}$$

The estimators (1.81) and (1.83) are then substituted into (1.80) to obtain $\widehat{w}$ and $\widehat{w}_j (j = 1, 2, \ldots, b)$. If in (1.83) $\widehat{\sigma}_\beta^2 \leq 0$ for a given data set, we take

$$\widehat{w}'_j = \widehat{w} = \frac{1}{\text{MS}(\boldsymbol{I} \mid \mathfrak{I}, \boldsymbol{X}_\beta, \boldsymbol{X}_\tau)}$$

In either case $\widehat{w}$ and $\widehat{w}'_j$ are substituted into (1.58) and (1.59) and hence into the solution (1.61). Also, $\text{var}\left(\widehat{\widehat{\tau}}_i - \widehat{\widehat{\tau}}_{i'}\right)$ is estimated by substituting $\widehat{w}, \widehat{w}'_j$ and $\widehat{\sigma}_e^2$ into (1.62) and (1.64).

For alternative estimation procedures see Section 1.11, and for a numerical example see Section 1.14.3.

1.10.2 Properties of Combined Estimators

As we have already pointed out, the fact that the unknown parameters in (1.61) are replaced by their estimators will have an effect on the properties of the estimators for treatment effects. The two properties we are concerned about here are unbiasedness and minimum variance.

Let $\boldsymbol{c}'\boldsymbol{\tau}$ be an estimable function of the treatment effects, let $t = \boldsymbol{c}'\widehat{\boldsymbol{\tau}}$ be its intrablock estimator, $t(\rho) = \boldsymbol{c}'\widehat{\widehat{\boldsymbol{\tau}}}$ its combined (Aitken) estimator with ρ known (for the present discussion we shall confine ourselves to proper designs), and $t(\widehat{\rho}) = \boldsymbol{c}'\widetilde{\boldsymbol{\tau}}$ the combined estimator when in (1.68) ρ is replaced by $\widehat{\rho} = \widehat{w}/\widehat{w}'$.

Roy and Shah (1962) have shown that for the general incomplete block design, although the Yates procedure leads to a biased estimator for ρ, the estimators for treatment contrasts obtained by the method just described are unbiased, that is,

$$E\left[t(\widehat{\rho})\right] = \boldsymbol{c}'\boldsymbol{\tau}$$

With regard to $\text{var}[t(\widehat{\rho})]$, it is clear that due to sampling fluctuations of $\widehat{\rho}$ we have

$$\text{var}[t(\rho)] < \text{var}[t(\widehat{\rho})]$$

that is, the combined estimators no longer have minimum variance. The crucial question in this context, however, is: When is $\text{var}[t(\widehat{\rho})] < \text{var}(t)$? In other words: When is the combined estimator more efficient than the intrablock estimator?

The answer to this question depends on several things such as (1) the true value of ρ, (2) the type of estimator for ρ, and (3) the number of treatments and blocks. It is therefore not surprising that so far no complete answer has been given.

The most general result for the Yates estimator (and a somewhat larger class of estimators) is that of Shah (1964) based upon some general results by Roy and Shah (1962) [see also Bhattacharya (1998) for proper designs]. It is shown there that the combined estimator for any treatment contrast in any (proper) incomplete block design has variance smaller than that of the corresponding intrablock estimator if ρ does not exceed 2, or, equivalently, if $\sigma_\beta^2 \leq \sigma_e^2/k$. This is a condition that, if blocking is effective at all, one would not in general expect to be satisfied. The problem therefore remains to find methods of constructing estimators for ρ such that the combined estimator for treatment contrasts is uniformly better than the corresponding intrablock estimator, in the sense of having smaller variances for all values of ρ. For certain incomplete block designs this goal has been achieved. We shall mention these results in the following chapters.

The only general advice that we give at this point in conjunction with the use of the Yates estimator is the somewhat imprecise advice to use the intrablock rather than the combined estimator if the number of treatments is "small." The reason for this is that in such a situation the degrees of freedom for $\mathrm{MS}(\boldsymbol{I}|\mathfrak{I}, \boldsymbol{X}_\beta, \boldsymbol{X}_\tau)$ and $\mathrm{MS}(\boldsymbol{X}_\beta|\mathfrak{I}, \boldsymbol{X}_\tau)$ are likely to be small also, which would imply that σ_e^2 and σ_β^2, and hence ρ, cannot be estimated very precisely.

1.11 MAXIMUM-LIKELIHOOD TYPE ESTIMATION

In this section we discuss alternatives to the Yates procedure (see Section 1.10) of estimating the variance components σ_β^2 and σ_e^2 for the combined analysis. These estimators are maximum-likelihood type estimators. This necessitates the assumption of normality, which is not in agreement with our underlying philosophy of finite population randomization analysis. The reason for discussing them, however, is the fact that they can easily be implemented in existing software, in particular SAS PROC MIXED (SAS, 1999–2000) (see Section 1.14).

1.11.1 Maximum-Likelihood Estimation

It is convenient to rewrite model (1.49) with its covariance structure (1.51) in matrix notation as follows:

$$\begin{aligned} \boldsymbol{y} &= \mu\mathfrak{I} + \boldsymbol{X}_\tau\boldsymbol{\tau} + \boldsymbol{X}_\beta\boldsymbol{\beta} + \boldsymbol{e} \\ &= \boldsymbol{X}\boldsymbol{\alpha} + \boldsymbol{U}\boldsymbol{\beta} + \boldsymbol{e} \end{aligned} \tag{1.84}$$

where $\boldsymbol{X\alpha}$ represents the fixed part, with $\boldsymbol{X} = (\mathfrak{I}\ \boldsymbol{X}_\tau)$, $\boldsymbol{\alpha}' = (\mu, \boldsymbol{\tau}')$, and $\boldsymbol{U\beta} + \boldsymbol{e}$ represents the random part. Thus

$$E(\boldsymbol{y}) = \boldsymbol{X\alpha}$$

and

$$\begin{aligned} \mathrm{var}(\boldsymbol{y}) &= \boldsymbol{U}\boldsymbol{U}'\sigma_\beta^2 + \boldsymbol{I}_n\sigma_e^2 \\ &= \boldsymbol{V}\sigma_e^2 \end{aligned} \tag{1.85}$$

with $V = \gamma UU' + I_n$ and $\gamma = \sigma_\beta^2/\sigma_e^2$ [see also (1.53), (1.54)]. We then assume that

$$y \sim N_n(X\alpha, V\sigma_e^2) \tag{1.86}$$

that is, y follows as a multivariate normal distribution (see I.4.17.1).

The logarithm of the likelihood function for $\mathbf{y}$ of (1.86) is then given by

$$\lambda = -\tfrac{1}{2}n \log \pi - \tfrac{1}{2}n\sigma_e^2 - \tfrac{1}{2} \log|V| - \tfrac{1}{2}(y - X\alpha)'V^{-1}(y - X\alpha)/\sigma_e^2 \tag{1.87}$$

Hartley and Rao (1967) show that the maximum-likelihood (ML) estimator of α, γ, and σ_e^2 are obtained by solving the following equations:

$$\frac{1}{\sigma_e^2}(X'V^{-1}y - X'V^{-1}X\alpha) = \mathbf{0} \tag{1.88}$$

$$-\frac{1}{2}\operatorname{tr}(V^{-1}UU') + \frac{1}{2\sigma_e^2}(y - X\alpha)'V^{-1}UU'V^{-1}(By - X\alpha) = 0$$

$$-\frac{n}{2\sigma_e^2} + \frac{1}{2\sigma_e^4}(y - X\alpha)'V^{-1}(y - X\alpha) = 0$$

where $\operatorname{tr}(A)$ represents the trace of the matrix A.

The basic feature of this method is that the fixed effects and the variance components associated with the random effects are estimated simultaneously in an iterative procedure. We shall not go into the details of the numerical implementation (see, e.g., Hemmerle and Hartley, 1973), but refer to the example given in Section 1.14.4 using SAS.

1.11.2 Restricted Maximum-Likelihood Estimation

Specifically for the estimation of weights for the purpose of recovery of interblock information, Patterson and Thompson (1971) introduced a modified maximum-likelihood procedure. The basic idea is to obtain estimators for the variance components that are free of the fixed effects in the sense that the likelihood does not contain the fixed effect. Operationally this is accomplished by dividing the likelihood function (1.87) into two parts, one being based on treatment contrasts and the other being based on error contrasts, that is, contrasts with expected value zero. Maximizing this second part will lead to estimates of functions of σ_β^2 and σ_e^2. Because of the procedure employed, these estimates are by some referred to as residual maximum-likelihood estimates (REML), by others as restricted maximum likelihood estimates (REML). The latter name is derived from the fact that maximizing the part of the likelihood free of the fixed effects can be thought of as maximizing the likelihood over a restricted parameter set, an idea first proposed by Thompson (1962) for random effects models and generalized for the general linear mixed model by Corbeil and Searle (1976), based on the work by Patterson and Thompson (1971). We shall give below a brief outline of the basic idea of REML estimation following Corbeil and Searle (1976).

Consider model (1.84) and assume that the observations are ordered by treatments, where the ith treatment is replicated r_i times ($i = 1, 2, \ldots, t$). Then the matrix $\boldsymbol{X}$ can simply be written as

$$\boldsymbol{X} = \begin{bmatrix} \mathfrak{I}_{r_1} & & & \\ & \mathfrak{I}_{r_2} & & 0 \\ & & \ddots & \\ 0 & & & \mathfrak{I}_{r_t} \end{bmatrix} \equiv \sum_{i=1}^{t}{}^{+} \mathfrak{I}_{r_i} \tag{1.89}$$

To separate the log-likelihood function (1.87) into the two parts mentioned above, we employ the transformation (as proposed by Patterson and Thompson, 1971)

$$\boldsymbol{y}'[\boldsymbol{S} \vdots \boldsymbol{V}^{-1}\boldsymbol{X}] \tag{1.90}$$

where

$$\begin{aligned} \boldsymbol{S} &= \boldsymbol{I} - \boldsymbol{X}(\boldsymbol{X}'\boldsymbol{X})^{-1}\boldsymbol{X}' \\ &= \sum_{i=1}^{t}{}^{+} \left(\boldsymbol{I}_{r_i} - \frac{1}{r_i}\mathfrak{I}_{r_i}\mathfrak{I}'_{r_i} \right) \end{aligned} \tag{1.91}$$

is symmetric and idempotent. Furthermore, $\boldsymbol{SX} = \boldsymbol{0}$, and hence $\boldsymbol{Sy}$ is distributed $N(\boldsymbol{0}, \boldsymbol{SVS}\sigma_e^2)$ independently of $\boldsymbol{X}'\boldsymbol{V}^{-1}\boldsymbol{y}$.

It follows from (1.91) that $\boldsymbol{S}$ is singular. Hence, instead of $\boldsymbol{S}$ we shall use in (1.90) a matrix, $\boldsymbol{T}$ say, which is derived from $\boldsymbol{S}$ by deleting its r_1th, $(r_1 + r_2)$th, $(r_1 + r_2 + r_3)$th, ..., $(r_1 + r_2 + \cdots + r_t)$th rows, thereby reducing an $n \times n$ matrix to an $(n - t) \times n$ matrix (with $n - t$ representing the number of linearly independent error contrasts). More explicitly, we can write $\boldsymbol{T}$ as

$$\begin{aligned} \boldsymbol{T} &= \sum_{i=1}^{t}{}^{+} \left[\boldsymbol{I}_{r_i-1} \ \vdots \ \boldsymbol{0}_{r_i-1} - \frac{1}{r_i}\mathfrak{I}_{r_i-1}\mathfrak{I}'_{r_i} \right] \\ &= \sum_{i=1}^{t}{}^{+} \left[\boldsymbol{I}_{r_i-1} - \frac{1}{r_i}\mathfrak{I}_{r_i-1}\mathfrak{I}'_{r_i-1} \ \vdots \ -\frac{1}{r_i}\mathfrak{I}_{r_i-1} \right] \end{aligned} \tag{1.92}$$

It follows from (1.89) and (1.92) that

$$\boldsymbol{TX} = \boldsymbol{0} \tag{1.93}$$

Considering now the transformation

$$z = \begin{pmatrix} \boldsymbol{T} \\ \boldsymbol{X}'\boldsymbol{V}^{-1} \end{pmatrix} \boldsymbol{y}$$

it follows from (1.86) and (1.93) that

$$z \sim N\left[\begin{pmatrix} \mathbf{0} \\ \boldsymbol{X}'\boldsymbol{V}^{-1}\boldsymbol{X}\boldsymbol{\alpha} \end{pmatrix}, \begin{pmatrix} \boldsymbol{T}\boldsymbol{V}\boldsymbol{T}'\sigma_e^2 & \mathbf{0} \\ \mathbf{0} & \boldsymbol{X}'\boldsymbol{V}^{-1}\boldsymbol{X}\sigma_e^2 \end{pmatrix}\right] \tag{1.94}$$

Clearly, the likelihood function of z consists of two parts, one for $\boldsymbol{T}\boldsymbol{y}$, which is free of fixed effects, and one for $\boldsymbol{X}'\boldsymbol{V}^{-1}\boldsymbol{y}$ pertaining to the fixed effects. In particular, the log likelihood of $\boldsymbol{T}\boldsymbol{y}$ then is

$$\lambda_1 = \tfrac{1}{2}(n-t)\log 2\pi - \tfrac{1}{2}(n-t)\log \sigma_e^2$$

$$-\tfrac{1}{2}\log|\boldsymbol{T}\boldsymbol{V}\boldsymbol{T}'| - \tfrac{1}{2}\boldsymbol{y}'\boldsymbol{T}'(\boldsymbol{T}\boldsymbol{V}\boldsymbol{T}')^{-1}\boldsymbol{y}/\sigma_e^2 \tag{1.95}$$

The REML estimators for $\gamma = \sigma_\beta^2/\sigma_e^2$ and σ_e^2 are obtained by solving the equations

$$\frac{\partial \lambda_1}{\partial \gamma} = 0 \tag{1.96}$$

$$\frac{\partial \lambda_1}{\partial \sigma_e^2} = 0 \tag{1.97}$$

The resulting equations have no analytic solutions and have to be solved iteratively. We denote the solutions, that is, estimates by $\widetilde{\gamma}$ and $\widetilde{\sigma}_e^2$, respectively.

The fixed effects, represented by $\boldsymbol{\alpha}$ in (1.84), can be estimated by considering the log likelihood of $\boldsymbol{X}'\boldsymbol{V}^{-1}\boldsymbol{y}$, which is given by

$$\lambda_2 = -\tfrac{1}{2}t\log 2\pi - \tfrac{1}{2}t\ \log \sigma_e^2$$

$$-\tfrac{1}{2}\log|\boldsymbol{X}'\boldsymbol{V}^{-1}\boldsymbol{X}|$$

$$-\tfrac{1}{2}\ (\boldsymbol{y} - \boldsymbol{X}\boldsymbol{\alpha})'\,\boldsymbol{V}^{-1}\boldsymbol{X}(\boldsymbol{X}'\boldsymbol{V}^{-1}\boldsymbol{X})^{-1}\boldsymbol{X}'\boldsymbol{V}^{-1}(\boldsymbol{y} - \boldsymbol{X}\boldsymbol{\alpha})/\sigma_e^2 \tag{1.98}$$

Solving

$$\frac{\partial \lambda_2}{\partial \boldsymbol{\alpha}} = 0$$

leads to the estimator

$$\widehat{\boldsymbol{\alpha}} = (\boldsymbol{X}'\boldsymbol{V}^{-1}\boldsymbol{X})^{-1}\boldsymbol{X}'\boldsymbol{V}^{-1}\boldsymbol{y} \tag{1.99}$$

This, of course, assumes that V is known. Since it is not, we substitute $\widetilde{\gamma}$ from (1.96) and (1.97) for γ in (1.99) and obtain the estimate

$$\widetilde{\boldsymbol{\alpha}} = (\boldsymbol{X}'\widetilde{\boldsymbol{V}}^{-1}\boldsymbol{X})^{-1}\boldsymbol{X}'\widetilde{\boldsymbol{V}}^{-1}\boldsymbol{y} \tag{1.100}$$

where $\widetilde{\boldsymbol{V}}$ denotes $\boldsymbol{V}$ with γ replaced by $\widetilde{\gamma}$.

An approximate estimate of the variance of $\widetilde{\alpha}$ is given by

$$\widetilde{\text{var}}(\widetilde{\boldsymbol{\alpha}}) \cong (\boldsymbol{X}'\widetilde{\boldsymbol{V}}^{-1}\boldsymbol{X})^{-1}\widetilde{\sigma}_e^2 \tag{1.101}$$

For a numerical example using REML see Section 1.14.4.

1.12 EFFICIENCY FACTOR OF AN INCOMPLETE BLOCK DESIGN

We have seen in Sections I.9.3 and I.10.2.9, for example, how we can compare different error control designs with each other by using the notion of relative efficiency. In this case, we compare two error control designs after we have performed the experiment using a particular error control design. For example, after we have used an RCBD we might ask: How would we have done with a corresponding CRD? In other cases, however, we may want to compare error control designs before we begin an experiment. In particular, we may want to compare an incomplete block design (IBD) with either a CRD or an RCBD, or we may want to compare competing IBDs with each other. For this purpose we shall use a quantity that is referred to as the *efficiency factor* of the IBD. It compares, apart from the residual variance, σ_e^2, the average variance of simple treatment comparisons for the two competing designs.

1.12.1 Average Variance for Treatment Comparisons for an IBD

Let us now consider

$$\underset{i \neq i'}{\text{av.}}\ \text{var}(\widehat{\tau}_i - \widehat{\tau}_{i'}) \tag{1.102}$$

for a connected IBD. We know, of course, that (1.102) is a function of $\boldsymbol{C}^-$, a generalized inverse of the $\boldsymbol{C}$ matrix. Suppose now that all the block sizes are equal to k. Then we have

$$\boldsymbol{C} = \boldsymbol{R} - \frac{1}{k}\boldsymbol{N}\boldsymbol{N}'$$

and we know that $\boldsymbol{C}$ has one zero root, $d_t = 0$ say, with associated normalized eigenvector $\boldsymbol{\xi}_t = (1/\sqrt{t})\mathfrak{I}$. Let the other roots be $d_1, d_2, \ldots, d_{t-1}$ with associated orthonormal eigenvectors $\boldsymbol{\xi}_1, \boldsymbol{\xi}_2, \ldots, \boldsymbol{\xi}_{t-1}$. Then

$$\boldsymbol{\xi}_i'\boldsymbol{C} = d_i\boldsymbol{\xi}_i' \qquad (i = 1, 2, \ldots, t-1)$$

and from

$$\boldsymbol{\xi}_i'\boldsymbol{C}\boldsymbol{\tau} = d_i\boldsymbol{\xi}_i'\boldsymbol{\tau}$$

it follows that

$$\widehat{\boldsymbol{\xi}_i'\boldsymbol{\tau}} = \frac{1}{d_i}\boldsymbol{\xi}_i'\boldsymbol{Q}$$

and

$$\text{var}(\boldsymbol{\xi}_i'\widehat{\boldsymbol{\tau}}) = \frac{1}{d_i^2}\boldsymbol{\xi}_i'\boldsymbol{C}\boldsymbol{\xi}_i\sigma_e^2 = \frac{1}{d_i}\sigma_e^2 \tag{1.103}$$

Using the fact that $\boldsymbol{\xi}_t = (1/\sqrt{t})\mathfrak{I}$, that $\boldsymbol{\xi}_1, \boldsymbol{\xi}_2, \ldots, \boldsymbol{\xi}_{t-1}$ are mutually perpendicular and perpendicular to $\boldsymbol{\xi}_1$, and that

$$\sum_{i=1}^{t} \boldsymbol{\xi}_i\,\boldsymbol{\xi}_i' = \mathbf{I}$$

we have with $\boldsymbol{z}' = (z_1, z_2, \ldots, z_t)$

$$\begin{aligned}
\sum_{i=1}^{t-1} (\boldsymbol{\xi}_i'\boldsymbol{z})^2 &= \boldsymbol{z}'\left(\sum_{i=1}^{t-1} \boldsymbol{\xi}_i\,\boldsymbol{\xi}_i'\right)\boldsymbol{z} \\
&= \boldsymbol{z}'(\boldsymbol{I} - \boldsymbol{\xi}_t\,\boldsymbol{\xi}_t')\boldsymbol{z} \\
&= \sum_{i=1}^{t} z_i^2 - \frac{1}{t}\left(\sum_{i=1}^{t} z_i\right)^2 \\
&= \sum_{i=1}^{t} (z_i - \overline{z})^2
\end{aligned} \tag{1.104}$$

It is also easy to verify that

$$\frac{1}{t(t-1)}\sum_{i \neq i'} (z_i - z_{i'})^2 = \frac{2}{t-1}\sum_{i=1}^{t} (z_i - \overline{z})^2 \tag{1.105}$$

Taking $z_i = \widehat{\tau}_i - \tau_i$, substituting into (1.105) using (1.104) and then taking expectations and using (1.103), yields for (1.102)

$$\underset{i \neq i'}{\text{av.}} \operatorname{var}(\widehat{\tau}_i - \widehat{\tau}_{i'}) = \frac{2}{t-1} \sum_{i=1}^{t-1} \frac{1}{d_i} \sigma_e^2 \tag{1.106}$$

1.12.2 Definition of Efficiency Factor

It is natural in attempting to evaluate the efficiency of an IBD to compare it with a CRD since this is always a possible competing design. For a CRD with r_i replications for treatment i, the average variance of treatment differences is

$$\underset{i \neq i'}{\text{av.}} \operatorname{var}(\widehat{\tau}_i - \widehat{\tau}_{i'}) = \underset{i \neq i'}{\text{av.}} \left(\frac{1}{r_i} + \frac{1}{r_i'} \right) \sigma^2_{e(\text{CRD})} = \frac{2}{\overline{r}_h} \sigma^2_{e(\text{CRD})} \tag{1.107}$$

where $\overline{r}_h$ is the harmonic mean of the r_i, that is,

$$\frac{1}{\overline{r}_h} = \frac{1}{t} \sum_i \frac{1}{r_i}$$

We shall digress here for a moment and show that the best CRD is the one with all $r_i = r$, and that is the design with which we shall compare the IBD. For this and later derivations we need the "old" result that the harmonic mean of a set of positive numbers is not greater than the arithmetic mean. It seems useful to give an elementary proof of this.

Let the set of numbers be $\{x_i, i = 1, 2, \ldots, m\}$. Consider the quadratic

$$q(\beta) = \sum_{i=1}^{m} \left(\sqrt{x_i} - \beta \frac{1}{\sqrt{x_i}} \right)^2$$

Clearly $q(\beta) \geq 0$ for all β. The minimizing value of β is obtained by using least squares which gives the NE

$$\sum_i \frac{1}{x_i} \widetilde{\beta} = m$$

The minimum sum of squares is

$$\sum_i x_i - \widetilde{\beta} m$$

Hence

$$\sum_i x_i - \frac{m^2}{\sum_i \frac{1}{x_i}} \geq 0$$

or

$$\left(\frac{1}{m}\sum_i x_i\right)\left(\frac{1}{m}\sum_i \frac{1}{x_i}\right) \geq 1$$

or

$$\frac{\overline{x}}{\overline{x}_h} \geq 1$$

with equality if and only if $x_i = x$ for all i.

This result implies that the best CRD will have $r_i = r$ and $r = n/t$ where n is the total numbers of EUs. This can happen, of course, only if n/t is an integer. If n/t is not an integer so that $n = pt + q$ $(0 < q < t)$, then the best CRD will have q treatments replicated $p + 1$ times.

Consider now the case of an IBD with b blocks of size k and r_i replications for the ith treatment. Then the total number of EUs is $n = bk = \sum r_i$. Suppose also that $n = rt$, so that an equireplicate CRD is possible. The average variance for such a design is $2\sigma^2_{e(\text{CRD})}/r$, whereas the average variance for the IBD is $2\sigma^2_{e(\text{IBD})}/c$ where, as shown in (1.106), c is the harmonic mean of the positive eigenvalues of $\left(\boldsymbol{R} - (1/k)\, \boldsymbol{N}\boldsymbol{N}'\right)$ (see Kempthorne, 1956). It is natural to write $c = rE$, so that with $\sigma^2_{e(\text{CRD})} = \sigma^2_{e(\text{IBD})}$ we have

$$\frac{\text{av. var}(\widehat{\tau}_i - \widehat{\tau}_{i'})_{\text{CRD}}}{\text{av. var}(\widehat{\tau}_i - \widehat{\tau}_{i'})_{\text{IBD}}} = \frac{2/r}{2/rE} = E \tag{1.108}$$

The quantity E thus defined is called the *efficiency factor* of the IBD. It is clearly a numerical property of the treatment-block configuration only and hence a characteristic of a given IBD.

We add the following remarks:

1. The same definition of E in (1.108) could have been obtained by using the average variance for an RCBD with $b = r$ blocks instead of the average variance for an equireplicate CRD assuming that $\sigma^2_{e(\text{RCBD})} = \sigma^2_{e(\text{IBD})}$.
2. Although E is a useful quantity to compare designs, it does not, of course, give the full story. It compares average variances only under the assumption of equality of residual variances, whereas we typically expect $\sigma^2_{e(\text{IBD})} < \sigma^2_{e(\text{CRD})}$ and $\sigma^2_{e(\text{IBD})} < \sigma^2_{e(\text{RCBD})}$.
3. The efficiency factor pertains only to the intrablock analysis and ignores the interblock information.
4. Each IBD will have associated with it an efficiency factor E. In order to compare two competing IBDs with the same n and with efficiency factors E_1 and E_2, respectively, we would typically choose the one with the higher E value.

1.12.3 Upper Bound for the Efficiency Factor

Using again the fact that the harmonic mean of positive numbers is not greater than the arithmetic mean, we have

$$\begin{aligned}(t-1)c \leq \sum_{i=1}^{t-1} d_i &= \text{trace}\left(\boldsymbol{R} - \frac{1}{k}\boldsymbol{N}\boldsymbol{N}'\right) \\ &= \sum_{i=1}^{t} r_i - \frac{1}{k}\sum_{ij} n_{ij}^2 \\ &= n - \frac{1}{k}\sum_{ij} n_{ij}^2 \qquad (1.109)\end{aligned}$$

The largest value of the right-hand side of (1.109) is obtained for the smallest value of $\sum_{ij} n_{ij}^2$. Since n_{ij} is one of the numbers $0, 1, 2, \ldots, k$, the minimum value of $\sum n_{ij}^2$ will be achieved when n of the n_{ij}'s are 1 and the remaining are zero. Since then $n_{ij}^2 = n_{ij}$ and $\sum_j n_{ij} = r_i$, it follows from (1.109) that

$$(t-1)c \leq \sum_i r_i - \frac{1}{k}\sum_i r_i = \frac{k-1}{k} t\bar{r}$$

or, since $c = rE$,

$$E \leq \frac{(k-1)t/(t-1)k}{\bar{r}/r}$$

But since $t\bar{r} = n = tr$, we have finally

$$E \leq \frac{(k-1)t}{(t-1)k} \qquad (1.110)$$

Since for an IBD $k < t$, we can write further

$$E \leq \frac{(k-1)t}{(t-1)k} < 1 \qquad (1.111)$$

We shall see later (see Chapter 2, also Section I.9.8.2) that the upper bound given in (1.110) will be achieved for the balanced incomplete block design.

Sharper upper bounds for certain classes of IBDs are given by Jacroux (1984), Jarrett (1983), Paterson (1983), and Tjur (1990); see also John and Williams (1995).

1.13 OPTIMAL DESIGNS

We have argued in the previous section that in order to compare two designs, d_1 and d_2 say, we may consider their efficiency factors E_1 and E_2, respectively, and choose the design with the higher efficiency factor. In particular, if the efficiency factor of one of those designs achieves the upper bound for that class of designs, we would consider that design to be optimal in some sense. Such considerations have led to the development of the notion of *optimal designs* and to various criteria for optimality. We shall describe briefly some of these criteria.

1.13.1 Information Function

Initial contributions to the formal discussion of optimal designs were made by Wald (1943) and Ehrenfeld (1953). Extending their results, Kiefer (1958) provided a systematic account of different optimality criteria. These can be discussed either in terms of maximizing a suitable function of the information matrix or minimizing a corresponding function of the dispersion matrix of a maximal set of orthonormal treatment contrast estimates.

In the context of our discussion the information matrix is given by $\boldsymbol{C}$ in (1.9). Let $\boldsymbol{P}'\boldsymbol{\tau}$ represent a set of $t-1$ orthonormal contrasts of the treatment effects. Using intrablock information from a connected design d, the estimator for $\boldsymbol{P}'\boldsymbol{\tau}$ is given by $\boldsymbol{P}'\widehat{\boldsymbol{\tau}}$ with $\widehat{\boldsymbol{\tau}}$ from (1.18). The dispersion matrix for $\boldsymbol{P}'\widehat{\boldsymbol{\tau}}$ is then given by

$$\boldsymbol{V}_d\, \sigma_e^2 = \boldsymbol{P}'\boldsymbol{C}_d^- \boldsymbol{P}\sigma_e^2 = (\boldsymbol{P}'\boldsymbol{C}_d\boldsymbol{P})^{-1}\sigma_e^2 \tag{1.112}$$

[see (1.20)], where $\boldsymbol{C}_d^-$ and hence $\boldsymbol{C}_d$ refer to the specific design d used. The information matrix for the design d is then defined as

$$\boldsymbol{C}_d^* = (\boldsymbol{P}'\boldsymbol{C}_d^-\boldsymbol{P})^{-1} \tag{1.113}$$

which shows, of course, the connection between $\boldsymbol{C}_d^*$ and $\boldsymbol{C}_d$.

An *information function* or *optimality criterion* is then a real-valued function ϕ that has the following properties (see Pukelsheim, 1993):

1. Function ϕ is a monotonic function; that is, an information matrix $\boldsymbol{C}^*$ is at least as good as another information matrix $\boldsymbol{D}^*$ if $\phi(\boldsymbol{C}^*) \geq \phi(\boldsymbol{D}^*)$;
2. Function ϕ is a concave function, that is, $\phi[(1-\alpha)\,\boldsymbol{C}^* + \alpha\boldsymbol{D}^*] = (1-\alpha)\phi(\boldsymbol{C}^*) + \alpha\,\phi(\boldsymbol{D}^*)$ for $\alpha\ \in (0, 1)$;
3. Function ϕ is positively homogeneous, that is, $\phi(\delta\boldsymbol{C}^*) = \delta\,\phi(\boldsymbol{C}^*)$.

Condition (2) says that information cannot be increased by interpolation. And condition (3) says that even if we define the information matrix to be directly proportional to the number of observations, n, and inversely proportional to σ_e^2, that is, the information matrix is of the form $(n/\sigma_e^2)\boldsymbol{C}^*$, we need to consider only $\boldsymbol{C}^*$.

Let $\mathcal{D}$ be the set of competing designs. The problem of finding an optimal design d in $\mathcal{D}$ can then be reduced to finding a design that maximizes $\phi(\boldsymbol{C}_d^*)$ over d in $\mathcal{D}$ (see Cheng, 1996). Such a design is called ϕ-optimal.

As indicated above, an alternative, and historically original, approach to finding an optimal design is to consider minimization of some convex and nonincreasing function Φ of dispersion matrices, as indicated by the relationship between (1.112) and (1.113). Accordingly, we shall then talk about a Φ-optimal design.

1.13.2 Optimality Criteria

Several optimality criteria, that is, several functions ϕ or Φ have been considered for studying optimal designs. These criteria can be expressed conveniently in terms of the eigenvalues of $\boldsymbol{C}_d^*$ or, equivalently, the nonzero eigenvalues of $\boldsymbol{C}_d$, say $\mu_{d1} \geq \mu_{d2} \geq \cdots \geq \mu_{d,t-1}$.

The most commonly used optimality criteria are D-, A-, and E-optimality, which maximize the following information functions:

1. *D-optimality*: Determinant criterion or

$$\phi(\boldsymbol{C}_d^*) = \prod_{i=1}^{t-1} \mu_{di}$$

2. *A-optimality*: Average variance criterion or

$$\phi(\boldsymbol{C}_d^*) = \left(\frac{1}{t-1} \sum_{i=1}^{t-1} \mu_{di}^{-1} \right)^{-1}$$

3. *E-optimality*: Smallest eigenvalue criterion or

$$\phi(\boldsymbol{C}_d^*) = \mu_{d,t-1}$$

In terms of the corresponding Φ function, these optimality criteria can be expressed as minimizing

1. $$\Phi(\boldsymbol{V}_d) = \det \boldsymbol{V}_d = \prod_{i=1}^{t-1} \mu_{di}^{-1} \tag{1.114}$$

2. $$\Phi(\boldsymbol{V}_d) = \operatorname{tr} \boldsymbol{V}_d = \sum_{i=1}^{t-1} \mu_{di}^{-1} \tag{1.115}$$

3. $$\Phi(\boldsymbol{V}_d) = \text{maximum eigenvalue of } \boldsymbol{V}_d = \mu_{d,t-1}^{-1} \tag{1.116}$$

The statistical meaning of these criteria is that minimizing (1.114) minimizes the generalized variance of $\boldsymbol{P}'\widehat{\boldsymbol{\tau}}$, (1.115) minimizes the average variance of the set $\boldsymbol{P}'\widehat{\boldsymbol{\tau}}$, and (1.116) minimizes the maximum variance of a single normalized contrast.

1.13.3 Optimal Symmetric Designs

There exist special classes of designs for which all the nonzero eigenvalues of the information matrix $\boldsymbol{C}_d$ are equal. Such designs are called *symmetric designs*. Examples of symmetric designs are balanced incomplete block designs (Chapter 2), Latin square designs (Section I.10.2), Youden squares (Section I.10.5), and so forth. The information matrix of a symmetric design is of the form $a\boldsymbol{I} + b\,\mathfrak{I}\mathfrak{I}'$, which is referred to as a *completely symmetric matrix*.

In general, if a design is Φ_1 optimal, it may not be Φ_2 optimal for two different optimality criteria Φ_1 and Φ_2. However, for symmetric designs Kiefer (1958) showed that they are A-, D-, and E-optimal. This led to the definition of *universal optimality* (Kiefer, 1975a) or *Kiefer optimality* (Pukelsheim, 1993).

Definition 1.1 (Kiefer, 1975a) Let $\mathcal{B}_{t,0}$ be the set of all $t \times t$ nonnegative definite matrices with zero row and column sums, and let Φ be a real-valued function on $\mathcal{B}_{t,0}$ such that

(a) Φ is convex,

(b) $\Phi(\delta \boldsymbol{C})$ is nonincreasing in the scalar $\delta \geq 0$, and

(c) Φ is invariant under each simultaneous permutation of rows and columns.

A design d^* is called *universally optimal* in $\mathcal{D}$ if d^* minimizes $\Phi(\boldsymbol{C}_d)$ for every Φ satisfying conditions (a), (b), and (c). □

To help identify universally optimal designs we have the following fundamental theorem.

Theorem 1.2 (Kiefer, 1975a) Suppose a class $\mathcal{C} = \{\boldsymbol{C}_d, d \in \mathcal{D}\}$ of matrices in $\mathcal{B}_{t,0}$ contains a $\boldsymbol{C}_{d^*}$ for which

(a) $\boldsymbol{C}_{d^*}$ is completely symmetric, and

(b) $\operatorname{tr} \boldsymbol{C}_{d^*} = \max_{d \in \mathcal{D}} \operatorname{tr} \boldsymbol{C}_d$,

Then d^* is universally optimal in $\mathcal{D}$.

1.13.4 Optimality and Research

We have just discussed the notion of design optimality and some of the available optimality criteria. Other criteria have been introduced in other contexts, in particular, in the area of regression or response surface designs (see, e.g., Atkinson and Donev, 1992). And thus optimality has become a powerful concept, but we need to remember that, although it has statistical meaning, it is a mathematical concept. It has a definite and restricted connotation and it may be difficult to apply it in the larger context of designing a "very good" experiment for a researcher who has a scientific or technological problem.

An immediate difficulty is that there is no simple classification of real problems. There are discovery problems, for example, finding the point in factor space at which yield is maximum. There are exploration problems, for example, to obtain a "good" representation of the nature of the dependence of a noisy dependent variable on a given set of independent variables. There is the mathematical problem in that same context that the dependence is known to be of a definite functional form with some specified but unknown parameters, which are to be determined from observations at some locations in the factor space. A common problem in technology and some scientific areas is what is called screening of factors. It is useful and important to think about this overall picture because there is a tendency to interpret the term "optimality of design" in a very limited context, a context that is very valuable, but misleading, in the sphere of total human investigation.

The moral of the situation is multifold: (1) Researchers have to make a choice about problems and often work on unrealistic ones as the closest workable approximation to real live problems and should not be criticized for so doing; (2) almost any optimality problem is to some extent artificial and limited because criteria of value of designs must be introduced, and in almost any investigative situation it is difficult to map the possible designs valuewise into the real line; and (3) a solution to a mathematically formulated problem may have limited value, so to promote one design that is optimal only with respect to a particular criterion of value, C_1, and to declare another design to be of poor value because it is not optimal may be unfair because that design may be better with respect to a different criterion of value, C_2 say. And for one researcher C_1 may be irrelevant, whereas C_2 may be more appropriate.

Considerations of optimality involve, of course, comparison of designs. But how does one do this when error reduction needs to be taken into account? For example, how does one compare the randomized complete block and the Latin square design? Or how does one compare designs when different aspects of statistical inference are involved? This was at the basis of heated discussion between Neyman, who was interested in hypothesis testing, and Fisher and Yates, who were interested in precision of estimation (see Neyman, Iwaszkiewicz, and Kolodziejczyk, 1935).

Informal optimality considerations early on gave probably rise to the heuristic (or perhaps mathematical) idea of symmetry and balancedness, and we shall encounter these characteristics throughout much of the book. Even though these properties do not always guarantee optimality in many cases they lead to near optimality. And from a practical point of view that may be good enough. On the other hand, if a balanced design is an optimal design, but we cannot use that design because of practical constraints and need to use instead a near-balanced design, then we have a way to evaluate the efficiency of the design we are going to use. For example, we know that a balanced incomplete block design (Chapter 2) is optimal. However, we cannot use the design and need to use a partially balanced incomplete block design (Chapter 4). We may then choose, if possible, a design with efficiency "close" to that of the optimal design.

Thus, the insistence on an optimal design may be frustrating for the user because practical reasons may dictate otherwise and because an experimenter rarely has one criterion of value. Rather, he has many criteria of value and is in a mathematical programming situation in which he wishes the design to have "reasonable" efficiencies with respect to criteria $C_1, C_2, \ldots, C_K$. The dilemma is often that the design that is optimal with respect to C_1 is completely nonoptimal with respect to C_2.

In summary, mathematical ideals and requirements of empirical science do not always meet, but it is worth trying to find common ground. In the end, practical considerations may dictate compromises in many instances of experimental research.

1.14 COMPUTATIONAL PROCEDURES

In this section we shall discuss some computational aspects of performing the intrablock analysis and the combined intra- and interblock analysis, mainly in the context of SAS procedures (SAS, 2000). For the intrablock analysis (see Section 1.3) we shall use SAS PROC GLM, and for the combined analysis (see Section 1.8) we shall use SAS PROC MIXED.

1.14.1 Intrablock Analysis Using SAS PROC GLM

Consider the following data set IBD (Table 1.7) with $t = 4$ treatments in $b = 5$ blocks of size $k = 2$, such that treatments 1 and 4 are replicated 3 times, and treatments 2 and 3 are replicated 2 times. An example might be 5 pairs of identical twins representing the blocks, each twin being an experimental unit to whom different drugs are assigned according to the given plan.

The SAS PROC GLM input statements for the intrablock analysis and the results are given in Table 1.8. We shall comment briefly on some aspects of the SAS output (see Table 1.8):

1. The coefficient matrix as well as the right-hand side (RHS) of (1.5) are given under the heading "The $\boldsymbol{X}'\boldsymbol{X}$ Matrix."
2. A generalized inverse for the coefficient matrix $\boldsymbol{X}'\boldsymbol{X}$ is obtained by first eliminating the rows and columns for β_5 and τ_4 from $\boldsymbol{X}'\boldsymbol{X}$ as a consequence of imposing the conditions $\beta_5^* = 0$ and $\tau_4^* = 0$ (we shall denote the SAS solutions to the NE by $\boldsymbol{\beta}^*$ and $\boldsymbol{\tau}^*$). The reduced matrix is of full rank and thus can be inverted. The inverted matrix is restored to the original dimension by inserting zeros in the rows and columns corresponding to β_5^* and τ_4^*. This matrix, together with $\widehat{\sigma}_e^2$ =MS(Error) from the ANOVA table, can be used to find the standard errors for the estimators of estimable functions for treatment effects.

Table 1.7 Data for Incomplete Block Design (t = 4, b = 5, k = 2, r1 = 3, r2 = r3 = 2, r4 = 3)

```
options pageno=1 nodate;
data IBD;
input TRT BLOCK Y @@;
datalines;
1 1 10  2 1 12
3 2 23  4 2 28
1 3 13  3 3 27
2 4 14  4 4 20
1 5 15  4 5 32
;
run;

proc print data=IBD;
title1 'TABLE 1.7';
title2 'DATA FOR INCOMPLETE BLOCK DESIGN';
title3 '(t=4, b=5, k=2, r1=3, r2=r3=2, r4=3)';
run;
```

Obs	TRT	BLOCK	Y
1	1	1	10
2	2	1	12
3	3	2	23
4	4	2	28
5	1	3	13
6	3	3	27
7	2	4	14
8	4	4	20
9	1	5	15
10	4	5	32

3. The general form of an estimable function for treatment effects is given by

$$L7\tau_1 + L8\tau_2 + L9\tau_3 - (L7 + L8 + L9)\tau_4$$

for any values of $L7, L8$, and $L9$; that is, only contrasts are estimable.

4. The general form of estimable functions can also be used to identify the solutions to the NE by putting sequentially (and in order) each $Li = 1$ and the remaining $Lj = 0, (j \neq i)$. For example, the "Estimate" of "Intercept" is actually the estimate of $\mu + \beta_5 + \tau_4$; that is, $L1 = 1, Lj = 0, (j \neq 1)$. Expressed in terms of the SAS solutions we thus have

$$\mu^* = \widehat{\mu + \beta_5 + \tau_4} = 31.125$$

Another example, putting $L7 = 1, Lj = 0 (j \neq 7)$, yields

$$\tau_1^* = \widehat{\tau_1 - \tau_4} = -15.25$$

Table 1.8 Intrablock Analysis with Post-hoc Comparisons

```
proc glm data=IBD;
class BLOCK TRT;
model Y = BLOCK TRT/XPX inverse solution e;

lsmeans TRT/stderr e;

estimate 'TRT1 - (TRT2+TRT3)/2' TRT 1 -.5 -.5 0;
estimate 'TRT1 - TRT4' TRT 1 0 0 -1;
estimate 'TRT2 - TRT3' TRT 0 1 -1 0;
title1 'TABLE 1.8';
title2 'INTRA-BLOCK ANALYSIS';
title3 'WITH POST-HOC COMPARISONS';
run;
```

The GLM Procedure

Class Level Information

Class	Levels	Values
BLOCK	5	1 2 3 4 5
TRT	4	1 2 3 4

Number of observations 10

The X'X Matrix

	Intercept	BLOCK 1	BLOCK 2	BLOCK 3	BLOCK 4	BLOCK 5
Intercept	10	2	2	2	2	2
BLOCK 1	2	2	0	0	0	0
BLOCK 2	2	0	2	0	0	0
BLOCK 3	2	0	0	2	0	0
BLOCK 4	2	0	0	0	2	0
BLOCK 5	2	0	0	0	0	2
TRT 1	3	1	0	1	0	1
TRT 2	2	1	0	0	1	0
TRT 3	2	0	1	1	0	0
TRT 4	3	0	1	0	1	1
Y	194	22	51	40	34	47

Table 1.8 (*Continued*)

The X′X Matrix

	TRT 1	TRT 2	TRT 3	TRT 4	Y
Intercept	3	2	2	3	194
BLOCK 1	1	1	0	0	22
BLOCK 2	0	0	1	1	51
BLOCK 3	1	0	1	0	40
BLOCK 4	0	1	0	1	34
BLOCK 5	1	0	0	1	47
TRT 1	3	0	0	0	38
TRT 2	0	2	0	0	26
TRT 3	0	0	2	0	50
TRT 4	0	0	0	3	80
Y	38	26	50	80	4300

X′X Generalized Inverse (g2)

	Intercept	BLOCK 1	BLOCK 2	BLOCK 3	BLOCK 4	BLOCK 5
Intercept	0.75	-0.375	-0.625	-0.375	-0.625	0
BLOCK 1	-0.375	1.3125	0.4375	0.5625	0.6875	0
BLOCK 2	-0.625	0.4375	1.3125	0.6875	0.5625	0
BLOCK 3	-0.375	0.5625	0.6875	1.3125	0.4375	0
BLOCK 4	-0.625	0.6875	0.5625	0.4375	1.3125	0
BLOCK 5	0	0	0	0	0	0
TRT 1	-0.5	-0.25	0.25	-0.25	0.25	0
TRT 2	-0.25	-0.625	0.125	-0.125	-0.375	0
TRT 3	-0.25	-0.125	-0.375	-0.625	0.125	0
TRT 4	0	0	0	0	0	0
Y	31.125	-7.6875	-4.0625	-1.9375	-9.3125	0

X′X Generalized Inverse (g2)

	TRT 1	TRT 2	TRT 3	TRT 4	Y
Intercept	-0.5	-0.25	-0.25	0	31.125
BLOCK 1	-0.25	-0.625	-0.125	0	-7.6875
BLOCK 2	0.25	0.125	-0.375	0	-4.0625
BLOCK 3	-0.25	-0.125	-0.625	0	-1.9375
BLOCK 4	0.25	-0.375	0.125	0	-9.3125
BLOCK 5	0	0	0	0	0
TRT 1	1	0.5	0.5	0	-15.25
TRT 2	0.5	1.25	0.25	0	-9.625
TRT 3	0.5	0.25	1.25	0	-3.125
TRT 4	0	0	0	0	0
Y	-15.25	-9.625	-3.125	0	18.1875

Table 1.8 (*Continued*)

General Form of Estimable Functions

Effect		Coefficients
Intercept		L1
BLOCK	1	L2
BLOCK	2	L3
BLOCK	3	L4
BLOCK	4	L5
BLOCK	5	L1-L2-L3-L4-L5
TRT	1	L7
TRT	2	L8
TRT	3	L9
TRT	4	L1-L7-L8-L9

The GLM Procedure

Dependent Variable: Y

Source	DF	Sum of Squares	Mean Square	F Value	Pr > F
Model	7	518.2125000	74.0303571	8.14	0.1137
Error	2	18.1875000	9.0937500		
Corrected Total	9	536.4000000			

R-Square	Coeff Var	Root MSE	Y Mean
0.966093	15.54425	3.015585	19.40000

Source	DF	Type I SS	Mean Square	F Value	Pr > F
BLOCK	4	261.4000000	65.3500000	7.19	0.1259
TRT	3	256.8125000	85.6041667	9.41	0.0975

Source	DF	Type III SS	Mean Square	F Value	Pr > F
BLOCK	4	79.1458333	19.7864583	2.18	0.3388
TRT	3	256.8125000	85.6041667	9.41	0.0975

Table 1.8 *(Continued)*

Parameter		Estimate	Standard Error	t Value	Pr > |t|
Intercept		31.12500000 B	2.61157280	11.92	0.0070
BLOCK	1	-7.68750000 B	3.45478608	-2.23	0.1560
BLOCK	2	-4.06250000 B	3.45478608	-1.18	0.3607
BLOCK	3	-1.93750000 B	3.45478608	-0.56	0.6314
BLOCK	4	-9.31250000 B	3.45478608	-2.70	0.1145
BLOCK	5	0.00000000 B	.	.	.
TRT	1	-15.25000000 B	3.01558452	-5.06	0.0369
TRT	2	-9.62500000 B	3.37152599	-2.85	0.1039
TRT	3	-3.12500000 B	3.37152599	-0.93	0.4518
TRT	4	0.00000000 B	.	.	.

NOTE: The X'X matrix has been found to be singular, and a generalized inverse was used to solve the normal equations. Terms whose estimates are followed by the letter 'B' are not uniquely estimable.

Least Squares Means

Coefficients for TRT Least Square Means

TRT Level

Effect		1	2	3	4
Intercept		1	1	1	1
BLOCK	1	0.2	0.2	0.2	0.2
BLOCK	2	0.2	0.2	0.2	0.2
BLOCK	3	0.2	0.2	0.2	0.2
BLOCK	4	0.2	0.2	0.2	0.2
BLOCK	5	0.2	0.2	0.2	0.2
TRT	1	1	0	0	0
TRT	2	0	1	0	0
TRT	3	0	0	1	0
TRT	4	0	0	0	1

TRT	Y LSMEAN	Standard Error	Pr > |t|
1	11.2750000	1.9774510	0.0294
2	16.9000000	2.6632921	0.0239
3	23.4000000	2.6632921	0.0127
4	26.5250000	1.9774510	0.0055

Dependent Variable: Y

Parameter	Estimate	Standard Error	t Value	Pr > |t|
TRT1 - (TRT2+TRT3)/2	-8.8750000	2.61157280	-3.40	0.0768
TRT1 - TRT4	-15.2500000	3.01558452	-5.06	0.0369
TRT2 - TRT3	-6.5000000	4.26468053	-1.52	0.2670

5. The top part of the ANOVA table provides the partition

$$\text{SS(MODEL)} + \text{SS(ERROR)} = \text{SS(TOTAL)}$$

and from that produces

$$\text{MS(ERROR)} = \widehat{\sigma}_e^2 = 9.094$$

6. The lower part of the ANOVA table provides type I SS (sequential SS for ordered model; see I.4.7.2) and type III SS (partial SS). From the latter we obtain the P value (.0975) for the test of

$$H_0: \quad \tau_1 = \tau_2 = \cdots = \tau_t$$

versus

$$H_1: \quad \text{not all } \tau_i \text{ are the same}$$

We stress that the P value for blocks should be ignored (see I.9.2).

7. The e-option of LSMEANS gives the coefficients for the solution vector to compute the treatment least squares means, for example,

$$\begin{aligned} \text{LSMEAN(TRT} \quad 1) &= \mu^* + .2\sum \beta_i^* + \tau_1^* \\ &= 31.125 + .2(-7.6875 - 4.0625 - 1.9375 \\ &\quad - 9.3125 + 0) - 15.25 \\ &= 11.275 \end{aligned}$$

The standard error is computed by making use of the G inverse (see item 2) and MS(ERROR).

8. The t tests are performed for the prespecified contrasts among the least-squares means; for example,

$$\begin{aligned} \text{TRT2-TRT3} &= \tau_2^* - \tau_3^* = 9.625 + 3.125 = -\ 6.5 \\ se(\tau_2^* - \tau_3^*) &= [(1.25 + 1.25 - 2 \times .25) \times 9.094]^{1/2} = 4.265 \\ t &= \frac{-\ 6.5}{4.265} = -1.52 \end{aligned}$$

1.14.2 Intrablock Analysis Using the Absorb Option in SAS PROC GLM

A computational method as described in Section 1.3 using the RNE can be implemented in SAS PROC GLM by using the ABSORB OPTION. This is illustrated in Table 1.9.

Table 1.9 Intra Block Analysis Using Reduced Normal Equations With Post-hoc Comparisons

```
proc glm data=IBD;
class BLOCK TRT;
absorb BLOCK;
model Y = TRT/XPX inverse solution e;

estimate 'TRT1 - (TRT2+TRT3)/2' TRT 1 -.5 -.5 0;
estimate 'TRT1 - TRT4' TRT 1 0 0 -1;
estimate 'TRT2 - TRT3' TRT 0 1 -1 0;
title1 'TABLE 1.9';
title2 'INTRA-BLOCK ANALYSIS USING REDUCED NORMAL EQUATIONS';
title3 'WITH POST-HOC COMPARISONS';
run;
```

The GLM Procedure

Class Level Information

Class	Levels	Values
BLOCK	5	1 2 3 4 5
TRT	4	1 2 3 4

Number of observations 10

The X'X Matrix

	TRT 1	TRT 2	TRT 3	TRT 4	Y
TRT 1	1.5	-0.5	-0.5	-0.5	-16.5
TRT 2	-0.5	1	0	-0.5	-2
TRT 3	-0.5	0	1	-0.5	4.5
TRT 4	-0.5	-0.5	-0.5	1.5	14
Y	-16.5	-2	4.5	14	275

The GLM Procedure

X'X Generalized Inverse (g2)

	TRT 1	TRT 2	TRT 3	TRT 4	Y
TRT 1	1	0.5	0.5	0	-15.25
TRT 2	0.5	1.25	0.25	0	-9.625
TRT 3	0.5	0.25	1.25	0	-3.125
TRT 4	0	0	0	0	0
Y	-15.25	-9.625	-3.125	0	18.1875

General Form of Estimable Functions

Given that the coefficients for all absorbed effects are zero

Table 1.9 *(Continued)*

Effect		Coefficients
TRT	1	L1
TRT	2	L2
TRT	3	L3
TRT	4	-L1-L2-L3

Dependent Variable: Y

Source	DF	Sum of Squares	Mean Square	F Value	Pr > F
Model	7	518.2125000	74.0303571	8.14	0.1137
Error	2	18.1875000	9.0937500		
Corrected Total	9	536.4000000			

R-Square	Coeff Var	Root MSE	Y Mean
0.966093	15.54425	3.015585	19.40000

Source	DF	Type I SS	Mean Square	F Value	Pr > F
BLOCK	4	261.4000000	65.3500000	7.19	0.1259
TRT	3	256.8125000	85.6041667	9.41	0.0975

Source	DF	Type III SS	Mean Square	F Value	Pr > F
TRT	3	256.8125000	85.6041667	9.41	0.0975

Parameter	Estimate	Standard Error	t Value	Pr > |t|
TRT1 - (TRT2+TRT3)/2	-8.8750000	2.61157280	-3.40	0.0768
TRT1 - TRT4	-15.2500000	3.01558452	-5.06	0.0369
TRT2 - TRT3	-6.5000000	4.26468053	-1.52	0.2670

Parameter		Estimate	Standard Error	t Value	Pr > |t|
TRT	1	-15.25000000 B	3.01558452	-5.06	0.0369
TRT	2	-9.62500000 B	3.37152599	-2.85	0.1039
TRT	3	-3.12500000 B	3.37152599	-0.93	0.4518
TRT	4	0.00000000 B	.	.	.

NOTE: The X'X matrix has been found to be singular, and a generalized inverse was used to solve the normal equations. Terms whose estimates are followed by the letter 'B' are not uniquely estimable.

We make the following comments about the SAS output:

1. The $X'X$ matrix is now the C-matrix of (1.9).
2. The $X'X$ generalized inverse is obtained by the SAS convention of setting $\tau_4^* = 0$. This g inverse is therefore different from $\widetilde{C}^{-1}$ of (1.16).
3. The ANOVA table provides the same information as in Table 1.8, except that it does not give solutions for the intercept and blocks. Hence, this analysis cannot be used to obtain treatment least-squares means and their standard error.

1.14.3 Combined Intra- and Interblock Analysis Using the Yates Procedure

Using SAS PROC VARCOMP illustrates the estimation of σ_β^2 according to the method described in Section 1.10.1. The result is presented in Table 1.10. The option type I produces Table 1.2 with

$$\text{E[MS(BLOCK)]} = \text{E}\left[\text{MS}(\boldsymbol{X}_{\boldsymbol{\beta}} | \mathfrak{I}, \boldsymbol{X}_{\boldsymbol{\tau}})\right]$$

as given in (1.82). This yields $\widehat{\sigma}_e^2 = 9.09$ (as in Table 1.8) and $\widehat{\sigma}_\beta^2 = 7.13$.

Substituting $\widehat{\rho} = \dfrac{9.09 + 2 \times 7.13}{9.09} = 2.57$ into (1.60) we obtain

$$\widetilde{\boldsymbol{A}} = \begin{pmatrix} 2.0814 & -0.3062 & -0.3062 & -0.3062 \\ -0.3062 & 1.3877 & 0.0000 & -0.3062 \\ -0.3062 & 0.0000 & 1.3877 & -0.3062 \\ -0.3062 & -0.3062 & -0.3062 & 2.0814 \end{pmatrix}$$

and

$$\widetilde{\boldsymbol{A}}^{-1} = \begin{pmatrix} 0.541654 & 0.146619 & 0.146619 & 0.122823 \\ 0.146619 & 0.785320 & 0.064704 & 0.146619 \\ 0.146619 & 0.064704 & 0.785320 & 0.146619 \\ 0.122823 & 0.146619 & 0.146619 & 0.541654 \end{pmatrix}$$

and

$$\widetilde{\boldsymbol{P}} = \begin{pmatrix} 4.6242 \\ 8.8528 \\ 22.1358 \\ 39.5816 \end{pmatrix}$$

Table 1.10 Estimation of Block Variance Component Using the Yates Procedure

```
proc varcomp data=IBD method=type1;
class BLOCK TRT;
model Y = TRT BLOCK/fixed=1;
title1 'TABLE 1.10';
title2 'ESTIMATION OF BLOCK VARIANCE COMPONENT';
title3 'USING THE YATES PROCEDURE';
run;
```

Variance Components Estimation Procedure

Class Level Information

Class	Levels	Values
BLOCK	5	1 2 3 4 5
TRT	4	1 2 3 4

Number of observations 10

Dependent Variable: Y

Type 1 Analysis of Variance

Source	DF	Sum of Squares	Mean Square	Expected Mean Square
TRT	3	439.066667	146.355556	Var(Error) + 0.6667 Var(BLOCK) + Q(TRT)
BLOCK	4	79.145833	19.786458	Var(Error) + 1.5 Var(BLOCK)
Error	2	18.187500	9.093750	Var(Error)
Corrected Total	9	536.400000	.	.

Type 1 Estimates

Variance Component	Estimate
Var(BLOCK)	7.12847
Var(Error)	9.09375

We then obtain

$$\widetilde{\boldsymbol{\tau}} = \widetilde{\boldsymbol{A}}^{-1}\widetilde{\boldsymbol{P}} = \begin{pmatrix} 11.9097 \\ 14.8659 \\ 24.4379 \\ 26.5510 \end{pmatrix}$$

We note that the elements of $\widetilde{\boldsymbol{\tau}}$ are actually the treatment least-squares means. Their estimated variances and the estimated variances for the treatment contrasts are obtained from $\widetilde{\boldsymbol{A}}^{-1} \times 9.09$ (see Tables 1.13 and 1.14).

1.14.4 Combined Intra- and Interblock Analysis Using SAS PROC MIXED

We illustrate here the numerical implementation of the ML and REML procedures as described in Sections 1.11.1 and 1.11.2, respectively, using SAS PROC MIXED. The results of the ML estimation are given in Table 1.11.

It takes four interations to obtain the solutions, yielding $\widehat{\sigma}_\beta^2 = 7.45$ and $\widehat{\sigma}_e^2 = 4.14$ and hence $\widehat{\gamma} = 1.80$ (Notice that these are quite different from the estimates obtained by the Yates procedure (Section 1.14.3) and the REML procedure as given below).

Since SAS uses a different parametrization than the one used in (1.84) it obtains "estimates" of μ and $\tau_i (i = 1, 2, 3, 4)$ separately. The type 3 coefficients indicate that the solutions for $\mu, \tau_i (i = 1, 2, 3, 4)$ are actually estimates of $\mu + \tau_4, \tau_1 - \tau_4, \tau_2 - \tau_4, \tau_3 - \tau_4$, respectively. From these solutions the least-squares means are then obtained as

$$\begin{aligned}
\text{LSMEAN(TRT 1)} = \widehat{\mu} + \widehat{\tau}_1 &= 11.65 = \widehat{\alpha}_1 \\
\text{LSMEAN(TRT 2)} = \widehat{\mu} + \widehat{\tau}_2 &= 15.63 = \widehat{\alpha}_2 \\
\text{LSMEAN(TRT 3)} = \widehat{\mu} + \widehat{\tau}_3 &= 24.09 = \widehat{\alpha}_3 \\
\text{LSMEAN(TRT 4)} = \widehat{\mu} &= 26.53 = \widehat{\alpha}_4
\end{aligned}$$

where the $\widehat{\alpha}_i$ denote the solutions to (1.87).

The REML procedure is illustrated in Table 1.12. It takes three iterations to obtain the estimates $\widetilde{\sigma}_\beta^2 = 6.35$ and $\widetilde{\sigma}_e^2 = 10.17$, and hence $\widetilde{\gamma} = 0.62$. We note that the REML and ML least-squares means are numerically quite similar even though $\widehat{\gamma}$ and $\widehat{\sigma}_e^2$ are substantially different from $\widetilde{\gamma}$ and $\widetilde{\sigma}_e^2$, respectively.

1.14.5 Comparison of Estimation Procedures

For a small proper incomplete block design we have employed the above four methods of estimating treatment least-squares means and treatment comparisons:

M1: Intrablock analysis

M2: Combined analysis: Yates

Table 1.11 Combined Analysis Using Maximum Likelihood With Post-hoc Comparisons

```
proc mixed data=IBD method=ML;

class BLOCK TRT;
model Y = TRT/ solution E3 ddfm=Satterth;

random BLOCK;
lsmeans TRT;
estimate 'TRT1 - (TRT2+TRT3)/2' TRT 1 -.5 -.5 0;
estimate 'TRT1 - TRT4' TRT 1 0 0 -1;
estimate 'TRT2 - TRT3' TRT 0 1 -1 0;
title1 'TABLE 1.11';
title2 'COMBINED ANALYSIS USING MAXIMUM LIKELIHOOD';
title3 'WITH POST-HOC COMPARISONS';
run;
                        The Mixed Procedure

                         Model Information

        Data Set                        WORK.IBD
        Dependent Variable              Y
        Covariance Structure            Variance Components
        Estimation Method               ML
        Residual Variance Method        Profile
        Fixed Effects SE Method         Model-Based
        Degrees of Freedom Method       Satterthwaite


                      Class Level Information

                 Class     Levels     Values

                 BLOCK          5     1 2 3 4 5
                 TRT            4     1 2 3 4


                            Dimensions

                  Covariance Parameters       2
                  Columns in X                5
                  Columns in Z                5
                  Subjects                    1
                  Max Obs Per Subject        10
                  Observations Used          10
                  Observations Not Used       0
                  Total Observations         10
```

Table 1.11 (*Continued*)

Iteration History

Iteration	Evaluations	-2 Log Like	Criterion
0	1	51.13433487	
1	2	50.29813032	0.00396161
2	1	50.22541829	0.00030283
3	1	50.22033455	0.00000230
4	1	50.22029773	0.00000000

Convergence criteria met.

Covariance Parameter Estimates

Cov Parm	Estimate
BLOCK	7.4528
Residual	4.1426

Fit Statistics

-2 Log Likelihood	50.2
AIC (smaller is better)	62.2
AICC (smaller is better)	90.2
BIC (smaller is better)	59.9

Solution for Fixed Effects

Effect	TRT	Estimate	Standard Error	DF	t Value	Pr > |t|
Intercept		26.5329	1.7767	8.87	14.93	<.0001
TRT	1	-14.8822	1.9330	4.75	-7.70	0.0007
TRT	2	-10.9029	2.1495	4.82	-5.07	0.0043
TRT	3	-2.4380	2.1495	4.82	-1.13	0.3099
TRT	4	0	.	.	.	.

Type 3 Coefficients for TRT

Effect	TRT	Row1	Row2	Row3
Intercept				
TRT	1	1		
TRT	2		1	
TRT	3			1
TRT	4	-1	-1	-1

Table 1.11 (*Continued*)

Type 3 Tests of Fixed Effects

Effect	Num DF	Den DF	F Value	Pr > F
TRT	3	4.76	23.37	0.0028

Estimates

Label	Estimate	Standard Error	DF	t Value	Pr > |t|
TRT1 - (TRT2+TRT3)/2	-8.2117	1.7085	4.3	-4.81	0.0072
TRT1 - TRT4	-14.8822	1.9330	4.75	-7.70	0.0007
TRT2 - TRT3	-8.4650	2.6087	5.86	-3.24	0.0182

Least Squares Means

Effect	TRT	Estimate	Standard Error	DF	t Value	Pr > |t|
TRT	1	11.6506	1.7767	8.87	6.56	0.0001
TRT	2	15.6299	2.0786	9.98	7.52	<.0001
TRT	3	24.0949	2.0786	9.98	11.59	<.0001
TRT	4	26.5329	1.7767	8.87	14.93	<.0001

M3: Combined analysis: ML

M4: Combined analysis: REML

In Tables 1.13 and 1.14 we present the estimates and their standard errors (exact or approximate) for these methods for purely numerical comparisons.

Based on the numerical results, we make the following observations, which should not necessarily be generalized:

1. For the least-squares means, M1 produces slightly smaller standard errors than M2, but the result is reversed for the contrast estimates.
2. The results for M2 and M4 are very similar, both with respect to estimates and standard errors.
3. In both tables M3 produces the smallest standard errors.

1.14.6 Testing of Hypotheses

To test the hypothesis

$$H_0: \tau_1 = \tau_2 = \cdots = \tau_t$$

Table 1.12 Combined Analysis Using Residual Maximum Likelihood With Post-hoc Comparisons

```
proc mixed data=IBD;

class BLOCK TRT;
model Y = TRT/ solution E3 ddfm=Satterth;
random BLOCK;
lsmeans TRT;
estimate 'TRT1 - (TRT2+TRT3)/2' TRT 1 -.5 -.5 0;
estimate 'TRT1 - TRT4' TRT 1 0 0 -1;
estimate 'TRT2 - TRT3' TRT 0 1 -1 0;
title1 'TABLE 1.12';
title2 'COMBINED ANALYSIS USING RESIDUAL MAXIMUM LIKELIHOOD';
title3 'WITH POST-HOC COMPARISONS';
run;
```

The Mixed Procedure

Model Information

Data Set	WORK.IBD
Dependent Variable	Y
Covariance Structure	Variance Components
Estimation Method	REML
Residual Variance Method	Profile
Fixed Effects SE Method	Model-Based
Degrees of Freedom Method	Satterthwaite

Class Level Information

Class	Levels	Values
BLOCK	5	1 2 3 4 5
TRT	4	1 2 3 4

Dimensions

Covariance Parameters	2
Columns in X	5
Columns in Z	5
Subjects	1
Max Obs Per Subject	10
Observations Used	10
Observations Not Used	0
Total Observations	10

Table 1.12 (*Continued*)

Iteration History

Iteration	Evaluations	-2 Res Log Like	Criterion
0	1	37.32907360	
1	2	37.14557447	0.00022052
2	1	37.14253025	0.00000186
3	1	37.14250585	0.00000000

Convergence criteria met.

Covariance Parameter Estimates

Cov Parm	Estimate
BLOCK	6.3546
Residual	10.1681

Fit Statistics

-2 Res Log Likelihood	37.1
AIC (smaller is better)	41.1
AICC (smaller is better)	45.1
BIC (smaller is better)	40.4

Solution for Fixed Effects

Effect	TRT	Estimate	Standard Error	DF	t Value	Pr > |t|
Intercept		26.5596	2.2615	5.93	11.74	<.0001
TRT	1	-14.5682	2.8843	2.68	-5.05	0.0196
TRT	2	-11.9152	3.2045	2.66	-3.72	0.0415
TRT	3	-2.0305	3.2045	2.66	-0.63	0.5766
TRT	4	0	.	.	.	.

Type 3 Coefficients for TRT

Effect	TRT	Row1	Row2	Row3
Intercept				
TRT	1	1		
TRT	2		1	
TRT	3			1
TRT	4	-1	-1	-1

Table 1.12 (*Continued*)

Type 3 Tests of Fixed Effects

Effect	Num DF	Den DF	F Value	Pr > F
TRT	3	2.42	10.82	0.0615

Estimates

Label	Estimate	Standard Error	DF	t Value	Pr > |t|
TRT1 - (TRT2+TRT3)/2	-7.5953	2.5979	2.2	-2.92	0.0891
TRT1 - TRT4	-14.5682	2.8843	2.68	-5.05	0.0196
TRT2 - TRT3	-9.8847	3.7522	3.82	-2.63	0.0607

Least Squares Means

Effect	TRT	Estimate	Standard Error	DF	t Value	Pr > |t|
TRT	1	11.9914	2.2615	5.93	5.30	0.0019
TRT	2	14.6444	2.7365	5.52	5.35	0.0023
TRT	3	24.5291	2.7365	5.52	8.96	0.0002
TRT	4	26.5596	2.2615	5.93	11.74	<.0001

versus

$$H_1\text{: not all } \tau_i \text{ are equal}$$

we consider a set of $t - 1$ linearly independent contrasts, say $\boldsymbol{C\tau}$, and test equivalently

$$H_0\text{: } \boldsymbol{C\tau} = \boldsymbol{0}$$

Table 1.13 Comparison of Least-Squares Means

TRT	M1		M2		M3		M4	
i	LSM(TRT$_i$)	SE	LSM(TRT$_i$)	SE	LSM(TRT$_i$)	SE	LSM(TRT$_i$)	SE
1	11.28	1.98	11.91	2.22	11.65	1.78	11.99	2.26
2	16.90	2.66	14.87	2.67	15.63	2.08	14.64	2.74
3	23.40	2.66	24.44	2.67	24.09	2.08	24.53	2.74
4	26.53	1.98	26.55	2.22	26.53	1.78	26.56	2.26

Table 1.14 Comparison of Contrast Estimates

	M1		M2		M3		M4	
CONTRAST[a]	$\widehat{C}$	SE	$\widetilde{C}$	SE	$\widetilde{C}$	SE	$\widetilde{C}$	SE
C1	−8.88	2.61	−7.75	2.47	−8.21	1.71	−7.59	2.60
C2	−15.25	3.02	−14.64	2.76	−14.88	1.93	−14.57	2.88
C3	−6.50	4.26	−9.57	3.62	−8.47	2.61	−9.88	3.75

[a]C1 = TRT1 − (TRT2 + TRT3)/2
C2 = TRT1 − TRT4
C3 = TRT2 − TRT3

Table 1.15 Comparison of Testing H_0:$\tau_1 = \tau_2 = \tau_3 = \tau_4$

Method	F Ratio	Denominator d.f.	P Value	Source
M1	9.41	2	.0975	Table 1.8
M2	11.73	2	.0796	See below
M3	23.37	4.76	.0028	Table 1.11
M4	10.82	2.42	.0615	Table 1.12

versus

$$H_1\colon \boldsymbol{C\tau} \neq \mathbf{0}$$

We then compute the test statistic

$$F = (\boldsymbol{C}\widetilde{\boldsymbol{\tau}})'[\boldsymbol{C}\widetilde{\boldsymbol{A}}^{-1}\boldsymbol{C}]^{-1}\boldsymbol{C}\widetilde{\boldsymbol{\tau}}/[(t-1)\ \mathrm{MS}(E)]$$

which follows approximately an F distribution with $t-1$ and $n-t-b+1$ d.f.
For the data set in Table 1.7, using

$$\boldsymbol{C} = \begin{pmatrix} 1 & -1 & 0 & 0 \\ 1 & 0 & -1 & 0 \\ 1 & 0 & 0 & -1 \end{pmatrix}$$

with the Yates procedure we obtain $F = 11.73$ with 3 and 2 d.f.

A comparison of the four methods of analysis [as described in (Section 1.13.5)] concerning the test of treatment effects is given in Table 1.15

It is interesting to note that the results for M1, M2, and M4 are in close agreement, whereas the result for M3 is quite different. This appears due to the fact that the estimate of σ_e^2 for M3, namely 4.1426, is quite different from the corresponding estimates using M2, namely, 9.09375, and M4, namely 10.1681.

CHAPTER 2

Balanced Incomplete Block Designs

2.1 INTRODUCTION

In the previous chapter we were concerned with incomplete block designs in very general terms, mainly with the analyses of these designs. But as mentioned earlier, there exist, among the totality of all incomplete block designs, several special types that have evolved in the process of refining the art of experimental design, these types being characterized by some common feature or property.

One such type of design is the balanced incomplete block (BIB) design introduced by Yates (1936a) (see also Section I.9.8.2). This is a proper, equireplicate, binary design such that each elementary treatment contrast, that is, the difference between two treatment effects, is estimated with the same variance.

2.2 DEFINITION OF THE BIB DESIGN

In this section we shall give a more precise and more formal definition of the BIB design. First we note that for a binary design the incidence matrix $\boldsymbol{N} = (n_{ij})$ has elements

$$n_{ij} = \begin{cases} 1 & \text{if treatment } i \text{ occurs in block } j \\ 0 & \text{otherwise} \end{cases}$$

Furthermore, for an equireplicate design

$$\sum_{j=1}^{b} n_{ij} = r \qquad \text{for all } i \tag{2.1}$$

Design and Analysis of Experiments. Volume 2: Advanced Experimental Design
By Klaus Hinkelmann and Oscar Kempthorne
ISBN 0-471-55177-5

and for a proper design

$$\sum_{i=1}^{t} n_{ij} = k \qquad \text{for all } j \tag{2.2}$$

where r is the number of replications for each treatment and k is the block size.

We denote the (i, i') element of $\boldsymbol{NN'}$ by $\lambda_{ii'}$; that is, $\sum_j n_{ij} n_{i'j} = \lambda_{ii'}$, where $\lambda_{ii'}$ denotes simply the number of blocks in which treatments i and i' appear together (it is for this reason that $\boldsymbol{NN'}$ is referred to as the *concordance* or *concurrence matrix*; see Pearce, 1963). Then, with (2.1) and (2.2) and the fact that all differences between treatment effects are estimated with the same variance, we must have $\lambda_{ii'} = \lambda$ for all i, $i'(i \neq i')$, as was shown by Thompson (1956) and Rao (1958). We thus have the following definition.

Definition 2.1 An incomplete block design is said to be a *balanced incomplete block* (BIB) design if it satisfies the following conditions:

i. The experimental material is divided into b blocks of k units each, different treatments being applied to the units in the same block.
ii. There are t treatments each of which occurs in r blocks.
iii. Any two treatments occur together in exactly λ blocks.

The quantities t, b, r, k, and λ are called the parameters of the BIB design. We note here that for a given set of parameters there may or may not exist a BIB design. □

2.3 PROPERTIES OF BIB DESIGNS

The following relations hold among the parameters, and even these are only necessary conditions for the existence of a BIB design:

$$rt = kb \tag{2.3}$$

$$\lambda(t-1) = r(k-1) \tag{2.4}$$

$$r > \lambda \tag{2.5}$$

$$b \geq t \tag{2.6}$$

Relationship (2.3) follows immediately from the fact that the total number of observation for the BIB design is

$$n = \sum_i \left(\sum_j n_{ij} \right) = \sum_j \left(\sum_i n_{ij} \right)$$

and to this we apply (2.1) and (2.2).

To establish (2.4) we consider a fixed treatment that is replicated r times and in each of these r blocks there are $k - 1$ other treatments. On the other hand, because of (iii) in Definition 2.1, each of the remaining $t - 1$ treatments must occur in these blocks exactly λ times. This leads immediately to (2.4).

Condition (2.5) follows from (2.4) and the fact that for an incomplete block design $k < t$.

Fisher (1940) first proved (2.6) to be a necessary condition for the existence of a balanced incomplete block design. The following proof is based on a property of the matrix $\boldsymbol{NN}'$. From the definition of the BIB design it follows that

$$\boldsymbol{NN}' = \begin{bmatrix} r & \lambda & & \lambda \\ \lambda & r & & \cdot \\ \cdot & \cdot & \cdot & \cdot \\ \cdot & & \cdot & \lambda \\ \cdot & \cdot & \cdot & \cdot \\ \lambda & & \lambda & r \end{bmatrix} = (r - \lambda)\boldsymbol{I} + \lambda \mathfrak{J}\mathfrak{J}' \tag{2.7}$$

Then

$$\begin{aligned} \boldsymbol{NN}'\mathfrak{J} &= (r - \lambda)\mathfrak{J} + t\lambda\mathfrak{J} \\ &= (r - \lambda + \lambda t)\mathfrak{J} \end{aligned}$$

So $(1/\sqrt{t})\mathfrak{J}$ is a normalized eigenvector with root $r - \lambda + \lambda t$. If $\boldsymbol{\xi}$ is another eigenvector of $\boldsymbol{NN}'$, which is orthogonal to $(1/\sqrt{t})\mathfrak{J}$, then we have from (2.7) that

$$\boldsymbol{NN}'\boldsymbol{\xi} = (r - \lambda)\boldsymbol{\xi}$$

and hence $(r - \lambda)$ is an eigenvalue with multiplicity $t - 1$. It follows then that

$$\det \boldsymbol{NN}' = (r - \lambda)^{t-1}(r - \lambda + \lambda t)$$

or, using (2.4),

$$\det \boldsymbol{NN}' = (r - \lambda)^{t-1} rk \tag{2.8}$$

Because of (2.5), this determinant is positive definite of rank t. Now $t =$ rank $\boldsymbol{NN}' =$ rank $\boldsymbol{N} \leq \min(t, b)$. Hence $b \geq t$.

In connection with the above necessary conditions for the existence of a BIB design, we shall mention briefly the special case when $t = b$ and hence $r = k$.

Definition 2.2 An incomplete block design (IBD) with the same number of treatments and blocks is called a *symmetrical incomplete block design*.

For a symmetrical BIB design, (2.8) will be of the form

$$\det NN' = (\det N)^2 = (r - \lambda)^{t-1} r^2$$

This implies, for example, that if t is even then $r - \lambda$ must be a perfect square for a symmetrical BIB design to exist. This is only one of several necessary conditions (Shrikhande, 1950) for the existence of a symmetrical BIB design.

Nonexistence theorems and methods of actually constructing BIB designs for a given set of parameters have become an important subject in itself, and we shall treat some aspects of this in Chapter 3.

Before we consider the analysis of BIB designs, we give as an example the design mentioned in Section 1.1. □

Example 2.1 We have $t = 7, b = 7, r = 4, k = 4, \lambda = 2$ and denoting the treatments now by $0, 1, \ldots, 6$ the design is listed as

(0, 3, 6, 5)

(2, 5, 4, 6)

(6, 0, 1, 4)

(0, 1, 2, 5)

(1, 6, 2, 3)

(4, 2, 3, 0)

(1, 3, 4, 5)

where each row represents a block, the treatments in each block being randomly assigned to the experimental units. □

2.4 ANALYSIS OF BIB DESIGNS

2.4.1 Intrablock Analysis

Using the results of Section 1.3.2 the RNE (1.7) take on the form

$$\begin{bmatrix} r\dfrac{k-1}{k} & -\dfrac{\lambda}{k} & & -\dfrac{\lambda}{k} \\ -\dfrac{\lambda}{k} & r\dfrac{k-1}{k} & & \vdots \\ \vdots & & \vdots & -\dfrac{\lambda}{k} \\ -\dfrac{\lambda}{k} & & -\dfrac{\lambda}{k} & r\dfrac{k-1}{k} \end{bmatrix} \begin{bmatrix} \widehat{\tau}_1 \\ \widehat{\tau}_2 \\ \vdots \\ \widehat{\tau}_t \end{bmatrix} = \begin{bmatrix} Q_1 \\ Q_2 \\ \vdots \\ Q_t \end{bmatrix}$$

or

$$\left[\left(r\frac{k-1}{k}+\frac{\lambda}{k}\right)\boldsymbol{I}-\frac{\lambda}{k}\mathfrak{I}\mathfrak{I}'\right]\widehat{\boldsymbol{\tau}}=\boldsymbol{Q} \tag{2.9}$$

A solution to (2.9) is obtained by using Theorem 1.1 with $a = \lambda/k$. Then

$$\widetilde{\boldsymbol{C}}^{-1}=\frac{1}{r\dfrac{k-1}{k}+\dfrac{\lambda}{k}}\boldsymbol{I} \tag{2.10}$$

Using (2.4) in (2.10) we can write

$$r\frac{k-1}{k}+\frac{\lambda}{k}=\frac{r}{k}\left(k-1+\frac{k-1}{t-1}\right)=r\frac{(k-1)t}{k(t-1)}$$

and hence

$$\widetilde{\boldsymbol{C}}^{-1}=\frac{1}{r\dfrac{(k-1)t}{k(t-1)}}\boldsymbol{I} \tag{2.11}$$

Recall that $\widetilde{\boldsymbol{C}}^{-1}\sigma_e^2$ is the variance–covariance matrix for estimable functions of treatment effects. Thus, from (2.11) we have

$$\text{var}\,(\widehat{\tau}_i-\widehat{\tau}_{i'})=\frac{2}{r\dfrac{(k-1)t}{k(t-1)}}\,\sigma_e^2$$

for all $i \neq i'$. Therefore also

$$\underset{i\neq i'}{\text{av.}}\ \text{var}(\widehat{\tau}_i-\widehat{\tau}_{i'})=\frac{2}{r\dfrac{(k-1)t}{k(t-1)}}\,\sigma_e^2$$

It follows then from (1.89) that $(k-1)t/k(t-1)$ is the efficiency factor E of the BIB design and according to (1.91) the upper bound for any IBD. We then write

$$\widetilde{\boldsymbol{C}}^{-1}=\frac{1}{rE}\boldsymbol{I} \tag{2.12}$$

Hence

$$\widehat{\boldsymbol{\tau}}=\frac{\boldsymbol{Q}}{rE} \tag{2.13}$$

is a solution to (2.9). Furthermore, it follows from (2.13) and (1.20) that any linear estimable function $\boldsymbol{c}'\boldsymbol{\tau} = \sum_i c_i \tau_i$ with $\sum c_i = 0$ is estimated by

$$\sum_i c_i \frac{Q_i}{rE} \tag{2.14}$$

with variance, using (2.12),

$$\sum_i c_i^2 \frac{\sigma_e^2}{rE} \tag{2.15}$$

The intrablock error variance σ_e^2 is estimated in the usual way by using the analysis of variance given in Table 1.2.

2.4.2 Combined Analysis

It follows from the definition of the BIB design and the form of $\boldsymbol{NN}'$ as given in (2.7) that the coefficient matrix, $\boldsymbol{A}$ say, in (1.60) is now of the form

$$\boldsymbol{A} = \left[r - (1-\rho^{-1})\left(\frac{r}{k} - \frac{\lambda}{k}\right)\right]\boldsymbol{I} - (1-\rho^{-1})\frac{\lambda}{k}\mathfrak{I}\mathfrak{I}' \tag{2.16}$$

By a familiar result in matrix algebra (e.g., Graybill, 1969) we find

$$\boldsymbol{A}^{-1} = \frac{1}{r - (1-\rho^{-1})\left(\dfrac{r}{k} - \dfrac{\lambda}{k}\right)}\left[\boldsymbol{I} + \frac{(1-\rho^{-1})\dfrac{\lambda}{k}}{r - (1-\rho^{-1})\left[\dfrac{r}{k} + (t-1)\dfrac{\lambda}{k}\right]}\mathfrak{I}\mathfrak{I}'\right] \tag{2.17}$$

Denoting the (i, i') element of $\boldsymbol{A}^{-1}$ by $a^{ii'}$, it follows from (1.62) and (1.63) that the linear function of treatment effects, $\boldsymbol{c}'\boldsymbol{\tau}$, with $\sum c_i = 0$ is estimated by $\boldsymbol{c}'\widehat{\widehat{\boldsymbol{\tau}}}$ with variance

$$\text{var}\left(\sum_i c_i \widehat{\widehat{\tau}}_i\right) = \sum_i c_i^2 a^{ii}\sigma_e^2 + \sum_{\substack{i,i' \\ i \neq i'}} c_i c_{i'} a^{ii'}\sigma_e^2 \tag{2.18}$$

Substituting the elements of (2.17) into (2.18) yields

$$\text{var}\left(\sum_i c_i \widehat{\widehat{\tau}}_i\right) = \sum_i c_i^2 \frac{\sigma_e^2}{r - (1-\rho^{-1})\left(\dfrac{r}{k} - \dfrac{\lambda}{k}\right)} \tag{2.19}$$

or, substituting (1.65) in (2.19),

$$\text{var}\left(\sum_i c_i \widehat{\widehat{\tau}}_i\right) = \sum_i c_i^2 \frac{\left(\sigma_e^2 + k\sigma_\beta^2\right)\sigma_e^2}{r\sigma_e^2 + (rk - r + \lambda)\sigma_\beta^2} \tag{2.20}$$

The following special familiar cases can be deduced easily from (2.19):

1. $r = \lambda$ (i.e., RCBD):

$$\text{var}\left(\sum_i c_i \widehat{\widehat{\tau}}_i\right) = \sum_i c_i^2 \frac{\sigma_e^2}{r}$$

2. $\rho^{-1} = 0$:

$$\text{var}\left(\sum_i c_i \widehat{\widehat{\tau}}_i\right) = \sum_i c_i^2 \frac{\sigma_e^2}{rE}$$

which is the same as (2.15), implying that no interblock information exists.

3. $\rho^{-1} = 1$:

$$\text{var}\left(\sum_i c_i \widehat{\widehat{\tau}}_i\right) = \sum_i c_i^2 \frac{\sigma_e^2}{r}$$

which is the same as the variance for a CRD. We note that σ_e^2 here refers to the error variance in a CRD, whereas σ_e^2 in case 1 refers to the within-block variance.

2.5 ESTIMATION OF ρ

As we already know from the discussion in the previous chapter and as is obvious from (2.17), the combined estimator $\boldsymbol{c}'\widehat{\widehat{\boldsymbol{\tau}}}$ for $\boldsymbol{c}'\boldsymbol{\tau}$ depends on $w = 1/\sigma_e^2$ and $w' = 1/\left(\sigma_e^2 + k\sigma_\beta^2\right)$ through $\rho = w/w'$. Since, in general, these parameters are not known, they will have to be estimated from the data; the estimates will have to be substituted into (2.17) in order to obtain the estimator for $\boldsymbol{c}'\boldsymbol{\tau}$, which we shall denote by $\boldsymbol{c}'\widetilde{\tau}$. Note that in our notation $c_i\widehat{\widehat{\tau}}$ depends on ρ and $\boldsymbol{c}'\widetilde{\tau}$ depends on $\widehat{\rho}$, where $\widehat{\rho}$ is an estimator for ρ.

In the previous chapter we described one method of estimating ρ, the Yates procedure, which estimates σ_e^2 and σ_β^2 from the intrablock analyses of variance with the model assuming the block effects to be random variables. This is a general procedure and as such applicable also to BIB designs. However, we also mentioned the problem that arises when using the Yates estimators, since one is not assured that the combined estimator for $\boldsymbol{c}'\boldsymbol{\tau}$ is better than the corresponding intrablock estimator.

The following theorem, which is a special case of a more general theorem given by Shah (1964), deals with this problem and gives a solution under a nonrestrictive condition.

Let

$$Z = \frac{t(t-1)k(k-1)}{(t-k)^2}\left[\mathrm{SS}(\boldsymbol{X}_\tau|\mathfrak{I}) - 2\boldsymbol{T}'\widehat{\boldsymbol{\tau}} + \frac{1}{E}\,\mathrm{SS}(\boldsymbol{X}_\tau|\mathfrak{I}, \boldsymbol{X}_\beta)\right] \tag{2.21}$$

where $\mathrm{SS}(\boldsymbol{X}_\tau|\mathfrak{I})$ and $\mathrm{SS}(\boldsymbol{X}_\tau|\mathfrak{I}, \boldsymbol{X}_\beta,)$ are given in Tables 1.2 and 1.1, respectively, $\boldsymbol{T}' = (T_1, T_2, \ldots, T_t)$ is the vector of the treatment totals and $\widehat{\boldsymbol{\tau}}$ is given by (2.13).

Theorem 2.1 For $t \geq 6$ and

$$\widehat{\rho} = \begin{cases} \dfrac{t-k}{t(k-1)}\left[\dfrac{Z}{(t-1)\mathrm{MS}_E} - 1\right] & \text{if} \quad \dfrac{Z}{\mathrm{MS}_E} > \dfrac{k(t-1)}{t-k} \\ 1 & \text{otherwise} \end{cases} \tag{2.22}$$

with Z defined by (2.21) and $\mathrm{MS}_E = \mathrm{MS}(\boldsymbol{I}|\mathfrak{I}, \boldsymbol{X}_\beta, \boldsymbol{X}_\tau)$ being the residual mean square as given in Table 1.1, the combined estimator, $\boldsymbol{c}'\widetilde{\boldsymbol{\tau}}$, for $\boldsymbol{c}'\boldsymbol{\tau}$ is uniformly better than the intrablock estimator, $\boldsymbol{c}'\widehat{\boldsymbol{\tau}}$, that is,

$$\mathrm{var}(\boldsymbol{c}'\widetilde{\boldsymbol{\tau}}) < \mathrm{var}(\boldsymbol{c}'\widehat{\boldsymbol{\tau}})$$

For the proof we refer to Shah (1964).

Similar estimators for ρ have been developed by Seshadri (1963) and still others by Graybill and Weeks (1959) and Graybill and Deal (1959).

We shall comment briefly on the nature of these estimators without going into any detail. Of particular interest is, of course, their relationship to the Yates estimator and the comparison of the performance of all the estimators with regard to estimating differences of treatment effects.

We note that the Yates estimator utilizes all $b-1$ d.f. between blocks as does the estimator proposed by Graybill and Weeks (1959), which, however, was shown by Seshadri (1963) to be not as good as the Yates estimator. All the other estimators utilize only part of the $b-1$ d.f., either $t-1$ or $b-t$. Under certain conditions these estimators have been shown to give rise to estimators for treatment differences uniformly better than the corresponding intrablock estimators. Such a property has not yet been proved for the Yates procedure, but the fact that it uses more information than the other estimators may lead one to suspect that it might actually be better than the estimators mentioned above that enjoy this property. To some extent this aspect was investigated by Shah (1970) along the following lines.

Let $\overline{V}$ be the average variance for treatment comparisons. Then $\overline{V}$ can be expressed as

$$\overline{V} = \frac{2\sigma_e^2}{r^*}$$

where r^* is defined as the *effective number of replications*. For example, r^* corresponding to the intrablock estimator is given by $r_I^* = rE$. The largest possible value for r^* is that corresponding to the combined estimator when ρ is known. Denote this value by r_c^*. Since usually ρ is not known and hence has to be estimated from the data, some loss of information is incurred; that is, $r^*(\widehat{\rho}) < r_c^*$, where $r^*(\widehat{\rho})$ is the effective number of replications when an estimator $\widehat{\rho}$ is used instead of ρ in the combined analysis. Shah (1970) shows that for his estimator (given in Theorem 2.1) the difference between r_c^* and $r^*(\widehat{\rho})$ is not appreciable for moderate values of ρ. This implies that the loss of information is not great and not much improvement, if any, can be expected from any other estimation procedure, including the Yates procedure. On the other hand we know that for Shah's estimator $r^*(\widehat{\rho}) > r_I^*$ when $t \geq 6$ (see Theorem 2.1), which is not guaranteed for the Yates procedure. For large values of ρ the difference between r_c^* and r_I^* is fairly small, so that not much is gained by using a combined estimator.

The conclusion from this discussion then is that the Shah procedure leads in general to quite satisfactory results and should therefore be used in the combined analysis.

2.6 SIGNIFICANCE TESTS

As mentioned in Section 1.13.4 significance tests concerning the treatment effects are performed as approximate F test by substituting $\widehat{\rho}$ for ρ in the coefficient matrix $\boldsymbol{A}$ of (1.60), using any of the previously described methods of estimating ρ. Such tests are mostly conveniently performed by choosing any option in SAS PROC MIXED.

An exact test, based, however, on the assumption of normality, for

$$H_0\colon \quad \tau_1 = \tau_2 = \cdots = \tau_t$$

against

$$H_1\colon \quad \text{not all } \tau_i \text{ equal}$$

was developed by Cohen and Sackrowitz (1989). We shall give a brief description here, but the reader should refer to the original article for details. The test is based on invariance properties and utilizes information from the intrablock analysis (as described in Sections 2.4.1 and 1.3) and the interblock analysis (as described for the general case in Section 1.7.2) by combining the P values from the two respective independent tests. In order to define the test we establish the following notation:

$$
\begin{aligned}
\widehat{\boldsymbol{\tau}} &= \text{solution to the intrablock normal equations, (2.13)} \\
\boldsymbol{\tau}^* &= \text{solution to the interblock normal equations, (1.48),} \\
&\quad\ \text{with } \boldsymbol{L} = \boldsymbol{I} \\
\boldsymbol{O} &= \text{matrix of } t-1 \text{ orthonormal contrasts for treatment effects} \\
\boldsymbol{U}_1 &= \boldsymbol{O}\widehat{\boldsymbol{\tau}} \\
\boldsymbol{U}_2 &= \boldsymbol{O}\boldsymbol{\tau}^* \\
\|\boldsymbol{U}_1\|^2 &= \boldsymbol{U}_1'\boldsymbol{U}_1 \\
\|\boldsymbol{U}_2\|^2 &= \boldsymbol{U}_2'\boldsymbol{U}_2 \\
\boldsymbol{V}_1 &= \boldsymbol{U}_1/\ \|\boldsymbol{U}_1\| \\
\boldsymbol{V}_2 &= \boldsymbol{U}_2/\ \|\boldsymbol{U}_2\| \\
R &= \boldsymbol{V}_1'\boldsymbol{V}_2 \\
S^2 &= \text{SS(Error from intrablock analysis)} \\
S^{*^2} &= \text{SS(Error from interblock analysis)} \\
&= \sum_{j=1}^{b}\left(B_j - k\mu^* - \sum_{i=1}^{t} n_{ij}\tau_i^*\right)^2 \\
T_1 &= S^2 + \frac{\lambda t}{k}\|\boldsymbol{U}_1\|^2 \\
T_2 &= S^{*^2} + \frac{r-\lambda}{k}\|\boldsymbol{U}_2\|^2 \\
a &= \min(T_1/T_2, 1) \\
\gamma &= 1/(a+1) \\
P &= P \text{ value for testing } H_0 \text{ using intrablock analysis} \\
P^* &= P \text{ value for testing } H_0 \text{ using interblock analysis, based on} \\
&\quad\ F \text{ statistic} \\
F^* &= \frac{(b-t)(r-\lambda)}{k(t-1)}\frac{\|\boldsymbol{U}_2\|^2}{S^{*^2}} \quad \text{with } t-1 \text{ and } b-t \text{ d.f.} \\
Z_1 &= -\ell_n P \\
Z_2 &= -\ell_n P^* \\
z &= \gamma Z_1 + (1-\gamma)Z_2
\end{aligned}
$$

Then, for $\gamma \neq \frac{1}{2}$, the P value for the exact test is given by

$$P^{**} = \left[\gamma e^{-z/\gamma} - (1-\gamma)e^{-z/(1-\gamma)}\right]/[(2\gamma - 1)(1+R)]$$

and for $\gamma = \frac{1}{2}$ by $\left[(2z+1)e^{-2z}\right]/(1+R)$. Cohen and Sackrowitz (1989) performed some power simulations and showed that the exact test is more powerful than the usual (approximate) F test. For a more general discussion see also Mathew, Sinha, and Zhou (1993) and Zhou and Mathew (1993).

We shall illustrate the test by using an example from Lentner and Bishop (1993).

Example 2.2 An experiment is concerned with studying the effect of six diets on weight gain of rabbits using 10 litters of three rabbits each. The data are given in Table 2.1A. The parameters of the BIB design are $t = 6, b = 10, k = 3, r = 5, \lambda = 2$.

The intrablock analysis using SAS PROC GLM is presented in Table 2.1A yielding $F = 3.16$ and $P = 0.0382$. In addition, Table 2.1A gives the SAS solutions to the normal equations, which are then used to compute the set of five orthonormal contrasts based on

$$\begin{pmatrix} -5/\sqrt{70} & -3/\sqrt{70} & -1/\sqrt{70} & 1/\sqrt{70} & 3/\sqrt{70} & 5/\sqrt{70} \\ 5/\sqrt{84} & -1/\sqrt{84} & -4/\sqrt{84} & -4/\sqrt{84} & -1/\sqrt{84} & 5/\sqrt{84} \\ -5/\sqrt{180} & 7/\sqrt{180} & 4/\sqrt{180} & -4/\sqrt{180} & -7/\sqrt{180} & 5/\sqrt{180} \\ 1/\sqrt{28} & -3/\sqrt{28} & 2/\sqrt{28} & 2/\sqrt{28} & -3/\sqrt{28} & 1/\sqrt{28} \\ -1/\sqrt{252} & 5/\sqrt{252} & -10/\sqrt{252} & 10/\sqrt{252} & -5/\sqrt{252} & 1/\sqrt{252} \end{pmatrix}$$

The linear, quadratic, cubic, quartic, and quintic parameter estimates represent the elements of $\boldsymbol{U}_1$, yielding $\| \boldsymbol{U}_1 \|^2 = 39.6817$.

A solution to the interblock normal equations [using $\boldsymbol{L} = \boldsymbol{I}$ in (1.48)] is given in Table 2.1B.

The estimates $\boldsymbol{\tau}^*$ are then used to obtain $\boldsymbol{U}_2$ and $\| \boldsymbol{U}_2 \|^2 = 537.5284$. We also obtain from Table 2.1B $F^* = 2.23$ and $P^* = 0.2282$.

Using $S^2 = 150.7728$ from Table 2.1A and $S^{*2} = 577.9133$ from Table 2.1B, all other values needed for the test can be computed. With $\gamma = 0.7828$ we obtain $Z = 2.8768$, and finally $P^{**} = 0.03$. This result is in good agreement with the P value of 0.0336 obtained from the combined analysis using SAS PROC MIXED with the REML option (see Table 2.1A). □

Table 2.1A Data, Intrablock, and Combined Analysis for BIB Design ($t = 6$, $b = 10$, $r = 5$, $k = 3$, LAMBDA=2)

```
options nodate pageno=1;
data rabbit1;
input B T Y @@;
datalines;
1 6 42.2 1 2 32.6 1 3 35.2
2 3 40.9 2 1 40.1 2 2 38.1
3 3 34.6 3 6 34.3 3 4 37.5
4 1 44.9 4 5 40.8 4 3 43.9
5 5 32.0 5 3 40.9 5 4 37.3
6 2 37.3 6 6 42.8 6 5 40.5
7 4 37.9 7 1 45.2 7 2 40.6
8 1 44.0 8 5 38.5 8 6 51.9
9 4 27.5 9 2 30.6 9 5 20.6
10 6 41.7 10 4 42.3 10 1 37.3
;
run;

proc print data=rabbit1;
title1 'TABLE 2.1A';
title2 'DATA FOR BIB DESIGN';
title3 '(t=6, b=10, r=5, k=3, LAMBDA=2)';
run;

proc glm data=rabbit1;
class B T;
model Y=B T/solution;
estimate 'linear' T -5 -3 -1 1 3 5/divisor=8.3667;
estimate 'quad' T 5 -1 -4 -4 -1 5/divisor=9.1652;
estimate 'cubic' T -5 7 4 -4 -7 5/divisor=13.4164;
estimate 'quartic' T 1 -3 2 2 -3 1/divisor=5.2915;
estimate 'quintic' T -1 5 -10 10 -5 1/divisor=15.8745;
title2 'INTRA-BLOCK ANALYSIS';
title3 'WITH ORTHONORMAL CONTRASTS';
run;

proc mixed data=rabbit1;
class B T;
model Y=T/solution;
random B;
lsmeans T;
title2 'COMBINED INTRA- AND INTER-BLOCK ANALYSIS';
title3 '(USING METHOD DESCRIBED IN SECTION 1.8)';
run;
```

Table 2.1A (*Continued*)

Obs	B	T	Y
1	1	6	42.2
2	1	2	32.6
3	1	3	35.2
4	2	3	40.9
5	2	1	40.1
6	2	2	38.1
7	3	3	34.6
8	3	6	34.3
9	3	4	37.5
10	4	1	44.9
11	4	5	40.8
12	4	3	43.9
13	5	5	32.0
14	5	3	40.9
15	5	4	37.3
16	6	2	37.3
17	6	6	42.8
18	6	5	40.5
19	7	4	37.9
20	7	1	45.2
21	7	2	40.6
22	8	1	44.0
23	8	5	38.5
24	8	6	51.9
25	9	4	27.5
26	9	2	30.6
27	9	5	20.6
28	10	6	41.7
29	10	4	42.3
30	10	1	37.3

The GLM Procedure

Class Level Information

Class	Levels	Values
B	10	1 2 3 4 5 6 7 8 9 10
T	6	1 2 3 4 5 6

Table 2.1A *(Continued)*

Number of observations 30

Dependent Variable: Y

Source	DF	Sum of Squares	Mean Square	F Value	Pr > F
Model	14	889.113889	63.508135	6.32	0.0005
Error	15	150.772778	10.051519		
Corrected Total	29	1039.886667			

R-Square	Coeff Var	Root MSE	Y Mean
0.855010	8.241975	3.170413	38.46667

Source	DF	Type I SS	Mean Square	F Value	Pr > F
B	9	730.3866667	81.1540741	8.07	0.0002
T	5	158.7272222	31.7454444	3.16	0.0382

Source	DF	Type III SS	Mean Square	F Value	Pr > F
B	9	595.7352222	66.1928025	6.59	0.0008
T	5	158.7272222	31.7454444	3.16	0.0382

Parameter	Estimate	Standard Error	t Value	Pr > |t|
linear	0.68326421	1.58518760	0.43	0.6726
quad	2.35674071	1.58519809	1.49	0.1578
cubic	3.14664639	1.58520742	1.99	0.0657
quartic	4.74975590	1.58520728	3.00	0.0090
quintic	1.09504761	1.58520728	0.69	0.5002

Parameter		Estimate		Standard Error	t Value	Pr > |t|
Intercept		42.61111111	B	2.24182052	19.01	<.0001
B	1	-3.29722222	B	2.79604144	-1.18	0.2567
B	2	0.83611111	B	2.79604144	0.30	0.7690

Table 2.1A (*Continued*)

Parameter		Estimate		Standard Error	t Value	Pr > |t|
B	3	-5.10000000	B	2.69433295	-1.89	0.0778
B	4	5.49722222	B	2.79604144	1.97	0.0681
B	5	-0.99166667	B	2.79604144	-0.35	0.7278
B	6	2.11111111	B	2.79604144	0.76	0.4619
B	7	2.48055556	B	2.69433295	0.92	0.3718
B	8	6.13055556	B	2.69433295	2.28	0.0380
B	9	-10.77777778	B	2.79604144	-3.85	0.0016
B	10	0.00000000	B	.	.	.
T	1	-3.30000000	B	2.24182052	-1.47	0.1617
T	2	-5.04166667	B	2.24182052	-2.25	0.0400
T	3	-2.90000000	B	2.24182052	-1.29	0.2154
T	4	-3.23333333	B	2.24182052	-1.44	0.1698
T	5	-8.52500000	B	2.24182052	-3.80	0.0017
T	6	0.00000000	B	.	.	.

NOTE: The X'X matrix has been found to be singular, and a generalized inverse was used to solve the normal equations. Terms whose estimates are followed by the letter 'B' are not uniquely estimable.

COMBINED INTRA- AND INTERBLOCK ANALYSIS
(USING METHOD DESCRIBED IN SECTION 1.8)

The Mixed Procedure

Model Information

Data Set	WORK.RABBIT1
Dependent Variable	Y
Covariance Structure	Variance Components
Estimation Method	REML
Residual Variance Method	Profile
Fixed Effects SE Method	Model-Based
Degrees of Freedom Method	Containment

Class Level Information

Class	Levels	Values
B	10	1 2 3 4 5 6 7 8 9 10
T	6	1 2 3 4 5 6

Table 2.1A (*Continued*)

Dimensions

Covariance Parameters	2
Columns in X	7
Columns in Z	10
Subjects	1
Max Obs Per Subject	30
Observations Used	30
Observations Not Used	0
Total Observations	30

Iteration History

Iteration	Evaluations	-2 Res Log Like	Criterion
0	1	160.26213715	
1	2	150.36288495	0.00010765
2	1	150.35691057	0.00000054
3	1	150.35688183	0.00000000

Convergence criteria met.

Covariance Parameter Estimates

Cov Parm	Estimate
B	21.6953
Residual	10.0840

Fit Statistics

-2 Res Log Likelihood	150.4
AIC (smaller is better)	154.4
AICC (smaller is better)	154.9
BIC (smaller is better)	155.0

Solution for Fixed Effects

Effect	T	Estimate	Standard Error	DF	t Value	Pr > |t|
Intercept		42.3454	2.1303	9	19.88	<.0001
T	1	-2.8100	2.2087	15	-1.27	0.2227

Table 2.1A (*Continued*)

Solution for Fixed Effects

Effect	T	Estimate	Standard Error	DF	t Value	Pr > |t|
T	2	-5.3172	2.2087	15	-2.41	0.0294
T	3	-2.9941	2.2087	15	-1.36	0.1953
T	4	-3.6952	2.2087	15	-1.67	0.1150
T	5	-8.4560	2.2087	15	-3.83	0.0016
T	6	0	.	.	.	.

Type 3 Tests of Fixed Effects

Effect	Num DF	Den DF	F Value	Pr > F
T	5	15	3.28	0.0336

Least Squares Means

Effect	T	Estimate	Standard Error	DF	t Value	Pr > |t|
T	1	39.5354	2.1303	15	18.56	<.0001
T	2	37.0282	2.1303	15	17.38	<.0001
T	3	39.3513	2.1303	15	18.47	<.0001
T	4	38.6502	2.1303	15	18.14	<.0001
T	5	33.8894	2.1303	15	15.91	<.0001
T	6	42.3454	2.1303	15	19.88	<.0001

Table 2.1B Data of Block Totals and Interblock Analysis

```
options nodate pageno=1;
data rabbit2;
input y x1 x2 x3 x4 x5 x6;
datalines;
110.0 0 1 1 0 0 1
119.1 1 1 1 0 0 0
106.4 0 0 1 1 0 1
129.6 1 0 1 0 1 0
110.2 0 0 1 1 1 0
120.6 0 1 0 0 1 1
123.7 1 1 0 1 0 0
134.4 1 0 0 0 1 1
```

Table 2.1B *(Continued)*

```
78.7 0 1 0 1 1 0
121.3 1 0 0 1 0 1
;
run;

proc print data=rabbit2;
title1 'TABLE 2.1 B';
title2 'DATA OF BLOCK TOTALS';

proc glm data=rabbit2;
model y=x1 x2 x3 x4 x5 x6;
title2 'INTER-BLOCK ANALYSIS';
run;
```

Obs	y	x1	x2	x3	x4	x5	x6
1	110.0	0	1	1	0	0	1
2	119.1	1	1	1	0	0	0
3	106.4	0	0	1	1	0	1
4	129.6	1	0	1	0	1	0
5	110.2	0	0	1	1	1	0
6	120.6	0	1	0	0	1	1
7	123.7	1	1	0	1	0	0
8	134.4	1	0	0	0	1	1
9	78.7	0	1	0	1	1	0
10	121.3	1	0	0	1	0	1

INTERBLOCK ANALYSIS

The GLM Procedure

Number of observations 10

The GLM Procedure

Dependent Variable: y

Source	DF	Sum of Squares	Mean Square	F Value	Pr > F
Model	5	1613.246667	322.649333	2.23	0.2282
Error	4	577.913333	144.478333		
Corrected Total	9	2191.160000			

Table 2.1B *(Continued)*

R-Square	Coeff Var	Root MSE	y Mean
0.736252	10.41587	12.01991	115.4000

Source	DF	Type I SS	Mean Square	F Value	Pr > F
x1	1	1044.484000	1044.484000	7.23	0.0547
x2	1	89.792667	89.792667	0.62	0.4746
x3	1	10.454444	10.454444	0.07	0.8012
x4	1	407.075556	407.075556	2.82	0.1685
x5	1	61.440000	61.440000	0.43	0.5499
x6	0	0.000000	.	.	.

Parameter	Estimate	Standard Error	t Value	Pr > |t|
Intercept	131.1000000 B	19.38152901	6.76	0.0025
x1	11.8000000 B	9.81421871	1.20	0.2955
x2	-13.5333333 B	9.81421871	-1.38	0.2400
x3	-5.8000000 B	9.81421871	-0.59	0.5863
x4	-17.4666667 B	9.81421871	-1.78	0.1497
x5	-6.4000000 B	9.81421871	-0.65	0.5499
x6	0.0000000 B	.	.	.

2.7 SOME SPECIAL ARRANGEMENTS

In some cases it is possible to arrange the treatments in the blocks of a BIB design in a special way that may lead to a reduction of either σ_e^2 or σ_β^2 or both. These arrangements lead therefore to more efficient designs.

2.7.1 Replication Groups Across Blocks

Designs of this type are characterized by the fact that the first s positions in the b blocks form a group of r' replications of all t treatments, the same being true for the next s positions, and so forth. If there are s' such groups, then $s'r' = r$.

We illustrate this type of arrangement in Example 2.3.

Example 2.3 Consider the BIB design with parameters $t = 10$, $r = 6$, $b = 15$, $k = 4$, $\lambda = 2$ given by the following plan:

Block	Replication Group I		Replication Group II	
1	1	2	3	4
2	1	5	2	6
3	1	3	7	8
4	4	9	1	10
5	5	7	1	9
6	6	8	1	10
7	3	9	2	6
8	2	4	7	10
9	5	10	2	8
10	2	7	8	9
11	3	10	5	9
12	6	10	3	7
13	4	8	3	5
14	6	7	4	5
15	8	9	4	6

It can be verified easily that for this design $s = 2$, $s' = 2$, $r' = 3$. □

The analysis is based on the following model:

$$\boldsymbol{y} = \mu\mathfrak{I} + \boldsymbol{X}_\tau\boldsymbol{\tau} + \boldsymbol{X}_\beta\boldsymbol{\beta} + \boldsymbol{X}_\gamma\boldsymbol{\gamma} + \boldsymbol{e} \tag{2.23}$$

or

$$y_{ij\ell} = \mu + \tau_i + \beta_j + \gamma_\ell + e_{ij\ell}$$

where τ_i and β_j are as defined previously and γ_ℓ is the (fixed) effect of the ℓth replication group ($\ell = 1, 2, \ldots, s'$). Since the replication groups are orthogonal to blocks and to treatments, the procedure for estimating linear contrasts of treatment effects is the same as before. More specifically, the RNE (2.9) are unchanged and the combined estimators are obtained by using (2.17).

The only change occurs with regard to the estimation of σ_e^2. This is reflected in the analysis of variance according to model (2.23) as given in Table 2.2. The form of this analysis of variance table shows a certain similarity to that of the Latin square design except that treatments and blocks are not orthogonal. But having sources of variation due to blocks and replication groups orthogonal to blocks achieves elimination of heterogeneity in two directions and hence leads to a reduction of the residual variance.

As can be seen from Example 2.3 and model (2.23), the replication groups play the same role as the columns in a Latin square type of design. It is for this reason that such designs have also been referred to as *Latinized incomplete block designs* (Harshbarger and Davis, 1952; John and Williams, 1995).

Table 2.2 ANOVA for BIB Design with Replication Groups Across Blocks

Source	d.f.	SS
$\boldsymbol{X}_\beta \mid \mathfrak{I}, \boldsymbol{X}_\gamma$	$b-1$	$\frac{1}{k}\sum_j B_j^2 - \frac{G^2}{n}$
$\boldsymbol{X}_\gamma \mid \mathfrak{I}, \boldsymbol{X}_\beta, \boldsymbol{X}_\tau$	$s'-1$	$\frac{1}{r't}\sum_\ell R_\ell^2 - \frac{G^2}{n}$
$\boldsymbol{X}_\tau \mid \mathfrak{I}, \boldsymbol{X}_\beta, \boldsymbol{X}_\gamma$	$t-1$	$\sum_i \widehat{\tau}_i Q_i$
$\boldsymbol{I} \mid \mathfrak{I}, \boldsymbol{X}_\beta, \boldsymbol{X}_\tau \boldsymbol{X}_\tau$	$n-t-b-s'+2$	Subtraction
Total	$n-1$	$\sum_{ij\ell} y_{ij\ell}^2 - \frac{G^2}{n}$

R_ℓ = total of ℓth replicate.

As a special case of designs described in this section we mention the following definition.

Definition 2.3 A BIB design that can be arranged in replication groups across blocks, each group containing a complete replicate (i.e., $r' = 1$, $s' = r = k$), is called a *Youden square design* (see I.10.5).

These designs, which for obvious reasons are also called *incomplete Latin square designs*, were first introduced by Youden (1937, 1940). They overcome the most severe objection to Latin square designs, namely that the number of replications (of each treatment) is the same as the number of treatments, t, which for large t is difficult to achieve from a practical point of view. Also, the block size in a Latin square design will then usually become too large.

It was shown by Hartley and Smith (1948) that for all incomplete block designs in which the number of treatments is equal to the number of blocks, such arrangements exist. Their proof provides, in fact, a general procedure of constructing a Youden square from a given symmetrical BIB design. This is accomplished by repeatedly interchanging the positions of pairs of treatments in the same block. A list of Youden squares is given by Cochran and Cox (1957, 1992). □

2.7.2 Grouped Blocks

Designs of this type are characterized by the fact that the first b' blocks form a group of α replicates of the t treatments, so do the next b' blocks, and so on, these groups of blocks being thereby orthogonal to treatments. If there are q of such groups, then $qb' = b$ and $q\alpha = r$.

Following Shrikhande and Raghavarao (1963) we formalize this in the following definition.

Definition 2.4 A BIB design with parameters t, b, k, r, λ that can be arranged in groups of blocks, each group containing α replicates of the t treatments, is said to be α-*resolvable*.

This concept of α-resolvability is an extension of the concept of resolvability given by Bose (1942), which refers to the case $\alpha = 1$.

The following design is an example of a 3-resolvable design. □

Example 2.4 Consider the following design with parameters $t = t, b = 10, k = 3, 4 = 6, \lambda = 3$:

Block	**Treatment**			**Replication Group**
1	1	2	3	
2	1	2	5	
3	1	4	5	I
4	2	3	4	
5	3	4	5	
6	1	2	4	
7	1	3	4	
8	1	3	5	II
9	2	3	5	
10	2	4	5	

We find $b' = 5, q = 2, \alpha = 3$. □

For α-resolvable BIB designs inequality (2.6) can be improved as follows.

Theorem 2.2 If a BIB design with parameters t, b, k, r, λ is α-resolvable, then the following inequality must hold:

$$b \geq t + q - 1 \tag{2.24}$$

Proof Since for an α-resolvable BIB design there exist $q - 1$ linearly independent relationships between the columns of $\boldsymbol{N}$, we have

$$t = \text{rank}(\boldsymbol{N}\boldsymbol{N}') = \text{rank}(\boldsymbol{N}) \leq b - q + 1$$

which implies (2.24). □

Definition 2.5 An α-resolvable BIB design is called *affine α-resolvable* if any pair of blocks in the same replication group have q_1 treatments in common and if any pair of blocks from two different replication groups have q_2 treatments in common. □

For affine α-resolvable BIB designs the following results are due to Shrikhande and Raghavarao (1963).

Theorem 2.3 In an affine α-resolvable BIB design k^2/t is an integer.

Proof Without loss of generality consider the first block in the first replication group, B_{11} say. By definition, B_{11} has q_2 treatments in common with every block in the remaining $q - 1$ replication groups, each group consisting of b/q blocks. Hence

$$(q-1)\,\frac{b}{q}\,q_2 = k(r-\alpha)$$

which, since $\alpha q = r$, yields $q_2 = k^2/t$. Since q_2 is an integer, k^2/t is an integer. □

Theorem 2.4 For an affine α-resolvable BIB design, the equality

$$b = t + q - 1 \tag{2.25}$$

holds.

Proof We shall give here the proof for $\alpha = 1$ following Bose (1942) and refer to Shrikhande and Raghavarao (1963) for the general case.

Consider a resolvable BIB design. We then have

$$t = b'k \qquad b = b'r \tag{2.26}$$

since the b blocks are divisible into r sets of b' blocks each, each set containing each treatment exactly once. Let the blocks belonging to the ith set, S_i, be denoted by $B_{i1}, B_{i2}, \ldots, B_{ib'}$ for $i = 1, 2, \ldots, r$. Consider now a particular block, B_{11} say. Let ℓ_{ij} be the number of treatments common to blocks B_{11} and $B_{ij} (i = 2, \ldots, r;\ j = 1, 2, \ldots, b')$. Further, let m denote the average and s^2 the variance of the $b'(r-1)$ numbers ℓ_{ij}.

Now each of the k treatments in B_{11} is replicated r times, and since the design is resolvable, that is, 1-resolvable, a given treatment in B_{11} will occur in exactly one block each in the $r - 1$ other sets, $S_2, S_3, \ldots, S_r$. This is true for all k treatments in B_{11}, and hence

$$\sum_{ij} \ell_{ij} = k(r-1) \tag{2.27}$$

and therefore

$$m = \frac{k}{b'} = \frac{k^2}{t} \tag{2.28}$$

using (2.26) for the last expression.

Further, the $k(k-1)/2$ pairs of treatments in B_{11} each occur $\lambda - 1$ times together in the sets $S_2, S_3, \ldots, S_r$. Hence

$$\frac{1}{2}\sum_{ij} \ell_{ij}(\ell_{ij} - 1) = \frac{1}{2}(\lambda - 1)k(k-1)$$

and therefore, using (2.27)

$$\sum_{ij} \ell_{ij}^2 = k[r - 1 + (\lambda - 1)(k - 1)] \tag{2.29}$$

Since

$$\lambda = \frac{r(k-1)}{t-1} = \frac{r(k-1)}{b'k-1}$$

we rewrite (2.29) as

$$\sum_{ij} \ell_{ij}^2 = \frac{k[(b'k-1)(r-k) + r(k-1)^2]}{b'k-1}$$

Then we find after some algebra, using (2.28), (2.29), and (2.26),

$$\begin{aligned} s^2 &= \frac{\sum_{ij}(\ell_{ij} - m)^2}{b'(r-1)} \\ &= \frac{\sum_{ij}\ell_{ij}^2}{b'(r-1)} - m^2 \\ &= \frac{k(t-k)(b-t-r+1)}{b'^2(r-1)(t-1)} \end{aligned} \tag{2.30}$$

Since $s^2 \geq 0$, (2.30) implies, of course, the earlier result (2.24), with $q = r$ for $\alpha = 1$.

We now consider an affine resolvable BIB design. Then, by definition, $\ell_{ij} = q_2$ for all i, j. Hence s^2 of (2.30) equals zero, which implies (2.25) with $q = r$ for a 1-resolvable BIB design. □

Corollary 2.1 If the parameters of a BIB design satisfy (2.25) and k^2/t is not an integer, then the design is not resolvable.

Example 2.5 An example of an affine resolvable BIB design is given by the following design with parameters $t = 8, b = 14, k = 4, r = 7, \lambda = 3, b' = 2, q = 7, q_1 = 0, q_2 = 2, \alpha = 1$:

Block	Treatments				Replication Group
1	1	2	3	4	I
2	5	6	7	8	
3	1	2	7	8	II
4	3	4	5	6	
5	1	3	6	8	III
6	2	4	5	7	
7	1	4	6	7	IV
8	2	3	5	8	
9	1	2	5	6	V
10	3	4	7	8	
11	1	3	5	7	VI
12	2	4	6	8	
13	1	4	5	8	VII
14	2	3	6	7	

A natural model for the analysis of such a design is

$$\boldsymbol{y} = \mu \mathfrak{I} + \boldsymbol{X}_\tau \boldsymbol{\tau} + \boldsymbol{X}_\eta \boldsymbol{\eta} + \boldsymbol{X}_{\beta^*} \boldsymbol{\beta}^* + \boldsymbol{e} \tag{2.31}$$

or

$$y_{ij\ell} = \mu + \tau_i + \eta_j + \beta^*_{j\ell} + e_{ij\ell}$$

where η_j is the (fixed) effect of the jth replication group ($j = 1, 2, \ldots, q$) and $\beta^*_{j\ell}$ is the effect of the ℓth block in the jth group ($\ell = 1, 2, \ldots, b'$). This brings out the point that the blocks are nested in the replication groups and, hence, in the case of the $\beta^*_{j\ell}$ being random variables (as assumed for the combined analysis), that $\sigma^2_{\beta^*}$ now measures the variability of the blocks within groups. Since $b' < b$, it can then usually be concluded that $\sigma^2_{\beta*}$ for this design will be smaller than the corresponding σ^2_β in the BIB design without grouping of blocks. Consequently, only the part of the analysis of variance that deals with the estimation of $\sigma^2_{\beta*}$ (to be used for the combined analysis, in particular, using the Yates estimator) is affected when using this design and model (2.31). This is shown in Table 2.3, where R_j refers to the total of the jth replication group, and $B_{j\ell}$ refers to the total of the ℓth block within the jth replication group. □

2.7.3 α-Resolvable BIB Designs with Replication Groups Across Blocks

The designs of this type combine the properties of the designs discussed in Sections 2.7.1 and 2.7.2 and hence lead to a possible reduction in both σ^2_e and σ^2_β.

Example 2.6 As an example we consider the following BIB design with parameters $t = 6$, $b = 15$, $k = 4$, $r = 10$, $\lambda = 6$:

	Replication Groups Across Blocks				
Block	**I**			**II**	**Replication Group**
1	1	2	3	4	
2	1	5	4	6	I
3	2	3	5	6	
4	1	2	3	5	
5	1	2	4	6	II
6	3	5	4	6	
7	1	3	2	6	
8	4	5	1	3	III
9	4	6	2	5	
10	2	4	1	5	
11	5	6	1	3	IV
12	3	6	2	4	
13	5	6	1	2	
14	4	6	1	3	V
15	3	4	2	5	

We find that $r' = 5$, $s' = 2$, $\alpha = 2$, $q = 5$ in the notation of Sections 2.7.1 and 2.7.2. □

A natural model for this type of design is

$$\boldsymbol{y} = \mu\mathfrak{I} + \boldsymbol{X}_\tau\boldsymbol{\tau} + \boldsymbol{X}_\gamma\boldsymbol{\gamma} + \boldsymbol{X}_\eta\boldsymbol{\eta} + \boldsymbol{X}_{\beta*}\boldsymbol{\beta}^* + \boldsymbol{e} \tag{2.32}$$

Table 2.3 Analysis of Variance for Resolvable BIB Design

Source	d.f.	SS	E(MS)
$\boldsymbol{X}_\tau \mid \mathfrak{I}, \boldsymbol{X}_\eta$	$t-1$	$\frac{1}{r}\sum_i T_i^2 - \frac{G^2}{n}$	
$\boldsymbol{X}_\eta \mid \mathfrak{I}, \boldsymbol{X}_\tau$	$q-1$	$\frac{1}{b'k}\sum_j R_j^2 - \frac{G^2}{n}$	
$\boldsymbol{X}_{\beta^*} \mid \mathfrak{I}, \boldsymbol{X}_\tau, \boldsymbol{X}_\eta$	$b-q$	Difference	$\sigma_e^2 + \frac{n-t-k(q-1)}{b-q}\sigma_{\beta^*}^2$
$\boldsymbol{I} \mid \mathfrak{I}, \boldsymbol{X}_\tau, \boldsymbol{X}_\eta, \boldsymbol{X}_{\beta^*}$	$n-t-b+1$	$\sum_{ij\ell} y_{ij\ell}^2 - \sum_i \widehat{\tau}_i Q_i - \frac{1}{k}\sum_{j\ell} B_{j\ell}^2$	σ_e^2
Total	$n-1$	$\sum_{ij\ell} y_{ij\ell}^2 - \frac{G^2}{n}$	

or

$$y_{ij\ell u} = \mu + \tau_i + \gamma_j + \eta_\ell + \beta_{\ell u}^* + e_{ij\ell u}$$

where all the parameters are as previously defined.

The analysis of variance associated with (2.32) follows easily from those presented in Tables 2.2 and 2.3 with the partitioning of the total d.f. as given in Table 2.4.

Table 2.4 Outline of Analysis of Variance for Model (2.32)

Source	d.f.
$\boldsymbol{X}_\tau \mid \mathfrak{I}, \mathbf{X}_\gamma, \mathbf{X}_\eta$	$t-1$
$\boldsymbol{X}_\gamma \mid \mathfrak{I}, \boldsymbol{X}_\tau, \boldsymbol{X}_\eta$	$s'-1$
$\boldsymbol{X}_\eta \mid \mathfrak{I}, \boldsymbol{X}_\tau, \boldsymbol{X}_\gamma$	$q-1$
$\boldsymbol{X}_{\beta^*} \mid \mathfrak{I}, \boldsymbol{X}_\tau, \boldsymbol{X}_\gamma, \boldsymbol{X}_\eta$	$b-q$
$\boldsymbol{I} \mid \mathfrak{I}, \boldsymbol{X}_\tau, \boldsymbol{X}_\gamma, \boldsymbol{X}_\eta, \boldsymbol{X}_{\beta^*}$	$n-t-b-s'+2$
Total	$n-1$

2.8 RESISTANT AND SUSCEPTIBLE BIB DESIGNS

2.8.1 Variance-Balanced Designs

The class of BIB designs belongs to the class of *variance-balanced designs*, that is, designs for which every normalized estimable linear function of the treatment effects is estimated with the same variance. It is easy to establish the following theorem.

Theorem 2.5 A necessary and sufficient condition for a connected design to be variance balanced is that its $\boldsymbol{C}$ matrix is of the form

$$\boldsymbol{C} = c_1 \boldsymbol{I} + c_2 \mathfrak{I}\mathfrak{I}' \tag{2.33}$$

Proof For an IBD with rank $\boldsymbol{C} = t - 1$ let $\boldsymbol{\xi}_1, \boldsymbol{\xi}_2, \ldots, \boldsymbol{\xi}_{t-1}, \boldsymbol{\xi}_t = (1/\sqrt{t})\mathfrak{I}$ be a set of orthonormal eigenvectors of $\boldsymbol{C}$. If we write

$$\boldsymbol{O} = \left(\boldsymbol{O}_1 \ \vdots\ \frac{1}{\sqrt{t}}\mathfrak{I}\right)$$

then

$$\boldsymbol{O}'\boldsymbol{C}\boldsymbol{O} = \begin{pmatrix} \boldsymbol{D} & \boldsymbol{0}' \\ \boldsymbol{0} & 0 \end{pmatrix}$$

where $\boldsymbol{D} = \text{diag}(d_i)$. Then the RNE

$$\boldsymbol{C}\widehat{\boldsymbol{\tau}} = \boldsymbol{Q}$$

can be written as

$$\boldsymbol{O}'\boldsymbol{C}\boldsymbol{O}\boldsymbol{O}'\widehat{\boldsymbol{\tau}} = \boldsymbol{O}'\boldsymbol{Q}$$

or

$$\begin{pmatrix} \boldsymbol{D} & \boldsymbol{0}' \\ \boldsymbol{0} & 0 \end{pmatrix} \begin{pmatrix} \boldsymbol{O}_1'\widehat{\boldsymbol{\tau}} \\ 0 \end{pmatrix} = \begin{pmatrix} \boldsymbol{O}_1\boldsymbol{Q} \\ 0 \end{pmatrix}$$

Hence

$$\boldsymbol{O}_1'\widehat{\boldsymbol{\tau}} = \boldsymbol{D}^{-1}\boldsymbol{O}_1\boldsymbol{Q}$$

with

$$\begin{aligned} \text{var}(\boldsymbol{O}_1'\widehat{\boldsymbol{\tau}}) &= \boldsymbol{D}^{-1}\boldsymbol{O}_1'\boldsymbol{C}\boldsymbol{O}_1\boldsymbol{D}^{-1}\sigma_e^2 \\ &= \boldsymbol{D}^{-1}\sigma_e^2 \end{aligned}$$

For a variance-balanced design we must then have $\boldsymbol{D} = d\boldsymbol{I}$. It then follows that

$$\boldsymbol{C} = \boldsymbol{O}\begin{pmatrix} d\boldsymbol{I} & \boldsymbol{0} \\ \boldsymbol{0}' & 0 \end{pmatrix}\boldsymbol{O}'$$

$$= d\boldsymbol{O}_1\boldsymbol{O}_1' - d\left(\boldsymbol{I} - \frac{1}{\sqrt{t}}\,\mathfrak{I}\mathfrak{I}'\right)$$

[see (1.17)] which we can write as

$$\boldsymbol{C} = c_1\boldsymbol{I} + c_2\mathfrak{I}\mathfrak{I}'$$

which is (2.33). Obviously, if $\boldsymbol{C}$ is of the form (2.33), the design is variance balanced. □

Further, among all binary, equireplicate, proper designs the BIB designs are the only variance-balanced designs. There do exist, however, variance-balanced designs with unequal block sizes, for example.

2.8.2 Definition of Resistant Designs

An interesting question then is: If from a BIB design one or several treatments are deleted, is the resulting design still variance balanced? In general the answer will be negative, but for some designs the answer will be, more or less, positive. Such designs have been called *resistant BIB designs* by Hedayat and John (1974) who have also given a characterization of such designs and have shown their existence. We shall give some results.

Let D be a BIB design with parameters t, b, k, r, λ and let D^* be the design obtained from D by deleting all experimental units that received a certain subset L of $\nu(\leq t-2)$ treatments. Following Hedayat and John (1974) we then consider the following definitions.

Definition 2.6 The design D is said to be *globally resistant* of degree ν if D^* is variance balanced when any subset L of ν treatments is deleted. □

Definition 2.7 The design D is said to be *locally resistant* of degree ν if D^* is variance balanced only with respect to certain subsets L of ν treatments. □

Definition 2.8 The design D is said to be *susceptible* if there exist no subsets L such that D^* is variance balanced. □

2.8.3 Characterization of Resistant Designs

Of particular interest, from a practical point of view, are resistant designs (locally and globally) of degree 1. It may happen that the researcher feels that a particular treatment, say a drug, may have to be discontinued before the termination of the experiment. If a BIB design had been chosen for this experiment to achieve equal precision for treatment comparisons, then such a course of action might destroy this feature, unless the BIB design is globally resistant of degree 1 or locally resistant of degree 1 with respect to treatment θ, say, and the treatment in question has been assigned the number θ.

To characterize suitable designs, let D denote a BIB design with parameters t, b, k, r, λ and let D be divided into two parts D_θ and $D_{\overline{\theta}}$. Design D_θ consists of all the blocks of D that contain the treatment θ, and $D_{\overline{\theta}}$ consists of the remaining blocks of D that do not contain the treatment θ. If $\boldsymbol{N}$, $\boldsymbol{N}_\theta$, $\boldsymbol{N}_{\overline{\theta}}$ are the incidence matrices of D, D_θ, and $D_{\overline{\theta}}$, respectively, then

$$\boldsymbol{N} = (\boldsymbol{N}_\theta, \boldsymbol{N}_{\overline{\theta}}) \tag{2.34}$$

with

$$\boldsymbol{N}\boldsymbol{N}' = \boldsymbol{N}_\theta \boldsymbol{N}'_\theta + \boldsymbol{N}_{\overline{\theta}}\boldsymbol{N}'_{\overline{\theta}} = (r - \lambda)\boldsymbol{I} + \lambda \mathfrak{I}\mathfrak{I}'$$

Now suppose that treatment θ has been deleted from D, that is, from D_θ. Call the resulting designs D^* and D^*_θ, respectively, and their corresponding incidence matrices $\boldsymbol{N}^*$ and $\boldsymbol{N}^*_\theta$. Also, let $\boldsymbol{N}^*_{\overline{\theta}}$ be the incidence matrix obtained from $\boldsymbol{N}_{\overline{\theta}}$ in (2.34) by deleting the row (of zeros) corresponding to θ, and call the corresponding design $D^*_{\overline{\theta}}$. We then have the following theorem due to Hedayat and John (1974).

Theorem 2.6 The BIB design D is locally resistant with respect to θ if and only if $D_{\overline{\theta}}$ is a BIB design.

Proof Let $\boldsymbol{C}^*$ denote the $\boldsymbol{C}$ matrix for D^*. Then, with

$$\boldsymbol{N}^* = (\boldsymbol{N}^*_\theta, \boldsymbol{N}^*_{\overline{\theta}})$$

a $(t-1) \times b$ incidence matrix, we have

$$\boldsymbol{C}^* = r\boldsymbol{I}_{t-1} - \frac{1}{k-1}\boldsymbol{N}^*_\theta \boldsymbol{N}^{*'}_\theta - \frac{1}{k}\boldsymbol{N}^*_{\overline{\theta}}\boldsymbol{N}^{*'}_{\overline{\theta}} \tag{2.35}$$

Since D^*_θ and $D^{*'}_{\overline{\theta}}$ are derived from the BIB design D, we know that if the general (off-diagonal) element in $\boldsymbol{N}^*_{\overline{\theta}}\boldsymbol{N}^{*'}_{\overline{\theta}}$ is λ_{ij}, the corresponding element in $\boldsymbol{N}^*_\theta \boldsymbol{N}^{*'}_\theta$ must be $\lambda - \lambda_{ij}$. Hence, the diagonal and off-diagonal elements of C^* are

$$c^*_{ii} = r - \frac{\lambda}{k-1} - \frac{r-\lambda}{k}$$

$$c_{ij}^{*} = -\frac{\lambda - \lambda_{ij}}{k-1} - \frac{\lambda_{ij}}{k}$$

$$= -\lambda_{ij}\left(\frac{1}{k} - \frac{1}{k-1}\right) - \frac{\lambda}{k}$$

Hence $\boldsymbol{C}^*$ of (2.35) is of the form (2.33) if and only if all λ_{ij} are equal, that is, if $D_{\overline{\theta}}^{*}$ and hence $D_{\overline{\theta}}$ is a BIB design (with parameters $t_1 = t-1, b_1 = b-r, k_1 = k, r_1 = r-\lambda, \lambda_1$). □

We illustrate this theorem with the following example.

Example 2.7 The following plan, D, is a BIB design with parameters $t = 8, b = 14, k = 4, r = 7, \lambda = 3$:

1	2	3	4
5	6	7	8
1	2	7	8
3	4	5	6
1	3	6	8
2	4	5	7
1	4	6	7
2	3	5	8
1	2	5	6
3	4	7	8
1	3	5	7
2	4	6	8
1	4	5	8
2	3	6	7

Let $\theta = 1$; then $D_{\overline{1}}$ is given by

5	6	7	8
3	4	5	6
2	4	5	7
2	3	5	8
3	4	7	8
2	4	6	8
2	3	6	7

This is a BIB design with $t_1 = 7, b_1 = 7, k_1 = 4, r_1 = 4, \lambda_1 = 2$. Hence D is resistant w.r.t. treatment 1.

The design D^* obtained from design D in Example 2.7 provides an illustration of the construction of an equireplicate, variance-balanced incomplete block design for $t^* = 7$ treatments with unequal block sizes. In particular, we have here $b^* = 14$ blocks, with 7 blocks of size $k_1^* = 3$ and 7 blocks of size $k_2^* = 4$. □

Corollary 2.2 The BIB design D is globally resistant of degree 1 if and only if $D_{\overline{\theta}}$ is a BIB design for all θ of D.

Corollary 2.3 The property of being a resistant BIB design depends not only on the parameters of D but also in the way D has been constructed.

An interesting example of this is provided by Hedayat and John (1974).

Corollary 2.4 The BIB design D is locally resistant w.r.t. θ if and only if every triple containing θ appears the same number of times in D.

Proof From the proof of Theorem 2.6 it follows that if $D_{\overline{\theta}}^*$ is a BIB design, then also D_θ^* is a BIB design, that is, every pair of treatments (other than θ) appears together the same number of times. Hence in D_θ (and therefore in D) every triple containing θ appears together the same number of times. □

This leads one to another characterization of globally resistant designs of degree 1. Before we state this we give the following definition.

Definition 2.9 An incomplete block design in which each triple of treatments occurs together the same number of times, δ, is referred to as a *doubly balanced* incomplete block (DBIB) design.

This definition is due to Calvin (1954), and such designs and their properties have been discussed by Calvin (1954) and Raghavarao and Thartare (1967, 1970). The design given in Example 2.7 is a DBIB design with $\delta = 1$. □

Based on Definition 2.9 and the proof of Corollary 2.4 we then have the following theorem.

Theorem 2.7 A BIB design D is globally resistant of degree 1 if and only if it is a DBIB design.

Existence, construction, and properties of resistant BIB designs have been discussed by Hedayat and John (1974) and John (1976), and we shall not go into this, except for the following theorem.

Theorem 2.8 Every symmetric BIB design with blocks of size k is locally resistant of degree k.

The proof follows immediately from the construction procedure of BIB designs described in Section 3.2.

An interesting follow-up to Theorem 2.8 is provided by Baksalary and Puri (1990). They show that the symmetry of the design is both necessary and sufficient for the local resistance of degree k with respect to L consisting of k treatments that occur in exactly one block of the design if we demand D^* to be a BIB design (rather than just a variance-balanced design).

2.8.4 Robustness and Connectedness

Another form of robustness of BIB designs is concerned with the question to what extent the unavailability of observations, such as missing or lost observations, affects the property of connectedness of the BIB design.

Following Ghosh (1982) we give the following definition.

Definition 2.10 A BIB design is said to be *robust* against unavailability of any s observations if the block design obtained by omitting any s observations remains a connected design. □

Using the notion of a bipartite graph, consisting of the subsets $\mathcal{T} = (1, 2, \ldots, t)$ and $\mathcal{B} = (1,\ 2,\ \ldots,\ b)$, and the notion of connectedness (see Section 1.5) Ghosh (1982) then proves the following theorems.

Theorem 2.9 A BIB design is robust against the unavailability of any $s \leq r - 1$ observations.

Theorem 2.10 A BIB design is robust against the availability of all observations in $b_0 \leq r - 1$ blocks.

With regard to Theorem 2.10, of particular interest is the case $b_0 = 1$. This case is considered by Srivastava, Gupta, and Dey (1990) who show that the efficiency factor for the resulting design relative to the original BIB design is given by

$$E = (\lambda t - k)(t - 1)/(\lambda t^2 - \lambda t - tk + k^2)$$

Considering many of the existing BIB designs, they found that for most designs $E > .80$.

CHAPTER 3

Construction of Balanced Incomplete Block Designs

3.1 INTRODUCTION

In Chapter 2 we discussed at great length the nature of BIB designs and their associated analyses. Although extensive lists of actual plans of such designs are available (see Cochran and Cox, 1957; Fisher and Yates, 1957; Beyer, 1991), these lists do not cover all presently existing BIB designs. Also, it is desirable to have some understanding of the combinatorics of design and of the resulting algebras. The actual structure has, naturally, an impact on the analysis of resulting data.

Several methods of constructing BIB designs have been introduced (for references, see Raghavarao, 1971) using such mathematical tools as finite Euclidian and projective geometries. We shall present here a very powerful and at the same time a quite simple method, the method of cyclical development of initial blocks, which is due to Bose (1939). With this method one can construct most of the existing BIB designs (Rao, 1946b) as one can see from the essentially complete listing of BIB designs given by Raghavarao (1971). Mainly from a historical point we shall also discuss some other methods.

3.2 DIFFERENCE METHODS

3.2.1 Cyclic Development of Difference Sets

The basic ideas of this method of constructing BIB designs are as follows:

1. We have a set of t treatments that we denote by $\mathcal{T} = \{t_0, t_1, \ldots, t_{t-1}\}$.

Design and Analysis of Experiments. Volume 2: Advanced Experimental Design
By Klaus Hinkelmann and Oscar Kempthorne
ISBN 0-471-55177-5

2. We have a set of s initial blocks, each containing k treatments. Suppose we denote these initial blocks by

$$\mathcal{B}_{j0} = (t_{j1}, t_{j2}, \ldots, t_{jk}) \qquad (j = 1, 2, \ldots, s)$$

where the t_{ji} $(i = 1, 2, \ldots, k)$ are elements in $\mathcal{T}$. The $\mathcal{B}_{j0}$ are chosen in a particular way such that they satisfy certain properties (which we shall discuss below).

3. Each initial block $\mathcal{B}_{j0}$ is then developed cyclically, that is, by forming

$$\mathcal{B}_{j\theta} = (t_{j1} + t_\theta, t_{j2} + t_\theta, \ldots, t_{jk} + t_\theta)$$

for $\theta = 1, 2, \ldots, t - 1$ assuming that we have an addition rule such that with $t_{ji}, t_\theta \in \mathcal{T}$ also $t_{ji} + t_\theta \in \mathcal{T}$ $(\theta = 0, 1, \ldots, t - 1)$ and $t_{ji} + t_0 = t_{ji}$.

4. The set of blocks $\{\mathcal{B}_{j\theta};\ j = 1, 2, \ldots, s; \theta = 0, 1, \ldots, t - 1\}$ forms then a BIB design with parameters $t, b = st, k, r = sk, \lambda$.

The crucial part of this method is obviously the choice of $\mathcal{T}$ and the initial blocks $\mathcal{B}_{j0}$ $(j = 1, 2, \ldots, s)$. To discuss their properties we shall confine ourselves at first to the case $s = 1$, that is, one initial block, $\mathcal{B}_0$ say. Extensions to other situations will then become fairly obvious.

We start with the following definition.

Definition 3.1 Let $\mathcal{G} = 0, 1, \ldots, t - 1$ be an Abelian group under addition, that is, for $g_1, g_2 \in \mathcal{G}$ define $g_1 + g_2$ to be $g_1 + g_2$ mod t and define $g_1 - g_2$ to be g, where $g_2 + g = g_1$ mod t. Let $\mathcal{A}$ be a subset (of $\mathcal{G}$) of k elements such that the $k(k - 1)$ differences (mod t) between members of $\mathcal{A}$ comprise all nonzero elements of $\mathcal{G}$ exactly λ times. Then $\mathcal{A}$ is called a *difference set* of size k. □

Example 3.1 Let $t = 7, \mathcal{G} = \{0, 1, 2, 3, 4, 5, 6\}, \mathcal{A} = \{0, 1, 3\}$. The differences between members in $\mathcal{A}$ are $0 - 1 \equiv 6, 0 - 3 \equiv 4, 1 - 0 \equiv 1, 1 - 3 \equiv -2 \equiv 5, 3 - 0 \equiv 3, 3 - 1 \equiv 2$, and hence $\lambda = 1$. □

Taking $\mathcal{T} = \mathcal{G}$, we now state the main result of this chapter in the form of the following theorem.

Theorem 3.1 Let $\mathcal{A}$ be a difference set of size k for a group $\mathcal{G}$ of t elements such that each nonzero difference occurs λ times. If the set $\mathcal{A}$ is taken as the initial block $\mathcal{B}_0$ and developed cyclically, it generates a symmetrical BIB design with parameters $t, b = t, r = k, \lambda$.

Proof Let $\mathcal{A} = \mathcal{B}_0 = \{g_1, g_2, \ldots, g_k\}$ with $g_i \in \mathcal{G}(i = 1, 2, \ldots, k)$. By cyclically developing $\mathcal{B}_0$, each element in $\mathcal{G}$ occurs exactly once in each of the k positions; hence each element occurs in the b sets (blocks) $\mathcal{B}_\theta(\theta = 0, 1, \ldots, t - 1)$ exactly k times. The only property of a BIB design that needs to be shown then to hold is that any two elements (treatments) occur together in λ blocks. To see this, consider without loss of generality $g_u, g_v, \in \mathcal{G}$. Suppose that for $g_{u'}, g_{v'} \in \mathcal{A}$ we have

$$\begin{aligned} g_u &= g_{u'} + \theta \\ g_v &= g_{v'} + \theta \end{aligned} \tag{3.1}$$

that is, g_u and g_v occur together in $\mathcal{B}_\theta$. Then

$$g_u - g_{u'} = g_v - g_{v'}$$

or

$$g_u - g_v = g_{u'} - g_{v'}$$

Now $g_u - g_v = c$ with $c \in \{1, 2, \ldots, t-1\}$. Since $\mathcal{A}$ is a difference set, we know that there exist λ differences of the form $g_{u'} - g_{v'}$ with the same constant c. Hence there exist exactly λ solutions to (3.1) as equations in θ. Call these solutions $\theta_1, \theta_2, \ldots, \theta_\lambda$. It then follows that g_u and g_v appear together in blocks $\mathcal{B}_{\theta_1}, \mathcal{B}_{\theta_2}, \ldots, \mathcal{B}_{\theta_\lambda}$, which proves the theorem. □

As an illustration reconsider the following example.

Example 3.2 Let $t = 7, \mathcal{G} = \{0, 1, 2, 3, 4, 5, 6, \}, \mathcal{A} = \mathcal{B}_0 = \{0, 1, 3\}, \lambda = 1$. The BIB design then is

(0, 1, 3)
(1, 2, 4)
(2, 3, 5)
(3, 4, 6)
(4, 5, 0)
(5, 6, 1)
(6, 0, 2)

where each row represents a block. □

It is now easy to generalize this procedure to the case of several initial blocks. We do this in the next theorem.

Theorem 3.2 Let $\mathcal{A} = \{\mathcal{B}_{10}, \mathcal{B}_{20}, \ldots, \mathcal{B}_{s0}\}$ consist of s sets of k elements each such that among the $sk(k-1)$ differences between the elements in each set each nonzero element of $\mathcal{G}$ occurs exactly λ times. Developing each set cyclically yields a BIB design with parameters $t, b = st, k, r = sk, \lambda$.

The proof follows along the same lines as that of Theorem 3.1.

To illustrate Theorem 3.2 we consider the following example.

Example 3.3 Let $t = 11, \mathcal{G} = \{0, 1, \ldots, 10\}$, and

$$\mathcal{B}_{10} = (0, 1, 10)$$

$$\mathcal{B}_{20} = (0, 2, 9)$$

$$\mathcal{B}_{30} = (0, 4, 7)$$

$$\mathcal{B}_{40} = (0, 8, 3)$$

$$\mathcal{B}_{50} = (0, 5, 6)$$

The differences resulting from

$$\mathcal{B}_{10} \text{ are } 10, 1, 1, 2, 10, 9$$

$$\mathcal{B}_{20} \text{ are } 9, 2, 2, 4, 9, 7$$

$$\mathcal{B}_{30} \text{ are } 7, 4, 4, 8, 7, 3$$

$$\mathcal{B}_{40} \text{ are } 3, 8, 8, 5, 3, 6$$

$$\mathcal{B}_{50} \text{ are } 6, 5, 5, 10, 6, 1$$

Inspection shows that $\lambda = 3$ and hence $\mathcal{A} = \{\mathcal{B}_{10}, \ldots, \mathcal{B}_{50}\}$ satisfies the condition of Theorem 3.2. The resulting BIB design with 11 treatments in 55 blocks of size 3 and 15 replicates per treatment is then as follows:

(0, 1, 10)	(0, 2, 9)	(0, 4, 7)	(0, 8, 3)	(0, 5, 6)
(1, 2, 0)	(1, 3, 10)	(1, 5, 8)	(1, 9, 4)	(1, 6, 7)
(2, 3, 1)	(2, 4, 0)	(2, 6, 9)	(2, 10, 5)	(2, 7, 8)
(3, 4, 2)	(3, 5, 1)	(3, 7, 10)	(3, 0, 6)	(3, 8, 9)
(4, 5, 3)	(4, 6, 2)	(4, 8, 0)	(4, 1, 7)	(4, 9, 10)
(5, 6, 4)	(5, 7, 3)	(5, 9, 1)	(5, 2, 8)	(5, 10, 0)
(6, 7, 5)	(6, 8, 4)	(6, 10, 2)	(6, 3, 9)	(6, 0, 1)
(7, 8, 6)	(7, 9, 5)	(7, 0, 3)	(7, 4, 10)	(7, 1, 2)
(8, 9, 7)	(8, 10, 6)	(8, 1, 4)	(8, 5, 0)	(8, 2, 3)
(9, 10, 8)	(9, 0, 7)	(9, 2, 5)	(9, 6, 1)	(9, 3, 4)
(10, 0, 9)	(10, 1, 8)	(10, 3, 6)	(10, 7, 2)	(10, 4, 5)

We note that this design could be used as a Youden square with 5 replication groups of 11 blocks each (see Section 2.6.1) □

3.2.2 Method of Symmetrically Repeated Differences

So far the procedure of cyclic development has led to BIB designs with the number of blocks being a multiple of the number of treatments. As exemplified in Example 3.3, this can mean a rather large number of blocks. This number can possibly be reduced, however, if the number t can be factored, say $t = pq$. For this situation the notion of a difference set and hence the construction procedure can be modified as follows (Bose, 1939).

Consider a finite additive group $\mathcal{G}$ containing p elements, $\mathcal{G} = \{a^{(0)}, a^{(1)}, \ldots, a^{(p-1)}\}$ say. To each element of the group $\mathcal{G}$ let there correspond q symbols, the symbols associated with $a^{(u)}$ being denoted by $a_1^{(u)}, a_2^{(u)}, \ldots, a_q^{(u)}$. Symbols with the subscript j are said to belong to the jth class, that is, $a_j^{(0)}, a_j^{(1)}, \ldots, a_j^{(p-1)}$ belong to the jth class. We consider differences of the form $a_i^{(u)} - a_j^{(v)}$, which are called differences of type (i, j), and their value is given by $a^{(u)} - a^{(v)} = d$, say, where $d \in \mathcal{G}$. When $i = j$, the differences are called pure, when $i \neq j$ the differences are called mixed, there being q types of pure differences and $q(q-1)$ types of mixed differences.

We then have the following definition.

Definition 3.2 Let $\mathcal{A} = \{\mathcal{B}_{10}, \mathcal{B}_{20}, \ldots, \mathcal{B}_{s0}\}$ be a collection of s sets satisfying the following conditions:

1. Each set contains k symbols with

$$k = n_{1\ell} + n_{2\ell} + \cdots + n_{q\ell} \qquad (\ell = 1, 2, \ldots, s)$$

   and $n_{j\ell}$ denoting the number of symbols of the jth class in set ℓ.
2. Among the $\sum_{\ell=1}^{s} n_{i\ell}(n_{i\ell} - 1)$ pure differences of type (i, i) arising from the s sets, every nonzero element of $\mathcal{G}$ is repeated exactly λ times for each $i = 1, 2, \ldots, q$.
3. Among the $\sum_{\ell=1}^{s} n_{i\ell} n_{j\ell}$ mixed differences of type (i, j) arising from the s sets, every element of $\mathcal{G}$ is repeated exactly λ times for every (i, j); $i, j = 1, 2, \ldots, q$; $i \neq j$.

If conditions 1–3 are satisfied, the differences are said to be *symmetrically repeated* in $\mathcal{G}$, with each difference occurring λ times. □

To illustrate Definition 3.2 we consider the following example.

Example 3.4 Let $t = 10$, $p = 5$, $q = 2$, $\mathcal{G} = \{0, 1, 2, 3, 4\}$,

$$\begin{aligned}
\mathcal{A}: \quad & \mathcal{B}_{10} = (0_2, 1_2, 2_2) \qquad && \mathcal{B}_{20} = (1_1, 4_1, 0_2) \\
& \mathcal{B}_{30} = (2_1, 3_1, 0_2) && \mathcal{B}_{40} = (1_1, 4_1, 2_2) \\
& \mathcal{B}_{50} = (2_1, 3_1, 2_2) && \mathcal{B}_{60} = (0_1, 0_2, 2_2)
\end{aligned}$$

that is,

$$\begin{aligned}
n_{11} &= 0 \qquad & n_{21} &= 3 \\
n_{12} &= 2 & n_{22} &= 1 \\
n_{13} &= 2 & n_{23} &= 1 \\
n_{14} &= 2 & n_{24} &= 1 \\
n_{15} &= 2 & n_{25} &= 1 \\
n_{16} &= 1 & n_{26} &= 2
\end{aligned}$$

Pure differences of type (1, 1):

$$\begin{aligned}
1-4 &= -3 \equiv 2\\
4-1 &= 3 \equiv 3\\
2-3 &= -1 \equiv 4\\
3-2 &= 1 \equiv 1\\
1-4 &= -3 \equiv 2\\
4-1 &= 3 \equiv 3\\
2-3 &= -1 \equiv 4\\
3-2 &= 1 \equiv 1 : \lambda = 2
\end{aligned}$$

Pure differences of type (2, 2):

$$\begin{aligned}
0-1 &= -1 \equiv 4\\
0-2 &= -2 \equiv 3\\
1-0 &= 1 \equiv 1\\
1-2 &= -1 \equiv 4\\
2-0 &= 2 \equiv 2\\
2-1 &= 1 \equiv 1\\
0-2 &= -2 \equiv 3\\
2-0 &= 2 \equiv 2 : \lambda = 2
\end{aligned}$$

Mixed differences of type (1, 2):

$$\begin{aligned}
1-0 &= 1 \equiv 1\\
4-0 &= 4 \equiv 4\\
2-0 &= 2 \equiv 2\\
3-0 &= 3 \equiv 3\\
1-2 &= -1 \equiv 4\\
4-2 &= 2 \equiv 2\\
2-2 &= 0 \equiv 0\\
3-2 &= 1 \equiv 1\\
0-0 &= 0 \equiv 0\\
0-2 &= -2 \equiv 3 : \lambda = 2
\end{aligned}$$

Mixed differences of type (2, 1) are the negative values of the differences of type (1, 2): $\lambda = 2$. Hence $\mathcal{A}$ satisfies conditions 1–3. □

Now we let the pq symbols $a_j^{(u)} (u = 1, 2, \ldots, p, j = 1, 2, \ldots, q)$ represent the t treatments. The method of cyclic development is then used again to generate BIB designs as described in the following theorem.

Theorem 3.3 Let it be possible to find a collection of s sets $\mathcal{A} = \{\mathcal{B}_{10}, \mathcal{B}_{20}, \ldots, \mathcal{B}_{s0}\}$ satisfying the following conditions:

i. Among the ks symbols (treatments) occurring in the s sets (blocks), exactly r symbols belong to the jth class $(j = 1, 2, \ldots, q)$, that is, $ks = qr$.

ii. The differences in $\mathcal{A}$ are symmetrically repeated, each occurring λ times.

Then, if each set in $\mathcal{A}$ is cyclically developed, obtaining $\mathcal{B}_{\ell\theta}$ by adding $\theta \in \mathcal{G}$ to each element in $\mathcal{B}_{\ell 0}(\ell = 1, 2, \ldots, s)$ and retaining the class number, the resulting ps sets $\mathcal{B}_{\ell\theta}(\ell = 1, 2, \ldots, s; \theta \in \mathcal{G})$ form a BIB design with parameters $t = pq, b = ps, k, r = ks/q, \lambda$.

The proof follows arguments similar to those used in the proof of Theorem 3.1 (for more details see Raghavarao, 1971).

We shall illustrate this procedure with a continuation of Example 3.4.

Example 3.4 (Continued) Let $t = 10 = 5 \times 2, \mathcal{G} = \{0, 1, 2, 3, 4\}$. Developing the initial blocks given earlier, we obtain the following BIB design with $t = 10, b = 30, k = 3, r = 9, \lambda = 2$:

$(0_2, 1_2, 2_2)$	$(1_1, 4_1, 0_2)$	$(2_1, 3_1, 0_2)$
$(1_2, 2_2, 3_2)$	$(2_1, 0_1, 1_2)$	$(3_1, 4_1, 1_2)$
$(2_2, 3_2, 4_2)$	$(3_1, 1_1, 2_2)$	$(4_1, 0_1, 2_2)$
$(3_2, 4_2, 0_2)$	$(4_1, 2_1, 3_2)$	$(0_1, 1_1, 3_2)$
$(4_2, 0_2, 1_2)$	$(0_1, 3_1, 4_2)$	$(1_1, 2_1, 4_2)$
$(1_1, 4_1, 2_2)$	$(2_1, 3_1, 2_2)$	$(0_1, 0_2, 2_2)$
$(2_1, 0_1, 3_2)$	$(3_1, 4_1, 3_2)$	$(1_1, 1_2, 3_2)$
$(3_1, 1_1, 4_2)$	$(4_1, 0_1, 4_2)$	$(2_1, 2_2, 4_2)$
$(4_1, 2_1, 0_2)$	$(0_1, 1_1, 0_2)$	$(3_1, 3_2, 0_2)$
$(0_1, 3_1, 1_2)$	$(1_1, 2_1, 1_2)$	$(4_1, 4_2, 1_2)$

We can then identify the symbols with the treatments in any way we want to, such as

Symbol	0_1	0_2	1_1	1_2	2_1	2_2	3_1	3_2	4_1	4_2
Treatment	1	2	3	4	5	6	7	8	9	10

□

Further modifications of the procedures discussed so far can be achieved by considering the set of treatments $\mathcal{T}_\infty = \{t_1, t_2, \ldots, t_v, \infty\}$ where $\mathcal{G} = \{t_1, t_2, \ldots, t_v\}$ represents an additive group and the element ∞ is such that $\infty + x = \infty$ for any $x \in \mathcal{G}$. As an extension of Theorem 3.2 we then have the following theorem.

Theorem 3.4 Let $\mathcal{G}$ be an additive Abelian group with v elements. Let $\mathcal{A}$ represent a collection of $s + u$ initial blocks $\mathcal{A} = \{\mathcal{B}_{10}, \mathcal{B}_{20}, \ldots, \mathcal{B}_{s0}, \mathcal{B}'_{10}, \mathcal{B}'_{20}, \ldots, \mathcal{B}'_{u0}\}$, each of size k, subject to the following conditions:

i. The blocks $\mathcal{B}_{i0}(i = 1, 2, \ldots, s)$ contain k elements of $\mathcal{G}$.
ii. The blocks $\mathcal{B}'_{j0}(j = 1, 2, \ldots, u)$ contain ∞ and $k - 1$ elements of $\mathcal{G}$.

iii. Ignoring ∞, the $sk(k-1)+u(k-1)(k-2)$ differences in $\mathcal{A}$ contain each nonzero member of $\mathcal{G}$ the same number, λ, of times.

iv. $\lambda = u(k-1)$.

Then developing the blocks in $\mathcal{A}$ cyclically by means of the elements in $\mathcal{G}$ (mod v) yields a BIB design with parameters $t = v+1, b = (s+u)v, k, r = uv$, and $\lambda = u(k-1)$.

We illustrate this theorem with the following example.

Example 3.5 Let $v = 11, \mathcal{T}_\infty = \{0, 1, \ldots, 10, \infty\}, \mathcal{G} = \{0, 1, \ldots, 10\}, s = 1, u = 1$, and

$$\mathcal{B}_{10} = (0, 1, 3, 7, 8, 10)$$

$$\mathcal{B}'_{10} = (\infty, 0, 5, 6, 8, 10)$$

The $30+20 = 50$ differences yield each nonzero element of $\mathcal{G}$ five times, that is, $\lambda = 5$. The resulting BIB design with parameters $t = 12, b = 22, k = 6, r = 11, \lambda = 5$ is given by

(0,	1,	3,	7,	8,	10)	(∞,	0,	5,	6,	8,	10)
(1,	2,	4,	8,	9,	0)	(∞,	1,	6,	7,	9,	0)
(2,	3,	5,	9,	10,	1)	(∞,	2,	7,	8,	10,	1)
(3,	4,	6,	10,	0,	2)	(∞,	3,	8,	9,	0,	2)
(4,	5,	7,	0,	1,	3)	(∞,	4,	9,	10,	1,	3)
(5,	6,	8,	1,	2,	4)	(∞,	5,	10,	0,	2,	4)
(6,	7,	9,	2,	3,	5)	(∞,	6,	0,	1,	3,	5)
(7,	8,	10,	3,	4,	6)	(∞,	7,	1,	2,	4,	6)
(8,	9,	0,	4,	5,	7)	(∞,	8,	2,	3,	5,	7)
(9,	10,	1,	5,	6,	8)	(∞,	9,	3,	4,	6,	8)
(10,	0,	2,	6,	7,	9)	(∞,	10,	4,	5,	7,	9)

□

Theorem 3.4 can be generalized in much the same way that Theorem 3.2 was generalized by defining pq symbols of the form $a_j^{(w)} (w = 1, 2, \ldots, p; j = 1, 2, \ldots, q)$ with the $a^{(w)}$ forming an additive group $\mathcal{G}$. These pq symbols are now referred to as finite symbols to which we adjoin the symbol ∞, so that $\mathcal{T}_\infty = \{a_j^{(w)} (w = 1, 2, \ldots, p; j = 1, 2, \ldots, q), \infty\}$. We then have the following theorem.

Theorem 3.5 Let $\mathcal{A} = \{\mathcal{B}_{10}, \mathcal{B}_{20}, \ldots, \mathcal{B}_{s0}, \mathcal{B}'_{10}, \mathcal{B}'_{20}, \ldots, \mathcal{B}'_{u0}\}$ satisfy the following conditions:

i. Each set $\mathcal{B}_{i0} (i = 1, 2, \ldots, s)$ contains k different finite symbols.

ii. Each set $\mathcal{B}'_{j0} (j = 1, 2, \ldots, u)$ contains ∞ and $(k-1)$ different finite symbols.

iii. Of the finite symbols of the ℓth class, $(pu - \lambda)$ occur in the sets $\{\mathcal{B}_{i0}, i = 1, 2, \ldots, s\}$ and λ occur in the sets $\{\mathcal{B}'_{j0}, j = 1, 2, \ldots, u\}$ for each $\ell = 1, 2, \ldots, q$, implying $ks = q(pu - \lambda)$, $(k - 1)u = q\lambda$.

iv. The differences from the finite symbols in $\mathcal{A}$ are symmetrically repeated, each occurring λ times.

Then developing each set (block) in $\mathcal{A}$ cyclically (as in Theorem 3.3) yields a BIB design with parameters $t = pq + 1, b = (s + u)p, k, r = up, \lambda$.

As an illustration consider the following example.

Example 3.6 Let $p = 7, q = 2, t = 15, s = 3, u = 2$, with

$$\mathcal{B}_{10} = (0_1, 1_1, 3_1, 0_2, 2_2, 6_2)$$

$$\mathcal{B}_{20} = (0_1, 1_1, 3_1, 1_2, 5_2, 6_2)$$

$$\mathcal{B}_{30} = (0_1, 4_1, 5_1, 0_2, 1_2, 3_2)$$

$$\mathcal{B}'_{10} = (\infty, 0_1, 0_2, 1_2, 2_2, 4_2)$$

$$\mathcal{B}'_{20} = (\infty, 0_1, 3_1, 5_1, 6_1, 0_2)$$

The reader can check easily that conditions (iii) and (iv) are met with $\lambda = 5$. Hence the resulting design is a BIB design with parameters $t = 15, b = 35, k = 6, r = 14, \lambda = 5$. □

3.2.3 Formulation in Terms of Galois Field Theory

In all the examples given so far, the additive group $\mathcal{G}$ has been written as $\mathcal{G} = \{0, 1, \ldots, t - 1\}$ or $\mathcal{G} = \{0, 1, \ldots, v - 1\}$, representing the residues mod t or mod v, respectively. If t or v is prime or prime power, the elements of $\mathcal{G}$ can be expressed also in terms of the powers of a primitive root of the Galois field GF(t) or GF(v), respectively. For a discussion of Galois fields we refer to Appendix A.

Since $\mathcal{G}$ can be expressed in terms of all powers of a primitive root of GF(t), then also $\mathcal{A}$ can be represented in terms of certain powers of such a primitive root. It is this fact that makes this representation useful. The reader will have noticed that the construction procedures presented in Theorems 3.1–3.5 depend upon the existence of difference sets, but we have not yet said how these can be obtained. One way, certainly, is by trial and error, but that is feasible only for small t and not very satisfactory in general. A more direct way is to make use of mathematical results that for certain forms of t give general explicit expressions for certain types of difference sets in the form of powers of primitive roots. For example, if $t = 10\ell + 1$ is a prime or prime power and x is a primitive root of GF$(10\ell + 1)$, then the ℓ sets $\mathcal{B}_{i0} = (x^i, x^{2\ell+i}, x^{4\ell+i}, x^{6\ell+i}, x^{8\ell+i})$, $i = 0, 1, \ldots, \ell - 1$ form a difference set $\mathcal{A}$ with $\lambda = 2$. As an illustration consider the following example.

Example 3.7 Let $t = 11$, $\ell = 1$, then $\mathcal{A} = \mathcal{B}_{00} = (x^0, x^2, x^4, x^6, x^8)$, which, with $x = 2$ being a primitive root of GF(11), translates into

$$\mathcal{B}_{00} = (1, 4, 5, 9, 3)$$

and that can then be used in Theorem 3.1 to construct a BIB design with $t = 11, b = 11, k = 5, r = 5, \lambda = 2$. □

We shall not pursue this any further here, but refer the reader to Raghavarao (1971) for an extensive list of such results and further references.

3.3 OTHER METHODS

To complete this chapter we mention briefly four other methods that give rise to quite a number of BIB designs. Of these methods, one is a simple direct construction procedure, two derive new BIB designs from existing BIB designs, and the last is based on notions of factorial experiments.

3.3.1 Irreducible BIB Designs

This method is applicable mainly for small t and consists of taking of the t symbols all possible combinations of k symbols. This leads to a BIB design with parameters $t, b = \binom{t}{k}, k, r = \binom{t-1}{k-1}, \lambda = \binom{t-2}{k-2}$.

3.3.2 Complement of BIB Designs

If the collection of blocks $D = \{\mathcal{B}_1, \mathcal{B}_2, \ldots, \mathcal{B}_b\}$ represents a BIB design with parameters t, b, k, r, and λ, then, if we let $\mathcal{B}_i^* = \mathcal{T} - \mathcal{B}_i$, the collection of blocks $D_1 = \{\mathcal{B}_1^*, \mathcal{B}_2^*, \ldots, \mathcal{B}_b^*\}$ represents a BIB design with parameters $t_1 = t, b_1 = b, k_1 = t - k, r_1 = b - r, \lambda_1 = b - 2r + \lambda$.

That this is true can be seen as follows: If $\boldsymbol{N}$ denotes the incidence matrix of the existing BIB design, and $\overline{\boldsymbol{N}}$ denotes the incidence matrix of its complement, then

$$\boldsymbol{N} + \overline{\boldsymbol{N}} = \mathfrak{I}_t \mathfrak{I}_b'$$

that is,

$$\overline{\boldsymbol{N}} = \mathfrak{I}_t \mathfrak{I}_b' - \boldsymbol{N}$$

and hence

$$\begin{aligned}
\overline{\boldsymbol{N}}\,\overline{\boldsymbol{N}}' &= (\mathfrak{I}_t \mathfrak{I}_b' - \boldsymbol{N})(\mathfrak{I}_t \mathfrak{I}_b' - \boldsymbol{N})' \\
&= b\mathfrak{I}_t \mathfrak{I}_t' - \mathfrak{I}_t \mathfrak{I}_b' \boldsymbol{N}' - \boldsymbol{N}\mathfrak{I}_b \mathfrak{I}_t' + \boldsymbol{N}\boldsymbol{N}' \\
&= (b - 2r)\mathfrak{I}_t \mathfrak{I}_t' + \boldsymbol{N}\boldsymbol{N}' \\
&= (b - 2r)\mathfrak{I}_t \mathfrak{I}_t' + (r - \lambda)\boldsymbol{I} + \lambda \mathfrak{I}_t \mathfrak{I}_t' \\
&= (r - \lambda)\boldsymbol{I} + (b - 2r + \lambda)\mathfrak{I}_t \mathfrak{I}_t'
\end{aligned}$$

which implies that $\overline{N}$ is the incidence matrix of a BIB design, and

$$r_1 = (r - \lambda) + (b - 2r + \lambda) = b - r$$
$$\lambda_1 = b - 2r + \lambda$$

3.3.3 Residual BIB Designs

If the collection of blocks $D = \{\mathcal{B}_1, \mathcal{B}_2, \ldots, \mathcal{B}_b\}$ represents a symmetrical BIB design with parameters $t, b = t, k, r = k, \lambda$, then one can consider the design D_2, which is obtained as follows: Select any block from D, say $\mathcal{B}_1$, and delete from each of the remaining blocks $\mathcal{B}_2, \mathcal{B}_3, \ldots, \mathcal{B}_b$ the λ elements it has in common with $\mathcal{B}_1$. Call those sets $\mathcal{B}_2^*, \mathcal{B}_3^*, \ldots, \mathcal{B}_t^*$. Then $D_2 = \{\mathcal{B}_2^*, \mathcal{B}_3^*, \ldots, \mathcal{B}_t^*\}$ is the residual BIB design with parameters $t_2 = t - k, b_2 = b - 1, k_2 = k - \lambda, r_2 = k, \lambda_2 = \lambda$.

To see that this is true we show that in a symmetrical BIB design any two blocks have λ treatments in common. This follows from

$$\begin{aligned} \boldsymbol{N}'\boldsymbol{N}\boldsymbol{N}' &= \boldsymbol{N}'[(r - \lambda)\boldsymbol{I} + \lambda \mathfrak{I}_t \mathfrak{I}_t'] \\ &= [(r - \lambda)\boldsymbol{I} + \lambda \mathfrak{I}_t \mathfrak{I}_t']\boldsymbol{N}' \end{aligned} \tag{3.2}$$

since $\boldsymbol{N}'\mathfrak{I}_t = \mathfrak{I}_b'\boldsymbol{N}' = r\mathfrak{I}_t$; and since $\boldsymbol{N}'$ is nonsingular, (3.2) implies

$$\boldsymbol{N}'\boldsymbol{N} = (r - \lambda)\boldsymbol{I} + \lambda \mathfrak{I}_t \mathfrak{I}_t'$$

which is the desired result. The rest is then obvious.

3.3.4 Orthogonal Series

This method is actually the oldest method. These designs are due to Yates (1936b) and are referred to as quasi-factorial or lattice designs, since their construction is based on concepts of factorial experiments. The designs generated by these methods have parameters $t = K^2, k = K, b = K(K + 1), r = K + 1, \lambda = 1$ with K prime or prime power. This series of designs is also called orthogonal series 1 (OS1). We shall deal with the construction of OS1 designs in Section 18.3.

An alternative method of constructing these designs is given by Khare and Federer (1981). This algorithm can be described as follows:

1. Write the treatment numbers $1, 2, \ldots, t$ consecutively in a square array of K rows and K columns to yield replicate 1 of the resolvable design, with rows constituting the blocks.
2. Transpose the rows and columns of replicate 1 to obtain replicate 2.
3. Take the main right diagonal of replicate 2 to form the first row of replicate 3, and write the remaining elements in each column of replicate 2 in a cyclic order in the same column for replicate 3.

4. Repeat step 3 for replicate 3 to generate replicate 4.
5. Continue this process on the just generated replicate until $K + 1$ replicates have been obtained.

As an illustration consider the following example.

Example 3.8 Let $t = 16 = 4^2, k = 4, b = 20, r = 5$:

Replicate 1			
1	2	3	4
5	6	7	8
9	10	11	12
13	14	15	16

Replicate 2			
1	5	9	13
2	6	10	14
3	7	11	15
4	8	12	16

Replicate 3			
1	6	11	16
2	7	12	13
3	8	9	14
4	5	10	15

Replicate 4			
1	7	9	15
2	8	10	16
3	5	11	13
4	6	12	14

Replicate 5			
1	8	11	14
2	5	12	15
3	6	9	16
4	7	10	13

Because of the repeated application of step 3 above, this method has been referred to as the *successive diagonalizing method* (Khare and Federer, 1981). □

3.4 LISTING OF EXISTING BIB DESIGNS

Raghavarao (1971) provides a complete list of existing BIB designs with parameters $t, b \leq 100, r, k \leq 15$ together with their method of construction. A similar table with difference sets is given by Takeuchi (1962). Another extensive list of parameters and references concerning construction of BIB designs for $r \leq 41$ and $k \leq t/2$ is given by Mathon and Rosa (1996). Table 3.1 gives a list of BIB designs with $t \leq 25$ and $k \leq 11$, which can be constructed using the methods discussed in this chapter. For each design we give the parameters t, b, r, k, λ and the method by which the design can be constructed. If t is a prime power the elements in the difference sets are given in terms of powers of the primitive element x for the primitive polynomials, $P(x)$, given in Table A2 (see Appendix A).

Table 3.1 BIB Designs with $t \leq 25, k \leq 11$

Design	t	b	r	k	λ	Method
1	3	3	2	2	1	Irreducible
2	4	6	3	2	1	Irreducible
3	4	4	3	3	2	Irreducible
4	5	10	4	2	1	Irreducible
5	5	5	4	4	3	Irreducible
6	5	10	6	3	3	Irreducible
7	6	15	5	2	1	Irreducible
8	6	10	5	3	2	Residual of design 25
9	6	6	5	5	4	Irreducible
10	6	15	10	4	6	Irreducible
11	7	21	6	2	1	Irreducible
12	7	7	6	6	5	Irreducible
13	7	21	15	5	10	Complement of design 11
14	8	28	7	2	1	Irreducible
15	8	14	7	4	3	Difference set: $(\infty, 1, 2, 4)$; $(0, 3, 6, 5)$ mod 7
16	8	8	7	7	6	Irreducible
17	9	36	8	2	1	Irreducible
18	9	18	8	4	3	Difference set: (x^0, x^2, x^4, x^6); (x, x^3, x^5, x^7); $x \in GF(3^2)$
19	9	9	8	8	7	Irreducible
20	9	18	10	5	5	Complement of design 18
21	10	45	9	2	1	Irreducible
22	10	30	9	3	2	Difference set: $(0_2, 1_2, 2_2,)$; $(1_1, 4_1, 0_2)(2_1, 3_1, 0_2)$; $(1_1, 4_1, 2_2,)$; $(2_1, 3_1, 2_2)$; $(0_1, 0_2, 2_2)$ mod 5
23	10	18	9	5	4	Residual of design 42
24	10	10	9	9	8	Irreducible
25	11	11	5	5	2	Difference set: $(1, 4, 5, 9, 3)$ mod 11
26	11	11	6	6	3	Complement of design 25
27	11	55	10	2	1	Irreducible
28	11	11	10	10	9	Irreducible

Table 3.1 (*Continued*)

Design	t	b	r	k	λ	Method
29	11	55	15	3	3	Difference set: (0, 1, 10); (0, 2, 9) (0, 4, 7) (0, 8, 3); (0, 5, 6) mod 11
30	12	44	11	3	2	Difference set: (0, 1, 3); (0, 1, 5); (0, 4, 6); $(\infty, 0, 3)$ mod 11
31	12	33	11	4	3	Difference set: (0, 1, 3, 7); (0, 2, 7, 8); $(\infty, 0, 1, 3)$ mod 11
32	12	22	11	6	5	Difference set: (0, 1, 3, 7, 8, 10); $(\infty, 0, 5, 6, 8, 10)$ mod 11
33	13	26	6	3	1	Difference set: (1, 3, 9); (2, 6, 5) mod 13
34	13	26	12	6	5	Difference set: (1, 4, 3, 12, 9, 10); (2, 8, 6, 11, 5, 7) mod 13
35	13	39	15	5	5	Difference set: (0, 1, 8, 12, 5); (0, 2, 3, 11, 10); (0, 4, 6, 9, 7) mod 13
36	15	35	7	3	1	Difference set: $(1_1, 4_1, 0_2,)$; $(2_1, 3_1, 0_2)$; $(1_2, 4_2, 0_3)$; $(2_2, 3_2, 0_3,)$; $(1_3, 4_3, 0_1)$; $(2_3, 3_3, 0_1)$; $(0_1, 0_2, 0_3)$ mod 5
37	15	35	14	6	5	Difference set: $(\infty, 0_1, 0_2, 1_2, 2_2, 4_2)$; $(\infty, 0_1, 3_1, 5_1, 6_1, 0_2)$; $(0_1, 1_1, 3_1, 0_2, 2_2, 6_2)$; $(0_1, 1_1, 3_1, 1_2, 5_2, 6_2)$; $(0_1, 4_1, 5_1, 0_2, 1_2, 3_2)$ mod 7
38	16	24	9	6	3	Residual of design 50
39	16	80	15	3	2	Different set: (x^0, x^5, x^{10}); (x^1, x^6, x^{11}); (x^2, x^7, x^{12}); (x^3, x^8, x^{13}); (x^4, x^9, x^{14}); $x \in GF(2^4)$
40	16	48	15	5	4	Difference set: $(x^0, x^3, x^6, x^9, x^{12})$; $(x^1, x^4, x^7, x^{10}, x^{13})$; $(x^2, x^5, x^8, x^{11}, x^{14})$; $x \in GF(2^4)$

Table 3.1 (*Continued*)

Design	t	b	r	k	λ	Method
41	19	57	9	3	1	Difference set: (1, 7, 11); (2, 14, 3); (4, 9, 6) mod 19
42	19	19	9	9	4	Difference set: (1, 4, 16, 7, 9, 17, 11, 6, 5) mod 19
43	19	19	10	10	5	Complement of design 42
44	19	57	12	4	2	Difference set: (0, 1, 7, 11); (0, 2, 14, 3); (0, 4, 9, 6) mod 19
45	21	70	10	3	1	Difference set: $(1_1, 6_1, 0_2)$; $(2_1, 5_1, 0_2)$; $(3_1, 4_1, 0_2)$; $(1_2, 6_2, 0_3)$; $(2_2, 5_2, 0_3)$; $(3_2, 4_2, 0_3)$; $(1_3, 6_3, 0_1)$; $(2_3, 5_3, 0_1)$; $(3_3, 4_3, 0_1)$; $(0_1, 0_2, 0_3)$ mod 7
46	21	42	12	6	3	Difference set: $(0_1, 5_1, 1_2, 4_2, 2_3, 3_3)$; $(0_1, 1_1, 3_1, 0_2, 1_2, 3_2)$; $(0_2, 5_2, 1_3, 4_3, 2_1, 3_1)$; $(0_2, 1_2, 3_2, 0_3, 1_3, 3_3)$; $(0_3, 5_3, 1_1, 4_1, 2_2, 3_2)$; $(0_3, 1_3, 3_3, 0_1, 1_1, 3_1)$ mod 7
47	22	77	14	4	2	Difference set: $(1_1, 6_1, 3_2, 4_2)$; $(3_1, 4_1, 2_2, 5_2)$; $(2_1, 5_1, 6_2, 1_2)$; $(1_2, 6_2, 3_3, 4_3)$; $(3_2, 4_2, 2_3, 5_3)$; $(2_2, 5_2, 6_3, 1_3)$; $(1_3, 6_3, 3_1, 4_1)$; $(3_3, 4_3, 2_1, 5_1)$; $(2_3, 5_3, 6_1, 1_1)$; $(\infty, 0_1, 0_2, 0_3)$; $(\infty, 0_1, 0_2, 0_3)$ mod 7
48	23	23	11	11	5	Difference set: (1, 2, 4, 8, 16, 9, 18, 13, 3, 6, 12) mod 23
49	25	50	8	4	1	Difference set: $(0, x^0, x^8, x^{16})$; $(0, x^2, x^{10}, x^{18})$; $x \in GF(5^2)$
50	25	25	9	9	3	See Fisher and Yates (1957)
51	25	100	12	3	1	Difference set: (x^0, x^8, x^{16}); (x^1, x^9, x^{17}); (x^2, x^{10}, x^{18}); (x^3, x^{11}, x^{19}); $x \in GF(5^2)$

CHAPTER 4

Partially Balanced Incomplete Block Designs

4.1 INTRODUCTION

It can be seen very easily from the list of existing BIB designs [see Raghavarao (1971), Mathon and Rosa (1996), and also Table 3.1] that such designs exist for certain parameters only and often with an inordinately large number of replicates, the main reason being that condition (2.4) has to be satisfied with λ being an integer. For example, using (2.4), with 8 treatments and blocks of 3 units, the lower limit for the number of replicates is 21, and with this number of replicates the blocks would consist of all combinations 3 at a time of the treatments. In fact, Yates himself recognized that BIB designs are "rare" and then developed different types of quasi-factorial or lattice designs (see Chapter 18). Although some of the lattice designs are special cases of BIB designs (see Section 3.3.4), others are special cases of a much larger class of designs, namely *partially balanced incomplete block* (PBIB) designs. These were introduced by Bose and Nair (1939).

Recall that BIB designs have the property that all treatment differences are estimated with the same accuracy. For PBIB designs this property will not be sacrificed completely, but only to the extent that, loosely speaking, pairs of treatments can be arranged in different sets such that the difference between the treatment effects of a pair, for all pairs in a set, is estimated with the same accuracy.

4.2 PRELIMINARIES

Before giving the definition of a PBIB design, we shall take a look at the reduced normal equations (1.7) for a general incomplete block design, investigate the

Design and Analysis of Experiments. Volume 2: Advanced Experimental Design
By Klaus Hinkelmann and Oscar Kempthorne
ISBN 0-471-55177-5

structure of these equations in light of the remarks made in the introduction, and see what that might imply.

4.2.1 Association Scheme

Recall that the RNE for a proper equireplicate incomplete block design are of the form [see (1.7)]

$$\left(r\boldsymbol{I} - \frac{1}{k}\boldsymbol{N}\boldsymbol{N}'\right)\widehat{\boldsymbol{\tau}} = \boldsymbol{T} - \frac{1}{k}\boldsymbol{N}\boldsymbol{B}$$

where

$$\boldsymbol{N}\boldsymbol{N}' = \left(\sum_j n_{ij}\, n_{i'j}\right) = (\lambda_{ii'}) \quad \text{(say)}$$

For the BIB design $\lambda_{ii'} = \lambda$ for all i, i' $(i \neq i')$ and $\lambda_{ii} = r$ for all i. The term $\lambda_{ii'}$ denotes the number of times that treatments i and i' appear together in the same block. We now wish to consider the case that for a given treatment i, the numbers $\lambda_{ii'}$ are equal within groups of the other treatments. Suppose that over all i and i' the $\lambda_{ii'}$ take on the values $\lambda_1, \lambda_2, \ldots, \lambda_m$. For the sake of argument we shall assume that the $\lambda_u (u = 1, 2, \ldots, m)$ are different although it will become clear later that this is not necessary. The set of treatments i' for which $\lambda_{ii'} = \lambda_1$ will be denoted by $S(i, 1)$ and called the *first associates* of treatment i. The set of treatments i' for which $\lambda_{ii'} = \lambda_2$ will be denoted by $S(i, 2)$ and called the second associates of treatment i, and so forth. For any given i then the remaining treatments fall into one of the sets $S(i, 1), S(i, 2), \ldots, S(i, m)$. The question now is: Under what conditions does this *association scheme* imply that, for example, treatment i and all the treatments i' in $S(i, u)$ are compared with the same precision and when is this the same for all pairs of uth associates independent of the treatment i?

4.2.2 Association Matrices

In order to investigate this question we shall use the concept of an *association matrix*, which was introduced by Thompson (1958) and investigated in more detail by Bose and Mesner (1959). Define the uth association matrix, a $t \times t$ matrix, as

$$\boldsymbol{B}_u = (b^u_{\alpha\beta}) \qquad (u = 1, 2, \ldots, m)$$

with

$$b^u_{\alpha\beta} = \begin{cases} 1 & \text{if treatment } \alpha \text{ and } \beta \text{ are } u\text{th associates} \\ 0 & \text{otherwise} \end{cases}$$

At this point it is convenient to introduce $\boldsymbol{B}_0 = \boldsymbol{I}$ and $\lambda_0 = r$. It follows then from what has been said earlier that

$$\sum_{u=0}^{m} \boldsymbol{B}_u = \mathfrak{I}\mathfrak{I}' \tag{4.1}$$

Furthermore,

$$\boldsymbol{N}\boldsymbol{N}' = \sum_{u=0}^{m} \lambda_u \boldsymbol{B}_u \tag{4.2}$$

and hence

$$\boldsymbol{C} = r\boldsymbol{I} - \frac{1}{k}\boldsymbol{N}\boldsymbol{N}' = r\,\frac{k-1}{k}\,\boldsymbol{B}_0 - \sum_{u=1}^{m} \frac{\lambda_u}{k}\,\boldsymbol{B}_u \tag{4.3}$$

or in words: The coefficient matrix $\boldsymbol{C}$ of the RNE is a linear combination of the association matrices.

4.2.3 Solving the RNE

To solve the RNE, we know from (1.17) that

$$\boldsymbol{C}\widetilde{\boldsymbol{C}}^{-1} = \boldsymbol{I} - \frac{1}{t}\,\mathfrak{I}\mathfrak{I}' \tag{4.4}$$

where $\widetilde{\boldsymbol{C}} = \boldsymbol{C} + a\mathfrak{I}\mathfrak{I}'$ and $a \neq 0$, real. Using (4.1) and (4.3) we can rewrite (4.4) in terms of association matrices as follows:

$$\left(\sum_{u=0}^{m} c_u \boldsymbol{B}_u\right)\left(\sum_{u=0}^{m} c_u^* \boldsymbol{B}_u\right)^{-1} = \frac{t-1}{t}\,\boldsymbol{B}_0 - \frac{1}{t}\sum_{u=1}^{m} \boldsymbol{B}_u \tag{4.5}$$

where

$$c_0 = r\frac{k-1}{k} \qquad c_u = -\frac{\lambda_u}{k} \quad (u = 1, 2, \ldots, m) \qquad c_u^* = c_u + a$$

Now, let

$$\left(\sum_{u=0}^{m} c_u^* \boldsymbol{B}_u\right)^{-1} = \boldsymbol{G} = (g_{ii'}) \tag{4.6}$$

Then we know that for any two pairs (i, i') and (i, i'')

$$\operatorname{var}(\widehat{\tau}_i - \widehat{\tau}_{i'}) = (g_{ii} + g_{i'i'} - 2g_{ii'})\sigma_e^2 \tag{4.7}$$

$$\operatorname{var}(\widehat{\tau}_i - \widehat{\tau}_{i''}) = (g_{ii} + g_{i''i''} - 2g_{ii''})\sigma_e^2 \tag{4.8}$$

Suppose that i' and i'' are uth associates of i. Then we would like to have that (4.7) and (4.8) are equal. One way that this can happen is if

$$g_{ii} = g_{i'i'} = g_{i''i''}$$

and

$$g_{ii'} = g_{ii''} \tag{4.9}$$

Furthermore, we would like $\text{var}(\widehat{\tau}_j - \widehat{\tau}_{j'})$ to be equal to the right-hand side of (4.7) for any pair (j, j') that are uth associates. It is obvious that this can be achieved if $\boldsymbol{G}$ is of the form

$$\boldsymbol{G} = \sum_{u=0}^{m} g_u \boldsymbol{B}_u \tag{4.10}$$

Substituting (4.10) into (4.5) yields

$$\sum_{u,v=0}^{m} c_u g_v \boldsymbol{B}_u \boldsymbol{B}_v = \frac{t-1}{t} \boldsymbol{B}_0 - \frac{1}{t} \sum_{u=1}^{m} \boldsymbol{B}_u \tag{4.11}$$

Since the right-hand side of (4.11) is a linear combination of the $\boldsymbol{B}_u$'s, the left-hand side also must be one. This would imply that $\boldsymbol{B}_u \boldsymbol{B}_v$ is of the form

$$\boldsymbol{B}_u \boldsymbol{B}_v = \sum_{k=0}^{m} p_{uv}^k \boldsymbol{B}_k \tag{4.12}$$

that is, $\boldsymbol{B}_u \boldsymbol{B}_v$ is itself a linear combination of the $\boldsymbol{B}_u$'s.

4.2.4 Parameters of the Second Kind

It will turn out that the coefficients p_{uv}^k in (4.12) play an important role in the definition of PBIB designs. It is for this reason that they are referred to as *parameters of the second kind.* To understand what they really are and to grasp the full meaning of (4.12), let us consider two treatments, α and β say. Consider then the (α, β) element of $\boldsymbol{B}_u \boldsymbol{B}_v$ and of $\sum_k p_{uv}^k \boldsymbol{B}_k$. The (α, β) element of $\boldsymbol{B}_u \boldsymbol{B}_v$ is of the form

$$\sum_{\gamma=1}^{t} b_{\alpha\gamma}^u b_{\gamma\beta}^v$$

Since $\boldsymbol{B}_u$ is symmetric (follows from the fact that if α is the ℓth associate of β, then β is also the ℓth associate of α), this is the same as

$$\sum_{\gamma=1}^{t} b_{\alpha\gamma}^u b_{\beta\gamma}^v \tag{4.13}$$

which equals the number of unity elements the αth row of $\boldsymbol{B}_u$ and βth row of $\boldsymbol{B}_v$ have in common, or expressed differently, which equals the number of treatments that $S(\alpha, u)$ and $S(\beta, v)$ have in common. The corresponding element of the right-hand side of (4.12) is

$$\sum_{k=0}^{m} p_{uv}^{k} b_{\alpha\beta}^{k} \tag{4.14}$$

Because of (4.1) there is only one element $b_{\alpha\beta}^{k}$ equal to one, namely $b_{\alpha\beta}^{\ell}$ if α and β are ℓth associates. Hence

$$p_{uv}^{\ell} = \sum_{\gamma=1}^{t} b_{\alpha\gamma}^{u} b_{\beta\gamma}^{v} \tag{4.15}$$

The left-hand side of (4.15) does not depend explicitly on α and β but only on the type of association between α and β and hence is the same for any two treatments that are ℓth associates. In particular for $\alpha = \beta$ and $u = v$ we obtain

$$p_{uu}^{0} = \sum_{\gamma=1}^{t} (b_{\alpha\gamma}^{u})^2 = \sum_{\gamma=1}^{t} b_{\alpha\gamma}^{u} = n_u \quad \text{(say)}$$

which is the number of elements in $S(\alpha, u)$. This implies that every treatment has n_u uth associates. In summary then, if there exists an association scheme such that (4.15) holds, then (4.10) and hence (4.9) can be achieved.

4.3 DEFINITION AND PROPERTIES OF PBIB DESIGNS

We now give the formal definition of a PBIB design based on the definition given by Bose and Shimamoto (1952) and derive some of its properties in terms of its various parameters.

4.3.1 Definition of PBIB Designs

With the motivation provided in Section 4.2 we now have the following definition.

Definition 4.1 An incomplete block design is said to be a PBIB design, if it satisfies the following conditions:

1. The experimental material is divided into b blocks of k units each, different treatments being applied to the units in the same block.
2. There are t treatments each of which occurs in r blocks.

3. Any two treatments are either 1st, 2nd, ..., *mth* associates; each treatment is the 0th associate of itself and of no other treatment.
4. Each treatment has n_u uth associates ($u = 0, 1, \ldots, m$).
5. If any two treatments are kth associates, then the number of treatments common to the uth associates of the first and the vth associates of the second is p_{uv}^k, independent of the pair of kth associates.
6. Any two treatments that are uth associates appear together in λ_u blocks.

The terms $t, b, r, k, \lambda_1, \lambda_2, \ldots, \lambda_m, n_1, n_2, \ldots, n_m$ are called the *parameters of the first kind*, and the p_{uv}^k are called *parameters of the second kind.*

The parameters of the second kind are generally exhibited in matrix form as $\boldsymbol{P}_k = (p_{uv}^k)(k = 0, 1, \ldots, m)$, and in our notation these are $(m + 1) \times (m + 1)$ matrices (this deviates from the original notation by Bose and Nair, 1939). □

As an illustration of a PBIB design we consider the following PBIB design with three associate classes.

Example 4.1 Let $t = 8, b = 6, r = 3, k = 4$

Block	Treatments
1	1, 2, 3, 4
2	1, 2, 5, 6
3	1, 3, 5, 7
4	2, 4, 6, 8
5	3, 4, 7, 8
6	5, 6, 7, 8

The association scheme is as follows:

0th Associate	1st Associates	2nd Associates	3rd Associates
1	2, 3, 5	4, 6, 7	8
2	1, 4, 6	3, 5, 8	7
3	1, 4, 7	2, 5, 8	6
4	2, 3, 8	1, 6, 7	5
5	1, 6, 7	2, 3, 8	4
6	2, 5, 8	1, 4, 7	3
7	3, 5, 8	1, 4, 6	2
8	4, 6, 7	2, 3, 5	1

Hence $n_1 = 3, n_2 = 3, n_3 = 1$, and, from inspection of the design, $\lambda_1 = 2, \lambda_2 = 1, \lambda_3 = 0$, and

$$P_0 = \begin{bmatrix} 1 & & & \\ & 3 & & 0 \\ 0 & & 3 & \\ & & & 1 \end{bmatrix} \qquad P_1 = \begin{bmatrix} 0 & 1 & 0 & 0 \\ 1 & 0 & 2 & 0 \\ 0 & 2 & 0 & 1 \\ 0 & 0 & 1 & 0 \end{bmatrix}$$

$$P_2 = \begin{bmatrix} 0 & 0 & 1 & 0 \\ 0 & 2 & 0 & 1 \\ 1 & 0 & 2 & 0 \\ 0 & 1 & 0 & 0 \end{bmatrix} \qquad P_3 = \begin{bmatrix} 0 & 0 & 0 & 1 \\ 0 & 0 & 3 & 0 \\ 0 & 3 & 0 & 0 \\ 1 & 0 & 0 & 0 \end{bmatrix}$$

□

4.3.2 Relationships Among Parameters of a PBIB Design

The following relationships exist among the various parameters of a PBIB design:

$$tr = bk \tag{4.16}$$

$$\sum_{u=0}^{m} n_u = t \tag{4.17}$$

$$\sum_{u=0}^{m} n_u \lambda_u = rk \tag{4.18}$$

$$p_{uv}^k = p_{vu}^k \tag{4.19}$$

$$\sum_{u=0}^{m} p_{uv}^k = n_v \tag{4.20}$$

$$n_k p_{uv}^k = n_u p_{kv}^u = n_v p_{uk}^v \tag{4.21}$$

Relationships (4.16), (4.17), and (4.18) follow immediately from the definition of a PBIB design.

To verify (4.19) we note that the relation of association is symmetric and that therefore the association matrices $\boldsymbol{B}_u$ are symmetric. Hence

$$\begin{aligned} \boldsymbol{B}_u \boldsymbol{B}_v = \boldsymbol{B}_u' \boldsymbol{B}_v' = (\boldsymbol{B}_v \boldsymbol{B}_u)' &= \left(\sum_k p_{vu}^k \boldsymbol{B}_k \right)' \\ &= \sum_k p_{vu}^k \boldsymbol{B}_k' = \sum_k p_{vu}^k \boldsymbol{B}_k = \boldsymbol{B}_v \boldsymbol{B}_u \end{aligned}$$

which implies (4.19).

To verify (4.20) consider

$$\left(\sum_{u=0}^{m} \boldsymbol{B}_u\right) \boldsymbol{B}_v = \mathfrak{I}\mathfrak{I}' \boldsymbol{B}_v = n_v \mathfrak{I}\mathfrak{I}' = n_v \sum_{k=0}^{m} \boldsymbol{B}_k$$

and

$$\sum_{u=0}^{m} \boldsymbol{B}_u \boldsymbol{B}_v = \sum_{u} \sum_{k} p_{uv}^{k} \boldsymbol{B}_k = \sum_{k} \left(\sum_{u} p_{uv}^{k}\right) \boldsymbol{B}_k$$

Comparing coefficients yields (4.20).

Finally, to derive (4.21) we use (4.12) twice and consider

$$(\boldsymbol{B}_u \boldsymbol{B}_v)\boldsymbol{B}_\ell = \sum_{s=0}^{m} \left(\sum_{k=0}^{m} p_{uv}^{k}\, p_{k\ell}^{s}\right) \boldsymbol{B}_s$$

and

$$(\boldsymbol{B}_\ell \boldsymbol{B}_v)\boldsymbol{B}_u = \sum_{s=0}^{m} \left(\sum_{k=0}^{m} p_{\ell v}^{k}\, p_{ku}^{s}\right) \boldsymbol{B}_s$$

Comparing coefficients yields

$$\sum_{k} p_{uv}^{k} p_{k\ell}^{s} = \sum_{k} p_{\ell v}^{k} p_{ku}^{s} \tag{4.22}$$

which for $s = 0$ reduces to

$$n_\ell p_{uv}^{\ell} = n_u p_{\ell v}^{u}$$

since $p_{k\ell}^{0} = n_\ell$ for $k = \ell$ and zero otherwise as follows easily from the, trivial, definition of 0th associates. Concerning the parameters of the second kind, we have already mentioned that

$$p_{uv}^{0} = \begin{cases} n_u & \text{for } u = v \\ 0 & \text{otherwise} \end{cases} \tag{4.23}$$

Also,

$$p_{0v}^{k} = \begin{cases} 1 & \text{for } k = v \\ 0 & \text{otherwise} \end{cases} \tag{4.24}$$

which follows immediately from

$$\boldsymbol{B}_v = \boldsymbol{B}_0 \boldsymbol{B}_v = \sum_k p_{0v}^k \, \boldsymbol{B}_k$$

Equations (4.23) and (4.24) hold for any PBIB design, whereas the remaining p_{uv}^k are determined by the particular association scheme for a given PBIB design, as will be shown later. For this reason we refer to a PBIB design as defined above as an m-associate class PBIB design, which henceforth we shall denote by PBIB(m) design.

4.4 ASSOCIATION SCHEMES AND LINEAR ASSOCIATIVE ALGEBRAS

In the motivation and derivations given in Sections 4.2 and 4.3 we have made use of some results given by Bose and Mesner (1959), in particular, relationship (4.12), which leads to the introduction of the parameters of the second kind. There exists, indeed, a much more mathematical relationship between the association matrices, $\boldsymbol{B}_u$, and certain matrices with elements p_{uv}^k. This relationship provides some further insights into the structures of PBIB designs. We shall give a brief account of the major results but refer the reader to Bose and Mesner (1959) for details.

4.4.1 Linear Associative Algebra of Association Matrices

Consider the association matrices $\boldsymbol{B}_u (u = 0, 1, 2, \ldots, m)$ defined in Section 4.2.2. We have seen [see (4.1)] that

$$\boldsymbol{B}_0 + \boldsymbol{B}_1 + \cdots + \boldsymbol{B}_m = \mathfrak{I}\mathfrak{I}'$$

It follows then that the linear form

$$c_0 \boldsymbol{B}_0 + c_1 \boldsymbol{B}_1 + \cdots + c_m \boldsymbol{B}_m \tag{4.25}$$

is equal to zero if and only if

$$c_0 = c_1 = \cdots = c_m = 0$$

Hence the linear functions (4.25) of the $\boldsymbol{B}_u$'s form a vector space with $\boldsymbol{B}_0, \boldsymbol{B}_1, \ldots, \boldsymbol{B}_m$ as basis.

We also know [see (4.12)] that

$$\boldsymbol{B}_u \boldsymbol{B}_v = \sum_{k=0}^{m} p_{uv}^k \, \boldsymbol{B}_k \tag{4.26}$$

If $\boldsymbol{A}$ and $\boldsymbol{B}$ are two matrices of the form (4.25), then the product $\boldsymbol{AB}$ is, because of (4.26), again of the form (4.25). The set of matrices of the form (4.25) is, therefore, closed under multiplication. It also forms an Abelian group under addition. Also, because of (4.19), multiplication is commutative. As a consequence, the set of matrices of the form (4.25) constitute a ring with unit element, $\boldsymbol{B}_0$. It will be a *linear associative algebra* if the coefficients c_u range over a field.

4.4.2 Linear Associative Algebra of $\mathcal{P}$ Matrices

We have introduced in Section 4.3.1 matrices $\boldsymbol{P}_k = (p^k_{uv})$ whose elements are the parameters of the second kind ($k = 0, 1, \ldots, m$). Following Bose and Mesner (1959) we introduce matrices

$$\mathcal{P}_v = \begin{bmatrix} p^0_{0v} & p^1_{0v} & \cdots & p^m_{0v} \\ p^0_{1v} & p^1_{1v} & \cdots & p^m_{1v} \\ & & \cdots & \\ p^0_{mv} & p^1_{mv} & \cdots & p^m_{mv} \end{bmatrix} \tag{4.27}$$

with $v = 0, 1, \ldots, m$. We note that in (4.27), the superscript k is the column index, and the first subscript u is the row index. Also note that although the $\boldsymbol{P}_k$ are symmetric, the $\mathcal{P}_v$ are not necessarily symmetric.

We now consider equation (4.22), that is,

$$\sum_k p^k_{uv} p^s_{k\ell} = \sum_k p^k_{\ell v} p^s_{ku}$$

which follows from the associative and commutative laws of multiplication for the $\boldsymbol{B}_u$'s, and that, because of (4.19), can be written as

$$\sum_k p^k_{uv} p^s_{k\ell} = \sum_k p^k_{v\ell} p^s_{uk} \tag{4.28}$$

The left-hand side of (4.28) is the element in the uth row and sth column of $\mathcal{P}_v\mathcal{P}_\ell$. Since the element in the uth row and sth column of $\mathcal{P}_k$ is p^s_{uk}, the right-hand side of (4.28) is the element in the uth row and sth column of

$$p^0_{v\ell}\mathcal{P}_0 + p^1_{v\ell}\,\mathcal{P}_1 + \cdots + p^m_{v\ell}\,\mathcal{P}_m$$

It follows then that we have the representation

$$\mathcal{P}_v\,\mathcal{P}_\ell = \sum_{k=0}^{m} p^k_{v\ell}\,\mathcal{P}_k \tag{4.29}$$

This corresponds to (4.12) and shows that the $\mathcal{P}$ matrices multiply in the same way as the $\boldsymbol{B}$ matrices.

Since $p_{0v}^k = 1$ for $k = v$ and 0 otherwise [see (4.24)], the 0th row of $\mathcal{P}_v$ contains a 1 in column v and 0's in the other positions. This implies immediately that the linear form

$$c_0\mathcal{P}_0 + c_1\mathcal{P}_1 + \cdots + c_m\mathcal{P}_m \tag{4.30}$$

is zero if and only if $c_0 = c_1 = \cdots = c_m = 0$. Thus the $\mathcal{P}$ matrices are linearly independent and hence form the basis of a vector space. Since they combine in exactly the same way as the $\boldsymbol{B}$ matrices under addition and multiplication, they provide a regular representation in terms of $(m+1) \times (m+1)$ matrices of the algebra of the $\boldsymbol{B}$ matrices, which are $t \times t$ matrices with $t > m + 1$. The $\mathcal{P}$ matrices behave in the same way as the $\boldsymbol{B}$ matrices. Hence, the set of matrices of the form (4.30) also constitute a linear associative algebra with unit element $\mathcal{P}_0 = \boldsymbol{I}_{m+1}$.

4.4.3 Applications of the Algebras

Bose and Mesner (1959) describe in detail applications of the algebras to combinatorial problems and algebraic properties. We have touched on some of these in Section 4.3. We shall mention here only one further application, namely the determination of the eigenvalues of the $\boldsymbol{C}$ matrix for PBIB(2) designs.

We have seen in our earlier discussion (see, e.g., Sections 1.3.3 and 1.10) that the eigenvalues of the $\boldsymbol{C}$ matrix of a given design or, alternatively, of its $\boldsymbol{NN}'$ matrix play a major role in describing properties, for example, the efficiency factor, of the design. We have seen that $\boldsymbol{NN}'$ has one eigenvalue equal to rk. To obtain the other eigenvalues we make use of the following arguments.

Let $\boldsymbol{B} = \sum c_u \boldsymbol{B}_u$ and let $f(\phi)$ be a polynomial. Then $f(\boldsymbol{B})$ can be expressed as

$$f(\boldsymbol{B}) = \sum_{u=0}^{m} \ell_u \boldsymbol{B}_u$$

If $\mathcal{P} = \sum c_u \mathcal{P}_u$ is a representation of $\boldsymbol{B}$, then

$$f(\mathcal{P}) = \sum_{u=0}^{m} \ell_u \mathcal{P}_u$$

If $f(\lambda)$ is the minimum function of $\boldsymbol{B}$ and $\Phi(\lambda)$ the minimum function of $\mathcal{P}$, it can be shown that

$$f(\lambda) = \Phi(\lambda)$$

which implies that $\boldsymbol{B}$ and $\mathcal{P}$ have the same eigenvalues.

We can apply this result to the specific case where

$$\boldsymbol{B} = \boldsymbol{N}\boldsymbol{N}' = \lambda_0 \boldsymbol{B}_0 + \lambda_1 \boldsymbol{B}_1 + \cdots + \lambda_m \boldsymbol{B}_m$$

and

$$\mathcal{P} = \lambda_0 \mathcal{P}_0 + \lambda_1 \mathcal{P}_1 + \cdots + \lambda_m \mathcal{P}_m$$

It follows then that $\boldsymbol{N}\boldsymbol{N}'$ has at most $m+1$ distinct eigenvalues. We know already that rk is one eigenvalue of $\boldsymbol{N}\boldsymbol{N}'$ with multiplicity 1, and it is the largest eigenvalue of $\boldsymbol{N}\boldsymbol{N}'$. The eigenvalues are determined from $|\,\mathcal{P} - \theta \boldsymbol{I}\,| = 0$. Specifically for $m = 2$ and after removal of $\theta = rk$, the remaining two eigenvalues are determined from

$$(\lambda_0 - \theta)^2 + [(\lambda_1 - \lambda_2)(p_{12}^2 - p_{12}^1) - (\lambda_1 + \lambda_2)]\,(\lambda_0 - \theta)$$
$$+ (\lambda_1 - \lambda_2)(\lambda_2 p_{12}^1 - \lambda_1 p_{12}^2) + \lambda_1 \lambda_2 = 0$$

which yields [see Bose and Mesner (1959), Connor and Clatworthy (1954), and Street and Street (1987) for an explicit derivation]

$$\theta_u = \lambda_0 - \tfrac{1}{2}\,[(\lambda_1 - \lambda_2)(-\lambda + (-1)^u \sqrt{\Delta} + (\lambda_1 + \lambda_2)] \tag{4.31}$$

for $u = 1, 2$, and

$$\gamma = p_{12}^2 - p_{12}^1 \qquad \Delta = \gamma^2 + 2\beta + 1 \qquad \beta = p_{12}^1 + p_{12}^2$$

and, of course, $\lambda_0 = r$. We then have

$$|\,\boldsymbol{N}\boldsymbol{N}' - \theta \boldsymbol{I}\,| = (rk - \theta)(\theta_1 - \theta)^{\alpha_1}(\theta_2 - \theta)^{\alpha_2}$$

where α_1, α_2 are the multiplicities of θ_1 and θ_2, respectively. To determine α_1 and α_2 we use the fact that

$$\text{trace } \boldsymbol{I} = 1 + \alpha_1 + \alpha_2 = t$$

and

$$\text{trace } \boldsymbol{N}\boldsymbol{N}' = rk + \alpha_1\theta_1 + \alpha_2\theta_2 = tr$$

Solving these equations and substituting (4.31) yields

$$\alpha_u = \frac{n_1 + n_2}{2} + (-1)^u \frac{(n_1 - n_2) + \gamma(n_1 + n_2)}{2\sqrt{\Delta}} \tag{4.32}$$

for $u = 1, 2$. These results imply that the matrix

$$C = rI - \frac{1}{k} NN'$$

has eigenvalues

$$d_u = r - \frac{\theta_u}{k} \tag{4.33}$$

with multiplicities $\alpha_u (u = 0, 1, 2)$, where $\theta_0 = rk, \alpha_0 = 1$, and θ_1, θ_2 and α_1, α_2 are given by (4.31) and (4.32), respectively.

Similar results can, of course, be obtained for PBIB designs with $m > 2$ associate classes.

The eigenvalues given in (4.33) together with their multiplicities for $u = 1, 2$ can be used to obtain the efficiency factor of a given design (see Section 1.11). We shall also make use of them in Section 4.7 in connection with the combined analysis of PBIB designs.

4.5 ANALYSIS OF PBIB DESIGNS

4.5.1 Intrablock Analysis

The analysis of PBIB(m) designs can be derived easily from the general analysis of incomplete block designs as presented in Section 1.3. In order to solve the RNE [see (1.8)]

$$C\widehat{\tau} = Q$$

where C is given in (4.3) and Q is the vector of adjusted treatment totals, we shall find a generalized inverse C^- for C satisfying (1.17); that is,

$$CC^- = I - \frac{1}{t} \mathfrak{I}\mathfrak{I}'$$

by utilizing (4.11) and (4.12) as outlined previously. This leads to the following system of $m + 1$ equations in $m + 1$ unknowns $g_0, g_1, \ldots, g_m$:

$$\begin{aligned} \sum_{u=0}^{m} \sum_{v=0}^{m} c_u p_{uv}^k g_v &= 1 - \frac{1}{t} \qquad \text{for} \quad k = 0 \\ &= -\frac{1}{t} \qquad \text{for} \quad k = 1, 2, \ldots, m \end{aligned} \tag{4.34}$$

Since

$$\sum_{u=0}^{m} \sum_{v=0}^{m} c_u p_{uv}^k = \sum_{u=0}^{m} c_u n_u = 0$$

it follows that this system of equations is not of full rank. Therefore any m of the equations in (4.34) can be taken and solved with an additional convenient restriction like $\sum g_u = 0$ or $\sum n_u g_u = 0$, or for some v, $g_v = 0$.

We shall illustrate this procedure for a PBIB(2) design. The system (4.34) is then of the form

$$\begin{aligned} c_0 g_0 + c_1 n_1 g_1 + c_2 n_2 g_2 &= 1 - \frac{1}{t} \\ c_1 g_0 + (c_0 + c_1 p_{11}^1 + c_2 p_{12}^1) g_1 + (c_1 p_{12}^1 + c_2 p_{22}^1) g_2 &= -\frac{1}{t} \\ c_2 g_0 + (c_1 p_{11}^2 + c_2 p_{12}^2) g_1 + (c_0 + c_1 p_{12}^2 + c_2 p_{22}^2) g_2 &= -\frac{1}{t} \end{aligned} \tag{4.35}$$

with

$$c_0 = \frac{r(k-1)}{k} \qquad c_1 = -\frac{\lambda_1}{k} \qquad c_2 = -\frac{\lambda_2}{k}$$

Letting $g_0 = 0$ and omitting the first equation then yields

$$\begin{aligned} (c_0 + c_1 p_{11}^1 + c_2 p_{12}^1) g_1 + (c_1 p_{12}^1 + c_2 p_{22}^1) g_2 &= -\frac{1}{t} \\ (c_1 p_{11}^2 + c_2 p_{12}^2) g_1 + (c_0 + c_1 p_{12}^2 + c_2 p_{22}^2) g_2 &= -\frac{1}{t} \end{aligned} \tag{4.36}$$

Example 4.2 Consider the following PBIB(2) design (SR 36 in Clatworthy, 1973) with parameters $t = 8, r = 4, k = 4, b = 8, \lambda_1 = 0, \lambda_2 = 2$:

Block	Treatments
1	1, 2, 3, 4
2	5, 6, 7, 8
3	2, 7, 8, 1
4	6, 3, 4, 5
5	3, 8, 1, 6
6	7, 4, 5, 2
7	4, 1, 6, 7
8	8, 5, 2, 3

which has the following association scheme:

0th Associate	1st Associates	2nd Associates
1	5	2, 3, 4, 6, 7, 8
2	6	1, 3, 4, 5, 7, 8
3	7	1, 2, 4, 5, 6, 8
4	8	1, 2, 3, 5, 6, 7
5	1	2, 3, 4, 6, 7, 8
6	2	1, 3, 4, 5, 7, 8
7	3	1, 2, 4, 5, 6, 8
8	4	1, 2, 3, 5, 6, 7

giving $n_1 = 1, n_2 = 6$, and

$$\boldsymbol{P}_0 = \begin{bmatrix} 1 & & \\ & 1 & \\ & & 6 \end{bmatrix} \qquad \boldsymbol{P}_1 = \begin{bmatrix} 0 & 1 & 0 \\ 1 & 0 & 0 \\ 0 & 0 & 6 \end{bmatrix} \qquad \boldsymbol{P}_2 = \begin{bmatrix} 0 & 0 & 1 \\ 0 & 0 & 1 \\ 1 & 1 & 4 \end{bmatrix}$$

Hence $c_0 = 3, c_1 = 0, c_2 = -\frac{1}{2}$, and (4.36) becomes

$$\begin{aligned} 3g_1 - 3g_2 &= -\tfrac{1}{8} \\ -\tfrac{1}{2}g_1 + g_2 &= -\tfrac{1}{8} \end{aligned}$$

yielding $g_1 = -0.3333$ and $g_2 = -0.2917$.

Having solved (4.34) or its reduced form for $g_0, g_1, \ldots, g_m$, we obtain the intrablock estimates of treatment effects as

$$\widehat{\boldsymbol{\tau}} = \left(\sum_{v=0}^{m} g_v \boldsymbol{B}_v \right) \boldsymbol{Q} \tag{4.37}$$

or, equivalently,

$$\widehat{\tau}_i = g_0 Q_i + g_1 \sum_{j \epsilon S(i,1)} Q_j + g_2 \sum_{k \epsilon S(i,2)} Q_k + \cdots + g_m \sum_{\ell \epsilon S(i,m)} Q_\ell \tag{4.38}$$

$(i = 1, 2, \ldots, t)$ where $S(i, u)$ is, as defined earlier, the set of uth associates of treatment i. It follows further from (4.37) and the definition of the association matrices that

$$\operatorname{var}(\widehat{\tau}_i - \widehat{\tau}_{i'}) = 2(g_0 - g_u)\sigma_e^2 \tag{4.39}$$

if treatments i and i' are uth associates. There are, therefore, m types of comparisons which, however, in certain cases may not all be distinct. Correspondingly, there are also at most m types of efficiency factors, E_u. Since

$$\text{var}(\widehat{\tau}_i - \widehat{\tau}_{i'}) = \frac{2\sigma_e^2}{r}$$

for a CRD or RCBD, we have

$$E_u = \frac{1}{r(g_0 - g_u)} \tag{4.40}$$

$(u = 1, 2, \ldots, m)$ (see Section 1.11). The overall efficiency factor of a PBIB(m) design relative to the CRD or the RCBD is then given by

$$E = \frac{1}{t-1} \sum_{u=1}^{m} n_u E_u \tag{4.41}$$

For the PBIB(2) design discussed in Example 4.2, we find (since $g_0 = 0$)

$$\text{var}(\widehat{\tau}_i - \widehat{\tau}_{i'}) = .6666\sigma_e^2$$

for 1st associates and

$$\text{var}(\widehat{\tau}_i - \widehat{\tau}_{i'}) = .5834\sigma_e^2$$

for 2nd associates. Also, from (4.40) and (4.41)

$$E_1 = -\frac{1}{rg_1} = .750$$

$$E_2 = -\frac{1}{rg_2} = .875$$

$$E = \tfrac{1}{7}(E_1 + 6E_2) = .842$$

□

4.5.2 Combined Analysis

As developed in Section 1.8.3, the set of normal equations for the combined analysis is of the form

$$\boldsymbol{A}\widehat{\widehat{\boldsymbol{\tau}}} = \boldsymbol{P} \tag{4.42}$$

with

$$\boldsymbol{A} = r\boldsymbol{I} - \frac{1-\rho^{-1}}{k}\boldsymbol{N}\boldsymbol{N}' \tag{4.43}$$

and

$$\boldsymbol{P} = \boldsymbol{T} - \frac{1-\rho^{-1}}{k}\boldsymbol{N}\boldsymbol{B} \tag{4.44}$$

In order to obtain $\widehat{\widehat{\boldsymbol{\tau}}}$ we use the same type of argument as in Section 4.2.1, that is, make use of the special form of $\boldsymbol{A}$ as expressed in terms of association matrices; that is,

$$\boldsymbol{A} = \sum_{u=0}^{m} a_u \boldsymbol{B}_u \tag{4.45}$$

with

$$a_0 = r\left(1 - \frac{1-\rho^{-1}}{k}\right) \tag{4.46}$$

$$a_u = -\frac{1-\rho^{-1}}{k}\lambda_u \qquad (u = 1, 2, \ldots, m) \tag{4.47}$$

Writing $\boldsymbol{A}^{-1}$ as

$$\boldsymbol{A}^{-1} = \sum_{u=0}^{m} a^u \boldsymbol{B}_u$$

we obtain, equivalent to the set of equations (4.34) for the intrablock analysis, the following system of equations:

$$\boldsymbol{A}\boldsymbol{A}^{-1} = \left(\sum_{u=0}^{m} a_u \boldsymbol{B}_u\right)\left(\sum_{v=0}^{m} a^v \boldsymbol{B}_v\right) = \sum_{uv}\sum_{k=0}^{m} a_u a^v\, p_{uv}^k \boldsymbol{B}_k = \boldsymbol{I} \tag{4.48}$$

or equivalently,

$$\sum_{u=0}^{m}\sum_{v=0}^{m} a_u p_{uv}^k a^v = \delta_{0k} \qquad (k = 0, 1, \ldots, m) \tag{4.49}$$

where δ_{0k} is the Kronecker symbol. This set of equations is of full rank and hence we have $(m+1)$ equations in $(m+1)$ unknowns.

It follows then that

$$\widehat{\widehat{\tau}}_i = a^0 P_i + a^1 \sum_{j \epsilon S(i,1)} P_j + a^2 \sum_{k \epsilon S(i,2)} P_k + \cdots + a^m \sum_{\ell \epsilon S(i,m)} P_\ell \tag{4.50}$$

where the $P_u (u = 1, 2, \ldots, t)$ are the elements of $\boldsymbol{P}$ in (4.44).

For any two treatments that are uth associates we have consequently

$$\operatorname{var}\left(\widehat{\widehat{\tau}}_i - \widehat{\widehat{\tau}}_{i'}\right) = 2(a^0 - a^u)\sigma_e^2 \tag{4.51}$$

We shall illustrate the procedure by using Example 4.2 given in Section 4.4.1.

Example 4.2 (Continued) We have $t = 8, r = \lambda_0 = 4, k = 4, b = 8, \lambda_1 = 0, \lambda_2 = 2$.

$$\boldsymbol{P}_0 = \begin{bmatrix} 1 & & \\ & 1 & \\ & & 6 \end{bmatrix} \qquad \boldsymbol{P}_1 = \begin{bmatrix} 0 & 1 & 0 \\ 1 & 0 & 0 \\ 0 & 0 & 6 \end{bmatrix}$$

$$\boldsymbol{P}_2 = \begin{bmatrix} 0 & 0 & 1 \\ 0 & 0 & 1 \\ 1 & 1 & 4 \end{bmatrix}$$

Then, from (4.46) and (4.47),

$$a_0 = 4\left(1 - \frac{1 - \rho^{-1}}{4}\right) = 3 + \rho^{-1}$$

$$a_1 = 0$$

$$a_2 = -(1 - \rho^{-1})\tfrac{1}{2}$$

and hence Eqs. (4.49) are given as

$$\begin{aligned} (3 + \rho^{-1})\, a^0 \qquad\qquad\qquad\qquad\qquad & - 3(1 - \rho^{-1})\, a^2 = 1 \\ (3 + \rho^{-1})\, a^1 \qquad & - 3(1 - \rho^{-1})\, a^2 = 0 \\ -\tfrac{1}{2}(1 - \rho^{-1})\, a^0 - \tfrac{1}{2}(1 - \rho^{-1})\, a^1 + [3 + \rho^{-1} & - 2\,(1 - \rho^{-1})]\, a^2 = 0 \end{aligned}$$

There remains, of course, the problem of estimating ρ or ρ^{-1}. Substituting such an estimator in the equations above, or more generally equations (4.49), will then yield solutions for a^0, a^1, a^2 which we denote by $\widehat{a}^0, \widehat{a}^1, \widehat{a}^2$ and which, in turn, are used in Eqs. (4.50) and (4.51). One such estimator is the Yates estimator as described in Section 1.10.1. For certain types of PBIB(2) designs we shall present in Section 4.7 other estimators for ρ similar to the one given for BIB designs (see Section 2.5), which will lead to uniformly better estimators (than the intrablock estimators) for treatment contrasts under certain conditions. □

4.6 CLASSIFICATION OF PBIB DESIGNS

Since their introduction by Bose and Nair (1939) and Bose and Shimamoto (1952) different types of PBIB designs have evolved. These can be classified according to the number of associate classes and type of association scheme. (We mention here parenthetically that the association scheme and the parameters of a PBIB design do not lead immediately to the actual experimental layout or field plan; see also Chapter 5). In this section we shall give a brief survey (without being complete) of types of PBIB designs.

The designs perhaps most often used are PBIB(2) designs. The reason, of course, is that they constitute the simplest extension of the BIB design. They have been tabulated (for certain ranges of parameters) extensively by Bose, Clatworthy, and Shrikhande (1954) and Clatworthy (1956, 1973) and are classified according to which of the following association schemes (to be explained in the following sections) is being used:

1. Group-divisible PBIB(2) designs
2. Triangular PBIB(2) designs
3. Latin square type PBIB(2) designs
4. Cyclic PBIB(2) designs

We then mention the association schemes and parameters for some PBIB(3) designs:

1. Rectangular PBIB(3) designs
2. Generalized group-divisible PBIB(3) designs
3. Generalized triangular PBIB(3) designs
4. Cubic PBIB(3) designs

Finally, we discuss some rather general classes of PBIB designs with $m > 3$ associate classes:

1. Extended group-divisible PBIB designs
2. Hypercubic PBIB designs
3. Right-angular PBIB(4) designs
4. Cyclic PBIB designs

4.6.1 Group-Divisible (GD) PBIB(2) Designs

Suppose that t can be written as $t = t_1 t_2$ and that the treatments are divided into t_1 groups of t_2 treatments each by arranging them in a rectangular array of t_1 rows and t_2 columns ($t_1, t_2 > 1$), the rows constituting the t_1 groups. Then two treatments are called first associates if they are in the same group and they are called second associates otherwise.

This association scheme implies $n_1 = t_2 - 1$ and $n_2 = (t_1 - 1)t_2$. Further

$$\boldsymbol{P}_1 = \begin{bmatrix} 0 & 1 & 0 \\ 1 & t_2 - 2 & 0 \\ 0 & 0 & (t_1 - 1)t_2 \end{bmatrix} \qquad \boldsymbol{P}_2 = \begin{bmatrix} 0 & 0 & 1 \\ 0 & 0 & t_2 - 1 \\ 1 & t_2 - 1 & (t_1 - 2)t_2 \end{bmatrix}$$

Example 4.3 Let $t = 8, t_1 = 2, t_2 = 4$.

Association scheme:

$$\begin{matrix} 1 & 3 & 5 & 7 \\ 2 & 4 & 6 & 8 \end{matrix}$$

0th Associate	1st Associates	2nd Associates
1	3, 5, 7	2, 4, 6, 8
2	4, 6, 8	1, 3, 5, 7
3	1, 5, 7	2, 4, 6, 8
4	2, 6, 8	1, 3, 5, 7
5	1, 3, 7	2, 4, 6, 8
6	2, 4, 8	1, 3, 5, 7
7	1, 3, 5	2, 4, 6, 8
8	2, 4, 6	1, 3, 5, 7

□

The factorization of t is, of course, not always unique. For Example 4.3 we could just as well have $t_1 = 4, t_2 = 2$, in which case the rectangular array is

$$\begin{matrix} 1 & 5 \\ 2 & 6 \\ 3 & 7 \\ 4 & 8 \end{matrix}$$

and hence $n_1 = 1, n_2 = 6$.

Bose and Connor (1952) have shown that for GD-PBIB(2) designs the following inequalities hold:

$$r \geq \lambda_1 \qquad rk - \lambda_2 t \geq 0$$

Accordingly, they have divided the GD designs further into three subclasses:

1. Singular if $r = \lambda_1$
2. Semiregular if $r > \lambda_1, rk - \lambda_2 t = 0$
3. Regular if $r > \lambda_1, rk - \lambda_2 t > 0$

4.6.2 Triangular PBIB(2) Designs

Suppose that t can be written as $t = q(q-1)/2$ and that the treatments are arranged in triangular form above the main diagonal of a $q \times q$ array and repeated symmetrically below the main diagonal leaving the main diagonal blank. Then two treatments are called first associates if they lie in the same row (or column) of the $q \times q$ array and are second associates otherwise.

This association scheme implies

$$n_1 = 2q - 4 \qquad n_2 = (q-2)(q-3)/2$$

$$\boldsymbol{P}_1 = \begin{bmatrix} 0 & 1 & 0 \\ 1 & q-2 & q-3 \\ 0 & q-3 & (q-3)(q-4)/2 \end{bmatrix}$$

$$\boldsymbol{P}_2 = \begin{bmatrix} 0 & 0 & 1 \\ 1 & 4 & 2q-8 \\ 1 & 2q-8 & (q-4)(q-5)/2 \end{bmatrix}$$

Example 4.4 Let $t = 10, q = 5$.

Association scheme:

*	1	2	3	4
1	*	5	6	7
2	5	*	8	9
3	6	8	*	10
4	7	9	10	*

0th Associate	**1st Associates**	**2nd Associates**
1	2, 3, 4, 5, 6, 7	8, 9, 10
2	1, 3, 4, 5, 8, 9	6, 7, 10
3	1, 2, 4, 6, 8, 10	5, 7, 9
4	1, 2, 3, 7, 9, 10	5, 6, 8
5	1, 2, 6, 7, 8, 9	3, 4, 10
6	1, 3, 5, 7, 8, 10	2, 4, 9
7	1, 4, 5, 6, 9, 10	2, 3, 8
8	2, 3, 5, 6, 9, 10	1, 4, 7
9	2, 4, 5, 7, 8, 10	1, 3, 6
10	3, 4, 6, 7, 8, 9	1, 2, 5

□

4.6.3 Latin Square Type PBIB(2) Designs

Suppose that t can be written as $t = q^2$ and that the treatments are arranged in a square array of size q.

For a Latin square type PBIB(2) design with two constraints [L_2-PBIB(2) design], two treatments are called first associates if they occur together in the same row or the same column of the square array and are second associates otherwise.

For the Latin square type design with i constraints ($i > 2$) [L_i-PBIB(2) design] the association scheme is used in conjunction with $i - 2$ mutually orthogonal Latin squares (if they exist) (see I.10.6.2). Then two treatments are called first associates if they both occur in the same row or column or correspond to the same letter of the superimposed Latin squares. Hence we obtain for the L_i-PBIB(2) design

$$n_1 = i(q-1) \qquad n_2 = (q-1)(q-i+1)$$

$$\boldsymbol{P}_1 = \begin{bmatrix} 0 & 1 & 0 \\ 1 & i^2 - 3i + q & (i-1)(q-i+1) \\ 0 & (i-1)(q-i+1) & (q-i)(q-i+1) \end{bmatrix}$$

$$\boldsymbol{P}_2 = \begin{bmatrix} 0 & 0 & 1 \\ 0 & i(i-1) & i(q-1) \\ 1 & i(q-i) & (q-i)^2 + i - 2 \end{bmatrix}$$

Example 4.5 Let $t = 9, q = 3$.

L_2-association scheme:

1	2	3
4	5	6
7	8	9

0th Associate	1st Associates	2nd Associates
1	2, 3, 4, 7	5, 6, 8, 9
2	1, 3, 5, 8	4, 6, 7, 9
3	1, 2, 6, 9	4, 5, 7, 8
4	1, 5, 6, 7	2, 3, 8, 9
5	2, 4, 6, 8	1, 3, 7, 9
6	3, 4, 5, 9	1, 2, 7, 8
7	1, 4, 8, 9	2, 3, 5, 6
8	2, 5, 7, 9	1, 3, 4, 6
9	3, 6, 7, 8	1, 2, 4, 5

□

Example 4.6 Let $t = 9, q = 3$.

L_3-association scheme:

$$\begin{array}{ccc} 1A & 2B & 3C \\ 4C & 5A & 6B \\ 7B & 8C & 9A \end{array}$$

0th Associate	**1st Associates**	**2nd Associates**
1	2, 3, 4, 5, 7, 9	6, 8
2	1, 3, 5, 6, 7, 8	4, 9
3	1, 2, 4, 6, 8, 9	5, 7
4	1, 3, 5, 6, 7, 8	2, 9
5	1, 2, 4, 6, 8, 9	3, 7
6	2, 3, 4, 5, 7, 9	1, 8
7	1, 2, 4, 6, 8, 9	3, 5
8	2, 3, 4, 5, 7, 9	1, 6
9	1, 3, 5, 6, 7, 8	2, 4

□

Other examples are mentioned in Section 18.7.

4.6.4 Cyclic PBIB(2) Designs

Suppose the treatments are denoted by $0, 1, 2, \ldots, t-1$. Then the first associates of treatment i are $i + d_1, i + d_2, \ldots, i + d_{n_1}$ (mod t) where the d_j are integers satisfying the following conditions:

1. The d_j are all different with $0 < d_j < t$ for each j.
2. Among the $n_1(n_1 - 1)$ differences $d_j - d_{j'}$ (mod t) each of the integers $d_1, d_2, \ldots, d_{n_1}$ occurs α times, and each of the integers $e_1, e_2, \ldots, e_{n_2}$ occurs β times, where $d_1, d_2, \ldots, d_{n_1}, e_1, e_2, \ldots, e_{n_2}$ are all the different integers $1, 2, \ldots, t-1$.

The second associates of treatment i obviously are $i + e_1, i + e_2, \ldots, i + e_{n_2}$ (mod t). We then have, as a consequence of condition 2,

$$n_1\alpha + n_2\beta = n_1(n_1 - 1)$$

and

$$\boldsymbol{P}_1 = \begin{bmatrix} 0 & 1 & 0 \\ 1 & \alpha & n_1 - \alpha - 1 \\ 0 & n_1 - \alpha - 1 & n_2 - n_1 + \alpha + 1 \end{bmatrix}$$

$$P_2 = \begin{bmatrix} 0 & 0 & 1 \\ 0 & \beta & n_1 - \beta \\ 0 & n_1 - \beta & n_2 - n_1 + \beta - 1 \end{bmatrix}$$

Example 4.7 Let $t = 5, d_1 = 2, d_2 = 3, \alpha = 0, \beta = 1$. With $n_1 = 2$, the possible differences $d_j - d_{j'}$ are $d_1 - d_2 = -1 = 4$ and $d_2 - d_1 = 1$; that is, d_1 and d_2 occur $\alpha = 0$ times among these differences and $e_1 = 1, e_2 = 4$ occur $\beta = 1$ time. Hence condition 2 is satisfied. We then have

0th Associate	1st Associates	2nd Associates
0	2, 3	1, 4
1	3, 4	0, 2
2	4, 0	1, 3
3	0, 1	2, 4
4	1, 2	0, 3

□

4.6.5 Rectangular PBIB(3) Designs

This association scheme was first proposed by Vartak (1959). Suppose that $t = t_1 t_2$ and that the treatments are arranged in a rectangular array with t_1 rows and t_2 columns. Two treatments are said to be first associates if they occur together in the same row of the array; they are said to be second associates if they occur together in the same column; they are third associates otherwise. Hence

$$n_1 = t_2 - 1, \qquad n_2 = t_1 - 1 \qquad n_3 = (t_1 - 1)(t_2 - 1)$$

and

$$P_1 = \begin{bmatrix} 0 & 1 & 0 & 0 \\ 1 & t_2 - 2 & 0 & 0 \\ 0 & 0 & 0 & t_1 - 1 \\ 0 & 0 & t_1 - 1 & (t_1 - 1)(t_2 - 2) \end{bmatrix}$$

$$P_2 = \begin{bmatrix} 0 & 0 & 1 & 0 \\ 0 & 0 & 0 & t_2 - 1 \\ 1 & 0 & t_1 - 2 & 0 \\ 0 & t_2 - 1 & 0 & (t_1 - 2)(t_2 - 1) \end{bmatrix}$$

$$P_3 = \begin{bmatrix} 0 & 0 & 0 & 1 \\ 0 & 0 & 1 & t_2 - 2 \\ 0 & 1 & 0 & t_1 - 2 \\ & t_2 - 2 & t_1 - 2 & (t_1 - 2)(t_2 - 2) \end{bmatrix}$$

Example 4.8 Let $t = 6, t_1 = 2, t_2 = 3$.

Association scheme:

$$\begin{matrix} 1 & 3 & 5 \\ 2 & 4 & 6 \end{matrix}$$

0th Associate	1st Associates	2nd Associates	3rd Associates
1	3, 5	2	4, 6
2	4, 6	1	3, 5
3	1, 5	4	2, 6
4	2, 6	3	1, 5
5	1, 3	6	2, 4
6	2, 4	5	1, 3

□

4.6.6 Generalized Group-Divisible (GGD) PBIB(3) Designs

In a more general way this association scheme was proposed by Roy (1953–1954) and described in more detail by Raghavarao (1960).

Suppose that $t = t_1 t_2 t_3$ and that the treatments are arranged in a three-dimensional array with t_1 rows, t_2 columns, and t_3 layers. Two treatments are said to be first associates if they occur together in the same row and column, but in different layers; they are said to be second associates if they occur in the same row but in different columns (and the same or different layers); they are third associates otherwise. Hence

$$n_1 = t_3 - 1 \qquad n_2 = (t_2 - 1)t_3 \qquad n_3 = (t_1 - 1)t_2 t_3$$

and

$$\boldsymbol{P}_1 = \begin{bmatrix} 0 & 1 & 0 & 0 \\ 1 & t_3 - 2 & 0 & 0 \\ 0 & 0 & (t_2 - 1)t_3 & 0 \\ 0 & 0 & 0 & (t_1 - 1)t_2 t_3 \end{bmatrix}$$

$$\boldsymbol{P}_2 = \begin{bmatrix} 0 & 0 & 1 & 0 \\ 0 & 0 & t_3 - 1 & 0 \\ 1 & t_3 - 1 & (t_2 - 2)t_3 & 0 \\ 0 & 0 & 0 & (t_1 - 1)t_2 t_3 \end{bmatrix}$$

$$\boldsymbol{P}_3 = \begin{bmatrix} 0 & 0 & 0 & 1 \\ 0 & 0 & 0 & t_3 - 1 \\ 0 & 0 & 0 & (t_2 - 1)t_3 \\ 1 & t_3 - 1 & (t_2 - 1)t_3 & (t_1 - 2)t_2 t_3 \end{bmatrix}$$

Example 4.9 Let $t = 12, t_1 = 2, t_2 = 3, t_3 = 2$.

Association scheme:

Layer 1			Layer 2		
1	3	5	7	9	11
2	4	6	8	10	12

0th Associate	1st Associates	2nd Associates	3rd Associates
1	7	3, 5, 9, 11	2, 4, 6, 8, 10, 12
2	8	4, 6, 10, 12	1, 3, 5, 7, 9, 11
3	9	1, 5, 7, 11	2, 4, 6, 8, 10, 12
4	10	2, 6, 8, 12	1, 3, 5, 7, 9, 11
5	11	1, 3, 7, 9	2, 4, 6, 8, 10, 12
6	12	2, 4, 8, 10	1, 3, 5, 7, 9, 11
7	1	3, 5, 9, 11	2, 4, 6, 8, 10, 12
8	2	4, 6, 10, 12	1, 3, 5, 7, 9, 11
9	3	1, 5, 7, 11	2, 4, 6, 8, 10, 12
10	4	2, 6, 8, 12	1, 3, 5, 7, 9, 11
11	5	1, 3, 7, 11	2, 4, 6, 8, 10, 12
12	6	2, 4, 8, 10	1, 3, 5, 7, 9, 11

□

4.6.7 Generalized Triangular PBIB(3) Designs

John (1966) has shown that the triangular association scheme discussed in Section 4.5.2 can be described equivalently by representing the treatments by ordered pairs (x, y) with $1 \leq x < y \leq q$ and calling two treatments first associates if they have one integer in common, second associates otherwise.

This can then be generalized (John, 1966) for the case that $t = q(q-1)(q-2)/6$ $(q > 3)$ and that the treatments are represented by ordered triplets (x, y, z) with $1 \leq x < y < z \leq q$. Two treatments are said to be first associates if they have two integers in common, second associates if they have one integer in common but not two, and third associates otherwise. Hence

$$n_1 = 3(q-3) \qquad n_2 = 3(q-3)(q-4)/2 \qquad n_3 = (q-3)(q-4)(q-5)/6$$

and

$$\boldsymbol{P}_1 = \begin{bmatrix} 0 & 1 & 0 & 0 \\ 1 & q-2 & 2(q-4) & 0 \\ 0 & 2(q-4) & (q-4)^2 & (q-4)(q-5)/2 \\ 0 & 0 & (q-4)(q-5)/2 & (q-4)(q-5)(q-6)/6 \end{bmatrix}$$

$$\boldsymbol{P}_2 = \begin{bmatrix} 0 & 0 & 1 & 0 \\ 1 & 4 & 2(q-4) & q-5 \\ 1 & 2(q-4) & (q+2)(q-5)/2 & (q-5)(q-6) \\ 0 & q-5 & (q-5)(q-6) & (q-5)(q-6)(q-7)/6 \end{bmatrix}$$

$$\boldsymbol{P}_3 = \begin{bmatrix} 0 & 0 & 0 & 1 \\ 0 & 0 & 9 & 3(q-6) \\ 0 & 9 & 9(q-6) & 3(q-6)(q-7)/2 \\ 1 & 3(q-6) & 3(q-6)(q-7)/2 & (q-6)(q-7)(q-8)/6 \end{bmatrix}$$

Example 4.10 Let $t = 20, q = 6$. The treatments are represented as

$$\begin{array}{llll}
1 \equiv (1,2,3) & 2 \equiv (1,2,4) & 3 \equiv (1,2,5) & 4 \equiv (1,2,6) \\
 & 5 \equiv (1,3,4) & 6 \equiv (1,3,5) & 7 \equiv (1,3,6) \\
 & & 8 \equiv (1,4,5) & 9 \equiv (1,4,6) \\
 & & & 10 \equiv (1,5,6) \\
 & 11 \equiv (2,3,4) & 12 \equiv (2,3,5) & 13 \equiv (2,3,6) \\
 & & 14 \equiv (2,4,5) & 15 \equiv (2,4,6) \\
 & & & 16 \equiv (2,5,6) \\
 & & 17 \equiv (3,4,5) & 18 \equiv (3,4,6) \\
 & & & 19 \equiv (3,5,6) \\
 & & & 20 \equiv (4,5,6)
\end{array}$$

and the association scheme follows:

0th Associate	1st Associates	2nd Associates	3rd Associates
1	2, 3, 4, 5, 6, 7, 11, 12, 13	8, 9, 10, 14, 15, 16, 17, 18, 19	20
2	1, 3, 4, 5, 8, 9, 11, 14, 15	6, 7, 10, 12, 13, 16, 17, 18, 20	19
3	1, 2, 4, 6, 8, 10, 12, 14, 16	5, 7, 9, 11, 13, 15, 17, 19, 20	18
4	1, 2, 3, 7, 9, 10, 13, 15, 16	5, 6, 8, 11, 12, 14, 18, 19, 20	17
5	1, 2, 6, 7, 8, 9, 11, 17, 18	3, 4, 10, 12, 13, 14, 15, 19, 20	16
6	1, 3, 5, 7, 8, 10, 12, 17, 19	2, 4, 9, 11, 13, 14, 16, 18, 20	15
7	1, 4, 5, 6, 9, 10, 13, 18, 19	2, 3, 8, 11, 12, 15, 16, 17, 20	14
8	2, 3, 5, 6, 9, 10, 14, 17, 20	1, 4, 7, 11, 12, 15, 16, 18, 19	13
9	2, 4, 5, 7, 8, 10, 15, 18, 20	1, 3, 6, 11, 13, 14, 16, 17, 19	12
10	3, 4, 6, 7, 8, 9, 16, 19, 20	1, 2, 5, 12, 13, 14, 15, 17, 18	11
11	1, 2, 5, 12, 13, 14, 15, 17, 18	3, 4, 6, 7, 8, 9, 16, 19, 20	10
12	1, 3, 6, 11, 13, 14, 16, 17, 19	2, 4, 5, 7, 8, 10, 15, 18, 20	9
13	1, 4, 7, 11, 12, 15, 16, 18, 19	2, 3, 5, 6, 9, 10, 14, 17, 20	8
14	2, 3, 8, 11, 12, 15, 16, 17, 20	1, 4, 5, 6, 9, 10, 13, 18, 19	7
15	2, 4, 9, 11, 13, 14, 16, 18, 20	1, 3, 5, 7, 8, 10, 12, 17, 19	6
16	3, 4, 10, 12, 13, 14, 15, 19, 20	1, 2, 6, 7, 8, 9, 11, 17, 18	5
17	5, 6, 8, 11, 12, 14, 18, 19, 20	1, 2, 3, 7, 9, 10, 13, 15, 16	4
18	5, 7, 9, 11, 13, 15, 17, 19, 20	1, 2, 4, 6, 8, 10, 12, 14, 16	3
19	6, 7, 10, 12, 13, 16, 17, 18, 20	1, 3, 4, 5, 8, 9, 11, 13, 15	2
20	8, 9, 10, 14, 15, 16, 17, 18, 19	2, 3, 4, 5, 6, 7, 11, 12, 13	1

□

4.6.8 Cubic PBIB(3) Designs

This association scheme, described by Raghavarao and Chandrasekhararao (1964), is an extension of the L_2-association scheme discussed in Section 4.5.3, which can be described equivalently as follows: Denote the $t = q^2$ treatments by pairs (x, y) with $1 \le x, y \le q$. Define the distance δ between two treatments (x, y) and (x', y') as the number of nonzero elements in $(x - x', y - y')$. For the L_2-association scheme then two treatments are said to be first associates if $\delta = 1$, and second associates if $\delta = 2$.

For the cubic association scheme we have $t = q^3$, each treatment being represented by a triplet (x, y, z) with $1 \le x, y, z \le q$. If the distance δ between any two treatments (x, y, z) and (x', y', z') is the number of nonzero elements in $(x - x', y - y', z - z')$, then two treatments are said to be first, second, or third associates if $\delta = 1, 2$, or 3, respectively.

This can be interpreted geometrically by realizing that the above representation of the treatments corresponds to an arrangement in a cube of side q. Two treatments are then first associates if they are lying on the same axis, second associates if they are lying on the same plane (but not on the same axis), and third associates otherwise. The design given in Example 4.1 has this association scheme with $q = 2$.

The cubic association scheme implies immediately

$$n_1 = 3(q-1) \qquad n_2 = 3(q-1)^2 \qquad n_3 = (q-1)^3$$

and

$$\boldsymbol{P}_1 = \begin{bmatrix} 0 & 1 & 0 & 0 \\ 1 & q-2 & 2(q-1) & 0 \\ 0 & 2(q-1) & 2(q-1)(q-2) & (q-1)^2 \\ 0 & 0 & (q-1)^2 & (q-1)^2(q-2) \end{bmatrix}$$

$$\boldsymbol{P}_2 = \begin{bmatrix} 0 & 0 & 1 & 0 \\ 0 & 2 & 2(q-2) & q-1 \\ 1 & 2(q-2) & 2(q-1)+(q-2)^2 & 2(q-1)(q-2) \\ 0 & q-1 & 2(q-1)(q-2) & (q-1)(q-2)^2 \end{bmatrix}$$

$$\boldsymbol{P}_3 = \begin{bmatrix} 0 & 0 & 0 & 1 \\ 0 & 0 & 3 & 3(q-2) \\ 0 & 3 & 6(q-2) & 3(q-2)^2 \\ 1 & 3(q-2) & 3(q-2)^2 & (q-2)^3 \end{bmatrix}$$

Example 4.11 Let $t = 8, q = 2$. The treatments are represented as

$$\begin{array}{ll} 1 \equiv (1, 1, 1) & 2 \equiv (1, 1, 2) \\ 3 \equiv (1, 2, 1) & 4 \equiv (1, 2, 2) \\ 5 \equiv (2, 1, 1) & 6 \equiv (2, 1, 2) \\ 7 \equiv (2, 2, 1) & 8 \equiv (2, 2, 2) \end{array}$$

Association scheme:

0th Associate	1st Associates	2nd Associates	3rd Associates
1	2, 3, 5	4, 6, 7	8
2	1, 4, 6	3, 5, 8	7
3	1, 4, 7	2, 5, 8	6
4	2, 3, 8	1, 6, 7	5
5	1, 6, 7	2, 3, 8	4
6	2, 5, 8	1, 4, 7	3
7	3, 5, 8	1, 4, 6	2
8	4, 6, 7	2, 3, 5	1

□

Other examples are mentioned in Section 18.7.

4.6.9 Extended Group-Divisible (EGD) PBIB Designs

These designs, which we shall abbreviate as EGD/$(2^\nu - 1)$-PBIB, were first introduced by Hinkelmann and Kempthorne (1963) as an extension of the rectangular PBIB(3) design of Vartak (1959), and described in more detail by Hinkelmann (1964). The association scheme for these designs is as follows:

Let there be $t = t_1 t_2, \ldots, t_\nu$ treatments denoted by $(i_1, i_2, \ldots, i_\nu)$, where $i_\ell = 1, 2, \ldots, t_\ell$ and $\ell = 1, 2, \ldots, \nu$, and i_ℓ is called the ℓth component of the treatment. Two treatments are said to be $\boldsymbol{\gamma}$th associates, $\boldsymbol{\gamma} = (\gamma_1, \gamma_2, \ldots, \gamma_\nu)$ and $\gamma_\ell = 0$ or $1 (\ell = 1, 2, \ldots, \nu)$, if the treatments differ only in the components that correspond to the unity components of $\boldsymbol{\gamma}$. We then have $m = 2^\nu - 1$ associate classes as each component in $\boldsymbol{\gamma}$ takes on the possible values 0, 1, and $\boldsymbol{\gamma} = (0, 0, \ldots, 0)$ represents the trivial association corresponding to the 0th associate class of other PBIB designs. The number of associates in the $\boldsymbol{\gamma}$th associate class is denoted by $n(\boldsymbol{\gamma})$ and

$$n(\boldsymbol{\gamma}) = n(\gamma_1, \gamma_2, \ldots, \gamma_\nu) = \prod_{\ell=1}^{\nu} (t_{\ell-1})^{\gamma_\ell} \tag{4.52}$$

If we write the associate classes in lexicographic order, for example, for $\nu = 3$: 000, 001, 010, 011, 100, 101, 110, 111, and if we denote by

$$\boldsymbol{P}_0^{(t_\ell)} = \begin{pmatrix} 1 & 0 \\ 0 & t_\ell - 1 \end{pmatrix}$$

$$\boldsymbol{P}_1^{(t_\ell)} = \begin{pmatrix} 0 & 1 \\ 1 & t_\ell - 2 \end{pmatrix}$$

the $\boldsymbol{P}$ matrices for a BIB design with t_ℓ treatments, then we can write the $\boldsymbol{P}$ matrices for the EGD/$(2^\nu - 1)$-PBIB design recursively as the following Kronecker product:

$$\boldsymbol{P}_{\gamma_1\gamma_2\ldots\gamma_{\nu-1}0} = \boldsymbol{P}_{\gamma_1\gamma_2\ldots\gamma_{\nu-1}} \times \boldsymbol{P}_0^{(t_\nu)} \tag{4.53}$$

$$\boldsymbol{P}_{\gamma_1\gamma_2\ldots\gamma_{\nu-1}1} = \boldsymbol{P}_{\gamma_1\gamma_2\ldots\gamma_{\nu-1}} \times \boldsymbol{P}_1^{(t_\nu)} \tag{4.54}$$

where $\boldsymbol{P}_{\gamma_1\gamma_2\ldots\gamma_{\nu-1}}$ are the $\boldsymbol{P}$ matrices for an EGD/$(2^{\nu-1} - 1)$ design with $t = t_1 t_2, \ldots, t_{\nu-1}$ treatments. In particular then for $\nu = 2$:

$$\boldsymbol{P}_{00} = \begin{bmatrix} 1 & & \emptyset & \\ & t_2 - 1 & & \\ & & t_1 - 1 & \\ & \emptyset & & (t_1-1)(t_2-1) \end{bmatrix}$$

$$\boldsymbol{P}_{01} = \begin{bmatrix} 0 & 1 & & \\ & & \emptyset & \\ 1 & t_2 - 2 & & \\ & & 0 & t_1 - 1 \\ & \emptyset & t_1 - 1 & (t_1-1)(t_2-2) \end{bmatrix}$$

$$\boldsymbol{P}_{10} = \begin{bmatrix} & & 1 & 0 \\ & \emptyset & & \\ & & 0 & t_2 - 1 \\ 1 & 0 & t_1 - 2 & 0 \\ 0 & t_2 - 1 & 0 & (t_1-2)(t_2-1) \end{bmatrix}$$

$$\boldsymbol{P}_{11} = \begin{bmatrix} & & 0 & 1 \\ & \emptyset & & \\ & & 1 & t_2 - 2 \\ 0 & 1 & 0 & t_1 - 2 \\ 1 & t_2 - 2 & t_1 - 2 & (t_1-2)(t_2-2) \end{bmatrix}$$

which are, apart from the order of rows and columns, the same as those given for the rectangular PBIB(3) design (see Section 4.5.5).

Example 4.12 Let $t = 36, t_1 = t_2 = 3, t_3 = 4$. The treatments are

$$\begin{array}{llll}
1 \equiv (1,1,1) & 2 \equiv (1,1,2) & 3 \equiv (1,1,3) & 4 \equiv (1,1,4) \\
5 \equiv (1,2,1) & 6 \equiv (1,2,2) & 7 \equiv (1,2,3) & 8 \equiv (1,2,4) \\
9 \equiv (1,3,1) & 10 \equiv (1,3,2) & 11 \equiv (1,3,3) & 12 \equiv (1,3,4) \\
13 \equiv (2,1,1) & 14 \equiv (2,1,2) & 15 \equiv (2,1,3) & 16 \equiv (2,1,4) \\
17 \equiv (2,2,1) & 18 \equiv (2,2,2) & 19 \equiv (2,2,3) & 20 \equiv (2,2,4) \\
21 \equiv (2,3,1) & 22 \equiv (2,3,2) & 23 \equiv (2,3,3) & 24 \equiv (2,3,4) \\
25 \equiv (3,1,1) & 26 \equiv (3,1,2) & 27 \equiv (3,1,3) & 28 \equiv (3,1,4) \\
29 \equiv (3,2,1) & 30 \equiv (3,2,2) & 31 \equiv (3,2,3) & 32 \equiv (3,2,4) \\
33 \equiv (3,3,1) & 34 \equiv (3,3,2) & 35 \equiv (3,3,3) & 36 \equiv (3,3,4)
\end{array}$$

The associate classes and the numbers of associates are

$$\begin{array}{ll}
0 \equiv (0,0,0) & n(0,0,0) \equiv 1 \\
1 \equiv (0,0,1) & n(0,0,1) \equiv 3 \\
2 \equiv (0,1,0) & n(0,1,0) \equiv 2 \\
3 \equiv (0,1,1) & n(0,1,1) \equiv 6 \\
4 \equiv (1,0,0) & n(1,0,0) \equiv 2 \\
5 \equiv (1,0,1) & n(1,0,1) \equiv 6 \\
6 \equiv (1,1,0) & n(1,1,0) \equiv 4 \\
7 \equiv (1,1,1) & n(1,1,1) \equiv 12
\end{array}$$

and the (partial) association scheme is given in Table 4.1 □

4.6.10 Hypercubic PBIB Designs

These designs, first indicated by Shah (1958) and more formally defined by Kusumoto (1965), are extensions of cubic PBIB designs (see Section 4.5.8). Suppose we have $t = q^\nu$ treatments, which are again denoted by $(i_1, i_2, \ldots, i_\nu)$ where $i_\ell = 1, 2, \ldots, q$ and $\ell = 1, 2, \ldots, \nu$. The association scheme for this design is then as follows: Two treatments are said to be jth associates if they differ in exactly j components. Hence we have $m = \nu$ associate classes, and the number of jth associates is

$$n_j = \binom{\nu}{j}(q-1) \qquad (j = 1, 2, \ldots, \nu)$$

Following Shah (1958), the general element of the $\boldsymbol{P}_k$ matrix $(k = 0, 1, \ldots, \nu)$ is given as

$$p_{ij}^k = \sum_u \binom{\nu-k}{u}\binom{k}{\nu-i-u}\binom{u+k+i-\nu}{\nu-j-u} \times (q-1)^{\nu-k-u}(q-2)^{k+i+j+2u-2\nu} \tag{4.55}$$

Table 4.1 Partial Association Scheme for EGD(7)-PBIB with $t = 36$ Treatments

	Associate Classes						
000	001	010	011	100	101	110	111
1	2,3,4	5,9	26,27,28,10,11,12	13,25	14,15,16,26,27,28	17,21,29,33	18,19,20,22,23,24 30,31,32,34,35,36
2	1,3,4	6,10	5,7,8,9,11,12	14,26	13,15,16,25,27,28	18,22,30,34	17,19,20,21,23,24 29,31,32,33,35,36
3	1,2,4	7,11	5,6,8,9,10,12	15,27	13,14,16,25,26,28	19,23,31,35	17,18,20,21,22,24 29,31,32,33,34,36
4	1,2,3	8,12	5,6,7,9,10,11	16,28	13,14,15, 25,26,27	20,24,32,36	17,18,19,21,22,23 29,30,31,33,34,35
5	6,7,8	1,9	2,3,4,10,11,12	17,29	18,19,20,30,31,32	13,21,25,33	14,15,16,22,23,24 26,27,28,34,35,36
⋮					⋮		
13	14,15,16	17,21	18,19,20,22,23,24	1,25	2,3,4,26,27,28	5,9,29,33	6,7,8,10,11,12 30,31,32,34,35,36
⋮					⋮		
36	33,34,35	28,32	25,26,27,29,30,31	12,24	9,10,11,21,22,23	4,8,16,20	1,2,3,5,6,7 13,14,15,17,18,19

where $i, j = 0, 1, \ldots, \nu$ and the summation extends over all integer values of u such that

$$u \leq \min(\nu - i, \nu - j, \nu - k)$$

and

$$u \geq \tfrac{1}{2}(2\nu - i - j - k)$$

and if for a given combination i, j, k no such u value exists, then $p_{ij}^k = 0$. Specifically, from (4.55) we obtain for $\nu = 3$ the $\boldsymbol{P}$ matrices for the cubic association scheme of Section 4.5.8.

Example 4.13 Let $t = 16, q = 2, \nu = 4$. The treatments are

$1 \equiv (1, 1, 1, 1)$	$2 \equiv (1, 1, 1, 2)$	$3 \equiv (1, 1, 2, 1)$	$4 \equiv (1, 1, 2, 2)$
$5 \equiv (1, 2, 1, 1)$	$6 \equiv (1, 2, 1, 2)$	$7 \equiv (1, 2, 2, 1)$	$8 \equiv (1, 2, 2, 2)$
$9 \equiv (2, 1, 1, 1)$	$10 \equiv (2, 1, 1, 2)$	$11 \equiv (2, 1, 2, 1)$	$12 \equiv (2, 1, 2, 2)$
$13 \equiv (2, 2, 1, 1)$	$14 \equiv (2, 2, 1, 2)$	$15 \equiv (2, 2, 2, 1)$	$16 \equiv (2, 2, 2, 2)$

and the association scheme follows:

0th Associate	1st Associates	2nd Associates	3rd Associates	4th Associates
1	2, 3, 5, 9	4, 6, 7, 10, 11, 13	8, 12, 14, 15	16
2	1, 4, 6, 10	3, 5, 8, 9, 12, 14	7, 11, 13, 16	15
3	1, 4, 7, 11	2, 5, 8, 9, 12, 15	6, 10, 13, 16	14
4	2, 3, 8, 12	1, 6, 7, 10, 11, 16	5, 9, 14, 15	13
5	1, 6, 7, 13	2, 3, 8, 9, 14, 15	4, 10, 11, 16	12
6	2, 5, 8, 14	1, 4, 7, 10, 13, 16	3, 9, 12, 15	11
7	3, 5, 8, 15	1, 4, 6, 11, 13, 16	2, 9, 12, 14	10
8	4, 6, 7, 16	2, 3, 5, 12, 14, 15	1, 10, 11, 13	9
9	1, 10, 11, 13	2, 3, 5, 12, 14, 15	4, 6, 7, 16	8
10	2, 9, 12, 14	1, 4, 6, 11, 13, 16	3, 5, 8, 15	7
11	3, 9, 12, 15	1, 4, 7, 10, 13, 16	2, 5, 8, 14	6
12	4, 10, 11, 16	2, 3, 8, 9, 14, 15	1, 6, 7, 13	5
13	5, 9, 14, 15	1, 6, 7, 10, 11, 16	2, 3, 8, 12	4
14	6, 10, 13, 16	2, 5, 8, 9, 12, 15	1, 4, 7, 11	3
15	7, 11, 13, 16	3, 5, 8, 9, 12, 14	1, 4, 6, 10	2
16	8, 12, 14, 15	4, 6, 7, 10, 11, 13	2, 3, 5, 9	1

□

4.6.11 Right-Angular PBIB(4) Designs

The right-angular association scheme was introduced by Tharthare (1963) for $t = 2sq$ treatments. The treatments are arranged in q right angles with arms of length s, keeping the angular positions of the right angles blank. The association scheme is then as follows: Any two treatments on the same arm are first associates; any two treatments on different arms of the same right angle are second associates;

any two treatments on the same (i.e., parallel) arm but in different right angles are third associates; any two treatments are fourth associates otherwise.

It follows then that

$$n_1 = s - 1 \qquad n_2 = s \qquad n_3 = (q-1)s = n_4$$

and

$$\boldsymbol{P}_1 = \begin{pmatrix} 0 & 1 & 0 & 0 & 0 \\ 1 & s-2 & 0 & 0 & 0 \\ 0 & 0 & s & 0 & 0 \\ 0 & 0 & 0 & s(q-1) & 0 \\ 0 & 0 & 0 & 0 & s(q-1) \end{pmatrix}$$

$$\boldsymbol{P}_2 = \begin{pmatrix} 0 & 0 & 1 & 0 & 0 \\ 0 & 0 & s-1 & 0 & 0 \\ 1 & s-1 & 0 & 0 & 0 \\ 0 & 0 & 0 & 0 & s(q-1) \\ 0 & 0 & 0 & s(q-1) & 0 \end{pmatrix}$$

$$\boldsymbol{P}_3 = \begin{pmatrix} 0 & 0 & 0 & 1 & 0 \\ 0 & 0 & 0 & s-1 & 0 \\ 0 & 0 & 0 & 0 & s \\ 1 & s-1 & 0 & s(q-2) & 0 \\ 0 & 0 & s & 0 & s(q-1) \end{pmatrix}$$

$$\boldsymbol{P}_4 = \begin{pmatrix} 0 & 0 & 0 & 0 & 1 \\ 0 & 0 & 0 & 0 & s-1 \\ 0 & 0 & 0 & s & 0 \\ 0 & 0 & s & 0 & s(q-2) \\ 1 & s-1 & 0 & s(q-2) & 0 \end{pmatrix}$$

Example 4.14 Let $t = 12, q = 3, s = 2$.

Treatments:

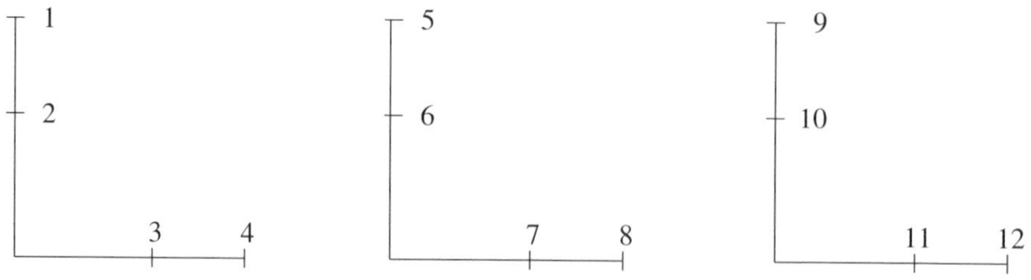

The association scheme follows:

0th Associate	1st Associates	2nd Associates	3rd Associates	4th Associates
1	2	3, 4	5, 6, 9, 10	7, 8, 11, 12
2	1	3, 4	5, 6, 9, 10	7, 8, 11, 12
3	4	1, 2	7, 8, 11, 12	5, 6, 9, 10
4	3	1, 2	7, 8, 11, 12	5, 6, 9, 10
5	6	7, 8	1, 2, 9, 10	3, 4, 11, 12
6	5	7, 8	1, 2, 9, 10	3, 4, 11, 12
7	8	5, 6	3, 4, 11, 12	1, 2, 9, 10
8	7	5, 6	3, 4, 11, 12	1, 2, 9, 10
9	10	11, 12	1, 2, 5, 6	3, 4, 7, 8
10	9	11, 12	1, 2, 5, 6	3, 4, 7, 8
11	12	9, 10	3, 4, 7, 8	1, 2, 5, 6
12	11	9, 10	3, 4, 7, 8	1, 2, 5, 6

□

Tharthare (1965) extended this association scheme to what he called the generalized right-angular association scheme for $t = pqs$ treatments with $p > 2$. To describe the association scheme succinctly, it is useful to denote the treatment by triplets (x_1, x_2, x_3) with $x_1 = 1, 2, \ldots, q$; $x_2 = 1, 2, \ldots, p$; $x_3 = 1, 2, \ldots, s$. Then, two treatments (x_1, x_2, x_3) and (y_1, y_2, y_3) are said to be

1st Associates: if $x_1 = y_1, x_2 = y_2, x_3 \neq y_3$

2nd Associates: if $x_1 = y_1, x_2 \neq y_2$

3rd Associates: if $x_1 \neq y_1, x_2 = y_2$

4th Associates: otherwise

This, obviously, leads to $n_1 = s - 1, n_2 = (p-1)s, n_3 = (q-1)s, n_4 = (q-1)(p-1)s$.

4.6.12 Cyclic PBIB Designs

Cyclic PBIB designs represent a rather large and flexible class of incomplete block designs. They were introduced by Kempthorne (1953) and Zoellner and Kempthorne (1954) for blocks of size $k = 2$. Further developments are due to for example, David (1963, 1965), David and Wolock (1965), John (1966, 1969), and Wolock (1964).

Let us denote the treatments by $0, 1, 2, \ldots, t-1$. The association scheme can then be defined as follows, where we need to distinguish between t even and t odd:

t even: For a fixed treatment θ, the uth associates are $(\theta + u, \theta - u)$mod $t(u = 1, 2, \ldots, t/2 - 1)$ and the $t/2$th associate is $\theta + t/2$ (mod t). Thus we have $m = t/2$ associate classes with $n_u = 2\ (u = 1, 2, \ldots, t/2 - 1)$ and $n_{t/2} = 1$.

t odd: The uth associates of θ are $(\theta + u, \theta - u)$mod t for $u = 1, 2, \ldots, (t - 1)/2$. Thus we have $m = (t - 1)/2$ associate classes with $n_u = 2\,[u = 1, 2, \ldots, (t - 1)/2]$.

Example 4.15 Let $t = 6$.

0th Associate	1st Associates	2nd Associates	3rd Associates
0	1, 5	2, 4	3
1	2, 0	3, 5	4
2	3, 1	4, 0	5
3	4, 2	5, 1	0
4	5, 3	0, 2	1
5	0, 4	1, 3	2

We draw the reader's attention to the cyclic development of the treatments in a given associate class, hence the name of the association scheme. The $\boldsymbol{P}$ matrices are as follows:

$$\boldsymbol{P}_1 = \begin{pmatrix} 0 & 1 & 0 & 0 \\ 1 & 0 & 1 & 0 \\ 0 & 1 & 0 & 1 \\ 0 & 0 & 1 & 0 \end{pmatrix}$$

$$\boldsymbol{P}_2 = \begin{pmatrix} 0 & 0 & 1 & 0 \\ 0 & 1 & 0 & 1 \\ 1 & 0 & 1 & 0 \\ 0 & 1 & 0 & 0 \end{pmatrix}$$

$$\boldsymbol{P}_3 = \begin{pmatrix} 0 & 0 & 0 & 1 \\ 0 & 0 & 2 & 0 \\ 0 & 2 & 0 & 0 \\ 1 & 0 & 0 & 0 \end{pmatrix}$$

□

4.6.13 Some Remarks

We conclude this section with the following remarks:

1. As we have mentioned earlier, the list of association schemes given here is not exhaustive. Other association schemes with $m > 2$ associate classes were given by for example, Roy (1953–1954), Raghavarao (1960), Ogasawara (1965), Yamamoto, Fuji, and Hamada (1965), and Adhikary (1966, 1967). In most cases they represent generalizations of existing association schemes.

2. We have chosen our list of association schemes because they lead to a rather large class of existing and practical PBIB designs [in particular the PBIB(2) designs and the cyclic PBIB designs are well documented (see Chapter 5)] and/or are particularly useful in the construction of systems of confounding for symmetrical and asymmetrical factorial designs (see Chapters 11 and 12).
3. We emphasize that an association scheme does not constitute a PBIB design, nor does it lead automatically to a PBIB design. Methods of constructing designs with a certain association scheme is the subject of Chapter 5. It will turn out that certain association schemes are connected with certain construction methods.

4.7 ESTIMATION OF ρ FOR PBIB(2) DESIGNS

Recall from Sections 1.10 and 2.5 that the general problem with utilizing the combined intra- and interblock information is that of finding an estimator for ρ, $\widehat{\rho}$, such that

$$\text{var}[t(\widehat{\rho})] \le \text{var}(t) \tag{4.56}$$

where t is the intrablock estimator for the estimable function $\boldsymbol{c}'\boldsymbol{\tau}$ and $t(\widehat{\rho})$ is the combined estimator for the same function using $\widehat{\rho}$ as the estimate of ρ.

Shah (1964) has derived as estimator for ρ for a certain class of designs such that (4.56) holds. We shall state his result without proof and then apply it to the various types of PBIB(2) designs as in Section 4.6.

4.7.1 Shah Estimator

Let D_1 be the class of proper, equireplicate, binary incomplete block designs for which the concordance matrix $\boldsymbol{NN}'$ has only one nonzero eigenvalue (other than rk). Let θ denote this eigenvalue with multiplicity $\alpha = \text{rank}(\boldsymbol{NN}') - 1$. Further, define

$$Z = \overline{c}\,(1 + \overline{c})\{SS(\boldsymbol{X}_\tau \mid \mathfrak{I}) - (2\boldsymbol{T}' - r\widehat{\boldsymbol{\tau}}')\widehat{\boldsymbol{\tau}}\} \tag{4.57}$$

where $\overline{c} = (rk/\theta) - 1$ and all the other terms are as defined in earlier sections.

We now state the following theorem.

Theorem 4.1 Consider an incomplete block design belonging to the class D_1. When

$$\widehat{\rho} = \begin{cases} \dfrac{\theta}{rk - \theta}\left(\dfrac{Z}{\alpha \text{MS}_E} - 1\right) & \text{if } \dfrac{Z}{\text{MS}_E} > \dfrac{rk\alpha}{\theta} \\ 1 & \text{otherwise} \end{cases}$$

with Z as given in (4.57), is used in the combined analysis, then for any treatment contrast $\mathbf{c}'\boldsymbol{\tau}$,

$$\text{var}[t(\widehat{\rho})] < \text{var}(t)$$

for all values of ρ provided that

$$(\alpha - 4)(n - t - b - 1) \geq 8 \tag{4.58}$$

4.7.2 Application to PBIB(2) Designs

We have already used the result of Theorem 4.1 in connection with BIB designs (see Section 2.5) that clearly belong to D_1. We have mentioned earlier that among all PBIB designs the PBIB(2) designs are quite numerous and most useful from a practical point of view. As we have seen in Section 4.4.3, they have two distinct eigenvalues other than rk and expressions for these and their multiplicities are given in (4.31) and (4.32), respectively. Using these results we shall now investigate which PBIB(2) designs belong to the class D_1 and give for them the corresponding condition (4.58).

1. *Group-Divisible Designs* Using the explicit expressions for $\boldsymbol{P}_1$ and $\boldsymbol{P}_2$ as given in Section 4.6.1, we find that

$$\theta_1 = r - \lambda_1 \qquad \alpha_1 = t_1(t_2 - 1)$$
$$\theta_2 = r - \lambda_1 + t_2(\lambda_1 - \lambda_2) \qquad \alpha_2 = t_1 - 1$$

   In particular, we have, following Bose and Connor (1952), for
   - **a.** *Singular GD designs*: $r - \lambda_1 = 0$. Hence the designs belong to D_1, with $\theta = \theta_2$ and $\alpha = \alpha_2$. Condition (4.58) then becomes $(t_1 - 5)(n - b - t - 1) \geq 8$.
   - **b.** *Semiregular GD designs*: $r - \lambda_1 + t_2(\lambda_1 - \lambda_2) = 0$. Hence these designs also belong to D_1, with $\theta = \theta_1$ and $\alpha = \alpha_1$. Condition (4.58) can, therefore, be written as $[t_1(t_2 - 1) - 4](n - b - t - 1) \geq 8$.
   - **c.** *Regular GD designs*: They do not belong to D_1 since $\theta_1 > 0$ and $\theta_2 > 0$.
2. *Triangular Designs* Using the results of Section 4.6.2 we obtain

$$\theta_1 = r + (q - 4)\lambda_1 - (q - 3)\lambda_2 \qquad \alpha_1 = q - 1$$
$$\theta_2 = r - 2\lambda_1 + \lambda_2 \qquad \alpha_2 = q(q - 3)/2$$

   They belong to D_1 if either one of these eigenvalues is zero, which is possible.

3. *Latin Square Type Designs* For the L_i-association scheme (see Section 4.6.3) we have

$$\theta_1 = r - (i - q)(\lambda_1 - \lambda_2) - \lambda_2 \qquad \alpha_1 = i(q - 1)$$

$$\theta_2 = r - i(\lambda_1 - \lambda_2) - \lambda_2 \qquad \alpha_2 = (q - 1)(q - i + 1)$$

they belong to D_1 if either θ_1 or θ_2 is zero, which is possible.

4. *Cyclic Designs* Neither the expressions for the eigenvalues or their multiplicities reduce to simple expressions. To check whether a given design belongs to D_1, we need to work out (4.31) and check whether either θ_1 or θ_2 equals zero.

CHAPTER 5

Construction of Partially Balanced Incomplete Block Designs

In the previous chapter we discussed several types of association schemes for PBIB designs with two or more associate classes. There exists a very large number of PBIB designs with these, and other, association schemes. But as we have pointed out earlier, the association schemes themselves do not constitute or generate the actual plan, that is, the assignment of treatment to blocks. To obtain such plans, different methods of construction have been developed and used. It is impossible to mention all of them as the literature in this area is immense. Rather, we shall discuss in some detail only a few methods with emphasis mainly on three types of PBIB designs, namely PBIB(2) designs, in particular, group-divisible PBIB(2) designs, cyclic PBIB designs, and EGD-PBIB designs. The first class is important because it constitutes the largest class of PBIB designs. The second class also represents a rather large class, and the designs are easy to construct and are widely applicable. The last class is important with respect to the construction of systems of confounding for asymmetrical factorial experiments (see Chapter 12).

5.1 GROUP-DIVISIBLE PBIB(2) DESIGNS

We shall give a brief discussion of the basic methods for constructing group-divisible designs. These methods are (i) duals of BIB designs, (ii) method of differences, (iii) finite geometries, and (iv) orthogonal arrays, as developed mainly by Bose, Shrikhande, and Bhattacharya (1953).

5.1.1 Duals of BIB Designs

Let $\boldsymbol{N}$ be the incidence matrix of a design D with parameters t, b, k, r. Then $\boldsymbol{N}'$ denotes the incidence matrix of a design D', with parameters t', b', k', r', which

Design and Analysis of Experiments. Volume 2: Advanced Experimental Design
By Klaus Hinkelmann and Oscar Kempthorne
ISBN 0-471-55177-5

has been obtained from D by interchanging treatments and blocks. This implies that D' has parameters $t' = b, b' = t, k' = r, r' = k$. The design D' is referred to as the *dual* of the design D (and, of course, vice versa). If the design D has a certain structure, such as being a BIB or PBIB design, then the dual design D' very often also has a particular recognizable structure. It is this relationship between D and D' that we shall now utilize to construct certain GD-PBIB(2) designs from existing BIB designs. Consider the following theorem.

Theorem 5.1 Let D be a BIB design with parameters $t = s^2, b = s(s+1)$, $k = s, r = s+1, \lambda = 1$, where s is prime or prime power (see Sections 3.34 and 18.7). Then the dual D' is a GD-PBIB(2) design with parameters $t' = (s+1)s, t_1 = s+1, t_2 = s, b' = s^2, k' = s+1, r' = s, \lambda_1 = 0, \lambda_2 = 1$.

Proof As can be seen from the construction in Section 3.7, D is a resolvable ($\alpha = 1$) BIB design, consisting of $s+1$ replication groups with s blocks each and containing each treatment once. Denote the blocks by $B_{j\ell}$ where j denotes the replication group ($j = 1, 2, \ldots, s+1$) and ℓ denotes the block within the jth replication group ($\ell = 1, 2, \ldots, s$). Now consider a treatment θ, say, and suppose it occurs in blocks $B_{1\ell_1}, B_{2\ell_2}, \ldots, B_{s+1,\ell_{s+1}}$. Then in D' block θ contains the treatments $1\ell_1, 2\ell_2, \ldots, \overline{s+1}\ell_{\overline{s+1}}$. Further, let $\theta_1, \theta_2, \ldots, \theta_{s-1}$ be the treatments that occur together with θ in $B_{1\ell_1}$ in D. Since $\{\theta, \theta_1, \theta_2, \ldots, \theta_{s-1}\}$ appear all in different blocks in the remaining s replication groups (since $\lambda = 1$), it follows then that in D' treatment $1\ell_1$ occurs together exactly once with all $j\ell (j = 2, \ldots, s+1; \ell = 1, 2, \ldots, s)$, but not at all with the $s-1$ treatments $1\ell (\ell \neq \ell_1)$. This implies the association scheme

$$
\begin{array}{cccc}
11 & 12 & \cdots & 1s \\
21 & 22 & \cdots & 2s \\
\overline{s+1}\ 1 & \overline{s+1}\ 2 & \cdots & \overline{s+1}\ s
\end{array}
$$

where treatments in the same row are 1st associates and 2nd associates otherwise, which is, in fact, the GD association scheme, and since the GD association is unique (Shrikhande, 1952), the resulting design is the GD-PBIB(2) design. □

Example 5.1 Consider the BIB design D with $t = 3^2, b = 12, k = 3, r = 4, \lambda = 1$, given by the plan [denoting the treatments by $(x_1, x_2), x_1, x_2 = 0, 1, 2$]:

Block	Treatments		
11	(0, 0)	(0, 1)	(0, 2)
12	(1, 0)	(1, 1)	(1, 2)
13	(2, 0)	(2, 1)	(2, 2)
21	(0, 0)	(1, 0)	(2, 0)
22	(0, 1)	(1, 1)	(2, 1)
23	(0, 2)	(1, 2)	(2, 2)

Block	Treatments		
31	(0, 0)	(1, 2)	(2, 1)
32	(1, 0)	(0, 1)	(2, 2)
33	(2, 0)	(0, 2)	(1, 1)
41	(0, 0)	(1, 1)	(2, 2)
42	(1, 0)	(0, 2)	(2, 1)
43	(2, 0)	(0, 1)	(1, 2)

The dual D' is then given by

Block	Treatments
(0, 0)	11, 21, 31, 41
(0, 1)	11, 22, 32, 43
(0, 2)	11, 23, 33, 42
(1, 0)	12, 21, 32, 42
(1, 1)	12, 22, 33, 41
(1, 2)	12, 23, 31, 43
(2, 0)	13, 21, 33, 43
(2, 1)	13, 22, 31, 42
(2, 2)	13, 23, 32, 41

The reader can verify easily that this is indeed the plan for a GD-PBIB(2) design with the parameters as indicated. It is isomorphic to plan SR41 given by Clatworthy (1973). □

5.1.2 Method of Differences

We denote and write the $t = t_1 t_2$ treatments now as follows:

$$\begin{array}{cccc} 0_1 & 0_2 & \cdots & 0_{t_2} \\ 1_1 & 1_2 & \cdots & 1_{t_2} \\ (t_1 - 1)_1 & (t_1 - 1)_2 & \cdots & (t_1 - 1)_{t_2} \end{array} \tag{5.1}$$

that is, the $(i + 1)$th group in the association scheme for the GD design consists of the treatments $(i_1, i_2, \ldots, i_{t_2})$ with $i = 0, 1, \ldots, t_1 - 1$, these treatments being 1st associates of each other. The treatments in the ℓth column are referred to as treatments of class ℓ.

The method of differences as described for BIB designs (Section 3.2) can now be extended for purposes of constructing GD-PBIB(2) designs as stated in the following theorem.

Theorem 5.2 Let it be possible to find s initial blocks $B_{10}, B_{20}, \ldots, B_{s0}$ such that

i. Each block contains k treatments.

ii. Treatments of class $\ell(\ell = 1, 2, \ldots, t_2)$ are represented r times among the s blocks.

iii. Among the differences of type (ℓ, ℓ) arising from the s blocks, each nonzero residue mod t_1 occurs the same number of times, λ_2, say, for all $\ell = 1, 2, \ldots, t_2$.

iv. Among the differences of type (ℓ, ℓ') arising from the s blocks, each nonzero residue mod t_1 occurs λ_2 times and 0 occurs λ_1 times, say, for all $(\ell, \ell')(\ell, \ell' = 1, 2, \ldots, t_2, \ell \neq \ell')$.

Then, developing the initial blocks $B_{10}, B_{20}, \ldots, B_{s0}$ cyclically, mod t_1 yields a GD-PBIB(2) design with parameters $t = t_1 t_2, b = t_1 s, k, r = ks/t_2, \lambda_1, \lambda_2$.

The proof is quite obvious as it is patterned after that of Theorem 3.1. By looking at array (5.1), it is clear that through the cyclic development of the initial blocks, any two treatments in the same row occur λ_1 times together in the same block, and any two treatments not in the same row occur λ_2 times together in the same block. This then satisfies the association scheme for the GD-PBIB(2) design.

We shall illustrate this with the following example.

Example 5.2 Let $t = 14, t_1 = 7, t_2 = 2$; that is, array (5.1) is

$$\begin{array}{cc} 0_1 & 0_2 \\ 1_1 & 1_2 \\ 2_1 & 2_2 \\ 3_1 & 3_2 \\ 4_1 & 4_2 \\ 5_1 & 5_2 \\ 6_1 & 6_2 \end{array}$$

Let the initial blocks of size 4 be

$$B_{10} = (0_1, 0_2, 1_1, 1_2)$$

$$B_{20} = (0_1, 0_2, 2_1, 2_2)$$

$$B_{30} = (0_1, 0_2, 3_1, 3_2)$$

To verify conditions 2–4 of Theorem 5.2 we see that treatments of class 1 and 2 occur $r = 6$ times. Further, the differences of type (1, 1) are

$$\begin{array}{lll} B_{10}: & 0_1 - 1_1 \equiv 6 & 1_1 - 0_1 \equiv 1 \\ B_{20}: & 0_1 - 2_1 \equiv 5 & 2_1 - 0_1 \equiv 2 \\ B_{30}: & 0_1 - 3_1 \equiv 4 & 3_1 - 0_1 \equiv 3 \end{array}$$

and the same holds for differences of type (2, 2); similarly, differences of type (1, 2) are

$$\begin{array}{lllll} B_{10}: & 0_1 - 0_2 \equiv 0 & 0_1 - 1_2 \equiv 6 & 1_1 - 0_2 \equiv 1 & 1_1 - 1_2 \equiv 0 \\ B_{20}: & 0_1 - 0_2 \equiv 0 & 0_1 - 2_2 \equiv 5 & 2_1 - 0_2 \equiv 2 & 2_1 - 2_2 \equiv 0 \\ B_{30}: & 0_1 - 0_2 \equiv 0 & 0_1 - 3_2 \equiv 4 & 3_1 - 0_2 \equiv 3 & 3_1 - 3_2 \equiv 0 \end{array}$$

and the same is true for differences of type (2, 1). Hence $\lambda_1 = 6, \lambda_2 = 1$. The plan for the GD-PBIB(2) design with $b = 21$ blocks is then

$$\begin{array}{lll} (0_1, 0_2, 1_1, 1_2) & (0_1, 0_2, 2_1, 2_2) & (0_1, 0_2, 3_1, 3_2) \\ (1_1, 1_2, 2_1, 2_2) & (1_1, 1_2, 3_1, 3_2) & (1_1, 1_2, 4_1, 4_2) \\ (2_1, 2_2, 3_1, 3_2) & (2_1, 2_2, 4_1, 4_2) & (2_1, 2_2, 5_1, 5_2) \\ (3_1, 3_2, 4_1, 4_2) & (3_1, 3_2, 5_1, 5_2) & (3_1, 3_2, 6_1, 6_2) \\ (4_1, 4_2, 5_1, 5_2) & (4_1, 4_2, 6_1, 6_2) & (4_1, 4_2, 0_1, 0_2) \\ (5_1, 5_2, 6_1, 6_2) & (5_1, 5_2, 0_1, 0_2) & (5_1, 5_2, 1_1, 1_2) \\ (6_1, 6_2, 0_1, 0_2) & (6_1, 6_2, 1_1, 1_2) & (6_1, 6_2, 2_1, 2_2) \end{array}$$

which is the same as plan S13 of Clatworthy (1973). We note that this PBIB design is, of course, resolvable and that it lends itself to two-way elimination of heterogeneity in that each treatment occurs exactly three times in the first two and the last two positions of the 21 blocks. □

5.1.3 Finite Geometries

We shall first consider a method of using a projective geometry, PG(K, p^n), to construct GD-PBIB(2) designs. The result can then be applied similarly to a Euclidean geometry, EG(K, p^n). (See Appendix B for details about projective and Euclidean geometries). The basic idea in both cases is to omit one point from a finite geometry and all the M-spaces containing that point. The remaining M-spaces are then taken as the blocks of a PBIB design. More specifically we have the following theorem.

Theorem 5.3 Omitting one point from a PG(K, p^n) and all the M-spaces containing that point, gives rise to a GD-PBIB(2) design, if one identifies the remaining points with the treatments and the remaining M-spaces with the blocks. Any two treatments are 1st associates if the line joining them goes through the omitted point, they are 2nd associates otherwise.

Proof The PG(K, p^n) has $1 + p^n + p^{2n} + \cdots + p^{Kn}$ points. Omitting one point leads to

$$\begin{aligned} t &= p^n + p^{2n} + \cdots + p^{Kn} \\ &= \left(1 + p^n + \cdots + p^{(K-1)n}\right) p^n \\ &\equiv t_1 t_2 \end{aligned} \tag{5.2}$$

treatments. In the PG(K, p^n) there are $\psi(M, K, p^n)$ M-spaces, of which $\varphi(0, M, K, p^n)$ contain the omitted point. Hence there are

$$b = \psi(M, K, p^n) - \varphi(0, M, K, p^n) \tag{5.3}$$

M-spaces left that constitute the blocks, each M-space (block) containing

$$k = \psi(0, M, p^n) = 1 + p^n + p^{2n} + \cdots + p^{Mn} \tag{5.4}$$

points (treatments). Since each point is contained in $\varphi(0, M, K, p^n)$ M-spaces and since each line joining a given point and the omitted point is contained in $\varphi(1, M, K, p^n)$ M-spaces, it follows that each retained point (treatment) occurs in

$$r = \varphi(0, M, K, p^n) - \varphi(1, M, K, p^n) \tag{5.5}$$

retained M-spaces (blocks). Further

$$n_1 = \psi(0, 1, p^n) - 2 = p^n - 1 = t_2 - 1 \tag{5.6}$$

$$\begin{aligned} n_2 &= t - n_1 - 1 = \left(p^n + p^{2n} + \cdots + p^{(K-1)n}\right) p^n \\ &= (t_1 - 1)t_2 \end{aligned} \tag{5.7}$$

and

$$\lambda_1 = 0, \lambda_2 = \varphi(1, M, K, p^n) \tag{5.8}$$

Thus, we are obviously led to the association scheme for the GD-PBIB(2) design. Because of the uniqueness of the association scheme, the design constructed in this manner is a GD-PBIB design. □

As an illustration we consider the following example.

Example 5.3 Let $p = 3$, $K = 2$, $n = 1$. The points of the PG(2, 2) are given by triplets (u_1, u_2, u_3) with $u_1, u_2, u_3 = 0, 1$ except $(u_1 u_2 u_3) = (000)$. Suppose we omit the point (100), then the remaining points are: 010, 110, 001, 101, 011, 111, that is, $t = 6$. The lines (i.e., $M = 1$) not passing through 100 constitute the blocks of the design and are given by

001	010	011	
110	001	111	
010	101	111	
110	101	011	(5.9)

that is, $b = 4, k = 3, r = 2$. The association scheme is obtained by determining for each point which other points lie on the line going through it and the omitted point. For example, the only point on the line going through (001) and (100) is obtained from $(\mu 0 + \nu 1, \mu 0 + \nu 0, \mu 1 + \nu 0)$ with $\mu = \nu = 1$, that is, the point (101). Doing this for all six points leads to the following association scheme:

0th Associate	1st Associates	2nd Associates
010	110	001, 101, 011, 111
110	010	001, 101, 011, 111
001	101	010, 110, 011, 111
101	001	010, 110, 011, 111
011	111	010, 110, 001, 101
111	011	010, 110, 001, 101

This is, of course, the GD-PBIB association scheme if we write the treatments in the following array:

010	110
001	101
011	111

Inspection of the plan (5.9) shows that $\lambda_1 = 0, \lambda_2 = 1$. With suitable labeling this is plan SR18 of Clatworthy (1973). □

5.1.4 Orthogonal Arrays

Orthogonal arrays (for a description see Appendix C) play an important role in the construction of experimental designs (see also Chapters 13 and 14). Of particular interest are orthogonal arrays of strength 2. This is the type we shall employ here to construct GD-PBIB designs. More specifically we shall employ orthogonal arrays OA$[N, K, p, 2; \lambda]$, where p is prime. Their relationship to the GD-PBIB(2) design is as follows: Replace any integer x appearing in the ith row of the array by the treatment $(i-1)p + x$. The ith row contains then the treatments

$$(i-1)p, (i-1)p + 1, \ldots, (i-1)p + p - 1$$

each occurring $r = N/p$ times since each symbol occurs equally often in each row. The total number of treatments is $t = Kp$ and we have $t_1 = K, t_2 = p$. The columns of this derived scheme form the blocks of the PBIB design, that is, there are $b = N$ blocks, each of size $k = K$. Treatments in the same row are 1st associates, that is, each treatment has $n_1 = p - 1$ 1st associates and since these treatments do not appear in any other row, it follows that $\lambda_1 = 0$. Treatments in different rows are 2nd associates, giving $n_2 = (K-1)p$. Since we have an OA$[b, k, p, 2; \lambda]$, that is, each possible 2×1 column vector appearing λ times, we have $\lambda_2 = \lambda$. This is, of course, the association scheme of a GD-PBIB(2) design with the parameters as stated above.

We illustrate this with the following example.

Example 5.4 Consider the following OA[8, 4, 2, 2, 2]:

$$\mathbf{A} = \begin{bmatrix} 0 & 1 & 0 & 1 & 0 & 1 & 0 & 1 \\ 0 & 0 & 1 & 1 & 0 & 0 & 1 & 1 \\ 0 & 0 & 0 & 0 & 1 & 1 & 1 & 1 \\ 0 & 1 & 1 & 0 & 1 & 0 & 0 & 1 \end{bmatrix}$$

This leads to the derived GD-PBIB(2) design (with columns representing blocks):

Block							
1	**2**	**3**	**4**	**5**	**6**	**7**	**8**
0	1	0	1	0	1	0	1
2	2	3	3	2	2	3	3
4	4	4	4	5	5	5	5
6	7	7	6	7	6	6	7

with $t = 8, b = 8, k = 4, r = 4, \lambda_1 = 0, \lambda_2 = 2$. This design is isomorphic to plan SR36 of Clatworthy (1973). □

For further specialized methods of constructing GD-PBIB(2) designs the reader is referred to Raghavarao (1971). An essentially complete list of existing plans for GD-PBIB(2) designs is given by Clatworthy (1973) together with references and methods concerning their construction.

5.2 CONSTRUCTION OF OTHER PBIB(2) DESIGNS

We shall mention here only a few methods but otherwise refer the reader to the pertinent literature.

5.2.1 Triangular PBIB(2) Designs

Recall that the number of treatments is $t = q(q-1)/2$ and that for the association scheme the treatments are arranged in a triangular array and its mirror image (see Section 4.6.2). One method of constructing triangular designs can then be described as follows.

If we let each row of the association scheme be a block, then the resulting design is a triangular PBIB design with parameters:

$$t = q(q-1)/2 \qquad b = q \qquad k = q-1 \qquad r = 2 \qquad \lambda_1 = 1 \qquad \lambda_2 = 0$$

Example 5.5 Let $t = 6 = 4 \times \frac{3}{2}$. The association scheme is

*	1	2	3
1	*	4	5
2	4	*	6
3	5	6	*

The design then is given by

Block	1:	1	2	3
	2:	1	4	5
	3:	2	4	6
	4:	3	5	6

that is, $b = 4, k = 3, r = 2, \lambda_1 = 1, \lambda_2 = 0$. □

Other methods given by Shrikhande (1960, 1965) and Chang, Liu, and Liu (1965) are based on the existence of certain BIB designs and considering, for example, the dual of the BIB design as omitting certain blocks from the BIB design. Still other methods are given by Clatworthy (1955, 1956), Masuyama (1965), and Ray-Chaudhuri (1965). A listing of practically useful designs is given by Clatworthy (1973).

5.2.2 Latin Square PBIB(2) Designs

Similar to the procedure described in the previous section, the following method of constructing L_2-PBIB designs is closely connected to the square treatment array for the L_2-association scheme.

We have $t = q^2$, and suppose q is prime or prime power. We know that there exist then $q - 1$ mutually orthogonal Latin squares (MOLS) of order q (see Section I.10.6.2). The languages for these $q - 1$ MOLS are then superimposed on the treatment array, that is, each treatment has associated with it $q - 1$ letters, one from each of the $q - 1$ MOLS. Collecting the treatments that have the same letter from each of the $q - 1$ languages and calling those sets blocks, leads to an L_2-PBIB design with the following parameters:

$$t = q^2 \qquad b = q(q-1) \qquad k = q \qquad r = q - 1 \qquad \lambda_1 = 0 \qquad \lambda_2 = 1$$

Example 5.6 Let $t = 16, q = 4$. The 4×4 association scheme with the superimposed three MOLS is as follows:

1	2	3	4
$A\,\alpha$ I	$B\,\beta$ II	$C\,\gamma$ III	$D\,\delta$ IV
5	6	7	8
$B\,\gamma$ IV	$A\,\delta$ III	$D\,\alpha$ II	$C\,\beta$ I
9	10	11	12
$C\,\delta$ II	$D\,\gamma$ I	$A\,\beta$ IV	$B\,\alpha$ III
13	14	15	16
$D\,\beta$ III	$C\,\alpha$ IV	$B\,\delta$ I	$A\,\gamma$ II

The blocks are then

Block	Treatments			
1(A)	1,	6,	11,	16
2(B)	2,	5,	12,	15
3(C)	3,	8,	9,	14
4(D)	4,	7,	10,	13
5(α)	1,	7,	12,	14
6(β)	2,	8,	11,	13
7(γ)	3,	5,	10,	16
8(δ)	4,	6,	9,	15
9(I)	1,	8,	10,	15
10(II)	2,	7,	9,	16
11(III)	3,	6,	12,	13
12(IV)	4,	5,	11,	14

It is, of course, obvious from the method of construction that this PBIB design is a resolvable design with the blocks from each language forming a complete replicate. □

Other methods of construction, some based on the existence of certain BIB designs, have been given by for example, Bose, Clatworthy, and Shrikhande (1954), Chang and Liu (1964), Clatworthy (1955, 1956, 1967), and, as lattice designs, by Yates (1936b) (see Section 18.7). A list of these designs can be found in Clatworthy (1973).

5.3 CYCLIC PBIB DESIGNS

We shall now turn to the rather large class of cyclic PBIB designs we introduced in Section 4.6.12. These designs are quite useful since (1) they are easy to construct, namely as the name suggests through cyclic development of initial blocks, (2) they exist for various combinations of design parameters, and (3) they are easy to analyze, that is, their analytical structure is easy to derive.

5.3.1 Construction of Cyclic Designs

As mentioned earlier, the construction of cyclic designs is based on the cyclic development of a set of initial blocks. We can distinguish basically between four types of cyclic designs according to whether the number of blocks b is (1) $b = t$, (2) $b = st$, (3) $b = t/d$, or (4) $b = t(s + 1/d)$, where d is a divisor of t. It is then obvious that

For (1) we need one initial block of size k;

For (2) we need s distinct (nonisomorphic) initial blocks of size k;

For (3) we need one initial block of size k such that, if cyclically developed, block $b + 1 = t/d + 1$ is the same as the initial block and each treatment is replicated r times. This means that if the initial block contains treatment i, it also must contain treatments $i + b, i + 2b, \ldots, i + (d - 1)b$. This implies that d is also a divisor of k, say $k/d = k'$. Then any k' treatments $i_1, i_2, \ldots, i_{k'}$ can be chosen and the remaining treatments in the initial block are determined;

For (4) we combine initial blocks from type (1) if $s = 1$ or (2) if $s > 1$ with an initial block from type (3).

The choice of initial blocks for (1), (2), and hence (4) above is quite arbitrary since any choice will lead to an appropriate design. There is, however, one other consideration that plays an important role in choosing initial blocks and that is the efficiency E of the resulting design. Efficiency here is defined in terms of the average variance of all treatment comparisons (see Section 1.10), and high efficiency is closely related to a small number of associate classes. Whereas the number of associate classes for a cyclic design is, in general, $m = t/2$ for t even and $m = (t - 1)/2$ for t odd, this number can sometimes be reduced by a proper choice of initial blocks, in some cases even to $m = 1$, that is, BIB designs, or $m = 2$, that is, PBIB(2) designs. For example, consider $t = 6, k = 3, r = 3, b = 6$: If the initial block is (1, 2, 3), then the plan is

1	2	3
2	3	4
3	4	5
4	5	6
5	6	1
6	1	2

and by inspection we can establish the following association scheme:

0th Associate	1st Associates	2nd Associates	3rd Associates
1	2, 6	3, 5	4
2	3, 1	4, 6	5
3	4, 2	5, 1	6
4	5, 3	6, 2	1
5	6, 4	1, 3	2
6	1, 5	2, 4	3

with $\lambda_1 = 2, \lambda_2 = 1, \lambda_3 = 0$, and

$$\boldsymbol{P}_1 = \begin{pmatrix} 0 & 1 & 0 & 0 \\ 1 & 0 & 1 & 0 \\ 0 & 1 & 0 & 1 \\ 0 & 0 & 1 & 0 \end{pmatrix} \qquad \boldsymbol{P}_2 = \begin{pmatrix} 0 & 0 & 1 & 0 \\ 0 & 1 & 0 & 1 \\ 1 & 0 & 1 & 0 \\ 0 & 1 & 0 & 0 \end{pmatrix} \qquad \boldsymbol{P}_3 = \begin{pmatrix} 0 & 0 & 0 & 1 \\ 0 & 0 & 2 & 0 \\ 0 & 2 & 0 & 0 \\ 1 & 0 & 0 & 0 \end{pmatrix}$$

Hence we have a PBIB(3) design. If, on the other hand, the initial block is (1, 2, 4), then the resulting plan is

1	2	4
2	3	5
3	4	6
4	5	1
5	6	2
6	1	3

and, by inspection, we can speculate on the association scheme as being

0th Associate	1st Associates	2nd Associates
1	2, 3, 5, 6	4
2	3, 4, 6, 1	5
3	4, 5, 1, 2	6
4	5, 6, 2, 3	1
5	6, 1, 3, 4	2
6	1, 2, 4, 5	3

with $\lambda_1 = 1, \lambda_2 = 2$, and

$$\boldsymbol{P}_1 = \begin{pmatrix} 0 & 1 & 0 \\ 1 & 2 & 1 \\ 0 & 1 & 0 \end{pmatrix} \qquad \boldsymbol{P}_2 = \begin{pmatrix} 0 & 0 & 1 \\ 0 & 4 & 0 \\ 1 & 0 & 0 \end{pmatrix}$$

Hence we have in fact a PBIB(2) design. For the first design we obtain $E = 0.743$ [see (5.14)], whereas for the second design we find $E = 0.784$. Hence the second design is slightly preferable from an efficiency point of view, and this design is therefore listed by John, Wolock, and David (1972). One can convince oneself that no better cyclic design for this combination of parameters exists.

The association scheme for the second design is, indeed, the association scheme for a cyclic PBIB(2) design as discussed in Section 4.6.4, except that we have to relabel the treatments as $0, 1, \ldots, 5$. We then have $d_1 = 1, d_2 = 2, d_3 = 4, d_4 = 5, e_1 = 3, \alpha = 2, \beta = 4$.

An extensive list of initial blocks for cyclic designs for various parameters with $6 \leq t \leq 30, k \leq 10, r \leq 10$, and fractional cyclic designs for $10 \leq t \leq 60, 3 \leq k \leq 10$ is given by John, Wolock, and David (1972). A special group in this collection of plans are those with $k = 2$, which are often referred to as *paired comparison designs*.

5.3.2 Analysis of Cyclic Designs

The general methods of analyzing incomplete block designs (see Chapter 1) apply, of course, also to the cyclic designs just described, or even more specifically, the methods of analyzing BIB designs (where appropriate) or PBIB(m)

designs can be used. However, as pointed out earlier, for many cyclic designs the number of associate classes, m, is rather large and hence the system of equations (4.34) becomes rather large. Instead, another method of solving the RNE

$$\boldsymbol{C}\hat{\boldsymbol{\tau}} = \boldsymbol{Q}$$

for cyclic designs can be used. This method is interesting in itself as it makes use of the particular construction process that leads to a circulant matrix $\boldsymbol{C}$ and hence to a circulant matrix $\boldsymbol{C}^-$, the elements of which can be written out explicitly (Kempthorne, 1953).

Let us denote the first row of $\boldsymbol{C}$ by $(c_1, c_2, \ldots, c_t)$. The second row is then obtained by shifting each element of the first row one position to the right (in a circulant manner), and so forth so that $\boldsymbol{C}$ is determined entirely by its first row. The same is true for $\boldsymbol{C}^-$, so all we need to know in order to solve the RNE and to obtain expressions for variances of estimable functions $\boldsymbol{\lambda}'\boldsymbol{\tau}$, is the first row of $\boldsymbol{C}^-$, which we denote by $c^1, c^2, \ldots, c^t$. Let $d_j (j = 1, 2, \ldots, t)$ denote the eigenvalues of $\boldsymbol{C}$ where (Kempthorne, 1953)

$$d_j = \sum_{\ell=1}^{t} c_\ell (\ell - 1)(j - 1)\theta \tag{5.10}$$

and $\theta = 2\pi/t$. Note that $d_1 = 0$; hence $d_2, d_3, \ldots, d_t$ are the nonzero eigenvalues of $\boldsymbol{C}$. Then, as shown by Kempthorne (1953), the elements $c^i (i = 1, 2, \ldots, t)$ are given as

$$\begin{aligned} c^1 &= \frac{1}{t} \sum_{j=2}^{t} \frac{1}{d_j} \\ c^i &= \frac{1}{t} \sum_{j=2}^{t} \frac{\cos(j-1)(i-1)\theta}{d_j} \qquad (i = 2, 3, \ldots, t) \end{aligned} \tag{5.11}$$

Expressions (5.10) and (5.11) can actually be simplified somewhat. Because of the construction of the designs and the resulting association scheme, we have for the $\lambda_{1i'}$ in $\boldsymbol{NN}'$

$$\lambda_{12} = \lambda_{1t}, \lambda_{13} = \lambda_{1,t-1},$$

$$\lambda_{14} = \lambda_{1,t-2}, \ldots, \begin{cases} \lambda_{1,t/2} = \lambda_{1,t/2+2} & \text{for } t \text{ even} \\ \lambda_{1,(t+1)/2} = \lambda_{1,(t+3)/2} & \text{for } t \text{ odd} \end{cases}$$

and hence we have for $\boldsymbol{C}$

$$c_1 = r\frac{k-1}{k}$$

$$c_2 = c_t = \frac{-\lambda_{12}}{k}$$

$$c_3 = c_{t-1} = \frac{-\lambda_{13}}{k}$$

$$c_4 = c_{t-2} = \frac{-\lambda_{14}}{k}$$

$$c_{(t+1)/2} = c_{(t+3)/2} = \frac{-\lambda_{1,(t+1)/2}}{k} \qquad \text{for } t \text{ odd}$$

or

$$c_{(t+2)/2} = \frac{-\lambda_{1,(t+2)/2}}{k} \qquad \text{for } t \text{ even}$$

As a consequence, also the maximum number of different c^i values is $(t+1)/2$ for t odd or $(t+2)/2$ for t even. With the c^i values given in (5.11) we can find the variances of treatment comparisons in general via comparisons with treatment 1 as follows:

$$\begin{aligned}\operatorname{var}(\widehat{\tau}_i - \widehat{\tau}_{i'}) &= \operatorname{var}(\widehat{\tau}_1 - \widehat{\tau}_{i'-i+1}) \\ &= 2(c^1 - c^{i'-i+1})\sigma_e^2 \qquad \text{for } i' > i \end{aligned} \tag{5.12}$$

This is, of course, a consequence of the circulant form of $\boldsymbol{C}^-$. The average variance of all such treatment comparisons then is

$$\begin{aligned}\text{av. var} &= \frac{1}{t(t-1)} \sum_{\substack{ii' \\ i \neq i'}} \operatorname{var}(\widehat{\tau}_i - \widehat{\tau}_{i'}) \\ &= \frac{1}{t-1} \sum_{i \neq 1} \operatorname{var}(\widehat{\tau}_1 - \widehat{\tau}_i)\end{aligned}$$

and using (5.12),

$$\begin{aligned}\text{av. var} &= \frac{2}{t-1} \sum_{i \neq 1} (c^1 - c^i)\sigma_e^2 \\ &= \frac{2}{t-1}\left(tc^1 - \sum_{i=1}^{t} c^i\right)\sigma_e^2 \\ &= \frac{2t}{t-1} c^1 \sigma_e^2 \end{aligned} \tag{5.13}$$

since $\sum_{i=1}^{t} c^i = 0$ as can be verified directly by using (5.11). From (5.12) it follows then that the efficiency factor for the treatment comparison between the ith and the i'th treatment is

$$E_{ii'} = \frac{2\sigma_e^2/r}{2(c^1 - c^{i'-i+1})\sigma_e^2} = \frac{1}{r(c^1 - c^{i'-i+1})}$$

and from (5.13) that the overall efficiency factor is

$$E = \frac{2\sigma_e^2/r}{2tc^1\sigma_e^2/(t-1)} = \frac{t-1}{rtc^1} \tag{5.14}$$

5.4 KRONECKER PRODUCT DESIGNS

In previous sections we have used the fact that incomplete block designs can sometimes be generated from some other existing incomplete block design, for example, the residual design of a BIB design is also a BIB design, the duals of certain BIB designs are PBIB(2) designs. In this section we shall use another widely applicable and useful technique to generate new incomplete block designs from existing designs. This method and hence the resulting designs are referred to as Kronecker product designs. These designs were first introduced by Vartak (1955) and later put on a more formal basis by Surendran (1968).

5.4.1 Definition of Kronecker Product Designs

Suppose we have two PBIB designs, D_1 and D_2, with m_1 and m_2 associate classes, respectively [in what follows we consider a BIB design as a PBIB(1) design] defined by their respective incidence matrices $\boldsymbol{N}_1$ and $\boldsymbol{N}_2$ with the following parameters:

$$\begin{aligned} &D_1\colon \quad t_1, r_1, k_1, b_1, n_u, \lambda_u \qquad \boldsymbol{P}_k = (p_{uv}^k) \qquad (u, v, k = 0, 1, \ldots, m_1) \\ &D_2\colon \quad t_2, r_2, k_2, b_2, n_f^*, \lambda_f^* \qquad \boldsymbol{P}_f^* = (q_{gh}^f) \qquad (f, g, h = 0, 1, \ldots, m_2) \end{aligned} \tag{5.15}$$

Consider now the Kronecker product

$$\boldsymbol{N} = \boldsymbol{N}_1 \times \boldsymbol{N}_2 \tag{5.16}$$

where $\boldsymbol{N}$ is obtained by multiplying each element in $\boldsymbol{N}_1$ by the matrix $\boldsymbol{N}_2$. It is immediately obvious that $\boldsymbol{N}$ is an incidence matrix of dimension $t \times b$ with $t = t_1t_2$ and $b = b_1b_2$. Furthermore, each row of $\boldsymbol{N}$ contains $r = r_1r_2$ unity elements and each column contains $k = k_1k_2$ unity elements. Hence the design D, defined by its incidence matrix (5.16), is an incomplete block design with t treatments, b blocks, r replicates per treatment, and k units per block. It remains for us to show that this *Kronecker product design* is indeed a PBIB design and to establish its association scheme and the remaining parameters.

5.4.2 Properties of Kronecker Product Designs

Let us denote the treatments of D_1 by the vector $\boldsymbol{\theta}' = (\theta_1, \theta_2, \ldots, \theta_{t_1})$ and the treatments of D_2 by the vector $\boldsymbol{\phi}' = (\phi_1, \phi_2, \ldots, \phi_{t_2})$. The treatments for design D can then be defined by what Kurkjian and Zelen (1962, 1963) have called the symbolic direct product between $\boldsymbol{\theta}$ and $\boldsymbol{\phi}$.

Definition 5.1 Let $\boldsymbol{\theta}' = (\theta_1, \theta_2, \ldots, \theta_q)$ and $\boldsymbol{\phi}' = (\phi_1, \phi_2, \ldots, \phi_s)$ be two arrays of size q and s, respectively. Then a new array $\boldsymbol{x}$ of size $q \cdot s$ is created by the *symbolic direct product* (SDP) of $\boldsymbol{\theta}$ and $\boldsymbol{\phi}$ as follows:

$$\boldsymbol{x} = \boldsymbol{\theta} \otimes \boldsymbol{\phi} = \begin{bmatrix} x_{11} \\ x_{12} \\ \vdots \\ x_{1s} \\ x_{21} \\ \vdots \\ x_{qs} \end{bmatrix} \tag{5.17}$$

□

It is convenient to enumerate the rows of the matrix $\boldsymbol{N}$ in (5.16) in the same way as the array in (5.17). We shall now prove the following theorem.

Theorem 5.4 Let D_1 and D_2 be two PBIB designs with parameters as given in (5.15). Then the Kronecker product design D as defined by its incidence matrix $\boldsymbol{N} = \boldsymbol{N}_1 \times \boldsymbol{N}_2$ is a PBIB design with $m = (m_1 + 1)(m_2 + 1) - 1$ associate classes and the following association scheme: If treatments θ_ν and $\theta_{\nu'}$ are uth associates in D_1 and if treatments ϕ_μ and $\phi_{\mu'}$ are gth associates in D_2 then the treatments $x_{\nu\mu}$ and $x_{\nu'\mu'}$ are said to be $(u : g)$th associates in D with $(u : g) = (0 : 0), (0 : 1), \ldots, (0 : m_2), (1 : 0), \ldots, (m_1 : m_2)$.

Proof We shall prove the theorem by establishing that the concordance matrix and the association matrices satisfy certain conditions as established by Bose and Mesner (1959) and as used in Chapter 4.

Let $\boldsymbol{B}_u (u = 0, 1, \ldots, m_1)$ and $\boldsymbol{B}_g^* (g = 0, 1, \ldots, m_2)$ be the association matrices for designs D_1 and D_2, respectively. We then have

$$\boldsymbol{N}_1\boldsymbol{N}_1' = \sum_{u=0}^{m_1} \lambda_u \boldsymbol{B}_u \tag{5.18}$$

and

$$\boldsymbol{N}_2\boldsymbol{N}_2' = \sum_{g=0}^{m_2} \lambda_g^* \boldsymbol{B}_g^* \tag{5.19}$$

Then, using (5.16), (5.18), and (5.19), we obtain

$$\begin{aligned} \boldsymbol{N}\boldsymbol{N}' &= (\boldsymbol{N}_1 \times \boldsymbol{N}_2)(\boldsymbol{N}_1 \times \boldsymbol{N}_2)' \\ &= (\boldsymbol{N}_1\boldsymbol{N}_1') \times (\boldsymbol{N}_2\boldsymbol{N}_2') \end{aligned}$$

$$= \left(\sum_{u=0}^{m_1} \lambda_u \boldsymbol{B}_u \right) \times \left(\sum_{g=0}^{m_2} \lambda_g^* \boldsymbol{B}_g^* \right)$$

$$= \sum_{u=0}^{m_1} \sum_{g=0}^{m_2} \lambda_u \lambda_g^* \left(\boldsymbol{B}_u \times \boldsymbol{B}_g^* \right) \tag{5.20}$$

Now let

$$\boldsymbol{B}_{u:g} = \boldsymbol{B}_u \times \boldsymbol{B}_g^* \tag{5.21}$$

It can be seen easily that the $\boldsymbol{B}_{u:g}$ are association matrices since

(i) $\boldsymbol{B}_{0:0} = \boldsymbol{B} \times \boldsymbol{B}_0^* = \boldsymbol{I}_t$

(ii) $$\sum_{u=0}^{m_1} \sum_{g=0}^{m_2} \boldsymbol{B}_{u:g} = \sum_u \sum_g \boldsymbol{B}_u \times \boldsymbol{B}_g^*$$

$$= \left(\sum_u \boldsymbol{B}_u \right) \times \left(\sum_g \boldsymbol{B}_g^* \right) = \mathfrak{I}_{t_1} \mathfrak{I}'_{t_1} \times \mathfrak{I}_{t_2} \mathfrak{I}'_{t_2} = \mathfrak{I}_t \mathfrak{I}'_t$$

(iii) $$\boldsymbol{B}_{u:g} \boldsymbol{B}_{v:h} = \left(\boldsymbol{B}_u \times \boldsymbol{B}_g^* \right) \left(\boldsymbol{B}_v \times \boldsymbol{B}_h^* \right) = (\boldsymbol{B}_u \boldsymbol{B}_v) \times \left(\boldsymbol{B}_g^* \boldsymbol{B}_h^* \right)$$

$$= \left(\sum_{k=0}^{m_1} p_{uv}^k \boldsymbol{B}_k \right) \times \left(\sum_{f=0}^{m_2} q_{gh}^f \boldsymbol{B}_f^* \right)$$

$$= \sum_{k=0}^{m_1} \sum_{f=0}^{m_2} p_{uv}^k q_{gh}^f \left(\boldsymbol{B}_k \times \boldsymbol{B}_f^* \right)$$

$$= \sum_{k=0}^{m_1} \sum_{f=0}^{m_2} p_{uv}^k q_{gh}^f \boldsymbol{B}_{k:f}$$

$$= \sum_{k=0}^{m_1} \sum_{f=0}^{m_2} p_{u:g,v:h}^{k:f} \boldsymbol{B}_{k:f}$$

with

$$p_{u:g,v:h}^{k:f} = p_{uv}^k q_{gh}^f$$

and

$$p_{u:g,v:h}^{k:f} = p_{u:h,u:g}^{k:f}$$

Further, the association scheme defined by the association matrices (5.21) is indeed the same as that stated in the theorem, which is immediately obvious from the way the treatments of $\boldsymbol{N}$ have been enumerated in (5.17).

This completes the proof of the theorem. □

By way of proving the theorem, we also obtain the remaining parameters of the Kronecker product design D as

$$\begin{aligned} n_{u:f} &= n_u n_f^* \\ \lambda_{u:f} &= \lambda_u \lambda_f^* \\ \boldsymbol{P}_{k:f} &= \left(p_{u:g,v:h}^{k:f}\right) \\ &= \left(p_{uv}^k \cdot p_{gh}^f\right) \\ &= \boldsymbol{P}_k \times \boldsymbol{P}_f^* \end{aligned}$$

with $(u : f) = (0 : 0), (0 : 1), \ldots, (0 : m_2), (1, 0), \ldots, (m_1 : m_2)$.

As a simple application we consider the following example.

Example 5.7 Let D_1 be a BIB design with parameters $t_1 = 4, b_1 = 4, k_1 = 3, r_1 = 3(= \lambda_0), \lambda_1 = 2$, and incidence matrix

$$\boldsymbol{N}_1 = \begin{pmatrix} 1 & 1 & 1 & 0 \\ 1 & 1 & 0 & 1 \\ 1 & 0 & 1 & 1 \\ 0 & 1 & 1 & 1 \end{pmatrix}$$

and let D_2 be a BIB design with $t_2 = 3, b_2 = 3, k_2 = 2, r_2 = 2(= \lambda_0^*), \lambda_1^* = 1$, and incidence matrix

$$\boldsymbol{N}_2 = \begin{pmatrix} 1 & 1 & 0 \\ 1 & 0 & 1 \\ 0 & 1 & 1 \end{pmatrix}$$

Then D has the parameters $t = 12, b = 12, k = 6, r = 6(= \lambda_{0:0}), \lambda_{0:1} = 3$; $\lambda_{1:0} = 4, \lambda_{1:1} = 2$, and incidence matrix

$$\boldsymbol{N} = \begin{pmatrix} \boldsymbol{N}_2 & \boldsymbol{N}_2 & \boldsymbol{N}_2 & 0 \\ \boldsymbol{N}_2 & \boldsymbol{N}_2 & 0 & \boldsymbol{N}_2 \\ \boldsymbol{N}_2 & 0 & \boldsymbol{N}_2 & \boldsymbol{N}_2 \\ 0 & \boldsymbol{N}_2 & \boldsymbol{N}_2 & \boldsymbol{N}_2 \end{pmatrix}$$

With

$$\begin{aligned} \boldsymbol{B}_0 &= \boldsymbol{I}_{t_1} & \boldsymbol{B}_1 &= \mathfrak{I}_{t_1}\mathfrak{I}'_{t_1} - \boldsymbol{I}_{t_1} \\ \boldsymbol{B}_0^* &= \boldsymbol{I}_{t_2} & \boldsymbol{B}_1^* &= \mathfrak{I}_{t_2}\mathfrak{I}'_{t_2} - \boldsymbol{I}_{t_2} \end{aligned}$$

as the association matrices for D_1 and D_2, respectively, the association matrices for D are

$$\begin{aligned} \boldsymbol{B}_{0:0} &= \boldsymbol{I}_{t_1} \times \boldsymbol{I}_{t_2} = \boldsymbol{I}_t \\ \boldsymbol{B}_{0:1} &= \boldsymbol{I}_{t_1} \times \left(\mathfrak{I}_{t_2}\mathfrak{I}'_{t_2} - \boldsymbol{I}_{t_1}\right) = \boldsymbol{I}_{t_1} \times \mathfrak{I}_{t_2}\mathfrak{I}'_{t_2} - \boldsymbol{I}_t \end{aligned}$$

$$\boldsymbol{B}_{1:0} = (\mathfrak{I}_{t_1}\mathfrak{I}'_{t_1} - \boldsymbol{I}_{t_1}) \times \boldsymbol{I}_{t_2} = \mathfrak{I}_{t_1}\mathfrak{I}'_{t_1} \times \boldsymbol{I}_{t_2} - \boldsymbol{I}_t$$

$$\boldsymbol{B}_{1:1} = (\mathfrak{I}_{t_1}\mathfrak{I}'_{t_1} - \boldsymbol{I}_{t_1}) \times (\mathfrak{I}_{t_2}\mathfrak{I}'_{t_2} - \boldsymbol{I}_{t_2})$$

$$= \mathfrak{I}_t\mathfrak{I}'_t - \mathfrak{I}_{t_1}\mathfrak{I}'_{t_1} \times \boldsymbol{I}_{t_2} - \boldsymbol{I}_{t_1} \times \mathfrak{I}_{t_2}\mathfrak{I}'_{t_2} + \boldsymbol{I}_t$$

Obviously

$$\boldsymbol{B}_{0:0} + \boldsymbol{B}_{0:1} + \boldsymbol{B}_{1:0} + \boldsymbol{B}_{1:1} = \mathfrak{I}_t\mathfrak{I}'_t$$

Any treatment in D is denoted, according to (5.17), by (i, j) where $i = 1, 2, 3, 4$, $j = 1, 2, 3$, and these treatments are arranged in the order as given by (5.17). The association scheme for this PBIB(3) design can then be expressed in two ways, both of which are useful:

1. Any two treatments (i, j) and (i', j') are

0:0 associates if	$i = i'$	$j = j'$
0:1 associates if	$i = i'$	$j \neq j'$
1:0 associates if	$i \neq i'$	$j = j'$
1:1 associates if	$i \neq i'$	$j \neq j'$

or

2. If the treatments are arranged in the 4×3 array

(1, 1)	(1, 2),	(1, 3)
(2, 1)	(2, 2),	(2, 3)
(3, 1)	(3, 2),	(3, 3)
(4, 1)	(4, 2),	(4, 3)

then each treatment is the 0:0th associate of itself, and any two treatments are

0:1 associates if they are in the same row
1:0 associates if they are in the same column
1:1 associates if they are in different rows and columns

This association scheme has been referred to as the extended group-divisible association scheme for an EGD(3)-PBIB design by Hinkelmann and Kempthorne (1963) using representation 1 above, and as the rectangular association scheme by Vartak (1955) using representation 2 (see Sections 4.6.9 and 4.6.5).

Finally, the $\boldsymbol{P}$ matrices for the design D are obtained from those for design D_1:

$$\boldsymbol{P}_0 = \begin{pmatrix} 1 & 0 \\ 0 & 3 \end{pmatrix} \qquad \boldsymbol{P}_1 = \begin{pmatrix} 0 & 1 \\ 1 & 2 \end{pmatrix}$$

and design D_2:

$$\boldsymbol{P}_0^* = \begin{pmatrix} 1 & 0 \\ 0 & 2 \end{pmatrix} \qquad \boldsymbol{P}_1^* = \begin{pmatrix} 0 & 1 \\ 1 & 1 \end{pmatrix}$$

as

$$\boldsymbol{P}_{0:0} = \begin{pmatrix} 1 & & & \\ & 2 & \phi & \\ & \phi & 3 & \\ & & & 6 \end{pmatrix} \qquad \boldsymbol{P}_{0:1} = \begin{pmatrix} 0 & 1 & & \phi \\ 1 & 1 & & \\ & & 0 & 3 \\ \phi & & 3 & 3 \end{pmatrix}$$

$$\boldsymbol{P}_{1:0} = \begin{pmatrix} & & 1 & 0 \\ & \phi & 0 & 2 \\ 1 & 0 & 2 & 0 \\ 0 & 2 & 0 & 4 \end{pmatrix} \qquad \boldsymbol{P}_{1:1} = \begin{pmatrix} & & 0 & 1 \\ & \phi & 1 & 1 \\ 0 & 1 & 0 & 2 \\ 1 & 1 & 2 & 2 \end{pmatrix}$$

□

The method of constructing PBIB designs as Kronecker product designs can, obviously, be extended to Kronecker products of incidence matrices for three or more existing PBIB designs, for example,

$$\boldsymbol{N} = \boldsymbol{N}_1 \times \boldsymbol{N}_2 \times \boldsymbol{N}_3$$

and the reader should have no difficulties working out the parameters of the resulting PBIB design with $(m_1 + 1)(m_2 + 1)(m_3 + 1) - 1$ associate classes, where m_1, m_2, m_3 are the numbers of associate classes of the component designs.

5.4.3 Usefulness of Kronecker Product Designs

It is, of course, clear that the usefulness of Kronecker product designs is somewhat limited as even for two component designs, that is, for $\boldsymbol{N} = \boldsymbol{N}_1 \times \boldsymbol{N}_2$ they may require rather large block sizes and/or large numbers of replications depending on the corresponding parameters for the underlying designs $\boldsymbol{N}_1$ and $\boldsymbol{N}_2$. In general, therefore, only those designs $\boldsymbol{N}_1$ and $\boldsymbol{N}_2$ with moderate values for k and/or r will be useful in this connection.

Also, the number of associate classes for Kronecker product PBIB designs may be quite large since the maximum number of associate classes for $\boldsymbol{N} = \boldsymbol{N}_1 \times \boldsymbol{N}_2$ is $m_1 m_2 + m_1 + m_2$. Vartak (1955) and Kageyama (1972) have shown, however, that under certain conditions this number can be reduced considerably, thereby actually inducing a different association scheme. Kageyama (1972), for example, proves that a necessary and sufficient condition for an EGD-$(2^\nu - 1)$-PBIB design (see Section 4.6.9) based on the ν-fold Kronecker product of BIB designs to reduce to PBIB(ν) designs having the hypercubic association scheme (see Section 4.6.10) is that the ν BIB designs have the same number of treatments and the same block size. This not only makes Kronecker product PBIB designs more attractive from a practical point of view but it also provides methods of constructing PBIB designs with different association schemes.

5.5 EXTENDED GROUP-DIVISIBLE PBIB DESIGNS

We have mentioned earlier that the EGD-PBIB designs play a major role in connection with systems of confounding for asymmetrical factorial experiments and balanced factorial experiments (see Chapter 12). Knowing how to construct EGD-PBIB designs therefore means knowing how to construct such systems of confounding. It is for this reason that we shall present a few methods of constructing EGD-PBIB designs.

5.5.1 EGD-PBIB Designs as Kronecker Product Designs

As illustrated already by Example 5.7, the EGD-PBIB(3) design can be constructed as the Kronecker product of two BIB designs. It is then quite obvious that a ν-fold Kronecker product of BIB designs will lead to an EGD-PBIB($2^\nu - 1$) design. More formally, we state the following theorem.

Theorem 5.5 Let there be ν designs D_i with parameters $t_i, b_i, k_i, r_i, \lambda_{(i)}$ and incidence matrix $N_i (i = 1, 2, \ldots, \nu)$, respectively. Then the Kronecker product design D as given by its incidence matrix

$$N = N_1 \times N_2 \times \cdots \times N_\nu$$

is an EGD/($2^\nu - 1$)-PBIB design with parameters

$$t = \prod_{i=1}^{\nu} t_i \qquad b = \prod_{i=1}^{\nu} b_i \qquad k = \prod_{i=1}^{\nu} k_i \qquad r = \prod_{i=1}^{\nu} r_i$$

$$\lambda_{\gamma_1 \gamma_2, \ldots, \gamma_\nu} = \prod_{i=1}^{\nu} r_i^{1-\gamma_i} \lambda_{(i)}^{\gamma_i}$$

where the power $\gamma_i = 0$ or 1.

Proof For the proof we simply refer to the definition of the Kronecker product design (Section 5.4) and the definition of the EGD/($2^\nu - 1$)-PBIB design (Section 4.6.9). □

5.5.2 Method of Balanced Arrays

The second method of constructing EGD-PBIB designs is due to Aggarwal (1974). It is based on balanced arrays. Before we give the corresponding theorem we need first the following definition.

Definition 5.2 Let $\boldsymbol{\alpha}' = (\alpha_1, \alpha_2, \ldots, \alpha_q)$ and $\boldsymbol{\beta}' = (\beta_1, \beta_2, \ldots, \beta_q)$ be two arrays of q elements, then the *symbolic inner product* (SIP), $\boldsymbol{\alpha}' \odot \boldsymbol{\beta}'$, of these arrays is the array defined as

$$\boldsymbol{\alpha}' \odot \boldsymbol{\beta}' = (\alpha_1 \beta_1, \alpha_2\beta_2, \ldots, \alpha_q\beta_q)$$

□

We then state the following theorem.

Theorem 5.6 The existence of an EGD/$(2^\nu - 1)$-PBIB design with parameters $t = \prod_{i=1}^{\nu} t_i$, $b = \prod_{i=2}^{\nu} t_i(t_i - 1)$, $k = t_i$, $r = \prod_{i=2}^{\nu}(t_i - 1)$, and

$$\lambda_{\gamma_1\gamma_2\ldots\gamma_\nu} = \begin{cases} 1 & \text{if } \gamma_1 = \gamma_2 = \ldots = \gamma_\nu = 1 \\ 0 & \text{otherwise} \end{cases}$$

where $t_1 \leq t_i (i = 2, 3, \ldots, \nu)$, is implied by the existence of $\nu - 1$ balanced arrays $\text{BA}_i[t_i(t_i - 1), t_1, t_i, 2](i = 2, 3, \ldots, \nu)$, that is, with $N_i = t_i(t_i - 1)$ assemblies, t_1 constraints, strength 2, t_i levels, and for each i

$$\lambda(x_1, x_2) = \begin{cases} 0 & \text{if } x_1 = x_2 \\ 1 & \text{otherwise} \end{cases}$$

[for a definition of balanced array (BA) see Appendix C].

Proof This proof actually gives the construction of the EGD-PBIB design, assuming that the BA_i exist. Denote the columns of BA_i by $\boldsymbol{a}_{i1}, \boldsymbol{a}_{i2}, \ldots, \boldsymbol{a}_{i,(t_i-1)}$ and the elements in BA_i by $0, 1, \ldots, t_i - 1$. Then construct successively the following EGD-PBIB designs:

1. Take the SIP of the array $[0, 1, 2, \ldots, (t_1 - 1)]$ and each $\boldsymbol{a}'_{2j}[j = 1, 2, \ldots, t_2(t_2 - 1)]$ of BA_2. If each SIP is taken as a block with treatments of the form (x_1x_2), then this yields an EGD(3)-PBIB design with the following parameters: $t = t_1t_2$, $b = t_2(t_2 - 1)$, $k = t_1$, $r = t_2 - 1$, $n_{01} = t_2 - 1$, $n_{10} = t_1 - 1$, $n_{11} = (t_1 - 1)(t_2 - 1)$, $\lambda_{01} = 0$, $\lambda_{10} = 0$, $\lambda_{11} = 1$. This can be verified easily by noting that (a) each element of BA_2 must occur $t_2 - 1$ times in each row of BA_2, hence each element in $(0, 1, \ldots, t_1 - 1)$ is combined with each element of $(0, 1, \ldots, t_2 - 1)$, implying $r = t_2 - 1$; (b) each SIP forms a block of size $k = t_1$ and hence $b = t_2(t_2 - 1)$; (c) because of the BA property concerning the $\lambda(x_1, x_2)$, no two pairs of elements, that is, no two treatments (i, j) and (i', j'), say, occur together in the same block if $i = i'$ or $j = j'$, implying $\lambda_{01} = \lambda_{10} = 0$, and two pairs of elements (i, j) and (i', j') occur together exactly once if $i \neq i'$ and $j \neq j'$, implying $\lambda_{11} = 1$. This establishes the EGD(3)-PBIB property of this design. We denote the blocks, that is, SIP arrays, of this EGD-PBIB design by $\boldsymbol{b}'_{21}, \boldsymbol{b}'_{22}, \ldots, \boldsymbol{b}'_{2,t_2(t_2-1)}$.

2. Then take the SIP of the $\boldsymbol{b}'_{2j}[j = 1, 2, \ldots, t_2(t_2 - 1)]$ with the $\boldsymbol{a}'_{3\ell}[\ell = 1, 2, \ldots, t_3(t_3 - 1)]$ of BA_3. This leads to an EGD/7-PBIB design with parameters $t = t_1 t_2 t_3, b = t_2(t_2 - 1)t_3(t_3 - 1), k = t_1, r = (t_2 - 1)(t_3 - 1), \lambda_{001} = \lambda_{010} = \lambda_{011} = \lambda_{100} = \lambda_{101} = \lambda_{110} = 0, \lambda_{111} = 1, n_{100} = t_1 - 1, \; n_{010} = t_2 - 1, \; n_{001} = t_3 - 1, \; n_{110} = (t_1 - 1)(t_2 - 1), \; n_{101} = (t_1 - 1)(t_3 - 1), \; n_{011} = (t_2 - 1)(t_3 - 1), n_{111} = (t_1 - 1)(t_2 - 1)(t_3 - 1)$. This can be verified by repeating the arguments given in item 1.
3. Continue the process of forming the SIP of the blocks of the EGD/$(2^q - 1)$-PBIB design with the columns of BA_{q+1} until $q = \nu - 1$. The resulting design is then an EGD/$(2^\nu - 1)$-PBIB design with parameters as stated in the theorem. □

Example 5.8 Let $t = 36 = 3 \times 3 \times 4$, that is, $t_1 = 3, t_2 = 3, t_3 = 4$. The two required BAs are (see Appendix C for construction)

$$\mathrm{BA}_2 = \begin{pmatrix} 0 & 1 & 2 & 0 & 1 & 2 \\ 2 & 0 & 1 & 1 & 2 & 0 \\ 1 & 2 & 0 & 2 & 0 & 1 \end{pmatrix}$$

$$\mathrm{BA}_3 = \begin{pmatrix} 0 & 1 & 2 & 3 & 0 & 1 & 2 & 3 & 0 & 1 & 2 & 3 \\ 1 & 0 & 3 & 2 & 2 & 3 & 0 & 1 & 3 & 2 & 1 & 0 \\ 2 & 3 & 0 & 1 & 3 & 2 & 1 & 0 & 1 & 0 & 3 & 2 \end{pmatrix}$$

The SIP of (0, 1, 2) with the columns of BA_2 yields

$$\boldsymbol{b}'_{21} = (00, 12, 21)$$
$$\boldsymbol{b}'_{22} = (01, 10, 22)$$
$$\boldsymbol{b}'_{23} = (02, 11, 20)$$
$$\boldsymbol{b}'_{24} = (00, 11, 22)$$
$$\boldsymbol{b}'_{25} = (01, 12, 20)$$
$$\boldsymbol{b}'_{26} = (02, 10, 21)$$

The SIP of the $\boldsymbol{b}'_{2j}$'s with the columns of BA_3 yields the final plan with 72 blocks of size 3 as given in Table 5.1. □

5.5.3 Direct Method

The methods described in Sections 5.5.1 and 5.5.2 depend both on the availability of appropriate BIB designs or BAs, respectively. The method to be described in this section, which is due to Chang and Hinkelmann (1987), is a direct method in the sense that it is based *ab initio* on simple combinatorial operations.

Table 5.1 EGD-PBIB Design for $t = 36, b = 72, k = 3$, and $r = 6$

(000, 121, 212)	(010, 101, 222)
(001, 120, 213)	(011, 100, 223)
(002, 123, 210)	(012, 103, 220)
(003, 122, 211)	(013, 102, 221)
(000, 122, 213)	(010, 102, 223)
(001, 123, 212)	(011, 103, 222)
(002, 120, 211)	(012, 100, 221)
(003, 121, 210)	(013, 101, 220)
(000, 123, 211)	(010, 103, 221)
(001, 122, 210)	(011, 102, 220)
(002, 121, 213)	(012, 101, 223)
(003, 120, 212)	(013, 100, 222)
(020, 111, 202)	(000, 111, 222)
(021, 110, 203)	(001, 110, 223)
(022, 113, 200)	(002, 113, 220)
(023, 112, 201)	(003, 112, 221)
(020, 112, 203)	(000, 112, 223)
(021, 113, 202)	(001, 113, 222)
(022, 110, 201)	(002, 110, 221)
(023, 111, 200)	(003, 111, 220)
(020, 113, 201)	(000, 113, 221)
(021, 112, 200)	(001, 112, 220)
(022, 111, 203)	(002, 111, 223)
(023, 110, 202)	(003, 110, 222)
(010, 121, 202)	(020, 101, 212)
(011, 120, 203)	(021, 100, 213)
(012, 123, 200)	(022, 103, 210)
(013, 122, 201)	(023, 102, 211)
(010, 122, 203)	(020, 102, 213)
(011, 123, 202)	(021, 103, 212)
(012, 120, 201)	(022, 100, 211)
(013, 121, 200)	(023, 101, 210)
(010, 123, 201)	(020, 103, 211)
(011, 122, 200)	(021, 102, 210)
(012, 121, 203)	(022, 101, 213)
(013, 120, 202)	(023, 100, 212)

Before we state the main result, we introduce, for ease of notation, the following definition.

Definition 5.3 For x and y intergers and $x > y$, let

$$x^{(y)} = \frac{x!}{(x-y)!} = x(x-1)\cdots(x-y+1) \tag{5.22}$$

□

We then have the following theorem.

Theorem 5.7 For $t = \prod_{i=1}^{\nu} t_i$ treatments with $t_1 \leq t_2 \leq \cdots \leq t_\nu$ there exists an EGD/$(2^\nu - 1)$-PBIB design with parameters

$$t = \prod_{i=1}^{\nu} t_i \qquad b = \prod_{i=2}^{\nu} t_i^{(t_1)}$$

$$r = \prod_{i=2}^{\nu} (t_i - 1)^{(t_1 - 1)} \qquad k = t_1$$

$$\lambda(0, 0, \ldots, 0) = r \qquad \lambda(1, \ldots, 1) = \prod_{i=2}^{\nu} (t_1 - 2)^{(t_1 - 2)}$$

$$\text{all other } \lambda(\boldsymbol{\gamma}) = 0$$

The proof is rather elementary but quite lengthy. We shall not go through all the details here and instead refer the reader to Chang and Hinkelmann (1987). Just as for Theorem 5.6, however, the proof is based on the actual construction of the design, and the individual steps of the method will be described now. To do so, it is useful to introduce some notation.

Let $(x_1, x_2, \ldots, x_\nu)$ denote a treatment with $x_i \epsilon \{1, 2, \ldots, t_i\}$ for $i = 1, 2, \ldots, \nu$ and x_i is called the ith component of the treatment (see Section 4.6.9). We then define a matrix

$$\boldsymbol{P} = \begin{bmatrix} \boldsymbol{P}^{(1)} \\ \boldsymbol{P}^{(2)} \\ \vdots \\ \boldsymbol{P}^{(t_1)} \end{bmatrix} \tag{5.23}$$

of dimension $\left(\prod_{i=1}^{\nu-1} t_i\right) \times t_\nu$ with the submatrix $\boldsymbol{P}^{(\ell)}$ of dimension $\left(\prod_{i=2}^{\nu-1} t_i\right) \times t_\nu$. (In what follows we shall abbreviate $\prod_{i=1}^{\nu-1}$ by $\prod$ and $\prod_{i=2}^{\nu-1}$ by $\prod^*$). The elements of $\boldsymbol{P}$ are the treatments $(x_1, x_2, \ldots, x_\nu)$, and $\boldsymbol{P}^{(\ell)}$ has the following properties:

1. The first component of any element is ℓ for $\ell = 1, 2, \ldots, t_1$.
2. The last component of any element in the sth column is s $(1 \leq s \leq t_\nu)$.
3. The remaining components of any element are changed from one row to the next by first changing the $(\nu - 1)$th component from 1 through $t_{\nu-1}$, then the $(\nu - 2)$th component from 1 through $t_{\nu-2}$, and so on.

For example, for $\nu = 3$ the submatrix $\boldsymbol{P}^{(\ell)}$ is of the form

$$\begin{bmatrix} (\ell, 1, 1) & (\ell, 1, 2) & \ldots & (\ell, 1, t_3) \\ (\ell, 2, 1) & (\ell, 2, 2) & \ldots & (\ell, 2, t_3) \\ & \ldots & \ldots & \\ (\ell, t_2, 1) & (\ell, t_2, 2) & \ldots & (\ell, t_2, t_3) \end{bmatrix} \tag{5.24}$$

for $\ell = 1, 2, \ldots, t_1$.

The next step is to define a matrix derived from $\boldsymbol{P}$,

$$\boldsymbol{Q}(\boldsymbol{P}) = \begin{bmatrix} \boldsymbol{q}_1 \\ \boldsymbol{q}_2 \\ \vdots \\ \boldsymbol{q}_{t_1} \end{bmatrix} \tag{5.25}$$

of order $t_1 \times t_\nu$ such that $\boldsymbol{q}_\ell$ is a row vector from $\boldsymbol{P}^{(\ell)}$ $(\ell = 1, 2, \ldots, t_1)$ and any two vectors, $\boldsymbol{q}_{(\ell)}$ and $\boldsymbol{q}_{(\ell')}$, do not have the same row number in $\boldsymbol{P}^{(\ell)}$ and $\boldsymbol{P}^{(\ell')}$, respectively. Denote the possible choices of $\boldsymbol{Q}(\boldsymbol{P})$ by $\boldsymbol{Q}_1, \boldsymbol{Q}_2, \ldots, \boldsymbol{Q}_N$ (suppressing $\boldsymbol{P}$), where

$$N = \Pi^* t_i^{(t_1)}$$

Example 5.9 Let $t = 24, t_1 = 2, t_2 = 3, t_3 = 4$. The $\boldsymbol{P}$ matrices as given by (5.24) are

$$\boldsymbol{P}^{(1)} = \begin{bmatrix} (1, 1, 1) & (1, 1, 2) & (1, 1, 3) & (1, 1, 4) \\ (1, 2, 1) & (1, 2, 2) & (1, 2, 3) & (1, 2, 4) \\ (1, 3, 1) & (1, 3, 2) & (1, 3, 3) & (1, 3, 4) \end{bmatrix}$$

$$\boldsymbol{P}^{(2)} = \begin{bmatrix} (2, 1, 1) & (2, 1, 2) & (2, 1, 3) & (2, 1, 4) \\ (2, 2, 1) & (2, 2, 2) & (2, 2, 3) & (2, 2, 4) \\ (2, 3, 1) & (2, 3, 2) & (2, 3, 3) & (2, 3, 4) \end{bmatrix}$$

With $N = 3^{(2)} = 6$ we then have

$$\boldsymbol{Q}_1 = \begin{bmatrix} (1, 1, 1) & (1, 1, 2) & (1, 1, 3) & (1, 1, 4) \\ (2, 2, 1) & (2, 2, 2) & (2, 2, 3) & (2, 2, 4) \end{bmatrix}$$

$$\boldsymbol{Q}_2 = \begin{bmatrix} (1, 1, 1) & (1, 1, 2) & (1, 1, 3) & (1, 1, 4) \\ (2, 3, 1) & (2, 3, 2) & (2, 3, 3) & (2, 3, 4) \end{bmatrix}$$

$$\boldsymbol{Q}_3 = \begin{bmatrix} (1, 2, 1) & (1, 2, 2) & (1, 2, 3) & (1, 2, 4) \\ (2, 1, 1) & (2, 1, 2) & (2, 1, 3) & (2, 1, 4) \end{bmatrix}$$

$$Q_4 = \begin{bmatrix} (1,2,1) & (1,2,2) & (1,2,3) & (1,2,4) \\ (2,3,1) & (2,3,2) & (2,3,3) & (2,3,4) \end{bmatrix}$$

$$Q_5 = \begin{bmatrix} (1,3,1) & (1,3,2) & (1,3,3) & (1,3,4) \\ (2,1,1) & (2,1,2) & (2,1,3) & (2,1,4) \end{bmatrix}$$

$$Q_6 = \begin{bmatrix} (1,3,1) & (1,3,2) & (1,3,3) & (1,3,4) \\ (2,2,1) & (2,2,2) & (2,2,3) & (2,2,4) \end{bmatrix}$$

The PBIB design is now obtained by forming, for each $Q_\ell (\ell = 1, 3, \ldots, N)$, arrays

$$B_{\ell_j} = \left(b_{\ell j1}, b_{\ell j2}, \ldots, b_{\ell j t_1}\right) \tag{5.26}$$

such that

1. $b_{\ell ji}$ is an element of the ith row of Q_ℓ.
2. For $i \neq i'$, $b_{\ell ji}$ and $b_{\ell ji'}$ are selected from different columns of Q_ℓ.

Thus for any given ℓ, there are $t_\nu^{(t_1)}$ such $B_{\ell j}$. The total number of arrays then is

$$b = N \times t_\nu^{(t_1)} = \left(\prod\nolimits^* t_i^{(t_1)}\right) \times t_\nu^{(t_1)} = \prod_{i=2}^{\nu} t_i^{(t_1)} \tag{5.27}$$

Each array (5.26) represents a block of size t_1 and (5.27) gives the total number of blocks. Hence the collection of all B_{ℓ_j}'s is the desired EGD/$(2^\nu - 1)$-PBIB design.

From Q_1 we obtain the following 12 B_{1j}:

B_{11}:	(1, 1, 1)	(2, 2, 2)	B_{14}:	(1, 1, 2)	(2, 2, 1)
B_{12}:	(1, 1, 1)	(2, 2, 3)	B_{15}:	(1, 1, 2)	(2, 2, 1)
B_{13}:	(1, 1, 1)	(2, 2, 4)	B_{16}:	(1, 1, 2)	(2, 2, 4)
B_{17}:	(1, 1, 3)	(2, 2, 1)	B_{10}:	(1, 4, 2)	(2, 2, 1)
B_{18}:	(1, 1, 3)	(2, 2, 2)	B_{11}:	(1, 1, 4)	(2, 2, 2)
B_{19}:	(1, 1, 3)	(2, 2, 4)	B_{12}:	(1, 1, 4)	(2, 2, 3)

The other blocks can be constructed in a similar way. □

It is, of course, obvious from the design above that only treatments that differ in all three components appear together in the same block. In general, this is assured by the definition of the P matrices in (5.24), the Q matrices in (5.25) and the B arrays in (5.26). Hence, only $\lambda(1, 1, \ldots, 1) > 0$.

Finally, we point out that if $t_1 = 2$ (as in Example 5.9) the method just described yields the same design as Aggarwal's (1974) method described in Section 5.5.2.

5.5.4 Generalization of the Direct Method

As stated in Theorem 5.7 the direct method leads to EGD-PBIB designs with block size $k = t_1$. Following Chang and Hinkelmann (1987) we shall now show that the theorem can be modified and used to construct certain EGD-PBIB designs with blocks of size $k = t_i$ where $2 \le i \le \nu$.

Suppose we have $t = t_1 \times t_2 \times \cdots \times t_\nu$ treatments with $t_1 \le t_2 \le \cdots \le t_\nu$. Let $A = \{i_1, i_2, \ldots, i_n\}$ be a subset of $1, 2, \ldots, \nu$ with $2 \le n \le \nu$. For ease of notation we take $A = \{1, 2, \ldots, n\}$, but the reader should have no difficulty replacing in actual applications $1, 2, \ldots, n$ by $i_1, i_2, \ldots, i_n$. The main result can then be stated as in the following theorem.

Theorem 5.8 Consider an experiment with $t = \prod_{i=1}^{\nu} t_i$ treatments with $t_1 \le t_2 \le \cdots \le t_\nu$. Let $A = \{1, 2, \ldots, n\}$. Then there exists an EGD/$(2^{\nu-1})$-PBIB design with parameters

$$t = \prod_{i=1}^{\nu} t_i \qquad b = \left(\prod_{i=2}^{n} t_i^{(t_1)}\right)\left(\prod_{j=n+1}^{\nu} t_j\right)$$

$$r = \prod_{i=2}^{n} (t_i - 1)^{(t_1 - 1)} \qquad k = t_1$$

$$\lambda(\underbrace{1, 1, \ldots, 1}_{n}, 0, \ldots, 0) = \prod_{i=2}^{n} (t_i - 2)^{(t_1 - 2)}$$

all other $\lambda(\boldsymbol{\gamma}) = 0$

Before we prove the theorem we introduce the notation in the following definition.

Definition 5.4 Let $\boldsymbol{a}_i' = (1, 2, \ldots, t_i)$ be $1 \times t_i$ vectors for $i = 1, 2, \ldots, \nu$. The *symbolic direct multiplication* (SDM) of $\boldsymbol{a}_p$ and $\boldsymbol{a}_q (1 \le p, q \le \nu)$, denoted by $\boldsymbol{a}_p \odot \boldsymbol{a}_q$ is given by an array of t_q vectors:

$$\boldsymbol{a}_p \odot \boldsymbol{a}_q = \left[\begin{pmatrix} (1,1) \\ (2,1) \\ \vdots \\ (t_p, 1) \end{pmatrix}, \begin{pmatrix} (1,2) \\ (2,2) \\ \vdots \\ (t_p, 2) \end{pmatrix}, \ldots, \begin{pmatrix} (1,t_q) \\ (2,t_q) \\ \vdots \\ (t_p, t_q) \end{pmatrix}\right]$$

where each vector has t_p elements and the second components in a given vector are identical and equal to $1, 2, \ldots, t_q$, respectively. □

Proof of Theorem 5.8 We first construct an EGD/$(2^n - 1)$-PBIB design for $t^* = \prod_{i=1}^{n} t_i$ treatments according to Theorem 5.7. Let the parameters of this design be $b^*, r^*, k^*, \boldsymbol{\lambda}^*$ say, as given in Theorem 5.7. Let $\boldsymbol{B}_j^*(j = 1, 2, \ldots, b^*)$ be a block in the EGD/$(2^n - 1)$-PBIB design. Further, let $\boldsymbol{d}_i' = (1, 2, \ldots, t_i)(i = 1, 2, \ldots, \nu)$, and let $\overline{A} = \{1, 2, \ldots, \nu\} - A = \{n+1, n+2, \ldots, \nu\}$ (because we have chosen $A = \{1, 2, \ldots, n\}$). We then define the SDP

$$\boldsymbol{D}_{\overline{A}} = \boldsymbol{d}_{n+1} \otimes \boldsymbol{d}_{n+2} \otimes \cdots \otimes \boldsymbol{d}_{\nu}$$

and consider the SDM

$$\boldsymbol{B}_j = \boldsymbol{B}_j^* \odot \boldsymbol{D}_{\overline{A}}$$

for $j = 1, 2, \ldots, b^*$. Each $\boldsymbol{B}_j^*$ consists of $\prod_{j=n+1}^{\nu} t_j$ vectors, each vector having t_1 elements, each element being an ν-tuple and representing a treatment. We thus have

$$b = b^* \times \prod_{j=n+1}^{\nu} t_j = \left(\prod_{i=2}^{n} t_i^{(t_1)}\right)\left(\prod_{j=n+1}^{\nu} t_j\right)$$

vectors and each such vector is taken as a block for the EGD/$(2^\nu - 1)$-PBIB design. From the construction it is obvious that any two elements in the same block (vector) of $\boldsymbol{B}_j$ are different in the first n positions and identical in the last $(\nu - n)$ positions. Hence $\lambda(1, \ldots, 1, 0, \ldots, 0) = \lambda^*(1, \ldots, 1)$ and all other $\lambda(\boldsymbol{\gamma})$ are zero. The other properties of the EGD/$(2^\nu - 1)$-PBIB design follow easily. □

We shall illustrate this procedure to construct a design with blocks of size 3 for $t = 24$ treatments.

Example 5.10 Let $t = 24$, $t_1 = 2$, $t_2 = 3$, $t_3 = 4$. To construct an EGD-PBIB design with blocks of size $k = 3$, we take $A = \{2, 3\}$ and construct an EGD/3-PBIB design for $t^* = 12 = 3 \times 4$ according to Theorem 5.7. For this case we define $\prod_{i=2}^{\nu-1} t_i$ (for $\nu = 2$) to be 1, so that the matrix $\boldsymbol{P}$ of (5.23) becomes

$$\boldsymbol{P} = \begin{bmatrix} (1,1) & (1,2) & (1,3) & (1,4) \\ (2,1) & (2,2) & (2,3) & (2,4) \\ (3,1) & (3,2) & (3,3) & (3,4) \end{bmatrix}$$

and $\boldsymbol{Q}(\boldsymbol{P}) = \boldsymbol{P}$. Then the $\boldsymbol{B}_j^*$ are of the form

$$\begin{bmatrix} (1,1) \\ (2,2) \\ (3,3) \end{bmatrix} \quad \begin{bmatrix} (1,1) \\ (2,2) \\ (3,4) \end{bmatrix} \quad \begin{bmatrix} (1,1) \\ (2,3) \\ (3,2) \end{bmatrix} \quad \begin{bmatrix} (1,1) \\ (2,3) \\ (3,4) \end{bmatrix} \quad \begin{bmatrix} (1,1) \\ (2,4) \\ (3,2) \end{bmatrix} \quad \begin{bmatrix} (1,1) \\ (2,4) \\ (3,3) \end{bmatrix}$$

plus 18 more blocks. Now $\overline{A} = \{1\}$ and hence

$$D_{\{1\}} = \begin{pmatrix} 1 \\ 2 \end{pmatrix}$$

Then, for any B_j^* above we obtain

$$B_j = B_j^* \odot D_{\{1\}}$$

that is, the B_j's are of the form

$$\begin{array}{ccccc} (1, 1, 1) & (1, 1, 2) & (1, 1, 1) & (1, 1, 2) & \\ (2, 2, 1) & (2, 2, 2) & (2, 2, 1) & (2, 2, 2) & \cdots \\ (3, 3, 1) & (3, 3, 2) & (3, 4, 1) & (3, 4, 2) & \end{array}$$

and we have a total of 48 blocks. All we need to do now to obtain the final result is to interchange the elements in each triplet above so that they are in the "right" order, that is,

$$\begin{array}{ccccc} (1, 1, 1) & (2, 1, 1) & (1, 1, 1) & (2, 1, 2) & \\ (1, 2, 2) & (2, 2, 2) & (1, 2, 2) & (2, 2, 2) & \cdots \\ (1, 3, 3) & (2, 3, 3) & (1, 3, 4) & (2, 3, 4) & \end{array}$$

The parameters for this design then are $t = 24, b = 48, k = 3, r = 6$, and $\lambda(0, 1, 1) = 1$. □

We conclude this section to point out that the EGD-PBIB designs thus constructed are disconnected except when $A = \{1, 2, \ldots, \nu\}$ which, of course, is the design of Theorem 5.7. Though disconnectedness is not desirable in general, it can be useful in connection with asymmetrical factorial experiments (see Chapter 12).

5.6 HYPERCUBIC PBIB DESIGNS

Using the result due to Kageyama (1972) that an EGD/$(2^\nu - 1)$-PBIB design constructed as a ν-fold Kronecker product of BIB designs reduces to a hypercubic PBIB design with ν associate classes if and only if the ν BIB designs have the same number of treatments and the same block size, we can state the following corollary to Theorem 5.5.

Corollary 5.1 Let there be ν BIB designs D_i with parameters $t_i, b_i, k_i, r_i, \lambda_{(i)}$ and incidence matrix N_i such that $t_i = q$ and $k_i = k^*(i = 1, 2, \ldots, \nu)$. Then the Kronecker product design D given by its incidence matrix

$$N = N_1 \times N_2 \times \cdots \times N_\nu$$

is a hypercubic PBIB(ν) design with parameters $t = q^\nu, b = \prod_{i=1}^{\nu} b_i, k = (k^*)^\nu, r = \prod_{i=1}^{\nu} r_i$, and $\lambda_j = \prod_{i=1}^{j} r_i \prod_{i=j+1}^{\nu} \lambda_{(i)}$ for $j = 1, 2, \ldots, \nu$.

Proof We refer simply to the definition of the hypercubic association scheme (Section 4.6.10). Concerning the expression for λ_j, we note that because the t_i's and k_i's are identical it follows that $r_i/\lambda_{(i)}$ = constant for each i and hence all the $\lambda(\gamma_1, \gamma_2, \ldots, \gamma_\nu)$ of the EDG-PBIB with the same number of unity components are identical. □

Similarly, for a second method of constructing hypercubic PBIB designs, we consider a special case of Theorem 5.6 and state the following corollary.

Corollary 5.2 (Aggarwal, 1974) A hypercubic PBIB design with the parameters $t = q^\nu$, $b = q^{\nu-1}(q-1)^{\nu-1}$, $k = q, r = (q-1)^{\nu-1}$, $\lambda_1 = \lambda_2 = \cdots = \lambda_{\nu-1} = 0, \lambda_\nu = 1$ can be constructed if q is a prime or a prime power.

For the proof we refer to the proof of Theorem 5.6 and the definition of the hypercubic association scheme (see Section 4.6.10).

For another method of constructing hypercubic designs, see Chang (1989).

CHAPTER 6

More Block Designs and Blocking Structures

6.1 INTRODUCTION

In Chapters 3 and 5 we described the construction of two rather large classes of incomplete block designs with desirable statistical and mathematical properties. Yet, these designs do not cover the need for incomplete block designs for many practical applications, such as variety, environmental, medical, sensory, and screening trials. It is impossible to create a catalogue with all possible designs because of the large number of combinations of design parameters. Hence there is a need for simple methods of constructing designs. Of particular interest often are resolvable designs (see Section 2.7.2) as they allow a certain amount of flexibility with respect to spatial and/or sequential experimentation. In this chapter we shall present three quite general algorithms to construct resolvable incomplete block designs allowing equal or unequal block sizes.

Although in general comparisons among all treatments are of primary interest, there are situations where this is only of secondary importance. In such situations the forms of the experiment is on comparing so-called *test treatments* versus a *control* (the word *control* is used here in a general sense with different meaning in different areas of practical applications). Any of the designs discussed in previous chapters can be used for that purpose by simply declaring one of the t treatments as the control. Because of the special nature of the desired inference, however, special designs have been developed, and we shall discuss them briefly here.

So far we have only considered designs with one blocking factor. In most practical situations this is clearly sufficient to develop an efficient experimental protocol. There are, however, situations where it is useful to take into account additional blocking factors to further reduce experimental error. We shall consider

Design and Analysis of Experiments. Volume 2: Advanced Experimental Design
By Klaus Hinkelmann and Oscar Kempthorne
ISBN 0-471-55177-5

here briefly one additional blocking factor, one that leads us to the notion of *row–column designs* as generalization of the Latin-square-type designs of Chapter I.10.

6.2 ALPHA DESIGNS

These designs, called *α-designs*, were introduced by Patterson and Williams (1976) and further developed by John and Williams (1995) to be used mainly in the setting of variety trials in agronomy.

6.2.1 Construction Method

The general ideas and steps of constructing an α-design are as follows:

1. Denote the t treatments by $0, 1, \ldots, t-1$.
2. Let $t = ks$, where k is the desired block size and s is the number of blocks in each replicate, there being r replicates in the final design.
3. Select as the generating array a $k \times r$ array α with elements $a(p, q)$ in the set of residues mod s $(p = 0, 1, 2, \ldots, k-1; q = 1, 2, \ldots, r)$.
4. Develop each column of α cyclically mod s to generate an intermediate $k \times rs$ array, α^*.
5. Add $i \cdot s$ to each element in the $(i+1)$th row of $\alpha^*(i = 1, 2, \ldots, k-1)$ to obtain the final array, α^{**}, containing elements $0, 1, 2, \ldots, t-1$.
6. Separate α^{**} into r sets, where the jth set is formed by the columns, $(j-1)s+1, (j-1)s+2, \ldots, (j-1)s+s (j = 1, 2, \ldots, r)$, each column representing a block.

Example 6.1 We shall illustrate this construction method by adapting an example from John and Williams (1995) for $t = 12 = 4 \times 3$; that is, $k = 4, s = 3$. In Table 6.1 we give the generating array α, for $r = 2$, the intermediate array α^*, the final array α^{**}, and the final design.

The generating array α in Table 6.1 is called a *reduced* array, having zeros in the first row and first column. All arrays can be represented in the reduced form by adding suitable elements to individual rows and columns, always reducing mod s. This is convenient for the search of optimal α-designs; that is, designs with the highest efficiency factor E [see (1.108)]. For more details see Williams and John (2000).

The efficiency factor itself is closely tied to the concurrence matrix NN'. It is easy to verify that NN' for the design in Table 6.1 has off-diagonal elements equal to $0, 1, 2$, implying that some treatment pairs never occur together in a block, other pairs occur together once, and still others occur together twice. To indicate this property the design will be called a $\alpha(0, 1, 2)$-design. Just as, in general, 2-associate class PBIB designs are more efficient than 3-associate class PBIB designs for the same parameters (t, b, k, r), so are 2-occurrence class

Table 6.1 Construction of Design for $t = 12$, $s = 3$, $k = 4$, and $r = 2$

α Array

0	0
0	1
0	2
0	0

α^* Array

0	1	2	0	1	2
0	1	2	1	2	0
0	1	2	2	0	1
0	1	2	0	1	2

α^{**} Array

0	1	2	0	1	2
3	4	5	4	5	3
6	7	8	8	6	7
9	10	11	9	10	11

Final Design

Replicate	1			2		
Block	1	2	3	1	2	3
	0	1	2	0	1	2
	3	4	5	4	5	3
	6	7	8	8	6	7
	9	10	11	9	10	11

α-designs preferred over 3-occurrence class α designs (Patterson and Williams, 1976). For example, the generating array

$$\begin{array}{cc} 0 & 0 \\ 0 & 2 \\ 0 & 1 \\ 0 & 1 \end{array}$$

leads to an $\alpha(0, 1)$-design. The efficiency factor for this design can be computed to be .71, which, in this case, is about the same as that for the $\alpha(0, 1, 2)$-design. For purposes of comparison we mention that the upper bound (1.110) for the efficiency factor is .82. □

6.2.2 Available Software

Williams and Talbot (1993) have provided a design generation software package, called ALPHA$^+$, which produces α-designs in the parameter ranges

$$2 \leq r \leq 10 \qquad 2 \leq k \leq 20 \qquad t \leq 500$$

A more general software package, called CyDesigN, has been provided by Whitaker, Williams, and John (1997). Both packages are able to construct designs with high efficiency factors relative to the upper bounds for such design thus leading to near-optimal designs.

6.2.3 Alpha Designs with Unequal Block Sizes

The construction of α-designs is quite flexible and able to accommodate most any desirable, practical block size for a given t, or to deal with the situation when t cannot be written as $t = ks(s > 1)$. In either case we may write

$$t = s_1 k_1 + s_2 k_2$$

and construct a design with s_1 blocks of size k_1 and s_2 blocks of size k_2. Of particular interest is the case where $k_2 = k_1 - 1$, that is, where the block sizes are not too different, differing only by one unit. For this situation an appropriate design can be derived from an α-design as follows:

1. Construct an α-design for $t + s_2$ treatments in $s = s_1 + s_2$ blocks of size k_1 in each replicate.
2. Delete from this design the treatments labeled $t, t+1, \ldots, t+s_2-1$.

Example 6.2 Consider

$$t = 11 = 2 \times 4 + 1 \times 3$$

that is, $s_1 = 2, s_2 = 1, k_1 = 4, k_2 = 3$. The α-design in Table 6.1 is a design for $t + s_2 = 11 + 1 = 12$ treatments in blocks of size 4. Deleting treatment 11 yields the derived design with two blocks of size 4 and one block of size 3. □

6.3 GENERALIZED CYCLIC INCOMPLETE BLOCK DESIGNS

We have seen in Section 5.3 that cyclic PBIB designs are easy to construct and encompass a large class of useful designs. Jarrett and Hall (1978) extended the idea underlying these designs and introduced the notion of *generalized cyclic incomplete block designs*. Basically, for $t = m \times n$, the designs are obtained by cyclic development of one or more initial blocks, but rather than adding 1 to the element in the initial blocks(s) we now add successively m to each element and reduce mod t.

Example 6.3 Consider $t = 12 = 3 \times 4$; that is, $m = 3, n = 4$. Let the initial blocks be

$$(0, 2, 3, 7)_3 \quad \text{and} \quad (1, 4, 5, 9)_3$$

where the subscript indicates the incrementing number. Cyclic development then leads to the following design with $b = 8, k = 4$:

$$\begin{array}{ll} (0,\ 2, 3,\ 7) & (\ 1,\ 4,\ 5, 9) \\ (3,\ 5, 6, 10) & (\ 4,\ 7,\ 8, 0) \\ (6,\ 8, 9,\ 1) & (\ 7, 10, 11, 3) \\ (9, 11, 0,\ 4) & (10,\ 1,\ 2, 6) \end{array}$$

Inspection shows that $r_0 = r_2 = r_3 = r_5 = r_6 = r_8 = r_9 = r_{11} = 3, r_1 = r_4 = r_7 = r_{10} = 2$. The reason for the two different number of replications can be explained as follows. The treatments can be divided into $m = 3$ groups of $n = 4$ elements by considering the residue classes mod m, that is,

$$S_i = \{i, i + m, \ldots, i + m(n - 1)\}$$

for $i = 0, 1, \ldots, m - 1$. For our example we have

$$\begin{aligned} S_0 &= \{0, 3, 6,\ 9\} \\ S_1 &= \{1, 4, 7, 10\} \\ S_2 &= \{2, 5, 8, 11\} \end{aligned}$$

It follows that when an initial block is developed the elements in each residue class are equally replicated as each initial block contributes n blocks to the design. Now, the two initial blocks contain three elements each from S_0 and S_1, and two elements from S_2.

From this argument it follows immediately that in order to construct an equireplicate design we need three initial blocks such that each residual class is represented the same number of times in these three blocks. One possible set of initial blocks is

$$(0, 2, 3, 7)_3 \qquad (1, 4, 5, 9)_3 \qquad (6, 8, 10, 11)_3$$

It is easy to select initial blocks to construct an incomplete block design with desired parameters. There is, however, no guarantee that this would lead to a design with a high efficiency factor. One would need a search algorithm to find the "best" design. In general, though, aiming for nearly equal numbers of concurrences of all pairs of treatments will yield a design with a high efficiency factor.

Cyclic development can also be used to construct incomplete block designs with unequal block sizes by specifying initial blocks of different sizes. □

6.4 DESIGNS BASED ON THE SUCCESSIVE DIAGONALIZING METHOD

Khare and Federer (1981) have presented several methods of constructing resolvable incomplete block designs for various factorizations of t, the number of treatments. In Section 3.3.4 we have discussed their method when $t = K^2$ and K is a prime or prime power. Here we shall discuss a few other cases.

6.4.1 Designs for $t = Kk$

For K prime or prime power and $k < K$, where k is the block size, we can use the method described in Section 3.3.4 for $t = K^2$. We then consider only replicates $2, 3, \ldots, K + 1$ and delete in each replicate the treatments numbered $Kk + 1, \ldots, K^2$. This leads to a binary design with $r = K$ replications and rK blocks of size k. These designs are also called rectangular lattice designs (see Section 18.11).

Example 6.4 For $t = 12 = 4 \times 3$ we use the design given in Example 3.8 and delete treatments 13, 14, 15, 16 from replicates 2, 3, 4, and 5. □

Example 6.5 For $t = 8 = 4 \times 2$ we can again use the design from Example 3.8 and delete treatments $9, 10, \ldots, 16$ from replicates 2, 3, 4, and 5 to obtain a design with $b = 16, k = 2$, and $r = 4$. □

6.4.2 Designs with $t = n^2$

We consider here the situation where t is a square but not a square of a prime number or prime power. The n can be written as $n = mp_s$, where p_s is the smallest prime in n. The first $p_s + 1$ replicates from the $n + 1$ replicates obtained by the successive diagonalizing method on n^2 treatments in blocks of size n yield an incomplete block design with (0, 1)-concurrences for treatment pairs. For additional replicates one can either repeat some of the original $p_s + 1$ replicates or develop further replicates continuing with the successive diagonalizing method. In the latter case the numbers of concurrences increase, however, in groups of p_s.

The above method can also be used to generate a design for $t = nk(k < n)$. We first obtain the design for $t = n^2$ and then delete from replicates $2, \ldots, p_s + 1$ the treatments numbered $nk + 1, \ldots, n^2$. The versatility of this method is illustrated by the following example.

Example 6.6 Fichtner (2000) considered the case of $t = 400$ varieties (treatments) in blocks of size $k = 10$ with $r = 2$ replicates for an agronomic trial. In the context discussed above, we have $t = 40 \times 10$. Rather than writing out the 40×40 array for replicate 2 (of the generating design) one only needs to write out the 40×10 array. Thus the first block would contain the varieties

$$(1, 41, 81, 121, 161, 201, 241, 281, 321, 361)$$

Adding $1, 2, \ldots, 39$ to each element produces the remaining 39 blocks. From this it is easy to see that the first block in replicate 2 contains the varieties

$$(1, 42, 83, 124, 165, 206, 247, 288, 329, 370)$$

The remaining blocks can then be filled in easily using replicate 1. We thus obtain the following replicate:

Block					**Treatments**					
1	1	42	83	124	165	206	247	288	329	370
2	2	43	84	125	166	207	248	289	330	371
3	3	44	85	126	167	208	249	290	331	372
⋮					⋮					
39	39	80	81	122	163	204	245	286	327	368
40	40	41	82	123	164	205	246	287	328	369

If a third replicate were desired, the method of successive diagonalizing would yield the initial block as

$$(1, 43, 85, 127, 169, 211, 253, 295, 337, 379)$$

The remaining blocks are obtained by columnwise cyclic substitution (as described in Section 3.3.4) using the replicate given above. □

For other cases of constructing (0, 1)-concurrence resolvable incomplete block designs, we refer the reader to Khare and Federer (1981), where methods are discussed for $t = p^{2k},\ \ t = n^3, t = n^{2k}$.

6.5 COMPARING TREATMENTS WITH A CONTROL

In some experimental situations one of the treatments, often referred to as the *control* or *standard*, may play a major role in that the focus is not so much the comparisons among all treatments but rather the comparisons of the remaining treatments, often referred to as the *test treatments*, with the control. The function of the controls might simply be to provide an indication whether the treatments are even worth consideration or to show their superiority to established procedures, such as comparing new drugs against an established medication, or new varieties against an old variety.

Obviously, any of the designs we have discussed so far would be suitable if we simply declare one of the t treatments as the control. However, because of the limitation on the inference space the designs may not be a good choice and more specialized designs may be called for that are more efficient for this particular purpose. Some of these designs will be described in the following sections.

6.5.1 Supplemented Balance

Suppose we have t test treatments and one control treatment. Let us denote the $t+1$ treatments by $T_0, T_1, T_2, \ldots, T_t$, where T_0 represents the control. In the context of block designs, a natural way to construct suitable designs for comparing $T_1, T_2, \ldots, T_t$ with T_0 is to consider a block design for $T_1, T_2, \ldots, T_t$ and augment or reinforce each block with $q(\geq 1)$ replications of T_0. This idea was suggested by Cox (1958) using a BIB design as the generating block design. Das (1958) referred to such designs as *reinforced BIB designs*.

A more general class of designs, referred to as *supplemented balance designs*, was introduced by Hoblyn, Pearce, and Freeman (1954) and more formally by Pearce (1960). These designs are characterized by the following form of the concurrence matrix:

$$NN' = \begin{pmatrix} s_0 & \lambda_0 & \cdots & & \lambda_0 \\ \lambda_0 & s & \cdots & & \vdots \\ & & & \lambda_1 & \\ \vdots & \lambda_1 & \ddots & & \vdots \\ \lambda_0 & & & & s \end{pmatrix} \tag{6.1}$$

implying that the self-concurrence is the same for all the test treatments. Further, T_0 occurs together λ_0 times with each of $T_1, T_2, \ldots, T_t$, and T_i and $T_j (i \neq j, i, j \neq 0)$ occur together λ_1 times.

The example below of a supplemented balance design was given by Pearce (1953) for $t = 4, b = 4, k = 7, r_0 = 8, r = 5, \lambda_0 = 10, \lambda_1 = 6$:

0	0	1	2	3	4	1
0	0	1	2	3	4	2
0	0	1	2	3	4	3
0	0	1	2	3	4	4

where each row represents a block. We note here in passing that the design above is an example of what we have called an extended block design (see Section I.9.8.5). And for the reinforced BIB design we have $\lambda_0 = qr$.

For incomplete blocks an obvious way to construct a supplemented balance design is to augment each block in a BIB design with t treatments by one or more replications of the control. Under certain conditions such designs are A-optimal (see Hedayat, Jacroux, and Majumdar, 1988; Stufken, 1987).

6.5.2 Efficiencies and Optimality Criteria

Clearly, there exist many ways to construct suitable augmented designs for given parameters t, b, and k. It is, therefore, important to distinguish between competing designs and choose the one that is "best" for the purpose at hand, which may not only depend on statistical but also on practical considerations.

From the statistical, that is, inferential, point of view we would like to obtain maximum information with respect to comparing T_0 versus T_i $(i = 1, 2, \ldots, t)$. To formulate this more precisely, we consider the usual model for observations from a block design, that is,

$$y_{ij} = \mu + \beta_i + \tau_j + e_{ij} \tag{6.2}$$

or, in matrix notation

$$\boldsymbol{y} = \mu \mathfrak{I} + \boldsymbol{X}_\beta \boldsymbol{\beta} + \boldsymbol{X}_\tau \boldsymbol{\tau}_j + \boldsymbol{e} \tag{6.3}$$

The information matrix [see (1.9)], assuming equal block sizes, is given by

$$\boldsymbol{C} = \boldsymbol{R} - \frac{1}{k}\boldsymbol{N}\boldsymbol{N}' \tag{6.4}$$

where $\boldsymbol{R} = \boldsymbol{X}_\tau' \boldsymbol{X}_\tau = \operatorname{diag}(r_0, r_1, \ldots, r_t)$ and $\boldsymbol{N} = \boldsymbol{X}_\tau' \boldsymbol{X}_\beta$. We know [see (1.20)] that for an estimable function $\boldsymbol{c}'\boldsymbol{\tau}$ of the treatment effects

$$\operatorname{var}(\boldsymbol{c}'\hat{\boldsymbol{\tau}}) = \boldsymbol{c}'\boldsymbol{C}^{-}\boldsymbol{c}\,\sigma_e^2$$

For the designs discussed in this section the estimable functions of primary interest are of the form $\tau_j - \tau_0 (j = 1, 2, \ldots, t)$. If we denote by

$$\boldsymbol{P} = \begin{pmatrix} -1 & 1 & 0 & \ldots & \ldots & \ldots & 0 \\ -1 & 0 & 1 & 0 & \ldots & \ldots & 0 \\ -1 & 0 & 0 & 1 & 0 & \ldots & 0 \\ & & \vdots & & & & \\ -1 & 0 & 0 & \ldots & \ldots & 0 & 1 \end{pmatrix} = (-1\mathfrak{I}, \boldsymbol{I}_t) \tag{6.5}$$

the coefficient matrix for the above contrasts, then the variance–covariance matrix for the estimators for those contrasts is given by $\boldsymbol{P}\boldsymbol{C}^{-}\boldsymbol{P}'\sigma_e^2$.

If we write $\boldsymbol{C}$ of (6.4) as

$$\boldsymbol{C} = \begin{pmatrix} c_0 & \boldsymbol{c}_1' \\ \boldsymbol{c}_1 & \boldsymbol{C}_{11} \end{pmatrix} \tag{6.6}$$

then a generalized inverse of $\boldsymbol{C}$ is given by

$$\boldsymbol{C}^- = \begin{pmatrix} 0 & \cdots & 0 \\ \vdots & & \\ 0 & \boldsymbol{C}_{11}^{-1} & \end{pmatrix} \tag{6.7}$$

which is equivalent to solving the reduced NE $\boldsymbol{C}\hat{\boldsymbol{\tau}} = \boldsymbol{Q}$ [see (1.8)] by imposing the condition $\hat{\tau}_0 = 0$ (recall that SAS PROC GLM, e.g., obtains a generalized inverse equivalently, by assuming $\hat{\tau}_t = 0$ rather than $\hat{\tau}_0 = 0$). Using $\boldsymbol{C}^-$ of the form (6.7) implies that

$$\boldsymbol{P}\boldsymbol{C}^-\boldsymbol{P}' = \boldsymbol{C}_{11}^{-1}$$

and further that (1) the information matrix for treatment control contrasts is given by $\boldsymbol{C}_{11}$ of (6.6) [see also Majumdar (1996)] and (2) with $\boldsymbol{C}_{11}^{-1} = (c^{ii})$

$$\text{av. var}\,(\hat{\tau}_j - \hat{\tau}_0) = \frac{1}{t}\sum_{j=1}^{t} c^{jj}\sigma_e^2 = \frac{1}{t}\sum_{j=1}^{t}\frac{1}{d_j}\sigma_e^2 \tag{6.8}$$

where $d_j (j = 1, 2, \ldots, t)$ are the eigenvalues of $\boldsymbol{C}_{11}$ [see (1.106) and Constantine (1983)].

For a CRD with $r_j (j = 0, 1, 2, \ldots, t)$ replications for the jth treatment we have, for $j \neq 0$,

$$\text{av. var}\,(\hat{\tau}_j - \hat{\tau}_0) = \left(\frac{1}{r_0} + \frac{1}{\overline{r}_h}\right)\sigma^2_{e(\text{CRD})} \tag{6.9}$$

where

$$\frac{1}{\overline{r}_h} = \frac{1}{t}\sum_{j=1}^{t}\frac{1}{r_j}$$

The ratio of (6.9) over (6.8) $\left(\text{assuming } \sigma_e^2 = \sigma^2_{e(\text{CRD})}\right)$ can then be defined as the efficiency factor E for a treatment control comparison design (see also Section 1.12.2), that is,

$$E = \frac{1/r_0 + 1/\overline{r}_h}{(1/t)\sum_{j=1}^{t}(1/d_j)} \tag{6.10}$$

can be used to compare competing designs.

Another way to compare designs is to make use of (6.8) and find, for a given set of parameters t, b, k, the design D^*, say, with minimum value of

$$\frac{1}{t}\sum_{j=1}^{t}\frac{1}{d_j}$$

where D^* is referred to as an A-optimal design (see Hedayat, Jacroux, and Majumdar, 1988, and Section 1.13).

An alternative criterion, called MV-optimality, is to minimize

$$\max_{1\le j\le t}(1/d_j)$$

for competing designs (see Hedayat, Jacroux and, Majumdar, 1988).

A different approach to optimal designs is suggested and explored by Bechhofer and Tamhane (1981, 1985). They propose to maximize the confidence coefficient for simultaneous confidence intervals for all $\tau_j - \tau_0 (j = 1, 2, \ldots, t)$, thereby linking the designs to Dunnett's procedures (see I.7.5.7) for comparing treatments with a control (Dunnett, 1955, 1964). In addition to assuming model (6.2), this method has to make distributional assumptions about the e_{ij}'s, which for practical purposes amounts to assuming normality (see Section 6.5.5).

6.5.3 Balanced Treatment Incomplete Block Designs

An important and useful class of treatment control designs was introduced by Bechhofer and Tamhane (1981), a class they refer to as *balanced treatment incomplete block designs* (BTIBD).

6.5.3.1 Definition and Properties

Definition 6.1 A treatment control design with t test treatments and one control treatment in b blocks of size $k < t + 1$ is called a *balanced treatment incomplete block design*, denoted by BTIBD($t, b, k; \lambda_0, \lambda_1$), if (1) each test treatment occurs together with the control λ_0 times in a block and (2) any two test treatments occur together λ_1 times in a block. □

The concurrence matrix of a BTIBD is then of the form

$$NN' = \begin{pmatrix} s_0 & \lambda_0 & \cdots & \cdots & \lambda_0 \\ \lambda_0 & s_1 & \lambda_1 & \cdots & \lambda_1 \\ \lambda_0 & \lambda_1 & s_2 & \cdots & \vdots \\ \vdots & \cdots & & \ddots & \lambda_1 \\ \lambda_0 & \lambda_1 & \cdots & \lambda_1 & s_t \end{pmatrix} \tag{6.11}$$

Comparing (6.11) and (6.1) shows that some designs with supplemented balance are special cases of BTIBDs.

Since the BTIBDs are, by definition, connected designs, it follows that $\text{rank}(\boldsymbol{C}) = t$, where $\boldsymbol{C}$ is given in (6.4). It follows then from (6.4), with $\boldsymbol{N} = (n_{ji})$, and (6.11) that

$$r_0 - \frac{1}{k}\sum_{i=1}^{b} n_{oi}^2 - \frac{t\lambda}{k} = 0$$

and

$$r_j - \frac{1}{k}\sum_{i=1}^{b} n_{ji}^2 - \frac{\lambda_0}{k} - \frac{(t-1)\lambda_1}{k} = 0 \qquad (j = 1, 2, \ldots, t) \tag{6.12}$$

It follows from (6.12) that

$$r_j - \frac{1}{k}\sum_{i} n_{ji}^2 = [\lambda_0 - (t-1)\lambda_1]/k$$

and hence $\boldsymbol{C}_{11}$ of (6.6) has the form

$$\boldsymbol{C}_{11} = a\boldsymbol{I} + b\mathfrak{I}\mathfrak{I}' \tag{6.13}$$

with

$$a = \frac{\lambda_0 + t\lambda_1}{k} \qquad b = \frac{-\lambda_1}{k} \tag{6.14}$$

Using (6.13) and (6.14) it is then easy to obtain $\boldsymbol{C}_{11}^{-1}$, which is of the same form as $\boldsymbol{C}_{11}$, say

$$\boldsymbol{C}_{11}^{-1} = c\boldsymbol{I} + d\mathfrak{I}\mathfrak{I}' \tag{6.15}$$

We find

$$c = \frac{k}{\lambda_0 + t\lambda_1} \quad \text{and} \quad d = \frac{k\lambda_1}{\lambda_0(\lambda_0 + t\lambda_1)} \tag{6.16}$$

and hence from our discussion above this implies that

$$\text{var}(\hat{\tau}_j - \hat{\tau}_0) = \frac{k(\lambda_0 + \lambda_1)}{\lambda_0(\lambda_0 + t\lambda_1)}\sigma_e^2 \qquad (j = 1, 2, \ldots, t) \tag{6.17}$$

and

$$\text{cov}(\hat{\tau}_j - \hat{\tau}_0, \hat{\tau}_{j'} - \hat{\tau}_0) = \frac{k\lambda_1}{\lambda_0(\lambda_0 + t\lambda_1)}\sigma_e^2 \qquad (j \neq j') \tag{6.18}$$

Expressions (6.17) and (6.18) explain, of course, in what sense these designs are balanced.

6.5.3.2 Construction Methods

There exist different methods of constructing BTIB designs, and we shall mention a few without going into all the details.

One obvious method, mentioned already in Section 6.5.1, is to augment each block of a BIBD(t, b, k, r, λ_1) with n copies of the control. This is a special case of a more general class of BTIB designs, which we denote (following Hedayat, Jacroux, and Majumdar, 1988) by BTIBD$(t, b, k^*; u, s; \lambda_0, \lambda_1)$ and which is characterized by the following form of the incidence matrix:

$$n_{ji} = 0 \text{ or } 1 \quad \text{for} \quad j = 1, 2, \ldots, t \qquad i = 1, 2, \ldots, b$$

and

$$n_{01} = n_{02} = \cdots = n_{0s} = u + 1$$

$$n_{0,s+1} = n_{0,s+2} = \cdots = n_{0b} = u$$

The special case mentioned above has $s = 0$ and hence $k^* = k + u$. Such a design is of rectangular or R-type, and the general form is depicted in Figure 6.1, where D_1 represents the BIBD(t, b, k, r, λ_1).

For $s > 0$ the design is said to be of step or S-type. Its general form is shown in Figure 6.1.

One method to construct an S-type BTIB$(t, b, k^*, \lambda_0, \lambda_1)$ is to choose BIB designs for D_2 and D_3, more specifically

$$D_2 = \text{BIBD}(t, s, k - 1, r_{(2)}; \lambda_{(2)}) \tag{6.19}$$

and

$$D_3 = \text{BIBD}(t, b - s, k, r_{(3)}; \lambda_{(3)}) \tag{6.20}$$

so that $k^* = k + u$, $\lambda_0 = (u + 1)r_{(2)} + ur_{(3)}$, and $\lambda_1 = \lambda_{(2)} + \lambda_{(3)}$.

For this particular method of construction we make the following comments:

1. Let us denote by D_2^* and D_3^* the two component designs of the S-type BTIB design (see Fig. 6.1). Then D_2^* and D_3^* with D_2 and D_3 as given in (6.19) and (6.20), respectively, are themselves BTIB designs. They constitute what Bechhofer and Tamhane (1981) have called *generator designs* (see Definition 6.2).
2. As a special case of S-type designs we can have $u = 0$; that is, the blocks in D_3^* do not contain the control, which means that $D_3^* = D_3$.
3. For $u = 0$ the design D_3 can be a RCBD, possibly with $b - s = 1$.

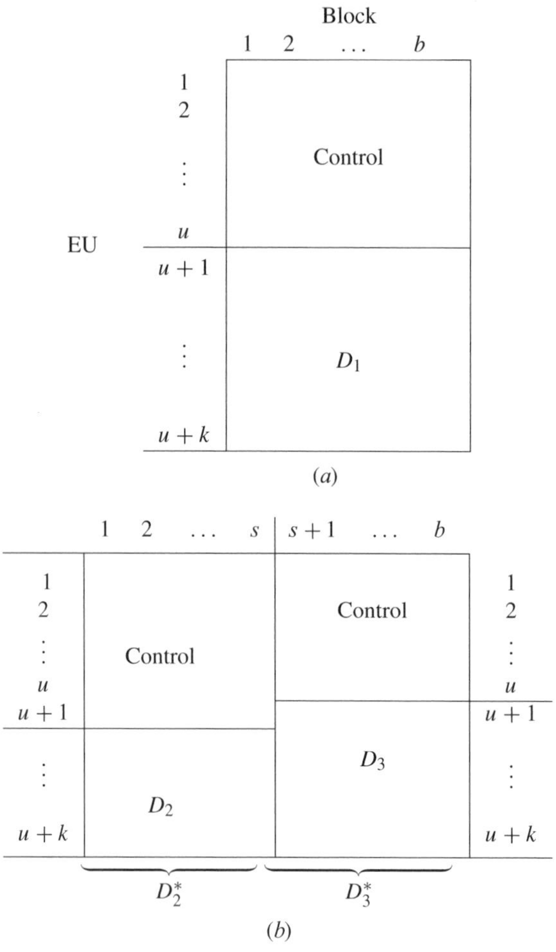

Figure 6.1 General form of (a) R- and (b) S-type BTIB designs.

4. As a trivial case (but only in the context of BTIB designs) the design D_2 may have blocks of size 1 with $s = mt$; that is, each treatment occurs m times. An important case in this context is $k^* = 2$, $m = 1$ with

$$D_2 = \{1, 2, \ldots, t\} \tag{6.21}$$

and

$$D_3 = \left\{ \begin{array}{cccc} 1 & 1 & & t-1 \\ 2 & 3 & \ldots & t \end{array} \right\} \tag{6.22}$$

Hedayat and Majumdar (1984) denote D_2 of (6.21) and D_3 of (6.22) by $t \sum 1$ and $t \sum 2$, respectively, where $t \sum p$ in general denotes all $\binom{t}{p}$ distinct blocks of size p for t treatments.

Let us now return to the notion of generator designs (GD) and their function in constructing BTIB designs. Bechhofer and Tamhane (1981) give the following definition.

Definition 6.2 For t test treatments and blocks of size k^* a *generator design* is a BTIB design no proper subset of whose blocks forms a BTIB design, and no block of which contains only one of the $t + 1$ treatments.

To illustrate this notion we use the following example. □

Example 6.7 For $t = 3$ and $k^* = 3$ consider

$$\text{GD}_1 = \begin{Bmatrix} 0 & 0 & 0 \\ 0 & 0 & 0 \\ 1 & 2 & 3 \end{Bmatrix} \qquad \lambda_0 = 2 \qquad \lambda_1 = 0$$

$$\text{GD}_2 = \begin{Bmatrix} 0 & 0 & 0 \\ 1 & 1 & 2 \\ 2 & 3 & 3 \end{Bmatrix} \qquad \lambda_0 = 2 \qquad \lambda_1 = 1$$

$$\text{GD}_3 = \begin{Bmatrix} 1 \\ 2 \\ 3 \end{Bmatrix} \qquad \lambda_0 = 0 \qquad \lambda_1 = 1$$

$$\text{GD}_4 = \begin{Bmatrix} 0 & 0 & 0 & 1 \\ 0 & 0 & 0 & 2 \\ 1 & 2 & 3 & 3 \end{Bmatrix} \qquad \lambda_0 = 2 \qquad \lambda_1 = 1$$

Of these four designs GD_1, GD_2, and GD_3 are generator designs, but GD_4 is not, because $\text{GD}_4 = \text{GD}_1 \bigcup \text{GD}_3$. □

The design GD_4 illustrates the construction of a BTIB design, D say, in terms of the generator designs:

$$D = f_1\text{GD}_1 \bigcup f_2\text{GD}_2 \bigcup f_3\text{GD}_3 \tag{6.23}$$

with $f_i \geq 0 (i = 1, 2, 3)$ and at least f_1 or $f_2 \geq 1$ for an implementable BTIB design. We note that the design D in (6.23) with $f_i \geq 1 (i = 1, 2, 3)$ represents another form of S-type BTIB design, which we might refer to as a S_2-type design because it has two steps. In general, we might have a S_q-type design

$$D = \bigcup_{i=1}^{q} f_i \text{ GD}_i$$

with GD_i having b_i blocks and $b = \sum_i f_i b_i$. Apart from the control the GD_i represent BIB/RCB designs with increasing block size $k (1 \leq k \leq k^*)$.

The above then represents a method of constructing generator designs. Another method is to start with a BIBD$(t^*, b, k, r; \lambda)$ with $t^* > t$ and identify the treatments $t + 1, t + 2, \ldots, t^*$ with the control, that is, treatment 0 (Bechhofer and Tamhane, 1981). Any block with 0's only would be deleted. We illustrate this method with the following example.

Example 6.8 Starting with the BIBD(7, 7, 4, 4; 2)

$$\begin{array}{ccccccc} 3 & 1 & 1 & 1 & 2 & 1 & 2 \\ 5 & 4 & 2 & 2 & 3 & 3 & 4 \\ 6 & 6 & 5 & 3 & 4 & 4 & 5 \\ 7 & 7 & 7 & 6 & 7 & 5 & 6 \end{array}$$

and replacing 7 by 0 leads to the BTIBD(6, 7, 4; 2, 2):

$$\begin{array}{ccccccc} 3 & 1 & 1 & 1 & 2 & 1 & 2 \\ 5 & 4 & 2 & 2 & 3 & 3 & 4 \\ 6 & 6 & 5 & 3 & 4 & 4 & 5 \\ 0 & 0 & 0 & 6 & 0 & 5 & 6 \end{array} \tag{6.24}$$

Replacing 6 and 7 by 0 leads to the BTIBD(5, 7, 4; 4, 2):

$$\begin{array}{ccccccc} 3 & 1 & 1 & 1 & 2 & 1 & 2 \\ 5 & 4 & 2 & 2 & 3 & 3 & 4 \\ 0 & 0 & 5 & 3 & 4 & 4 & 5 \\ 0 & 0 & 0 & 0 & 0 & 5 & 0 \end{array} \tag{6.25}$$

Replacing 5, 6, and 7 by 0 leads to the BTIBD(4, 7, 4; 6, 2):

$$\begin{array}{ccccccc} 3 & 1 & 1 & 1 & 2 & 1 & 2 \\ 0 & 4 & 2 & 2 & 3 & 3 & 4 \\ 0 & 0 & 0 & 3 & 4 & 4 & 0 \\ 0 & 0 & 0 & 0 & 0 & 0 & 0 \end{array} \tag{6.26}$$

Suppose we want to construct a BTIBD(4, 8, 4; λ_0, λ_1). We could adjoin to the GD_1 given by (6.26) the $GD_2 = 4 \sum 4$ and obtain the following BTIBD(4, 8, 4; 6, 3):

$$\begin{array}{cccccccc} 3 & 1 & 1 & 1 & 2 & 1 & 2 & 1 \\ 0 & 4 & 2 & 2 & 3 & 3 & 4 & 2 \\ 0 & 0 & 0 & 3 & 4 & 4 & 0 & 3 \\ 0 & 0 & 0 & 0 & 0 & 0 & 0 & 4 \end{array} \tag{6.27}$$

We should mention that instead of replacing 5, 6, 7, by 0 we could have replaced, for example, 6, 7, and 2 by 0, yielding

$$\begin{array}{ccccccc} 0 & 0 & 0 & 0 & 0 & 0 & 1 \\ 0 & 0 & 0 & 0 & 0 & 0 & 3 \\ 3 & 1 & 1 & 1 & 3 & 4 & 4 \\ 5 & 4 & 5 & 3 & 4 & 5 & 5 \end{array}$$

Other methods of constructing generator designs are given by Bechhofer and Tamhane (1981). □

6.5.4 Partially Balanced Treatment Incomplete Block Designs

To further enrich the class of treatment control designs, an obvious step is to generalize BTIB designs to PBTIB, or *partially balanced treatment incomplete block designs*, just as BIB designs were generalized to PBIB designs. Rashed (1984) introduced two types of PBTIB designs, and we shall give some of his results below. Another form of PBTIB design was defined by Jacroux (1987) (see also Stufken, 1991).

6.5.4.1 Definitions and Structures

Definition 6.3 (Rashed, 1984) A treatment control design for t test treatments and one control is said to be a PBTIB type I design if it satisfies the following conditions:

1. The experimental units are divided into b blocks of size $k (k \leq t)$.
2. Each test treatment occurs in r blocks, but at most once in a block, and the control is replicated r_0 times with possibly multiple applications in a block.
3. The set of test treatments $T = \{1, 2, \ldots, t\}$ is divided into two disjoint subsets, T_1 and T_2, with q and $t - q$ treatments, respectively. Each test treatment occurs together in a block with the control λ_{01} or λ_{02} times, where

$$\lambda_{01} = \sum_{i=1}^{b} n_{ji} n_{0i} \quad \text{for} \quad j \in T_1$$

$$\lambda_{02} = \sum_{i=1}^{b} n_{ji} n_{0i} \quad \text{for} \quad j \in T_2$$

4. Any two treatments in T_ℓ occur together in a block $\lambda_{\ell\ell}$ times ($\ell = 1, 2$).
5. Any two treatments, one from T_1 and the other from T_2, occur together in a block λ_{12} times.

It follows then that the matrix $\boldsymbol{C}_{11}$ of (6.6) is of the form

$$\boldsymbol{C}_{11} = \begin{pmatrix} a\boldsymbol{I}_q + b\mathfrak{I}_q\mathfrak{I}'_q & e\mathfrak{I}_q\mathfrak{I}'_{\overline{q}} \\ e\mathfrak{I}'_{\overline{q}}\mathfrak{I}'_q & c\boldsymbol{I}'_{\overline{q}} + d\mathfrak{I}'_{\overline{q}}\mathfrak{I}'_{\overline{q}} \end{pmatrix} \tag{6.28}$$

with $\overline{q} = t - q$ and $a = r - r/k + \lambda_{11}/k, b = -\lambda_{11}/k, c = r - r/k + \lambda_{22}/k,$ $d = -\lambda_{11}/k$, and $e = -\lambda_{12}/k$. □

Since $\boldsymbol{C}_{11}^{-1}$ is of the same form as C_{11} of (6.28) we can state the following theorem.

Theorem 6.1 For a PBTIB type I design the variance–covariance structure of the estimators $\widehat{\tau}_j - \widehat{\tau}_0$ for $\tau_j - \tau_0$ is given by

$$\text{var}(\widehat{\tau}_j - \widehat{\tau}_0) = \begin{cases} \alpha_1^2\sigma_e^2 & \text{for} \quad j \in T_1 \\ \alpha_2^2\sigma_e^2 & \text{for} \quad j \in T_2 \end{cases}$$

$$\text{cov}(\widehat{\tau}_j - \widehat{\tau}_0, \widehat{\tau}_{j'} - \widehat{\tau}_0) = \begin{cases} \rho_1\sigma_e^2 & \text{for} \quad j, i' \in T_1 \\ \rho_2\sigma_e^2 & \text{for} \quad j, i' \in T_2 \\ \rho_3\sigma_e^2 & \text{for} \quad j' \in T_1, j' \in T_2 \end{cases}$$

where $\alpha_1^2, \alpha_2^2, \rho_1, \rho_2, \rho_3$ are functions of the design parameters (see Rashed, 1984).

An illustration of a PBTIB type I design is given in the following example.

Example 6.9 For $t = 5, b = 8, k = 3, r = 3, r_0 = 9$ the two groups are $T_1 = \{2, 3, 4\}, T_2 = \{1, 5\}$ with $\lambda_{01} = 2, \lambda_{02} = 3, \lambda_{11} = 1, \lambda_{22} = 0, \lambda_{12} = 1$. The design is given by

$$D = \begin{bmatrix} 0 & 0 & 0 & 0 & 0 & 0 & 1 & 2 \\ 0 & 0 & 0 & 1 & 3 & 4 & 2 & 3 \\ 1 & 2 & 5 & 3 & 4 & 5 & 4 & 5 \end{bmatrix}$$

and

$$\boldsymbol{C}_{11}^{-1} = \begin{pmatrix} .619 & .19 & .19 & .166 & .166 \\ .19 & .619 & .19 & .166 & .166 \\ .19 & .19 & .619 & .166 & .166 \\ .166 & .166 & .166 & .583 & .083 \\ .166 & .166 & .166 & .083 & .583 \end{pmatrix}$$

that is, $\alpha_1^2 = .619, \alpha_2^2 = .583, \rho_1 = .19, \rho_2 = .083$, and $\rho_3 = .166$. □

Another type of PBTIB design that retains the feature of two variances but allows for up to six different covariances for comparison estimators is described in the following definition.

Definition 6.4 (Rashed, 1984) A treatment control design for t test treatments and one control is said to be a PBTIB type II design if it satisfies the following conditions. Conditions 1, 2, and 3 are as in Definition 6.3:

4. Each group T_i considered by itself has either a BIB or a PBIB association structure, with at least one of these groups having PBIB design properties; that is, any two treatments in T_i occur together in either $\lambda_{ii}^{(1)}$ or $\lambda_{ii}^{(2)}$ blocks with $\lambda ii^{(1)} \neq \lambda_{ii}^{(2)}$ for at least one $i = 1, 2$.

5. Any two treatments, one from T_1 and the other from T_2, occur together in λ_{12} blocks with $\lambda_{12} = \lambda_{ii}^{(1)}$ or $\lambda_{ii}^{(2)} (i = 1, 2)$. □

We note that, if in condition 4 above both groups have BIB association structure, then the PBTIB type II design reduces to a PBTIB type I design. This gives an indication how much more general type II designs are compared to type I designs.

For type II designs Rashed (1984) proved the following theorem.

Theorem 6.2 For a PBTIB type II design the variance–covariance structure of the estimators $\widehat{\tau}_j - \widehat{\tau}_0$ for $\tau_j - \tau_0$ is given by

$$\operatorname{var}(\widehat{\tau}_j - \widehat{\tau}_0) = \begin{cases} \alpha_1^2\sigma_e^2 & \text{for} \quad j \in T_1 \\ \\ \alpha_2^2\sigma_e^2 & \text{for} \quad j \in T_2 \end{cases}$$

$$\operatorname{cov}(\widehat{\tau}_j - \widehat{\tau}_0, \widehat{\tau}_{j'} - \widehat{\tau}_0) = \begin{cases} \rho_{11}\sigma_e^2 & \text{or} \quad \rho_{12}\sigma_e^2 & \text{for} \quad j, j' \in T_1 \\ \rho_{21}\sigma_e^2 & \text{or} \quad \rho_{22}\sigma_e^2 & \text{for} \quad j, j' \in T_2 \\ \rho_{31}\sigma_e^2 & \text{or} \quad \rho_{32}\sigma_e^2 & \text{for} \quad j \in T_1, j' \in T_2 \end{cases}$$

We shall illustrate this structure with the following example.

Example 6.10 For $t = 6, b = 8, k = 3, r = 3, r = 3, r_0 = 6$, the two groups are $T_1 = \{1, 2, 5, 6\}$ and $T_2 = \{3, 4\}$, and the design is given by

$$D = \begin{bmatrix} 1 & 2 & 3 & 0 & 0 & 0 & 0 & 0 \\ 2 & 3 & 4 & 4 & 5 & 1 & 0 & 1 \\ 4 & 5 & 6 & 5 & 6 & 6 & 2 & 3 \end{bmatrix}$$

with $\lambda_{01} = 2, \lambda_{02} = 1, \lambda_{11}^{(1)} = 1, \lambda_{11}^{(2)} = 0$ (for a PBIB association structure):

0th Associate	1st Associates	2nd Associates
1	2, 6	5
2	1, 5	6
5	2, 6	1
6	1, 5	2

and $\lambda_{22}^{(1)} = 1$ (for a BIB association structure) and $\lambda_{12} = 1$. The variance–covariance matrix (if the treatments are written in the order 1, 2, 5, 6, 3, 4) is found to be

$$\boldsymbol{C}_{11}^{-1} = \begin{pmatrix} .66 & .22 & .16 & .22 & .25 & .25 \\ .22 & .66 & .22 & .16 & .25 & .25 \\ .16 & .22 & .66 & .22 & .25 & .25 \\ .22 & .16 & .22 & .66 & .25 & .25 \\ .25 & .25 & .25 & .25 & .71 & .29 \\ .25 & .25 & .25 & .25 & .29 & .71 \end{pmatrix}$$

that is, $\alpha_1^2 = .66, \alpha_2^2 = .71, \rho_{11} = .22, \rho_{12} = .16, \rho_{21} = .20, \rho_{31} = .25$. □

6.5.4.2 Construction of PBTIB Designs

We shall describe briefly one method of constructing PBTIB designs as developed by Rashed (1984). It is an extension of one of the methods described by Bechhofer and Tamhane (1981) to construct BTIB designs (see Section 6.5.3.2). Generally speaking, in order to construct a PBTIB design with t treatments in b blocks of size $k(\leq t)$, we start with a PBIB design with $t^*(>t)$ treatments in $b^*(\leq b)$ blocks of size $k^* = k$, and then replace the $t^* - t$ "excess" treatments by 0, deleting any resulting blocks which contain only the control. This is a very general recipe and may not necessarily lead to a PBTIB design unless we impose further conditions. Therefore, to be more precise, we start with a GD-PBIB design (see Section 4.6.1) and proceed as follows:

1. Write out the GD association scheme for $t^* = t_1^* \times t_2^*$ treatments, that is, an array of t_1^* rows and t_2^* columns, which we denote by T.
2. For

$$t = (t_{11} \times t_{12}) + (t_{21} \times t_{22})$$

   form two nonoverlapping subarrays $T_1(t_{11} \times t_{12})$ and $T_2(t_{21} \times t_{22})$ within T such that T_1 and T_2 each form a GD association scheme.
3. Denote the remaining set of treatments by T_0 and replace those treatments by 0 (these are the excess treatments mentioned above).

Schematically this procedure can be depicted as follows:

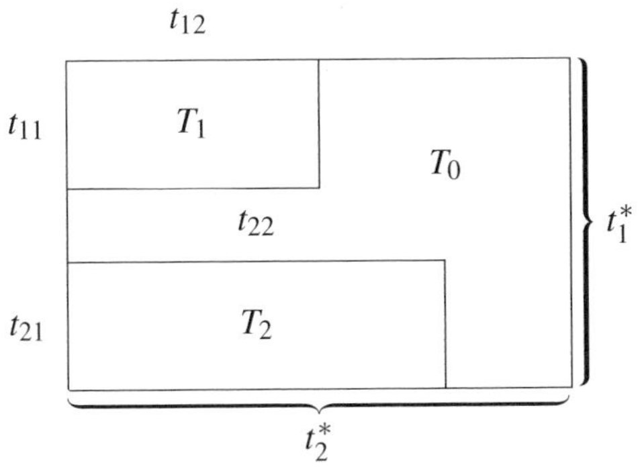

This implies that

$$t_{11} + t_{21} \leq t_1^*$$
$$t_{12} \leq t_2^*$$
$$t_{22} \leq t_2^*$$

Following steps 1–3 and returning to the starting PBIB design yields the required PBTIB design in which each treatment is replicated r times and the control is replicated r_0 times where

$$r_0 = (t_1^* t_2^* - t_{11}t_{12} - t_{21}t_{22})r = (t^* - t)r$$

Furthermore, the remaining parameters of the PBTIB design are related to those of the PBIB design as follows:

$$\lambda_{ii}^{(1)} = \lambda_1 \qquad \lambda_{ii}^{(2)} = \lambda_2 \qquad (i = 1, 2)$$

($\lambda_{ii}^{(1)}$ exists only if T_i has more than one treatment in the same row of T, and $\lambda_{ii}^{(2)}$ exists only if T_i has treatments in more than one row of T),

$$\lambda_{0i} = a_i\lambda_1 + b_i\lambda_2 \qquad (i = 1, 2)$$

where a_i equals the number of treatments that are replaced by zeros and are 1st associates of a treatment in T_i, that is, $a_i = t_2^* - t_{i2}$, and b_i equals the number of treatments that are replaced by zeros and are 2nd associates of a treatment in T_i, that is, $b_i = t^* - t - (t_2^* - t_{i2}) = t^* - t - a_i (i = 1, 2)$, and

$$\lambda_{12} = \lambda_2$$

We shall illustrate the construction of a PBTIB design with the following examples.

Example 6.11 Suppose $t = 8, k = 3$. We start with design SR 23 from Clatworthy (1973) with $t^* = 9, k = 3, r = 3, b^* = 9, \lambda_1 = 0, \lambda_2 = 1$:

$$\begin{bmatrix} 1 & 4 & 7 & 1 & 2 & 3 & 1 & 2 & 3 \\ 2 & 5 & 8 & 5 & 6 & 4 & 6 & 4 & 5 \\ 3 & 6 & 9 & 9 & 7 & 8 & 8 & 9 & 7 \end{bmatrix}$$

and association structure T:

$$\begin{matrix} 1 & 4 & 7 \\ 2 & 5 & 8 \\ 3 & 6 & 9 \end{matrix}$$

If we replace 9 by 0, that is, $T_0 = \{9\}$, then

$$T_1 = \begin{Bmatrix} 1 & 4 & 7 \\ 2 & 5 & 8 \end{Bmatrix} \qquad T_2 = \begin{Bmatrix} 3 & 6 \end{Bmatrix}$$

and the PBTIB design is given by

$$\begin{bmatrix} 1 & 4 & 7 & 1 & 2 & 3 & 1 & 2 & 3 \\ 2 & 5 & 8 & 5 & 6 & 4 & 6 & 4 & 5 \\ 3 & 6 & 0 & 0 & 7 & 8 & 8 & 0 & 7 \end{bmatrix} \tag{6.29}$$

The association schemes for T_1 and T_2 follow from the association scheme for T as

	0th Associate	1st Associates	2nd Associates
T_1	1	4, 7	2, 3, 5, 6, 8
	2	5, 8	1, 3, 4, 6, 7
	4	1, 7	2, 3, 5, 6, 8
	5	2, 8	1, 3, 4, 6, 7
	7	1, 4	2, 3, 5, 6, 8
	8	2, 5	1, 3, 4, 6, 7
T_2	3	6	1, 2, 4, 5, 7, 8
	6	3	1, 2, 4, 5, 7, 8

It follows, therefore, that the design (6.29) is a PBTIB type II design with parameters $r = r_0 = 3, \lambda_{01} = 1, \lambda_{02} = 0, \lambda_{11}^{(1)} = 0, \lambda_{11}^{(2)} = 1, \lambda_{22}^{(1)} = 0, \lambda_{12} = 1$. □

Example 6.12 Suppose $t = 7, k = 3$. We start again with design SR 23 as given in Example 6.11. We now replace 8 and 9 by 0, that is, $T_0 = \{8, 9\}$ and then have

$$T_1 = \begin{Bmatrix} 1 & 4 & 7 \end{Bmatrix} \qquad T_2 = \begin{Bmatrix} 2 & 5 \\ 3 & 6 \end{Bmatrix}$$

or, schematically,

$$
\begin{array}{ccc|cl}
\multicolumn{1}{c}{1} & \multicolumn{1}{c}{4} & \multicolumn{1}{c}{7} & \longleftarrow & T_1 \\
\hline
2 & 5 & \multicolumn{1}{|c}{8} & & \\
 & & \multicolumn{1}{|c}{} & \longleftarrow & T_0 \\
3 & 6 & \multicolumn{1}{|c}{9} & & \\
\multicolumn{2}{c}{\uparrow} & & & \\
\multicolumn{2}{c}{T_2} & & &
\end{array}
$$

The resulting PBTIB design is

$$
\begin{bmatrix}
1 & 4 & 7 & 1 & 2 & 3 & 1 & 2 & 3 \\
2 & 5 & 0 & 5 & 6 & 4 & 6 & 4 & 5 \\
3 & 6 & 0 & 0 & 7 & 0 & 0 & 0 & 7
\end{bmatrix} \tag{6.30}
$$

with parameters $r_0 = 6, \lambda_{01} = 2, \lambda_{02} = 1, \lambda_{11}^{(1)} = 0, \lambda_{22}^{(1)} = 0, \lambda_{22}^{(2)} = 1, \lambda_{12} = 1$. It is easy to verify that the design (6.30) is a PBTIB type II design.

Since there exist many GD-PBIB designs (see Clatworthy, 1973) the method described above will yield a rather rich class of PBTIB designs, most of them being of type II. □

6.5.5 Optimal Designs

In the preceding sections we have described the construction of various types of treatment control designs. For a prespecified value of the number of test treatments t and block size k, different methods may lead to different designs, some more efficient than others. In deciding which design to choose among competing designs and to use in a practical application, one consideration may be to compare their efficiencies and look, if possible, for the optimal design. In this context different optimality criteria have been used (see Section 6.5.2), based on different philosophies [for a discussion see Hedayat, Jacroux, and Majumdar (1988), and the comments following their article]. Unfortunately, different optimality criteria may lead to different optimal designs, but often the optimal design using one criterion is near optimal under another criterion.

Using the A-optimality criterion Hedayat and Majumdar (1984) provide for the class of BTIB designs a catalogue of optimal R- and S-type designs for block sizes $2 \le k^* \le 8$. Optimal R-type designs are also given by Stufken (1987). These results are based on a theorem by Majumdar and Notz (1983), which can be stated as in the following theorem (Hedayat and Majumdar, 1984).

Theorem 6.3 For given (t, b, k^*) a BTIBD$(t, b, k^*, u, s; \lambda_0, \lambda_1)$ (see Section 6.5.3.2) is A-optimal if for $x = u$ and $z = s$ the following function is minimized:

$$
\begin{aligned}
g(x, z) = (t-1)^2[btk^*(k^*-1) - (bx+z)(k^*t - t + k) + (bx^2 + 2xz + z)]^{-1} \\
+ [k^*(bx+x) - (bx^2 + 2xz + z)]^{-1}
\end{aligned}
$$

among integers $x = 0, 1, \ldots, [k^*/2]$, $z = 0, 1, \ldots, b-1$, with $[k^*/2]$ the largest integer not greater than $k^*/2$, and z positive when $x = 0$ and $z = 0$ when $x = [k^*/2]$.

A different approach to constructing optimal BTIB designs has been provided by Bechhofer and Tamhane (1983a, 1983b, 1985). Their criterion is maximization of the confidence coefficient for the joint confidence intervals of prescribed allowance (or length) for $\tau_j - \tau_0 (j = 1, 2, \ldots, t)$, namely, for $\ell > 0$,

$$P_1 = P\{\tau_j - \tau_0 \geq \widehat{\tau}_j - \widehat{\tau}_0 - \ell;\ j = 1, 2, \ldots, t\} \tag{6.31}$$

for one-sided confidence intervals, and

$$P_2 = P\{\widehat{\tau}_j - \widehat{\tau}_0 - \ell \leq \tau_j - \tau_0 \leq \widehat{\tau}_j - \widehat{\tau}_0 + \ell;\ j = 1, 2, \ldots, t\} \tag{6.32}$$

for two-sided confidence intervals. This requires that additional assumptions have to be made about the error e_{ij} in model (6.2). Assuming that the e_{ij} are i.i.d. $N(0, \sigma_e^2)$ and denoting the right-hand sides of (6.17) and (6.18) by $\alpha^2\sigma_e^2$ and $\rho\alpha^2\sigma_e^2$, respectively, then expressions for P_1 in (6.31) and P_2 in (6.32) can be written as

$$P_1 = \int_{-\infty}^{\infty} \left[\Phi\left(\frac{x\sqrt{\rho} + \ell/\alpha\sigma_e}{\sqrt{1-\rho}}\right)\right]^t d\Phi(x) \tag{6.33}$$

and

$$P_2 = \int_{-\infty}^{\infty} \left[\Phi\left(\frac{x\sqrt{\rho} + \ell/\alpha\sigma_e}{\sqrt{1-\rho}}\right) - \Phi\left(\frac{x\sqrt{\rho} - \ell/\alpha\sigma_e}{\sqrt{1-\rho}}\right)\right]^t d\Phi(x) \tag{6.34}$$

where $\Phi\ (\cdot)$ denotes the standard normal cumulative distribution function (cdf).

We note here, parenthetically, that the assumption about the e_{ij} does not agree with our underlying framework of randomization theory (see Sections 1.6 and I.10). We, therefore, consider the confidence coefficients in (6.33) and (6.34) to be approximate confidence coefficients. Nevertheless, the proposed procedures of finding optimal designs based on maximizing P_1 or P_2 are useful. We note also that for design purposes σ_e^2 is assumed to be known, based on prior experiences or theoretical subject matter considerations. But as Bechhofer and Tamhane (1985) suggest, for analysis purposes an estimate of σ_e^2, based on the analysis of variance, should be used in connection with Dunnett's procedure (Dunnett, 1955; see also I.7).

Tables of optimal BTIB designs for one-sided and two-sided comparisons for $2 \leq t \leq 6$), $k^* = 2, 3$ and a wide range of values for b and ℓ/σ_e are given by Bechhofer and Tamhane (1985). The procedure is, for a desired confidence coefficient, to combine one or more replicates from a table of minimal complete sets of generator designs.

6.6 ROW–COLUMN DESIGNS

6.6.1 Introduction

For purposes of error control or error reduction the principle of blocking is of primary importance. So far we have considered only one blocking factor, and for most situations this is quite satisfactory. There are, however, situations where blocking in more than one dimension may be needed. For example, in an agronomic or forestry trial it is not uncommon to compare a large number of hybrid varieties (which represent the treatments), which leads to a large experimental area, and it may be important to account for or eliminate the effects of fertility trends in the land in two directions. This implies that one needs to block or eliminate heterogeneity in two directions. The two blocking systems are referred to generally (borrowing terminology from field experiments) as row blocking and column blocking, and the resulting designs are referred to as *row–column designs*. Examples of such designs are the Latin square type designs of Chapter I.10, but other designs may be needed with incomplete blocks in both directions. As an example consider the following design given by John and Williams (1995) with $t = 12$ treatments in $r = 4$ rows and $c = 9$ columns:

		Column								
		1	2	3	4	5	6	7	8	9
	1	4	8	5	9	11	3	2	1	7
	2	12	11	2	3	5	10	4	8	6
Row	3	7	10	9	4	6	5	1	12	8
	4	11	9	12	6	1	7	10	3	2

Both rows and columns represent incomplete blocks, and the row design and column design, which we refer to as the component designs, are characterized by their respective incidence matrices, $\boldsymbol{N}_\rho$ and $\boldsymbol{N}_\gamma$ say. We shall see later (see Section 6.6.2) that the properties of the row–column design depend on $\boldsymbol{N}_\rho$ and $\boldsymbol{N}_\gamma$, but that it is not straightforward how to combine the component designs to produce a "good" final design. We shall return to this point in Section 6.6.3 after discussing the analysis of a row–column design in Section 6.6.2.

6.6.2 Model and Normal Equations

Consider a row–column design with r rows, c columns, and t treatments where treatment k is replicated r_k times ($k = 1, 2, \ldots, t$). Let y_{ijk} denote the observation for the experimental unit in the ith row and jth column to which treatment k has been applied. Based on the assumption of unit treatment additivity (see Section I.10.2.3), a model for y_{ijk} can be derived [despite criticism by Srivastava (1993, 1996), and Srivastava and Wang (1998)] and written as

$$y_{ijk} = \mu + \rho_i + \gamma_j + \tau_k + e_{ijk} \tag{6.35}$$

where $\rho_i (i = 1, 2, \ldots, r)$, $\gamma_j (j = 1, 2, \ldots, c)$, $\tau_k (k = 1, 2, \ldots, t)$ represent the row, column, and treatment effects, respectively, which are defined such that $\sum \rho_i = 0, \sum \gamma_j = 0, \sum \tau_k = 0$ (see Section I.10.2.3). The e_{ijk} represent the combined experimental and observational error. Let us rewrite (6.35) in the usual fashion in matrix notation as follows:

$$\boldsymbol{y} = \mu \mathfrak{I} + \boldsymbol{X}_\rho \boldsymbol{\rho} + \boldsymbol{X}_\gamma \boldsymbol{\gamma} + \boldsymbol{X}_\tau \boldsymbol{\tau} + \boldsymbol{e} \tag{6.36}$$

where $\boldsymbol{\rho}, \boldsymbol{\gamma}, \boldsymbol{\tau}$ represent the row, column, and treatment effect vectors, respectively, and $\boldsymbol{X}_\rho, \boldsymbol{X}_\gamma, \boldsymbol{X}_\tau$ are known matrices linking the row, column, and treatment effects to the observations. If we write, for example, the observation $\boldsymbol{y}$ such that the first c components are the observation in row 1, the next c from row 2, and so forth, then we have

$$\boldsymbol{X}_\rho = \begin{pmatrix} \mathfrak{I}_c & & & \\ & \mathfrak{I}_c & & \\ & & \ddots & \\ & & & \mathfrak{I}_c \end{pmatrix} \tag{6.37}$$

of dimension $rc \times r$, and

$$\boldsymbol{X}_\gamma = \begin{pmatrix} \boldsymbol{I}_c \\ \boldsymbol{I}_c \\ \cdots \\ \boldsymbol{I}_c \end{pmatrix} \tag{6.38}$$

of dimension $rc \times c$. Further, the relationships between $\boldsymbol{X}_\rho$, $\boldsymbol{X}_\gamma$, and $\boldsymbol{X}_\tau$ are

$$\boldsymbol{X}'_\tau \boldsymbol{X}_\rho = \boldsymbol{N}_\rho \qquad \boldsymbol{X}'_\tau \boldsymbol{X}_\gamma = \boldsymbol{N}_\gamma \tag{6.39}$$

Using (6.37), (6.38), and (6.39), it is then easy to write the NE for model (6.36) as

$$\begin{pmatrix} rc & c\mathfrak{I}'_r & r\mathfrak{I}'_c & \mathbf{r}' \\ c\mathfrak{I}_r & c\boldsymbol{I}_r & \mathfrak{I}_r\mathfrak{I}'_c & \boldsymbol{N}'_\rho \\ r\mathfrak{I}_c & \mathfrak{I}_c\mathfrak{I}'_r & r\boldsymbol{I}_c & \boldsymbol{N}'_\gamma \\ \boldsymbol{r} & \boldsymbol{N}_\rho & \boldsymbol{N}_\gamma & \boldsymbol{r}^\delta \end{pmatrix} \begin{pmatrix} \widehat{\mu} \\ \widehat{\boldsymbol{\rho}} \\ \widehat{\boldsymbol{\gamma}} \\ \widehat{\boldsymbol{\tau}} \end{pmatrix} = \begin{pmatrix} \mathfrak{I}'_{rc}\boldsymbol{y} \\ \boldsymbol{X}'_\rho \boldsymbol{y} \\ \boldsymbol{X}'_\gamma \boldsymbol{y} \\ \boldsymbol{X}'_\tau \boldsymbol{y} \end{pmatrix} \tag{6.40}$$

where $\boldsymbol{r}' = (r_1, r_2, \ldots, r_t)$ and $\boldsymbol{r}^\delta = \text{diag}(r_1, r_2, \ldots, r_t)$.

Similar to Section 1.3.2 we shall derive from (6.40) the reduced NE as follows: Using $\sum \widehat{\rho}_i = 0$ and $\sum \widehat{\gamma}_j = 0$ (as suggested by $\sum \rho_i = 0$ and $\sum \gamma_j = 0$), we obtain from the first equation of (6.40)

$$\widehat{\mu} = \frac{G}{n} - \frac{\boldsymbol{r}'\widehat{\boldsymbol{\tau}}}{n} \tag{6.41}$$

where we are using the following obvious notation for the right-hand side of (6.40):

$$\begin{pmatrix} \mathfrak{I}'_{rc}\boldsymbol{y} \\ \boldsymbol{X}'_{\rho}\boldsymbol{y} \\ \boldsymbol{X}'_{\gamma}\boldsymbol{y} \\ \boldsymbol{X}'_{\tau}\boldsymbol{y} \end{pmatrix} = \begin{pmatrix} G \\ \boldsymbol{P} \\ \boldsymbol{\Gamma} \\ \boldsymbol{T} \end{pmatrix} \tag{6.42}$$

and $rc = n$. From the second and third set of the NE (6.40) we obtain

$$c\widehat{\boldsymbol{\rho}} = \boldsymbol{P} - c\widehat{\mu}\mathfrak{I}_r - \boldsymbol{N}'_{\rho}\widehat{\boldsymbol{\tau}} \tag{6.43}$$

and

$$r\widehat{\boldsymbol{\gamma}} = \boldsymbol{\Gamma} - r\widehat{\mu}\mathfrak{I}_c - \boldsymbol{N}'_{\gamma}\widehat{\boldsymbol{\tau}} \tag{6.44}$$

Substituting (6.41), (6.43), and (6.44) into the fourth set of equations in (6.40) and simplifying, we obtain the RNE for $\boldsymbol{\tau}$ as

$$\left(\boldsymbol{r}^{\delta} - \frac{1}{c}\boldsymbol{N}_{\rho}\boldsymbol{N}'_{\rho} - \frac{1}{r}\boldsymbol{N}_{\gamma}\boldsymbol{N}\gamma' + \frac{\boldsymbol{r}\boldsymbol{r}'}{n}\right)\widehat{\boldsymbol{\tau}} = \left(\boldsymbol{T} - \frac{1}{c}\boldsymbol{N}_{\rho}\boldsymbol{P} - \frac{1}{r}\boldsymbol{N}_{\gamma}\boldsymbol{\Gamma} + \boldsymbol{r}\frac{G}{n}\right) \tag{6.45}$$

which we write in abbreviated form as

$$\boldsymbol{C}\widehat{\tau} = \boldsymbol{Q} \tag{6.46}$$

with $\boldsymbol{C}$ and $\boldsymbol{Q}$ as defined in (6.45). A solution to (6.46) can be written as

$$\widehat{\boldsymbol{\tau}} = \boldsymbol{C}^{-}\boldsymbol{Q} \tag{6.47}$$

where $\boldsymbol{C}^{-}$ is a generalized inverse of $\boldsymbol{C}$.

Any statistical inference about the treatment effects is then derived, similar to the developments in Chapter 1, from (6.46) and (6.47), in conjunction with information from the analysis of variance (see Section 6.6.3). For example, rank $(\boldsymbol{C})(\leq t-1)$ determines the number of linearly independent estimable functions, $\boldsymbol{c}'\boldsymbol{\tau}$, of the treatment effects. For any estimable function $\boldsymbol{c}'\boldsymbol{\tau}$ we have $\widehat{\boldsymbol{c}'\boldsymbol{\tau}} = \boldsymbol{c}'\widehat{\boldsymbol{\tau}}$, with $\widehat{\boldsymbol{\tau}}$ from (6.47), and

$$\mathrm{var}(\boldsymbol{c}'\widehat{\boldsymbol{\tau}}) = \boldsymbol{c}'\boldsymbol{C}^{-}\boldsymbol{c}\sigma_e^2$$

6.6.3 Analysis of Variance

Following general principles (see, e.g., Sections I.4.11 and 1.3.6), it is easy to write out the analysis of variance table for a row–column design that will enable

Table 6.2 Analysis of Variance for Row–Column Design

Source	d.f.	SS
$X_\rho \mid \mathfrak{I}$	$r-1$	$c\sum_{i=1}^{r}(\overline{y}_{i..}-\overline{y}_{...})^2$
$X_\gamma \mid \mathfrak{I}, X_\rho$	$c-1$	$r\sum_{j=1}^{c}(\overline{y}_{.j.}-\overline{y}_{...})^2$
$X_\tau \mid \mathfrak{I}, X_\rho, X_\gamma$	$t-1$[a]	$\widehat{\boldsymbol{\tau}}'\boldsymbol{Q}$
$\boldsymbol{I} \mid \mathfrak{I}, X_\rho, X_\gamma, X_\tau$	Difference $\equiv \nu_E$	Difference $\equiv$ SS(E)
Total	$rc-1$	$\sum_{i,j,k}(y_{ijk}-\overline{y}_{...})^2$

[a] Assuming a connected design.

us to (i) estimate σ_e^2 and (ii) obtain an (approximate) F test for testing H_0: $\tau_1 = \tau_2 = \cdots = \tau_t = 0$. Such an ANOVA table is given in Table 6.2, using an earlier established notation. We then have

$$\widehat{\sigma}_e^2 = \text{MS}(E) = \frac{\text{SS}(E)}{\nu_E}$$

and

$$F = \frac{\widehat{\boldsymbol{\tau}}'\boldsymbol{Q}/(t-1)}{\text{MS}(E)}$$

provides an approximate test for testing H_0: $\tau_1 = \tau_2 = \cdots = \tau_t = 0$.

It should be mentioned (as is, of course, obvious from the notation used) that the ANOVA in Table 6.2 is a sequential (or type I in SAS terminology) ANOVA. The nature of the design implies immediately that $\text{SS}(X_\gamma \mid \mathfrak{I}, X_\rho) = \text{SS}(X_\gamma \mid \mathfrak{I})$, and hence the simple expression for $\text{SS}(X_\gamma \mid \mathfrak{I}, X_\rho)$ given in Table 6.2.

6.6.4 An Example

We shall illustrate the analysis of data from a row–column design, using the design of Section 6.1 and using SAS PROC GLM. The data and analysis are given in Table 6.3. We make the following comments about the analysis and the SAS output in Table 6.3:

1. The inv-option in the model statement provides a generalized inverse to the coefficient matrix of the NE (6.40), and the 12×12 submatrix given by the rows and columns $T_1, T_2, \ldots, T_{12}$ serve as a $\boldsymbol{C}^-$ for the RNE (6.46).

Table 6.3 Data and Analysis of Row–Column Design ($t = 12, r = 4, c = 9$)

```
options nodate pageno=1;
data rowcoll;
input R C T y @@;
datalines;
1 1 4 21  1 2 8 18  1 3 5 14  1 4 9 15
1 5 11 9  1 6 3 17  1 7 2 24  1 8 1 22
1 9 7 13  2 1 12 8  2 2 11 11  2 3 2 20
2 4 3 25  2 5 5 17  2 6 10 24  2 7 4 14
2 8 8 19  2 9 6 25  3 1 7 30  3 2 10 31
3 3 9 36  3 4 4 29  3 5 6 30  3 6 5 35
3 7 1 29  3 8 12 28  3 9 8 25  4 1 11 40
4 2 9 41  4 3 12 43  4 4 6 39  4 5 1 42
4 6 7 50  4 7 10 47  4 8 3 44  4 9 2 39
;
run;

proc print data=rowcoll;
title1 'TABLE 6.3';
title2 'DATA ROW-COLUMN DESIGN';
title3 'WITH t=12 TREATMENTS, r=4 ROWS, c=9 COLUMNS';
run;

proc glm data=rowcoll;
class R C T;
model y = R C T/inv e;
lsmeans T/stderr;
estimate 'T1-T2' T 1 -1;
estimate 'T1-T3' T 1 0 -1;
estimate 'T1-T4' T 1 0 0 -1;
```

Table 6.3 (*Continued*)

```
estimate 'T1-T5' T 1 0 0 0 -1;
estimate 'T1-T6' T 1 0 0 0 0 -1;
estimate 'T1-T7' T 1 0 0 0 0 0 -1;
estimate 'T1-T8' T 1 0 0 0 0 0 0 -1;
estimate 'T1-T9' T 1 0 0 0 0 0 0 0 -1;
estimate 'T1-T10' T 1 0 0 0 0 0 0 0 0 -1;
estimate 'T1-T11' T 1 0 0 0 0 0 0 0 0 0 -1;
estimate 'T1-T12' T 1 0 0 0 0 0 0 0 0 0 0 -1;
title2 'ANALYSIS OF VARIANCE';
title3 'FOR ROW-COLUMN DESIGN';
run;
```

Obs	R	C	T	Y
1	1	1	4	21
2	1	2	8	18
3	1	3	5	14
4	1	4	9	15
5	1	5	11	9
6	1	6	3	17
7	1	7	2	24
8	1	8	1	22
9	1	9	7	13
10	2	1	12	8
11	2	2	11	11
12	2	3	2	20
13	2	4	3	25
14	2	5	5	17

15	2	6	10	24
16	2	7	4	14
17	2	8	8	19
18	2	9	6	25
19	3	1	7	30
20	3	2	10	31
21	3	3	9	36
22	3	4	4	29
23	3	5	6	30
24	3	6	5	35
25	3	7	1	29
26	3	8	12	28
27	3	9	8	25
28	4	1	11	40
29	4	2	9	41
30	4	3	12	43
31	4	4	6	39
32	4	5	1	42
33	4	6	7	50
34	4	7	10	47
35	4	8	3	44
36	4	9	2	39

ANALYSIS OF VARIANCE FOR ROW-COLUMN DESIGN
The GLM Procedure
Class Level Information

Class	Levels	Values
R	4	1 2 3 4
C	9	1 2 3 4 5 6 7 8 9
T	12	1 2 3 4 5 6 7 8 9 10 11 12

Number of observations 36

X'X Generalized Inverse (g2)

	Intercept	R 1	R 2	R 3	R 4	C 1
Intercept	0.8611111111	-0.083333333	-0.125	-0.125	0	-0.444444444
R 1	-0.083333333	0.25	0.125	0.125	0	-5.68289E-19
R 2	-0.125	0.125	0.25	0.125	0	-9.73435E-17
R 3	-0.125	0.125	0.125	0.25	0	-8.46741E-17
R 4	0	0	0	0	0	0
C 1	-0.444444444	-5.68289E-19	-9.73435E-17	-8.46741E-17	0	0.6666666667
C 2	-0.333333333	-6.06814E-18	-6.94448E-17	-6.61156E-17	0	0.3333333333
C 3	-0.444444444	-3.65495E-19	-9.88873E-17	-8.6178E-17	0	0.3333333333
C 4	-0.333333333	1.010807E-17	-5.45625E-17	-5.73645E-17	0	0.3333333333
C 5	-0.333333333	-8.63974E-19	-7.38351E-17	-5.91767E-17	0	0.3333333333
C 6	-0.333333333	5.078598E-18	-6.80636E-17	-6.29509E-17	0	0.3333333333
C 7	-0.333333333	5.223935E-18	-6.84973E-17	-6.39636E-17	0	0.3333333333
C 8	-0.444444444	-1.12406E-17	-1.07621E-16	-8.74456E-17	0	0.3333333333
C 9	0	0	0	0	0	0

T 1	-0.421296296	-0.041666667	0.0416666667	-8.68327E-17	0	0.1111111111
T 2	-0.532407407	-0.041666667	-1.56588E-16	0.0416666667	0	0.2222222222
T 3	-0.421296296	-0.041666667	-1.31958E-16	0.0416666667	0	0.1111111111
T 4	-0.37962963	-0.083333333	-0.041666667	-0.041666667	0	-8.59604E-16
T 5	-0.37962963	-0.083333333	-0.041666667	-0.041666667	0	0.1111111111
T 6	-0.555555556	-2.16298E-17	-1.63766E-16	-1.12908E-16	0	0.2222222222
T 7	-0.532407407	-0.041666667	0.0416666667	-1.05476E-16	0	0.1111111111
T 8	-0.490740741	-0.083333333	-0.041666667	-0.041666667	0	0.2222222222
T 9	-0.421296296	-0.041666667	0.0416666667	-8.40883E-17	0	0.1111111111
T 10	-0.444444444	-2.56719E-17	-1.38442E-16	-8.45796E-17	0	0.1111111111
T 11	-0.421296296	-0.041666667	-1.28995E-16	0.0416666667	0	-7.20447E-16
T 12	0	0	0	0	0	0
Y	34.208333333	-25.54166667	-24.16666667	-12.45833333	0	3

X'X Generalized Inverse (g2)

	C 2	C 3	C 4	C 5	C 6	C 7
Intercept	-0.333333333	-0.444444444	-0.333333333	-0.333333333	-0.333333333	-0.333333333
R 1	-6.06814E-18	-3.65495E-19	1.010807E-17	-8.63974E-19	5.078598E-18	5.223935E-18
R 2	-6.94448E-17	-9.88873E-17	-5.45625E-17	-7.38351E-17	-6.80636E-17	-6.84973E-17
R 3	-6.61156E-17	-8.6178E-17	-5.73645E-17	-5.91767E-17	-6.29509E-17	-6.39636E-17
R 4	0	0	0	0	0	0
C 1	0.3333333333	0.3333333333	0.3333333333	0.3333333333	0.3333333333	0.3333333333
C 2	0.6666666667	0.3333333333	0.3333333333	0.3333333333	0.3333333333	0.3333333333
C 3	0.3333333333	0.6666666667	0.3333333333	0.3333333333	0.3333333333	0.3333333333
C 4	0.3333333333	0.3333333333	0.6666666667	0.3333333333	0.3333333333	0.3333333333
C 5	0.3333333333	0.3333333333	0.3333333333	0.6666666667	0.3333333333	0.3333333333
C 6	0.3333333333	0.3333333333	0.3333333333	0.3333333333	0.6666666667	0.3333333333
C 7	0.3333333333	0.3333333333	0.3333333333	0.3333333333	0.3333333333	0.6666666667

Table 6.3 (*Continued*)

X'X Generalized Inverse (g2)

	C 2	C 3	C 4	C 5	C 6	C 7
C 8	0.3333333333	0.3333333333	0.3333333333	0.3333333333	0.3333333333	0.3333333333
C 9	0	0	0	0	0	0
T 1	-6.6489E-16	0.1111111111	-7.25482E-16	-0.111111111	-7.19612E-16	-0.111111111
T 2	0.1111111111	0.1111111111	0.1111111111	0.1111111111	0.1111111111	-6.41164E-16
T 3	-6.59544E-16	0.1111111111	-0.111111111	-6.80361E-16	-0.111111111	-7.03875E-16
T 4	-7.08048E-16	0.1111111111	-0.111111111	-7.21926E-16	-7.52338E-16	-0.111111111
T 5	-6.10521E-16	-7.265E-16	-6.70087E-16	-0.111111111	-0.111111111	-6.53118E-16
T 6	0.1111111111	0.2222222222	-7.08466E-16	-6.35922E-16	0.1111111111	0.1111111111
T 7	0.1111111111	0.2222222222	0.1111111111	0.1111111111	-7.09929E-16	0.1111111111
T 8	-6.29051E-16	0.2222222222	0.1111111111	0.1111111111	0.1111111111	0.1111111111
T 9	-0.111111111	-7.48079E-16	-0.111111111	-6.44453E-16	-6.91857E-16	-6.73674E-16
T 10	-0.111111111	0.1111111111	-7.945E-16	-7.4153E-16	-0.111111111	-0.111111111
T 11	-0.111111111	0.1111111111	-6.75091E-16	-0.111111111	-6.70031E-16	-6.46005E-16
T 12	0	0	0	0	0	0
Y	1.4444444444	5	1.5555555556	0.7777777778	6.3333333333	2.8888888889

X'X Generalized Inverse (g2)

	C 8	C 9	T 1	T 2	T 3	T 4
Intercept	-0.444444444	0	-0.421296296	-0.532407407	-0.421296296	-0.37962963
R 1	-1.12406E-17	0	-0.041666667	-0.041666667	-0.041666667	-0.083333333
R 2	-1.07621E-16	0	0.0416666667	-1.56588E-16	-1.31958E-16	-0.041666667
R 3	-8.74456E-17	0	-8.68327E-17	0.0416666667	0.0416666667	-0.041666667

R 4	0	0	0	0	0	0
C 1	0.3333333333	0	0.1111111111	0.2222222222	0.1111111111	-8.59604E-16
C 2	0.3333333333	0	-6.6489E-16	0.1111111111	-6.59544E-16	-7.08048E-16
C 3	0.3333333333	0	0.1111111111	0.1111111111	0.1111111111	0.1111111111
C 4	0.3333333333	0	-7.25482E-16	0.1111111111	-0.111111111	-0.111111111
C 5	0.3333333333	0	-0.1111111111	0.1111111111	-6.80361E-16	-7.21926E-16
C 6	0.3333333333	0	-7.19612E-16	0.1111111111	-0.111111111	-7.52338E-16
C 7	0.3333333333	0	-0.111111111	-6.41164E-16	-7.03875E-16	-0.111111111
C 8	0.6666666667	0	-7.15563E-16	0.2222222222	-7.50593E-16	0.1111111111
C 9	0	0	0	0	0	0
T 1	-7.15563E-16	0	0.8425925926	0.4212962963	0.4212962963	0.4212962963
T 2	0.2222222222	0	0.4212962963	0.8425925926	0.3981481481	0.4212962963
T 3	-7.50593E-16	0	0.4212962963	0.3981481481	0.8425925926	0.4212962963
T 4	0.1111111111	0	0.4212962963	0.4212962963	0.4212962963	0.8425925926
T 5	0.1111111111	0	0.4212962963	0.4212962963	0.4212962963	0.3981481481
T 6	0.2222222222	0	0.4444444444	0.4444444444	0.4444444444	0.4444444444
T 7	0.2222222222	0	0.3981481481	0.4212962963	0.4212962963	0.4212962963
T 8	0.1111111111	0	0.4212962963	0.4212962963	0.4212962963	0.3981481481
T 9	0.1111111111	0	0.3981481481	0.4212962963	0.4212962963	0.4212962963
T 10	0.1111111111	0	0.4444444444	0.4444444444	0.4444444444	0.4444444444

X'X Generalized Inverse (g2)

	C 8	C 9	T 1	T 2	T 3	T4
T 11	0.1111111111	0	0.4212962963	0.3981481481	0.3981481481	0.4212962963
T 12	0	0	0	0	0	0
Y	5	0	6.5694444444	7.3981481481	6.7314814815	5.3657407407

Table 6.3 *(Continued)*

X'X Generalized Inverse (g2)

	T 5	T 6	T 7	T 8	T 9	T 10
Intercept	-0.37962963	-0.555555556	-0.532407407	-0.490740741	-0.421296296	-0.444444444
R 1	-0.083333333	-2.16298E-17	-0.041666667	-0.083333333	-0.041666667	-2.56719E-17
R 2	-0.041666667	-1.63766E-16	0.0416666667	-0.041666667	0.0416666667	-1.38442E-16
R 3	-0.041666667	-1.12908E-16	-1.05476E-16	-0.041666667	-8.40883E-17	-8.45796E-17
R 4	0	0	0	0	0	0
C 1	0.1111111111	0.2222222222	0.1111111111	0.2222222222	0.1111111111	0.1111111111
C 2	-6.10521E-16	0.1111111111	0.1111111111	-6.29051E-16	-0.111111111	-0.1111111111
C 3	-7.265E-16	0.2222222222	0.2222222222	0.2222222222	-7.48079E-16	0.1111111111
C 4	-6.70087E-16	-7.08466E-16	0.1111111111	0.1111111111	-0.1111111111	-7.945E-16
C 5	-0.111111111	-6.35922E-16	0.1111111111	0.1111111111	-6.44453E-16	-7.4153E-16
C 6	-0.111111111	0.1111111111	-7.09929E-16	0.1111111111	-6.91857E-16	-0.111111111
C 7	-6.53118E-16	0.1111111111	0.1111111111	0.1111111111	-6.73674E-16	-0.111111111
C 8	0.1111111111	0.2222222222	0.2222222222	0.1111111111	0.1111111111	0.1111111111
C 9	0	0	0	0	0	0
T 1	0.4212962963	0.4444444444	0.3981481481	0.4212962963	0.3981481481	0.4444444444
T 2	0.4212962963	0.4444444444	0.4212962963	0.4212962963	0.4212962963	0.4444444444
T 3	0.4212962963	0.4444444444	0.4212962963	0.4212962963	0.4212962963	0.4444444444
T 4	0.3981481481	0.4444444444	0.4212962963	0.3981481481	0.4212962963	0.4444444444
T 5	0.8425925926	0.4444444444	0.4212962963	0.3981481481	0.4212962963	0.4444444444
T 6	0.4444444444	0.8888888889	0.4444444444	0.4444444444	0.4444444444	0.4444444444
T 7	0.4212962963	0.4444444444	0.8425925926	0.4212962963	0.3981481481	0.4444444444
T 8	0.3981481481	0.4444444444	0.4212962963	0.8425925926	0.4212962963	0.4444444444
T 9	0.4212962963	0.4444444444	0.3981481481	0.4212962963	0.8425925926	0.4444444444
T 10	0.4444444444	0.4444444444	0.4444444444	0.4444444444	0.4444444444	0.8888888889

T 11	0.4212962963	0.4444444444	0.4212962963	0.4212962963	0.4212962963	0.4444444444
T 12	0	0	0	0	0	0
y	4.4768518519	8.5555555556	6.3472222222	5.0324074074	6.4583333333	8.4444444444

X'X Generalized Inverse (g2)

	T 11	T 12	y
Intercept	-0.421296296	0	34.208333333
R 1	-0.041666667	0	-25.54166667
R 2	-1.28995E-16	0	-24.16666667
R 3	0.0416666667	0	-12.45833333
R 4	0	0	0
C 1	-7.20447E-16	0	3
C 2	-0.111111111	0	1.4444444444
C 3	0.1111111111	0	5
C 4	-6.75091E-16	0	1.5555555556
C 5	-0.111111111	0	0.7777777778
C 6	-6.70031E-16	0	6.3333333333
C 7	-6.46005E-16	0	2.8888888889
C 8	0.1111111111	0	5
C 9	0	0	0
T 1	0.4212962963	0	6.5694444444
T 2	0.3981481481	0	7.3981481481
T 3	0.3981481481	0	6.7314814815
T 4	0.4212962963	0	5.3657407407

Table 6.3 *(Continued)*

T 5	0.4212962963	0	4.4768518519
T 6	0.4444444444	0	8.5555555556
T 7	0.4212962963	0	6.3472222222
T 8	0.4212962963	0	5.0324074074
T 9	0.4212962963	0	6.4583333333
T 10	0.4444444444	0	8.4444444444
T 11	0.8425925926	0	0.6203703704
T 12	0	0	0
y	0.6203703704	0	338.76851852

General Form of Estimable Functions

Effect		Coefficients
Intercept		L1
R	1	L2
R	2	L3
R	3	L4
R	4	L1-L2-L3-L4
C	1	L6
C	2	L7
C	3	L8
C	4	L9
C	5	L10
C	6	L11
C	7	L12

```
      C     8      L13
      C     9      L1-L6-L7-L8-L9-L10-L11-L12-L13

      T     1      L15
      T     2      L16
      T     3      L17
      T     4      L18
      T     5      L19
      T     6      L20
      T     7      L21
      T     8      L22
      T     9      L23
      T     10     L24
      T     11     L25
      T     12     L1-L15-L16-L17-L18-L19-L20-L21-L22-L23-L24-L25

Dependent Variable: y

Source              DF    Sum of Squares    Mean Square    F Value    Pr > F

Model               22       4305.120370     195.687290       7.51    0.0003
Error               13        338.768519      26.059117
Corrected Total     35       4643.888889

               R-Square     Coeff Var     Root MSE      y Mean

               0.927051      18.86789     5.104813    27.05556
```

Table 6.3 *(Continued)*

Source	DF	Type I SS	Mean Square	F Value	Pr > F
R	3	3951.444444	1317.148148	50.54	<.0001
C	8	168.888889	21.111111	0.81	0.6062
T	11	184.787037	16.798822	0.64	0.7638

Source	DF	Type III SS	Mean Square	F Value	Pr > F
R	3	3403.527778	1134.509259	43.54	<.0001
C	8	112.370370	14.046296	0.54	0.8077
T	11	184.787037	16.798822	0.64	0.7638

Least Squares Means

T	y LSMEAN	Standard Error	Pr > |t|
1	28.1250000	3.2997161	<.0001
2	28.9537037	3.2997161	<.0001
3	28.2870370	3.2997161	<.0001
4	26.9212963	3.2997161	<.0001
5	26.0324074	3.2997161	<.0001
6	30.1111111	3.2997161	<.0001
7	27.9027778	3.2997161	<.0001
8	26.5879630	3.2997161	<.0001

9	28.0138889	3.2997161	<.0001
10	30.0000000	3.2997161	<.0001
11	22.1759259	3.2997161	<.0001
12	21.5555556	3.2997161	<.0001

Dependent Variable: y

Parameter	Estimate	Standard Error	t Value	Pr > |t|
T1-T2	-0.82870370	4.68585305	-0.18	0.8623
T1-T3	-0.16203704	4.68585305	-0.03	0.9729
T1-T4	1.20370370	4.68585305	0.26	0.8013
T1-T5	2.09259259	4.68585305	0.45	0.6625
T1-T6	-1.98611111	4.68585305	-0.42	0.6786
T1-T7	0.22222222	4.81286395	0.05	0.9639
T1-T8	1.53703704	4.68585305	0.33	0.7481
T1-T9	0.11111111	4.81286395	0.02	0.9819
T1-T10	-1.87500000	4.68585305	-0.40	0.6956
T1-T11	5.94907407	4.68585305	1.27	0.2265
T1-T12	6.56944444	4.68585305	1.40	0.1843

2. The e-option in the model statement provides a list of linearly independent estimable functions. The output shows that this design is a connected design, that is, all treatment comparisons are estimable.
3. The P-values for R and C should be ignored (see Section I.10.2.5).
4. All treatment least-squares means have the same standard error because of the same number of replications.
5. Not all simple treatment comparisons have the same standard error. This has to do with the fact that $\boldsymbol{N}_\rho$ and $\boldsymbol{N}_\gamma$ represent incomplete block designs. For example, treatment 1 occurs together with treatments 7 and 9 three times in the same row, whereas it occurs together twice in the same row with all other treatments; but treatment 1 never occurs together with treatments 7 and 9 in the same column, whereas it occurs together once in the same column with all other treatments. Hence there are two different variances and, hence, two different standard errors for simple treatment comparisons, for example, 4.81 and 4.68. We add here, however, that, in spite of their appearance, neither $\boldsymbol{N}_\rho$ nor $\boldsymbol{N}_\gamma$ represent PBIB(2) designs.

6.6.5 Regular Row–Column Designs

As we have already seen in the previous sections, a row–column design consists of the amalgamation of two component designs, characterized by $\boldsymbol{N}_\rho$ and $\boldsymbol{N}_\gamma$, the row and column designs, respectively. For example, the row–column designs in Table 6.4 can be characterized (prior to an appropriate rearrangement of the treatments) easily in terms of well-defined error-control designs. We might refer to these designs as *regular* row–column designs.

6.6.6 Doubly Incomplete Row–Column Designs

One feature that all the designs in Table 6.4 have in common is that the number of columns, that is, the number of experimental units in each row, is equal to a multiple of the number of treatments. Of considerable interest, however, are designs for which the number of rows and columns is less than the number of treatments. A natural way to construct such designs would be to amalgamate two PBIB designs.

Example 6.13 For $t = 8, r = 4, c = 4$ Eccleston and Russell (1975) give the following design:

		Column 1	2	3	4
Row	1	1	2	5	6
	2	3	4	7	8
	3	7	6	2	3
	4	5	8	4	1

Table 6.4 Regular Row–Column Designs

Row Design	Column Design	Row–Column Design	Reference
RCBD	RCBD	Latin square	I.10.2
RCBD	GRBD	Latin square	I.10.4
GRBD	GRBD	Frequency square	Hedayat and Seiden (1970); Hedayat, Raghavarao, and Seiden (1975)
RCBD	BIBD	Youden square	I. 10.5
RCBD	PBIBD	Youden square	I.10.5
RCBD	Extended block design[a]	Incomplete Latin square	Cochran and Cox (1957)
BBD	BBD[b]	Generalized Youden design	Kiefer (1975a, b)

[a] See I.9.8.5.
[b] See Definition 19.9.

By inspection we find that both row and column designs are PBIB(3) designs with the following association schemes:

Row design:

0th Associate	**1st Associates**	**2nd Associates**	**3rd Associates**
1	2, 4, 6, 8	5	3, 7
2	3, 5, 7, 1	6	4, 8
3	4, 6, 8, 2	7	5, 1
4	5, 7, 1, 3	8	6, 2
5	6, 8, 2, 4	1	7, 3
6	7, 1, 3, 5	2	8, 4
7	8, 2, 4, 6	3	1, 5
8	1, 3, 5, 7	4	2, 6

with $\lambda_1 = 1, \lambda_2 = 2, \lambda_3 = 0$, and

$$\boldsymbol{P}_1 = \begin{pmatrix} 0 & 1 & 0 & 0 \\ 1 & 0 & 1 & 2 \\ 0 & 1 & 0 & 0 \\ 0 & 2 & 0 & 0 \end{pmatrix} \qquad \boldsymbol{P}_2 = \begin{pmatrix} 0 & 0 & 1 & 0 \\ 0 & 4 & 0 & 0 \\ 1 & 0 & 0 & 0 \\ 0 & 0 & 0 & 2 \end{pmatrix} \qquad \boldsymbol{P}_3 = \begin{pmatrix} 0 & 0 & 0 & 1 \\ 0 & 4 & 0 & 0 \\ 0 & 0 & 0 & 1 \\ 1 & 0 & 1 & 0 \end{pmatrix}$$

Column design:

0th Associate	1st Associates	2nd Associates	3rd Associates
1	5, 6, 7, 8	3	2, 4
2	5, 6, 7, 8	4	1, 3
3	5, 6, 7, 8	1	2, 4
4	5, 6, 7, 8	2	1, 3
5	1, 2, 3, 4	7	6, 8
6	1, 2, 3, 4	8	5, 7
7	1, 2, 3, 4	5	6, 8
8	1, 2, 3, 4	6	5, 7

with $\lambda_1 = 1, \lambda_2 = 2, \lambda_3 = 0$, and $\boldsymbol{P}$ matrices the same as for the row design. □

6.6.7 Properties of Row–Column Designs

We shall restrict ourselves to designs for which each treatment is replicated the same number of times, say r_0. Moreover, we shall consider a subclass of designs, denoted by D_0 by Shah (1977), with the property that each row of the design has r_0 treatments in common with each column (this holds for all the designs discussed so far). In terms of the component design incidence matrices, this condition can be expressed as

$$\boldsymbol{N}'_\rho \boldsymbol{N}_\gamma = r_0 \mathfrak{I}_r \mathfrak{I}'_c \tag{6.48}$$

The condition (6.48) is equivalent to the property of *adjusted orthogonality* for the row–column design (John and Eccleston, 1986). For the type of design considered here Eccleston and Russell (1975, 1977) define adjusted orthogonality to mean that using model (6.35) the contrast estimators for row effects adjusted for treatment effects are uncorrelated with the contrast estimators for column effects adjusted for treatment effects. This property is important because it has implications with respect to the connectedness of a row–column design. Thus Eccleston and Russell (1975) show that a row–column design is connected if it satisfies the property of adjusted orthogonality and if both the row and column designs are connected. This, in turn, means that for any design in D_0 its properties are determined by the properties of the component designs.

One important characteristic of an incomplete block design is its efficiency factor, which is related to the properties of the information matrix $\boldsymbol{C}$ (see Section 1.12.2). For the row–column design the information matrix as given in (6.37) can be written in terms of the information matrices for the component designs as follows:

$$\begin{aligned}\boldsymbol{C} &= \left(\boldsymbol{r}^\delta - \frac{1}{c}\boldsymbol{N}_\rho \boldsymbol{N}'_\rho\right) + \left(\boldsymbol{r}^\delta - \frac{1}{r}\boldsymbol{N}_\gamma \boldsymbol{N}'_\gamma\right) - \left(\boldsymbol{r}^\delta - \frac{\boldsymbol{r}\boldsymbol{r}'}{n}\right) \\ &= \boldsymbol{C}_\rho + \boldsymbol{C}_\gamma - \left(\boldsymbol{r}^\delta - \frac{\boldsymbol{r}\boldsymbol{r}'}{n}\right)\end{aligned} \tag{6.49}$$

where $\boldsymbol{C}_\rho$ and $\boldsymbol{C}_\gamma$ denote the information matrices for the row and column designs, respectively. More specifically, for the class D_0 (6.49) can be written as

$$\boldsymbol{C} = \left(r_0\boldsymbol{I} - \frac{1}{c}\boldsymbol{N}_\rho\boldsymbol{N}'_\rho\right) + \left(r_0\boldsymbol{I} - \frac{1}{r}\boldsymbol{N}_\gamma\boldsymbol{N}'_\gamma\right) - r_0\left(\boldsymbol{I} - \frac{\mathfrak{I}\mathfrak{I}'}{t}\right) \tag{6.50}$$

In addition, for any design in D_0 $\boldsymbol{C}$ satisfies the condition

$$\boldsymbol{C} = \boldsymbol{C}_\rho\boldsymbol{C}_\gamma/r_0 \tag{6.51}$$

(John and Eccleston, 1986). The relationships (6.50) and (6.51), together with the fact that $\boldsymbol{C}$, $\boldsymbol{C}_\rho$, and $\boldsymbol{C}_\gamma$ have the same eigenvectors and that the eigenvalues of $\boldsymbol{C}$ are those of $\boldsymbol{C}_\rho$ or $\boldsymbol{C}_\gamma$ or 0 (John and Eccleston, 1986), it is possible to express the efficiency factor $E_{\rho\gamma}$ of a row–column design in D_0 in terms of the efficiency factors E_ρ and E_γ of the row and column component designs, respectively, as

$$E_{\rho\gamma} = \frac{E_\rho E_\gamma}{E_\rho + E_\gamma - E_\rho E_\gamma} \tag{6.52}$$

(see Eccleston and McGilchrist, 1986; John and Eccleston, 1986). For designs not in D_0 the right-hand side of (6.52) represents an upper bound for $E_{\rho\gamma}$ (Eccleston and McGilchrist, 1986).

We shall illustrate the relationship (6.52) numerically in terms of the design of Example 6.13.

Example 6.13 (Continued) Inspection shows that the design belongs to D_0 with $r_0 = 2$. The SAS output in Table 6.5 confirms that the design is connected.

To obtain E_ρ and E_γ (which in this case are identical) we use the system of equations (4.34) and solve for g_0, g_1, g_2, g_3 to be used in

$$\text{var}(\hat{\tau}_k - \hat{\tau}_{k'}) = 2(g_0 - g_u)\sigma_e^2$$

[see (4.39)] for the three types of associates. We find $g_0 = 0, g_1 = -\frac{5}{8}, g_2 = -\frac{1}{2}, g_3 = -\frac{3}{4}$ with $n_1 = 4, n_2 = 1, n_3 = 2$, and hence

$$\text{av. var}(\hat{\tau}_k - \hat{\tau}_{k'}) = \tfrac{9}{7}\,\sigma_e^2 = 1.2857\,\sigma_e^2$$

Thus, from (1.108) $E_\rho = E_\gamma = 1/1.2857 = .7778$ [we should point out that this value is slightly different from what would have been obtained from (4.41), which is defined as the weighted average of the three efficiencies E_1, E_2, E_3]. Using (6.52) we obtain

$$E_{\rho\gamma} = \frac{(.7778)^2}{2 \times .7778 - (.7778)^2} = .6364$$

Table 6.5 Data and Analysis for Row-Column Design With PBIB (3) Row and Column Designs ($t = 8, r = 4, c = 4$)

```
options nodate pageno=1;
data rcpbib;
input R C T Y @@;
datalines;
1 1 1 20  1 2 2 23  1 3 5 25  1 4 6 21
2 1 3 30  2 2 4 29  2 3 7 25  2 4 8 31
3 1 7 34  3 2 6 35  3 3 2 30  3 4 3 35
4 1 5 33  4 2 8 38  4 3 4 34  4 4 1 31
;
run;

proc print data=rcpbib;
title1 'TABLE 6.5';
title2 'DATA FOR ROW-COLUMN DESIGN';
title3 'W/PBIB(3) ROW AND COLUMN DESIGNS';
title4 't=8, r=4, c=4';
run;

proc glm data=rcpbib;
class R C T;
model Y=C R T/e;
estimate '1-2' T 1 -1;
estimate '1-3' T 1 0 -1;
estimate '1-4' T 1 0 0 -1;
estimate '1-5' T 1 0 0 0 -1;
estimate '1-6' T 1 0 0 0 0 -1;
estimate '1-7' T 1 0 0 0 0 0 -1;
estimate '1-8' T 1 0 0 0 0 0 0 -1;
title2 'ANALYSIS OF DATA';
title3 'FROM ROW-COLUMN DESIGN';
title4 'W/SIMPLE TREATMENT CONTRASTS';
run;
```

Obs	R	C	T	Y
1	1	1	1	20
2	1	2	2	23
3	1	3	5	25
4	1	4	6	21
5	2	1	3	30
6	2	2	4	29
7	2	3	7	25
8	2	4	8	31
9	3	1	7	34
10	3	2	6	35
11	3	3	2	30
12	3	4	3	35

Table 6.5 (*Continued*)

```
            13     4     1     5     33
            14     4     2     8     38
            15     4     3     4     34
            16     4     4     1     31

           ANALYSIS OF DATA FROM ROW-COLUMN
          DESIGN W/SIMPLE TREATMENT CONTRACTS

                 The GLM Procedure

              Class Level Information

        Class     Levels     Values

        R              4     1 2 3 4
        C              4     1 2 3 4
        T              8     1 2 3 4 5 6 7 8

             Number of observations    16

          General Form of Estimable Functions

    Effect              Coefficients

    Intercept           L1
    C             1     L2
    C             2     L3
    C             3     L4
    C             4     L1-L2-L3-L4
    R             1     L6
    R             2     L7
    R             3     L8
    R             4     L1-L6-L7-L8
    T             1     L10
    T             2     L11
    T             3     L12
    T             4     L13
    T             5     L14
    T             6     L15
    T             7     L16
    T             8     L1-L10-L11-L12-L13-L14-L15-L16

Dependent Variable: Y
```

Table 6.5 (*Continued*)

Source	DF	Sum of Squares	Mean Square	F Value	Pr > F
Model	13	426.5000000	32.8076923	7.09	0.1302
Error	2	9.2500000	4.6250000		
Corrected Total	15	435.7500000			

R-Square	Coeff Var	Root MSE	Y Mean
0.978772	7.259346	2.150581	29.62500

Source	DF	Type I SS	Mean Square	F Value	Pr > F
C	3	16.2500000	5.4166667	1.17	0.4913
R	3	357.2500000	119.0833333	25.75	0.0376
T	7	53.0000000	7.5714286	1.64	0.4305

Source	DF	Type III SS	Mean Square	F Value	Pr > F
C	3	10.2500000	3.4166667	0.74	0.6189
R	3	285.5000000	95.1666667	20.58	0.0467
T	7	53.0000000	7.5714286	1.64	0.4305

Parameter	Estimate	Standard Error	t Value	Pr > |t|
1-2	0.50000000	2.84495167	0.18	0.8767
1-3	-4.75000000	2.63391344	-1.80	0.2131
1-4	-3.25000000	2.84495167	-1.14	0.3716
1-5	-3.50000000	2.40442301	-1.46	0.2828
1-6	-1.00000000	2.63391344	-0.38	0.7407
1-7	-1.75000000	2.84495167	-0.62	0.6011
1-8	-6.25000000	2.63391344	-2.37	0.1410

This value can be confirmed by making use of the SAS output in Table 6.5, in particular the standard errors of simple treatment contrasts and the fact that, because of the combinatorial structure of the design, there are three different standard errors with frequency 3, 3, and 1, respectively. Thus, using the general definition of E given in (1.108),

$$E_{\rho\gamma} = \frac{2/2}{[3 \times (2.8449)^2 + 3 \times (2.6339)^2 + (2.4044)^2]/(7 \times 4.625)} = 0.6364$$

It follows, of course, from (6.52) that if, say, $E_\rho = 1$, then $E_{\rho\gamma} = E_\gamma$. □

6.6.8 Construction

From our discussion in the previous sections concerning the information matrix and the efficiency factor of a row–column design, it comes as no surprise that the construction of an efficient row–column design starts with efficient row and column component designs. This is, of course, easy to do for regular row–column designs for which at least one of the component designs has efficiency equal to 1. There are, however, no general rules for amalgamating two PBIB designs. Rather, we shall describe one construction method due to John and Eccleston (1986), which is based on the generalized cyclic incomplete block designs of Jarrett and Hall (1978) (see Section 6.3).

To construct a design for $t = r \times s$ treatments such that each treatment is replicated r_0 times (with $r_0 \leq r$) in r rows and $c = r_0 s$ columns, we consider the r residue classes $S_i = \{i, i + r, \ldots, i + r(s - 1)\}$ for $i = 0, 1, \ldots, r - 1$, and form an $r \times r_0$ array where each column contains one element from each of the r residue classes S_i. We shall refer to this array as the generator array. From each such column we generate $s - 1$ further columns by adding successively r mod t to each element in that column. Thus each initial column generates a complete replicate of the t treatments, and the columns of the $r \times r_0 s$ array represent the blocks of the generalized cyclic incomplete block design with r_0 replications for each treatment.

In order to obtain a design in D_0 we need to ensure that each row in the generator array contains at most one element from each residue class. Then each row will consist of elements from r_0 residue classes and hence each row and each column will have exactly r_0 elements in common. Thus the final design satisfies the condition of adjusted orthogonality.

To illustrate this method we consider the following example.

Example 6.14 Suppose $t = 12 = 3 \times 4, r = 3, s = 4, r_0 = 2, c = 8$. The residue classes are

$$S_0 = \{0, 3, 6, 9\} \qquad S_1 = \{1, 4, 7, 10\} \qquad S_2 = \{2, 5, 8, 11\}$$

Choose the generator array as

0	7
1	11
2	3

Then the final row–column design is

		Column 1	2	3	4	5	6	7	8
	1	0	3	6	9	7	10	1	4
Row	2	1	4	7	10	11	2	5	8
	3	2	5	8	11	3	6	9	0

Closer inspection shows that both row and column designs are actually PBIB(2) designs. □

As pointed out by John and Williams (1995), the method described above may not always lead to the most efficient row–column designs. Alternatively, computer algorithms have been proposed that rely on iterative improvements of a starting design or a starting component design by using exchange algorithms. The aim is to construct a design with as high an efficiency factor as possible. The criterion used most often in this context is that of (M, S) optimality since it is most amenable to computational procedures. We shall not go into details, but rather refer the reader to the appropriate literature, such as Russell, Eccleston, and Knudsen (1981), Nguyen and Williams (1993), Venables and Eccleston (1993), and John and Williams (1995).

6.6.9 Resolvable Row–Column Designs

It is sometimes desirable, especially when the number of treatments is large or when the experiment is done sequentially, to have the treatments arranged in a resolvable design. This means that the design consists of several replicates such that each replicate constitutes a row–column design and each treatment occurs exactly once in each replicate. An example of a resolvable design for $t = c^2$ are the lattice square design of Yates (1940b) (see Chapter 18).

For a resolvable row–column design with R distinct replications model (6.1) needs to be changed to

$$y_{ijk\ell} = \mu + R_i + \rho_{ij} + \gamma_{ik} + \tau_\ell + e_{ijk\ell} \tag{6.53}$$

with $i = 1, 2, \ldots, R$; $j = 1, 2, \ldots, r$; $k = 1, 2, \ldots, c$; $\ell = 1, 2, \ldots, t$, and where R_i are the replicate effects, ρ_{ij}, γ_{ik} are the row and column effects, respectively, nested within replicates, and τ_ℓ are the treatment effects. Because of the nesting structure these designs are a special case of what has been referred to as *nested row–column designs* (see, e.g., Morgan, 1996; John and Williams, 1995).

In matrix notation model (6.53) can be written as

$$\boldsymbol{y} = \mu \mathfrak{I} + \boldsymbol{X}_R \boldsymbol{R} + \boldsymbol{X}_\rho \boldsymbol{\rho} + \boldsymbol{X}_\gamma \boldsymbol{\gamma} + \boldsymbol{X}_\tau \boldsymbol{\tau} + \boldsymbol{e} \tag{6.54}$$

where, in contrast to (6.2), $\boldsymbol{X}_\rho$ is of dimension $n \times Rr$, and $\boldsymbol{X}_\gamma$ is of dimension $n \times Rc$, where $n = Rrc = Rt$. Using (6.54) we can write out the normal equations and then the reduced normal equations. In terms of the design matrices in (6.54) the information matrix can be written as (see Morgan and Bailey, 2000)

$$\boldsymbol{C} = \boldsymbol{X}'_\tau \left(\boldsymbol{I} - \frac{1}{c} \boldsymbol{X}_\rho \boldsymbol{X}'_\rho - \frac{1}{r} \boldsymbol{X}_\gamma \boldsymbol{X}'_\gamma + \frac{1}{rc} \boldsymbol{X}_R \boldsymbol{X}'_R \right) \boldsymbol{X}_\tau \tag{6.55}$$

With $X'_\tau X_\rho = N_\rho$, $X_\tau X_\gamma = N_\gamma$, and $X'_\tau X_R = N_R$ we can rewrite (6.55) as

$$C = RI - \frac{1}{c} N_\rho N'_\rho - \frac{1}{r} N_\gamma N'_\gamma + \frac{1}{rc} N_R N'_R$$

or, since the replicate design is an RCBD,

$$C = RI - \frac{1}{c} N_\rho N'_\rho - \frac{1}{r} N_\gamma N'_\gamma + \frac{R}{rc} \mathfrak{I}\mathfrak{I}' \tag{6.56}$$

When constructing resolvable row–column designs care must be taken to ensure that rank $(C) = t - 1$. This holds for the method of constructing lattice square designs (see Chapter 18). The method described in Section 6.6.7 can be used here also in conjunction with the interchange algorithm of Nguyen and Williams (1993) to produce connected and efficient designs.

Variations of the model (6.53) may be considered as dictated by the experimental conditions. For example, if in an industrial experiment the rows are different batches of raw material, the columns are different machines, the replicates are different times, and the experiment is conducted in the same factory, then the machines may be the same for the different replicates. Or, in an agriculture the replicates may be contiguous pieces of land so that, for example, the columns are the same for all the replicates. In those cases the columns are no longer nested in replicates but rather crossed with replicates, and hence model (6.53) changes to

$$y_{ijk\ell} = \mu + R_i + \rho_{ij} + \gamma_k + \tau_\ell + e_{ijk\ell} \tag{6.57}$$

(this is similar to the various forms of replicated Latin squares in I.10.3).

Finally, a further modification, or restriction, of the above design is considered by John and Williams (1995). In addition to having each treatment occur once in each replicate, each treatment should not occur more than once (or nearly equally often) in each column as, for example, in the design given below for $t = 12 = 3 \times 4$, $R = 3$, $r = 3$, $c = 4$:

Replicate				
1	0	3	6	9
	1	4	7	10
	2	5	8	11
2	4	8	9	0
	5	6	10	1
	3	7	11	2
3	7	10	0	5
	8	11	1	3
	4	9	2	4

This arrangement may be useful in, for example, an agronomic experiment to provide some protection against the loss of an entire column. This design is referred to as a *Latinized row–column design* (using terminology of Harshbarger and Davis, 1952). John and Williams (1995) recommend to modify model (6.57) to

$$y_{ijk\ell} = \mu + R_i + \rho_{ij} + \gamma_k + (R\gamma)_{ik} + \tau_\ell + e_{ijk\ell} \tag{6.58}$$

so that the effects R_i and $(R\gamma)_{ik}$ account for the $R(c-1)$ d.f. of the γ_{ij} in model (6.53). Obviously, subject matter knowledge may determine whether model (6.53) or (6.57) is appropriate.

From the above discussion it should be clear that the original row–column design may have to be modified in various ways to accommodate practical considerations and requirements, leading to still other forms of row–column designs.

CHAPTER 7

Two-Level Factorial Designs

7.1 INTRODUCTION

In the preceding chapters we have discussed various aspects of error control designs. Another component of an experimental design is, however, the treatment design (see Section I.2.2.3), in particular, designs with a factorial treatment structure (see I.11.1 and I.11.2).

In this chapter we shall consider in great detail the situation where an experiment involves several factors, each having two levels. Such experiments are generally referred to as 2^n factorials, where n denotes the number of factors, and hence 2^n denotes the total number of level combinations or treatment combinations.

Although the 2^n factorial represents only a special case of the general p^n factorial experiment, that is, n factors with p levels each, it deserves special consideration because of its practical importance and its special algebraic and combinatorial representation. At the same time, however, we shall use the 2^n case to lay the foundation for the discussion of the more general case by introducing appropriate notation and mathematics suitable for generalization in later chapters.

7.2 CASE OF TWO FACTORS

7.2.1 Definition of Main Effects and Interaction

Let us denote the two factors by A and B, where factor A has levels a_0 and a_1, and factor B has levels b_0 and b_1. We shall refer to a_0 and b_0 as 0 levels and to a_1 and b_1 as 1 levels of the two factors, respectively. The four possible treatment

---

Design and Analysis of Experiments. Volume 2: Advanced Experimental Design
By Klaus Hinkelmann and Oscar Kempthorne
ISBN 0-471-55177-5

combinations may be represented as

$$\begin{aligned} &a_0b_0 \\ &a_1b_0 \\ &a_0b_1 \\ &a_1b_1 \end{aligned} \tag{7.1}$$

We can then define the effect of factor A, that is, the effect of changing A from a_0 to a_1, at each of the two levels of factor B. Denoting these effects by $A(b_0)$ and $A(b_1)$, respectively, we define

$$\begin{aligned} A(b_0) &= a_1b_0 - a_0b_0 \\ A(b_1) &= a_1b_1 - a_0b_1 \end{aligned} \tag{7.2}$$

where a_ib_j now denotes the true response of the treatment combination a_ib_j. We shall use this dual meaning of a_ib_j throughout, but from the context it should always be clear whether we mean the treatment combination or the response of that treatment combination. The effect of A, also referred to as the *main effect* A, is now defined as the average of the so-called *simple effects* $A(b_0)$ and $A(b_1)$ of (7.2), denoted by A, that is,

$$A = \tfrac{1}{2}[A(b_0) + A(b_1)] \tag{7.3}$$

or, by using (7.2),

$$A = \tfrac{1}{2}[a_1b_0 - a_0b_0 + a_1b_1 - a_0b_1] \tag{7.4}$$

or, symbolically,

$$A = \tfrac{1}{2}(a_1 - a_0)(b_1 + b_0) \tag{7.5}$$

where the expression is to be expanded algebraically.

If the two factors are acting independently, we would expect $A(b_0)$ and $A(b_1)$ of (7.2) to be equal, but, in general they will be different, and their difference is a measure of the extent to which the two factors A and B interact. This *interaction* is denoted by AB (or $A \times B$), and defined as

$$\begin{aligned} AB &= \tfrac{1}{2}[A(b_1) - A(b_0)] \\ AB &= \tfrac{1}{2}[a_1b_1 - a_0b_1 - a_1b_0 + a_0b_0] \end{aligned} \tag{7.6}$$

or, symbolically,

$$AB = \tfrac{1}{2}(a_1 - a_0)(b_1 - b_0) \tag{7.7}$$

The factor $\frac{1}{2}$ here is a matter of convention and is used to express the interaction effect on a per-unit-difference basis, there being two differences in (7.6). We shall use this convention throughout.

Along the same lines we may also obtain the effect of factor B by first defining the simple effects:

$$\begin{aligned} B(a_0) &= a_0b_1 - a_0b_0 \\ B(a_1) &= a_1b_1 - a_1b_0 \end{aligned} \tag{7.8}$$

and then

$$\begin{aligned} B &= \tfrac{1}{2}[B(a_0) + B(a_1)] \\ &= \tfrac{1}{2}[a_0b_1 - a_0b_0 + a_1b_1 - a_1b_0] \\ &= \tfrac{1}{2}(a_1 + a_0)(b_1 - b_0) \end{aligned} \tag{7.9}$$

$B(a_0)$ and $B(a_1)$ from (7.8) can also be used to define the interaction between factors B and A as

$$\begin{aligned} BA &= \tfrac{1}{2}[B(a_1) - B(a_0)] \\ &= \tfrac{1}{2}[a_1b_1 - a_1b_0 - a_0b_1 + a_0b_0] \\ &= \tfrac{1}{2}(a_1 - a_0)(b_1 - b_0) = AB \end{aligned}$$

which shows that in defining the interaction we need not bother with the order in which we write down the letters.

7.2.2 Orthogonal Contrasts

We note that the effects A, B and the interaction AB are three mutually orthogonal contrasts of the true responses of the four treatment combinations:

	a_0b_0	a_1b_0	a_0b_1	a_1b_1
$2A$	−	+	−	+
$2B$	−	−	+	+
$2AB$	+	−	−	+

which follows immediately from (7.4), (7.9), and (7.6). If we denote the mean response of the four treatment combinations by M, we have the following transformation of the response vector $(a_0b_0, a_1b_0, a_0b_1, a_1b_1)'$:

$$\begin{pmatrix} 4M \\ 2A \\ 2B \\ 2AB \end{pmatrix} = \begin{pmatrix} 1 & 1 & 1 & 1 \\ -1 & 1 & -1 & 1 \\ -1 & -1 & 1 & 1 \\ 1 & -1 & -1 & 1 \end{pmatrix} \begin{pmatrix} a_0b_0 \\ a_1b_0 \\ a_0b_1 \\ a_1b_1 \end{pmatrix} \tag{7.10}$$

It is interesting to note that the transformation matrix in (7.10) can be written as the Kronecker product

$$\begin{pmatrix} 1 & 1 \\ -1 & 1 \end{pmatrix} \times \begin{pmatrix} 1 & 1 \\ -1 & 1 \end{pmatrix}$$

where the matrix

$$\boldsymbol{T} = \begin{pmatrix} 1 & 1 \\ -1 & 1 \end{pmatrix} \tag{7.11}$$

can be considered as the transformation matrix for the one factor situation, that is,

$$\begin{pmatrix} 2M \\ A \end{pmatrix} = \begin{pmatrix} 1 & 1 \\ -1 & 1 \end{pmatrix} \begin{pmatrix} a_0 \\ a_1 \end{pmatrix}$$

Hence, by using (7.11), we can write (7.10) as

$$\begin{pmatrix} 4M \\ 2A \\ 2B \\ 2AB \end{pmatrix} = (\boldsymbol{T} \times \boldsymbol{T}) \begin{pmatrix} a_0 b_0 \\ a_1 b_0 \\ a_0 b_1 \\ a_1 b_1 \end{pmatrix} \tag{7.12}$$

7.2.3 Parameterizations of Treatment Responses

Relationship (7.10) and hence (7.12) may be inverted, that is,

$$\begin{pmatrix} a_0 b_0 \\ a_1 b_0 \\ a_0 b_1 \\ a_1 b_1 \end{pmatrix} = (\boldsymbol{T} \times \boldsymbol{T})^{-1} \begin{pmatrix} 4M \\ 2A \\ 2B \\ 2AB \end{pmatrix}$$

and since $(\boldsymbol{T} \times \boldsymbol{T})^{-1} = \boldsymbol{T}^{-1} \times \boldsymbol{T}^{-1}$ and

$$\boldsymbol{T}^{-1} = \frac{1}{2} \begin{pmatrix} 1 & -1 \\ 1 & 1 \end{pmatrix}$$

we obtain

$$\begin{pmatrix} a_0 b_0 \\ a_1 b_0 \\ a_0 b_1 \\ a_1 b_1 \end{pmatrix} = \frac{1}{4} \begin{pmatrix} 1 & -1 & -1 & 1 \\ 1 & 1 & -1 & -1 \\ 1 & -1 & 1 & -1 \\ 1 & 1 & 1 & 1 \end{pmatrix} \begin{pmatrix} 4M \\ 2A \\ 2B \\ 2AB \end{pmatrix} \tag{7.13}$$

that is,

$$
\begin{aligned}
a_0b_0 &= M - \tfrac{1}{2}A - \tfrac{1}{2}B + \tfrac{1}{2}AB \\
a_1b_0 &= M + \tfrac{1}{2}A - \tfrac{1}{2}B - \tfrac{1}{2}AB \\
a_0b_1 &= M - \tfrac{1}{2}A + \tfrac{1}{2}B - \tfrac{1}{2}AB \\
a_1b_1 &= M + \tfrac{1}{2}A + \tfrac{1}{2}B + \tfrac{1}{2}AB
\end{aligned}
\tag{7.14}
$$

Parameterizations of type (7.14) will become very useful in later discussions of the 2^n factorial. At this point it may be of interest to relate representation (7.14) to the usual parameterization of a factorial structure in linear model terms. If μ_{ij} denotes the true response of the treatment combination a_ib_j, the usual parametric model (see Section I.4.13.2) is given as

$$\mu_{ij} = \mu + \alpha_i + \beta_j + (\alpha\beta)_{ij} \tag{7.15}$$

It is obvious that we cannot relate M, A, B, AB uniquely to the parameters in (7.15) without defining them more precisely. For that reason we write (7.15) as the identity

$$\mu_{ij} = \overline{\mu}_{..} + (\overline{\mu}_{i.} - \overline{\mu}_{..}) + (\overline{\mu}_{.j} - \overline{\mu}_{..}) + (\mu_{ij} - \overline{\mu}_{i.} - \overline{\mu}_{.j} + \overline{\mu}_{..}) \tag{7.16}$$

where, in the present context with $i, j = 0, 1$,

$$
\begin{aligned}
\overline{\mu}_{i.} &= \tfrac{1}{2}(\mu_{i0} + \mu_{i1}) \qquad (i = 0, 1) \\
\overline{\mu}_{.j} &= \tfrac{1}{2}(\mu_{0j} + \mu_{1j}) \qquad (j = 0, 1) \\
\overline{\mu}_{..} &= \tfrac{1}{4}\sum_{ij} \mu_{ij} = \mu
\end{aligned}
$$

so that

$$
\begin{aligned}
\alpha_i &= \overline{\mu}_{i.} - \overline{\mu}_{..} \\
\beta_j &= \overline{\mu}_{.j} - \overline{\mu}_{..} \\
(\alpha\beta)_{ij} &= \mu_{ij} - \overline{\mu}_{i.} - \overline{\mu}_{.j} + \overline{\mu}_{..}
\end{aligned}
$$

with

$$
\begin{aligned}
\alpha_0 + \alpha_1 &= 0 \\
\beta_0 + \beta_1 &= 0 \\
(\alpha\beta)_{00} + (\alpha\beta)_{10} &= 0 \\
(\alpha\beta)_{01} + (\alpha\beta)_{11} &= 0 \\
(\alpha\beta)_{00} + (\alpha\beta)_{01} &= 0 \\
(\alpha\beta)_{10} + (\alpha\beta)_{11} &= 0
\end{aligned}
\tag{7.17}
$$

It follows then from (7.10) that

$$\begin{aligned} M &= \mu \\ A &= \alpha_1 - \alpha_0 \\ B &= \beta_1 - \beta_0 \\ AB &= \tfrac{1}{2}\left[(\alpha\beta)_{00} - (\alpha\beta)_{10} - (\alpha\beta)_{01} + (\alpha\beta)_{11}\right] \end{aligned}$$

or, using the relationships (7.17),

$$\begin{aligned} M &= \mu \\ A &= 2\alpha_1 \\ B &= 2\beta_1 \\ AB &= 2(\alpha\beta)_{11} \end{aligned}$$

that is,

$$\begin{pmatrix} 2M \\ A \\ B \\ AB \end{pmatrix} = 2 \begin{pmatrix} \mu \\ \alpha_1 \\ \beta_1 \\ (\alpha\beta)_{11} \end{pmatrix}$$

Hence we obtain from (7.14)

$$\begin{aligned} a_0 b_0 &= \mu - \alpha_1 - \beta_1 + (\alpha\beta)_{11} \\ a_1 b_0 &= \mu + \alpha_1 - \beta_1 - (\alpha\beta)_{11} \\ a_0 b_1 &= \mu - \alpha_1 + \beta_1 - (\alpha\beta)_{11} \\ a_1 b_1 &= \mu + \alpha_1 + \beta_1 + (\alpha\beta)_{11} \end{aligned}$$

This again can be expressed as a linear transformation using the matrix $\boldsymbol{T}$ of (7.11) as

$$\begin{pmatrix} a_0 b_0 \\ a_1 b_0 \\ a_0 b_1 \\ a_1 b_1 \end{pmatrix} = 4(\boldsymbol{T} \times \boldsymbol{T})^{-1} \begin{pmatrix} \mu \\ \alpha_1 \\ \beta_1 \\ (\alpha\beta)_{11} \end{pmatrix}$$

Hence

$$\begin{pmatrix} \mu \\ \alpha_1 \\ \beta_1 \\ (\alpha\beta)_{11} \end{pmatrix} = \tfrac{1}{4}(\boldsymbol{T} \times \boldsymbol{T}) \begin{pmatrix} a_0 b_0 \\ a_1 b_0 \\ a_0 b_1 \\ a_1 b_1 \end{pmatrix}$$

We can, therefore, easily relate the parameters customarily used in factorial experiments to those commonly used in classificatory linear models and relate both, in turn, to the true responses.

7.2.4 Alternative Representation of Treatment Combinations, Main Effects, and Interaction

We now return to Eqs. (7.10) and (7.14) using two alternative ways of representing the treatment combinations and hence the responses, one that is peculiar to the 2^n factorial and simplifies the previous notation in a very useful way, and one that leads to a more formal, mathematical definition of main effects and interaction, which can be generalized for other factorials as well.

Instead of writing the treatment combinations as in (7.1), we shall now write them as (1) a, b, ab, respectively, that is, in each combination the letter corresponding to a 0 level is replaced by 1 and the subscript is dropped in a letter corresponding to a 1 level. With this notation (7.5), (7.9), and (7.7) can be written formally as

$$\begin{aligned} A &= \tfrac{1}{2}(a-1)(b+1) \\ B &= \tfrac{1}{2}(a+1)(b-1) \\ AB &= \tfrac{1}{2}(a-1)(b-1) \end{aligned} \tag{7.18}$$

each of which is then multiplied out as if these were mathematical quantities. We note that in the expressions above terms like $(a-1)$ do not have any meaning in themselves, only the quantities (1) a, b, ab after multiplication are meaningful. A simple rule for writing down the expressions in (7.18) is as follows: A minus sign appears in any factor on the right if the corresponding letter is present on the left, otherwise a plus sign appears. It is crucial in this notation to adhere to the rule that we have used, that effects and interactions are denoted by capital letters and treatment combinations and responses are denoted by small letters.

Still another way of representing the treatment combinations is by pairs (x_1, x_2), where x_1 represents the level of factor A and x_2 represents the level of factor B, with $x_i = 0, 1 (i = 1, 2)$. The four treatment combinations are then written as (0, 0), (1, 0), (0, 1), (1, 1). By looking at (7.10) we notice that, apart from a constant, main effect A is a comparison of treatment combinations with $x_1 = 1$ versus those with $x_1 = 0$. Similarly, B is a comparison of treatment combinations with $x_2 = 1$ versus those with $x_2 = 0$. And finally, AB is a comparison of treatment combinations satisfying $x_1 + x_2 = 0 \bmod 2$, that is, (0, 0) and (1, 1), versus those satisfying $x_1 + x_2 = 1$, that is, (1, 0) and (0, 1). Each of the three basic equations for

$$A\colon x_1 = i$$

$$B\colon x_2 = j$$

$$AB\colon x_1 + x_2 = k$$

where i, j, k take on the values 0 and 1, and where the equations are always reduced mod 2, represents a partitioning of the set of four treatment combinations into two sets of two treatment combinations each, that is,

$$\begin{aligned} x_1 &= 0\colon (0,0), (0,1) \\ &= 1\colon (1,0), (1,1) \\ x_2 &= 0\colon (0,0), (1,0) \\ &= 1\colon (0,1), (1,1) \\ x_1 + x_2 &= 0\colon (0,0), (1,1) \\ &= 1\colon (1,0), (0,1) \end{aligned}$$

A unified way to express these partitions is to write them as

$$\alpha_1 x_1 + \alpha_2 x_2 = 0, 1 \mod 2$$

where $\alpha_1, \alpha_2 = 0, 1$ but $(\alpha_1, \alpha_2) \neq (0, 0)$. These partitions are orthogonal in the sense that for a given $\alpha_1 x_1 + \alpha_2 x_2 = i$ the set of treatment combinations contains one from $\alpha_1' x_1 + \alpha_2' x_2 = 0$ and one from $\alpha_1' x_1 + \alpha_2' x_2 = 1$ where $(\alpha_1', \alpha_2') \neq (\alpha_1, \alpha_2)$. This, of course, is equivalent to our earlier statement that A, B and AB represent orthogonal comparisons of the treatment combination responses.

7.3 CASE OF THREE FACTORS

We now consider three factors A, B, and C, say, with levels $a_0, a_1, b_0, b_1, c_0, c_1$, respectively. The eight possible treatment combinations can be represented, in the standard order, in the following three ways:

$a_0b_0c_0$		(1)		(0, 0, 0)
$a_1b_0c_0$		a		(1, 0, 0)
$a_0b_1c_0$		b		(0, 1, 0)
$a_1b_1c_0$		ab		(1, 1, 0)
	or		or	
$a_0b_0c_1$		c		(0, 0, 1)
$a_1b_0c_1$		ac		(1, 0, 1)
$a_0b_1c_1$		bc		(0, 1, 1)
$a_1b_1c_1$		abc		(1, 1, 1)

7.3.1 Definition of Main Effects and Interactions

Denoting the true response by the same symbol as the treatment combination, we define now four simple effects associated with factor A:

$$\begin{aligned} A(b_0, c_0) &= a_1b_0c_0 - a_0b_0c_0 \\ A(b_1, c_0) &= a_1b_1c_0 - a_0b_1c_0 \\ A(b_0, c_1) &= a_1b_0c_1 - a_0b_0c_1 \\ A(b_1, c_1) &= a_1b_1c_1 - a_0b_1c_1 \end{aligned} \tag{7.19}$$

where $A(b_i, c_j)$ is the effect of factor A with factor B at level $b_i(i = 0, 1)$ and factor C at level $c_j(j = 0, 1)$. The average of the four simple effects of (7.19) is then defined as *the* effect of factor A, denoted by A, that is,

$$\begin{aligned} A &= \tfrac{1}{4}\sum_{ij} A(b_i, c_j) \\ &= \tfrac{1}{4}(a_1b_0c_0 + a_1b_1c_0 + a_1b_0c_1 + a_1b_1c_1 \\ &\qquad -a_0b_0c_0 - a_0b_1c_0 - a_0b_0c_1 - a_0b_1c_1) \end{aligned}$$

or, symbolically,

$$A = \tfrac{1}{4}(a_1 - a_0)(b_1 + b_0)(c_1 + c_0) \tag{7.20}$$

and using the alternative notation

$$\begin{aligned} A &= \tfrac{1}{4}(a + ab + ac + abc - (1) - b - c - bc) \\ &= \tfrac{1}{4}(a - 1)(b + 1)(c + 1) \end{aligned} \tag{7.21}$$

Next we define the interaction between factors A and B, denoted by AB, by considering the difference between the following average effects:

$$A(b_0, \bar{c}) = \tfrac{1}{2}[A(b_0, c_0) + A(b_0, c_1)]$$

and

$$A(b_1, \bar{c}) = \tfrac{1}{2}[A(b_1, c_0) + A(b_1, c_1)]$$

Then

$$\begin{aligned} AB &= \tfrac{1}{2}[A(b_1, \bar{c}) - A(b_0, \bar{c})] \\ &= \tfrac{1}{4}(a_1b_1c_0 - a_0b_1c_0 + a_1b_1c_1 - a_0b_1c_1 \\ &\qquad -a_1b_0c_0 + a_0b_0c_0 - a_1b_0c_1 + a_0b_0c_1) \\ &= \tfrac{1}{4}(a_1 - a_0)(b_1 - b_0)(c_1 + c_0) \end{aligned} \tag{7.22}$$

or, using the alternative notation

$$\begin{aligned} AB &= \tfrac{1}{4}[ab + abc + (1) + c - a - b - ac - bc] \\ &= \tfrac{1}{4}(a-1)(b-1)(c+1) \end{aligned} \tag{7.23}$$

We can also evaluate the *simple* interaction between factors A and B for each level of factor C, using an obvious modification of (7.6), as

$$AB(c_0) = \tfrac{1}{2}(a_1b_1c_0 + a_0b_0c_0 - a_1b_0c_0 - a_0b_1c_0) \tag{7.24}$$

and

$$AB(c_1) = \tfrac{1}{2}(a_1b_1c_1 + a_0b_0c_1 - a_1b_0c_1 - a_0b_1c_1) \tag{7.25}$$

Obviously,

$$AB = \tfrac{1}{2}[AB(c_0) + AB(c_1)]$$

The two interactions given by (7.24) and (7.25) may be different, and, as a measure of the extent to which they are different, we define the three-factor interaction of factors A, B, and C, denoted by ABC, as

$$\begin{aligned} ABC &= \tfrac{1}{4}[AB(c_1) - AB(c_0)] \\ &= \tfrac{1}{4}(a_1b_1c_1 + a_0b_0c_1 - a_1b_0c_1 - a_0b_1c_1 \\ &\qquad - a_1b_1c_0 - a_0b_0c_0 + a_1b_0c_0 + a_0b_1c_0) \\ &= \tfrac{1}{4}(a_1 - a_0)(b_1 - b_0)(c_1 - c_0) \end{aligned} \tag{7.26}$$

or

$$\begin{aligned} ABC &= \tfrac{1}{4}[abc + a + b + c - (1) - ab - ac - bc] \\ &= \tfrac{1}{4}(a-1)(b-1)(c-1) \end{aligned} \tag{7.27}$$

Similar arguments can be used to define the remaining main effects and two-factor interactions. We summarize them below:

$$\begin{aligned} B &= \tfrac{1}{4}(a_1 + a_0)(b_1 - b_0)(c_1 + c_0) \\ C &= \tfrac{1}{4}(a_1 + a_0)(b_1 + b_0)(c_1 - c_0) \\ AC &= \tfrac{1}{4}(a_1 - a_0)(b_1 + b_0)(c_1 - c_0) \\ BC &= \tfrac{1}{4}(a_1 + a_0)(b_1 - b_0)(c_1 - c_0) \end{aligned}$$

or alternatively,

$$B = \tfrac{1}{4}(a+1)(b-1)(c+1)$$

$$C = \tfrac{1}{4}(a+1)(b+1)(c-1)$$

$$AC = \tfrac{1}{4}(a-1)(b+1)(c-1)$$

$$BC = \tfrac{1}{4}(a+1)(b-1)(c-1)$$

Just as in the two-factor case we can express the effects and interactions defined above as a transformation of the treatment responses:

$$\begin{bmatrix} 8M \\ 4A \\ 4B \\ 4AB \\ 4C \\ 4AC \\ 4BC \\ 4ABC \end{bmatrix} = \begin{bmatrix} 1 & 1 & 1 & 1 & 1 & 1 & 1 & 1 \\ -1 & 1 & -1 & 1 & -1 & 1 & -1 & 1 \\ -1 & -1 & 1 & 1 & -1 & -1 & 1 & 1 \\ 1 & -1 & -1 & 1 & 1 & -1 & -1 & 1 \\ -1 & -1 & -1 & -1 & 1 & 1 & 1 & 1 \\ 1 & -1 & 1 & -1 & -1 & 1 & -1 & 1 \\ 1 & 1 & -1 & -1 & -1 & -1 & 1 & 1 \\ -1 & 1 & 1 & -1 & 1 & -1 & -1 & 1 \end{bmatrix} \begin{bmatrix} a_0b_0c_0 \\ a_1b_0c_0 \\ a_0b_1c_0 \\ a_1b_1c_0 \\ a_0b_0c_1 \\ a_1b_0c_1 \\ a_0b_1c_1 \\ a_1b_1c_1 \end{bmatrix} \tag{7.28}$$

where M is the mean response of the 8 treatment combinations. We shall comment briefly on the transformation matrix in (7.28): (i) It can be verified easily that, with the standard order of writing the symbols for the treatment combinations and the effects (main effects and interactions) that we have adopted, the matrix can be expressed as the threefold Kronecker product $\boldsymbol{T} \times \boldsymbol{T} \times \boldsymbol{T}$, where $\boldsymbol{T}$ is given by (7.11); (ii) the rows of the transformation matrix are orthogonal to each other; (iii) except for the first row, the rows of the transformation matrix represent comparisons of the treatment responses; (iv) it is easy to write down the elements in any given row: for A, B, and C we have a $+1$ if the corresponding factor appears at the 1 level in a treatment combination, and a -1 if the corresponding factor appears at the 0 level; for interactions we just multiply the corresponding elements for the main effects involved; for the alternate representation of treatment combinations this rule can be formulated as follows: Let X be any effect symbol; if X consists of an odd (even) number of factor symbols, then an element in the row corresponding to X is $+1$ if the associated treatment combination has an odd (even) number of letters in common with X, and it is -1 if the associated treatment combination has an even (odd) number of letters in common with X, where zero is considered to be an even number.

7.3.2 Parameterization of Treatment Responses

Inverting (7.28) leads to a representation of treatment responses in terms of the overall mean, main effects, and interactions:

$$\begin{bmatrix} a_0b_0c_0 \\ a_1b_0c_0 \\ a_0b_1c_0 \\ a_1b_1c_0 \\ a_0b_0c_1 \\ a_1b_0c_1 \\ a_0b_1c_1 \\ a_1b_1c_1 \end{bmatrix} = \begin{bmatrix} 1 & -1 & -1 & 1 & -1 & 1 & 1 & -1 \\ 1 & 1 & -1 & -1 & -1 & -1 & 1 & 1 \\ 1 & -1 & 1 & -1 & -1 & 1 & -1 & 1 \\ 1 & 1 & 1 & 1 & -1 & -1 & -1 & -1 \\ 1 & -1 & -1 & 1 & 1 & -1 & -1 & 1 \\ 1 & 1 & -1 & -1 & 1 & 1 & -1 & -1 \\ 1 & -1 & 1 & -1 & 1 & -1 & 1 & -1 \\ 1 & 1 & 1 & 1 & 1 & 1 & 1 & 1 \end{bmatrix} \begin{bmatrix} 8M \\ 4A \\ 4B \\ 4AB \\ 4C \\ 4AC \\ 4BC \\ 4ABC \end{bmatrix} \tag{7.29}$$

The rule for writing down the elements in the matrix in (7.29) corresponding to the treatment combination $a_ib_jc_k$ is: The coefficient for M is $+1$; the coefficient for the main effects is $+1$ if the corresponding factor appears at the 1 level and -1 if it appears at the 0 level; the coefficient for an interaction is the product of the coefficients for the main effects involved. An alternative way of writing (7.29) then is

$$\begin{aligned} 2a_ib_jc_k = {} & 2M + (-1)^{1-i}A + (-1)^{1-j}B + (-1)^{[(1-i)+(1-j)]}AB \\ & + (-1)^{1-k}C + (-1)^{[(1-i)+(1-k)]}AC \\ & + (-1)^{[(1-j)+(1-k)]}BC + (-1)^{[(1-i)+(1-j)+(1-k)]}ABC \end{aligned} \tag{7.30}$$

where $i, j, k = 0, 1$.

7.3.3 The x-Representation

To conclude this section, we shall discuss briefly the formal definition of main effects and interactions in terms of orthogonal partitions, using what we shall henceforth call the $\boldsymbol{x}$ representation of treatment combinations. A treatment combination is given by $\boldsymbol{x}' = (x_1, x_2, x_3)$ with $x_i = 0, 1 (i = 1, 2, 3)$. We then consider all partitions of the form

$$\alpha_1x_1 + \alpha_2x_2 + \alpha_3x_3 = 0, 1 \mod 2$$

where $\alpha_i = 0, 1 (i = 1, 2, 3)$ but $(\alpha_1, \alpha_2, \alpha_3) \neq (0, 0, 0)$. Each partition divides the set of treatment combinations into two sets of four treatment combinations each. The comparison of the average response for these two sets defines then a main effect or interaction. Specifically these sets are

$$\begin{array}{rll}
A\colon & \alpha_1 = 1, \alpha_2 = \alpha_3 = 0 & \text{i.e. } x_1 = 0 \\
& & \text{vs. } x_1 = 1 \\[1ex]
B\colon & \alpha_2 = 1, \alpha_1 = \alpha_3 = 0 & \text{i.e. } x_2 = 0 \\
& & \text{vs. } x_2 = 1 \\[1ex]
AB\colon & \alpha_1 = \alpha_2 = 1, \alpha_3 = 0 & \text{i.e. } x_1 + x_2 = 0 \\
& & \text{vs. } x_1 + x_2 = 1 \\[1ex]
C\colon & \alpha_3 = 1, \alpha_1 = \alpha_2 = 0 & \text{i.e. } x_3 = 0 \\
& & \text{vs. } x_3 = 1 \\[1ex]
AC\colon & \alpha_1 = \alpha_3 = 1, \alpha_2 = 0 & \text{i.e. } x_1 + x_3 = 0 \\
& & \text{vs. } x_1 + x_3 = 1 \\[1ex]
BC\colon & \alpha_1 = 0, \alpha_2 = \alpha_3 = 1 & \text{i.e. } x_2 + x_3 = 0 \\
& & \text{vs. } x_2 + x_3 = 1 \\[1ex]
ABC\colon & \alpha_1 = \alpha_2 = \alpha_3 = 1 & \text{i.e. } x_1 + x_2 + x_3 = 0 \\
& & \text{vs. } x_1 + x_2 + x_3 = 1
\end{array}$$

where every equation is reduced mod 2. We refer to each $\boldsymbol{\alpha} = (\alpha_1, \alpha_2, \alpha_3)'$ as a partition of the factor space $\mathcal{X} = \{\boldsymbol{x} = (x_1, x_2, x_3)'; x_i = 0, 1; i = 1, 2, 3\}$.

7.4 GENERAL CASE

We now have n factors and hence 2^n treatment combinations. It is convenient to denote the factors by $A_1, A_2, \ldots, A_n$ where factor A_i has levels a_{i0} and $a_{i1} (i = 1, 2, \ldots, n)$, or simply $x_i = 0$ and 1. Then a treatment combination can be represented again in three different forms: (1) the explicit form:

$$a_{1_{j_1}} a_{2_{j_2}} \cdots a_{nj_n}$$

where $j_i = 0$ or $1 (i = 1, 2, \ldots, n)$; (2) the condensed form:

$$a_1^{x_i} a_2^{x_2} \cdots a_n^{x_n} \tag{7.31}$$

where $x_i = 0$ or 1 with $a_i^0 = 1$ and $a_i^1 = a_i$ and all unity terms in (7.31) are ignored; (3) the $\boldsymbol{x}$ representation

$$(x_1, x_2, \ldots, x_n)$$

where, as in (2), $x_i = 0$ or 1.

7.4.1 Definition of Main Effects and Interactions

Conforming to the general expression for a partition of the 2^n treatment combinations into two sets of 2^{n-1} treatment combinations each, that is,

$$\alpha_1 x_1 + \alpha_2 x_2 + \cdots + \alpha_n x_n = 0, 1 \mod 2 \tag{7.32}$$

with $\alpha_i = 0, 1, x_i = 0, 1 (i = 1, 2, \ldots, n)$, we write the effects and interactions in the general form

$$A_1^{\alpha_1} A_2^{\alpha_2} \cdots A_n^{\alpha_n} \tag{7.33}$$

with $A_i^0 = 1$, $A_i^1 = A_i$, and we ignore all unity elements in this expression. If all $\alpha_i = 0$, then (7.33) is written as M, the mean of the 2^n treatment combinations.

Using (7.31) and (7.33) we now generalize the symbolic expressions given in the previous sections for main effects and interactions and write

$$A_1^{\alpha_1} A_2^{\alpha_2} \cdots A_n^{\alpha_n} = \frac{1}{2^{n-1}} \prod_{i=1}^{n} [a_i + (-1)^{\alpha_i}] \tag{7.34}$$

and

$$M = \frac{1}{2^n} \prod_{i=1}^{n} (a_i + 1)$$

We know that in (7.34) 2^{n-1} treatment combinations enter positively into $A_1^{\alpha_1} A_2^{\alpha_2} \cdots A_n^{\alpha_n}$ and the remaining 2^{n-1} treatment combinations enter negatively. For the case $n = 3$ we have given an even–odd rule to determine which treatment combinations enter positively and which enter negatively, a rule due to Fisher (1949). We shall use (7.34) to derive such a rule for the general case.

First we rewrite (7.34) as

$$A_1^{\alpha_1} A_2^{\alpha_2} \cdots A_n^{\alpha_n} = \frac{1}{2^{n-1}} \prod_{i=1}^{n} [a_i^1 + (-1)^{\alpha_i} a_i^0] \tag{7.35}$$

We then note that each treatment combination contains either a_i^1 or $a_i^0 (i = 1, 2, \ldots, n)$. To determine the sign of $a_1^{x_1} a_2^{x_2} \cdots a_n^{x_n}$ we therefore have to determine the sign with which $a_i^{x_i}$ enters into this "product": If $\alpha_i = 0$, then the sign for $a_i^{x_i}$ is $+1$ whether $x_i = 0$ or 1; if $\alpha_i = 1$, then the sign for $a_i^{x_i}$ is $+1$ for $x_i = 1$ and -1 for $x_i = 0$ as is obvious from (7.35). Both statements can be combined into the statement that the sign for $a_i^{x_i}$ is $(-1)^{\alpha_i(1-x_i)}$. Hence the sign of $a_1^{x_1} a_2^{x_2} \cdots a_n^{x_n}$ is

$$\prod_{i=1}^{n} [(-1)^{\alpha_i(1-x_1)}] = (-1)^{\sum_i \alpha_i(1-x_i)} = (-1)^{\sum \alpha_i - \sum \alpha_i x_i} \tag{7.36}$$

Now $\sum_i \alpha_i$ is the number of letters in $A_1^{\alpha_1} A_2^{\alpha_2} \cdots A_n^{\alpha_n}$ after deletion of terms of the form A_j^0. Also, $\alpha_i x_i = 1$ if a_i and A_i are present in (7.31) and (7.33), respectively. With (7.36) this establishes then the following rules.

Rule 1 The treatment combination $a_1^{x_1} a_2^{x_2} \cdots a_n^{x_n}$ enters positively into $A_1^{\alpha_1} A_2^{\alpha_2} \cdots A_n^{\alpha_n}$ if $\delta =$ [number of letters in $A_1^{\alpha_1} A_2^{\alpha_2} \cdots A_n^{\alpha_n} -$ number of letters in common between $A_1^{\alpha_1} A_2^{\alpha_2} \cdots A_n^{\alpha_n}$ and $a_1^{x_1} a_2^{x_2} \cdots a_n^{x_n}$] is even, and it enters negatively if δ is odd.

This leads immediately to the alternative formulation in rule 1′.

Rule 1′ If $A_1^{\alpha_1} A_2^{\alpha_2} \cdots A_n^{\alpha_n}$ has an odd (even) number of letters, then the treatment combination $a_1^{x_1} a_2^{x_2} \cdots a_n^{x_n}$ enters positively into $A_1^{\alpha_1} A_2^{\alpha_2} \cdots A_n^{\alpha_n}$ if it has an odd (even) number of letters in common with $A_1^{\alpha_1} A_2^{\alpha_2} \cdots A_n^{\alpha_n}$, and it enters negatively if it has an even (odd) number of letters in common with $A_1^{\alpha_1} A_2^{\alpha_2} \cdots A_n^{\alpha_n}$.

Based on the method of orthogonal partitioning as given in (7.32) and rule 1′, we can give yet another expression for $A_1^{\alpha_1} A_2^{\alpha_2} \cdots A_n^{\alpha_n}$ that will prove to be useful for further discussion of the general factorial experiment. We know that $A_1^{\alpha_1} A_2^{\alpha_2} \cdots A_n^{\alpha_n}$ is, apart from a constant, a comparison of responses of treatment combinations satisfying the equation

$$\alpha_1 x_1 + \alpha_2 x_2 + \cdots + \alpha_n x_n = 0 \mod 2 \tag{7.37}$$

versus those satisfying the equation

$$\alpha_1 x_1 + \alpha_2 x_2 + \cdots + \alpha_n x_n = 1 \mod 2 \tag{7.38}$$

Now, if in $A_1^{\alpha_1} A_2^{\alpha_2} \cdots A_n^{\alpha_n}$ the sum $\sum_i \alpha_i$ is odd, then all treatment combinations with $\sum_i \alpha_i x_i$ odd, that is, satisfying Eq. (7.38), enter positively into $A_1^{\alpha_1} A_2^{\alpha_2} \cdots A_n^{\alpha_n}$ and those with $\sum_i \alpha_i x_i$ even, that is, satisfying Eq. (7.37), enter negatively into $A_1^{\alpha_1} A_2^{\alpha_2} \cdots A_n^{\alpha_n}$. If on the other hand $\sum_i \alpha_i$ is even, then all treatment combinations satisfying (7.37) enter positively, and those satisfying (7.38) enter negatively. Let

$$(A_1^{\alpha_1} A_2^{\alpha_2} \cdots A_n^{\alpha_n})_0 = [\text{mean response of all treatment combinations satisfying } \sum_i \alpha_i x_i = 0] - M \tag{7.39}$$

and

$$(A_1^{\alpha_1} A_2^{\alpha_2} \cdots A_n^{\alpha_n})_1 = [\text{mean response of all treatment combinations satisfying } \sum_i \alpha_i x_i = 1] - M \tag{7.40}$$

We can then write

$$\begin{aligned} A_1^{\alpha_1} A_2^{\alpha_2} \cdots A_n^{\alpha_n} &= (A_1^{\alpha_1} A_2^{\alpha_2} \cdots A_n^{\alpha_n})_1 - (A_1^{\alpha_1} A_2^{\alpha_2} \cdots A_n^{\alpha_n})_0 \qquad \text{if } \sum \alpha_i \text{ is odd} \\ &= (A_1^{\alpha_1} A_2^{\alpha_2} \cdots A_n^{\alpha_n})_0 - (A_1^{\alpha_1} A_2^{\alpha_2} \cdots A_n^{\alpha_n})_1 \qquad \text{if } \sum \alpha_i \text{ is even} \end{aligned}$$

or, combining both cases,

$$A_1^{\alpha_1} A_1^{\alpha_2} \cdots A_n^{\alpha_n} = (-1)^{\sum \alpha_i} (A_1^{\alpha_1} \ A_2^{\alpha_2} \cdots A_n^{\alpha_n})_0 - (A_1^{\alpha_1} A_2^{\alpha_2} \cdots A_n^{\alpha_n})_1] \qquad (7.41)$$

7.4.2 Parameterization of Treatment Responses

We shall use this result now to derive an expression, due to Kempthorne (1952), for the parameterization of treatment combination responses in terms of main effects and interactions, that is, the generalization of (7.30) for the 2^n case. To abbreviate the notation, let $\boldsymbol{\alpha}' = (\alpha_1, \alpha_2, \ldots, \alpha_n)$, $\boldsymbol{x}' = (x_1, x_2, \ldots, x_n)$, $a(\boldsymbol{x}) = a_1^{x_1} a_2^{x_2}, \ldots, a_n^{x_n}$, $E^{\boldsymbol{\alpha}} = A_1^{\alpha_1} A_2^{\alpha_2} \cdots A_n^{\alpha_n}$, $E_i^{\boldsymbol{\alpha}} = (A_1^{\alpha_1} A_2^{\alpha_2} \cdots A_n^{\alpha_n})_i$ $(i = 0, 1)$.

First, we establish the following identity: For a fixed $\boldsymbol{x}$

$$a(\boldsymbol{x}) = M + \sum_{\alpha} E^{\boldsymbol{\alpha}}_{\boldsymbol{\alpha}'\boldsymbol{x}} \qquad (7.42)$$

where summation is over all $\boldsymbol{\alpha} \neq (0, 0, \ldots, 0)$, and $\boldsymbol{\alpha}'\boldsymbol{x}$ is reduced modulo 2. Expression (7.42) in itself is useful as a parameterization of $a(\boldsymbol{x})$, and we shall return to it in connection with the general factorial experiment (see Chapter 11). For the 2^n case, however, (7.42) can be rewritten in terms of effects and interactions as we shall see.

Now, to prove (7.42), we make use of (7.39) and (7.40), the definitions of $E_0^{\boldsymbol{\alpha}}$ and $E_1^{\boldsymbol{\alpha}}$, and we observe that

1. $\sum_{\boldsymbol{\alpha}} E^{\boldsymbol{\alpha}}_{\boldsymbol{\alpha}'\boldsymbol{x}}$ contains $2^n - 1$ terms.
2. Each $E^{\boldsymbol{\alpha}}_{\boldsymbol{\alpha}'\boldsymbol{x}}$ contains the term $-M$.
3. Each $E^{\boldsymbol{\alpha}}_{\boldsymbol{\alpha}'\boldsymbol{x}}$ contains the sum of 2^{n-1} treatment combinations divided by 2^{n-1}.
4. Each $E^{\boldsymbol{\alpha}}_{\boldsymbol{\alpha}'\boldsymbol{x}}$ contains $a(\boldsymbol{x})$.
5. If $\boldsymbol{y}' = (y_1, y_2, \ldots, y_n)$, with $\boldsymbol{y} \neq \boldsymbol{x}$, satisfies the same equation as $\boldsymbol{x}$, then $\boldsymbol{\alpha}'(\boldsymbol{x} - \boldsymbol{y}) = 0$, and there are exactly $2^{n-1} - 1$ distinct solutions $\boldsymbol{\alpha}$, that is, each $a(\boldsymbol{y}) \neq a(\boldsymbol{x})$ occurs in exactly $2^{n-1} - 1$ different $E^{\boldsymbol{\alpha}}_{\boldsymbol{\alpha}'\boldsymbol{x}}$.

Collecting terms on the right-hand side of (7.42) we then obtain

$$\begin{aligned} &M - (2^n - 1)M + \frac{2^n - 1}{2^{n-1}} a(\boldsymbol{x}) + \frac{2^{n-1} - 1}{2^{n-1}} \sum_{\boldsymbol{y} \neq \boldsymbol{x}} a(\boldsymbol{y}) \\ &\quad = 2(2^{n-1} - 1)M + \frac{2^n - 1}{2^{n-1}} a(\boldsymbol{x}) + \frac{2^{n-1} - 1}{2^{n-1}} [2^n M - a(\boldsymbol{x})] = a(\boldsymbol{x}) \end{aligned}$$

which proves (7.42).

We now use (7.42) to derive an expression of the form

$$a(\boldsymbol{x}) = M + \sum_{\boldsymbol{\alpha}} \phi(\boldsymbol{\alpha}, \boldsymbol{x}) E^{\boldsymbol{\alpha}} \tag{7.43}$$

where the coefficients $\phi(\boldsymbol{\alpha}, \boldsymbol{x})$ are either $+\frac{1}{2}$ or $-\frac{1}{2}$. Let $\delta_i^{\boldsymbol{\alpha}'\boldsymbol{x}} = 1$ if $\boldsymbol{\alpha}'\boldsymbol{x} = i$ and $= 0$ if $\boldsymbol{\alpha}'\boldsymbol{x} \neq i (i = 0, 1)$. We can then write (7.42) as

$$a(\boldsymbol{x}) = M + \sum_{\boldsymbol{\alpha}} [\delta_0^{\boldsymbol{\alpha}'\boldsymbol{x}} E_0^{\boldsymbol{\alpha}} + \delta_1^{\boldsymbol{\alpha}'\boldsymbol{x}} E_1^{\boldsymbol{\alpha}}] \tag{7.44}$$

From (7.41) we have

$$E^{\boldsymbol{\alpha}} = (-1)^{\sum \alpha_i} (E_0^{\boldsymbol{\alpha}} - E_1^{\boldsymbol{\alpha}})$$

and hence

$$E_0^{\boldsymbol{\alpha}} - E_1^{\boldsymbol{\alpha}} = (-1)^{\sum \alpha_i} E^{\boldsymbol{\alpha}} \tag{7.45}$$

Also, from (7.39) and (7.40), we have

$$E_0^{\boldsymbol{\alpha}} + E_1^{\boldsymbol{\alpha}} = 0 \tag{7.46}$$

so that from (7.45) and (7.46) it follows that

$$E_0^{\boldsymbol{\alpha}} = \tfrac{1}{2}(-1)^{\sum \alpha_i} E^{\boldsymbol{\alpha}} \qquad E_1^{\boldsymbol{\alpha}} = -\left(\tfrac{1}{2}\right)(-1)^{\sum \alpha_i} E^{\boldsymbol{\alpha}} \tag{7.47}$$

Substituting (7.47) into (7.44), we obtain

$$\begin{aligned} a(\boldsymbol{x}) &= M + \tfrac{1}{2} \sum_{\boldsymbol{\alpha}} \left[\delta_0^{\boldsymbol{\alpha}'\boldsymbol{x}} (-1)^{\sum \alpha_i} - \delta_1^{\boldsymbol{\alpha}'\boldsymbol{x}} (-1)^{\sum \alpha_i}\right] E^{\boldsymbol{\alpha}} \\ &= M + \tfrac{1}{2} \sum_{\boldsymbol{\alpha}} (-1)^{\sum \alpha_i} \left(\delta_0^{\boldsymbol{\alpha}'\boldsymbol{x}} - \delta_1^{\boldsymbol{\alpha}'\boldsymbol{x}}\right) E^{\boldsymbol{\alpha}} \end{aligned} \tag{7.48}$$

Now

$$\begin{aligned} \delta_0^{\boldsymbol{\alpha}'\boldsymbol{x}} - \delta_1^{\boldsymbol{\alpha}'\boldsymbol{x}} &= 1 \quad \text{if } \boldsymbol{\alpha}'\boldsymbol{x} = 0 \\ &= -1 \quad \text{if } \boldsymbol{\alpha}'\boldsymbol{x} = 1 \end{aligned}$$

that is,

$$\delta_0^{\boldsymbol{\alpha}'\boldsymbol{x}} - \delta_1^{\boldsymbol{\alpha}'\boldsymbol{x}} = (-1)^{\boldsymbol{\alpha}'\boldsymbol{x}} = (-1)^{-\boldsymbol{\alpha}'\boldsymbol{x}}$$

and hence (7.48) becomes

$$a(\boldsymbol{x}) = M + \tfrac{1}{2} \sum_{\boldsymbol{\alpha}} (-1)^{\sum_i \alpha_i (1 - x_i)} E^{\boldsymbol{\alpha}} \tag{7.49}$$

With regard to the contribution of $E^{\boldsymbol{\alpha}}$ to $a(\boldsymbol{x})$, we can state the following rules.

RULE 2 The effect or interaction $A_1^{\alpha_1} A_2^{\alpha_2} \cdots A_n^{\alpha_n}$ enters positively into the treatment combination response $a_1^{x_1} a_2^{x_2} \cdots a_n^{x_n}$ if δ (defined in rule 1) is even and it enters negatively if δ is odd.

The reader may verify that for $n = 3$ this rule agrees with the expressions given in (7.29) and that (7.49) does indeed agree with (7.30). Also, the reader will notice that rule 2 is an exact analog to rule 1. Hence rule 2 can be reformulated as rule 2′.

RULE 2′ If $A_1^{\alpha_1} A_2^{\alpha_2} \cdots A_n^{\alpha_n}$ has an odd (even) number of letters, then it will enter positively into $a_1^{x_1} a_2^{x_2} \cdots a_n^{x_n}$ if both have an odd (even) number of letters in common, and it will enter negatively into $a_1^{x_1} a_2^{x_2} \cdots a_n^{x_n}$ if both have an even (odd) number of letters in common.

This rule will become important in our discussion of systems of confounding (see Section 8.6) and fractional replication (see Section 13.3).

7.4.3 Generalized Interactions

To conclude this section we shall introduce the concept of a generalized interaction, a concept that also will take on great importance in connection with systems of confounding and fractional replication. To motivate this concept we return briefly to the 2^2 case and arrange the treatment combinations in the following 2×2 table:

		B	
		b_0	b_1
A	a_0	a_0b_0 $\boxed{1}$	a_0b_1 $\boxed{2}$
	a_1	a_1b_0 $\boxed{3}$	a_1b_1 $\boxed{4}$

The contrast defining main effects A, B and the interaction AB can then be written down symbolically in terms of the cell labels $\boxed{1}, \ldots, \boxed{4}$ as

$$A:\ \boxed{3} + \boxed{4} - \boxed{1} - \boxed{2}$$

$$B:\ \boxed{2} + \boxed{4} - \boxed{1} - \boxed{3}$$

$$AB:\ \boxed{1} + \boxed{4} - \boxed{2} - \boxed{3}$$

Returning to the 2^n case, let $\boldsymbol{X}$ and $\boldsymbol{Y}$ denote two interactions, that is,

$$X = A_1^{\alpha_1} A_2^{\alpha_2} \cdots A_n^{\alpha_n}$$

and

$$Y = A_1^{\beta_1} A_2^{\beta_2} \cdots A_n^{\beta_n}$$

which are defined by the partitions $\sum_i \alpha_i x_i = 0, 1$ and $\sum_i \beta_i x_i = 0, 1$, respectively. Now, any pair of equations

$$\sum_i \alpha_i x_i = k$$

$$\sum_i \beta_i x_i = \ell$$

with $k, \ell = 0, 1$ is satisfied by 2^{n-2} treatment combinations. Accordingly, we can arrange the 2^n treatment combinations in a 2×2 table as follows:

		$\sum \beta_i x_i =$ 0	1
$\sum \alpha_i x_i =$	0	$\boxed{1}$	$\boxed{2}$
	1	$\boxed{3}$	$\boxed{4}$

There are three orthogonal comparisons among the four cells, namely

$$X\text{: } \boxed{3} + \boxed{4} - \boxed{1} - \boxed{2}$$

$$Y\text{: } \boxed{2} + \boxed{4} - \boxed{1} - \boxed{3}$$

The remaining contrast

$$\boxed{1} + \boxed{4} - \boxed{2} - \boxed{3} \tag{7.50}$$

corresponds to that of the AB interaction in the 2^2 case above. We shall therefore denote the contrast (7.50) by XY and call it the *generalized interaction* (GI) of X and Y. From the table above it follows immediately that the treatment combinations in the four cells satisfy the following equations:

$$\boxed{1}\text{: } \sum_i \alpha_i x_i = 0 \quad \sum_i \beta_i x_i = 0 \quad \text{hence also} \quad \sum_i (\alpha_i + \beta_i) x_i = 0$$

$$\boxed{2}\text{: } \sum_i \alpha_i x_i = 0 \quad \sum_i \beta_i x_i = 1 \quad \text{hence also} \quad \sum_i (\alpha_i + \beta_i) x_i = 1$$

$$\boxed{3}\text{: } \sum_i \alpha_i x_i = 1 \quad \sum_i \beta_i x_i = 0 \quad \text{hence also} \quad \sum_i (\alpha_i + \beta_i) x_i = 1$$

$$\boxed{4}\text{: } \sum_i \alpha_i x_i = 1 \quad \sum_i \beta_i x_i = 1 \quad \text{hence also} \quad \sum_i (\alpha_i + \beta_i) x_i = 0$$

According to (7.50), the generalized interaction XY can then also be defined in terms of the partition

$$\sum_i (\alpha_i + \beta_i)x_i = 0, 1 \mod 2$$

This is the reason why we shall write the GI between $X = A_1^{\alpha_1} A_2^{\alpha_2} \cdots A_n^{\alpha_n}$ and $Y = A_1^{\beta_1} A_2^{\beta_2} \cdots A_n^{\beta_n}$ formally as

$$XY = A_1^{\alpha_1+\beta_1} A_2^{\alpha_2+\beta_2} \cdots A_n^{\alpha_n+\beta_n} \tag{7.51}$$

where the superscripts in (7.51) are reduced modulo 2, letters with superscript 0 being deleted and letters with superscript 1 being retained following previously established rules.

7.5 INTERPRETATION OF EFFECTS AND INTERACTIONS

The main effects and interactions are the building blocks of the theory of factorial experiments, a fact that will become more and more obvious in subsequent chapters. We have introduced the notion of main effects and interactions in such a way that they can be readily interpreted (see Sections 7.2 and 7.3). For example, the main effect A is the effect of changing factor A from the 0 level to the 1 level, averaging over all other factors, or the interaction AB is a measure of the extent that the simple effects of A at both levels of B, averaged over all other factors, are different from each other.

In addition to main effects and interactions, however, there may be other effects that are of interest to the experimenter. For example, in the presence of interaction AB, the simple effects $A(b_0)$ and $A(b_1)$ may be of interest, or we may ask: What is the effect of factor A when factor C is at the 0 level and we average over the levels of factor B? Using the previously established notation, we denote this effect by $A(\overline{b}, c_0)$, and it is given by

$$\begin{aligned} A(\overline{b}, c_0) &= \tfrac{1}{2}(a_1b_1c_0 - a_0b_1c_0 + a_1b_0c_0 - a_0b_0c_0) \\ &= \tfrac{1}{2}[ab - b + a - (1)] \end{aligned} \tag{7.52}$$

Alternatively, (7.52) can be expressed as

$$A(\overline{b}, c_0) = A - AC \tag{7.53}$$

which follows immediately from the definition of A and AC, namely

$$\begin{aligned} A &= \tfrac{1}{4}(a - 1)(b + 1)(c + 1) \\ AC &= \tfrac{1}{4}(a - 1)(b + 1)(c - 1) \end{aligned}$$

so that

$$A - AC = \tfrac{1}{4}(a-1)(b+1)$$

which is (7.52). More formally, we write (7.53) as the symbolic product

$$A(\overline{b}, c_0) = A(1 - C)$$

As another example we may ask: What is the interaction between factors A and B if factor C is at the 1 level and we average over the levels of factor D? The answer, of course, is

$$\begin{aligned} AB(c_1, \overline{d}) &= \tfrac{1}{4}(a_1b_1c_1d_0 - a_1b_0c_1d_0 - a_0b_1c_1d_0 + a_0b_0c_1d_0) \\ &\quad + a_1b_1c_1d_1 - a_1b_0c_1d_1 - a_0b_1c_1d_1 + a_0b_0c_1d_1 \\ &= \tfrac{1}{4}(abc - ac - bc + c + abcd - acd - bcd + cd) \\ &= \tfrac{1}{4}(a-1)(b-1)c(d+1) \\ &= AB + ABC \\ &= AB(1+C) \end{aligned} \tag{7.54}$$

We shall refer to effects and interactions such as (7.52) and (7.54) as *partial interactions*. For notational purposes in the 2^n case we partition the set $\mathcal{N} = (A_1, A_2, \ldots, A_n)$ of n factors into four disjoint sets:

$$\begin{aligned} \mathcal{N}_1 &= A_{i_1}, A_{i_2}, \ldots, A_{i_{n_1}} \\ \mathcal{N}_2 &= A_{j_1}, A_{j_2}, \ldots, A_{j_{n_2}} \\ \mathcal{N}_3 &= A_{k_1}, A_{k_2}, \ldots, A_{k_{n_3}} \\ \mathcal{N}_4 &= A_{\ell_1}, A_{\ell_2}, \ldots, A_{\ell_{n_4}} \end{aligned}$$

with $\mathcal{N}_1 \bigcup \mathcal{N}_2 \bigcup \mathcal{N}_3 \bigcup \mathcal{N}_4 = \mathcal{N}, n_1 + n_2 + n_3 + n_4 = n$ and $\mathcal{N}_1$ is always nonempty. We then write the partial interaction of all factors in $\mathcal{N}_1$ with all factors in $\mathcal{N}_2$ at the 0 level, all factors in $\mathcal{N}_3$ at the 1 level, and averaging over both levels of all factors in $\mathcal{N}_4$ as

$$A_{i_1}A_{i_2}\cdots A_{i_{n_1}}\left(a_{j_1,0}, \ldots, a_{j_{n_2},0}; a_{k_1,1}, \ldots, a_{k_{n_3},1}; \overline{a}_{\ell_1}, \ldots, \overline{a}_{\ell_{n_4}}\right) \tag{7.55}$$

By definition, (7.55) is given in terms of treatment combinations as

$$\frac{1}{2^{n_1+n_4-1}} \prod_{i \in \mathcal{N}_1} (a_i - 1) \prod_{k \in \mathcal{N}_3} a_k \prod_{\ell \in \mathcal{N}_4} (a_\ell + 1) \tag{7.56}$$

For purposes of expressing (7.55) in terms of effects and interactions, we rewrite (7.56) as

$$\frac{1}{2^{n_1+n_4-1}} \prod_{i \in \mathcal{N}_1} (a_i - 1) \prod_{j \in \mathcal{N}_2} \frac{1}{2}[(a_j + 1) - (a_j - 1)]$$

$$\prod_{k \in \mathcal{N}_3} \frac{1}{2}[(a_k + 1) + (a_k - 1)] \prod_{\ell \in \mathcal{N}_4} (a_\ell + 1)$$

Using (7.34) this product can then be written as the symbolic product

$$\prod_{i \in \mathcal{N}_1} A_i \prod_{j \in \mathcal{N}_2} (1 - A_j) \prod_{k \in \mathcal{N}_3} (1 + A_k) \tag{7.57}$$

In other words: After expansion (7.57) provides a representation of the partial interaction (7.55) in terms of effects and interactions. This representation rather than (7.56) may prove to be useful if (estimates of) effects and interactions are available, but the individual responses are not (see also Section 13.8).

We illustrate this result by the following example.

Example 7.1 Let $n = 7$, that is, we have factors $A_1, A_2, \ldots, A_7$. Suppose $\mathcal{N}_1 = (A_1, A_2)$, $\mathcal{N}_2 = (A_3, A_4)$, $\mathcal{N}_3 = (A_5)$, and $\mathcal{N}_4 = (A_6, A_7)$. Then

$$\begin{aligned} A_1A_2(a_{30}, a_{40}, a_{51}, \overline{a}_6, \overline{a}_7) &= A_1A_2 - A_1A_2A_3 - A_1A_2A_4 + A_1A_2A_3A_4 \\ &\quad + A_1A_2A_5 - A_1A_2A_3A_5 - A_1A_2A_4A_5 \\ &\quad + A_1A_2A_3A_4A_5 \end{aligned}$$

□

7.6 ANALYSIS OF FACTORIAL EXPERIMENTS

So far we have considered main effects and interactions as contrasts among the true treatment combination responses. These true responses are, of course, not available, but rather estimates of the true responses as the treatment combinations are used with a particular error control design. Consequently, we obtain estimates of the true effects and interactions.

7.6.1 Linear Models

In order to obtain such estimates, we utilize the fact that under additivity the observed yields will be given by a general linear model that characterizes the error control design that has been used. In its most general form such a model can be written as (using the notation of Section I.4.2)

$$\boldsymbol{y} \doteq \mu \boldsymbol{\mathfrak{I}} + \boldsymbol{X}_\delta \boldsymbol{\delta} + \boldsymbol{X}_\tau \boldsymbol{\tau} \tag{7.58}$$

where y is the vector of observations, δ is the vector of blocking effects, τ is the vector of treatment combination effects (by which we mean the true treatment combination response expressed as deviation from the true mean response), and X_δ and X_τ are known design-model matrices. For example, for the completely randomized design (see I.6) (7.58) is of the form

$$y \doteq \mu \mathfrak{I} + X_\tau \tau \tag{7.59}$$

For a randomized complete block design (see I.9) it is of the form

$$y \doteq \mu \mathfrak{I} + X_\beta \beta + X_\tau \tau \tag{7.60}$$

or for a Latin square design (see I.10) it is of the form

$$y \doteq \mu \mathfrak{I} + X_\rho \rho + X_\gamma \gamma + X_\tau \tau \tag{7.61}$$

The effects and interactions, as defined earlier, are then expressible as linear contrasts among the treatment combination effects, and we know that under the GMNLM (see I.4.17) the best estimate of such a contrast is the same contrast of the estimates of the treatment effects. For models (7.59), (7.60), and (7.61) with the same number, r say, of replications for each treatment combination, the best estimate of any treatment contrast is given by the same contrast of the treatment means, that is, to obtain the estimate of the interaction $A_1^{\alpha_1} A_2^{\alpha_2} \cdots A_n^{\alpha_n}$ we substitute for $a(x) = A_1^{x_1} A_2^{x_2} \cdots A_n^{x_n}$ its observed mean, $\overline{a}(x)$ say, into (7.34), using rule 1 of Section 7.4.1.

7.6.2 Yates Algorithm

A convenient way of obtaining those estimates is to apply the Yates algorithm (Yates, 1937b) which consists of adding and subtracting the treatment means in an appropriate way through several steps. Mainly for historical reasons we shall illustrate this algorithm for the 2^3 factorial. The steps are as follows:

Step 0: Write down the observed treatment means in standard order. (This is important since otherwise the algorithm will not work.)

Step 1: Starting from the top, add adjacent pairs of treatment means and record them in the first 4 $(= 2^{n-1})$ positions; for the same pairs, subtract the first from the second treatment mean and record those differences in the next 4 $(= 2^{n-1})$ positions.

Step 2: Repeat step 1 with the numbers recorded in step 1 (column 1).

Step 3: Repeat step 1 with the numbers obtained in step 2 (column 2); the numbers obtained are $8\widehat{M}, 4\widehat{A}, 4\widehat{B}, \ldots, 4\widehat{ABC}$, that is, apart from a constant the estimates of the mean response and of effects and interactions.

The three steps in terms of the formal expressions of the treatment means (leaving off the bar for convenience) are indicated below:

Step 0	1	2	3	Estimate
(1)	$a+1$	$(a+1)(b+1)$	$(a+1)(b+1)(c+1)$	$8\widehat{M}$
a	$b(a+1)$	$(a+1)(b+1)c$	$(a-1)(b+1)(c+1)$	$4\widehat{A}$
b	$ac(a+1)$	$(a-1)(b+1)$	$(a+1)(b-1)(c+1)$	$4\widehat{B}$
ab	$bc(a+1)$	$(a-1)(b+1)c$	$(a-1)(b-1)(c+1)$	$4\widehat{AB}$
c	$a-1$	$(a+1)(b-1)$	$(a+1)(b+1)(c-1)$	$4\widehat{C}$
ac	$b(a-1)$	$(a+1)(b-1)c$	$(a-1)(b+1)(c-1)$	$4\widehat{AC}$
bc	$ac(a-1)$	$(a-1)(b-1)$	$(a+1)(b-1)(c-1)$	$4\widehat{BC}$
abc	$bc(a-1)$	$(a-1)(b-1)c$	$(a-1)(b-1)(c-1)$	$4\widehat{ABC}$

For the 2^n factorial this algorithm requires n steps and at the last step we obtain an estimate of $2^{n-1}A_1^{\alpha_1}A_2^{\alpha_2}\cdots A_n^{\alpha_n}$ in standard order.

An inverse algorithm (Good, 1958) can be used to express the treatment combination responses in terms of effects and interactions. The mean, effects, and interactions are arranged in reverse standard order, and then we proceed exactly as in the Yates algorithm. We shall illustrate this again for the 2^3 factorial:

Step 0	1	2	3
$(a-1)(b-1)(c-1) = 4ABC$	$2a(b-1)(c-1)$	$4ab(c-1)$	$8abc$
$(a+1)(b-1)(c-1) = 4BC$	$2a(b+1)(c-1)$	$4ab(c+1)$	$8bc$
$(a-1)(b+1)(c-1) = 4AC$	$2a(b-1)(c+1)$	$4b(c-1)$	$8ac$
$(a+1)(b+1)(c-1) = 4C$	$2a(b+1)(c+1)$	$4b(c+1)$	$8c$
$(a-1)(b-1)(c+1) = 4AB$	$2(b-1)(c-1)$	$4a(c-1)$	$8ab$
$(a+1)(b-1)(c+1) = 4B$	$2(b+1)(c-1)$	$4a(c+1)$	$8b$
$(a-1)(b+1)(c+1) = 4A$	$2(b-1)(c+1)$	$4(c-1)$	$8a$
$(a+1)(b+1)(c+1) = 8M$	$2(b+1)(c+1)$	$4(c+1)$	$8(1)$

This shows that if we start at step 0 with $\frac{1}{2}ABC$, $\frac{1}{2}BC, \ldots, \frac{1}{2}A$, M we obtain the treatment combination response averages in reverse standard order. This method thus provides an easy way of estimating the residuals.

A modification of the Yates algorithm has been given by Riedwyl (1998) with a view toward applications for systems of confounding (Chapters 8 and 9) and fractional factorials (Chapter 13).

7.6.3 Variances of Estimators

Returning now to the estimation of effects and interactions, we know that each effect and interaction is estimated as

$$\frac{1}{2^{n-1}}[(\text{sum of } 2^{n-1} \text{ response means}) \\ - (\text{sum of } 2^{n-1} \text{ response means})]$$

where each response mean is based on r replications. Since each observation has variance σ_e^2 and since all the observations can be treated as being uncorrelated (see the argument in I.6.3.4), we obtain

$$\text{var}(\widehat{E^{\alpha}}) = \frac{1}{r2^{n-2}}\sigma_e^2 \tag{7.62}$$

for all $E^{\alpha} = A_1^{\alpha_1} A_2^{\alpha_2} \cdots A_n^{\alpha_n}$. Since all the effects and interactions are orthogonal to each other, so are their estimates, and hence the variance of a partial interaction (see Section 7.5) consisting of a linear combination of m interactions is $m\, \sigma_e^2/r2^{n-2}$.

7.6.4 Analysis of Variance

To estimate the variance (7.62) and that of partial interactions we need to estimate σ_e^2. This is accomplished through the ANOVA associated with the basic model (7.58) as outlined in Chapter I.4. Such an ANOVA can, of course, be written out explicitly for models (7.59), (7.60), and (7.61) according to the general rules for such error control designs discussed in Chapters I.6, I.9, and I.10, respectively. In any case, MS(Error) is an estimate of σ_e^2.

We shall comment briefly on the sum of squares due to treatments, SS(Treatments) say, which in the present situation has $t - 1 = 2^n - 1$ d.f. Using the fact that effects and interactions represent $2^n - 1$ orthogonal comparisons of the treatment effects, we can partition SS(Treatments) into $2^n - 1$ orthogonal 1-d.f. sums of squares associated with these comparisons (see Chapter I.7). We denote the sum of squares associated with $E^{\alpha} = A_1^{\alpha_1} A_2^{\alpha_2} \cdots A_n^{\alpha_n}$ by SS(E^{α}), and according to the general rules (see I.7.2.3)

$$\begin{aligned} \text{SS}\left(E^{\alpha}\right) &= \frac{\left(\widehat{E^{\alpha}}\right)^2}{\text{var}\left(\hat{E}^{\alpha}\right)/\sigma_e^2} \\ &= r2^{n-2}\left[\hat{E}^{\alpha}\right]^2 \end{aligned} \tag{7.63}$$

using (7.62).

Another way of writing SS(E^α), a generalization of which will be used in later chapters, is as follows. We recall from (7.41) that

$$E^\alpha = (-1)^{\sum \alpha_i} \left(E_0^\alpha - E_1^\alpha\right)$$

In terms of observed quantities we then write

$$\widehat{E}^\alpha = (-1)^{\sum \alpha_i} \left(\widehat{E}_0^\alpha - \widehat{E}_1^\alpha\right)$$

It follows then from (7.63) that

$$\text{SS}\left(E^\alpha\right) = r2^{n-2} \left(\widehat{E}_0^\alpha - \widehat{E}_1^\alpha\right)^2 \tag{7.64}$$

Using the fact that $\widehat{E}_0^\alpha + \widehat{E}_1^\alpha = 0$ we can rewrite (7.64) as

$$\begin{aligned} \text{SS}\left(E^\alpha\right) &= r2^{n-2} \left[\widehat{E}_0^\alpha - \widehat{E}_1^\alpha + \widehat{E}_0^\alpha + \widehat{E}_1^\alpha\right]^2 \\ &= r2^n \left(\widehat{E}_0^\alpha\right)^2 \end{aligned}$$

or

$$\begin{aligned} \text{SS}\left(E^\alpha\right) &= r2^{n-2} \left[\widehat{E}_0^\alpha - \widehat{E}_1^\alpha - \widehat{E}_0^\alpha - \widehat{E}_1^\alpha\right]^2 \\ &= r2^n \left(\widehat{E}_1^\alpha\right)^2 \end{aligned}$$

or

$$\begin{aligned} \text{SS}\left(E^\alpha\right) &= \tfrac{1}{2} r2^n \left[\left(\widehat{E}_0^\alpha\right)^2 + \left(\widehat{E}_1^\alpha\right)^2\right] \\ &= r2^{n-1} \left[\left(\widehat{E}_0^\alpha\right)^2 + \left(\widehat{E}_1^\alpha\right)^2\right] \end{aligned} \tag{7.65}$$

Now

$$\begin{aligned} E\left[\widehat{E}^\alpha\right]^2 &= \text{var}\left(\widehat{E}^\alpha\right) + \left[E\left(\widehat{E}^\alpha\right)\right]^2 \\ &= \frac{1}{r2^{n-2}} \sigma_e^2 + r2^{n-2} \left[E^\alpha\right]^2 \end{aligned} \tag{7.66}$$

so that, using (7.63) and (7.66),

$$E\left[\text{SS}\left(E^\alpha\right)\right] = \sigma_e^2 + r2^{n-2} \left[E^\alpha\right]^2 \tag{7.67}$$

Under the GMNLM we can then test

$$H_0\colon\ E^\alpha = 0$$

by means of the F test (as an approximation to the randomization test)

$$F = \frac{\text{SS}(E^\alpha)}{\text{MS(Error)}}$$

Table 7.1 ANOVA for 2^3 Factorial in RCBD

Source	d.f.	SS	E(MS)
Blocks	$r-1$	$8\sum_i(\overline{y}_{i\ldots}-\overline{y}_{\ldots.})^2$	
Treatments	7	$r\sum_{jk\ell}(\overline{y}_{\cdot jk\ell}-\overline{y}_{\ldots.})^2$	
A	1	$2r[\widehat{A}]^2$	$\sigma_e^2+2r[A]^2$
B	1	$2r[\widehat{B}]^2$	$\sigma_e^2+2r[B]^2$
AB	1	$2r[\widehat{AB}]^2$	$\sigma_e^2+2r[AB]^2$
C	1	$2r[\widehat{C}]^2$	$\sigma_e^2+2r[C]^2$
AC	1	$2r[\widehat{AC}]^2$	$\sigma_e^2+2r[AC]^2$
BC	1	$2r[\widehat{BC}]^2$	$\sigma_e^2+2r[BC]^2$
ABC	1	$2r[\widehat{ABC}]^2$	$\sigma_e^2+2r[ABC]^2$
Error	$7(r-1)$	$\sum_{ijk\ell}(y_{ijk\ell}-\overline{y}_{i\ldots}-\overline{y}_{\cdot jk\ell}+\overline{y}_{\ldots.})^2$	σ_e^2
Total	$8r-1$	$\sum_{ijk\ell}(\overline{y}_{ijk\ell}-\overline{y}_{\ldots})^2$	

As an example of the present discussion we give in Table 7.1 the ANOVA of a 2^3 experiment in a RCB design with r blocks, using the model equation

$$y_{ijk\ell}=\mu+\beta_i+\tau_{jk\ell}+e_{ijk\ell} \tag{7.68}$$

7.6.5 Numerical Examples

To illustrate some aspects of the theoretical developments in the previous sections, we shall consider two numerical examples. The analysis will be performed by using SAS PROC GLM and PROC MIXED (SAS Institute, 1999–2001), and we shall provide some comments on the output.

Example 7.2 Consider a 2^3 factorial in an RCBD with $r=2$ blocks. The data and the analysis, using the model equation (7.68), are given in Table 7.2.

Although most of the output is self-explanatory we shall make the following comments:

1. Estimates of the main and interaction effects can be obtained by using the ESTIMATE option. The coefficients for the contrasts are obtained by utilizing the relationship between the various parameterizations, as explained in Section 7.2.3.
2. The estimates of the effects are, of course, obtained as linear contrasts among the LSMEANS using the coefficients as specified in the ESTIMATE

Table 7.2 Data and Analysis for 2^3 Factorial in a RCBD

```
options pageno=1 nodate;
data example1;
input block A B C y @@;
datalines;
1 0 0 0 13  1 1 0 0 20  1 0 1 0 14  1 1 1 0 17
1 0 0 1 21  1 1 0 1 25  1 0 1 1 19  1 1 1 1 22
2 0 0 0 24  2 1 0 0 27  2 0 1 0 28  2 1 1 0 31
2 0 0 1 35  2 1 0 1 39  2 0 1 1 34  2 1 1 1 40
;
run;

proc print data=example1;
title1 'TABLE 7.2';
title2 'DATA FOR 2**3 FACTORIAL';
title3 'IN A RCBD';
run;

proc glm data=example1;
class block A B C;
model y=block A|B|C;
lsmeans A B C A*B A*C B*C A*B*C/stderr;
contrast 'Main effect A' A -1 1;
estimate 'Main effect A' A -1 1;
contrast 'Interaction AB' A*B .5 -.5 -.5 .5;
estimate 'Interaction AB' A*B .5 -.5 -.5 .5;
contrast 'Interaction ABC' A*B*C -.25 .25 .25 -.25 .25 -.25 -.25 .25;
estimate 'Interaction ABC' A*B*C -.25 .25 .25 -.25 .25 -.25 -.25 .25;
title1 'ANALYSIS OF 2**3 FACTORIAL';
title2 'IN A RCBD';
run;
```

Obs	block	A	B	C	y
1	1	0	0	0	13
2	1	1	0	0	20
3	1	0	1	0	14
4	1	1	1	0	17
5	1	0	0	1	21
6	1	1	0	1	25
7	1	0	1	1	19
8	1	1	1	1	22
9	2	0	0	0	24
10	2	1	0	0	27
11	2	0	1	0	28
12	2	1	1	0	31
13	2	0	0	1	35
14	2	1	0	1	39

Table 7.2 (*Continued*)

15	2	0	1	1	34
16	2	1	1	1	40

```
                    ANALYSIS OF 2**3 FACTORIAL
                             IN A RCBD

                         The GLM Procedure

                      Class Level Information

                     Class       Levels      Values

                     block            2      1 2

                     A                2      0 1

                     B                2      0 1

                     C                2      0 1

                    Number of observations 16

                         The GLM Procedure

Dependent Variable: y

                                  Sum of
Source                  DF       Squares     Mean Square    F Value    Pr > F

Model                    8   1026.000000      128.250000      24.98    0.0002

Error                    7     35.937500        5.133929

Corrected Total         15   1061.937500


             R-Square     Coeff Var      Root MSE       y Mean

             0.966159      8.863833      2.265817     25.56250

Source          DF        Type I SS     Mean Square     F Value     Pr > F

block            1      715.5625000     715.5625000      139.38     <.0001
A                1       68.0625000      68.0625000       13.26     0.0083
B                1        0.0625000       0.0625000        0.01     0.9152
A*B              1        0.5625000       0.5625000        0.11     0.7503
C                1      232.5625000     232.5625000       45.30     0.0003
A*C              1        0.0625000       0.0625000        0.01     0.9152
B*C              1        7.5625000       7.5625000        1.47     0.2642
A*B*C            1        1.5625000       1.5625000        0.30     0.5983
```

Table 7.2 (*Continued*)

Source	DF	Type III SS	Mean Square	F Value	Pr > F
block	1	715.5625000	715.5625000	139.38	<.0001
A	1	68.0625000	68.0625000	13.26	0.0083
B	1	0.0625000	0.0625000	0.01	0.9152
A*B	1	0.5625000	0.5625000	0.11	0.7503
C	1	232.5625000	232.5625000	45.30	0.0003
A*C	1	0.0625000	0.0625000	0.01	0.9152
B*C	1	7.5625000	7.5625000	1.47	0.2642
A*B*C	1	1.5625000	1.5625000	0.30	0.5983

Least Squares Means

A	y LSMEAN	Standard Error	Pr > |t|
0	23.5000000	0.8010874	<.0001
1	27.6250000	0.8010874	<.0001

B	y LSMEAN	Standard Error	Pr > |t|
0	25.5000000	0.8010874	<.0001
1	25.6250000	0.8010874	<.0001

C	y LSMEAN	Standard Error	Pr > |t|
0	21.7500000	0.8010874	<.0001
1	29.3750000	0.8010874	<.0001

A	B	y LSMEAN	Standard Error	Pr > |t|
0	0	23.2500000	1.1329087	<.0001
0	1	23.7500000	1.1329087	<.0001
1	0	27.7500000	1.1329087	<.0001
1	1	27.5000000	1.1329087	<.0001

A	C	y LSMEAN	Standard Error	Pr > |t|
0	0	19.7500000	1.1329087	<.0001
0	1	27.2500000	1.1329087	<.0001
1	0	23.7500000	1.1329087	<.0001
1	1	31.5000000	1.1329087	<.0001

Table 7.2 (*Continued*)

B	C	y LSMEAN	Standard Error	Pr > |t|
0	0	21.0000000	1.1329087	<.0001
0	1	30.0000000	1.1329087	<.0001
1	0	22.5000000	1.1329087	<.0001
1	1	28.7500000	1.1329087	<.0001

A	B	C	y LSMEAN	Standard Error	Pr > |t|
0	0	0	18.5000000	1.6021749	<.0001
0	0	1	28.0000000	1.6021749	<.0001
0	1	0	21.0000000	1.6021749	<.0001
0	1	1	26.5000000	1.6021749	<.0001
1	0	0	23.5000000	1.6021749	<.0001
1	0	1	32.0000000	1.6021749	<.0001
1	1	0	24.0000000	1.6021749	<.0001
1	1	1	31.0000000	1.6021749	<.0001

Dependent Variable: y

Contrast	DF	Contrast SS	Mean Square	F Value	Pr > F
Main effect A	1	68.06250000	68.06250000	13.26	0.0083
Interaction AB	1	0.56250000	0.56250000	0.11	0.7503
Interaction ABC	1	1.56250000	1.56250000	0.30	0.5983

Parameter	Estimate	Standard Error	t Value	Pr > |t|
Main effect A	4.12500000	1.13290871	3.64	0.0083
Interaction AB	-0.37500000	1.13290871	-0.33	0.7503
Interaction ABC	0.62500000	1.13290871	0.55	0.5983

statements. Recall that for an orthogonal design the least-squares means are just the ordinary means. Thus, we have, for example,

$$\begin{aligned}\widehat{AB} &= \tfrac{1}{2}[\text{LSMEAN}(a_0b_0) - \text{LSMEAN}(a_0b_1) \\ &\qquad - \text{LSMEAN}(a_1b_0) + \text{LSMEAN}(a_1b_1)] \\ &= \tfrac{1}{2}(\overline{y}_{\cdot 00\cdot} - \overline{y}_{\cdot 01\cdot} - \overline{y}_{\cdot 10\cdot} + \overline{y}_{\cdot 11\cdot}) \\ &= \tfrac{1}{2}(23.25 - 23.75 - 27.75 + 27.5) \\ &= -.375 \end{aligned} \tag{7.69}$$

3. The $\widehat{E}_i^{\alpha}$ terms, as introduced in Section 7.4.2, can easily be obtained from the LSMEANS. For example,

$$\begin{aligned} E_0^{(110)} &= \tfrac{1}{2}\,[\text{LSMEAN}(a_0b_0) + \text{LSMEAN}(a_1b_1)] \\ &\quad - \text{overall mean} \\ &= \tfrac{1}{2}\,(\overline{y}_{\cdot 00\cdot} + \overline{y}_{\cdot 11\cdot}) - \overline{y}_{\cdots\cdot} \\ &= \tfrac{1}{2}\,(23.25 + 27.50) - 25.5625 \\ &= -.1875 \end{aligned}$$

and

$$E_1^{(110)} = .1875$$

4. From the estimate in comment 2 and (7.63)we obtain

$$\text{SS}(AB) = 2 \cdot 2 \cdot (-.375)^2 = .5625$$

The same value is, of course, obtained in the ANOVA table and through the CONTRAST statement.

5. The standard error (SE) for the estimate of an effect is provided through the ESTIMATE statement. It can also be obtained directly from the information provided for the LSMEANS. For example, making use of (7.69), we obtain

$$\begin{aligned} \widehat{\text{var}}(\widehat{AB}) &= \tfrac{1}{4} \cdot 4\,\widehat{\text{var}}\,[\text{LSMEAN}(a_0b_0)] \\ &= (1.133)^2 = 1.283 \end{aligned}$$

and hence

$$\text{SE}(\widehat{AB}) = \sqrt{\widehat{\text{var}}(\widehat{AB})} = 1.133$$

The same result can be obtained by substituting in (7.62) the estimate for σ_e^2, namely $\widehat{\sigma}_e^2 =$ MS(Error) = 5.134 from the ANOVA table,

$$\text{SE}(\widehat{AB}) = \sqrt{\tfrac{1}{4} \cdot 5.134} = 1.133 \qquad \square$$

Example 7.3 The following is an example of combining all three elements of an experimental design, namely error control design, treatment design, and sampling design (see I.2.23 and I.3). More specifically, we consider a 2^3 factorial in an RCBD with $r' = 2$ blocks and subsampling ($n = 2$). The data and the analysis are given in Table 7.3.

Table 7.3 Data and Analysis for 2^3 Factorial in a RCBD With Subsampling

```
options pageno=1 nodate;
data example2;
input block A B C y @@;
datalines;
1 0 0 0 13  1 0 0 0 14  1 1 0 0 20  1 1 0 0 22
1 0 1 0 14  1 0 1 0 11  1 1 1 0 17  1 1 1 0 17
1 0 0 1 21  1 0 0 1 22  1 1 0 1 25  1 1 0 1 23
1 0 1 1 19  1 0 1 1 20  1 1 1 1 22  1 1 1 1 23
2 0 0 0 24  2 0 0 0 26  2 1 0 0 27  2 1 0 0 26
2 0 1 0 28  2 0 1 0 27  2 1 1 0 31  2 1 1 0 32
2 0 0 1 35  2 0 0 1 37  2 1 0 1 39  2 1 0 1 39
2 0 1 1 34  2 0 1 1 35  2 1 1 1 40  2 1 1 1 42
;
run;

proc print data=example2;
title1 'TABLE 7.3';
title2 'DATA FOR 2**3 FACTORIAL';
title3 'IN A RCBD WITH SUBSAMPLING';
run;

proc glm data=example2;
class block A B C;
model y=block A|B|C block*A*B*C;
test H=A B C A*B A*C B*C A*B*C E=block*A*B*C;
title1 'ANALYSIS OF 2**3 FACTORIAL';
title2 'IN A RCBD WITH SUBSAMPLING';
run;

proc mixed data=example2;
class block A B C;
model y=block A|B|C;
random block*A*B*C;
estimate 'Main effect A' A -1 1;
estimate 'Interaction AB' A*B .5 -.5 -.5 .5;
estimate 'Interaction ABC' A*B*C -.25 .25 .25 -.25 .25 -.25 -.25 .25;
run;
```

Obs	block	A	B	C	y
1	1	0	0	0	13
2	1	0	0	0	14
3	1	1	0	0	20
4	1	1	0	0	22
5	1	0	1	0	14
6	1	0	1	0	11
7	1	1	1	0	17

Table 7.2 (*Continued*)

8	1	1	1	0	17
9	1	0	0	1	21
10	1	0	0	1	22
11	1	1	0	1	25
12	1	1	0	1	23
13	1	0	1	1	19
14	1	0	1	1	20
15	1	1	1	1	22
16	1	1	1	1	23
17	2	0	0	0	24
18	2	0	0	0	26
19	2	1	0	0	27
20	2	1	0	0	26
21	2	0	1	0	28
22	2	0	1	0	27
23	2	1	1	0	31
24	2	1	1	0	32
25	2	0	0	1	35
26	2	0	0	1	37
27	2	1	0	1	39
28	2	1	0	1	39
29	2	0	1	1	34
30	2	0	1	1	35
31	2	1	1	1	40
32	2	1	1	1	42

ANALYSIS OF 2**3 FACTORIAL
IN A RCBD WITH SUBSAMPLING

The GLM Procedure

Class Level Information

Class	Levels	Values
block	2	1 2
A	2	0 1
B	2	0 1
C	2	0 1

Number of observations 32

Dependent Variable: y

Source	DF	Sum of Squares	Mean Square	F Value	Pr > F
Model	15	2244.968750	149.664583	129.44	<.0001

Table 7.3 *(Continued)*

Error	16	18.500000	1.156250		
Corrected Total	31	2263.468750			

R-Square	Coeff Var	Root MSE	y Mean
0.991827	4.170824	1.075291	25.78125

Source	DF	Type I SS	Mean Square	F Value	Pr > F
block	1	1498.781250	1498.781250	1296.24	<.0001
A	1	132.031250	132.031250	114.19	<.0001
B	1	0.031250	0.031250	0.03	0.8715
A*B	1	1.531250	1.531250	1.32	0.2667
C	1	504.031250	504.031250	435.92	<.0001
A*C	1	0.781250	0.781250	0.68	0.4232
B*C	1	3.781250	3.781250	3.27	0.0894
A*B*C	1	2.531250	2.531250	2.19	0.1584
block*A*B*C	7	101.468750	14.495536	12.54	<.0001

Source	DF	Type III SS	Mean Square	F Value	Pr > F
block	1	1498.781250	1498.781250	1296.24	<.0001
A	1	132.031250	132.031250	114.19	<.0001
B	1	0.031250	0.031250	0.03	0.8715
A*B	1	1.531250	1.531250	1.32	0.2667
C	1	504.031250	504.031250	435.92	<.0001
A*C	1	0.781250	0.781250	0.68	0.4232
B*C	1	3.781250	3.781250	3.27	0.0894
A*B*C	1	2.531250	2.531250	2.19	0.1584
block*A*B*C	7	101.468750	14.495536	12.54	<.0001

Tests of Hypotheses Using the Type III MS for block*A*B*C as an Error Term

Source	DF	Type III SS	Mean Square	F Value	Pr > F
A	1	132.0312500	132.0312500	9.11	0.0194
B	1	0.0312500	0.0312500	0.00	0.9643
C	1	504.0312500	504.0312500	34.77	0.0006
A*B	1	1.5312500	1.5312500	0.11	0.7547
A*C	1	0.7812500	0.7812500	0.05	0.8231
B*C	1	3.7812500	3.7812500	0.26	0.6252
A*B*C	1	2.5312500	2.5312500	0.17	0.6886

Table 7.3 (*Continued*)

```
                 The Mixed Procedure

                  Model Information

Data Set                        WORK.EXAMPLE2
Dependent Variable              y
Covariance Structure            Variance Components
Estimation Method               REML
Residual Variance Method        Profile
Fixed Effects SE Method         Model-Based
Degrees of Freedom Method       Containment

              Class Level Information

           Class     Levels     Values

           block          2     1 2
           A              2     0 1
           B              2     0 1
           C              2     0 1

                    Dimensions

           Covariance Parameters       2
           Columns in X               29
           Columns in Z               16
           Subjects                    1
           Max Obs Per Subject        32
           Observations Used          32
           Observations Not Used       0
           Total Observations         32

                 Iteration History

Iteration    Evaluations    -2 Res Log Like       Criterion

        0              1       116.43092171
        1              1        99.48076616      0.00000000

              Convergence criteria met.

               Covariance Parameter
                    Estimates

           Cov Parm               Estimate

           block*A*B*C              6.6696
           Residual                 1.1563
```

Table 7.3 (*Continued*)

Fit Statistics

-2 Res Log Likelihood	99.5
AIC (smaller is better)	103.5
AICC (smaller is better)	104.1
BIC (smaller is better)	105.0

Type 3 Tests of Fixed Effects

Effect	Num DF	Den DF	F Value	Pr > F
block	1	7	103.40	<.0001
A	1	7	9.11	0.0194
B	1	7	0.00	0.9643
A*B	1	7	0.11	0.7547
C	1	7	34.77	0.0006
A*C	1	7	0.05	0.8231
B*C	1	7	0.26	0.6252
A*B*C	1	7	0.17	0.6886

Estimates

Label	Estimate	Standard Error	DF	t Value	Pr > |t|
Main effect A	4.0625	1.3461	7	3.02	0.0194
Interaction AB	0.4375	1.3461	7	0.33	0.7547
Interaction ABC	0.5625	1.3461	7	0.42	0.6886

The analysis is based on the model

$$y_{ijk\ell m} = \mu + \beta_i + \tau_{jk\ell} + \epsilon_{ijk\ell} + \eta_{ijk\ell m}$$

(see I.6.9.1).

The following comments are intended to relate the output to the theoretical developments:

1. SAS PROC GLM is being used only to obtain an ANOVA table. Because of subsampling it is necessary to explicitly define the experimental error as block $* A * B * C$ in order to perform the correct tests about all effects, using MS(X)/MS(block $* A * B * C$), where X represents any of the seven effects and MS(Experimental error) = MS(block $* A * B * C$) represents

technically speaking, the 7 d.f. associated with the seven 1-d.f. interactions block $* X$.

2. Specifying the correct error term is not necessary when using SAS PROC MIXED because the random term block $* A * B * C$ automatically assumes that role. This can be verified by comparing the tests performed by PROC GLM and PROC MIXED (note, however, that the test about block effects should be ignored; see I.9.2.6).
3. PROC MIXED provides REML estimates of σ_ϵ^2 and σ_η^2 (PROC GLM provides an estimate of σ_η^2 only). We obtain

$$\widehat{\sigma}_\epsilon^2 = 6.6696 \qquad \widehat{\sigma}_\eta^2 = 1.1563$$

4. In estimating the various effects, and their standard errors as illustrated here for A, AB, and ABC, by means of the ESTIMATE statement, PROC MIXED uses the correct error term, namely MS(Experimental error).
5. The above statement is true also for the LSMEANS, and their standard error. □

7.6.6 Use of Only One Replicate

In the previous section we have assumed that each treatment combination is replicated r times in a certain error control design such as a CRD or a RCBD. In any case we usually have $r > 1$, according to the general principles of experimental design. One can visualize, however, that in a 2^n factorial experiment (particularly an exploratory-type experiment) the number of treatment combinations can be rather large, in fact so large that the experimenter cannot afford to replicate the treatments. How should we deal with such a situation?

Even with only one replicate, that is, $r = 1$, we can, of course, estimate all effects and interactions in the usual way. We cannot, however, estimate their variances as there are no d.f. left for error in the ANOVA. We then recall (see Chapter I.11) that many-factor, that is, higher order, interactions are often negligible. Assuming that they are actually zero, it follows then from (7.67) that the expected value of their sums of squares is σ_e^2. The idea then is to pool all sums of squares associated with interactions assumed (a priori) to be negligible and treat the resulting sum of squares as SS(Error) for analysis purposes. Even if the interactions are not zero (and they seldom are), this procedure will result in a conservative F test for the remaining effects and interactions.

CHAPTER 8

Confounding in 2^n Factorial Designs

8.1 INTRODUCTION

It is well known that the experimental error variance is related to the size of the block, increasing to a greater or lesser degree as the block size increases. If we were testing 5 factors each at 2 levels, we would have 32 treatment combinations and hence, if we were to use a randomized complete block design, we would require blocks of size 32 in order to compare them. In field experiments it is generally acknowledged that, if at all possible, the size of the block should not be greater than 16 and if possible 8, though, of course, there are no hard and fast rules, and one experiment with blocks of 16 EUs may well give a lower experimental error than another on similar material with blocks of 8 EUs. In other situations, as, for example, in industrial experimentation, it may be impossible to have blocks as large as this without randomizing over factors that contribute considerable variation, thereby causing large experimental error. Or in experiments with animals (e.g. mice), the litter size determines and hence limits the block size. It is desirable, therefore, to have some means of reducing the block size or of making use of blocks of smaller size (i.e., incomplete blocks), and for this purpose the device of *confounding* has been introduced. It is a fundamental concept in the theory and application of factorial experiments.

8.1.1 A Simple Example

We have introduced the concept of confounding certain interactions (or effects, in general) with block effects in Section I.11.5. The basic idea is that information about unimportant, that is, usually negligible interactions is sacrificed in the sense that only biased estimates of the confounded interactions are available, and the bias is a linear function (contrast) of block effects. For example, for the 2^3 factorial (with factors A, B, C) with blocks of size 4, we might confound the

Design and Analysis of Experiments. Volume 2: Advanced Experimental Design
By Klaus Hinkelmann and Oscar Kempthorne
ISBN 0-471-55177-5

3-factor interaction, ABC, with blocks that lead to the arrangement

$$\begin{array}{ll} \text{Block 1} & (1), ab, ac, bc \\ \text{Block 2} & a, b, c, abc \end{array} \tag{8.1}$$

Since for each observation we have derived a linear model of the form

$$\text{Observation} = \text{mean} + \text{block effect} + \text{treatment effect} + \text{error} \tag{8.2}$$

(see Sections I.9.2 and 1.3) and since block 1 contains all the treatment combinations that enter negatively into ABC (see Section 7.3) and block 2 contains those entering positively, we have

$$E(\widehat{ABC}) = ABC + (\beta_2 - \beta_1) \tag{8.3}$$

where β_1, β_2 are the block effects, in other words, the interaction ABC and the difference of block effects, $\beta_2 - \beta_1$, cannot be separated. We say, for short, that ABC is *confounded with blocks*.

8.1.2 Comparison of Information

As we have mentioned earlier (see Section I.11.2) the main value of the factorial structure lies in the fact that in many situations the high-order interactions are of inappreciable magnitude. If in the situation described above the 3-factor interaction can be assumed to be negligible, then the basic arrangement (8.1) is quite satisfactory since it allows the unbiased estimation of all main effects and 2-factor interactions. For practical purposes, however, one would have to repeat the basic arrangement (8.1), say r times (using, of course, a different randomization every time). Thus, with r replications for each treatment combination, we can compare the information on each effect and interaction obtained from the present design with that obtained from a design without confounding, that is, the RCBD. Using the reciprocal of the true variance of each estimator as a measure of the information on each effect and interaction, and denoting by σ_k^2 the true error variance [of the error in model (8.2)] with blocks of size k, we have the results of Table 8.1. Since generally $\sigma_4^2 < \sigma_8^2$, we have increased the information on main effects and 2-factor interactions in the ratio σ_8^2/σ_4^2 by using blocks of size 4 at the expense of obtaining zero information on the 3-factor interaction instead of $2r/\sigma_8^2$ units.

8.1.3 Basic Analysis

Before discussing the problem of confounding and appropriate systems of confounding more generally and more formally, we shall conclude this introductory section by outlining briefly the analysis of the experiment discussed above. As indicated earlier, all main effects and 2-factor interactions are estimated in the

Table 8.1 Comparison of Information

Effect or Interaction	Information with Blocks of 8 Units	Information with Blocks of 4 Units
A	$\dfrac{2r}{\sigma_8^2}$	$\dfrac{2r}{\sigma_4^2}$
B	$\dfrac{2r}{\sigma_8^2}$	$\dfrac{2r}{\sigma_4^2}$
AB	$\dfrac{2r}{\sigma_8^2}$	$\dfrac{2r}{\sigma_4^2}$
C	$\dfrac{2r}{\sigma_8^2}$	$\dfrac{2r}{\sigma_4^2}$
AC	$\dfrac{2r}{\sigma_8^2}$	$\dfrac{2r}{\sigma_4^2}$
BC	$\dfrac{2r}{\sigma_8^2}$	$\dfrac{2r}{\sigma_4^2}$
ABC	$\dfrac{2r}{\sigma_8^2}$	Zero

usual way based upon the average responses from the r replicates. We can, for example, use the Yates algorithm and obtain the estimates $\widehat{A}, \widehat{B}, \widehat{AB}, \widehat{C}, \widehat{AC}, \widehat{BC}$. We also obtain $\widehat{ABC}$, but we know already that it is an estimate of ABC plus a function of block effect differences. In order to distinguish it from unbiased estimates we shall denote it by $\widetilde{ABC}$ rather than $\widehat{ABC}$.

For purposes of writing out the analysis of variance we rewrite model (8.2) more explicitly as

$$y_{ijk\ell m} = \mu + \beta_{ij} + \tau_{k\ell m} + e_{ijk\ell m} \tag{8.4}$$

where β_{ij} is the effect of the jth block in the ith replicate ($i = 1, 2, \ldots, r$; $j = 1, 2$), $\tau_{k\ell m}$ is the effect of the treatment combination $a_k b_\ell c_m (k, \ell, m = 0, 1)$, and $e_{ijk\ell m}$ is the usual error, that is, experimental and observational error.

We then have the usual breakdown of the total sum of squares into the block sum of squares, the treatment sum of squares, and the error sum of squares as exhibited in Table 8.2. The treatment sum of squares is partitioned in the usual way (as explained in Chapter 7), except that there is no SS_{ABC}.

We close this section with a few remarks about some of the features of Table 8.2:

1. Due to the confounding of ABC we have only 6 d.f. for treatments, which means that this incomplete block design is a disconnected design (see Section 1.5). It is, of course, disconnected by choice.
2. The sums of squares for estimable (and unconfounded) main effects and interactions are obtained as if an RCBD had been used, that is, by linear combinations of treatment means. This is contrary to what would have happened had we assigned the treatment combinations in other ways to existing incomplete block designs, for example, BIB or PBIB designs. In that case the method described in Section 1.3 would have to be used, that is, treatment means would have to be replaced by least-squares means.
3. The statement above implies that the estimable effects are orthogonal to block effects; that is, we have, for example, that SS(A|Blocks) = SS(A).

Table 8.2 Analysis of Variance Table for 2^3 Factorial in Blocks of Size 4

Source of Variation	d.f.	Sum of Squares	Expected Mean Squares
Blocks	$2r - 1$	$\frac{1}{4}\sum_{ij} y_{ij\cdots}^2 - \frac{y_{\cdots\cdots}^2}{8r}$	
Replicates	$r - 1$	$\frac{1}{8}\sum_{i} y_{i\cdots\cdot}^2 - \frac{y_{\cdots\cdots}^2}{8r}$	
ABC	1	$2r[\widehat{ABC}]^2$	
Remainder	$r - 1$	Difference	
Treatments	6		
A	1	$2r[\widehat{A}]^2$	$\sigma_e^2 + 2r[A]^2$
B	1	$2r[\widehat{B}]^2$	$\sigma_e^2 + 2r[B]^2$
AB	1	$2r[\widehat{AB}]^2$	$\sigma_e^2 + 2r[AB]^2$
C	1	$2r[\widehat{C}]^2$	$\sigma_e^2 + 2r[C]^2$
AC	1	$2r[\widehat{AC}]^2$	$\sigma_e^2 + 2r[AC]^2$
BC	1	$2r[\widehat{BC}]^2$	$\sigma_e^2 + 2r[BC]^2$
Error	$6(r - 1)$	Difference	σ_e^2
Total	$8r - 1$	$\sum_{ijk\ell m} y_{ijk\ell m}^2 - \frac{y_{\cdots\cdots}^2}{8r}$	

4. Although it is not important at this point but will become important later (see Section 8.7), SS(Blocks) can be partitioned in various ways. Recall that we have r replicates, each consisting of two blocks. We therefore have

$$\text{SS(Blocks)} = \text{SS(Replicates)} + \text{SS(Blocks/Replicates)}$$

with $r - 1$ and r d.f., respectively. Now, the makeup of the two blocks in each replicate is the same, that is, one block contains the treatment combinations (1), ab, ac, bc and the other block contains a, b, c, abc. Apart from block effects, the comparison between the two blocks is an "estimate" of the (confounded) interaction ABC. Hence we can write

$$\text{SS(Blocks/Replicates)} = \text{SS}(ABC) + \text{SS(Remainder)}$$

where SS(ABC) is obtained in the usual way (see remark 2 above).

5. Although we indicate that SS(Error) and SS(Remainder) can be obtained by subtraction, there is also a direct way of computing them. Each main effect and 2-factor interaction can be estimated (unbiasedly) from the r replicates and comparisons among the r estimates of a particular effect provide a measure of the interaction between replicates and the effect. These interactions are, at least for our current discussion, assumed to be negligible. Hence, these comparisons "belong to error" (see I.4.17.1) and the corresponding sums of squares, for example, SS($A \times$ reps), are part of SS(Error). Each such "interaction" sum of squares has $r - 1$ d.f. Since we have six unconfounded effects we have six such sums of squares and hence $6(r - 1)$ d.f. for SS(Error). The same comment applies to SS(Remainder), which, technically speaking, is equal to SS($ABC \times$ reps). For a more precise discussion see Section 8.5.3.

8.2 SYSTEMS OF CONFOUNDING

In this section we present a formal method of constructing systems of confounding when the block sizes are equal to a power of 2. Such systems will be obtained by considering different partitions of the factor space as represented by corresponding systems of equations.

8.2.1 Blocks of Size 2^{n-1}

The general feature of the example discussed in the previous section is that the block size is $\frac{1}{2}$ the number of treatment combinations. As a consequence two blocks are necessary to accommodate all possible treatment combinations, and hence not all treatment contrasts are estimable. Through proper arrangement of the treatment combinations in the blocks, it can be achieved that the nonestimable contrast (i.e., the lost degree of freedom) is of minor or no importance.

For the 2^n factorial with blocks of size $k = 2^{n-1}$ this means that we can confound any interaction $A_1^{\alpha_1} A_2^{\alpha_2} \cdots A_n^{\alpha_n}$ with blocks, and the basic arrangement is obtained by assigning the set of treatment combinations satisfying

$$\alpha_1 x_1 + \alpha_2 x_2 + \cdots + \alpha_n x_n = 0 \mod 2$$

to one block and the set satisfying

$$\alpha_1 x_1 + \alpha_2 x_2 + \cdots + \alpha_n x_n = 1 \mod 2$$

to the other block. Typically, one would choose the interaction with all $\alpha_i = 1 (i = 1, 2, \ldots, n)$, that is, the highest-order interaction.

8.2.2 Blocks of Size 2^{n-2}

In many practical situations blocks of size 2^{n-1} may still be too large or may not be available. From a factorial point of view the next smaller block size is then $k = 2^{n-2}$. In this case four blocks are needed for the 2^n treatment combinations. It is obvious then that the 3 d.f. due to block comparisons are "lost" as far as treatment comparisons are concerned. With a proper allocation of the treatment combinations to the blocks this means that three interactions will not be estimable, that is, will have to be confounded with blocks. To obtain an appropriate arrangement, we assume first that the interaction $E^{\alpha} = A_1^{\alpha_1} A_2^{\alpha_2} \cdots A_n^{\alpha_n}$ is negligible, that is, can be confounded with blocks. Based on the equations

$$\alpha_1 x_1 + \alpha_2 x_2 + \cdots + \alpha_n x_n = 0 \mod 2 \tag{8.5}$$

$$\alpha_1 x_1 + \alpha_2 x_2 + \cdots + \alpha_n x_n = 1 \mod 2 \tag{8.6}$$

we divide the 2^n treatment combinations into two sets of 2^{n-1} treatment combinations each. We now assume further that the interaction $E^{\beta} = A_1^{\beta_1} A_2^{\beta_2} \cdots A_n^{\beta_n}$ is also negligible. We can then subdivide the set of treatment combinations into two sets of 2^{n-1} based on the equations

$$\beta_1 x_1 + \beta_2 x_2 + \cdots + \beta_n x_n = 0 \mod 2 \tag{8.7}$$

$$\beta_1 x_1 + \beta_2 x_2 + \cdots + \beta_n x_n = 1 \mod 2 \tag{8.8}$$

Because of orthogonality, we know that among the 2^{n-1} treatment combinations that satisfy (8.5), exactly 2^{n-2} satisfy also (8.7) and the remaining 2^{n-2} satisfy (8.8). The same statement can be made for the treatment combinations satisfying (8.6). This then gives us a method of dividing the treatment combinations into 4 sets of 2^{n-2} treatment combinations, which are then assigned to the four blocks, that is, we consider

Block 1: $\sum_i \alpha_i x_i = 0$

$\sum_i \beta_i x_i = 0$

Block 2: $\sum_i \alpha_i x_i = 0$

$\sum_i \beta_i x_i = 1$

Block 3: $\sum_i \alpha_i x_i = 1$

$\sum_i \beta_i x_i = 0$

Block 4: $\sum_i \alpha_i x_i = 1$

$\sum_i \beta_i x_i = 1$

where all equations are reduced modulo 2 as usual. In terms of block comparisons E^{α} and E^{β} are defined (apart from a constant) as follows:

	Block 1	2	3	4
$A_1^{\alpha_1} A_2^{\alpha_2} \cdots A_n^{\alpha_n}$	+	+	−	−
$A_1^{\beta_1} A_2^{\beta_2} \cdots A_n^{\beta_n}$	+	−	+	−

The third orthogonal comparison is

	Block 1	2	3	4
$A_1^{\alpha_1+\beta_1} A_2^{\alpha_2+\beta_2} \cdots A_n^{\alpha_n+\beta_n}$	+	−	−	+

which we recognize immediately as the GI of E^{α} and E^{β} satisfying the equations

$$\sum_i (\alpha_i + \beta_i) x_i = 0$$

and

$$\sum_i (\alpha_i + \beta_i) x_i = 1$$

The block contrasts given above also indicate at the same time with which linear function of block effects each of the three interactions is confounded.

The main result of this discussion, however, is that if two interactions E^{α} and E^{β} are each confounded with blocks then also their GI $E^{\alpha+\beta}$ is confounded with blocks. This imposes a certain restriction on the choice of the design, that is, the allocation of the treatment combinations to the blocks. It may very well be the case that, although E^{α} and E^{β} are negligible, $E^{\alpha+\beta}$ is not; in fact it may be a main effect or a low-order interaction. Hence care has to be exercised in choosing E^{α} and E^{β}.

8.2.3 General Case

The problem described above becomes even more acute if we have to resort to still smaller block sizes. Suppose then that we have available blocks of size $k = 2^p (p < n)$ that require 2^{n-p} blocks for the basic arrangement and hence the confounding of $2^{n-p} - 1$ interactions with blocks. To find a suitable system of confounding, we choose $n - p$ interactions E^{α_s} represented by the n-tuples

$$\boldsymbol{\alpha}_s' = (\alpha_{s1}, \alpha_{s2}, \ldots, \alpha_{sn}) \tag{8.9}$$

with $s = 1, 2, \ldots, n - p$, such that no $\boldsymbol{\alpha}_s'$ can be obtained as a linear combination of the remaining n-tuples. We refer to this set of interactions as *independent interactions*. Each set of equations associated with these $n - p$ interactions, namely

$$\begin{aligned} \alpha_{11}x_1 + \alpha_{12}x_2 + \cdots + \alpha_{1n}x_n &= \delta_1 \\ \alpha_{21}x_1 + \alpha_{22}x_2 + \cdots + \alpha_{2n}x_n &= \delta_2 \\ &\vdots \\ \alpha_{n-p,1}x_1 + \alpha_{n-p,2}x_2 + \cdots + \alpha_{n-p,n}x_n &= \delta_{n-p} \end{aligned} \tag{8.10}$$

for a fixed right-hand side with $\delta_s = 0, 1 (s = 1, 2, \ldots, n - p)$ is satisfied by 2^p treatment combinations $(x_1, x_2, \ldots, x_n)$. Since each δ_s can take on two different values, there are altogether 2^{n-p} equations of the form (8.10) (with the $\boldsymbol{\alpha}'_s$, $s = 1, 2, \ldots, n - p$, remaining constant, of course). Any treatment combination satisfying one set of equations cannot satisfy another set. Hence the 2^{n-p} sets of 2^p treatment combinations comprise together all 2^n treatment combinations and, moreover, each set makes up the treatment combinations for one block.

Because of the nature of the equations (8.10), the interactions $E^{\boldsymbol{\alpha}_s}$ are obviously confounded with blocks. We have shown earlier that if two interactions are confounded with blocks then so is their GI. Similarly if three interactions, $E^{\boldsymbol{\alpha}_1}$, $E^{\boldsymbol{\alpha}_2}$, $E^{\boldsymbol{\alpha}_3}$ say, are confounded with blocks, then any treatment combination $(x_1, x_2, \ldots, x_n)$ that satisfies the equations

$$\sum_i \alpha_{1i}x_i = \delta_1$$

$$\sum_i \alpha_{2i}x_i = \delta_2$$

$$\sum_i \alpha_{3i}x_i = \delta_3$$

also satisfies the equations

$$\sum_i (\alpha_{1i} + \alpha_{2i})x_i = \delta_1 + \delta_2$$

$$\sum_i (\alpha_{1i} + \alpha_{3i})x_i = \delta_1 + \delta_3$$

$$\sum_i (\alpha_{2i} + \alpha_{3i})x_i = \delta_2 + \delta_3$$

$$\sum_i (\alpha_{1i} + \alpha_{2i} + \alpha_{3i})x_i = \delta_1 + \delta_2 + \delta_3$$

which define the GIs $E^{\alpha_1+\alpha_2}$, $E^{\alpha_1+\alpha_3}$, $E^{\alpha_2+\alpha_3}$, and $E^{\alpha_1+\alpha_2+\alpha_3}$, respectively, which are then also confounded with blocks. By induction it is then easy to see that if $n - p$ interactions are confounded with blocks so are their

$$\binom{n-p}{2} + \binom{n-p}{3} + \cdots + \binom{n-p}{n-p} = 2^{n-p} - \binom{n-1}{1} - \binom{n-p}{0} = 2^{n-p} - (n-p) - 1 \tag{8.11}$$

GIs. It follows then from (8.11) that altogether $2^{n-p} - 1$ interactions (including the set of $n - p$ independent ones) are confounded with blocks, which is equal to the number of linearly independent comparisons among the 2^{n-p} blocks.

This result emphasizes the important role played by the original set of $n - p$ independent interactions. They must be chosen so that their GIs do not result in main effects and/or low-order interactions that are possibly nonnegligible.

We shall illustrate this procedure with the following example.

Example 8.1 Suppose we have six factors A, B, C, D, E, F and we consider blocks of size 8, that is, we have $n = 6, 2^6 = 64$ treatment combinations, $2^p = 8, n - p = 3, 2^{n-p} = 8$. A possible set of three independent interactions is

$$ABCD, CDE, BDF$$

Their GIs are

$$(ABCD)(CDE) = ABC^2D^2E = ABE$$

$$(ABCD)(BDF) = AB^2CD^2F = ACF$$

$$(CDE)(BDF) = BCD^2EF = BCEF$$

$$(ABCD)(CDE) = AB^2C^2D^3EF = ADEF$$

Hence the seven interactions $ABCD, CDE, BDF, ABE, ACF, BCEF$, and $ADEF$ will be confounded with blocks. □

We give in Table 8.3 typical examples of systems of confounding for up to 8 factors. Only those systems are given that result in a block size of 16 or less. The allocation of letters to the factors is a matter for the experimenter, and the type exemplified by ABC^*, ABD^*, CD for 4 factors in blocks of 4 experimental units includes 5 additional systems, namely, AB^*, ACD^*, BCD; AC^*, ABD^*, BCD; AD^*, ABC^*, BCD; BC^*, ABD^*, ACD; BD^*, ABC^*, ACD. If the experimenter wishes to use this type of confounding, he should decide which of the six 2-factor interactions is least important and choose the system involving that one.

Table 8.3 Types of Confounding for 2^n Factorial Systems in Blocks of Size k

n	k	Confounded Effects[a]
2	2	Any one effect or interaction
3	4	Any one effect or interaction
	2	AB^*, AC^*, BC
		A^*, BC^*, ABC
4	8	Any one effect or interaction
	4	$A^*, BCD^*, ABCD$
		$AB^*, CD^*, ABCD$
		AB^*, ACD^*, BCD
	2	$A^*, B^*, AB, C^*, AC, BC, ABC$
		$A^*, BC^*, ABC, BD^*, ABD, CD, ACD$
		$AB^*, AC^*, BC, AD^*, BD, CD, ABCD$
5	16	Any one effect or interaction
	8	$A^*, BCDE^*, ABCDE$
		$AB^*, CDE^*, ABCDE$
		$ABC^*, CDE^*, ABDE$
	4	$A^*, B^*, AB, CDE^*, ACDE, BCDE, ABCDE$
		$A^*, BC^*, ABC, DE^*, ADE, BCDE, ABCDE$
		$A^*, BC^*, ABC, CDE^*, ACDE, BDE, ABDE$
		$AB^*, AC^*, BC, DE^*, ABDE, ACDE, BCDE$
		$AB^*, CD^*, ABCD, BDE^*, ADE, BCE, ACE$
	2	AB^*, AC^*, AD^*, AE^*, and all GIs
		A^*, BC^*, AD^*, BE^*, and all GIs
6	16	$ABCD^*, CDEF^*, ABEF$
	8	$ACE^*, BDE^*, ABCD, BCF^*, ABEF, CDEF, ADF$
	4	$AB^*, CD^*, ABCD, EF^*, ABEF, CDEF, ABCDEF$
		$ACE^*, BCE^*, ADE, BDE^*, ACF, BCF, ADF, BDF$
	2	$AB^*, AC^*, AD^*, AE^*, AF^*$, and all GIs [not particularly useful except with partial confounding (see Chapter 9)]
		$A^*, BC^*, BD^*, BE^*, BF^*, BG^*$, and all GIs
7	16	$ACEG^*, BDE^*, ABCDG, BCF^*, ABEFG, CDEF, ADFG$
	8	$ABC^*, ADE^*, BCDE, BDF^*, ACDF, ABEF, CEF, ABCDEFG^*$ $DEFG, BCFG, AFG, ACEG, BEG, CDG, ABDG$
	4	$AB^*, AC^*, AD^*, AE^*, AF^*, AG^*$, and all GIs [again not particularly useful]
		$A^*, BC^*, BD^*, BE^*, BF^*, BG^*$, and all GIs
8	16	As for 7 factors in blocks of 8
		$ABC^*, ADE^*, AFG^*, BDG^*, CH^*$, and all their GIs; CH is the only 2-factor interaction confounded.

[a]Effects with an asterisk (*) are the independent effects of (8.10).

In general one would like to avoid confounding of main effects and 2-factor interactions. There is a remarkable theorem obtained by Fisher (1942) that this is possible only if the number of units in a block is larger than the number of factors. We shall return to this theorem in Section 11.7.

8.3 COMPOSITION OF BLOCKS FOR A PARTICULAR SYSTEM OF CONFOUNDING

Once it has been decided which effects and interactions are to be confounded with blocks, that is, once the set of independent interactions has been chosen, the composition of the individual blocks is (apart from randomization) completely determined. In fact, varying the right-hand side of the system of equations (8.10) in all possible ways yields the sets of treatment combinations that are to be allocated to the various blocks. We shall now describe a more systematic way of determining these sets.

8.3.1 Intrablock Subgroup

We consider the general case of a 2^n factorial in blocks of size 2^p. Suppose we confound the interactions $E^{\boldsymbol{\alpha}_s} (s = 1, 2, \ldots, n-p)$ with $\boldsymbol{\alpha}_s$ defined in (8.9). Out of the 2^{n-p} possible sets of equations (8.10) we then consider the set

$$\begin{array}{ccccccc} \alpha_{11}x_1 & + & \alpha_{12}x_2 & + \cdots + & \alpha_{1n}x_n = 0 \\ \alpha_{21}x_1 & + & \alpha_{22}x_2 & + \cdots + & \alpha_{2n}x_n = 0 \\ & & & \vdots & \\ \alpha_{n-p,1}x_1 & + & \alpha_{n-p,2}x_2 & + \cdots + & \alpha_{n-p,n}x_n = 0 \end{array} \tag{8.12}$$

As pointed out earlier, Eqs. (8.12) are satisfied by 2^p treatment combinations $\boldsymbol{x}'_j = (x_{j1}, x_{j2}, \ldots, x_{jn}) (j = 1, 2, \ldots, 2^p)$. We shall now exhibit the structure of this set, which will prove to be useful for the construction of all blocks.

We note that (i) the treatment combination $\boldsymbol{\phi}' = (0, 0, \ldots, 0)$, often referred to as the control, satisfies (8.12); and (ii) if two treatment combinations, $\boldsymbol{x}_k$ and $\boldsymbol{x}_\ell$ say, satisfy (8.12), then the treatment combination

$$\boldsymbol{x}'_k + \boldsymbol{x}'_\ell = (x_{k1} + x_{\ell 1}, x_{k2} + x_{\ell 2}, \ldots, x_{kn} + x_{\ell n}) \tag{8.13}$$

also satisfies (8.12), where each component in (8.13) is reduced mod 2. Hence, if we define "addition" of treatment combinations by (8.13), then the 2^p treatment combinations satisfying (8.12) form an Abelian group with respect to this addition operation and with $\boldsymbol{\phi}'$ as the zero element. More precisely, it forms a subgroup of the group of all 2^n treatment combinations. In the context of systems of confounding this subgroup is referred to as the *intrablock subgroup* (IBSG).

In terms of the reduced representation of treatment combinations, that is, (1), $a, b, ab, c, \ldots$, the operation for this subgroup is the ordinary algebraic multiplication with the square of any letter replaced by unity. The symbol (1) acts as the unity element of the group.

If we examine Eq. (8.12), we see immediately that associated with the nonzero α_{si} for each $s = 1, 2, \ldots, n - p$ there can be only an even number of $x_i (i = 1, 2, \ldots, n)$ equal to unity. This implies that each treatment combination in the IBSG has an even number of letters in common with each confounded interaction. Because of the group property of the IBSG, one only needs to find, in addition to the control, p "independent" treatment combinations that satisfy (8.12). The remaining ones are then determined automatically.

We shall illustrate this procedure by continuing Example 8.1.

Example 8.1 (Continued) Equations (8.12) are of the form

$$\begin{aligned} x_1 + x_2 + x_3 + x_4 \qquad\qquad &= 0 \\ x_3 + x_4 + x_5 \qquad &= 0 \\ x_2 \qquad + x_4 \qquad + x_6 &= 0 \ . \end{aligned}$$

The following treatment combinations satisfy these equations; that is, constitute the IBSG:

1. (0, 0, 0, 0, 0, 0)
2. (1, 1, 1, 1, 0, 0)
3. (1, 0, 1, 0, 1, 0)
4. (0, 1, 0, 1, 1, 0) (by adding 2 and 3)
5. (1, 0, 0, 1, 1, 1)
6. (0, 1, 1, 0, 1, 1) (by adding 2 and 5)
7. (0, 0, 1, 1, 0, 1) (by adding 3 and 5)
8. (1, 1, 0, 0, 0, 1) (by adding 4 and 5)

In terms of the reduced representation these treatment combinations are (1), $abcd, ace$, $bde(= abcd \times ace)$, $adef$, $bcef(= abcd \times adef)$, $cdf(= ace \times adef)$, $abf(= bde \times adef)$.

8.3.2 Remaining Blocks

So far we have described the composition of one block. Rather than change the right-hand side of (8.10) and solve the resulting equations to obtain the remaining $2^{n-p} - 1$ blocks, we make use of the IBSG and construct the blocks in the following way. Suppose the treatment combinations in the IBSG are denoted by

$$\boldsymbol{x}'_j = (x_{j1}, x_{j2}, \ldots, x_{jn}) \qquad (j = 1, 2, \ldots, 2^p)$$

and suppose further that $\boldsymbol{y}' = (y_1, y_2, \ldots, y_n)$ is a treatment combination not in the IBSG. Then the treatment combinations

$$\{\boldsymbol{x}'_j + \boldsymbol{y}';\ j = 1, 2, \ldots, 2^p\} \tag{8.14}$$

form a new set of 2^p treatment combinations. Obviously, these treatment combinations satisfy some set of equations (8.10) with $(\delta_1, \delta_2, \ldots, \delta_n) \neq (0, 0, \ldots, 0)$ and hence are different from the treatment combinations in the IBSG. Thus these treatment combinations form another block.

This process is now repeated, replacing $\boldsymbol{y}'$ in (8.14) by a treatment combination z', where z' is neither in the IBSG nor in a previously constructed set (block). This gives an easy and systematic way of constructing all blocks. We illustrate this again with a continuation of Example 8.1.

Example 8.1 (Continued) Let block 1 be the IBSG. The first treatment combination of each subsequent block takes the role of $\boldsymbol{y}'$ in (8.14). The eight blocks then are given in Table 8.4.

In terms of the reduced representation for a treatment combination the method just described means that the treatment combinations in any one block may be obtained from those in the IBSG by multiplying the symbols of the treatment combinations in the IBSG by the symbol of a treatment combination not contained in any previous block, replacing the square of any letter by unity.

Also, as a consequence of the procedure, we have that when an effect or interaction is confounded with blocks, *all* treatment combinations in a block have *either* an even number of letters in common with the effect or interaction *or* an odd number. No two treatment combinations in the same block can have an odd number of letters and the other an even number of letters in common with any effect or interaction that is confounded. This serves as a convenient check.

8.4 DETECTING A SYSTEM OF CONFOUNDING

Occasionally we may be confronted with the problem of analyzing an experiment that we have not designed. It may then be necessary to discover the system of confounding on which the plan for the experiment is based. This can be done by reversing the procedure outlined in the previous section.

To illustrate, consider the plan given for Example 8.1. Take the IBSG, that is, the block containing the control. We then know that each treatment combination in that block must have an even number of letters in common with each confounded interaction. Inspection shows that no 2-factor interaction meets that requirement; hence no 2-factor interaction can be confounded with blocks. Next going through the 3-factor interactions in a systematic fashion one finds that *ABE* and *ACF* meet the requirement. Hence they and their GI, *BCEF*, are confounded with blocks. Finally, *BDF* is found to be confounded with blocks, and so then are all GIs of *BDF* with the previously identified interactions. Thus the system of confounding has been determined completely.

Table 8.4 Basic Plan for 2^6 Factorial in Blocks of Size 8

Block 1	Block 2	Block 3
(0, 0, 0, 0, 0, 0)	(1, 0, 0, 0, 0, 0)	(0, 1, 0, 0, 0, 0)
(1, 1, 1, 1, 0, 0)	(0, 1, 1, 1, 0, 0)	(1, 0, 1, 1, 0, 0)
(1, 0, 1, 0, 1, 0)	(0, 0, 1, 0, 1, 0)	(1, 1, 1, 0, 1, 0)
(0, 1, 0, 1, 1, 0)	(1, 1, 0, 1, 1, 0)	(0, 0, 0, 1, 1, 0)
(1, 0, 0, 1, 1, 1)	(0, 0, 0, 1, 1, 1)	(1, 1, 0, 1, 1, 1)
(0, 1, 1, 0, 1, 1)	(1, 1, 1, 0, 1, 1)	(0, 0, 1, 0, 1, 1)
(0, 0, 1, 1, 0, 1)	(1, 0, 1, 1, 0, 1)	(0, 1, 1, 1, 0, 1)
(1, 1, 0, 0, 0, 1)	(0, 1, 0, 0, 0, 1)	(1, 0, 0, 0, 0,1)
Block 4	**Block 5**	**Block 6**
(0, 0, 1, 0, 0, 0)	(0, 0, 0, 1, 0, 0)	(0, 0, 0, 0, 1, 0)
(1, 1, 0, 1, 0, 0)	(1, 1, 1, 0, 0, 0)	(1, 1, 1, 1, 1, 0)
(1, 0, 0, 0, 1, 0)	(1, 0, 1, 1, 1, 0)	(1, 0, 1, 0, 0, 0)
(0, 1, 1, 1, 1, 0)	(0, 1, 0, 0, 1, 0)	(0, 1, 0, 1, 0, 0)
(1, 0, 1, 1, 1, 1)	(1, 0, 0, 0, 1, 1)	(1, 0, 0, 1, 0, 1)
(0, 1, 0, 0, 1, 1)	(0, 1, 1, 1, 1, 1)	(0, 1, 1, 0, 0, 1)
(0, 0, 0, 1, 0, 1)	(0, 0, 1, 0, 0, 1)	(0, 0, 1, 1, 1, 1)
(1, 1, 1, 0, 0, 1)	(1, 1, 0, 1, 0, 1)	(1, 1, 0, 0, 1, 1)
Block 7	**Block 8**	
(0, 0, 0, 0, 0, 1)	(1, 1, 1, 1, 1, 1)	
(1, 1, 1, 1, 0, 1)	(0, 0, 0, 0, 1, 1)	
(1, 0, 1, 0, 1, 1)	(0, 1, 0, 1, 0, 1)	
(0, 1, 0, 1, 1, 1)	(1, 0, 1, 0, 0, 1)	
(1, 0, 0, 1, 1, 0)	(0, 1, 1, 0, 0, 0)	
(0, 1, 1, 0, 1, 0)	(1, 0, 0, 1, 0, 0)	
(0, 0, 1, 1, 0, 0)	(1, 1, 0, 0, 1, 0)	
(1, 1, 0, 0, 0, 0)	(0, 0, 1, 1, 1, 0)	

This then exemplifies the general procedure that should be used in a case like this: Start with main effects, then 2-factor interactions, and so forth and find $n - p$ independent interactions that satisfy the even-or-odd-number-of-letters-in-common requirement for a given block. This presupposes, however, that a proper system of confounding has been used except that one does not know which. If one is not sure, one should check other blocks to see whether they too conform to the even-or-odd-numbers-of-letters-in-common requirement.

8.5 USING SAS FOR CONSTRUCTING SYSTEMS OF CONFOUNDING

In this section we shall illustrate briefly how the SAS procedure PROC FACTEX can be used to generate systems of confounding.

Example 8.2 In Table 8.5 we present the SAS input and output for the 2^3 factorial in blocks of size 4, assuming that the 3-factor interaction is negligible. The keys to generating the system are (1) specification of the block size and (2) specification of the effects to be estimated, in this example A, B, C, AB, AC, BC. The "Block Pseudo-factor Confounding Rules" give then the effect(s) confounded with blocks, in this example ABC. □

Example 8.3 This example is to illustrate how, in the 2^3 factorial with blocks of size 4, any effect can be confounded with blocks. For example, we may know already something about the A effect from a previous experiment, but we are mainly interested in the other main effects and all interactions. Table 8.6 gives such a plan. □

Example 8.4 Consider the 2^4 factorial in blocks of size 4. It is intended to estimate all the main effects and as many 2-factor interactions as possible, and then as many 3-factor interactions as possible. From Table 8.3 we see that design 3 for $n = 4, k = 4$ seems an appropriate choice, since it sacrifices information on one 2-factor interaction, AB, and two 3-factor interactions, ACD and BCD. This knowledge determines the model statement in Table 8.7 and yields the basic arrangement given in Table 8.7 The two independent interactions to be confounded with blocks are listed as BCD and ACD, which implies, of course, that the generalized interaction, AB, is also confounded with blocks. □

8.6 ANALYSIS OF EXPERIMENTS WITH CONFOUNDING

8.6.1 Estimation of Effects and Interactions

We consider the 2^n experiment with blocks of size 2^p, replicating the basic block arrangement r times. A linear model appropriate for this design is of the form (7.60), which we write more specifically as

$$y_{ij}(\boldsymbol{x}) = \mu + \beta_{ij} + \tau(\boldsymbol{x}) + e_{ij}(\boldsymbol{x}) \tag{8.15}$$

or, equivalently,

$$y_{ij}(\boldsymbol{x}) = \mu + \rho_i + \beta^*_{ij} + \tau(\boldsymbol{x}) + e_{ij}(\boldsymbol{x}) \tag{8.16}$$

Table 8.5 2^3 Factorial in Blocks of Size 4 (Confounding ABC)

```
options nodate pageno=1;
proc factex;
factors A B C;
size design=8;
blocks size=4;
model estimate=(A|B|C @2);
examine design confounding;
output out=design blockname=block nvals=(1 2);
title1 'TABLE 8.5';
title2 '2**3 FACTORIAL IN BLOCKS OF SIZE 4';
title3 'CONFOUNDING ABC';
run;

proc print data=design;
run;
```

```
                    The FACTEX Procedure

                       Design Points

          Experiment
           Number       A       B       C     Block
          ---------------------------------------
              1        -1      -1      -1        1
              2        -1      -1       1        2
              3        -1       1      -1        2
              4        -1       1       1        1
              5         1      -1      -1        2
              6         1      -1       1        1
              7         1       1      -1        1
              8         1       1       1        2

            Block Pseudofactor Confounding Rules

                      [B1] = A*B*C

             Obs     block      A       B       C

              1        1       -1      -1      -1
              2        1       -1       1       1
              3        1        1      -1       1
              4        1        1       1      -1
              5        2       -1      -1       1
              6        2       -1       1      -1
              7        2        1      -1      -1
              8        2        1       1       1
```

Table 8.6 2^3 Factorial in Blocks of Size 4 (Confounding Main Effect A)

```
options nodate pageno=1;
proc factex;
factors A B C;
blocks size=4;
model est=(B A*B C A*C B*C A*B*C);
examine design confounding;
output out=design blockname=block nvals=(1 2);
title1 'TABLE 8.6';
title2 '2**3 FACTORIAL IN BLOCKS OF SIZE 4';
title3 'CONFOUNDING MAIN EFFECT A';
run;

proc print data=design;
run;
```

```
                    The FACTEX Procedure

                       Design Points

          Experiment
           Number      A      B      C     Block
          ----------------------------------------
              1       -1     -1     -1       1
              2       -1     -1      1       1
              3       -1      1     -1       1
              4       -1      1      1       1
              5        1     -1     -1       2
              6        1     -1      1       2
              7        1      1     -1       2
              8        1      1      1       2

            Block Pseudofactor Confounding Rules

                        [B1] = A

             Obs     block     A      B      C

              1        1      -1     -1     -1
              2        1      -1     -1      1
              3        1      -1      1     -1
              4        1      -1      1      1
              5        2       1     -1     -1
              6        2       1     -1      1
              7        2       1      1     -1
              8        2       1      1      1
```

Table 8.7 2^4 Factorial in Blocks of Size 4

```
options nodate pageno=1;
proc factex;
factors A B C D;
model est=(A B C A*C B*C D A*D B*D C*D A*B*C A*B*D);
blocks size=4;
examine design confounding;
output out=design blockname=block nvals=(1 2 3 4);
title1 'TABLE 8.7';
title2 '2**4 FACTORIAL IN BLOCKS OF SIZE 4';
run;

proc print data=design;
run;
```

```
                      The FACTEX Procedure

                          Design Points

     Experiment
         Number      A      B      C      D     Block
     -----------------------------------------------
              1     -1     -1     -1     -1         1
              2     -1     -1     -1      1         4
              3     -1     -1      1     -1         4
              4     -1     -1      1      1         1
              5     -1      1     -1     -1         2
              6     -1      1     -1      1         3
              7     -1      1      1     -1         3
              8     -1      1      1      1         2
              9      1     -1     -1     -1         3
             10      1     -1     -1      1         2
             11      1     -1      1     -1         2
             12      1     -1      1      1         3
             13      1      1     -1     -1         4
             14      1      1     -1      1         1
             15      1      1      1     -1         1
             16      1      1      1      1         4

  Block Pseudofactor Confounding Rules

              [B1] = B*C*D
              [B2] = A*C*D
```

Table 8.7 (*Continued*)

Obs	block	A	B	C	D
1	1	-1	-1	-1	-1
2	1	-1	-1	1	1
3	1	1	1	-1	1
4	1	1	1	1	-1
5	2	-1	1	-1	-1
6	2	-1	1	1	1
7	2	1	-1	-1	1
8	2	1	-1	1	-1
9	3	-1	1	-1	1
10	3	-1	1	1	-1
11	3	1	-1	-1	-1
12	3	1	-1	1	1
13	4	-1	-1	-1	1
14	4	-1	-1	1	-1
15	4	1	1	-1	-1
16	4	1	1	1	1

where ρ_i = effect of the ith replicate ($i = 1, 2, \ldots, r$)

β^*_{ij} or β_{ij} = effect of jth block in ith replicate ($j = 1, 2, \ldots, 2^{n-p}$)

$\tau(\boldsymbol{x})$ = effect of treatment combination $\boldsymbol{x}' = (x_1, x_2, \ldots, x_n)$ ($x_\ell = 0, 1$; $\ell = 1, 2, \ldots, n$)

$e_{ij}(\boldsymbol{x})$ = error associated with treatment combination $\boldsymbol{x}$ in the jth block of the ith replicate,

or in matrix notation

$$\boldsymbol{y} = \mu\mathfrak{I} + \mathbf{X}_\rho\boldsymbol{\rho} + \mathbf{X}_{\beta^*}\boldsymbol{\beta}^* + \mathbf{X}_\tau\boldsymbol{\tau} + \mathbf{e} \tag{8.17}$$

with the obvious definition of all terms in this model equation. Let $\mathcal{E}_1 = \{E^{\alpha_\ell}, \ell = 1, 2, \ldots, q(= 2^{n-p} - 1)\}$ be the set of interactions confounded with blocks, and let $\mathcal{E}_2 = \{E^{\gamma_m}, m = 1, 2, \ldots, s[= 2^{n-p}(2^p - 1)]\}$ be the set of interactions not confounded with blocks. (We note that we use here the word *interaction* in a general sense that includes also main effects as 1-factor interactions.) We then obtain in the usual way, using the Yates algorithm for example, estimates $\widehat{E}^{\gamma_m}$ for each $E^{\gamma_m} \in \mathcal{E}_2$ and $\widetilde{E}^{\alpha_\ell}$ for each $E^{\alpha_\ell} \in \mathcal{E}_1$, such that $\widehat{E}^{\gamma_m}$is BLUE and $\widetilde{E}^{\alpha_\ell}$ is a biased estimate. From the general discussion in Chapter 7 it also follows that

$$\text{var}(\widehat{E}^{\gamma_m}) = \frac{1}{r2^{n-2}}\sigma_e^2 \tag{8.18}$$

8.6.2 Parameterization of Treatment Responses

It is clear from the discussion above that only the $E^{\gamma_m} \in \mathcal{E}_2$ are estimable. It is also obvious from the construction of a confounded design that certain treatment combinations never occur together in the same block and hence cannot, ordinarily, be compared with each other. We thus have a disconnected design, which generally is an undesirable situation. In the factorial setting this negative feature may be overcome, however. If we can assume that all $E^{\alpha_\ell} \in \mathcal{E}_1$ are negligible, then the reparameterization (7.42) and (7.43) may be rewritten as

$$a(\boldsymbol{x}) = M + \sum_{\boldsymbol{\gamma}_m} E^{\boldsymbol{x}}_{\boldsymbol{\gamma}'_m \boldsymbol{x}}$$

or

$$a(\boldsymbol{x}) = M + \sum_{\boldsymbol{\gamma}_m} \phi(\boldsymbol{\gamma}_m, \boldsymbol{x}) E^{\boldsymbol{\gamma}_m}$$

respectively, where $\sum_{\boldsymbol{\gamma}_m}$ is summation over all $\boldsymbol{\gamma}_m$ such that $E^{\boldsymbol{\gamma}_m} \in \mathcal{E}_2$. For any two treatment combinations $\boldsymbol{x}$ and $\boldsymbol{z}$, we can then obtain

$$\widehat{a}(\boldsymbol{x}) - \widehat{a}(\boldsymbol{z}) = \sum_{\boldsymbol{\gamma}_m} [\phi(\boldsymbol{\gamma}_m, \boldsymbol{x}) - \phi(\boldsymbol{\gamma}_m, \boldsymbol{z})]\, \widehat{E}^{\boldsymbol{\gamma}_m}$$

where $\phi(\boldsymbol{\gamma}_m, \boldsymbol{x}) - \phi(\boldsymbol{\gamma}_m, \boldsymbol{z})$ is either 0, 1, or -1. Using (8.18) we can then evaluate $\text{var}\,[\widehat{a}(\boldsymbol{x}) - \widehat{a}(\boldsymbol{z})]$.

As an illustration of this procedure consider the following example.

Example 8.5 Suppose we have a 2^3 experiment in blocks of size 4, confounding ABC. We have the basic arrangement

Block 1: $(1), ab, ac, bc$

Block 2: a, b, c, abc

which is replicated r times. Suppose $a(\boldsymbol{x}) = abc, a(\boldsymbol{z}) = (1)$. Then, assuming ABC to be negligible, we can write

$$\widehat{abc} = \widehat{M} + \tfrac{1}{2}\widehat{A} + \tfrac{1}{2}\widehat{B} + \tfrac{1}{2}\widehat{AB} + \tfrac{1}{2}\widehat{C} + \tfrac{1}{2}\widehat{AC} + \tfrac{1}{2}\widehat{BC}$$

and

$$(\widehat{1}) = \widehat{M} - \tfrac{1}{2}\widehat{A} - \tfrac{1}{2}\widehat{B} + \tfrac{1}{2}\widehat{AB} - \tfrac{1}{2}\widehat{C} + \tfrac{1}{2}\widehat{AC} + \tfrac{1}{2}\widehat{BC}$$

and hence

$$\widehat{abc} - (\widehat{1}) = \widehat{A} + \widehat{B} + \widehat{C}$$

and

$$\text{var}\left[\widehat{abc} - (\widehat{1})\right] = \frac{3}{r2^{n-2}}\sigma_e^2$$

□

8.6.3 ANOVA Tables

The ANOVA table for a system of confounding as described in this section follows the basic outline of the ANOVA table for incomplete block designs (see Table 1.3) and is as given in Table 8.8a. A partitioning of the various sums of

Table 8.8 ANOVA Tables for System of Confounding

Source	d.f.	SS
	a. Basic T|B-ANOVA	
$\boldsymbol{X}_\beta \mid \mathfrak{I}$	$r2^{n-p}-1$	$\frac{1}{2^p}\sum_{ij} y_{ij}^2(\cdot) - \frac{1}{r2^n} y_{..}^2(\cdot)$
$\boldsymbol{X}_\tau \mid \mathfrak{I}, \boldsymbol{X}_\beta$	$2^{n-p}(2^p-1)$	$r2^{n-2}\sum_m [\widehat{E}^{\boldsymbol{\gamma}_m}]^2$
$\boldsymbol{I} \mid \mathfrak{I}, \boldsymbol{X}_\beta, \boldsymbol{X}_\tau$	$(r-1)\,2^{n-p}\,(2^p-1)$	Subtraction
Total	2^n-1	$\sum_{ij}\sum_{\boldsymbol{x}} y_{ij}^2(\boldsymbol{x}) - \frac{1}{r2^n} y_{..}^2(\cdot)$
	b. Detailed T|B-ANOVA	
$\boldsymbol{X}_\beta \mid \mathfrak{I}$	$r2^{n-p}-1$	$\frac{1}{2^p}\sum_i\sum_j y_{ij}^2(\cdot) - \frac{1}{r2^n} y_{..}^2(\cdot)$
$\boldsymbol{X}_\rho \mid \mathfrak{I}$	$r-1$	$\frac{1}{2^n}\sum_i y_{i.}^2(\cdot) - \frac{1}{r2^n} y_{..}^2(\cdot)$
$\boldsymbol{X}_{\beta^*} \mid \mathfrak{I}, \boldsymbol{X}_\rho$	$r(2^{n-p}-1)=rq$	$\frac{1}{2^p}\sum_i\sum_j y_{ij}^2(\cdot) - \frac{1}{2^n}\sum_i y_{i.}^2(\cdot)$
$E^{\boldsymbol{\alpha}_1}$	1	$r2^{n-2}\,[\widetilde{E}^{\alpha_1}]^2$
⋮	⋮	⋮
$E^{\boldsymbol{\alpha}_q}$	1	$r2^{n-2}\,[\widetilde{E}^{\alpha_q}]^2$
Remainder	$(r-1)(2^{n-p}-1)=(r-1)\,q$	Subtraction
$\boldsymbol{X}_\tau \mid \mathfrak{I}, \boldsymbol{X}_\beta$	$2^{n-p}(2^p-1)=s$	
$E^{\boldsymbol{\gamma}_1}$	1	$r2^{n-2}\,[\widehat{E}^{\boldsymbol{\gamma}_1}]^2$
$E^{\boldsymbol{\gamma}_2}$	1	$r2^{n-2}\,[\widehat{E}^{\boldsymbol{\gamma}_2}]^2$
⋮	⋮	⋮
$E^{\boldsymbol{\gamma}_s}$	1	$r2^{n-2}\,[\widehat{E}^{\boldsymbol{\gamma}_s}]^2$
$\boldsymbol{I} \mid \mathfrak{I}, \boldsymbol{X}_\beta, \boldsymbol{X}_\tau$	$(r-1)\,2^{n-p}\,(2^p-1)=(r-1)s$	Subtraction
Total	$r2^n-1$	$\sum_i\sum_j\sum_{\boldsymbol{x}} y_{ij}^2(\boldsymbol{x}) - \frac{1}{r2^n} y_{..}^2(\cdot)$

squares is then given in Table 8.8b. Although quite self-explanatory we shall comment on these partitionings briefly.

The partitioning

$$SS(X_\beta|\mathfrak{I}) = SS(\text{Replicates}) + SS(\text{Blocks/reps})$$

follows immediately from model (8.16) and can be written more precisely as

$$SS(X_\beta \mid \mathfrak{I}) = SS(X_\rho \mid \mathfrak{I}) + SS(X_{\beta^*} \mid \mathfrak{I}, X_\rho) \tag{8.19}$$

Since each treatment combination occurs exactly once in each replicate, $SS(X_\rho|\mathfrak{I})$ is orthogonal to treatments in the same sense that blocks and treatments are orthogonal to each other in a randomized complete block design. However, $SS(X_{\beta^*}|\mathfrak{I}, X_\rho)$ is not orthogonal to treatments. This fact becomes evident from the partitioning

$$SS(X_{\beta^*} \mid \mathfrak{I}, X_\rho) = \sum_{\ell=1}^{q} SS(E^{\alpha_\ell}) + SS(\text{Remainder}) \tag{8.20}$$

where

$$SS(E^{\alpha_\ell}) = r2^{n-2}\,[\widetilde{E}^{\alpha_\ell}]^2 \tag{8.21}$$

with

$$ESS(E^{\alpha_\ell}) = r2^{n-2}\,[\widetilde{E}^{\alpha_\ell}]^2 + Q_\ell(\beta_{ij}^*) + \sigma_e^2 \tag{8.22}$$

and $Q_\ell(\beta_{ij}^*)$ a quadratic function of block effects corresponding to the block contrast defining E^{α_ℓ} in the chosen system of confounding.

There is another way of writing (8.20) that is informative. Let $\widetilde{E}_i^{\alpha_\ell}$ be the estimate of E^{α_ℓ} in the ith replicate ($i = 1, 2, \ldots, r$) and let

$$SS(E_i^{\alpha_\ell}) = 2^{n-2}\,[\widetilde{E}_i^{\alpha_\ell}]^2 \tag{8.23}$$

be the sum of squares associated with that estimate. It is clear then that

$$\begin{aligned} SS(X_{\beta^*} \mid \mathfrak{I}, X_\rho) &= \sum_{i=1}^{r}\sum_{\ell=1}^{q} SS(E_i^{\alpha_\ell}) \\ &= \sum_{\ell=1}^{q} SS(E^{\alpha_\ell}) \\ &\quad + \sum_{\ell=1}^{q}\left[\sum_{i=1}^{r} SS(E_i^{\alpha_\ell}) - SS(E^{\alpha_\ell})\right] \end{aligned} \tag{8.24}$$

Hence, in (8.20)

$$SS(\text{Remainder}) = \sum_{\ell=1}^{q}\left[\sum_{i=1}^{r} SS(E_i^{\alpha_\ell}) - SS(E^{\alpha_\ell})\right] \tag{8.25}$$

The purpose of writing $SS(\boldsymbol{X}_{\beta^*} \mid \mathfrak{I}, \boldsymbol{X}_\rho)$ in the form (8.24) is to partition it into what is sometimes referred to as a *treatment component* and a *pure block component*. Obviously, $\sum_\ell SS(E^{\boldsymbol{\alpha}_\ell})$ depends both on treatment effects, namely the $E^{\boldsymbol{\alpha}_\ell}$, and on block effects. As we shall show below SS(Remainder) does not depend on treatment effects, hence $\sum_\ell SS(E^{\boldsymbol{\alpha}_\ell})$ is the treatment component of the block sum of squares mentioned above.

We now consider SS(Remainder) and derive (8.25) in a different way, which will show that SS(Remainder) is algebraically equivalent to a sum of squares associated with the replicate $\times\, E^{\boldsymbol{\alpha}_\ell}\,(\ell = 1, 2, \ldots, q)$ interactions. Let

$$X_{\delta i}^{\boldsymbol{\alpha}_\ell} = \{\text{average observed response in replicate } i \text{ of all treatment combinations satisfying } \boldsymbol{\alpha}_\ell' \boldsymbol{x} = \delta\ (\delta = 0, 1)\}$$

We note that $X_{\delta i}^{\boldsymbol{\alpha}_\ell}$ is the mean of 2^{n-1} observations and that $E_{\delta i}^{\boldsymbol{\alpha}_\ell} = X_{\delta i}^{\boldsymbol{\alpha}_\ell} - \overline{y}_{i}.(\cdot)$. We then consider the following two-way table:

		Replicate (i) 1	2	$\ldots$	r	
$\delta =$	0	$X_{01}^{\boldsymbol{\alpha}_\ell}$	$X_{02}^{\boldsymbol{\alpha}_\ell}$	$\ldots$	$X_{0r}^{\boldsymbol{\alpha}_\ell}$	$X_{0\cdot}^{\boldsymbol{\alpha}_\ell}$
	1	$X_{11}^{\boldsymbol{\alpha}_\ell}$	$X_{12}^{\boldsymbol{\alpha}_\ell}$	$\ldots$	$X_{1r}^{\boldsymbol{\alpha}_\ell}$	$X_{1\cdot}^{\boldsymbol{\alpha}_\ell}$
		$X_{\cdot 1}^{\boldsymbol{\alpha}_\ell}$	$X_{\cdot 2}^{\boldsymbol{\alpha}_\ell}$	$\ldots$	$X_{\cdot r}^{\boldsymbol{\alpha}_\ell}$	$X_{\cdot\cdot}^{\boldsymbol{\alpha}_\ell}$

The interaction sum of squares in this table (on a per-observation basis) is

$$\begin{aligned} SS(I)_\ell &= 2^{n-1} \sum_i \sum_\delta \left[X_{\delta i}^{\boldsymbol{\alpha}_\ell} - \overline{X}_{\cdot i}^{\boldsymbol{\alpha}_\ell} - \overline{X}_{\delta\cdot}^{\boldsymbol{\alpha}_\ell} + \overline{X}_{\cdot\cdot}^{\boldsymbol{\alpha}_\ell}\right]^2 \\ &= 2^{n-1} \sum_i \sum_\delta \left[\widetilde{E}_{\delta i}^{\boldsymbol{\alpha}_\ell} - \widetilde{E}_{\delta}^{\boldsymbol{\alpha}_\ell}\right]^2 \end{aligned}$$

since $\overline{X}_{\cdot i}^{\boldsymbol{\alpha}_\ell} = \overline{y}_i^2(\cdot)$ and $\overline{X}_{\cdot\cdot}^{\boldsymbol{\alpha}_\ell} = \overline{y}.(\cdot)$. The expression above for $SS(I)_\ell$ shows that it is free of treatment effects as those cancel within each square bracket. Further,

$$\begin{aligned} SS(I)_\ell &= 2^{n-1} \left\{\sum_i \sum_\delta \left[\widetilde{E}_{\delta i}^{\boldsymbol{\alpha}_i}\right] + r \sum_\delta \left[\widetilde{E}_{\delta}^{\boldsymbol{\alpha}_\ell}\right] - 2 \sum_i \sum_\delta \widetilde{E}_{\delta i}^{\boldsymbol{\alpha}_\ell}\, \widetilde{E}_{\delta}^{\boldsymbol{\alpha}_\ell}\right\} \\ &= 2^{n-1} \left\{\sum_i \sum_\delta \left[\widetilde{E}_{\delta i}^{\boldsymbol{\alpha}_i}\right] - r \sum_\delta \left[\widetilde{E}_{\delta}^{\boldsymbol{\alpha}_\ell}\right]^2\right\} \qquad (8.26) \\ &= \sum_i \left(E_i^{\boldsymbol{\alpha}_\ell}\right) - SS\left(E^{\boldsymbol{\alpha}_\ell}\right) \end{aligned}$$

using (7.65). Hence

$$\sum_{\ell=1}^{q} \text{SS}(I)_\ell = \text{SS(Remainder)}$$

as given in (8.25).

Turning now to $\text{SS}(\boldsymbol{X}_\tau \mid \mathfrak{I}, \boldsymbol{X}_\beta)$ the treatment sum of squares adjusted for block effects with $s = 2^{n-p}(2^p - 1)$ d.f., we can partition it into s single-d.f. sums of squares associated with the unconfounded interactions $E^{\gamma_m} (m = 1, 2, \ldots, s)$, given by

$$\text{SS}(E^{\gamma_m}) = r2^{n-2}\left[\widehat{E}^{\gamma_m}\right]^2 \tag{8.27}$$

We then use the F test (as an approximation to the randomization test)

$$F = \frac{\text{SS}(E^{\gamma_m})}{\text{MS(Residual)}}$$

to test the hypothesis $H_0{:}E^{\gamma_m} = 0$.

Finally, corresponding to SS(Remainder) in (8.25) we can partition SS(Error) as s sums of squares with $r - 1$ d.f. each, that is,

$$\text{SS(Error)} = \sum_{m=1}^{s} \left[\sum_{i=1}^{r} \text{SS}\left(E_i^{\gamma_m}\right) - \text{SS}\left(E^{\gamma_m}\right)\right] \tag{8.28}$$

where

$$\text{SS}\left(E_i^{\gamma_m}\right) = 2^{n-2}\left[\widehat{E}_i^{\gamma_m}\right]^2$$

and $\widehat{E}_i^{\gamma_m}$ is the BLUE of E^{γ_m} in the ith replicate. Obviously, then, (8.28) is a function of the $e_{ij}(\boldsymbol{x})$ only and

$$E\left[\sum_{i=1}^{r} \text{SS}\left(E_i^{\gamma_m}\right) - \text{SS}\left(E^{\gamma_m}\right)\right] = (r-1)\sigma_e^2 \tag{8.29}$$

for $m = 1, 2, \ldots, s$. Hence we have s independent estimates of σ_e^2, a fact that may be useful for testing homogeneity of error.

To complete this discussion about the various sums of squares in Table 8.8b, we mention that just as

$$\text{SS}(I)_\ell = \sum_i \text{SS}(E_i^{\alpha_\ell}) - \text{SS}(E^{\alpha_\ell})$$

in (8.25) is algebraically the (replicate $\times$ E^{α_ℓ}) interaction sum of squares with $r - 1$ d.f. $(\ell = 1, 2, \ldots, q)$,

$$\text{SS}(I)_m = \sum_i \text{SS}(E_i^{\gamma_m}) - \text{SS}(E^{\gamma_m})$$

in (8.28) is algebraically the (replicate $\times E^{\gamma_m}$) interaction sum of squares with $r-1$ d.f. ($m = 1, 2, \ldots, s$).

8.7 INTERBLOCK INFORMATION IN CONFOUNDED EXPERIMENTS

In discussing the problem of confounding and the analysis of confounded experiments we have assumed that the block effects are fixed effects, as is evident from Eq. (8.3), for example. If, however, we consider block effects as random effects, then it becomes clear that, again referring to (8.3),

$$\tfrac{1}{4}E[\text{block 2} - \text{ block 1}] = ABC$$

that is, the block difference is an unbiased estimate of ABC. The only difference between this estimator, $\widetilde{ABC}$ say, and the estimators for the other effects and interactions, $\widehat{A}, \ldots, \widehat{BC}$ is that with r replications

$$\text{var}(\widehat{A}) = \cdots = \text{var}(\widehat{BC}) = \frac{1}{2r}\,\sigma_e^2 \tag{8.30}$$

whereas

$$\text{var}(\widetilde{ABC}) = \frac{1}{2r}\,(\sigma_e^2 + 4\sigma_\beta^2) \tag{8.31}$$

From the nature of the estimators it is clear that $\widehat{A}, \ldots, \widehat{BC}$ are intrablock estimates and that $\widetilde{ABC}$ is an interblock estimate as it is a linear function of block totals only (see Section 1.7).

For the general case of a 2^n factorial in blocks of size $k = 2^p$, using an appropriate system of confounding, we are then able to estimate all interactions as

Table 8.9 ANOVA for Interblock Analysis

Source	d.f.	SS
Replicates	$r-1$	
E^{α_1}	1	$\sigma_e^2 + 2^p\sigma_\beta^2 + r2^{n-2}[E^{\alpha_1}]^2$
E^{α_2}	1	$\sigma_e^2 + 2^p\sigma_\beta^2 + r2^{n-2}[E^{\alpha_2}]^2$
$\vdots$	$\vdots$	$\vdots$
E^{α_q}	1	$\sigma_e^2 + 2^p\sigma_\beta^2 + r2^{n-2}[E^{\alpha_q}]^2$
Remainder	$(r-1)(2^{n-p}-1)$	$\sigma_e^2 + 2^p\sigma_\beta^2$

Table 8.10 Data and Intrablock Analysis for 2^3 Factorial in Blocks of Size 4 With ABC Confounded with Blocks

```
options nodate pageno=1;
data example;
input rep block A B C y @@;
datalines;
1 1 0 0 0 14  1 1 0 1 1 25
1 1 1 0 1 17  1 1 1 1 0 20
1 2 0 0 1 13  1 2 0 1 0 16
1 2 1 0 0 14  1 2 1 1 1 26
2 3 0 0 0 24  2 3 0 1 1 34
2 3 1 0 1 27  2 3 1 1 0 29
2 4 0 0 1 24  2 4 0 1 0 26
2 4 1 0 0 25  2 4 1 1 1 35
;
run;

proc print data=example;
title1 'TABLE 8.10';
title2 'DATA FOR 2**3 FACTORIAL IN BLOCKS OF SIZE 4';
title3 'WITH ABC CONFOUNDED WITH BLOCKS';
run;

proc glm data=example;
class rep block A B C;
model y=rep block(rep) A|B|C@2;
estimate 'A' A -1 1;
estimate 'A*B' A*B 1 -1 -1 1/divisor=2;
title2 'INTRA-BLOCK ANALYSIS OF 2**3 FACTORIAL';
title3 'IN BLOCKS OF SIZE 4';
title4 'WITH ABC CONFOUNDED WITH BLOCKS';
run;
```

Obs	rep	block	A	B	C	y
1	1	1	0	0	0	14
2	1	1	0	1	1	25
3	1	1	1	0	1	17
4	1	1	1	1	0	20
5	1	2	0	0	1	13
6	1	2	0	1	0	16
7	1	2	1	0	0	14
8	1	2	1	1	1	26
9	2	3	0	0	0	24
10	2	3	0	1	1	34
11	2	3	1	0	1	27
12	2	3	1	1	0	29
13	2	4	0	0	1	24
14	2	4	0	1	0	26
15	2	4	1	0	0	25
16	2	4	1	1	1	35

Table 8.10 (*Continued*)

```
                 INTRABLOCK ANALYSIS OF 2**3 FACTORIAL
                          IN BLOCKS OF SIZE 4
                    WITH ABC CONFOUNDED WITH BLOCKS

                          The GLM Procedure

                       Class Level Information

                      Class     Levels    Values
                      rep            2    1 2
                      block          4    1 2 3 4
                      A              2    0 1
                      B              2    0 1
                      C              2    0 1

                      Number of observations 16

Dependent Variable: y

  Source             DF   Sum of Squares   Mean Square   F Value   Pr > F
  Model               9      699.0625000    77.6736111    248.56   <.0001
  Error               6        1.8750000     0.3125000
  Corrected Total    15      700.9375000

              R-Square    Coeff Var    Root MSE      y Mean
              0.997325     2.423922    0.559017    23.06250

Source          DF       Type I SS     Mean Square      F Value      Pr > F

rep              1     390.0625000     390.0625000      1248.20      <.0001
block(rep)       2       8.1250000       4.0625000        13.00      0.0066
A                1      18.0625000      18.0625000        57.80      0.0003
B                1     175.5625000     175.5625000       561.80      <.0001
A*B              1       0.0625000       0.0625000         0.20      0.6704
C                1      68.0625000      68.0625000       217.80      <.0001
A*C              1       0.0625000       0.0625000         0.20      0.6704
B*C              1      39.0625000      39.0625000       125.00      <.0001

Source          DF     Type III SS     Mean Square      F Value      Pr > F

rep              1     390.0625000     390.0625000      1248.20      <.0001
block(rep)       2       8.1250000       4.0625000        13.00      0.0066
A                1      18.0625000      18.0625000        57.80      0.0003
B                1     175.5625000     175.5625000       561.80      <.0001
A*B              1       0.0625000       0.0625000         0.20      0.6704
C                1      68.0625000      68.0625000       217.80      <.0001
A*C              1       0.0625000       0.0625000         0.20      0.6704
B*C              1      39.0625000      39.0625000       125.00      <.0001

                                             Standard
     Parameter          Estimate                Error     t Value     Pr > |t|
     A                2.12500000           0.27950850        7.60       0.0003
     A*B              0.12500000           0.27950850        0.45       0.6704
```

Table 8.11 Data and Combined Analysis for 2^3 Factorial in Blocks of Size 4 With ABC Confounded with Blocks

```
options nodate pageno=1;
data example;
input rep block A B C y @@;
datalines;
1 1 0 0 0 14  1 1 0 1 1 25
1 1 1 0 1 17  1 1 1 1 0 20
1 2 0 0 1 13  1 2 0 1 0 16
1 2 1 0 0 14  1 2 1 1 1 26
2 3 0 0 0 24  2 3 0 1 1 34
2 3 1 0 1 27  2 3 1 1 0 29
2 4 0 0 1 24  2 4 0 1 0 26
2 4 1 0 0 25  2 4 1 1 1 35
;
run;
proc print data=example;
title1 'TABLE 8.11';
title2 'DATA FOR 2**3 FACTORIAL IN BLOCKS OF SIZE 4';
title3 'WITH ABC CONFOUNDED WITH BLOCKS';

proc mixed data=example;
class rep block A B C;
model y=rep A|B|C/ddfm=satterth;
random block(rep);
estimate 'A' A -1 1;
estimate 'A*B' A*B 1 -1 -1 1/divisor=2;
estimate 'A*B*C' A*B*C -1 1 1 -1 1 -1 -1 1/divisor=4;
title1 'TABLE 8.11';
title2 'INTRA- AND INTER-BLOCK ANALYSIS';
title3 'OF 2**3 FACTORIAL IN BLOCKS OF SIZE 4';
title4 'WITH ABC CONFOUNDED WITH BLOCKS';
run;
```

Obs	rep	block	A	B	C	y
1	1	1	0	0	0	14
2	1	1	0	1	1	25
3	1	1	1	0	1	17
4	1	1	1	1	0	20
5	1	2	0	0	1	13
6	1	2	0	1	0	16
7	1	2	1	0	0	14
8	1	2	1	1	1	26
9	2	3	0	0	0	24
10	2	3	0	1	1	34
11	2	3	1	0	1	27
12	2	3	1	1	0	29
13	2	4	0	0	1	24
14	2	4	0	1	0	26
15	2	4	1	0	0	25
16	2	4	1	1	1	35

Table 8.11 *(Continued)*

```
              INTRA- AND INTERBLOCK ANALYSIS
           OF 2**3 FACTORIAL IN BLOCKS OF SIZE 4
              WITH ABC CONFOUNDED WITH BLOCKS

                    The Mixed Procedure

                     Model Information

      Data Set                       WORK.EXAMPLE
      Dependent Variable             y
      Covariance Structure           Variance Components
      Estimation Method              REML
      Residual Variance Method       Profile
      Fixed Effects SE Method        Model-Based
      Degrees of Freedom Method      Satterthwaite

                  Class Level Information

              Class     Levels     Values

              rep            2     1 2
              block          4     1 2 3 4
              A              2     0 1
              B              2     0 1
              C              2     0 1

                        Dimensions

                Covariance Parameters      2
                Columns in X              29
                Columns in Z               4
                Subjects                   1
                Max Obs Per Subject       16
                Observations Used         16
                Observations Not Used      0
                Total Observations        16

                     Iteration History

   Iteration    Evaluations    -2 Res Log Like    Criterion

           0              1        19.41205069
           1              1        19.24234227    0.00000000
```

Table 8.11 ***(Continued)***

```
                Convergence Criteria met.

                  Covariance Parameter
                       Estimates

                 Cov Parm           Estimate

                 block(rep)          0.06250
                 Residual             0.3125


                         Fit Statistics

                 -2 Res Log Likelihood           19.2
                 AIC (smaller is better)         23.2
                 AICC (smaller is better)        26.2
                 BIC (smaller is better)         22.0


                Type 3 Tests of Fixed Effects

    Effect      Num DF      Den DF      F Value      Pr > F

    rep              1           1       693.44      0.0242
    A                1           6        57.80      0.0003
    B                1           6       561.80      <.0001
    A*B              1           6         0.20      0.6704
    C                1           6       217.80      <.0001
    A*C              1           6         0.20      0.6704
    B*C              1           6       125.00      <.0001
    A*B*C            1           1        13.44      0.1695


                           Estimates

                          Standard
  Label      Estimate        Error     DF    t Value     Pr > |t|

  A            2.1250       0.2795      6       7.60       0.0003
  A*B          0.1250       0.2795      6       0.45       0.6704
  A*B*C       -1.3750       0.3750      1      -3.67       0.1695
```

follows (using the notation from the previous section): E^{α_ℓ} ($\ell = 1, 2, \ldots, 2^{n-p} - 1$) from interblock information only and with variance $1/r2^{n-2}$ $(\sigma_e^2 + 2^p\sigma_\beta^2)$, $E^{\gamma_m}[m = 1, 2, \ldots, 2^{n-p}(2^p - 1)]$ from intrablock information only and with variance $(1/r2^{n-2})\,\sigma_e^2$. The fact that either intrablock or interblock information only is available for a particular interaction is a major difference between the block arrangements we have been discussing in this chapter and the block

Table 8.12 Auxiliary Analysis of 2^3 Factorial With Blocks of Size 4 and ABC Confounded with Blocks

```
options nodate pageno=1;
proc glm data=example;
class rep block A B C;
model y=rep|A|B|C;
title1 'TABLE 8.12';
title2 'AUXILIARY ANALYSIS OF 2**3 FACTORIAL';
title3 'WITH BLOCKS OF SIZE4 AND ABC CONFOUNDED WITH BLOCKS';
run;
```

```
                    The GLM Procedure

                 Class Level Information

               Class     Levels     Values

               rep            2     1 2

               block          4     1 2 3 4

               A              2     0 1

               B              2     0 1

               C              2     0 1
                Number of observations 16


Dependent Variable: y

                              Sum of
Source                DF     Squares    Mean Square   F Value   Pr > F

Model                 15  700.9375000    46.7291667       .        .

Error                  0    0.0000000          .

Corrected Total       15  700.9375000

        R-Square     Coeff Var     Root MSE       y Mean

        1.000000          .             .        23.06250
```

Table 8.12 *(Continued)*

Source	DF	Type I SS	Mean Square	F Value	Pr > F
rep	1	390.0625000	390.0625000	.	.
A	1	18.0625000	18.0625000	.	.
rep*A	1	0.0625000	0.0625000	.	.
B	1	175.5625000	175.5625000	.	.
rep*B	1	1.5625000	1.5625000	.	.
A*B	1	0.0625000	0.0625000	.	.
rep*A*B	1	0.0625000	0.0625000	.	.
C	1	68.0625000	68.0625000	.	.
rep*C	1	0.0625000	0.0625000	.	.
A*C	1	0.0625000	0.0625000	.	.
rep*A*C	1	0.0625000	0.0625000	.	.
B*C	1	39.0625000	39.0625000	.	.
rep*B*C	1	0.0625000	0.0625000	.	.
A*B*C	1	7.5625000	7.5625000	.	.
rep*A*B*C	1	0.5625000	0.5625000	.	.

Source	DF	Type III SS	Mean Square	F Value	Pr > F
rep	1	390.0625000	390.0625000	.	.
A	1	18.0625000	18.0625000	.	.
rep*A	1	0.0625000	0.0625000	.	.
B	1	175.5625000	175.5625000	.	.
rep*B	1	1.5625000	1.5625000	.	.
A*B	1	0.0625000	0.0625000	.	.
rep*A*B	1	0.0625000	0.0625000	.	.
C	1	68.0625000	68.0625000	.	.
rep*C	1	0.0625000	0.0625000	.	.
A*C	1	0.0625000	0.0625000	.	.
rep*A*C	1	0.0625000	0.0625000	.	.
B*C	1	39.0625000	39.0625000	.	.
rep*B*C	1	0.0625000	0.0625000	.	.
A*B*C	1	7.5625000	7.5625000	.	.
rep*A*B*C	1	0.5625000	0.5625000	.	.

arrangements we have discussed in connection with incomplete block designs in Chapters 1, 2, and 4.

Although the choice of a particular system of confounding is no longer as crucial as it is in the case of fixed block effects, it is important nevertheless as is evident from the statement about the variances of the different estimators, interblock estimators having generally larger variance than the intrablock estimators. This is emphasized also by the way in which hypotheses about the E^{α_ℓ} can be tested in the ANOVA. To show this we give in Table 8.9 the appropriate

E(MS) in the ANOVA table pertaining to the interblock analysis. The hypothesis H_0: $E^{\alpha_\ell} = 0$ is then tested using an approximate F test by

$$F = \frac{\text{SS}(E^{\alpha_\ell})}{\text{MS(Remainder)}}$$

with $(r-1)(2^{n-p}-1)$ d.f. in the denominator, as compared to $(r-1)2^{n-p}(2^p - 1)$ for testing H_0: $E^{\gamma_m} = 0$ in the intrablock analysis.

8.8 NUMERICAL EXAMPLE USING SAS

The following example illustrates how SAS PROC GLM and PROC MIXED can be used to analyze the data from systems of confounding.

Consider again the 2^3 factorial in blocks of size 4 and with ABC confounded with blocks (see Example 8.2). The data are given in Table 8.10 together with the intrablock analysis, using PROC GLM.

The combined analysis, using intrablock information for A, B, AB, C, AC, BC and interblock information for ABC, is obtained with PROC MIXED and is given in Table 8.11.

We note here that in order to obtain the correct interblock information; that is, test of ABC, it is not sufficient to declare the block effects as random effects, but also amend the model statement by the option DDFM = SATTERTH (Note: It is always useful to check the d.f. of each F test as at least a partial assurance that the correct test has been performed). A check that the correct test for ABC has been performed is provided in Table 8.12

As explained in Section 8.7 (see also Table 8.8b), the denominator for testing H_0: $ABC = 0$ is given by MS(Remainder) [see (8.25)], which algebraically is given by MS(Rep $\times$ ABC) with $r - 1 = 1$ d.f. From Table 8.12 we obtain

$$\text{SS}(ABC) = 7.5625 \quad \text{and} \quad \text{SS(rep} \times ABC) = 0.5625$$

and hence

$$F = \frac{7.5625}{0.5625} = 13.38$$

which is approximately the same as the F value given in Table 8.11, namely 13.44.

As a final note we mention that estimates for σ_e^2 and $\sigma_{\beta*}^2$ are given in Table 8.11 as

$$\widehat{\sigma}_e^2 = 0.3125 \quad \text{and} \quad \widehat{\sigma}_{\beta*}^2 = 0.0625$$

from which we can also reconstruct MS(Remainder) as

$$\text{MS(Remainder)} = \widehat{\sigma}_e^2 + 4\,\widehat{\sigma}_{\beta*}^2 = 0.5625$$

which agrees with the value given in Table 8.12.

CHAPTER 9

Partial Confounding in 2^n Factorial Designs

9.1 INTRODUCTION

In the previous chapter we have seen how the treatment combinations of a 2^n experiment can be accommodated in blocks of size 2^p, where $p < n$, and what consequences this has with regard to the estimation of effects and interactions. The price we have to pay for being able to reduce the error variance is the loss of information on certain interactions and possibly main effects. As long as only high-order interactions are confounded, the price may not be too high, but as is evident from Table 8.3, this may not always be possible; that is, in certain cases we may have to confound low-order interactions and/or main effects. This is clearly undesirable. We would like to obtain at least partial information on all essential effects and interactions. Since most experiments are replicated, it seems quite reasonable to use different systems of confounding in different replicates, which may achieve the objective just stated. This is known as *partial confounding* (as compared to the *complete confounding* of Chapter 8).

9.2 SIMPLE CASE OF PARTIAL CONFOUNDING

9.2.1 Basic Plan

Consider the simplest possible factorial scheme, that involving two factors A, B, each at two levels, and suppose that it is necessary to use blocks of two experimental units. The necessity of blocks of this size might arise, for example, in an experiment on young cattle, because it is possible to obtain a number of identical twins, that is, twins of the same genetic constitution and each pair of twins forming a block. Or it may be that the experimenter can handle only two

Design and Analysis of Experiments. Volume 2: Advanced Experimental Design
By Klaus Hinkelmann and Oscar Kempthorne
ISBN 0-471-55177-5

	Replicate I			Replicate II			Replicate III	
Block 1	a	ab	Block 1	b	ab	Block 1	(1)	ab
Block 2	(1)	b	Block 2	(1)	a	Block 2	a	b

Figure 9.1 Partial confounding of a 2^2 experiment.

experimental units at a time and that, if units were treated in random order, the experimental error introduced by randomizing over "times" would be considerable. Another possible use of blocks of two EUs could arise in plant pathological work, the block being a leaf and the EUs the two halves. Other examples in other fields of experimental research (e.g., medicine, engineering, psychology) can be found easily.

Given such a situation, suppose also that the experimenter wishes to obtain equal information on both main effects and the 2-factor interaction. This implies immediately that A, B, and AB should be confounded equally often with blocks. The basic pattern of the design would then consist of three replicates arranged as in Figure 9.1. The main effect A is confounded in replicate I, B in replicate II, and the interaction AB in replicate III. Each effect and interaction is then "partially confounded" with blocks. In view of this, we shall estimate the effects and interaction from the replicates in which they are unconfounded with blocks, namely

Effect A	estimated from replicates II, III
Effect B	estimated from replicates I, III
Effect AB	estimated from replicates I, II.

These estimates will be subject to an error based on the variance of units within blocks of two units treated alike, and, in order to obtain a reasonably precise estimate of this error variance, we shall need several repetitions of the basic pattern given in Figure 9.1. Suppose we have q repetitions (i.e., $3q$ replicates in all); then the ANOVA will have the structure shown in Table 9.1.

9.2.2 Analysis

The ANOVA is based on the three-part model

$$y_{ijk\ell} = \mu + \rho_i + \beta^*_{ij} + \tau_{k\ell} + e_{ijk\ell} \tag{9.1}$$

where ρ_i is the effect of the ith replicate ($i = 1, 2, \ldots, 3q$), β^*_{ij} is the effect of the jth block ($j = 1, 2$) in the ith replicate, and $\tau_{k\ell}$ is the effect of the treatment combination $a_k b_\ell (k, \ell = 0, 1)$. The reader will recognize that the basic pattern of this design, given in Figure 9.1, is actually a BIB design (see Chapter 2). Hence

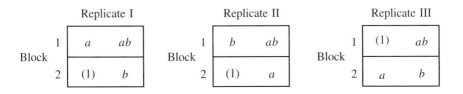

Table 9.1 Structure of T|B-ANOVA for Partially Confounded Design

Source	d.f.	SS
$X_\rho \mid \mathfrak{I}$	$3q-1$	$4\sum_i (\overline{y}_{i\ldots} - \overline{y}_{\ldots.})^2$
$X_{\beta^*} \mid \mathfrak{I},\ X_\rho$	$3q$	$2\sum_{ij} (\overline{y}_{ij..} - \overline{y}_{i\ldots})^2$
$X_\tau \mid \mathfrak{I},\ \mathbf{X}_\rho, \mathbf{X}_{\beta^*}$	3	
A	1	$2q\left[\widehat{A}_{\text{II, III}}\right]^2$
B	1	$2q\left[\widehat{B}_{\text{I, III}}\right]^2$
AB	1	$2q\left[\widehat{AB}_{\text{II, II}}\right]^2$
$\boldsymbol{I} \mid \mathfrak{I},\ X_\rho,\ X_{\beta^*},\ X_\tau$	$3(2q-1)$	Difference
Total	$12q-1$	$\sum_{ijk\ell} (y_{ijk\ell} - \overline{y}_{\ldots.})^2$

the complete design is a resolved BIB design and could be analyzed as such (see Section 9.3). It is, however, much simpler to use the factorial structure of the treatments and the particular system of confounding that has been employed.

As mentioned earlier, the effects and the interaction can be estimated (unbiasedly) only from those types of replicates in which they are not confounded. To emphasize this point, we write $\widehat{A}_{\text{II,III}}$, $\widehat{B}_{\text{I,III}}$, $\widehat{AB}_{\text{I,II}}$ indicating that A is estimated from replicates of types II and III, that is, from $r = 2q$ replicates; B is estimated from replicates of types I and III; and AB is estimated from replicates of types I and II. Consequently, the associated sums of squares are based on these estimates, as given in Table 9.1. Also, using the general formula (7.62), with $r = 2q, n = 2$, we have

$$\text{var}(\widehat{A}_{\text{II,III}}) = \text{var}(\widehat{B}_{\text{I,III}}) = \text{var}(\widehat{AB}_{\text{I,II}}) = \frac{1}{2q}\sigma_e^2$$

Following the procedure outlined in Section 8.5 we can partition SS(Blocks/replicates) = SS($X_{\beta^*} \mid \mathfrak{I}, X_\rho$) and SS(Residual)=SS($\boldsymbol{I} \mid \mathfrak{I},\ X_\rho,\ X_{\beta^*},\ X_\tau$) of Table 9.1 as given in Table 9.2. The SS(Blocks/replicates) is further partitioned into a treatment component, associated with the estimates of A, B, AB from those types of replicates in which they are confounded, and SS(Remainder), which is algebraically equal to the interactions between the effects and the replicates in which the effects are confounded. The actual computation of the various sums of squares is indicated explicitly in Table 9.2a. Here, for example, $\widetilde{A}_\text{I}$ is the estimate of A obtained from observations in the q replicates of type I in which A

Table 9.2 Partitioning of Sums of Squares

Source		d.f.	SS
		a. SS(Blocks/Replicates)	
A		1	$q\left[\widetilde{A}_{\text{I}}\right]^2$
B		1	$q\left[\widetilde{B}_{\text{II}}\right]^2$
AB		1	$q\left[\widetilde{AB}_{\text{III}}\right]^2$
$A \times \text{reps}_{\text{I}}$	Remainder	$q-1$	$\sum_{\ell=1}^{q}\left[\widetilde{A}_{\ell\text{I}}\right]^2 - q\left[\widetilde{A}_{\text{I}}\right]^2$
$B \times \text{reps}_{\text{II}}$	Remainder	$q-1$	$\sum_{\ell=1}^{q}\left[\widetilde{B}_{\ell\text{II}}\right]^2 - q\left[\widetilde{B}_{\text{II}}\right]^2$
$AB \times \text{reps}_{\text{III}}$	Remainder	$q-1$	$\sum_{\ell=1}^{q}\left[\widetilde{AB}_{\ell\text{III}}\right] - q\left[\widetilde{AB}_{\text{III}}\right]^2$
		b. SS(Residual)	
$A \times \text{reps}_{\text{II, III}}$		$2q-1$	$\sum_{m=1}^{2q}[\widehat{A}_{m,\text{ II, III}}]^2 - 2q[\widehat{A}_{\text{II, III}}]^2$
$B \times \text{reps}_{\text{I, III}}$		$2q-1$	$\sum_{m=1}^{2q}[\widehat{B}_{m,\text{ I, III}}]^2 - 2q[\widehat{B}_{\text{I, III}}]^2$
$AB \times \text{reps}_{\text{I, II}}$		$2q-1$	$\sum_{m=1}^{2q}[\widehat{AB}_{m,\text{ I, II}}]^2 - 2q[\widehat{AB}_{\text{I, II}}]^2$

is confounded. Similarly, $\widetilde{A}_{\ell\text{I}}$ is the estimate of A obtained from observations in the ℓth replicate of type I.

With regard to SS(Residual) it can be partitioned into three components each one algebraically equal to the interaction between an effect and replicates, in this case the replicates in which the corresponding effect is not confounded. The form of the sums of squares is given in Table 9.2b. The notation should be obvious now.

9.2.3 Use of Intra- and Interblock Information

In concluding the discussion of this introductory example of partial confounding, we point out that in this case we have both intra- and interblock information about A, B, and AB. If the block effects in model (9.1) were considered to be random effects, we could obtain combined estimates of the effects and interaction. We know from our previous discussion that

$$\text{var}(\widehat{A}_{\text{II,III}}) = \text{var}(\widehat{B}_{\text{I,III}}) = \text{var}(\widehat{AB}_{\text{I,II}}) = \frac{1}{2q}\sigma_e^2$$

and, as follows from Section 8.6,

$$\text{var}(\widetilde{A}_{\text{I}}) = \text{var}(\widetilde{B}_{\text{II}}) = \text{var}(\widetilde{AB}_{\text{III}}) = \frac{1}{q}(\sigma_e^2 + 2\sigma_\beta^2)$$

Hence the combined estimate of A, for example, is the weighted average of $\widehat{A}_{\text{II, III}}$ and $\widetilde{A}_I$, that is, letting $w = 1/\sigma_e^2$ and $w' = 1/(\sigma_e^2 + 2\sigma_\beta^2)$,

$$\widehat{\widehat{A}} = \frac{2qw\widehat{A}_{\text{II, III}} + qw'\widetilde{A}_{\text{I}}}{2qw + qw'} \tag{9.2}$$

with similar expressions for B and AB. (For a proof of this result we refer to Section 11.6.) Since w and w' or $\rho = w/w'$ are usually not known, the quantities have to be estimated and then substituted into (9.2). We know that

$$\widehat{\sigma}_e^2 = \text{MS(Residual)}$$

and we also know that (see Section 8.6)

$$\widehat{\sigma_e^2 + 2\sigma_\beta^2} = \text{MS(Remainder)}$$

There is, however, another way of estimating $\sigma_e^2 + 2\sigma_\beta^2$. This is accomplished through the B|T-ANOVA as given in Table 9.3 by utilizing all $3q$ d.f. for $\text{SS}(\boldsymbol{X}_{\beta^*}|\mathfrak{I}, \boldsymbol{X}_\rho, \boldsymbol{X}_\tau)$ and not just the $3(q-1)$ d.f. for SS(Remainder). We comment briefly on how $\text{SS}(\boldsymbol{X}_{\beta^*}|\mathfrak{I}, \boldsymbol{X}_\rho, \boldsymbol{X}_\tau)$ is obtained in a way other than the usual as indicated in Section 1.3.

As mentioned above, SS(Remainder) from the T|B-ANOVA is part of $\text{SS}(\boldsymbol{X}_{\beta^*}|\mathfrak{I}, \boldsymbol{X}_\rho, \boldsymbol{X}_\tau)$ because it is free of treatment effects. Since this SS accounts for $3(q-1)$ d.f., there remain 3 d.f. to be accounted for. These are obtained by realizing that for each effect and interaction we have two estimates, namely one from replicates in which the effect is not confounded and one from replicates in which the effect is confounded. Specifically, we have $\widehat{A}_{\text{II,III}}$ and $\widetilde{A}_{\text{I}}$, $\widehat{B}_{\text{I,III}}$ and $\widetilde{B}_{\text{II}}$, and $\widehat{AB}_{\text{I,II}}$ and $\widetilde{AB}_{\text{III}}$. Obviously, the comparison of these two types of estimates is a function of block effects and error only, and hence the associated sum of squares belongs to $\text{SS}(\boldsymbol{X}_{\beta^*}|\mathfrak{I}, \boldsymbol{X}_\rho, \boldsymbol{X}_\tau)$. Since, for example, $\widetilde{A}_{\text{II,III}}$ is obtained from $2q$ replicates and $\widetilde{A}_{\text{I}}$ from q replicates and these estimates are uncorrelated, we have

$$\text{SS}(\widehat{A}_{\text{II,III}} - \widetilde{A}_{\text{I}}) = \frac{[\widehat{A}_{\text{II,III}} - \widetilde{A}_{\text{I}}]^2}{1/2q + 1/q} = \frac{2q}{3}[\widehat{A}_{\text{II,III}} - \widetilde{A}_{\text{I}}]^2 \tag{9.3}$$

Table 9.3 Structure of B|T-ANOVA for Partially Confounded Design

Source	d.f.	SS	E(MS)
$X_\rho \mid \mathfrak{I}$	$3q-1$	From Table 9.1	
$X_\tau \mid \mathfrak{I}, \boldsymbol{X}_\rho$	3	$3q \sum_{k\ell} (\overline{y}_{..k\ell} - \overline{y}_{....})^2$	
$X_{\beta^*} \mid \mathfrak{I}, \boldsymbol{X}_\rho, \boldsymbol{X}_\tau$	$3q$		
$\widehat{A}_{\mathrm{II,III}}$ vs. $\widetilde{A}_{\mathrm{I}}$	1	$\frac{2q}{3}\left(\widehat{A}_{\mathrm{II,III}} - \widetilde{A}_{\mathrm{I}}\right)^2$	$\sigma_e^2 + \frac{4}{3}\sigma_\beta^2$
$\widehat{B}_{\mathrm{I,\ III}}$ vs. $\widetilde{B}_{\mathrm{II}}$	1	$\frac{2q}{3}\left(\widehat{B}_{\mathrm{I,III}} - \widetilde{B}_{\mathrm{II}}\right)^2$	$\sigma_e^2 + \frac{4}{3}\sigma_\beta^2$
$\widehat{AB}_{\mathrm{I,II}}$ vs. $\widetilde{A}_{\mathrm{III}}$	1	$\frac{2q}{3}\left(\widehat{AB}_{\mathrm{I,II}} - \widetilde{AB}_{\mathrm{III}}\right)^2$	$\sigma_e^2 + \frac{4}{3}\sigma_\beta^2$
Remainder	$3(q-1)$	From Table 9.2a	$\sigma_e^2 + 2\sigma_\beta^2$
Residual	$3(2q-1)$	From Table 9.2b	σ_e^2
Total	$12q-1$	From Table 9.1	

Similarly

$$\mathrm{SS}(\widehat{B}_{\mathrm{I,III}} - \widetilde{B}_{\mathrm{II}}) = \frac{2q}{3}[\widehat{B}_{\mathrm{I,III}} - \widetilde{B}_{\mathrm{II}}]^2 \tag{9.4}$$

$$\mathrm{SS}(\widehat{AB}_{\mathrm{I,II}} - \widetilde{AB}_{\mathrm{III}}) = \frac{2q}{3}[\widehat{AB}_{\mathrm{I,II}} - \widetilde{AB}_{\mathrm{III}}]^2 \tag{9.5}$$

Since the comparisons $\widehat{A}_{\mathrm{II,III}} - \widetilde{A}_{\mathrm{I}}$, $\widehat{B}_{\mathrm{I,III}} - \widetilde{B}_{\mathrm{III}}$, and $\widehat{AB}_{\mathrm{I,II}} - \widetilde{AB}_{\mathrm{III}}$ are orthogonal to each other, the SS (9.3), (9.4), and (9.5) are orthogonal. Also, the 3 d.f. associated with (9.3), (9.4), and (9.5) are not accounted for in SS(Remainder) since the comparisons leading to SS(Remainder) involve only comparisons among blocks from those replicates in which the respective effects are confounded, as is evident from Table 9.2a.

Returning now to the estimation of $\sigma_e^2 + 2\sigma_\beta^2$, to be used in (9.2), we evaluate as usual $E[\mathrm{MS}(\boldsymbol{X}_{\beta^*} \mid \mathfrak{I}, \boldsymbol{X}_\rho, \mathbf{X}_\tau)]$ assuming that the β_{ij}^* are i.i.d. random variables with mean zero and variance σ_β^2. We do this by obtaining the expected value of each component of $\mathrm{SS}(\boldsymbol{X}_{\beta^*} \mid \mathfrak{I}, \boldsymbol{X}_\rho, \boldsymbol{X}_\tau)$ as given in Table 9.3. We already know that

$$E[\mathrm{SS(Remainder)}] = 3(q-1)(\sigma_e^2 + 2\sigma_\beta^2) \tag{9.6}$$

Now

$$E[\mathrm{SS}(\widehat{A}_{\mathrm{II,III}} - \widetilde{A}_{\mathrm{I}})] = \frac{2q}{3}\mathrm{var}(\widehat{A}_{\mathrm{II,III}} - \widetilde{A}_{\mathrm{I}}) \tag{9.7}$$

$$= \sigma_e^2 + \frac{4}{3}\sigma_\beta^2$$

since

$$\text{var}(\widehat{A}_{\text{II,III}}) = \frac{1}{2q}\sigma_e^2$$

and

$$\text{var}(\widetilde{A}_{\text{I}}) = \frac{1}{q}(\sigma_e^2 + 2\sigma_\beta^2)$$

The same result (9.7) is, of course, obtained also for $\text{SS}(\widehat{B}_{\text{I,III}} - \widetilde{B}_{\text{II}})$ and $\text{SS}(\widehat{AB}_{\text{I,II}} - \widetilde{A}_{\text{III}})$. Hence,

$$\begin{aligned} E[(\text{MS}(\boldsymbol{X}_{\beta^*}|\boldsymbol{\mathfrak{I}}, \boldsymbol{X}_\rho, \boldsymbol{X}_\tau)] &= \frac{1}{3q}\left[3\left(\sigma_e^2 + \frac{4}{3}\sigma_\beta^2\right) + 3(q-1)(\sigma_e^2 + 2\sigma_\beta^2)\right] \\ &= \sigma_e^2 + 2\frac{3q-1}{3q}\sigma_\beta^2 \end{aligned} \tag{9.8}$$

An estimator for $\sigma_e^2 + 2\sigma_\beta^2$ is then

$$\widehat{\sigma_e^2 + 2\sigma_\beta^2} = \frac{1}{3q-1}[3q\ \text{MS}(\boldsymbol{X}_{\beta^*}|\boldsymbol{\mathfrak{I}}, \boldsymbol{X}_\rho, \boldsymbol{X}_\tau) - \text{MS(Residual)}] \tag{9.9}$$

We note here that if q is large, MS(Remainder) may actually be a quite satisfactory estimator for $\sigma_e^2 + 2\sigma_\beta^2$.

9.3 PARTIAL CONFOUNDING AS AN INCOMPLETE BLOCK DESIGN

In the previous section we have presented a simple example of partial confounding together with the analysis based entirely on reasoning suggested by the factorial structure of the treatments and the corresponding allocation of the treatments to blocks. The design given in Figure 9.1, replicated q times, is, of course, an example of an incomplete block design, in fact a resolvable BIB design in this case as we pointed out earlier. As such, data from this design can be analyzed according to the general principles discussed in Chapter 1, or, more specifically, as in Chapter 2. We shall do this now and show that the resulting analysis agrees with that given in the previous section.

9.3.1 Two Models

Without loss of generality and for purpose of ease of notation only, we assume that the arrangement of treatments in blocks is the same for each repetition of the basic design of Figure 9.1. Using the model for an incomplete block design, we have

$$\boldsymbol{y} = \mu\boldsymbol{\mathfrak{I}} + \boldsymbol{X}_\beta\boldsymbol{\beta} + \boldsymbol{X}_\tau\boldsymbol{\tau} + \boldsymbol{e} \tag{9.10}$$

with $\boldsymbol{y} = (y_{ijk\ell m})$, $i = 1, 2, \ldots, q$ indicating the repetition, $j = 1, 2, 3$ indicating the replication within repetition (as labeled in Fig. 9.1), $k = 1, 2$ denoting the block within replication, and $\ell, m = 0, 1$ denoting the levels of factors A and B. The observation vector $\boldsymbol{y}$ consists then of q segments $(y_{i1110}, y_{i1111}, y_{i1200}, y_{i1201}, y_{i2101}, y_{i2111}, y_{i2200}, y_{i2210}, y_{i3100}, y_{i3111}, y_{i3210}, y_{i3201})'$ $(i = 1, 2, \ldots, q)$ following Figure 9.1. Further, it can be deduced easily that

$$\boldsymbol{X}_\beta = \boldsymbol{I}_{6q} \times \begin{pmatrix} 1 \\ 1 \end{pmatrix} \tag{9.11}$$

$$\boldsymbol{X}_\tau = \mathfrak{I}_q \times \begin{pmatrix} 0 & 1 & 0 & 0 \\ 0 & 0 & 0 & 1 \\ 1 & 0 & 0 & 0 \\ 0 & 0 & 1 & 0 \\ 0 & 0 & 1 & 0 \\ 0 & 0 & 0 & 1 \\ 1 & 0 & 0 & 0 \\ 0 & 1 & 0 & 0 \\ 1 & 0 & 0 & 0 \\ 0 & 0 & 0 & 1 \\ 0 & 1 & 0 & 0 \\ 0 & 0 & 1 & 0 \end{pmatrix}$$

and $\boldsymbol{\tau}' = (\tau_{00}, \tau_{10}, \tau_{01}, \tau_{11})$.

Alternatively, we may write (9.1) as

$$\boldsymbol{y} = \mu\mathfrak{I} + \boldsymbol{X}_\beta\boldsymbol{\beta} + \boldsymbol{X}_{\tau^*}\boldsymbol{\tau}^* + \boldsymbol{e} \tag{9.12}$$

where

$$\boldsymbol{X}_{\tau^*} = \mathfrak{I}_q \times \begin{pmatrix} 1 & 1 & -1 & -1 \\ 1 & 1 & 1 & 1 \\ 1 & -1 & -1 & 1 \\ 1 & -1 & 1 & -1 \\ 1 & -1 & 1 & -1 \\ 1 & 1 & 1 & 1 \\ 1 & -1 & -1 & 1 \\ 1 & 1 & -1 & -1 \\ 1 & -1 & -1 & 1 \\ 1 & 1 & 1 & 1 \\ 1 & 1 & -1 & -1 \\ 1 & -1 & 1 & -1 \end{pmatrix} \tag{9.13}$$

and $\boldsymbol{\tau}^{*'} = \left(M, \frac{1}{2}A, \frac{1}{2}B, \frac{1}{2}AB\right)$. The "factorial incidence matrix" $\boldsymbol{N}_F = \boldsymbol{X}'_{\tau^*}\boldsymbol{X}_\beta$ is then of the form

$$\boldsymbol{N}_F = \mathfrak{I}'_q \times \begin{pmatrix} 2 & 2 & 2 & 2 & 2 & 2 \\ 2 & -2 & 0 & 0 & 0 & 0 \\ 0 & 0 & 2 & -2 & 0 & 0 \\ 0 & 0 & 0 & 0 & 2 & -2 \end{pmatrix}$$

or

$$\boldsymbol{N}_F = 2\mathfrak{I}'_q \times \begin{pmatrix} 1 & 1 & 1 & 1 & 1 & 1 \\ 1 & -1 & 0 & 0 & 0 & 0 \\ 0 & 0 & 1 & -1 & 0 & 0 \\ 0 & 0 & 0 & 0 & 1 & -1 \end{pmatrix} \tag{9.14}$$

9.3.2 Normal Equations

We consider now the RNE for $\boldsymbol{\tau}^*$, that is,

$$(\boldsymbol{X}'_{\tau^*}\boldsymbol{X}_{\tau^*} - \boldsymbol{N}_F(\boldsymbol{X}'_\beta\boldsymbol{X}_\beta)^{-1}\boldsymbol{N}'_F)\boldsymbol{\tau}^* = \boldsymbol{X}'_{\tau^*}\boldsymbol{y} - \boldsymbol{N}_F(\boldsymbol{X}'_\beta\boldsymbol{X}_\beta)^{-1}\boldsymbol{X}'_\beta\boldsymbol{y} \tag{9.15}$$

or

$$\left(12q\boldsymbol{I}_4 - \tfrac{1}{2}\boldsymbol{N}_F\boldsymbol{N}'_F\right)\boldsymbol{\tau}^* = \boldsymbol{T}^* - \tfrac{1}{2}\boldsymbol{N}_F\boldsymbol{B} \tag{9.16}$$

where

$$\boldsymbol{T}^* = \begin{pmatrix} 12q\widehat{M} \\ 6q\widetilde{A} \\ 6q\widetilde{B} \\ 6q\widetilde{AB} \end{pmatrix}, \qquad \boldsymbol{B} = \begin{pmatrix} B_{111} \\ B_{112} \\ \vdots \\ B_{q32} \end{pmatrix} \tag{9.17}$$

are the effect totals and block totals, respectively. With

$$\boldsymbol{N}_F\boldsymbol{N}'_F = 4q \begin{pmatrix} 6 & 0 & 0 & 0 \\ 0 & 2 & 0 & 0 \\ 0 & 0 & 2 & 0 \\ 0 & 0 & 0 & 2 \end{pmatrix}$$

and

$$\boldsymbol{N}_F\boldsymbol{B} = 2 \begin{pmatrix} B_{\dots} \\ B_{\cdot 11} - B_{\cdot 12} \\ B_{\cdot 21} - B_{\cdot 22} \\ B_{\cdot 31} - B_{\cdot 32} \end{pmatrix}$$

Eq. (9.16) reduces to

$$\begin{pmatrix} 8q \cdot \frac{1}{2}\widehat{A} \\ 8q \cdot \frac{1}{2}\widehat{B} \\ 8q \cdot \frac{1}{2}\widehat{AB} \end{pmatrix} = \begin{pmatrix} 6q\widetilde{A} - (B_{\cdot 11} - B_{\cdot 12}) \\ 6q\widetilde{B} - (B_{\cdot 21} - B_{\cdot 22}) \\ 6q\widetilde{AB} - (B_{\cdot 31} - B_{\cdot 32}) \end{pmatrix} = \begin{pmatrix} 4q\widehat{A}_{\text{II,III}} \\ 4q\widehat{B}_{\text{I,III}} \\ 4q\widehat{AB}_{\text{I,II}} \end{pmatrix}$$

and hence to

$$\begin{pmatrix} \widehat{A} \\ \widehat{B} \\ \widehat{AB} \end{pmatrix} = \begin{pmatrix} \widehat{A}_{\text{II,III}} \\ \widehat{B}_{\text{I,III}} \\ \widehat{AB}_{\text{I,II}} \end{pmatrix}$$

since $B_{\cdot 11} - B_{\cdot 12} = 2q\widetilde{A}_{\text{I}}$, $B_{\cdot 21} - B_{\cdot 22} = 2q\widetilde{B}_{\text{II}}$, $B_{\cdot 31} - B_{\cdot 32} = 2q\widetilde{AB}_{\text{III}}$, $\widetilde{A} = \frac{1}{3}(\widetilde{A}_{\text{I}} + 2\widehat{A}_{\text{II,III}})$, $\widetilde{B} = \frac{1}{3}(\widetilde{B}_{\text{II}} + 2\widehat{B}_{\text{I,III}})$, $\widetilde{AB} = \frac{1}{3}(\widetilde{AB}_{\text{III}} + 2\widehat{AB}_{\text{I,II}})$.

Turning to the RNE for $\boldsymbol{\beta}$, we consider

$$(\boldsymbol{X}'_\beta \boldsymbol{X}_\beta - \boldsymbol{N}'_F(\boldsymbol{X}'_{\tau^*}\boldsymbol{X}_{\tau^*})^{-1}\boldsymbol{N}_F)\boldsymbol{\beta} = \boldsymbol{X}'_\beta \boldsymbol{y} - \boldsymbol{N}'_F(\boldsymbol{X}'_{\tau^*}\boldsymbol{X}_{\tau^*})^{-1}\boldsymbol{X}'_{\tau^*}\boldsymbol{y} \tag{9.18}$$

or

$$\left(2\boldsymbol{I}_{6q} - \frac{1}{12q}\boldsymbol{N}'_F\boldsymbol{N}_F\right)\boldsymbol{\beta} = \boldsymbol{B} - \frac{1}{12q}\boldsymbol{N}'_F\boldsymbol{T}^* \tag{9.19}$$

with $\boldsymbol{B}$ and $\boldsymbol{T}^*$ as defined in (9.17). With

$$\begin{aligned} \boldsymbol{N}'_F\boldsymbol{N}_F &= 4(\mathfrak{I}_q\mathfrak{I}'_q) \times \begin{bmatrix} 2 & 0 & 1 & 1 & 1 & 1 \\ 0 & 2 & 1 & 1 & 1 & 1 \\ 1 & 1 & 2 & 0 & 1 & 1 \\ 1 & 1 & 0 & 2 & 1 & 1 \\ 1 & 1 & 1 & 1 & 2 & 0 \\ 1 & 1 & 1 & 1 & 0 & 2 \end{bmatrix} \\ &= 4(\mathfrak{I}_q\mathfrak{I}'_q) \times \boldsymbol{H} \text{ say} \end{aligned}$$

we see that the coefficient matrix for $\boldsymbol{\beta}$, that is,

$$\frac{1}{3q}\left(6q\boldsymbol{I} - \mathfrak{I}_q\mathfrak{I}'_q \times \boldsymbol{H}\right) \tag{9.20}$$

is of rank $6q - 1$.

9.3.3 Block Contrasts

From the form of (9.20) we can write out $6q - 1$ identifiable functions of $\boldsymbol{\beta}$, namely $3q - 1$ of the form

$$S_1\colon \{2(\beta_{ij1} + \beta_{ij2} - \beta_{q31} - \beta_{q32});\ i = 1, 2, \ldots, q;\ j = 1, 2, 3;\ (ij) \neq (q3)\}$$

$3(q - 1)$ of the form

$$S_2\colon \{2(\beta_{ij1} - \beta_{ij2} - \beta_{qj1} + \beta_{qj2});\ i = 1, 2, \ldots, q - 1;\ j = 1, 2, 3\}$$

and 3 of the form

$$S_3\colon \left\{\frac{4}{3}\sum_{i=1}^{q}(\beta_{ij1} - \beta_{ij2});\ j = 1, 2, 3\right\}$$

These three sets of identifiable functions are orthogonal to each other in the sense that any function in S_ν is orthogonal to any function in $S_{\nu'}(\nu, \nu' = 1, 2, 3;\ \nu \neq \nu')$.

The corresponding RHSs are obtained by writing the RHS of (9.19) in the form

$$\boldsymbol{B} - \mathfrak{I}_q \times \begin{bmatrix} 2\widehat{M} + \widetilde{A} \\ 2\widehat{M} - \widetilde{A} \\ 2\widehat{M} + \widetilde{B} \\ 2\widehat{M} - \widetilde{B} \\ 2\widehat{M} + \widetilde{AB} \\ 2\widehat{M} - \widetilde{AB} \end{bmatrix}$$

We then find the following RHSs for

$$S_1\colon\ \left\{B_{ij\cdot} - B_{q3\cdot};\ i = 1, 2\ldots, q;\ j = 1, 2, 3;\ (ij) \neq (3q)\right\}$$

$$S_2\colon\ \left\{B_{ij1} - B_{ij2} - B_{qj1} + B_{qj2};\ i = 1, 2\ldots, q - 1;\ j = 1, 2, 3\right\}$$

$$S_3\colon\ \left\{B_{\cdot 11} - B_{\cdot 12} - 2q\widetilde{A},\ B_{\cdot 21} - B_{\cdot 22} - 2q\widetilde{B},\ B_{\cdot 31} - B_{\cdot 32} - 2q\widetilde{AB}\right\}$$

Within the context of the factorial calculus we can rewrite these for S_2 and S_3 as

$$\begin{aligned} S_2\colon\ \Big\{&2(\widetilde{A}_i - \widetilde{A}_q) \text{ (for } j = 1) \\ &2(\widetilde{B}_i - \widetilde{B}_q) \text{ (for } j = 2) \\ &2(\widetilde{AB}_i - \widetilde{AB}_q) \text{ (for } j = 3);\ i = 1, 2\ldots, q - 1\Big\} \end{aligned}$$

$$S_3\colon\ \left\{\frac{4q}{3}(\widetilde{A}_{\mathrm{I}} - \widetilde{A}_{\mathrm{II,III}}),\ \frac{4q}{3}(\widetilde{B}_{\mathrm{II}} - \widetilde{B}_{\mathrm{I,III}}),\ \frac{4q}{3}(\widetilde{AB}_{\mathrm{III}} - \widetilde{AB}_{\mathrm{I,II}})\right\}$$

since $B_{i11} - B_{i12} = 2\widetilde{A}_{i\mathrm{I}}$, $B_{i21} - B_{i22} = 2\widetilde{B}_{i\mathrm{II}}$, $B_{i21} - B_{i32} = 2\widetilde{AB}_{i\mathrm{III}}$.

We recognize that the comparisons S_1 represent comparisons among replicates and the associated sum of squares is SS(Replicate) in Table 9.1. The comparisons of S_2 and S_3 are comparisons of block effects within replicates and hence belong to $(\boldsymbol{X}_{\beta*}|\mathfrak{I}, \boldsymbol{X}_\rho, \boldsymbol{X}_\tau)$. The sum of squares associated with S_2 is SS(Remainder) of Table 9.2, each set for $j = 1, 2, 3$ leading to a sum of squares with $q - 1$ d.f. Finally, the three functions of S_3 give rise to the remaining three single d.f. sums of squares belonging to $(\boldsymbol{X}_{\beta*}|\mathfrak{I}, \boldsymbol{X}_\rho, \boldsymbol{X}_\tau)$ as given in Table 9.3.

We have thus shown that a system of partial confounding can be analyzed using the general principles of incomplete block designs. Although we have only considered the intrablock analysis of a particular case, this line of argument can, of course, be carried over to the combined analysis and to the general case. It is hoped, however, that the reader realizes that the analysis can be described and performed much easier via the concepts of factorial experiments as the factorial structure of the treatments determines the structure of the incomplete block design.

9.4 EFFICIENCY OF PARTIAL CONFOUNDING

In general, reduction of block size will lead to a reduction of experimental error variance. On the other hand, this forces the experimenter to use a system of partial confounding, which may offset this gain. Therefore, if the experimenter has a choice of using blocks either of size 4 or of size 2, he or she needs a criterion on which to base the choice. For the simple case discussed so far, this is obtained by comparing the information provided by both designs (using the same number of replicates) as given in Table 9.4.

The information of the partially confounded scheme relative to that of the scheme with no confounding is

$$\frac{2q/\sigma_2^2}{3q/\sigma_4^2} = \frac{2}{3}\frac{\sigma_4^2}{\sigma_2^2} \tag{9.21}$$

where the subscript on σ^2 denotes the number of units per block. If σ_4^2 is greater than $3/2\sigma_2^2$ (or σ_2^2 less than $2/3\sigma_4^2$), the information is greater with the partially confounded design. Equivalently, we say then that the efficiency of the partially confounded design relative to the unconfounded design, as given by (9.21), is

Table 9.4 Information on Effects with Equal Confounding

	No Confounding	Partial Confounding of Section 9.2
A, B, AB	$\dfrac{3q}{\sigma_4^2}$	$\dfrac{2q}{\sigma_2^2}$

Table 9.5 Information on Effects with Unequal Confounding

	No Confounding	Partial Confounding
A	$\dfrac{4q}{\sigma_4^2}$	$\dfrac{3q}{\sigma_2^2}$
B	$\dfrac{4q}{\sigma_4^2}$	$\dfrac{3q}{\sigma_2^2}$
AB	$\dfrac{4q}{\sigma_4^2}$	$\dfrac{2q}{\sigma_2^2}$

larger than 1. In general, σ_4^2 will be greater than σ_2^2, but whether it will be sufficiently greater to give the advantage to the partially confounded design depends on the experimental material. In many cases this may prove to be a difficult question.

We have considered a scheme of partial confounding that results in equal information on main effects and the 2-factor interaction. In some cases it may be more appropriate to obtain greater information on main effects. This would entail greater representation of replicates of type III, to an extent depending on the relative amounts of information required. Suppose that a basic repetition consists of one replicate each of types I and II and 2 replicates of type III and that there are q such repetitions, that is, $4q$ replicates all together. The amounts of information to be compared then are as given in Table 9.5.

Since the emphasis is on main effects, it seems reasonable to use the partial confounding scheme if $\sigma_2^2 < \frac{3}{4}\sigma_4^2$.

The design utilizing partial confounding that we have just mentioned is obviously only one of many choices. The number of choices becomes even larger as the number of factors increases. Some such cases will be discussed in the following sections.

9.5 PARTIAL CONFOUNDING IN A 2^3 EXPERIMENT

As we have mentioned earlier, the arrangements utilizing partial confounding that are best for any given situation depend on the information the experimenter wishes to obtain. Suppose, for example, with an experiment of three factors A, B, C, each at two levels, the experimenter desires maximum possible accuracy on main effects and equal information on the 2-factor and 3-factor interactions.

9.5.1 Blocks of Size 2

A suitable system of partial confounding will consist of a number of repetitions, q say, of the following types of replicates:

Type I: Confound AB, AC, BC
Type II: Same as type I

Type III: Confound A, BC, ABC
Type IV: Confound B, AC, ABC
Type V: Confound C, AB, ABC

This design requires $5q$ replicates, that is, $20q$ blocks, which may not always be feasible. The information that this design yields as compared to the unconfounded design is given in Table 9.6.

For an example using SAS PROC FACTEX see Table 9.14, Section 9.10.1.

9.5.2 Blocks of Size 4

In this case the experimenter will use a number of repetitions of the following basic pattern (with each column representing a block):

I		II		III		IV	
(1)	a	(1)	b	(1)	b	(1)	a
ab	b	ac	c	bc	c	ab	b
c	ac	b	ab	a	ab	ac	c
abc	bc	abc	bc	abc	ac	bc	abc

confounding AB, AC, BC, ABC in replicates of types I, II, III, IV, respectively.

Suppose we use q repetitions, the positions of replicates, and blocks within replicates and treatment combinations within blocks being randomized. The information from this design and the corresponding unconfounded design is given in Table 9.7. It follows then that the partially confounded design will yield more information on interactions than the unconfounded design and substantially more information on main effects if σ_4^2 is less than $\frac{3}{4}\sigma_8^2$.

Table 9.6 Information Given by Design for 2^3 System in Blocks of Size 2

	No Confounding		Partial Confounding of Section 9.5.1	
	Information	Estimate From	Information	Estimate from Replicate of Types
A	$\frac{10q}{\sigma_8^2}$	All replicates	$8q/\sigma_2^2$	I, II, IV, V
B			$8q/\sigma_2^2$	I, II, III, V
AB			$4q/\sigma_2^2$	III, IV
C			$8q/\sigma_2^2$	I, II, III, IV
AC			$4q/\sigma_2^2$	III, V
BC			$4q/\sigma_2^2$	IV, V
ABC			$4q/\sigma_2^2$	I, II

Table 9.7 Information Given by Design for 2^3 System in Blocks of 4

	No Confounding		Partial Confounding of Section 9.5.2	
	Information	Estimate From	Information	Estimate from Replicate of Types
A			$8q/\sigma_4^2$	All
B			$8q/\sigma_4^2$	All
AB			$6q/\sigma_4^2$	II, III, V
C	$\dfrac{8q}{\sigma_8^2}$	All replicates	$8q/\sigma_4^2$	All
AC			$6q/\sigma_4^2$	I, III, V
BC			$6q/\sigma_4^2$	I, II, V
ABC			$6q/\sigma_4^2$	I, II, III

Table 9.8 Structure of T|B-ANOVA

	Source of Variation	d.f.
a.	Replicates	$4q - 1$
	Blocks/replicates	$4q$
	A	1
	B	1
	AB	1
	C	1
	AC	1
	BC	1
	ABC	1
	Residual	$24q - 7$
	Total	$32q - 1$
b.	Blocks/replicates	$4q$
	AB	1
	AC	1
	BC	1
	ABC	1
	$AB \times \text{reps}_{\text{I}}$	$q - 1$
	$AC \times \text{reps}_{\text{II}}$	$q - 1$
	$BC \times \text{reps}_{\text{III}}$	$q - 1$
	$ABC \times \text{reps}_{\text{IV}}$	$q - 1$

Table 9.8 **(*Continued*)**

	Source of Variation	d.f.
c.	Residual	$24q - 7$
	$A \times$ reps (all)	$4q - 1$
	$B \times$ reps (all)	$4q - 1$
	$AB \times$ reps (II, III, IV)	$3q - 1$
	$C \times$ reps (all)	$4q - 1$
	$AC \times$ reps (I, III, IV)	$3q - 1$
	$BC \times$ reps (I, II, IV)	$3q - 1$
	$ABC \times$ reps (I, II, III)	$3q - 1$

The partition of the degrees of freedom in the analysis of variance is given in Table 9.8a with a breakdown of the degrees of freedom for blocks in Table 9.8b and residual in Table 9.8c.

The computation of the various sums of squares follows the general procedure as indicated in Tables 9.1 and 9.2 using the observations from the respective replicates according to Tables 9.7 and 9.8.

9.6 PARTIAL CONFOUNDING IN A 2^4 EXPERIMENT

9.6.1 Blocks of Size 2

In this situation each replicate will consist of eight blocks. With s types of replicates, each utilizing a different system of confounding, there will be $8s - 1$ d.f. for blocks, 15 d.f. for treatments if no effect is completely confounded, and therefore $8s - 15$ d.f. for residual. In order to have an error based on a reasonable number of d.f., it seems that we need at least $s = 4$ replicates. It follows from Table 8.3 that under these circumstances the best system of confounding appears to be as follows:

Replicate	**Effects and Interactions Confounded**						
I	A,	BC,	ABC,	BD,	ABD,	CD,	ACD
II	B,	AC,	ABC,	AD,	ABD,	CD,	BCD
III	C,	AB,	ABC,	AD,	ACD,	BD,	BCD
IV	D,	AB,	ABD,	AC,	ACD,	BC,	BCD

These four replicates give $\frac{3}{4}$ relative information on main effects (in the sense that all main effects are estimated from three out of four replicates), $\frac{1}{2}$ relative information on 2-factor interactions, $\frac{1}{4}$ relative information on 3-factor interactions, and full information on the 4-factor interaction.

Since the number of observations is already quite large for one set of replicates, it does not seem to be practical to have repetitions of the basic pattern given above.

9.6.2 Blocks of Size 4

Referring again to Table 8.3, we find that a reasonable pattern of confounding is obtained by combining two systems of confounding as follows:

Replicate	Interactions Confounded		
I	AB,	CD,	$ABCD$
II	AC,	BD,	$ABCD$
III	AD,	ABC,	BCD
IV	BC,	ABD,	ACD

This yields full information on main effects, $\frac{3}{4}$ relative information on 2- and 3-factor interactions, and $\frac{1}{2}$ relative information on the 4-factor interaction. It also provides a sufficient number of degrees of freedom for residual.

For a design using replicates of types I and II see Section 9.10.2.

Another pattern of confounding, yielding equal information on all effects and interactions, is based on the completely orthogonalized 4×4 square given in Table 9.9. If we insert into each cell of this square the treatment combinations obtained by multiplying the treatment combinations listed for the corresponding row and column and then make up blocks of size 4 according to columns, rows, Latin letters, Greek letters, and numerals, we find the following pattern of confounding:

Replicate	Blocks Obtained From	Effects and Interactions Confounded
I	Columns	A, B, AB
II	Rows	C, D, CD
III	Latin letters	$AC, BD, ABCD$
IV	Greek letters	AD, ABC, BCD
V	Numerals	BC, ABD, ACD

The reader will realize that this design is in fact a resolvable BIB design with parameters $t = 16, b = 20, r = 5, k = 4, \lambda = 1$.

9.6.3 Blocks of Size 8

Most likely one would use here a system of complete confounding, namely confounding $ABCD$. If some information is wanted on all interactions, a reasonable

Table 9.9 Completely Orthogonalized 4 × 4 Square

	(1)	a	b	ab
(1)	$A\alpha 1$	$B\beta 2$	$C\gamma 3$	$D\delta 4$
c	$B\gamma 4$	$A\delta 3$	$D\alpha 2$	$C\beta 1$
d	$C\delta 2$	$D\gamma 1$	$A\beta 4$	$B\alpha 3$
cd	$D\beta 3$	$C\alpha 4$	$B\delta 1$	$A\gamma 2$

system of confounding would be to confound some or all of ABC, ABD, ACD, BCD, and $ABCD$, one in each replicate.

For all the examples discussed in this section it should be clear what the structure of the analysis is. Also, these examples should serve as an illustration for more general systems of partial confounding for 2^n experiments.

9.7 GENERAL CASE

The discussion in the preceding sections should have given an indication of how the concept of partial confounding can be utilized very effectively for obtaining information about most or all effects and interactions using blocks of small size. It also has pointed out that there exist many different designs, that is, patterns of confounding, for a given situation. There does not seem to be a reasonable way of discussing the various designs in general. A careful choice must be made for each specific experiment based on the objectives of the experiment and prior knowledge or expectations about the experimental situation. All designs, however, have certain properties in common, and we shall comment on these as they affect various parts of the analysis.

Suppose we have a 2^n factorial in blocks of size 2^p. We then have 2^{n-p} blocks in each replicate; that is, $2^{n-p} - 1$ interactions are confounded in each replicate. Let a basic pattern of partial confounding consist of s types of replicates and let there be q repetitions of the basic pattern. We can then divide the totality of the $2^n - 1$ interactions into three mutually exclusive sets as follows:

$$\mathcal{E}_1 = \{E^{\alpha_k}, k = 1, 2, \ldots, n_1;\ \text{completely confounded}\}$$

$$\mathcal{E}_2 = \{E^{\gamma_\ell}, \ell = 1, 2, \ldots, n_2;\ \text{partially confounded}\}$$

$$\mathcal{E}_3 = \{E^{\delta_m}, m = 1, 2, \ldots, n_3;\ \text{not confounded}\}$$

with $n_1 + n_2 + n_3 = 2^n - 1$. With regard to $\mathcal{E}_2$, let E^{γ_ℓ} be confounded in $c(\gamma_\ell)$ replicates and not confounded in $u(\gamma_\ell) = s - c(\gamma_\ell)$ replicates. Finally, let $N = 2^n sq$ denote the total number of observations.

9.7.1 Intrablock Information

Each $E^{\delta_m} \in \mathcal{E}_3$ is estimated from all replicates, and

$$\text{var}\left(\widehat{E}^{\delta_m}\right) = \frac{1}{sq2^{n-2}}\sigma_e^2$$

and hence

$$\text{SS}\left(E^{\delta_m}\right) = sq2^{n-2}\left[\widehat{E}^{\delta_m}\right]^2 \tag{9.22}$$

$E^{\gamma_\ell} \in \mathcal{E}_2$ is estimated from $qu(\gamma_\ell)$ replicates with

$$\text{var}\left(\widehat{E}^{\gamma_\ell}\right) = \frac{1}{qu(\gamma_\ell)2^{n-2}}\sigma_e^2 \tag{9.23}$$

and

$$\text{SS}\left(E^{\gamma_\ell}\right) = qu(\gamma_\ell)2^{n-2}\left[\widehat{E}^{\gamma_\ell}\right]^2 \tag{9.24}$$

9.7.2 The ANOVAs

The basic partitioning of the total sum of squares and its d.f. in the T|B-ANOVA is given in Table 9.10a, with a further partitioning of the block and residual sum of squares indicated symbolically in Tables 9.10b and 9.10c. The ANOVA is based on model (9.1).

In Table 9.10a SS(E^{γ_ℓ}) and SS(E^{δ_m}) for $E^{\gamma_\ell} \in \mathcal{E}_2$ and $E^{\delta_m} \in \mathcal{E}_3$ are obtained from (9.22) and (9.24), respectively, and ν_R is obtained by subtraction. In Table 9.10b SS$(E^{\alpha_k})_c$ and SS$(E^{\gamma_\ell})_c$ for $E^{\alpha_k} \in \mathcal{E}_1$ and $E^{\gamma_\ell} \in \mathcal{E}_2$ are obtained from the replicates in which these interactions are confounded, that is,

$$\text{SS}\left(E^{\alpha_k}\right)_c = sq2^{n-2}\left[\widetilde{E}^{\alpha_k}\right]^2$$

and

$$\text{SS}\left(E^{\gamma_\ell}\right)_c = qc(\gamma_\ell 2^{n-2})\left[\widetilde{E}^{\gamma_\ell}\right]^2$$

Similarly

$$\text{SS}\left(E_{ij}^{\alpha_k}\right)_c = 2^{n-2}\left[\widetilde{E}_{ij}^{\alpha_k}\right]^2$$

and

$$\text{SS}\left(E_{ij}^{\gamma_\ell}\right)_c = 2^{n-2}\left[\widetilde{E}_{ij}^{\gamma_\ell}\right]^2$$

where (ij) denotes the replicate j in repetition i and $\sum'_{ij}$ denotes summation over all replicates in which a given E^{γ_ℓ} is confounded, that is, $qc(\gamma_\ell)$ replicates.

Also, in Table 9.10c SS$(E_{ij}^{\gamma_\ell})$ and SS$(E_{ij}^{\delta_m})$ are obtained from those replicates in which the corresponding effects are not confounded, that is,

$$\text{SS}\left(E_{ij}^{\gamma_\ell}\right) = 2^{n-2}\left[\widehat{E}_{ij}^{\gamma_\ell}\right]^2$$

Table 9.10 T|B-ANOVA for Partial Confounding

Source	d.f.	SS
	a. Basic Partitioning	
$X_\rho \mid \mathfrak{I}$	$sq - 1$	Usual
$X_{\beta^*} \mid \mathfrak{I}, X_\rho$	$sq(2^{n-p} - 1)$	Usual
$X_\tau \mid \mathfrak{I}, X_\rho, X_{\beta^*}$	$n_2 + n_3$	$\sum_\ell \text{SS}(E^{\gamma_\ell}) + \sum_m \text{SS}(E^{\delta_m})$
$I \mid \mathfrak{I}, X_\rho, X_{\beta^*}, X_\tau$	ν_R	Difference
Total	$N - 1$	Usual
	b. Partitioning of Block Sum of Squares	
$X_{\beta^*} \mid \mathfrak{I}, X_\rho$	$sq(2^{n-p} - 1)$	
$\{E^{\alpha_k}\}$	n_1	$\sum_k \text{SS}(E^{\alpha_k})_c$
$\{E^{\gamma_\ell}\}$	n_2	$\sum_\ell \text{SS}(E^{\gamma_\ell})_c$
$\{E^{\alpha_k} \times \text{ reps}\}$	$n_1(sq - 1)$	$\sum_k \sum_{ij} \text{SS}(E_{ij}^{\alpha_k})_c - \sum_k \text{SS}(E^{\alpha_k})_c$
$\{E^{\gamma_\ell} \times \text{ reps}\}$	$q \sum_\ell c(\gamma_\ell) - n_2$	$\sum_\ell {\sum_{ij}}' \text{SS}(E_{ij}^{\gamma_\ell})_c - \sum_\ell \text{SS}(E^{\gamma_\ell})_c$
	c. Partitioning of Residual Sum of Squares	
$I \mid \mathfrak{I}, X_\rho, X_{\beta^*}, X_\tau$	ν_R	
$\{E^{\gamma_\ell} \times \text{ reps}\}$	$q \sum_\ell u(\gamma_\ell) - n_2$	$\sum_\ell {\sum_{ij}}'' \text{SS}(E_{ij}^{\gamma_\ell}) - \sum_\ell \text{SS}(E^{\gamma_\ell})$
$\{E^{\delta_m} \times \text{ reps}\}$	$n_3(sq - 1)$	$\sum_m \sum_{ij} \text{SS}(E_{ij}^{\delta_m}) - \sum_m \text{SS}(E^{\delta_m})$

and

$$\text{SS}\left(E_{ij}^{\delta_m}\right) = 2^{n-2}\left[\widehat{E}_{ij}^{\delta_m}\right]^2$$

and $\sum''$ denotes summation over all replicates (ij) in which a given E^{γ_ℓ} is not confounded, there being $qu(\gamma_\ell)$ such replicates.

Table 9.11 B|T-ANOVA for Partial Confounding

Source	d.f.	SS
	a. Basic Partitioning	
$\boldsymbol{X}_\rho \mid \mathfrak{I}$	$sq-1$	Usual
$\boldsymbol{X}_\tau \mid \mathfrak{I}, \boldsymbol{X}_\rho$	2^n-1	Usual
$\boldsymbol{X}_{\beta^*} \mid \mathfrak{I}, \boldsymbol{X}_\rho, \boldsymbol{X}_\tau$	$sq(2^{n-p}-1)-n_1$	Difference
$\boldsymbol{I} \mid \mathfrak{I}, \boldsymbol{X}_\rho, \boldsymbol{X}_\tau, \boldsymbol{X}_{\beta^*}$	ν_R	From Table 9.10
Total	$N-1$	Usual
	b. Partitioning of Block Sum of Squares	
$\boldsymbol{X}_{\beta^*} \mid \mathfrak{I}, \boldsymbol{X}_\rho, \boldsymbol{X}_\tau$	$sq(2^{n-p}-1)-n_1$	
$\widehat{E}^{\boldsymbol{\gamma}_\ell}$ vs. $\widetilde{E}^{\boldsymbol{\gamma}_\ell}$	n_2	$\dfrac{q2^{n-2}}{s}\sum_\ell c(\boldsymbol{\gamma}_\ell)u(\boldsymbol{\gamma}_\ell)\left[\widehat{E}^{\boldsymbol{\gamma}_\ell}-\widetilde{E}^{\boldsymbol{\gamma}_\ell}\right]^2$
$\{E^{\boldsymbol{\alpha}_k} \times \text{ reps}\}$	$n_1(sq-1)$	From Table 9.10b
$\{E^{\boldsymbol{\gamma}_\ell} \times \text{ reps}\}$	$q\sum_\ell c(\boldsymbol{\gamma}_\ell)-n_2$	From Table 9.10b

The B|T-ANOVA in its basic form is given in Table 9.11a and a partitioning of the block sum of squares is given in Table 9.11b. Here

$$\text{SS}\left(\widehat{E}^{\boldsymbol{\gamma}_\ell} \text{ vs. } \widetilde{E}^{\boldsymbol{\gamma}_\ell}\right)=\frac{\left(\widehat{E}^{\boldsymbol{\gamma}_\ell}-\widetilde{E}^{\boldsymbol{\gamma}_\ell}\right)^2}{\operatorname{var}\left(\widehat{E}^{\boldsymbol{\gamma}_\ell}-\widetilde{E}^{\boldsymbol{\gamma}_\ell}\right)/\sigma_e^2}$$

with

$$\operatorname{var}\left(\widehat{E}^{\boldsymbol{\gamma}_\ell}-\widetilde{E}^{\boldsymbol{\gamma}_\ell}\right)=\left(\frac{1}{qu(\boldsymbol{\gamma}_\ell)2^{n-2}}+\frac{1}{qc(\boldsymbol{\gamma}_\ell)2^{n-2}}\right)\sigma_e^2$$

$$=\frac{s}{qu(\boldsymbol{\gamma}_\ell)c(\boldsymbol{\gamma}_\ell)2^{n-2}}\sigma_e^2$$

9.7.3 Interblock Information

Assuming $\boldsymbol{\beta}^*$ to be a vector of i.i.d. random variables with $E(\boldsymbol{\beta}^*)=\boldsymbol{\phi}$, $E(\boldsymbol{\beta}^*\boldsymbol{\beta}^{*\prime})=\sigma_\beta^2\boldsymbol{I}$, we can obtain interblock information on the $E^{\boldsymbol{\alpha}_k}\in\mathcal{E}_1$ and $E^{\boldsymbol{\gamma}_\ell}\in\mathcal{E}_2$, namely

$$E\left(\widetilde{E}^{\boldsymbol{\alpha}_k}\right)=E^{\boldsymbol{\alpha}_k} \text{ with } \operatorname{var}\left(\widetilde{E}^{\boldsymbol{\alpha}_k}\right)=\frac{1}{sq2^{n-2}}\left(\sigma_e^2+2^p\sigma_\beta^2\right)$$

and

$$E\left(\widetilde{E}^{\boldsymbol{\gamma}_\ell}\right)=E^{\boldsymbol{\gamma}_\ell} \text{ with } \operatorname{var}\left(\widetilde{E}^{\boldsymbol{\gamma}_\ell}\right)=\frac{1}{qc(\boldsymbol{\gamma}_\ell)2^{n-2}}\left(\sigma_e^2+2^p\sigma_\beta^2\right) \tag{9.25}$$

9.7.4 Combined Intra- and Interblock Information

Combined information is available for $E^{\gamma_\ell} \in \mathcal{E}_2$. Using (9.23) and (9.25), we obtain the combined estimator

$$\widehat{\widehat{E}}^{\gamma_\ell} = \frac{u(\gamma_\ell)w\widehat{E}^{\gamma_\ell} + c(\gamma_\ell)w'\widetilde{E}^{\gamma_\ell}}{u(\gamma_\ell)w + c(\gamma_\ell)w'} \tag{9.26}$$

where, as usual, $w = 1/\sigma_e^2$, $w' = 1/(\sigma_e^2 + 2^p\sigma_\beta^2)$. Letting $w'/w = \rho$, we have

$$\begin{aligned} \text{var}\left(\widehat{\widehat{E}}^{\gamma_\ell}\right) &= \frac{1}{\left[u(\gamma_\ell)w + c(\gamma_\ell)w'\right]q2^{n-2}} \\ &= \frac{\sigma_e^2}{\left[u(\gamma_\ell) + c(\gamma_\ell)\rho\right]q2^{n-2}} \end{aligned} \tag{9.27}$$

9.7.5 Estimation of Weights

As always, we estimate σ_e^2 and hence w from the T|B-ANOVA, and using the Yates procedure, we estimate σ_β^2 and hence w' from the B|T-ANOVA. From Table 9.10 we obtain

$$\widehat{\sigma}_e^2 = \frac{\text{SS}(\boldsymbol{I}, \boldsymbol{X}_\rho, \mathbf{X}_{\beta_*}, \boldsymbol{X}_\tau)}{\nu_R} = \text{MS}(\boldsymbol{I}|\mathfrak{I}, \boldsymbol{X}_\rho, \boldsymbol{X}_{\beta^*}, \boldsymbol{X}_\tau) \tag{9.28}$$

and hence $\widehat{w} = 1/\widehat{\sigma}_e^2$.

In order to estimate w' we consider the expected values of the sums of squares that form $\text{SS}(\boldsymbol{X}_{\beta^*}|\mathfrak{I}, \boldsymbol{X}_\rho, \boldsymbol{X}_\tau)$ as given in Table 9.11b. Under the mixed model assumption given in Section 9.7.3 above we find, for each $E^{\gamma_\ell} \in \mathcal{E}_2$,

$$\begin{aligned} E\left(\widehat{E}^{\gamma_\ell} - \widetilde{E}^{\gamma_\ell}\right)^2 &= \text{var}\left(\widehat{E}^{\gamma_\ell} - \widetilde{E}^{\gamma_\ell}\right) \\ &= \frac{1}{q2^{n-2}}\left[\frac{1}{u(\gamma_\ell)}\sigma_e^2 + \frac{1}{c(\gamma_\ell)}\left(\sigma_e^2 + 2^p\sigma_\beta^2\right)\right] \\ &= \frac{s}{q2^{n-2}u(\gamma_\ell)c(\gamma_\ell)}\left[\sigma_e^2 + \frac{u(\gamma_\ell)2^p}{s}\sigma_\beta^2\right] \end{aligned} \tag{9.29}$$

and hence, from (9.29) and Table 9.11b,

$$ESS\left(\widehat{E}^{\gamma_\ell} \text{ vs. } \widetilde{E}^{\gamma_\ell}\right) = \sigma_e^2 + \frac{u(\gamma_\ell)2^p}{s}\sigma_\beta^2 \tag{9.30}$$

Since

$$ESS\left(E^{\alpha_k} \times \text{ reps}\right) = (sq - 1)\left(\sigma_e^2 + 2^p\sigma_\beta^2\right) \tag{9.31}$$

and

$$ESS\left(E^{\gamma_\ell} \times \text{ reps}\right) = \left[qc(\gamma_\ell) - 1\right]\left(\sigma_e^2 + 2^p\sigma_\beta^2\right) \tag{9.32}$$

we obtain, by using (9.30), (9.31), and (9.32) in Table 9.11b,

$$EMS\left(\boldsymbol{X}_{\beta*}|\mathfrak{I}, \boldsymbol{X}_\rho, \boldsymbol{X}_\tau\right) = \sigma_e^2 + \Delta 2^p \sigma_\beta^2 \tag{9.33}$$

where

$$\Delta = \frac{(1/s)\sum_\ell u(\gamma_\ell) + n_1(sq-1) + q\sum_\ell c(\gamma_\ell) - n_2}{sq(2^{n-p}-1) - n_1} \tag{9.34}$$

From (9.33), (9.34), and (9.28) we then obtain

$$\widehat{\sigma}_e^2 + 2^p\widehat{\sigma}_\beta^2 = \frac{1}{\widehat{w}'} = \left(1 - \frac{1}{\Delta}\right)\text{MS}(\boldsymbol{I}|\mathfrak{I}, \boldsymbol{X}_\rho, \boldsymbol{X}_{\beta*}, \boldsymbol{X}_\tau)$$
$$+\frac{1}{\Delta}\text{MS}(\boldsymbol{X}_{\beta*}|\mathfrak{I}, \boldsymbol{X}_\rho, \boldsymbol{X}_\tau) \tag{9.35}$$

An interesting consequence of our derivation of $E\text{MS}(\boldsymbol{X}_{\beta*}|\mathfrak{I}, \boldsymbol{X}_\rho, \boldsymbol{X}_\tau)$ is that because of (9.30), that is, the heterogeneous expected values of the components of $\text{SS}(\boldsymbol{X}_{\beta*}|\mathfrak{I}, \boldsymbol{X}_\rho, \boldsymbol{X}_\tau)$, it follows that even under the assumption of normality $\text{SS}(\boldsymbol{X}_{\beta*}|\mathfrak{I}, \boldsymbol{X}_\rho, \boldsymbol{X}_\tau)$ does not follow a scaled χ^2 distribution but rather is distributed as a linear combination of scaled χ^2 distributions.

Of course, as we have discussed earlier (see Section 1.11), other estimation procedures can be used to obtain estimates of σ_e^2 and σ_β^2 (see Section 9.10 for numerical examples using such procedures).

9.7.6 Efficiencies

Under a system of no confounding with $r = qs$ replicates per treatment combination (= number of blocks of size 2^n) and a particular pattern of partial confounding, we can compare the following amounts of information:

	No Confounding	Partial Confounding
$E^{\alpha_k} \in \mathcal{E}_1$	$sq2^{n-2}/\sigma_n^2$	0
$E^{\gamma_\ell} \in \mathcal{E}_2$	$sq2^{n-2}/\sigma_n^2$	$u(\gamma_\ell)q2^{n-2}/\sigma_p^2$
$E^{\delta_m} \in \mathcal{E}_3$	$sq2^{n-2}/\sigma_n^2$	$sq2^{n-2}/\sigma_p^2$

where σ_n^2 denotes σ_e^2 for blocks of size 2^n and σ_p^2 denotes σ_e^2 for blocks of size 2^p. From this we can compute the efficiency for each interaction and determine when the information on interactions in $\mathcal{E}_2$ and $\mathcal{E}_3$ under confounding is at least as good as that without confounding depending on the ratio σ_p^2 / σ_e^2. Excepting interactions in $\mathcal{E}_1$ we can express the average efficiency as

$$\frac{\left[\sum_\ell u(\gamma_\ell) + n_3 s\right] \Big/ (n_2 + n_3)s}{\sigma_n^2/\sigma_p^2}$$

9.8 DOUBLE CONFOUNDING

In certain experimental situations it may be necessary to use more sophisticated systems of confounding than we have discussed up to this point. Consider, for example, a 2^5 experiment, where each treatment is applied to a certain material that is then processed by a machine. Homogeneous batches of material, sufficient for four runs (treatment combinations), are available and, in order to complete the experiment in a reasonable time, four machines will be used; that is, each machine will handle eight pieces of material from eight different batches, each piece of material being treated differently (by one of the 2^5 treatment combinations). In order to eliminate batch-to-batch variation we might consider each batch as a block of size 4. Elimination of possible systematic differences among machines suggests to consider each machine as a block of size 8. Combining these two types of blocking and their corresponding systems of confounding leads to an arrangement of the treatment combinations in a rectangular array that is referred to as a system of *double confounding*. For the present example this is illustrated in Table 9.12.

This arrangement is obtained by confounding ABC, CDE, $ABDE$ with rows (machines) and AB, CD, $ABCD$, BDE, ADE, BCE, ACE with columns (batches) (see Table 8.3) and then constructing the intrablock subgroup for each system of confounding. These subgroups represent the first row and first column, respectively, in Table 9.12. The remaining treatment combinations are then filled in by taking products of appropriate treatment combinations according to the usual rules. In the actual experiment rows and columns are, of course, randomized.

Table 9.12 2^5 Experiment in 4 x 8 Pattern

		Batches							
		1	2	3	4	5	6	7	8
Machines	1	(1)	*ab*	*ace*	*bce*	*abde*	*de*	*bcd*	*acd*
	2	*abe*	*e*	*bc*	*ac*	*d*	*abd*	*acde*	*bcde*
	3	*cde*	*abcde*	*ad*	*bd*	*abc*	*c*	*be*	*ae*
	4	*abcd*	*cd*	*bde*	*ade*	*ce*	*abce*	*a*	*b*

Table 9.13 Analysis of Variance for Double Confounding Scheme

Source of Variation	d.f.
Replicates	1
Rows/replicates	6
Columns/replicates	14
$A, B, AB_{\text{II}}, C, AC_{\text{I}}, BC,$ $D, AD, BD, CD_{\text{II}}, E, AE,$ BE_{I}, CE, DE	15
Residual	27

The observation in the jth row and kth column receiving the ℓth treatment is represented by the model

$$y_{jk\ell} = \mu + \rho_j + \gamma_k + \tau_\ell + e_{jk\ell} \tag{9.36}$$

where the terms have the obvious meaning. However, this particular design is of little value unless repeated, for the only means of estimating the error variance is to use unconfounded interactions involving three or more factors, so that only 8 d.f. are available. In replications it would be well to use different schemes of confounding (i.e., partial confounding) so that some within-row-and-column information could be obtained on all important effects and interactions. If different sets of machines are used in different replicates, model (9.36) will be extended as follows:

$$y_{ijk\ell} = \mu + \alpha_i + \rho_{ij} + \gamma_{ik} + \tau_\ell + e_{ijk\ell} \tag{9.37}$$

where α_i is the effect of the ith replicate, ρ_{ij} the effect of the jth row (machine) in the ith replicate, γ_{ik} the effect of the kth column (batch) in the ith replicate, τ_ℓ the effect of the ℓth treatment combination, and $e_{ijk\ell}$ the error associated with the unit in the jth row and kth column of the ith replicate receiving the ℓth treatment. The structure of the ANOVA for model (9.37) is exhibited in Table 9.13 using two replicates and confounding in the second replicate ACD, BCE, $ABDE$ with rows, AC, BE, $ABCE$, BCD, ABD, CDE, ADE with columns. We assume that interactions involving three or more factors are negligible. The d.f. for error in Table 9.13 can be broken down into 12 d.f. from unconfounded or partially confounded higher order interactions and 15 d.f. from comparisons of unconfounded effects and interactions in the two replicates, for example, $\widehat{A}_{\text{I}}$ versus $\widehat{A}_{\text{II}}$.

9.9 CONFOUNDING IN SQUARES

In the previous section we have discussed systems of confounding in a rectangular array utilizing in general columns as blocks of size 2^p and rows as blocks of size

2^{n-p}. A special case, $p = n/2$, would lead to a square array. More generally, however, confounding in squares refers to any type of square array with blocks of size 2^p in both directions. Then every treatment combination appears 2^{2p-n} times in the array. We shall consider a few specific examples illustrating the procedure (Kempthorne, 1952, Chapter 15).

9.9.1 2^3 Factorial in Two 4 x 4 Squares

We need to confound one interaction with columns and one interaction with rows, and since two replicates are necessary to get an adequate estimate of the error variance, we may completely confound these two interactions, or we may partially confound four interactions, or we may partially confound two interactions (say, two 2-factor interactions) and completely confound one interaction (say, the 3-factor interaction). Examples of each follow.

1. If ABC is confounded with columns and BC with rows, both replicates will be randomizations of one of the squares

$$\begin{array}{cccc} (1) & a & bc & abc \\ ab & b & ac & c \\ ac & c & ab & b \\ bc & abc & (1) & a \end{array} \quad \text{or} \quad \begin{array}{cccc} (1) & a & bc & abc \\ ab & c & ac & b \\ ac & b & bc & c \\ bc & abc & (1) & a \end{array}$$

   Note that the first square is constructed by the method described in Section 9.8. Also, the ANOVA has a structure similar to that given in Table 9.13.

2. If ABC and BC are confounded with columns and rows, respectively, in one square, and AB and AC are confounded with columns and rows, respectively, in the second square, we obtain the following result:

$$\begin{array}{cccc} (1) & a & bc & abc \\ ab & b & ac & c \\ ac & c & ab & b \\ bc & abc & (1) & a \end{array} \quad \text{and} \quad \begin{array}{cccc} (1) & a & abc & ac \\ ab & b & c & bc \\ abc & bc & (1) & b \\ c & ac & ab & a \end{array}$$

   Here we obtain full information on main effects and $\frac{1}{2}$ relative information on interactions. As a result, however, the error d.f. are reduced from 13 to 11 compared to the previous design.

3. If we confound ABC with columns in both squares, BC with rows in one square, and AB with rows in the second square, we obtain

$$\begin{array}{cccc} (1) & a & bc & abc \\ ab & b & ac & c \\ ac & c & ab & b \\ bc & abc & (1) & a \end{array} \quad \text{and} \quad \begin{array}{cccc} (1) & c & ab & abc \\ ab & abc & (1) & c \\ ac & a & bc & b \\ bc & b & ac & a \end{array}$$

9.9.2 2^4 Factorial in 8 x 8 Squares

An interesting alternative to completely confounding two interactions with rows and columns is the following: Partially confound four 3-factor interactions with row pairs and completely confound the 4-factor interaction with the four column pairs. This leads to the following design, where ABC, ABD, ACD, BCD are confounded with row pairs I, II, III, IV, respectively:

I	(1)	d	ab	abd	ac	acd	bc	bcd
	ad	a	bd	b	cd	abc	$abcd$	c
II	ab	c	(1)	abc	ad	bcd	bd	acd
	ac	b	bc	a	$abcd$	abd	cd	d
III	cd	abc	ac	bcd	(1)	b	ad	abd
	bc	acd	$abcd$	c	bd	d	ab	a
IV	bd	abd	cd	acd	bc	a	(1)	abc
	$abcd$	bcd	ad	d	ab	c	ac	b

This design will provide $\frac{3}{4}$ relative information on the 3-factor interactions ABC, ABD, ACD, and BCD, no information on $ABCD$, and full information on main effects and 2-factor interactions. If replications can be done, other systems of partial confounding may be used. The structure of the ANOVA is similar to that given in Table 9.13.

Yates (1937a) has given arrangements for the 2^5 and 2^6 experiment in 8×8 squares. Through the process of randomizing rows and columns, these arrangements may lead (with considerable frequency) to undesirable configurations. Grundy and Healy (1950) have provided solutions to this problem by using the concept of restricted randomization.

9.10 NUMERICAL EXAMPLES USING SAS

In this section we shall illustrate the use of PROC FACTEX to construct designs with partial confounding and PROC GLM and PROC MIXED to analyze data from such designs.

9.10.1 2^3 Factorial in Blocks of Size 2

Using one replicate each of types III, IV, and V as defined in Section 9.5.1 leads to a design with partial confounding of A, B, C, AB, AC, BC and complete confounding of ABC. The SAS input and output are given in Table 9.14.

Table 9.14 2^3 Factorial in Blocks of Size 2 Using Partial Confounding

```
options nodate pageno=1;
proc factex;
factors A B C;
blocks size=2;
model est=(A B A*C B*C);
examine design confounding;
output out=replicate1 blockname=block nvals=(1 2 3 4);
title1 'TABLE 9.14';
title2 '2**3 FACTORIAL IN BLOCKS OF SIZE 2';
title3 'USING PARTIAL CONFOUNDING';
title4 '(CONFOUNDING ABC, AB AND C IN REP 1)';
run;

proc factex;
factors A B C;
blocks size=2;
model est=(A C A*B B*C);
examine design confounding;
output out=replicate2 blockname=block nvals=(5 6 7 8);
title4 '(CONFOUNDING ABC, AC AND B IN REP 2';
run;

proc factex;
factors A B C;
blocks size=2;
model est=(B C A*B A*C);
examine design confounding;
output out=replicate3 blockname=block nvals=(9 10 11 12);
title4 '(CONFOUNDING ABC, BC, AND A IN REP 3)';
run;

data combine;
set replicate1 replicate2 replicate3;
run;

proc print data=combine;
title4 '(PARTIALLY CONFOUNDING A, B, AB, C AC, BC';
title5 'AND COMPLETELY CONFOUNDING ABC)';
run;
```

Table 9.14 (*Continued*)

```
                2**3 FACTORIAL IN BLOCKS OF SIZE 2
                    USING PARTIAL CONFOUNDING
               (CONFOUNDING ABC, AB AND C IN REP 1)

                      The FACTEX Procedure
                         Design Points

            Experiment
               Number       A       B       C      Block
            -------------------------------------------
                    1      -1      -1      -1          1
                    2      -1      -1       1          4
                    3      -1       1      -1          2
                    4      -1       1       1          3
                    5       1      -1      -1          2
                    6       1      -1       1          3
                    7       1       1      -1          1
                    8       1       1       1          4

               Block Pseudofactor Confounding Rules

                         [B1] = A*B*C
                         [B2] = C

                2**3 FACTORIAL IN BLOCKS OF SIZE 2
                    USING PARTIAL CONFOUNDING
               (CONFOUNDING ABC, AC AND B IN REP 2)

                         Design Points

            Experiment
               Number       A       B       C      Block
            -------------------------------------------
                    1      -1      -1      -1          3
                    2      -1      -1       1          2
                    3      -1       1      -1          4
                    4      -1       1       1          1
                    5       1      -1      -1          2
                    6       1      -1       1          3
                    7       1       1      -1          1
                    8       1       1       1          4

               Block Pseudofactor Confounding Rules

                         [B1] = A*B*C
                         [B2] = A*C
```

Table 9.14 *(Continued)*

```
          2**3 FACTORIAL IN BLOCKS OF SIZE 2
              USING PARTIAL CONFOUNDING
         (CONFOUNDING ABC, BC, AND A IN REP 3)

                    Design Points

  Experiment
      Number       A        B        C        Block
  ---------------------------------------------------

           1      -1       -1       -1            3
           2      -1       -1        1            2
           3      -1        1       -1            2
           4      -1        1        1            3
           5       1       -1       -1            4
           6       1       -1        1            1
           7       1        1       -1            1
           8       1        1        1            4

        Block Pseudofactor Confounding Rules

                    [B1] = A*B*C
                    [B2] = B*C

          2**3 FACTORIAL IN BLOCKS OF SIZE 2
              USING PARTIAL CONFOUNDING
     (PARTIALLY CONFOUNDING A, B, AB, C, AC, BC
         AND COMPLETELY CONFOUNDING ABC)

      Obs     block       A        B        C

        1         1      -1       -1       -1
        2         1       1        1       -1
        3         2      -1        1       -1
        4         2       1       -1       -1
        5         3      -1        1        1
        6         3       1       -1        1
        7         4      -1       -1        1
        8         4       1        1        1
        9         5      -1        1        1
       10         5       1        1       -1
       11         6      -1       -1        1
       12         6       1       -1       -1
       13         7      -1       -1       -1
       14         7       1       -1        1
       15         8      -1        1       -1
       16         8       1        1        1
       17         9       1       -1        1
       18         9       1        1       -1
       19        10      -1       -1        1
       20        10      -1        1       -1
       21        11      -1       -1       -1
       22        11      -1        1        1
       23        12       1       -1       -1
       24        12       1        1        1
```

Note that there are three input statements, one for each replicate, and that the ESTIMATE statement specifies which effects should be estimable and hence which effects should be confounded in a given replicate.

Various aspects of the analysis of data from the design of Table 9.14 are illustrated in Table 9.15

The intrablock analysis is performed with PROC GLM and the combined analysis with PROC MIXED. The output is fairly self-explanatory, but we shall make the following comments, mainly to relate the numerical output to the theoretical developments in this chapter:

1. For the intrablock analysis, that is, the GLM procedure, estimates of all effects are obtained from the replicates in which they are not confounded with blocks. For example, the main effect A is estimated from replicates 1 and 2, that is, blocks 1–8, yielding $\widehat{A} = 3.0$.
2. The standard error for $\widehat{A}$ is obtained from

$$\widehat{\text{var}}(\widehat{A}) = \frac{1}{2 \cdot 2^{3-2}}\widehat{\sigma}_e^2 = \frac{1}{4} \cdot 7.5833 = 1.8958$$

as

$$\text{se}(\widehat{A}) = \sqrt{\widehat{\text{var}}(\widehat{A})} = 1.3769$$

3. Concerning the combined analysis using PROC MIXED, we notice that the estimate for σ_e^2 obtained here, namely $\widehat{\sigma}_e^2 = 8.0297$, is different from the estimate obtained from PROC GLM. The reason for this is that σ_e^2 and σ_β^2 are estimated jointly using the REML procedure (see Section 1.11.2) rather than the ANOVA procedure as outlined in Section 9.7.
4. With $\widehat{\sigma}_e^2 = 8.027$ and $\widehat{\sigma}_\beta^2 = 2.4656$ the weights w and w' (see Section 9.7.5) are estimated as

$$\widehat{w} = \frac{1}{\widehat{\sigma}_e^2} = .1245$$

$$\widehat{w}' = \frac{1}{\widehat{\sigma}_e^2 + 2\widehat{\sigma}_\beta^2} = \frac{1}{12.96} = .0772$$

5. To illustrate the combining of intra- and interblock information, we consider estimation of A. From the intrablock analysis we have $\widehat{A}_{\text{I,II}} = 3.0$. From replicate 3 we find $\widetilde{A}_{\text{III}} = 5.50$. Then [see (9.26)]

$$\widehat{\widehat{A}} = \frac{2 \times .1245 \times 3.0 + .0772 \times 5.50}{2 \times .1245 + .0772} = 3.5916$$

which is (apart from rounding error) the value given in the output.

Table 9.15 Analysis of 2^3 Factorial With Partial Confounding

```
options nodate pageno=1;
data confound;
input rep block A B C y @@;
datalines;
1 1 0 0 0 8   2 5 0 1 1 22   3 9 1 0 1 39
1 1 1 1 0 13   2 5 1 1 0 25   3 9 1 1 0 32
1 2 0 1 0 15   2 6 0 0 1 20   3 10 0 0 1 28
1 2 1 0 0 9   2 6 1 0 0 18   3 10 0 1 0 26
1 3 0 1 1 11   2 7 0 0 0 20   3 11 0 0 0 30
1 3 1 0 1 18   2 7 1 0 1 21   3 11 0 1 1 33
1 4 0 0 1 7   2 8 0 1 0 17   3 12 1 0 0 27
1 4 1 1 1 19   2 8 1 1 1 21   3 12 1 1 1 41
;
run;

proc print data=confound;
title1 'TABLE 9.15';
title2 '2**3 FACTORIAL WITH PARTIAL CONFOUNDING';
run;

proc glm data=confound;
class rep block A B C;
model y= rep block(rep) A|B|C@2;
estimate 'Main effect A' A -1 1;
estimate 'Interaction AB' A*B 1 -1 -1 1/divisor=2;
title3 'INTRA-BLOCK ANALYSIS';
run;

proc mixed data=confound;
class rep block A B C;
model y= rep A|B|C/ddfm=satterth;
random block(rep);
estimate 'Main effect A' A -1 1;
estimate 'Interaction AB' A*B 1 -1 -1 1/divisor=2;
estimate 'Interaction ABC' A*B*C 1 -1 -1 1 -1 1 1 -1/divisor=4;
title3 'COMBINED INTRA- AND INTER-BLOCK ANALYSIS';
run;

proc glm data=confound;
class rep block A B C;
model y= rep A|B|C block(rep);
title3 'AUXILIARY ANALYSIS';
run;
```

Table 9.15 *(Continued)*

Obs	rep	block	A	B	C	y
1	1	1	0	0	0	8
2	2	5	0	1	1	22
3	3	9	1	0	1	39
4	1	1	1	1	0	13
5	2	5	1	1	0	25
6	3	9	1	1	0	32
7	1	2	0	1	0	15
8	2	6	0	0	1	20
9	3	10	0	0	1	28
10	1	2	1	0	0	9
11	2	6	1	0	0	18
12	3	10	0	1	0	26
13	1	3	0	1	1	11
14	2	7	0	0	0	20
15	3	11	0	0	0	30
16	1	3	1	0	1	18
17	2	7	1	0	1	21
18	3	11	0	1	1	33
19	1	4	0	0	1	7
20	2	8	0	1	0	17
21	3	12	1	0	0	27
22	1	4	1	1	1	19
23	2	8	1	1	1	21
24	3	12	1	1	1	41

INTRABLOCK ANALYSIS

The GLM Procedure

Class Level Information

Class	Levels	Values
rep	3	1 2 3
block	12	1 2 3 4 5 6 7 8 9 10 11 12
A	2	0 1
B	2	0 1
C	2	0 1

Table 9.15 (*Continued*)

Number of observations 24

Dependent Variable: y

Source	DF	Sum of Squares	Mean Square	F Value	Pr > F
Model	17	1889.833333	111.166667	14.66	0.0016
Error	6	45.500000	7.583333		
Corrected Total	23	1935.333333			

R-Square	Coeff Var	Root MSE	y Mean
0.976490	12.70978	2.753785	21.66667

Source	DF	Type I SS	Mean Square	F Value	Pr > F
rep	2	1537.333333	768.666667	101.36	<.0001
block(rep)	9	127.000000	14.111111	1.86	0.2316
A	1	36.000000	36.000000	4.75	0.0722
B	1	36.000000	36.000000	4.75	0.0722
A*B	1	12.250000	12.250000	1.62	0.2508
C	1	56.250000	56.250000	7.42	0.0345
A*C	1	81.000000	81.000000	10.68	0.0171
B*C	1	4.000000	4.000000	0.53	0.4950

Source	DF	Type III SS	Mean Square	F Value	Pr > F
rep	2	1537.333333	768.666667	101.36	<.0001
block(rep)	9	119.833333	13.314815	1.76	0.2539
A	1	36.000000	36.000000	4.75	0.0722
B	1	36.000000	36.000000	4.75	0.0722
A*B	1	12.250000	12.250000	1.62	0.2508
C	1	56.250000	56.250000	7.42	0.0345
A*C	1	81.000000	81.000000	10.68	0.0171
B*C	1	4.000000	4.000000	0.53	0.4950

Parameter	Estimate	Standard Error	t Value	Pr > |t|
Main effect A	3.00000000	1.37689264	2.18	0.0722
Interaction AB	1.75000000	1.37689264	1.27	0.2508

Table 9.15 (*Continued*)

```
           2**3 FACTORIAL WITH PARTIAL CONFOUNDING
           COMBINED INTRA- AND INTERBLOCK ANALYSIS

                    The Mixed Procedure

                     Model Information

       Data Set                        WORK.CONFOUND
       Dependent Variable              y
       Covariance Structure            Variance Components
       Estimation Method               REML
       Residual Variance Method        Profile
       Fixed Effects SE Method         Model-Based
       Degrees of Freedom Method       Satterthwaite

                  Class Level Information

        Class     Levels     Values

        rep            3     1 2 3
        block         12     1 2 3 4 5 6 7 8 9 10 11 12
        A              2     0 1
        B              2     0 1
        C              2     0 1

                        Dimensions

               Covariance Parameters      2
               Columns in X              30
               Columns in Z              12
               Subjects                   1
               Max Obs Per Subject       24
               Observations Used         24
               Observations Not Used      0
               Total Observations        24

                     Iteration History

   Iteration     Evaluations     -2 Res Log Like       Criterion

           0               1         83.94834175
           1               2         83.76036493      0.00000969
           2               1         83.76008026      0.00000000
```

Table 9.15 *(Continued)*

Convergence criteria met.

Covariance Parameter Estimates

Cov Parm	Estimate
block(rep)	2.4656
Residual	8.0297

Fit Statistics

−2 Res Log Likelihood	83.8
AIC (smaller is better)	87.8
AICC (smaller is better)	88.9
BIC (smaller is better)	88.7

Type 3 Tests of Fixed Effects

Effect	Num DF	Den DF	F Value	Pr > F
rep	2	4.56	59.31	0.0005
A	1	9.98	8.41	0.0158
B	1	9.98	4.57	0.0584
A*B	1	9.98	0.63	0.4464
C	1	9.98	7.79	0.0191
A*C	1	9.98	6.19	0.0321
B*C	1	9.98	0.11	0.7481
A*B*C	1	4.56	1.85	0.2370

Estimates

Label	Estimate	Standard Error	DF	t Value	Pr > |t|
Main effect A	3.5913	1.2380	9.98	2.90	0.0158
Interaction AB	0.9814	1.2380	9.98	0.79	0.4464
Interaction ABC	2.0000	1.4697	4.56	1.36	0.2370

Table 9.15 (*Continued*)

```
              2**3 FACTORIAL WITH PARTIAL CONFOUNDING
                        AUXILIARY ANALYSIS

                         The GLM Procedure

                      Class Level Information

        Class       Levels     Values

        rep              3     1 2 3
        block           12     1 2 3 4 5 6 7 8 9 10 11 12
        A                2     0 1
        B                2     0 1
        C                2     0 1


                    Number of observations 24

Dependent Variable: y

                                 Sum of
Source                DF        Squares   Mean Square   F Value   Pr > F

Model                 17    1889.833333    111.166667     14.66   0.0016
Error                  6      45.500000      7.583333
Corrected Total       23    1935.333333


        R-Square      Coeff Var      Root MSE        y Mean

        0.976490       12.70978      2.753785      21.66667


Source            DF      Type I SS   Mean Square   F Value   Pr > F

rep                2    1537.333333    768.666667    101.36   <.0001
A                  1      88.166667     88.166667     11.63   0.0143
B                  1      37.500000     37.500000      4.95   0.0678
A*B                1       2.666667      2.666667      0.35   0.5748
C                  1      66.666667     66.666667      8.79   0.0251
A*C                1      37.500000     37.500000      4.95   0.0678
B*C                1       0.166667      0.166667      0.02   0.8870
A*B*C              1      24.000000     24.000000      3.16   0.1256
block(rep)         8      95.833333     11.979167      1.58   0.2973
```

Table 9.15 (*Continued*)

Source	DF	Type III SS	Mean Square	F Value	Pr > F
rep	2	1537.333333	768.666667	101.36	<.0001
A	1	36.000000	36.000000	4.75	0.0722
B	1	36.000000	36.000000	4.75	0.0722
A*B	1	12.250000	12.250000	1.62	0.2508
C	1	56.250000	56.250000	7.42	0.0345
A*C	1	81.000000	81.000000	10.68	0.0171
B*C	1	4.000000	4.000000	0.53	0.4950
A*B*C	0	0.000000	.	.	.
block(rep)	8	95.833333	11.979167	1.58	0.2973

Using (9.27), we obtain

$$\widehat{\text{var}}\left(\widehat{\widehat{A}}\right) = \frac{8.0297}{(2 + 0.6201)\,2} = 1.5323$$

and hence

$$\text{se}\left(\widehat{\widehat{A}}\right) = 1.2379$$

which is (apart from rounding error) the value given in the output.

6. Using all three replicates, we find

$$\widetilde{ABC} = 2.0$$

with

$$\widehat{\text{var}}\left(\widetilde{ABC}\right) = \frac{1}{3 \times 2} 12.96 = 2.16$$

and

$$\text{se}\left(\widetilde{ABC}\right) = 1.4697$$

7. We use the results of Table 9.15—auxiliary analysis—to illustrate the numerical implementation of some of the results of Sections 9.7.4 and 9.7.5 The reader will recognize that the type I SS are those of the B|T-ANOVA in Table 9.11. Its main purpose is the estimation of w and w'. Obviously,

$$\widehat{w} = \frac{1}{7.5833} = .1319$$

and with $\Delta = 0.75$ in (9.34), we obtain

$$\widehat{w}' = (-0.3333 \times 7.5833 + 1.3333 \times 11.9792)^{-1} = 0.0744.$$

The values for $\widehat{w}$ and $\widehat{w}'$ are used to obtain, for example,

$$\widehat{\widehat{A}} = \frac{0.1319 \times 3.0 + 0.0744 \times 5.5}{0.1319 + 0.0744} = 3.9016$$

[see (9.26) and from (9.27)],

$$\widehat{\text{var}}\left(\widehat{\widehat{A}}\right) = \frac{7.5833}{(2 + 0.5641)\,2} = 1.4787$$

$$\text{se}\left(\widehat{\widehat{A}}\right) = 1.2160$$

which is in good agreement with the results obtained in 5.

9.10.2 2^4 Factorial in Blocks of Size 4

Assuming that the 4-factor into action $ABCD$ is negligible and that we want full information on all main effects, using eight blocks of size 4, a suitable design (from inspection of Table 8.3) is to confound AB and CD and, hence, $ABCD$ in one replicate and AC and BD and, hence, $ABCD$ in the other replicate. Such a design is given in Table 9.16, generated by PROC FACTEX. This design results in the following three sets (see Section 9.7):

$$\mathcal{E}_1 = \{ABCD\} \qquad n_1 = 1$$

$$\mathcal{E}_2 = \{AB, AC, BD, CD\} \qquad n_2 = 4$$

$$\mathcal{E}_3 = \{A, B, C, D, AD, BC, ABC, ABD, ACD, BCD\} \qquad n_3 = 10$$

with zero information on $\mathcal{E}_1$, $\frac{1}{2}$ information on $\mathcal{E}_2$, and full information on $\mathcal{E}_3$.

The analysis of a data set using the design of Table 9.16 is given in Table 9.17 using PROC GLM for intrablock analysis and PROC MIXED for the combined analysis.

The same comments made in Section 9.10.1 apply here also, and the reader is encouraged to perform the computations using previously described methodology in order to verify agreement with the computer output. We only make one additional comment concerning unconfounded effects, for example, main effect A. Since A is unconfounded, only intrablock information exists. This is reflected in the fact that the estimates for A are the same, namely $\widehat{A} = 4.6250$, in the PROC GLM and PROC MIXED outputs. However, the standard errors are slightly different, 0.8705 for PROC GLM versus 0.8722 for PROC MIXED, due to the fact that the estimation procedures for variance components and hence the estimates are different (due to unbalancedness) for the two procedures, $\widehat{\sigma}_e^2 = 6.0625$ for PROC GLM and $\widehat{\sigma}_e^2 = 6.0857$ for PROC MIXED. The difference in this case is, of course, negligible.

Table 9.16 2^4 Factorial in Blocks of Size 4 Using Partial Confounding

```
options nodate pageno=1;
proc factex;
factors A B C D;
blocks size=4;
model est=(A B C A*C B*C D A*D B*D A*B*C A*B*D A*C*D B*C*D);
examine design confounding;
output out=replicate1 blockname=block nvals=(1 2 3 4);
title1 'TABLE 9.16';
title2 '2**4 FACTORIAL IN BLOCKS OF SIZE 4';
title3 'USING PARTIAL CONFOUNDING';
run;

proc factex;
factors A B C D;
blocks size=4;
model est=(A B A*B C B*C A*B*C D A*D C*D A*B*D A*C*D B*C*D);
examine design confounding;
output out=replicate2 blockname=block nvals=(5 6 7 8);
run;

proc print data=replicate1;
title3 '(PARTIALLY CONFOUNDING AB AC BD CD';
title4 'AND COMPLETELY CONFOUNDING ABCD)';
run;

proc print data=replicate2;
run;
```

The FACTEX Procedure

Design Points

Experiment Number	A	B	C	D	Block
1	-1	-1	-1	-1	4
2	-1	-1	-1	1	1
3	-1	-1	1	-1	1
4	-1	-1	1	1	4
5	-1	1	-1	-1	3
6	-1	1	-1	1	2
7	-1	1	1	-1	2
8	-1	1	1	1	3
9	1	-1	-1	-1	3
10	1	-1	-1	1	2
11	1	-1	1	-1	2
12	1	-1	1	1	3
13	1	1	-1	-1	4

Table 9.16 (*Continued*)

14	1	1	-1	1	1
15	1	1	1	-1	1
16	1	1	1	1	4

Block Pseudo-factor Confounding Rules

[B1] = A*B*C*D
[B2] = C*D

Design Points

Experiment Number	A	B	C	D	Block
1	-1	-1	-1	-1	4
2	-1	-1	-1	1	1
3	-1	-1	1	-1	3
4	-1	-1	1	1	2
5	-1	1	-1	-1	1
6	-1	1	-1	1	4
7	-1	1	1	-1	2
8	-1	1	1	1	3
9	1	-1	-1	-1	3
10	1	-1	-1	1	2
11	1	-1	1	-1	4
12	1	-1	1	1	1
13	1	1	-1	-1	2
14	1	1	-1	1	3
15	1	1	1	-1	1
16	1	1	1	1	4

Block Pseudo-factor Confounding Rules

[B1] = A*B*C*D
[B2] = B*D

2**4 FACTORIAL IN BLOCKS OF SIZE 4
(PARTIALLY CONFOUNDING AB AC BD CD
AND COMPLETELY CONFOUNDING ABCD)

Obs	block	A	B	C	D
1	1	-1	-1	-1	1
2	1	-1	-1	1	-1
3	1	1	1	-1	1
4	1	1	1	1	-1
5	2	-1	1	-1	1
6	2	-1	1	1	-1

Table 9.16 (*Continued*)

7	2	1	-1	-1	1
8	2	1	-1	1	-1
9	3	-1	1	-1	-1
10	3	-1	1	1	1
11	3	1	-1	-1	-1
12	3	1	-1	1	1
13	4	-1	-1	-1	-1
14	4	-1	-1	1	1
15	4	1	1	-1	-1
16	4	1	1	1	1
1	5	-1	-1	-1	1
2	5	-1	1	-1	-1
3	5	1	-1	1	1
4	5	1	1	1	-1
5	6	-1	-1	1	1
6	6	-1	1	1	-1
7	6	1	-1	-1	1
8	6	1	1	-1	-1
9	7	-1	-1	1	-1
10	7	-1	1	1	1
11	7	1	-1	-1	-1
12	7	1	1	-1	1
13	8	-1	-1	-1	-1
14	8	-1	1	-1	1
15	8	1	-1	1	-1
16	8	1	1	1	1

Table 9.17 Analysis of 2^4 Factorial With Partial Confounding

```
options nodate pageno=1;
data partial;
input rep block A B C D y @@;
datalines;
1 1 0 0 0 1 15   2 5 0 0 0 1 40
1 1 0 0 1 0 19   2 5 0 1 0 0 43
1 1 1 1 0 1 12   2 5 1 0 1 1 47
1 1 1 1 1 0 22   2 5 1 1 1 0 45
1 2 0 1 0 1 21   2 6 0 0 1 1 45
1 2 0 1 1 0 25   2 6 0 1 1 0 49
1 2 1 0 0 1 28   2 6 1 0 0 1 52
1 2 1 0 1 0 30   2 6 1 1 0 0 53
1 3 0 1 0 0 27   2 7 0 0 1 0 55
1 3 0 1 1 1 31   2 7 0 1 1 1 59
1 3 1 0 0 0 30   2 7 1 0 0 0 54
1 3 1 0 1 1 34   2 7 1 1 0 1 58
```

Table 9.17 (*Continued*)

```
1 4 0 0 0 0 14  2 8 0 0 0 0 45
1 4 0 0 1 1 20  2 8 0 1 0 1 50
1 4 1 1 0 0 22  2 8 1 0 1 0 55
1 4 1 1 1 1 27  2 8 1 1 1 1 63
;
run;

proc print data=partial;
title1 'TABLE 9.17';
title2 '2**4 FACTORIAL WITH PARTIAL CONFOUNDING';
run;

proc glm data=partial;
class rep block A B C D;
model y=rep block(rep)A|B|C|D@3;
estimate 'Main effect A' A -1 1;
estimate 'Interaction AB' A*B 1 -1 -1 1/divisor=2;
title3 'INTRA-BLOCK ANALYSIS';
run;

proc mixed data=partial;
class rep block A B C D;
model y= rep A|B|C|D/ddfm=satterth;
random block(rep);
estimate 'Main effect A' A -1 1;
estimate 'Interaction AB' A*B 1 -1 -1 1/divisor=2;

title3 'COMBINED INTRA- AND INTER-BLOCK ANALYSIS';
run;
```

Obs	rep	block	A	B	C	D	y
1	1	1	0	0	0	1	15
2	1	1	0	0	1	0	19
3	1	1	1	1	0	1	12
4	1	1	1	1	1	0	22
5	1	2	0	1	0	1	21
6	1	2	0	1	1	0	25
7	1	2	1	0	0	1	28
8	1	2	1	0	1	0	30
9	1	3	0	1	0	0	27
10	1	3	0	1	1	1	31
11	1	3	1	0	0	0	30
12	1	3	1	0	1	1	34
13	1	4	0	0	0	0	14
14	1	4	0	0	1	1	20
15	1	4	1	1	0	0	22

Table 9.17 (*Continued*)

16	1	4	1	1	1	1	27
17	2	5	0	0	0	1	40
18	2	5	0	1	0	0	43
19	2	5	1	0	1	1	47
20	2	5	1	1	1	0	45
21	2	6	0	0	1	1	45
22	2	6	0	1	1	0	49
23	2	6	1	0	0	1	52
24	2	6	1	1	0	0	53
25	2	7	0	0	1	0	55
26	2	7	0	1	1	1	59
27	2	7	1	0	0	0	54
28	2	7	1	1	0	1	58
29	2	8	0	0	0	0	45
30	2	8	0	1	0	1	50
31	2	8	1	0	1	0	55
32	2	8	1	1	1	1	63

2**4 FACTORIAL WITH PARTIAL CONFOUNDING
INTRABLOCK ANALYSIS

The GLM Procedure

Class Level Information

Class	Levels	Values
rep	2	1 2
block	8	1 2 3 4 5 6 7 8
A	2	0 1
B	2	0 1
C	2	0 1
D	2	0 1

Number of observations 32

Dependent Variable: y

Source	DF	Sum of Squares	Mean Square	F Value	Pr > F
Model	21	7132.250000	339.630952	56.02	<.0001
Error	10	60.625000	6.062500		
Corrected Total	31	7192.875000			

Table 9.17 *(Continued)*

R-Square	Coeff Var	Root MSE	y Mean
0.991572	6.621081	2.462214	37.18750

Source	DF	Type I SS	Mean Square	F Value	Pr > F
rep	1	5940.500000	5940.500000	979.88	<.0001
block(rep)	6	777.375000	129.562500	21.37	<.0001
A	1	171.125000	171.125000	28.23	0.0003
B	1	18.000000	18.000000	2.97	0.1156
A*B	1	1.562500	1.562500	0.26	0.6227
C	1	120.125000	120.125000	19.81	0.0012
A*C	1	0.562500	0.562500	0.09	0.7669
B*C	1	2.000000	2.000000	0.33	0.5784
A*B*C	1	4.500000	4.500000	0.74	0.4091
D	1	6.125000	6.125000	1.01	0.3385
A*D	1	1.125000	1.125000	0.19	0.6758
B*D	1	5.062500	5.062500	0.84	0.3823
A*B*D	1	0.500000	0.500000	0.08	0.7798
C*D	1	1.562500	1.562500	0.26	0.6227
A*C*D	1	10.125000	10.125000	1.67	0.2253
B*C*D	1	72.000000	72.000000	11.88	0.0063

Source	DF	Type III SS	Mean Square	F Value	Pr > F
rep	1	5940.500000	5940.500000	979.88	<.0001
block(rep)	6	406.875000	67.812500	11.19	0.0006
A	1	171.125000	171.125000	28.23	0.0003
B	1	18.000000	18.000000	2.97	0.1156
A*B	1	1.562500	1.562500	0.26	0.6227
C	1	120.125000	120.125000	19.81	0.0012
A*C	1	0.562500	0.562500	0.09	0.7669
B*C	1	2.000000	2.000000	0.33	0.5784
A*B*C	1	4.500000	4.500000	0.74	0.4091
D	1	6.125000	6.125000	1.01	0.3385
A*D	1	1.125000	1.125000	0.19	0.6758
B*D	1	5.062500	5.062500	0.84	0.3823
A*B*D	1	0.500000	0.500000	0.08	0.7798
C*D	1	1.562500	1.562500	0.26	0.6227
A*C*D	1	10.125000	10.125000	1.67	0.2253
B*C*D	1	72.000000	72.000000	11.88	0.0063

Parameter	Estimate	Standard Error	t Value	Pr > |t|
Main effect A	4.62500000	0.87052427	5.31	0.0003
Interaction AB	−0.62500000	1.23110723	−0.51	0.6227

Table 9.17 (*Continued*)

```
          2**4 FACTORIAL WITH PARTIAL CONFOUNDING
          COMBINED INTRA- AND INTERBLOCK ANALYSIS

                  The Mixed Procedure

                   Model Information

     Data Set                        WORK.PARTIAL
     Dependent Variable              y
     Covariance Structure            Variance Components
     Estimation Method               REML
     Residual Variance Method        Profile
     Fixed Effects SE Method         Model-Based
     Degrees of Freedom Method       Satterthwaite

                Class Level Information

            Class     Levels     Values

            rep            2     1 2
            block          8     1 2 3 4 5 6 7 8
            A              2     0 1
            B              2     0 1
            C              2     0 1
            D              2     0 1

                     Dimensions

              Covariance Parameters      2
              Columns in X              83
              Columns in Z               8
              Subjects                   1
              Max Obs Per Subject       32
              Observations Used         32
              Observations Not Used      0
              Total Observations        32

                   Iteration History

 Iteration     Evaluations     -2 Res Log Like      Criterion

         0               1        107.26388086
         1               2         97.51296909     0.01846419
         2               1         96.69150852     0.00679576
         3               1         96.40491364     0.00138264
         4               1         96.35086623     0.00008239
         5               1         96.34791369     0.00000036
         6               1         96.34790114     0.00000000
```

Table 9.17 *(Continued)*

Convergence criteria met.

Covariance Parameter Estimates

Cov Parm	Estimate
block(rep)	36.8465
Residual	6.0857

Fit Statistics

-2 Res Log Likelihood	96.3
AIC (smaller is better)	100.3
AICC (smaller is better)	101.3
BIC (smaller is better)	100.5

Type 3 Tests of Fixed Effects

Effect	Num DF	Den DF	F Value	Pr > F
rep	1	4.62	38.71	0.0021
A	1	9.92	28.12	0.0004
B	1	9.92	2.96	0.1164
A*B	1	10.7	0.63	0.4452
C	1	9.92	19.74	0.0013
A*C	1	10.7	0.02	0.8817
B*C	1	9.92	0.33	0.5792
A*B*C	1	9.92	0.74	0.4101
D	1	9.92	1.01	0.3396
A*D	1	9.92	0.18	0.6764
B*D	1	10.7	0.41	0.5366
A*B*D	1	9.92	0.08	0.7803
C*D	1	10.7	0.39	0.5438
A*C*D	1	9.92	1.66	0.2264
B*C*D	1	9.92	11.83	0.0064
A*B*C*D	1	4.62	0.01	0.9139

Estimates

Label	Estimate	Standard Error	DF	t Value	Pr > |t|
Main effect A	4.6250	0.8722	9.92	5.30	0.0004
Interaction AB	-0.9587	1.2097	10.7	-0.79	0.4452

CHAPTER 10

Designs with Factors at Three Levels

10.1 INTRODUCTION

In discussing the 2^n factorial design in Chapter 7 we saw that main effects and interactions can be defined simply as linear combinations of the true responses, more specifically as the average response of one set of 2^{n-1} treatment combinations minus the average response of the complementary set of 2^{n-1} treatment combinations. And even more specifically, the main effect of a certain factor is the average response with that factor at the 1 level minus the average response with that factor at the 0 level. Turning now to the situation where each factor has three levels, which we shall refer to as 0 level, 1 level, and 2 level, such a simple definition of main effects and interactions no longer exists. We can no longer talk about *the* main effect of a factor or *the* interaction between two or more factors but shall talk instead about main effect components or comparisons belonging to a certain factor and about interaction components. We shall see how all this can be developed as a generalization of the formal approach described for the 2^n experiment in Section 7.4.

10.2 DEFINITION OF MAIN EFFECTS AND INTERACTIONS

10.2.1 The 3^2 Case

To introduce the concepts we shall consider first the simplest case, namely that of two factors, A and B say, each having three levels, denoted by 0, 1, 2. A treatment combination of this 3^2 factorial is then represented by $\boldsymbol{x}' = (x_1, x_2)$ where $x_i = 0, 1, 2 (i = 1, 2)$, with x_1 referring to factor A and x_2 to factor B.

Design and Analysis of Experiments. Volume 2: Advanced Experimental Design
By Klaus Hinkelmann and Oscar Kempthorne
ISBN 0-471-55177-5

We now partition the set of nine treatment combinations into three sets of three treatment combinations each according to the levels of factor A:

$$\begin{aligned} \text{set I:} \quad & \{(0,0),\ (0,1),\ (0,2)\} \\ \text{set II:} \quad & \{(1,0),\ (1,1),\ (1,2)\} \\ \text{set III:} \quad & \{(2,0),\ (2,1),\ (2,2)\} \end{aligned}$$

More formally, we can define these three sets by the three equations:

$$\begin{aligned} \text{set I:} \quad & x_1 = 0 \\ \text{set II:} \quad & x_1 = 1 \\ \text{set III:} \quad & x_1 = 2 \end{aligned} \tag{10.1}$$

Comparisons among the mean true responses for these three sets are then said to *belong to main effect A*. Since there are three sets, there are two linearly independent comparisons among these three sets (i.e., their mean responses), and these comparisons represent the 2 d.f. for main effect A. For example, the comparisons could be (set I − set II) and (set I − III), or (set I − set II) and (set I + set II − 2 set III).

Similarly, we can divide the nine treatment combinations into three sets corresponding to the levels of factor B or, equivalently, corresponding to the equations:

$$\begin{aligned} x_2 = 0: \quad & \{(0,0),\ (1,0),\ (2,0)\} \\ x_2 = 1: \quad & \{(0,1),\ (1,1),\ (2,1)\} \\ x_2 = 2: \quad & \{(0,2),\ (1,2),\ (2,2)\} \end{aligned} \tag{10.2}$$

Comparisons among the mean responses of these three sets then constitute main effect B.

As in the 2^n case, the interaction between factors A and B will be defined in terms of comparisons of sets (of treatment combinations), which are determined by equations involving both x_1 and x_2. One such partitioning is given by

$$\begin{aligned} \text{set I:} \quad & x_1 + x_2 = 0 \quad \text{mod } 3: \quad \{(0,0),\ (1,2),\ (2,1)\} \\ \text{set II:} \quad & x_1 + x_2 = 1 \quad \text{mod } 3: \quad \{(1,0),\ (0,1),\ (2,2)\} \\ \text{set III:} \quad & x_1 + x_2 = 2 \quad \text{mod } 3: \quad \{(2,0),\ (0,2),\ (1,1)\} \end{aligned} \tag{10.3}$$

Comparisons among these three sets account for 2 of the 4 d.f. for the $A \times B$ interaction. The remaining 2 d.f. are accounted for by comparisons among the sets based on the following partition:

$$\begin{aligned} \text{set I:} \quad & x_1 + 2x_2 = 0 \quad \text{mod } 3: \quad \{(0,0),\ (1,1),\ (2,2)\} \\ \text{set II:} \quad & x_1 + 2x_2 = 1 \quad \text{mod } 3: \quad \{(1,0),\ (0,2),\ (2,1)\} \\ \text{set III:} \quad & x_1 + 2x_2 = 2 \quad \text{mod } 3: \quad \{(2,0),\ (0,1),\ (1,2)\} \end{aligned} \tag{10.4}$$

To see what our development so far means with respect to the usual factorial representation, we consider (see also Section I.11.8.1)

$$\tau_{ij} = \mu + A_i + B_j + (AB)_{ij} \tag{10.5}$$

with

$$\sum_{i=0}^{2} A_i = 0 \qquad \sum_{j=0}^{2} B_j = 0 \qquad \sum_{i=0}^{2} (AB)_{ij} = 0$$

for each j,

$$\sum_{j=0}^{2} (AB)_{ij} = 0$$

for each i, where τ_{ij} is the true response for the treatment combination $(x_1 = i, x_2 = j)$. With the model (10.5), a contrast among sets (10.1), that is,

$$\sum_i c_i \overline{\tau}_{i.} = \sum_i c_i A_i \qquad \left(\sum c_i = 0\right)$$

is the corresponding contrast among A main effects. A contrast among sets (10.2), that is,

$$\sum_j c_j \overline{\tau}_{.j} = \sum_j c_j B_j \qquad \left(\sum c_j = 0\right)$$

it is the corresponding contrast among B main effects. A contrast among sets (10.3) can be written as

$$c_1(\tau_{00} + \tau_{12} + \tau_{21}) + c_2(\tau_{10} + \tau_{01} + \tau_{22}) + c_3(\tau_{20} + \tau_{02} + \tau_{11}) \qquad (c_1 + c_2 + c_3 = 0)$$

which, using (10.5), reduces to the same contrast in the $(AB)_{ij}$'s. The same is true for comparisons among sets (10.4), that is,

$$c_1(\tau_{00} + \tau_{11} + \tau_{22}) + c_2(\tau_{10} + \tau_{02} + \tau_{21}) + c_3(\tau_{20} + \tau_{01} + \tau_{12}) \qquad (c_1 + c_2 + c_3 = 0)$$

The reader will notice that the last two comparisons have no particular meaning or interpretation for any choice of the c_i's, except that they each *belong to the 2-factor interaction* $A \times B$, and that each represents 2 d.f. of that interaction. This is in contrast to the parameterization given in Section I.11.8.1 in terms of orthogonal polynomials. One difference, of course, is that the parameterization given there is in terms of single degree of freedom parameters, and a second difference is that it is meaningful only for quantitative factors, whereas the definitions in terms of the partitions as summarized in (10.6) below hold for quantitative and qualitative factors. But the most important point is that the current definitions of effects and interactions will prove to be important in the context of systems of confounding (see Section 10.5) and fractional factorials (see Section 13.4).

To sum up our discussion so far, the effects and interactions for a 3^2 experiment are given in pairs of degrees of freedom by comparisons among three sets of treatment combinations as follows:

$$\left.\begin{array}{ll} A: & x_1 = 0, 1, 2 \\ B: & x_2 = 0, 1, 2 \\ A \times B: & \begin{cases} x_1 + x_2 = 0, 1, 2 \\ x_1 + 2x_2 = 0, 1, 2 \end{cases} \end{array}\right\} \text{mod } 3 \tag{10.6}$$

It is convenient to denote the pairs of degrees of freedom corresponding to $x_1 + x_2 = 0, 1, 2$ by the symbol AB and the pair corresponding to $x_1 + 2x_2 = 0, 1, 2$ by AB^2.

It is easy to see that the groups given by the symbols AB^2 and A^2B are the same. It is, therefore, convenient, in order to obtain a complete and unique enumeration of the pairs of degrees of freedom, to adopt the rule that an order of the letters is to be chosen in advance and that the power of the first letter in a symbol must be unity. This latter is obtained by taking the square of the symbol with the rule that the cube of any letter is to be replaced by unity, that is, if the initial letter of the symbol occurs raised to the power 2, for example, A^2B, we then obtain

$$A^2B \equiv (A^2B)^2 \equiv A^4B^2 \equiv AB^2$$

This procedure follows from the fact that the partitioning produced by

$$2x_1 + x_2 = 0, 1, 2$$

is the same as that produced by

$$2 \cdot 2x_1 + 2x_2 = 0, 2, 1$$

that is,

$$x_1 + 2x_2 = 0, 2, 1$$

which is the partitioning denoted by AB^2.

Table 10.1 Effects and Interactions for 3^3 Experiment

Effect/Interaction		Left-Hand Side of Defining Equation
A		x_1
B		x_2
$A \times B$	AB	$x_1 + x_2$
	AB^2	$x_1 + 2x_2$
C		x_3
$A \times C$	AC	$x_1 + x_3$
	AC^2	$x_1 + 2x_3$
$B \times C$	BC	$x_2 + x_3$
	BC^2	$x_2 + 2x_3$
$A \times B \times C$	ABC	$x_1 + x_2 + x_3$
	ABC^2	$x_1 + x_2 + 2x_3$
	AB^2C	$x_1 + 2x_2 + x_3$
	AB^2C^2	$x_1 + 2x_2 + 2x_3$

10.2.2 General Case

The procedure of formally defining effects and interactions, illustrated for the 3^2 experiment, can be extended easily to the 3^n case. We shall have then $(3^n - 1)/2$ symbols, each representing 2 d.f. For example, for the 3^3 experiment there will be 13 symbols as given in Table 10.1 together with their defining equations of the form $\alpha_1 x_1 + \alpha_2 x_2 + \alpha_3 x_3 = 0, 1, 2 \bmod 3$.

For the general case of the 3^n experiment, denoting the factors by $A_1, A_2, \ldots, A_n$, the $(3^n - 1)/2$ symbols can be written as $A_1^{\alpha_1}, A_2^{\alpha_2}, \ldots, A_n^{\alpha_n}$ with $\alpha_i = 0, 1, 2\ (i = 1, 2, \ldots, n)$ and the convention that (1) any letter A_i with $\alpha_i = 0$ is dropped from the expression, (2) the first nonzero α is equal to one (this can always be achieved by multiplying each α_i by 2), and (3) any $\alpha_i = 1$ is not written explicitly in the expression. (This is illustrated in Table 10.1 by replacing A by A_1, B by A_2, and C by A_3.)

The n-tuple $\boldsymbol{\alpha}' = (\alpha_1, \alpha_2, \cdots, \alpha_n)$ associated with $A_1^{\alpha_1} A_2^{\alpha_2} \cdots A_n^{\alpha_n}$ is referred to as a *partition* of the 3^n treatment combinations into three sets according to the equations

$$\alpha_1 x_1 + \alpha_2 x_2 + \cdots + \alpha_n x_n = 0, 1, 2 \mod 3 \tag{10.7}$$

We now list some properties of such partitions:

1. Each partition leads to three sets of 3^{n-1} treatment combinations each as is evident from Eqs. (10.7).
2. If $\boldsymbol{\alpha}' = (\alpha_1, \alpha_2, \ldots, \alpha_n)$ and $\boldsymbol{\beta}' = (\beta_1, \beta_2, \ldots, \beta_n)$ are two distinct partitions, then the two equations

$$\alpha_1 x_1 + \alpha_2 x_2 + \cdots + \alpha_n x_n = \delta_1 \mod 3 \tag{10.8}$$

and

$$\beta_1 x_1 + \beta_2 x_2 + \cdots + \beta_n x_n = \delta_2 \mod 3 \tag{10.9}$$

are satisfied by exactly 3^{n-2} treatment combinations $\boldsymbol{x}' = (x_1, x_2, \ldots, x_n)$. This implies that the set of treatment combinations determined by $\boldsymbol{\alpha}'\boldsymbol{x} = \delta_1$ has exactly 3^{n-2} treatment combinations in common with each of the three sets determined by the equations $\boldsymbol{\beta}'\boldsymbol{x} = 0, 1, 2 \mod 3$, respectively. It is in this sense that the two partitions $\boldsymbol{\alpha}$ and $\boldsymbol{\beta}$ are *orthogonal* to each other.
3. If a treatment combination $\boldsymbol{x}' = (x_1, x_2, \ldots, x_n)$ satisfies both Eqs. (10.8) and (10.9) for a particular choice of δ_1, δ_2, then $\boldsymbol{x}$ also satisfies the equation

$$(\alpha_1 + \beta_1)x_1 + (\alpha_2 + \beta_2)x_2 + \cdots + (\alpha_n + \beta_n)x_n = \delta_1 + \delta_2 \mod 3 \tag{10.10}$$

Equation (10.10) is, of course, one of the three equations associated with the partition $\boldsymbol{\alpha}' + \boldsymbol{\beta}' = (\alpha_1 + \beta_1, \alpha_2 + \beta_2, \ldots, \alpha_n + \beta_n)$, in which each component is reduced mod 3, and hence with the interaction $A_1^{\alpha_1+\beta_1} A_2^{\alpha_2+\beta_2} \cdots A_n^{\alpha_n+\beta_n}$. In agreement with the definition in Section 7.4 we refer to $E^{\boldsymbol{\alpha}+\boldsymbol{\beta}} = A_1^{\alpha_1+\beta_1} A_2^{\alpha_2+\beta_2} \cdots A_n^{\alpha_n+\beta_n}$ as generalized interaction (GI) of $E^{\boldsymbol{\alpha}} = A_1^{\alpha_1} A_2^{\alpha_2} \cdots A_n^{\alpha_n}$ and $E^{\boldsymbol{\beta}} = A_1^{\beta_1} A_2^{\beta_2} \cdots A_n^{\beta_n}$.
In addition to satisfying (10.10), the treatment combination $\boldsymbol{x}$, which satisfies (10.8) and (10.9), also satisfies the equation

$$(\alpha_1 + 2\beta_1)x_1 + (\alpha_2 + 2\beta_2)x_2 + \cdots + (\alpha_n + 2\beta_n)x_n = \delta_1 + 2\delta_2 \mod 3 \tag{10.11}$$

which is associated with the partition $\boldsymbol{\alpha}' + 2\boldsymbol{\beta}'$ and hence the interaction $E^{\boldsymbol{\alpha}+2\boldsymbol{\beta}} = A_1^{\alpha_1+2\beta_1} A_2^{\alpha_2+2\beta_2} \cdots A_n^{\alpha_n+2\beta_n}$. This interaction is therefore another GI of $E^{\boldsymbol{\alpha}}$ and $E^{\boldsymbol{\beta}}$. To summarize then, any two interactions $E^{\boldsymbol{\alpha}}$ and $E^{\boldsymbol{\beta}}$ have *two* GIs $E^{\boldsymbol{\alpha}+\boldsymbol{\beta}}$ and $E^{\boldsymbol{\alpha}+2\boldsymbol{\beta}}$, where $\boldsymbol{\alpha} + \boldsymbol{\beta}$ and $\boldsymbol{\alpha} + 2\boldsymbol{\beta}$ are formed mod 3 and are subject to the rules stated earlier. We illustrate this by the following example.

Example 10.1 Consider AB and ABC^2 in the 3^3 case, that is, $\boldsymbol{\alpha}' = (1, 1, 0)$ and $\boldsymbol{\beta}' = (1, 1, 2)$. Then

$$(\boldsymbol{\alpha} + \boldsymbol{\beta})' = (2, 2, 2) \equiv (1, 1, 1)$$

and

$$(\boldsymbol{\alpha} + 2\boldsymbol{\beta})' = (3, 3, 4) \equiv (0, 0, 1)$$

and hence the GIs of AB and ABC^2 are ABC and C. Another way of obtaining this result is through formal multiplication and reduction mod 3, that is,

$$(AB)(ABC^2) = A^2B^2C^2 = (A^2B^2C^2)^2 = ABC$$

and

$$(AB)(ABC^2)^2 = A^3B^3C^4 = C \qquad \square$$

10.3 PARAMETERIZATION IN TERMS OF MAIN EFFECTS AND INTERACTIONS

The symbols used in the previous section to denote pairs of degrees of freedom will also be used to denote the magnitude of effects and interactions in the following way (see also Section 10.4). Each symbol represents a division of the set of 3^n treatment combinations into three sets of $3^n - 1$ treatment combinations each. The symbol, with a subscript that is the right-hand side of the equation determining the particular one of the three sets in which the treatment combinations lie, will denote the mean response of that set as a deviation from the overall mean, M. If $E^{\boldsymbol{\alpha}} = A_1^{\alpha_1} A_2^{\alpha_2} \cdots A_n^{\alpha_n}$ represents an interaction, then

$$E_i^{\boldsymbol{\alpha}} = \left(A_1^{\alpha_1} A_2^{\alpha_2} \cdots A_n^{\alpha_n}\right)_i = \begin{pmatrix} \text{mean of treatment combinations} \\ \text{satisfying } \boldsymbol{\alpha}'\boldsymbol{x} = i \mod 3 \end{pmatrix} - M \quad (10.12)$$

We shall also use the notation $E_{\boldsymbol{\alpha}'\boldsymbol{x}}^{\boldsymbol{\alpha}}$ for given $\boldsymbol{\alpha}$ and $\boldsymbol{x}$ to denote one of the quantities $E_0^{\boldsymbol{\alpha}}, E_1^{\boldsymbol{\alpha}}, E_2^{\boldsymbol{\alpha}}$ depending on whether $\boldsymbol{\alpha}'\boldsymbol{x} = 0, 1, 2$ mod 3, respectively.

We note that a comparison belonging to $E^{\boldsymbol{\alpha}}$ is, of course, given by

$$c_0E_0^{\boldsymbol{\alpha}} + c_1E_1^{\boldsymbol{\alpha}} + c_2E_2^{\boldsymbol{\alpha}} \qquad (c_0 + c_1 + c_2 = 0) \qquad (10.13)$$

Also, it follows from (10.12) that

$$E_0^{\boldsymbol{\alpha}} + E_1^{\boldsymbol{\alpha}} + E_2^{\boldsymbol{\alpha}} = 0 \qquad (10.14)$$

so that any comparison of the form (10.13) could be expressed in terms of only two $E_i^{\boldsymbol{\alpha}}$. Such a procedure was, in fact, adopted for the 2^n factorial, but as we shall see below, in the present situation this would only lead to a certain amount of asymmetry.

As an extension of (7.42) we can now state and prove the following result, which expresses the response $a(\boldsymbol{x})$ of a treatment combination $\boldsymbol{x}$ as a linear combination of interaction components. This parameterization of $a(\boldsymbol{x})$ is given by

$$a(\boldsymbol{x}) = M + \sum_{\boldsymbol{\alpha}} E^{\boldsymbol{\alpha}}_{\boldsymbol{\alpha}'\boldsymbol{x}} \tag{10.15}$$

where summation is over all $\boldsymbol{\alpha}' = (\alpha_1, \alpha_2, \ldots, \alpha_n) \neq (0, 0, \ldots, 0)$, subject to the rule that the first nonzero α_i equals one, and $\boldsymbol{\alpha}'\boldsymbol{x}$ is reduced mod 3. The proof of (10.15) follows that of (7.42) and will be given for the general case in Section 11.5.

We illustrate (10.15) with the following example.

Example 10.2 Consider the 3^3 factorial with factors A, B, C and denote the true response of the treatment combination (i, j, k) by $a_i b_j c_k$. Then (10.15) can be written as

$$\begin{aligned} a_i b_j c_k = M &+ A_i + B_j + AB_{i+j} + AB^2_{i+2j} \\ &+ C_k + AC_{i+k} + AC^2_{i+2k} + BC_{j+k} \\ &+ BC^2_{j+2k} + ABC_{i+j+k} + ABC^2_{i+j+2k} \\ &+ AB^2C_{i+2j+k} + AB^2C^2_{i+2j+2k} \end{aligned}$$

For $i = 1, j = 1, k = 2$, for example, this becomes

$$\begin{aligned} a_1 b_1 c_2 = M &+ A_1 + B_1 + AB_2 + AB^2_0 + C_2 + AC_0 \\ &+ AC^2_2 + BC_0 + BC^2_2 + ABC_1 + ABC^2_0 \\ &+ AB^2C_2 + AB^2C^2_1 \end{aligned}$$ □

We emphasize again that the parameterization (10.15), which because of (10.14) is a non-full-rank parameterization, becomes important in connection with systems of confounding (Section 10.5) and fractional factorials (Section 13.4).

10.4 ANALYSIS OF 3^n EXPERIMENTS

Suppose that each treatment combination is replicated r times in an appropriate error control design, such as a CRD or a RCBD. Comparisons of treatments are then achieved by simply comparing the observed treatment means, and tests for main effects and interactions are done in an appropriate ANOVA.

Table 10.2 ANOVA for 3^3 Experiment in Randomized Complete Block Design

Source	d.f.	SS
Blocks	$r-1$	$27\sum_{i=1}^{r}(\overline{y}_{i\cdots} - \overline{y}_{\cdots\cdot})^2$
Treatments	$3^3 - 1 = 26$	
A	2	$9r\sum_{j=0}^{2}(\overline{y}_{\cdot j\cdot\cdot} - \overline{y}_{\cdots\cdot})^2$
B	2	$9r\sum_{k=0}^{2}(\overline{y}_{\cdot\cdot k\cdot} - \overline{y}_{\cdots\cdot})^2$
$A \times B$	4	$3r\sum_{jk}(\overline{y}_{\cdot jk\cdot} - \overline{y}_{\cdot j\cdot\cdot} - \overline{y}_{\cdot\cdot k\cdot} + \overline{y}_{\cdots\cdot})^2$
C	2	$9r\sum_{\ell=0}^{2}(\overline{y}_{\cdots\ell} - \overline{y}_{\cdots\cdot})^2$
$A \times C$	4	$3r\sum_{j\ell}(\overline{y}_{\cdot j\cdot\ell} - \overline{y}_{\cdot j\cdot\cdot} - \overline{y}_{\cdots\ell} + \overline{y}_{\cdots\cdot})^2$
$B \times C$	4	$3r\sum_{k\ell}(\overline{y}_{\cdot\cdot k\ell} - \overline{y}_{\cdot\cdot k\cdot} - \overline{y}_{\cdots\ell} + \overline{y}_{\cdots\cdot})^2$
$A \times B \times C$	8	$r\sum_{jk\ell}(\overline{y}_{\cdot jk\ell} - \overline{y}_{\cdot jk\cdot} - \overline{y}_{\cdot j\cdot\ell} - \overline{y}_{\cdot\cdot k\ell} + \overline{y}_{\cdot j\cdot\cdot}$ $+\overline{y}_{\cdot\cdot k\cdot} + \overline{y}_{\cdots\ell} - \overline{y}_{\cdots\cdot})^2$
Residual	$26(r-1)$	$\sum_{ijk\ell}(\overline{y}_{ijk\ell} - \overline{y}_{i\cdots} - \overline{y}_{\cdot jk\ell} + \overline{y}_{\cdots\cdot})^2$

For purposes of illustration we consider a 3^3 experiment in an RCBD with r blocks. With the usual model

$$y_{ijk\ell} = \mu + \beta_i + \tau_{jk\ell} + e_{ijk\ell}$$

or

$$\boldsymbol{y} = \mu\mathfrak{I} + \boldsymbol{X}_\beta\boldsymbol{\beta} + \boldsymbol{X}_\tau\boldsymbol{\tau} + \boldsymbol{e}$$

where $(jk\ell)$ denotes the level combinations for the three factors A, B, C, and with the factorial structure of the treatments, we obtain the usual ANOVA given in Table 10.2.

An alternative way of computing the various components of the treatment sum of squares is based upon the definition of the 2-d.f. components of any interaction and the corresponding symbols defined in Section 10.3. Let E^α denote any such interaction component, such as AB or AB^2C, and let $\widehat{E}_0^\alpha$, $\widehat{E}_1^\alpha$, $\widehat{E}_2^\alpha$ be the mean

Table 10.3 Partitioning of 3-Factor Interaction Sum of Squares

Source	d.f.	SS
ABC	2	$9r\left\{[\widehat{ABC_0}]^2+[\widehat{ABC_1}]^2+[\widehat{ABC_2}]^2\right\}$
ABC^2	2	$9r\left\{[\widehat{ABC_0^2}]^2+[\widehat{ABC_1^2}]^2+[\widehat{ABC_2^2}]^2\right\}$
AB^2C	2	$9r\left\{[\widehat{AB^2C_0}]^2+[\widehat{AB^2C_1}]^2+[\widehat{AB^2C_2}]^2\right\}$
AB^2C^2	2	$9r\left\{[\widehat{AB^2C_0^2}]^2+[\widehat{AB^2C_1^2}]^2+[\widehat{AB^2C_2^2}]^2\right\}$

observed responses (as a deviation from the overall mean) of the three sets defining E^α. The sum of squares associated with E^α, accounting for 2 d.f., is then given by

$$\mathrm{SS}(E^\alpha) = r3^{n-1}\left\{\left[\widehat{E}_0^\alpha\right]^2+\left[\widehat{E}_1^\alpha\right]^2+\left[\widehat{E}_2^\alpha\right]^2\right\} \tag{10.16a}$$

with

$$E\left[\mathrm{SS}(E^\alpha)\right] = 2\sigma_e^2 + r3^{n-1}\left\{\left[E_0^\alpha\right]^2+\left[E_1^\alpha\right]^2+\left[E_2^\alpha\right]^2\right\} \tag{10.16b}$$

Specifically, for the 3^3 experiment the sum of squares due to 3-factor interaction, for example, can be broken down as given in Table 10.3. The usefulness of this procedure will become apparent when we consider systems of confounding in Section 10.5. The SS given in (10.16a) is simply the SS (among sets) for the sets defined by $\boldsymbol{\alpha}'x = 0, 1, 2$. Since the various partitions are orthogonal to each other, so are their associated SSs. It is not difficult to show that the sum of the four SSs in Table 10.3 is the same as $\mathrm{SS}(A \times B \times C)$ in Table 10.2.

Generally, it is also useful to list the quantities $\widehat{E}_0^\alpha$, $\widehat{E}_1^\alpha$, $\widehat{E}_2^\alpha$ for main effects and interactions as they can be used to estimate the yield of any treatment combination or comparisons among treatment combinations (see Section 10.3).

10.5 CONFOUNDING IN A 3^n FACTORIAL

The necessity for using incomplete blocks for a 3^n factorial is even more obvious than for a 2^n factorial. Even for small n the number of treatment combinations is often too large for complete blocks, either because such large blocks are not available or they are no longer homogeneous. The use of smaller blocks will then lead to confounding of effects and interactions with blocks, a notion that we have discussed extensively for the 2^n factorials. We shall see that the basic ideas encountered there can be modified easily for 3^n factorials. We shall illustrate this with a simple example before discussing useful systems of confounding more generally.

10.5.1 The 3^3 Experiment in Blocks of Size 3

Suppose we wish to arrange a 3^3 experiment in blocks of size 3. This will result in nine blocks with 8 d.f. among them. Hence, we need to confound with blocks four pairs of degrees of freedom, each pair representing a main effect or an interaction component. Suppose we first choose to confound AB and AC^2. Recall that AB is represented by comparisons among the sets satisfying the equations

$$x_1 + x_2 = 0, 1, 2 \quad \text{mod } 3$$

and that AC^2 is similarly represented by the equations

$$x_1 + 2x_3 = 0, 1, 2 \quad \text{mod } 3$$

It is obvious then that considering jointly

$$\begin{aligned} x_1 + x_2 &= k \quad \text{mod } 3 \\ x_1 + 2x_3 &= \ell \quad \text{mod } 3 \end{aligned} \tag{10.17}$$

for all possible combinations of $k, \ell, = 0, 1, 2$, we partition the 27 treatment combinations into 9 sets of 3 treatment combinations each, these sets being the blocks for the desired system of confounding. Now, any treatment combination which satisfies (10.17) for given (k, ℓ), also satisfies

$$2x_1 + x_2 + 2x_3 = k + \ell \quad \text{mod } 3$$

or, equivalently,

$$x_1 + 2x_2 + x_3 = 2(k + \ell) \quad \text{mod } 3$$

and since $2(k + \ell) \equiv 0, 1, 2$ mod 3 it follows that there are three sets of three blocks which satisfy the equation

$$x_1 + 2x_2 + x_3 = 0, 1, 2 \quad \text{mod } 3$$

respectively. Comparisons among these sets, however, define the interaction component AB^2C. Hence AB^2C is also confounded with blocks, and we recognize immediately that AB^2C is a GI of AB and AC^2; that is,

$$(AB) \times (AC^2) = A^2BC^2 = A^4B^2C^4 = AB^2C$$

Similarly, any treatment combination that satisfies (10.17) also satisfies the equation

$$(x_1 + x_2) + 2(x_1 + 2x_3) = k + 2\ell \quad \text{mod } 3$$

Table 10.4 Plan for 3^3 Experiment in Blocks of Size 3

1	2	3	4	5	6	7	8	9
000	002	001	010	012	011	020	022	021
121	120	122	101	100	102	111	110	112
212	211	210	222	221	220	202	201	200

or

$$x_2 + x_3 = k + 2\ell \mod 3$$

Hence the other GI

$$(AB)(AC^2)^2 = A^3BC^4 = BC$$

is also confounded with blocks. These four interactions, AB, AC^2, AB^2C, and BC, then account for the 8 d.f. for comparisons among blocks.

The composition of the blocks for the above system of confounding can be obtained from Eqs. (10.17) with (k, ℓ) assuming all possible values. Alternatively, we can construct first the intrablock subgroup (IBSG) from

$$x_1 + x_2 = 0 \mod 3$$

$$x_1 + 2x_3 = 0 \mod 3$$

and then, using the x representation for treatment combinations, add (componentwise and mod 3) a treatment combination, not already contained in the IBSG, to each element in the IBSG. This process is continued as described in Section 8.3, until all blocks have been constructed in this manner as given in Table 10.4.

10.5.2 Using SAS PROC FACTEX

A similar design can be obtained by using SAS PROC FACTEX and is given in Table 10.5.

We shall comment briefly here on some aspects of the SAS output and how it relates to our discussion in this chapter:

1. We note that rather than using 0, 1, 2 for the factor levels, SAS uses -1, 0, 1, respectively, as commonly used in response surface and regression methodology.
2. The single degree of freedom associated with the main effects and interactions are listed formally akin to the linear-quadratic effects representation for quantitative factors (see I.11.8.1), for example,

$$A \longrightarrow A \text{ linear}$$

$$2 * A \longrightarrow A \text{ quadratic}$$

Table 10.5 3^3 Factorial in Blocks of Size 3

```
options nodate pageno=1;
proc factex;
factors A B C/nlev=3;
blocks size=3;
model est=(A B C);
examine design confounding aliasing;
output out=design blockname=block nvals=(1 2 3 4 5 6 7 8 9);
title1 'TABLE 10.5';
title2 '3**3 FACTORIAL IN BLOCKS OF SIZE 3';
run;

proc print data=design;
run;
```

The FACTEX Procedure

Design Points

Experiment Number	A	B	C	Block
1	-1	-1	-1	1
2	-1	-1	0	9
3	-1	-1	1	5
4	-1	0	-1	6
5	-1	0	0	2
6	-1	0	1	7
7	-1	1	-1	8
8	-1	1	0	4
9	-1	1	1	3
10	0	-1	-1	3
11	0	-1	0	8
12	0	-1	1	4
13	0	0	-1	5
14	0	0	0	1
15	0	0	1	9
16	0	1	-1	7
17	0	1	0	6
18	0	1	1	2
19	1	-1	-1	2
20	1	-1	0	7
21	1	-1	1	6
22	1	0	-1	4
23	1	0	0	3
24	1	0	1	8
25	1	1	-1	9
26	1	1	0	5
27	1	1	1	1

Block Pseudofactor Confounding Rules

```
[B1] = (2*A)+(2*B)+(2*C)
[B2] = B+(2*C)
```

Table 10.5 *(Continued)*

```
Aliasing Structure

           A
         (2*A)
           B
         (2*B)
           C
         (2*C)
           A  +   B
[B]  =  (2*A) +   B
[B]  =     A  +(2*B)
         (2*A) +(2*B)
           A  +   C
[B]  =  (2*A) +   C
[B]  =     A  +(2*C)
         (2*A) +(2*C)
           B  +   C
[B]  =  (2*B) +   C
[B]  =     B  +(2*C)
         (2*B) +(2*C)
```

Obs	block	A	B	C
1	1	-1	-1	-1
2	1	0	0	0
3	1	1	1	1
4	2	-1	0	0
5	2	0	1	1
6	2	1	-1	-1
7	3	-1	1	1
8	3	0	-1	-1
9	3	1	0	0
10	4	-1	1	0
11	4	0	-1	1
12	4	1	0	-1
13	5	-1	-1	1
14	5	0	0	-1
15	5	1	1	0
16	6	-1	0	-1
17	6	0	1	0
18	6	1	-1	1
19	7	-1	0	1
20	7	0	1	-1
21	7	1	-1	0
22	8	-1	1	-1
23	8	0	-1	0
24	8	1	0	1
25	9	-1	-1	0
26	9	0	0	1
27	9	1	1	-1

and

$$A + B \longrightarrow A \text{ linear} \times B \text{ linear}$$

$$2 * A + B \longrightarrow A \text{ quadratic} \times B \text{ linear}$$

$$A + 2 * B \longrightarrow A \text{ linear} \times B \text{ quadratic}$$

$$2 * A + 2 * B \longrightarrow A \text{ quadratic} \times B \text{ quadratic}$$

We should point out, however, that these representations are *not* identical [see also (4) below].

3. The confounding rules are essentially the same as those we have explained earlier. In this example the block compositions are obtained by satisfying the equations

$$2 * A + 2 * B + 2 * C = \delta_1$$

$$B + 2 * C = \delta_2$$

for some $\delta_1, \delta_2 (= 0, 1, 2 \bmod 3)$, where A, B, and C are the levels of those factors. In our notation this is equivalent to satisfying the equations

$$2x_1 + 2x_2 + 2x_3 = \gamma_1$$

$$x_2 + 2x_3 = \gamma_2$$

or

$$x_1 + x_2 + x_3 = \gamma_1^*$$

$$x_2 + 2x_3 = \gamma_2$$

Hence, in this example we confound ABC and BC^2 and, hence, AB^2 and AC^2 with blocks. We only need to remember that $-1 \equiv 2 \bmod 3$ and $-2 \equiv 1 \bmod 3$.

4. The aliasing structure gives a list of the main effects and 2-factor interactions that are either estimable or confounded with blocks, the latter being identified by $[B]$. More precisely, we should really say that the aliasing structure represents a list of the number of degrees of freedom associated with estimable and confounded effects, respectively. For example, the output identifies $A + B$ and $2 * A + 2 * B$ as estimable. This does not mean, however, that A linear $\times B$ linear, or A quadratic $\times B$ quadratic are estimable since there is no relationship between these components and the 2-d.f. component AB.

10.5.3 General Case

Following the example from Section 10.5.1 it is easy to see that for the general case of a 3^n factorial in blocks of size 3^p we then have the following:

1. $3^{n-p} - 1$ d.f. are confounded with blocks, that is, $(3^{n-p} - 1)/2$ main effects and/or interactions are confounded with blocks.
2. If two interactions E^{α} and E^{β} are confounded with blocks, then their GIs $E^{\alpha+\beta}$ and $E^{\alpha+2\beta}$ are also confounded with blocks.
3. To find a system of confounding, one needs to specify only $q = n - p$ independent main effects and/or interactions $E^{\alpha_1}, E^{\alpha_2}, \ldots, E^{\alpha_q}$ since the q interactions have

$$2\binom{q}{2} + 2^2\binom{q}{3} + 2^3\binom{q}{4} + \cdots + 2^{q-1}\binom{q}{q} = \frac{1}{2}3^q - \frac{1}{2} - q$$

GIs among them. Hence altogether $(3^q - 1)/2$ interactions are confounded with blocks.
4. The composition of the blocks is obtained by means of the IBSG, which is composed of the treatment combinations satisfying the equations

$$\alpha_{j1}x_1 + \alpha_{j2}x_2 + \cdots + \alpha_{jn}x_n = 0$$

$(j = 1, 2, \ldots, q = n - p)$ as determined by the independent confounded interactions $E^{\alpha_1}, E^{\alpha_2}, \ldots, E^{\alpha_q}$ in (3). The remaining blocks are then obtained as described in Section 10.5.1.

10.6 USEFUL SYSTEMS OF CONFOUNDING

As we have mentioned earlier, the number of treatment combinations is quite large even for a moderate number of factors. This would call in most cases for incomplete blocks and hence for a system of confounding. But even this may lead to certain difficulties since at this time we are only considering blocks the size of which is a power of 3, so that the choice is quite limited. (For other block sizes we refer to Section 11.14.4.) To complicate matters, according to Fisher's (1942, 1945) theorem (see Section 11.7) confounding of main effects and/or 2-factor interactions can be avoided only if the block size is larger than twice the number of factors, that is, $k > 2n$. For purposes of reference we list in Table 10.6 possible types of confounding involving up to five factors and various block sizes. Further systems can be obtained from this list by permuting the letters.

Table 10.6 Types of Systems of Confounding for 3^n Experiments

Number of Factors	Size of Blocks	Confounded Effects/Interactions[a]
2	3	Any main effect or interaction
3	3	A^*, B^*, AB, AB^2
		A^*, BC^*, ABC, AB^2C^2
		$A^*, BC^{2^*}, AB^2C, ABC^2$
		$AB^*, AC^*, BC^2, AB^2C^2$
		$AB^*, AC^{2^*}, BC, AB^2C$
		$AB^{2^*}, AC^*, BC, ABC^2$
		$AB^{2^*}, AC^{2^*}, BC^2, ABC$
	9	Any main effect or interaction
4	3	$A^*, BC^*, ABC, AB^2C^2, BD^*, ABD, AB^2D^2, BC^2D^2, CD^2, AB^2CD, ACD^2, AC^2D, ABC^2D^2$
		$B^*, AC^*, ABC, AB^2C, AD^{2^*}, ABD^2, AB^2D^2, AC^2D, CD, AB^2C^2D, BCD, ABC^2D, BC^2D^2$
		$C^*, AB^*, ABC, ABC^2, AD^*, ACD, AC^2D, AB^2D^2, BD^2, AB^2C^2D^2, BCD^2, AB^2CD^2, BC^2D^2$
		$D^*, AB^{2^*}, AB^2D, AB^2D^2, AC^{2^*}, AC^2D, AC^2D^2, ABC, BC^2, ABCD^2, BC^2D^2, ABCD, BC^2D$
	9	A^*, B^*, AB, AB^2
		A^*, BC^*, ABC, AB^2C^2
		$A^*, BCD^*, ABCD, AB^2C^2D^2$
		$AB^*, AC^*, BC^2, AB^2C^2$
		$AB^*, CD^*, ABCD, ABC^2D^2$
		$AB^*, ACD^*, BC^2D^2, AB^2C^2D^2$
		$ABD^*, ACD^{2^*}, AB^2C^2, BC^2D^2$
	27	Any main effect or interaction
5	9	$BE^*, ABC^*, AB^2CE, ACE^2, CDE^*$ $BCDE^2, BC^2D^2, ABC^2DE, ABD^2E^2$ $AB^2C^2DE^2, AB^2D^2, AC^2D, AD^2E$
	27	$ABC^*, AB^2DE^*, AC^2D^2E^2, BC^2DE$

[a]Effects with an asterisk (*) are the independent effects.

Generally not all components of a particular interaction are confounded with blocks and, hence, limited intrablock information on that interaction is still available (see also Sections 10.7.2 and 10.7.3). Even so, in most practical cases it will be useful to resort to partial confounding. These can be obtained easily from the systems provided in Table 10.6. In the following we shall comment briefly on some such systems.

10.6.1 Two Factors

Since we can confound any pair of effect or interaction degrees of freedom, the following systems suggest themselves: (1) Use a basic pattern of two replicates, confounding AB and AB^2, giving full information on main effects and $\frac{1}{2}$ information on the interaction. (2) Use a basic pattern of four replicates, confounding A, B, AB, and AB^2, each in one of the four replicates, giving equal information on main effects and interactions. The arrangement of the treatment combinations in blocks and the structure of the analysis of variance with q repetitions of the basic pattern are given in Table 10.7.

10.6.2 Three Factors

1. *Blocks of Size 3* Consulting Table 10.6 suggests that a suitable system of confounding consists of a basic pattern of four replicates of the following types:

Type I	confound	AB, AC, BC^2, AB^2C^2
Type II	confound	AB, AC^2, BC, AB^2C
Type III	confound	AB^2, AC, BC, ABC^2
Type IV	confound	AB^2, AC^2, BC^2, ABC

   This will yield full information on main effects, $\frac{1}{2}$ relative information on 2-factor interactions, and $\frac{3}{4}$ relative information on 3-factor interactions.
2. *Blocks of Size 9* The most useful design consists of one or more repetitions of a basic pattern of four replicates confounding ABC, ABC^2, AB^2C, and AB^2C^2, respectively. This will result in full information on all main effects and 2-factor interactions and $\frac{3}{4}$ relative information on the 3-factor interaction components. The arrangement of the blocks is given in Table 10.8, each column (level of C) combined with the levels of A and B giving a block and each set of three columns a replicate.

10.6.3 Treatment Comparisons

At this point we comment briefly on evaluating the amount of information on any treatment comparison provided by the confounded design relative to the unconfounded design. To do so we make use of the fact that the yield of a treatment combination can be represented in terms of main effect and interaction components (see Section 10.3). For purposes of illustration suppose that we are interested in the comparison $(a_0b_0c_0 - a_0b_0c_1)$, using q repetitions of the basic pattern given in Table 10.8. Notice that $a_0b_0c_0$ and $a_0b_0c_1$ never occur together in the same block, so that a simple comparison among their mean yields

Table 10.7 3^2 Experiment in Blocks of Size 3

Replicate Type	Effect/Interaction Confounded	Blocks
	a. Arrangement in Blocks	
I	A	00, 01, 02; 10, 11, 12; 20, 21, 22
II	B	00, 10, 20; 01, 11, 21; 02, 12, 22
III	AB	00, 12, 21; 01, 10, 22; 02, 20, 11
IV	AB^2	00, 11, 22; 10, 02, 21; 01, 12, 20

Source	d.f.
b. Analysis of Variance	
Replicates	$4q - 1$
Blocks/reps	$8q$
A	2
B	2
AB	2
AB^2	2
Residual	$24q - 8$
$A \times$ reps II, III, IV	$2(3q - 1)$
$B \times$ reps I, III, IV	$2(3q - 1)$
$AB \times$ reps I, II, IV	$2(3q - 1)$
$AB^2 \times$ reps I, II, III	$2(3q - 1)$

Table 10.8 Design for 3^3 Experiment in Blocks of Size 9

Level of A	Level of B	ABC	ABC^2	AB^2C	AB^2C^2
		Level of C			
0	0	0 1 2	0 2 1	0 1 2	0 2 1
1	0	2 0 1	1 0 2	2 0 1	1 0 2
2	0	1 2 0	2 1 0	1 2 0	2 1 0
0	1	2 0 1	1 0 2	1 2 0	2 1 0
1	1	1 2 0	2 1 0	0 1 2	0 2 1
2	1	0 1 2	0 2 1	2 0 1	1 0 2
0	2	1 2 0	2 1 0	2 0 1	1 0 2
1	2	0 1 2	0 2 1	1 2 0	2 1 0
2	2	2 0 1	1 0 2	0 1 2	0 2 1

(Column group header: Confounded Interaction — ABC, ABC^2, AB^2C, AB^2C^2.)

will not provide any useful information. Instead, we make use of the fact that [see (10.15)]

$$\begin{aligned} a_0b_0c_0 = {} & M + A_0 + B_0 + AB_0 + AB_0^2 + C_0 \\ & + AC_0 + AC_0^2 + BC_0 + BC_0^2 + ABC_0 \\ & + ABC_0^2 + AB^2C_0 + AB^2C_0^2 \end{aligned}$$

and

$$\begin{aligned} a_0b_0c_1 = {} & M + A_0 + B_0 + AB_0 + AB_0^2 + C_1 \\ & + AC_1 + AC_2^2 + BC_1 + BC_2^2 + ABC_1 \\ & + ABC_2^2 + AB^2C_1 + AB^2C_2^2 \end{aligned}$$

so that the estimator for $(a_0b_0c_0 - a_0b_0c_1)$ is

$$\begin{aligned} \widehat{a_0b_0c_0} - \widehat{a_0b_0c_1} = {} & \left(\widehat{C}_0 - \widehat{C}_1\right) + \left(\widehat{AC}_0 - \widehat{AC}_1\right) \\ & + \left(\widehat{AC}_0 - \widehat{AC_2^2}\right) + \left(\widehat{BC}_0 - \widehat{BC}_1\right) \\ & + \left(\widehat{BC_0^2} - \widehat{BC_2^2}\right) + \left(\widehat{ABC}_0 - \widehat{ABC}_1\right) \\ & + \left(\widehat{ABC_0^2} - \widehat{ABC_2^2}\right) + \left(\widehat{AB^2C}_0 - \widehat{AB^2C}_1\right) \\ & + \left(\widehat{AB^2C_0^2} - \widehat{AB^2C_2^2}\right) \end{aligned}$$

Each quantity in parentheses is statistically independent of the others (because of the orthogonality of the partitions), with a variance depending on the system of confounding used. In the present case any difference among main effect components (like $C_0 - C_1$) and among 2-factor interaction components (like $AC_0 - AC_1$) is estimated with variance

$$2\frac{\sigma_e^2}{4q \cdot 9} = \frac{\sigma_e^2}{18q}$$

and any difference among 3-factor interaction components (like $ABC_0 - ABC_1$) is estimated with variance

$$2\frac{\sigma_e^2}{3q \cdot 9} = \frac{2\sigma_e^2}{27q}$$

Hence

$$\text{var}\left(\widehat{a_0b_0c_0} - \widehat{a_0b_0c_1}\right) = 5 \cdot \frac{\sigma_e^2}{18q} + 4 \cdot \frac{2\sigma_e^2}{27q} = \frac{31}{54q}\sigma_e^2$$

With no confounding and the same error variance, the variance of the comparison would have been $\sigma_e^2/2q$, so that the relative information on this comparison is $\frac{54}{62} = \frac{27}{31}$.

10.6.4 Four Factors

1. *Blocks of Size 3* A suitable system consists of a basic pattern of four replicates, using the four systems of confounding given in Table 10.6. This design will result in $\frac{3}{4}$ information on main effects and all 2-factor interaction components.
2. *Blocks of Size 9* The most useful type of confounding is obviously the last one given in Table 10.6 since it confounds only 3-factor interaction components, one from each type of 3-factor interaction. Altogether there exist eight such systems of confounding.
3. *Blocks of Size 27* In general, the experimenter will wish to avoid blocks of size as large as 27, though, in some fields of experimentation and with some types of experimental material, the effect on error variance of reducing block size from 27 to 9 may be so small as not to offset any loss in relative information that results from confounding. If blocks of size 27 are being used, any of the eight 4-factor interaction components may be confounded.

10.6.5 Five Factors

1. *Blocks of Size 9* It is not possible to avoid confounding a main effect or 2-factor interactions. Under these circumstances, the system of confounding given in Table 10.6 and permutation of that set would be most useful.
2. *Blocks of Size 27* One can show that it is not possible to find a design confounding only 4- and 5-factor interactions. The most useful system of confounding is then one of the form given in Table 10.6.

10.6.6 Double Confounding

Occasionally it may be desirable to impose a double restriction on the pattern of a 3^n experiment. This leads to systems of double confounding. Suitable systems can be found by consulting Table 10.6. The actual arrangement of the treatment combinations can be obtained by first constructing the IBSGs for confounding with "rows" and "columns," respectively, and then adding the elements of the first row and first column termwise mod 3. As an example consider the 3^3 experiment with rows of size 9, confounding ABC^2, and columns of size 3, confounding AB, AC, BC^2, AB^2C^2; that is, construct row blocks from

$$x_1 + x_2 + 2x_3 = 0 \mod 3$$

and column blocks from

$$x_1 + x_2 = 0 \mod 3$$
$$x_1 + x_3 = 0 \mod 3$$

The final arrangement (apart from randomization of rows and columns) then is as follows:

	Column 1	2	3	4	5	6	7	8	9
Row 1	000	101	202	011	112	210	022	120	221
Row 2	122	220	021	100	201	002	111	212	010
Row 3	211	012	110	222	020	121	200	001	102

Obviously, one will need replications of this or a similar arrangement, using different systems of confounding.

10.7 ANALYSIS OF CONFOUNDED 3^n FACTORIALS

In the course of our discussion of 3^n factorial experiments we have already commented on some aspects of the analysis for particular situations. We shall now make some remarks about the general 3^n factorial experiment in blocks of size 3^p using some system of complete or partial confounding. Since the development parallels that of Section 9.7, we shall not repeat all the details, but only those that are specific to the 3^n case.

The basic model underlying the analysis is as before:

$$\boldsymbol{y} = \mu\mathfrak{I} + \boldsymbol{X}_\rho\boldsymbol{\rho} + \boldsymbol{X}_{\beta*}\boldsymbol{\beta}^* + \boldsymbol{X}_\tau\boldsymbol{\tau} + \boldsymbol{e}$$

where $\boldsymbol{\rho}$ represents the replicate effects, $\boldsymbol{\beta}^*$ the block within replicate effects, and $\boldsymbol{\tau}$ the treatment effects, or in its reparameterized form

$$\boldsymbol{y} = \mu\mathfrak{I} + \boldsymbol{X}_\rho\boldsymbol{\rho} + \boldsymbol{X}_{\beta*}\boldsymbol{\beta}^* + \boldsymbol{X}_{\tau*}\boldsymbol{\tau}^* + \boldsymbol{e}$$

where $\boldsymbol{\tau}^*$ represents the interaction components $E_i^{\boldsymbol{\alpha}}$ for all admissible $\boldsymbol{\alpha}' = (\alpha_1, \alpha_2, \ldots, \alpha_n)$ and $i = 0, 1, 2$. A basic pattern of partial confounding consists of s types of replicates, each replicate consisting of 3^{n-p} blocks, and the basic pattern is repeated q times.

We divide the $(3^n - 1)/2$ interactions E^{α} into three mutually exclusive sets based on the amount of confounding:

$$\mathcal{E}_1 = \left\{E^{\alpha_k}, k = 1, 2, \ldots, n_1\text{: completely confounded}\right\}$$

$$\mathcal{E}_2 = \left\{E^{\gamma_\ell}, \ell = 1, 2, \ldots, n_2\text{: partially confounded}\right\}$$

$$\mathcal{E}_3 = \left\{E^{\delta_m}, k = 1, 2, \ldots, n_3\text{: not confounded}\right\}$$

and $E^{\gamma_\ell} \in \mathcal{E}_2$ is confounded in $c(\gamma_\ell)$ replicates and not confounded in $u(\gamma_\ell)$ replicates of the basic pattern. We denote by $N = 3^n sq$ the total number of observations.

The following statements concerning the analysis are then obvious extensions of those made in Section 9.7.

10.7.1 Intrablock Information

1. Each $E_h^{\delta_m} \in \mathcal{E}_3 (h = 0, 1, 2)$ is estimated from all replicates, which implies

$$\operatorname{var}\left(\widehat{E}_h^{\delta_m} - \widehat{E}_{h'}^{\delta_m}\right) = \frac{2}{sq3^{n-1}} \sigma_e^2$$

and

$$\text{SS}\left(E^{\delta_m}\right) = sq3^{n-1} \left\{\left[\widehat{E}_0^{\delta_m}\right]^2 + \left[\widehat{E}_1^{\delta_m}\right]^2 + \left[\widehat{E}_2^{\delta_m}\right]^2\right\} \tag{10.18}$$

2. Each $E_h^{\gamma_\ell} \in \mathcal{E}_2$ is estimated from $qu(\gamma_\ell)$ replicates and hence

$$\operatorname{var}\left(\widehat{E}_h^{\gamma_\ell} - \widehat{E}_{h'}^{\gamma_\ell}\right) = \frac{2}{qu(\gamma_\ell)3^{n-1}} \sigma_e^2 \tag{10.19}$$

and

$$\text{SS}\left(E^{\gamma_\ell}\right) = qu(\gamma_\ell)3^{n-1} \left\{\left[\widehat{E}_0^{\gamma_\ell}\right]^2 + \left[\widehat{E}_1^{\gamma_\ell}\right]^2 + \left[\widehat{E}_2^{\gamma_\ell}\right]^2\right\} \tag{10.20}$$

10.7.2 The ANOVAs

The partitioning of the total sum of squares and its degree of freedom in the T|B-ANOVA is given in Table 10.9a with a further partitioning of the block and residual sums of squares in Tables 10.9b and 10.9c. All sums of squares associated with interactions are obtained in the usual way, following (10.18) and (10.20), except that $\text{SS}(\cdot)_c$ is obtained from only those replicates in which the corresponding interaction is confounded. $\text{SS}(E_{ij}^*)$ is obtained only from the jth replicate in the ith repetition, and $\sum_{ij}'$ and $\sum_{ij}''$ denote summation over

Table 10.9 T|B-ANOVA for Partial Confounding

Source	d.f.	SS
	a. Basic Partitioning	
$X_\rho \mid \mathfrak{I}$	$sq - 1$	Usual
$X_{\beta^*} \mid \mathfrak{I}, X_\rho$	$sq(3^{n-p} - 1)$	Usual
$X_{\tau^*} \mid \mathfrak{I}, X_\rho, X_{\beta^*}$	$2n_2 + 2n_3$	$\sum_\ell \text{SS}(E^{\gamma_\ell}) + \sum_m \text{SS}(E^{\delta_m})$
$I \mid \mathfrak{I}, X_\rho, X_{\beta^*}, X_{\tau^*}$	ν_R	Difference
Total	$N - 1$	Usual
	b. Partitioning of Block Sum of Squares	
$X_{\beta^*} \mid \mathfrak{I}, X_\rho$	$sq(3^{n-p} - 1)$	
$\{E^{\alpha_k}\}$	$2n_1$	$\sum_k \text{SS}(E^{\alpha_k})_c$
$\{E^{\gamma_\ell}\}$	$2n_2$	$\sum_\ell \text{SS}(E^{\gamma_\ell})_c$
$\{E^{\alpha_k} \times \text{reps}\}$	$2n_1(sq - 1)$	$\sum_k \sum_{ij} \text{SS}(E_{ij}^{\alpha_k})_c - \sum_k \text{SS}(E^{\alpha_k})_c$
$\{E^{\alpha_\ell} \times \text{reps}\}$	$2q \sum_\ell c(\gamma_\ell) - 2n_2$	$\sum_\ell \sum_{ij}' \text{SS}(E_{ij}^{\gamma_\ell})_c - \sum_\ell \text{SS}(E^{\gamma_\ell})_c$
	c. Partitioning of Residual Sum of Squares	
$I \mid \mathfrak{I}, X_\rho, X_{\beta^*}, X_{\tau^*}$	ν_R	
$\{E^{\gamma_\ell} \times \text{reps}\}$	$2q \sum_\ell u(\gamma_\ell) - 2n_2$	$\sum_\ell \sum_{ij}'' \text{SS}(E_{ij}^{\gamma_\ell}) - \sum_\ell \text{SS}(E^{\gamma_\ell})$
$\{E^{\delta_m} \times \text{reps}\}$	$2n_3(sq - 1)$	$\sum_m \sum_{ij} \text{SS}(E_{ij}^{\delta_m}) - \sum_m \text{SS}(E^{\delta_m})$

all replicates in which E^{γ_ℓ} is confounded and not confounded, respectively (see Section 9.7).

The B|T-ANOVA in its basic form is given in Table 10.10a with a partitioning of the block sum of squares in Table 10.10b. The only sum of squares that needs to be explained is that associated with $\{\widehat{E}^{\gamma_\ell} \text{ vs. } \widetilde{E}^{\gamma_\ell}\}$. More specifically this sum of squares is associated with the comparisons

$$\left(\widehat{E}_0^{\gamma_\ell} - \widetilde{E}_0^{\gamma_\ell}\right) \quad \text{vs.} \quad \left(\widehat{E}_1^{\gamma_\ell} - \widetilde{E}_1^{\gamma_\ell}\right) \quad \text{vs.} \quad \left(\widehat{E}_2^{\gamma_\ell} - \widetilde{E}_2^{\gamma_\ell}\right) \tag{10.21}$$

where $\widehat{E}_h^{\gamma_\ell}$ is obtained from the $qu(\gamma_\ell)$ replicates in which $E^{\gamma_\ell} \in \mathcal{E}_2$ is not confounded and $\widetilde{E}_h^{\gamma_\ell}$ is obtained from the $qc(\gamma_\ell)$ replicates in which E^{γ_ℓ} is confounded. Since $\widehat{E}_0^{\gamma_\ell} + \widehat{E}_1^{\gamma_\ell} + \widehat{E}_2^{\gamma_\ell} = 0$ and $\widetilde{E}_0^{\gamma_\ell} + \widetilde{E}_1^{\gamma_\ell} + \widetilde{E}_2^{\gamma_\ell} = 0$, it follows

Table 10.10 B|T-ANOVA for Partial Confounding

Source	d.f.	SS
	a. Basic Partitioning	
$\boldsymbol{X}_\rho \mid \mathfrak{I}$	$sq - 1$	Usual
$\boldsymbol{X}_{\tau^*} \mid \mathfrak{I}, \boldsymbol{X}_\rho$	$3^n - 1$	Usual
$\boldsymbol{X}_{\beta^*} \mid \mathfrak{I}, \boldsymbol{X}_\rho, \boldsymbol{X}_{\tau^*}$	$sq(3^{n-p} - 1) - 2n_1$	Difference
$\boldsymbol{I} \mid \mathfrak{I}, \boldsymbol{X}_\rho, \boldsymbol{X}_{\tau^*}, \boldsymbol{X}_{\beta^*}$	ν_R	From Table 10.9
Total	$N - 1$	Usual
	b. Partitioning of Block Sum of Squares	
$\boldsymbol{X}_{\beta^*} \mid \mathfrak{I}, \boldsymbol{X}_\rho, \boldsymbol{X}_{\tau^*}$	$sq(3^{n-p} - 1) - 2n_1$	
$\left\{\widehat{E}^{\gamma_\ell} \text{ vs. } \widetilde{E}^{\gamma_\ell}\right\}$	$2n_2$	See (10.24)
$\{E^{\alpha_k} \times \text{reps}\}$	$2n_1(sq - 1)$	From Table 10.9b
$\{E^{\gamma_\ell} \times \text{reps}\}$	$2q \sum_\ell c(\gamma_\ell) - 2n_2$	From Table 10.9b

that the comparisons (10.21) and hence the associated sum of squares carry 2 d.f. This sum of squares is obtained in the usual way. Define

$$\widehat{E}_h^{\gamma_\ell} - \widetilde{E}_h^{\gamma_\ell} = X_h^{\gamma_\ell} \qquad (h = 0, 1, 2)$$

Then

$$\text{SS}\left(\widehat{E}^{\gamma_\ell} \quad \text{vs.} \quad \widetilde{E}^{\gamma_\ell}\right) = \begin{pmatrix} X_0^{\gamma_\ell} - X_2^{\gamma_\ell} \\ X_1^{\gamma_\ell} - X_2^{\gamma_\ell} \end{pmatrix}' \left[\frac{\boldsymbol{V}_{\gamma_\ell}}{\sigma_e^2}\right]^{-1} \begin{pmatrix} X_0^{\gamma_\ell} - X_2^{\gamma_\ell} \\ X_1^{\gamma_\ell} - X_2^{\gamma_\ell} \end{pmatrix} \tag{10.22}$$

where $\boldsymbol{V}_{\gamma_\ell}$ is the variance–covariance matrix of

$$\begin{pmatrix} X_0^{\gamma_\ell} - X_2^{\gamma_\ell} \\ X_1^{\gamma_\ell} - X_2^{\gamma_\ell} \end{pmatrix}$$

It is easy to see that

$$\boldsymbol{V}_{\gamma_\ell} = \frac{s}{qu(\gamma_\ell)\, c(\gamma_\ell)\, 3^{n-1}} \begin{pmatrix} 2 & 1 \\ 1 & 2 \end{pmatrix} \sigma_e^2$$

and hence

$$\left[\frac{\boldsymbol{V}_{\gamma\ell}}{\sigma_e^2}\right]^{-1} = \frac{qu(\gamma_\ell)\, c(\gamma_\ell)\, 3^{n-1}}{s} \begin{pmatrix} 2 & -1 \\ -1 & 2 \end{pmatrix}$$

Then (10.22) reduces to

$$\text{SS}\left(\widehat{E}^{\gamma_\ell} \quad \text{vs.} \quad \widetilde{E}^{\gamma_\ell}\right) = \frac{qu(\gamma_\ell)\, c(\gamma_\ell)\, 3^{n-1}}{s} \sum_h \left[X_h^{\gamma_\ell}\right]^2 \tag{10.23}$$

Summing (10.23) over all $E^{\gamma_\ell} \in \mathcal{E}_2$ yields

$$\text{SS}\left(\{\widehat{E}^{\gamma_\ell} \quad \text{vs.} \quad \widetilde{E}^{\gamma_\ell}\}\right) = \frac{qu3^{n-1}}{s} \sum_\ell \sum_h u(\gamma_\ell)\, c(\gamma_\ell) \left[X_h^{\gamma_\ell}\right]^2 \tag{10.24}$$

which is a sum of squares with $2n_2$ d.f. and which depends only on block effects and error.

10.7.3 Tests of Hypotheses

We can test the hypotheses

$$H_0\colon\ E_0^{\gamma_\ell} = E_1^{\gamma_\ell} = E_2^{\gamma_\ell} = 0 \qquad (E^{\gamma_\ell} \in \mathcal{E}_2)$$

and

$$H_0\colon\ E_0^{\delta_m} = E_1^{\delta_m} = E_2^{\delta_m} = 0 \qquad (E^{\delta_m} \in \mathcal{E}_3)$$

by using, in the T|B-ANOVA, the F tests as approximations to the randomization tests (see, e.g., I.6.6 and I.9.2):

$$F = \frac{\text{MS}\,(E^{\gamma_\ell})}{\text{MS}\left(\boldsymbol{I}|\mathfrak{I}, \boldsymbol{X}_\rho, \boldsymbol{X}_{\beta^*}, \boldsymbol{X}_{\tau^*}\right)} \tag{10.25}$$

and

$$F = \frac{\text{MS}\left(E^{\delta_m}\right)}{\text{MS}\left(\boldsymbol{I}|\mathfrak{I}, \boldsymbol{X}_\rho, \boldsymbol{X}_{\beta^*}, \boldsymbol{X}_{\tau^*}\right)} \tag{10.26}$$

respectively, each with 2 and ν_R d.f. These are tests for no main effects or interactions. With regard to interactions, however, we must keep in mind that the total interaction among s factors has 2^s d.f. and hence has $2^s - 1$ components of the form E^α. For example, for $s = 3$ and factors A, B, C, the components are ABC, ABC^2, AB^2C, AB^2C^2. To test then the hypothesis that there is no $A \times B \times C$ interaction we use the F test:

$$F = \frac{\left[\text{SS}(ABC) + \text{SS}\left(ABC^2\right) + \text{SS}\left(AB^2C\right) + \text{SS}\left(AB^2C^2\right)\right]/8}{\text{MS}\left(\boldsymbol{I}|\mathfrak{I}, \boldsymbol{X}_\rho, \boldsymbol{X}_{\beta^*}, \boldsymbol{X}_{\tau^*}\right)}$$

with 8 and ν_R d.f. In this case an F test of the form (10.25) or (10.26) may not tell us very much about the 3-factor interaction, except that when one or the other of those tests is significant then we can conclude that $A \times B \times C$ interaction is possibly present.

For the general case of an s-factor interaction the following situations can occur for a particular system of confounding:

1. All 2^{s-1} components belong to $\mathcal{E}_2$ and/or $\mathcal{E}_3$. In this case we can pool the associated sums of squares in Table 10.9a and obtain an F test with 2^s and ν_R d.f.
2. Some components belong to $\mathcal{E}_1$. In this case we can only use the sums of squares associated with the components in $\mathcal{E}_2$ and/or $\mathcal{E}_3$. Great care has to be exercised in interpreting such a test.
3. All components belong to $\mathcal{E}_1$. In this case no test exists in the context of the T|B-ANOVA (but see Section 10.7.4).

10.7.4 Interblock Information

Assuming $\boldsymbol{\beta}^*$ to be a vector of i.i.d. random variables with $E(\boldsymbol{\beta}^*) = \boldsymbol{\phi}$ and $E(\boldsymbol{\beta}^*\boldsymbol{\beta}^{*\prime}) = \sigma_\beta^2 \boldsymbol{I}$, we can obtain interblock information regarding $E^{\alpha_k} \in \mathcal{E}_1$ and $E^{\gamma_\ell} \in \mathcal{E}_2$ by considering $\widetilde{E}_h^{\alpha_k}$ and $\widetilde{E}_h^{\gamma_\ell}$ $(h = 0, 1, 2)$ obtained from those replicates in which these interactions are confounded. We then have

$$E\left(\widetilde{E}_h^{\alpha_k} - \widetilde{E}_{h'}^{\alpha_k}\right) = E_h^{\alpha_k} - E_{h'}^{\alpha_k}$$

with

$$\operatorname{var}\left(\widetilde{E}_h^{\alpha_k} - \widetilde{E}_{h'}^{\alpha_k}\right) = \frac{2}{qs3^{n-1}}\left(\sigma_e^2 + 3^p\,\sigma_\beta^2\right) \tag{10.27}$$

and

$$E\left(\widetilde{E}_h^{\gamma_\ell} - \widetilde{E}_{h'}^{\gamma_\ell}\right) = \widetilde{E}_h^{\gamma_\ell} - \widetilde{E}_{h'}^{\gamma_\ell}$$

with

$$\operatorname{var}\left(\widetilde{E}_h^{\gamma_\ell} - \widetilde{E}_{h'}^{\gamma_\ell}\right) = \frac{2}{qc(\gamma_\ell)3^{n-1}}\left(\sigma_e^2 + 3^p\,\sigma_\beta^2\right) \tag{10.28}$$

If we define the pooled sum of squares associated with $\{E^{\alpha_k} \times \text{reps}\}$ and $\{E^{\gamma_\ell} \times \text{reps}\}$ in Table 10.9b as SS(Remainder) with $\nu = 2n_1(sq-1) + 2q\sum_\ell c(\gamma_\ell) - 2n_2$ d.f., then the hypotheses

$$H_0\colon E_0^{\alpha_k} = E_1^{\alpha_k} = E_2^{\alpha_k} = 0 \qquad (E^{\alpha_k} \in \mathcal{E}_1)$$

can be tested by using

$$F = \frac{\text{MS}\,(E^{\alpha_k})_c}{\text{MS(Remainder)}} \tag{10.29}$$

with 2 and ν d.f.

10.7.5 Combined Intra- and Interblock Information

Combined information is available and may be of interest for quantities of the form $E_h^{\gamma_\ell} - E_{h'}^{\gamma_\ell}$ for $E^{\gamma_\ell} \in \mathcal{E}_2$, or more generally for $c_0 E_0^{\gamma_\ell} + c_1 E_1^{\gamma_\ell} + c_2 E_2^{\gamma_\ell}$ with $c_0 + c_1 + c_2 = 0$. Using (10.19) and (10.28) we obtain

$$\widehat{\widehat{E}}_h^{\gamma_\ell} - \widehat{\widehat{E}}_{h'}^{\gamma_\ell} = \frac{u(\gamma_\ell)\, w(\widehat{E}_h^{\gamma_\ell} - \widehat{E}_{h'}^{\gamma_\ell}) + c\,(\gamma_\ell)\; w' \left(\widetilde{E}_h^{\gamma_\ell} - \widetilde{E}_{h'}^{\gamma_\ell}\right)}{u(\gamma_\ell)\, w + c(\gamma_\ell)\, w'} \tag{10.30}$$

with $w = 1/\sigma_e^2$, $w' = 1/(\sigma_e^2 + 3^p\, \sigma_\beta^2)$. Letting $w/w' = \rho$, we obtain

$$\mathrm{var}\left(\widehat{\widehat{E}}_h^{\gamma_\ell} - \widehat{\widehat{E}}_{h'}^{\gamma_\ell}\right) = \frac{2\sigma_e^2}{\left[(u(\gamma_\ell) + c(\gamma_\ell)\rho^{-1}\right] q 3^{n-1}} \tag{10.31}$$

10.7.6 Estimation of Weights

For the practical use of (10.30) and (10.31) in connection with, for example, the combined estimate of $a(\boldsymbol{x}) - a(\boldsymbol{z})$ for two treatment combinations $\boldsymbol{x}'$ and $\boldsymbol{z}'$, using (10.15), we usually need to estimate w and w' (or ρ). As always

$$\widehat{\sigma}_e^2 = \frac{1}{\widehat{w}} = \mathrm{MS}(\boldsymbol{I} \mid \mathfrak{I}, \boldsymbol{X}_\rho, \boldsymbol{X}_{\beta^*}, \boldsymbol{X}_{\boldsymbol{\tau}^*}) \tag{10.32}$$

which is obtained from the T|B-ANOVA of Table 10.10.

For the estimation of w' by the Yates procedure we use the B|T-ANOVA of Table 10.9. The two components of $\mathrm{SS}(\boldsymbol{X}_{\beta^*} \mid \mathfrak{I}, \boldsymbol{X}_\rho, \boldsymbol{X}_{\tau_*})$ are SS(Remainder) as defined in Section 10.7.4 above and $\mathrm{SS}\left(\widehat{E}^{\gamma_\ell} \quad \text{vs.} \quad \widetilde{E}^{\gamma_\ell}\right)$ given by (10.24). Now

$$E\ \mathrm{SS(Remainder)} = \nu(\sigma_e^2 + 3^p\, \sigma_\beta^2) \tag{10.33}$$

and, after straightforward derivation,

$$E\ \mathrm{SS}\left(\widehat{E}^{\gamma_\ell} \quad \text{vs.} \quad \widetilde{E}^{\gamma_\ell}\right) = 2\left[\sigma_e^2 + \frac{u(\gamma_\ell)}{s}\, 3^p\, \sigma_\beta^2\right] \tag{10.34}$$

Hence, from (10.33), (10.34) and Table 10.10,

$$E\ \mathrm{MS}(\boldsymbol{X}_{\beta_*} \mid \mathfrak{I}, \boldsymbol{X}_\rho, \boldsymbol{X}_{\tau^*}) = \sigma_e^2 + \Delta\, 3^p\, \sigma_\beta^2 \tag{10.35}$$

with

$$\Delta = \frac{\frac{2}{s} \sum_\ell u\,(\gamma_\ell) + \nu}{sq(3^{n-p} - 1) - 2n_1} \tag{10.36}$$

and ν as defined in Section 10.7.4. It then follows that

$$\widehat{\sigma_e^2 + 3^p\,\sigma_\beta^2} = \frac{1}{\widehat{w}'} = \left(1 - \frac{1}{\Delta}\right) \text{MS}(\boldsymbol{I}\,|\,\mathfrak{I}, \boldsymbol{X}_\rho, \boldsymbol{X}_{\beta^*}, \boldsymbol{X}_{\tau^*}$$
$$+ \frac{1}{\Delta}\text{MS}(\boldsymbol{X}_{\beta^*}\,|\,\mathfrak{I}, \boldsymbol{X}_\rho, \boldsymbol{X}_{\tau^*}) \tag{10.37}$$

with Δ given by (10.36).

There are, of course, other methods of estimating the weights as described in Section 1.11. In any case, Satterthwaite's procedure (Satterthwaite, 1946; see also I.9.7.7) must be used to obtain the degrees of freedom associated with the estimator given in (10.37).

10.8 NUMERICAL EXAMPLE

We consider the 3^2 factorial in blocks of size 3, confounding the 2-factor interaction component AB with blocks. The data as well as the intrablock analysis (using SAS PROC GLM) and combined intra- and interblock analysis (using SAS PROC MIXED) are given in Table 10.11. We shall comment briefly on these analyses.

10.8.1 Intrablock Analysis

We find $\widehat{\sigma}_e^2 = \text{MS}(\boldsymbol{I}\,|\,\mathfrak{I}, \boldsymbol{X}_\rho, \boldsymbol{X}_{\beta^*}, \boldsymbol{X}_{\tau^*}) = \text{MS}(E) = .8333$. MS($E$) (with 6 d.f.) is used to test hypotheses about A, B, and $A \times B$ by forming F ratios with the respective type III mean squares in the numerator. Concerning the $A \times B$ interaction, we know, of course, that it has only 2 d.f., which are associated with the interaction component AB^2.

10.8.2 Combined Analysis

In this example AB is confounded in both replicates, that is, AB belongs to $\mathcal{E}_1$, whereas A, B, and AB^2 belong to $\mathcal{E}_3$. This means that only interblock information is available for AB and only intrablock information is available for A, B, and AB^2. Thus, the combined analysis for A and B yields the same results as the intrablock analysis. We can verify this easily by comparing the F ratios and P values using the type III MS in PROC GLM and PROC MIXED, respectively. In both cases we find for A: $F = 126.87$, $P = .0001$, and for B: $F = 2.07$, $P = .2076$.

To test the hypothesis that there is no $A \times B$ interaction we obtain

$$\text{SS}(A \times B) = \text{SS}(AB) + \text{SS}(AB^2) \tag{10.38}$$

Table 10.11 Analysis of 3^2 Factorial in Blocks of Size 3 (With AB Completely Confounded)

```
data three;
input rep block A B y @@;
datalines;
1 1 0 0 11   1 1 1 2 14   1 1 2 1 20
1 2 0 2 20   1 2 1 1 23   1 2 2 0 27
1 3 0 1 20   1 3 1 0 27   1 3 2 2 30
2 4 0 0 20   2 4 1 2 25   2 4 2 1 29
2 5 0 2 27   2 5 1 1 30   2 5 2 0 32
2 6 0 1 31   2 6 1 0 41   2 6 2 2 43
;
run;

options nodate pageno=1;
proc print data=three;
title1 'TABLE 10.11';
title2 '3**2 FACTORIAL IN BLOCKS OF SIZE 3';
title3 '(WITH AB COMPLETELY CONFOUNDED)';
run;

proc glm data=three;
class rep block A B;
model y=rep A|B block(rep);
title3 'INTRA-BLOCK ANALYSIS';
run;

proc mixed data=three;
class rep block A B;
model y=rep A|B/ddfm=satterth;
random block(rep);
title3 'COMBINED INTRA- AND INTER-BLOCK ANALYSIS';
run;
```

Obs	rep	block	A	B	y
1	1	1	0	0	11
2	1	1	1	2	14
3	1	1	2	1	20
4	1	2	0	2	20
5	1	2	1	1	23
6	1	2	2	0	27
7	1	3	0	1	20
8	1	3	1	0	27
9	1	3	2	2	30
10	2	4	0	0	20
11	2	4	1	2	25
12	2	4	2	1	29
13	2	5	0	2	27
14	2	5	1	1	30
15	2	5	2	0	32
16	2	6	0	1	31
17	2	6	1	0	41
18	2	6	2	2	43

Table 10.11 *(Continued)*

INTRABLOCK ANALYSIS

The GLM Procedure

Class Level Information

Class	Levels	Values
rep	2	1 2
block	6	1 2 3 4 5 6
A	3	0 1 2
B	3	0 1 2

Number of observations 18

Dependent Variable: y

Source	DF	Sum of Squares	Mean Square	F Value	Pr > F
Model	11	1136.777778	103.343434	124.01	<.0001
Error	6	5.000000	0.833333		
Corrected Total	17	1141.777778			

R-Square	Coeff Var	Root MSE	y Mean
0.995621	3.496101	0.912871	26.11111

Source	DF	Type I SS	Mean Square	F Value	Pr > F
rep	1	410.8888889	410.8888889	493.07	<.0001
A	2	228.1111111	114.0555556	136.87	<.0001
B	2	3.4444444	1.7222222	2.07	0.2076
A*B	4	464.2222222	116.0555556	139.27	<.0001
block(rep)	2	30.1111111	15.0555556	18.07	0.0029

Source	DF	Type III SS	Mean Square	F Value	Pr > F
rep	1	410.8888889	410.8888889	493.07	<.0001
A	2	228.1111111	114.0555556	136.87	<.0001
B	2	3.4444444	1.7222222	2.07	0.2076
A*B	2	18.7777778	9.3888889	11.27	0.0093
block(rep)	2	30.1111111	15.0555556	18.07	0.0029

Table 10.11 *(Continued)*

```
              COMBINED INTRA- AND INTERBLOCK ANALYSIS

                       The Mixed Procedure

                        Model Information

          Data Set                        WORK.THREE
          Dependent Variable              y
          Covariance Structure            Variance Components
          Estimation Method               REML
          Residual Variance Method        Profile
          Fixed Effects SE Method         Model-Based
          Degrees of Freedom Method       Satterthwaite

                     Class Level Information

                  Class     Levels     Values

                  rep            2     1 2
                  block          6     1 2 3 4 5 6
                  A              3     0 1 2
                  B              3     0 1 2

                           Dimensions

                   Covariance Parameters      2
                   Columns in X              18
                   Columns in Z               6
                   Subjects                   1
                   Max Obs Per Subject       18
                   Observations Used         18
                   Observations Not Used      0
                   Total Observations        18

                        Iteration History

    Iteration     Evaluations     -2 Res Log Like      Criterion

            0               1         42.27802731
            1               1         34.77498334     0.00000000

                    Convergence criteria met.

                      Covariance Parameter
                            Estimates

                  Cov Parm              Estimate

                  block(rep)              4.7407
                  Residual                0.8333
```

Table 10.11 *(Continued)*

Fit Statistics

-2 Res Log Likelihood	34.8
AIC (smaller is better)	38.8
AICC (smaller is better)	41.2
BIC (smaller is better)	38.4

Type 3 Tests of Fixed Effects

Effect	Num DF	Den DF	F Value	Pr > F
rep	1	2	27.29	0.0347
A	2	6	136.87	<.0001
B	2	6	2.07	0.2076
A*B	4	2	13.03	0.0725

from the type I SS in PROC GLM as

$$\mathrm{SS}(A \times B) = 464.22$$

with 4 d.f. Now we know from (10.16b) and its extension to completely confounded effects that

$$E\left\{\mathrm{SS}(AB)\right\} = 2\sigma_e^2 + 2 \cdot 3\left\{[AB_0]^2 + [AB_1]^2 + [AB_2]^2\right\} \tag{10.39}$$

and

$$E\left\{\mathrm{SS}(AB^2)\right\} = 2(\sigma_e^2 + 3\sigma_\beta^2) + 2 \cdot 3\left\{[AB_0^2]^2 + [AB_1^2]^2 + [AB_2^2]^2\right\} \tag{10.40}$$

Hence, using (10.38), (10.39), and (10.40),

$$E\left\{\mathrm{MS}(A \times B)\right\} = \sigma_e^2 + \frac{3}{2}\sigma_\beta^2 + \frac{3}{2}\left\{\sum_i [AB_i]^2 + \sum_i [AB_i^2]^2\right\} \tag{10.41}$$

Under H_0 (10.41) reduces to $\sigma_e^2 + \frac{3}{2}\sigma_\beta^2$. We therefore need to obtain an estimator for this quantity that will then provide the denominator for the F ratio to test H_0. Using (10.37) and (10.32) it is easy to see that

$$\widehat{\sigma}_e^2 + \frac{3}{2}\widehat{\sigma}_\beta^2 = \left(1 - \frac{1}{2\Delta}\right)\mathrm{MS}(\boldsymbol{I}|\mathfrak{I}, \boldsymbol{X}_\rho, \boldsymbol{X}_{\beta^*}, \boldsymbol{X}_{\tau^*})$$

$$+ 1 - \frac{1}{2\Delta}\mathrm{MS}(\boldsymbol{X}_{\beta^*}|\mathfrak{I}, \boldsymbol{X}_\rho, \boldsymbol{X}_{\tau^*})$$

where Δ is obtained from (10.36), recognizing that $u(\gamma_\ell) = 0$ since $\mathcal{E}_2 = \phi$, $\nu = 2$, $s = 2$, $q = 1$, $n = 2$, $p = 1$, $n_1 = 1$. Thus, $\Delta = 1$ and

$$\widehat{\sigma}_e^2 + \tfrac{3}{2}\widehat{\sigma}_\beta^2 = \tfrac{1}{2}\,\mathrm{MS}(\boldsymbol{I}|\mathfrak{I}, \boldsymbol{X}_\rho, \boldsymbol{X}_{\beta^*}, \boldsymbol{X}_{\tau^*}) + \tfrac{1}{2}\,\mathrm{MS}(\boldsymbol{X}_{\beta^*}|\mathfrak{I}, \boldsymbol{X}_\rho, \boldsymbol{X}_{\tau^*})$$

$$= \tfrac{1}{2}\,(.8333 + 15.0556) = 7.95$$

using results from the ANOVA table. Finally,

$$F = \frac{\mathrm{MS}(A \times B)}{\widehat{\sigma}_e^2 + \frac{3}{2}\widehat{\sigma}_\beta^2} = \frac{116.06}{7.95} = 14.60$$

This value is comparable to the corresponding value 13.03 obtained with PROC MIXED, which uses REML to estimate the variance components σ_e^2 and σ_β^2.

The degrees of freedom for the denominator of the F ratio are obtained by Satterthwaite's procedure as

$$\text{d.f.} = \frac{\left[(.8333 + 15.0556)/2\right]^2}{\left(\frac{.8333}{2}\right)^2 \Big/ 6 + \left(\frac{15.0556}{2}\right)^2 \Big/ 2} = 2.22$$

which is comparable to Den DF = 2 in the PROC MIXED output.

CHAPTER 11

General Symmetrical Factorial Design

11.1 INTRODUCTION

In the preceding chapters we have discussed at great length the design and analysis of factorial experiments with two and three qualitative levels. In both cases the development is based upon orthogonal partitions of the complete set of treatment combinations and comparisons among the resulting subsets. These partitions are based on solving certain equations or sets of equations using only elements from the (mathematical) field of residue classes mod 2 and mod 3, respectively and elementary facts about ordinary arithmetic mod 2 and mod 3, respectively. The question then is whether this can be extended.

Consider, say, arithmetic mod 4. Addition with an additive identity and multiplication with a multiplicative identity are easily defined, for example,

$$2 + 3 = 1$$

and

$$2 \times 3 = 2$$

However, this is not enough. We want to set up families of hyperplanes defined by, for example,

$$x_1 + 2x_2 = 0, 1, 2, 3$$

and in order to achieve orthogonal partitions of the 4^n treatment combinations, we need the result that the equation in the unknown x

$$ax = b \qquad (a \neq 0)$$

Design and Analysis of Experiments. Volume 2: Advanced Experimental Design
By Klaus Hinkelmann and Oscar Kempthorne
ISBN 0-471-55177-5

has a unique solution. We see immediately that arithmetic mod 4 does not have this property; for example,

$$2x = 2$$

is satisfied by $x = 1$ and $x = 3$ and

$$2x = 1$$

is not satisfied by any x. So we see that the simple arithmetic mod k does not extend immediately to any $k > 3$. It does so, however, if k is a prime number, p say.

The addition and multiplication properties mod p of the set $S = \{0, 1, 2, \ldots, p - 1\}$ are obvious. It only remains to show that

$$ax = b \qquad (a \neq 0)$$

has a unique solution mod p. Suppose we have two solutions x_1, x_2 with $x_1 \neq x_2$. Then

$$a(x_1 - x_2) = 0$$

Since $x_1 - x_2 = d$, say, $\in S$, we have

$$ad = 0$$

but

$$ad = 0 \longleftrightarrow ad = cp$$

with $c \in S$. Since neither a nor d are divisible by p, this relationship cannot hold unless $d = 0$, and hence $x_1 = x_2$. So, division is unique, and hence S forms a field. We denote this field of p elements and arithmetic mod p by GF(p). Further generalizations can be made to $k = p^m$, a prime power, using the Galois field GF(p^m) (see Appendix A).

In this chapter we shall discuss the p^n factorial as a generalization of the 2^n and 3^n factorials and indicate how this, in turn, can be generalized to the $s^n = (p^m)^n$ factorial. Much of the development in this chapter is due to Bose (1947b), Bose and Kishen (1940), Fisher (1942, 1945), Kempthorne (1947, 1952), Rao (1946a, 1947b), and Yates (1937b).

11.2 REPRESENTATION OF EFFECTS AND INTERACTIONS

We now have n factors $A_1, A_2, \ldots, A_n$ each at p levels; hence we have p^n treatment combinations and $p^n - 1$ d.f. among treatments. A treatment combination is denoted by the n-tuple $\boldsymbol{x}' = (x_1, x_2, \ldots, x_n)$ with $x_i \in \text{GF}(p)$. We then consider a partition $\boldsymbol{\alpha}' = (\alpha_1, \alpha_2, \ldots, \alpha_n)$ with $\alpha_i \in \text{GF}(p)$, but not all $\alpha_i = 0$, which partitions the set of p^n treatment combinations into p sets of p^{n-1} treatment combinations each through the equations

$$\alpha_1 x_1 + \alpha_2 x_2 + \cdots + \alpha_n x_n = \delta \tag{11.1}$$

where δ takes on all values in GF(p). That this is true can be seen as follows: Suppose for fixed $\boldsymbol{\alpha}$ and δ we have $\alpha_i \neq 0$ for some i. We can then choose all $x_j (j \neq i)$ freely ($x_j = x_j^*$ say) and obtain, from (11.1),

$$\alpha_i x_i = \delta - \sum_{j \neq i} \alpha_j x_j^* = \delta^*, \quad \text{say}$$

Because of uniqueness of division in GF(p), x_i is then determined uniquely. Since each $x_j (j \neq i)$ can take on p different values x_j^*, we have p^{n-1} solutions to (11.1). Comparisons among the p sets of treatment combinations generated by (11.1) define the $p - 1$ d.f. associated with the effect or interaction $E^{\boldsymbol{\alpha}} = A_1^{\alpha_1} A_2^{\alpha_2} \cdots A_n^{\alpha_n}$.

Let $S(\boldsymbol{x}; \boldsymbol{\alpha}, \delta)$ denote the set of treatment combinations $\boldsymbol{x}' = (x_1, x_2, \ldots, x_n)$ satisfying (11.1). Then a contrast belonging to $E^{\boldsymbol{\alpha}}$ can be defined formally as $\sum_\delta c_\delta S(\boldsymbol{x}; \boldsymbol{\alpha}, \delta)$ with $\delta \in \text{GF}(p)$ and $\sum_\delta c_\delta = 0$.

Excluding $\boldsymbol{\alpha}' = (0, 0, \ldots, 0)$ there exist $p^n - 1$ partitions $\boldsymbol{\alpha}$. However, for any given $\boldsymbol{\alpha}$ the partitions $\boldsymbol{\alpha}, 2\boldsymbol{\alpha}, \ldots, (p-1)\boldsymbol{\alpha}$ are identical since they lead to the same equations of the form (11.1). Hence we have $(p^n - 1)/(p - 1)$ distinct partitions, each accounting for $p - 1$ d.f., and consequently, these $(p^n - 1)/(p - 1)$ partitions account for the $p^n - 1$ d.f. among treatments. In order to have a unique enumeration, we restrict the first nonzero α_i in a partition $\boldsymbol{\alpha}$ to be equal to 1. The partitions then define the main effects, 2-factor interactions, …, n-factor interaction, designated in general by $E^{\boldsymbol{\alpha}} = A_1^{\alpha_1} A_2^{\alpha_2} \cdots A_n^{\alpha_n}$ with the convention that a letter A_i with $\alpha_i = 0$ is dropped from the expression.

As a consequence of partitioning the totality of $p^n - 1$ d.f. into $(p^n - 1)/(p - 1)$ sets of $p - 1$ d.f. each, a k-factor interaction, for example, $A_1 \times A_2 \times \cdots \times A_k$, consists of $(p-1)^{k-1}$ components denoted by, for example, $A_1 A_2^{\alpha_2} \cdots A_k^{\alpha_k}$, where $\alpha_2, \alpha_3, \ldots, \alpha_k$ take on all nonzero values in GF(p). As an example, in a 5^n design the 2-factor interaction $A \times B$ consists of AB, AB^2, AB^3, AB^4.

We also note that for two distinct partitions $\boldsymbol{\alpha}' = (\alpha_1, \alpha_2, \ldots, \alpha_n)$ and $\boldsymbol{\beta}' = (\beta_1, \beta_2, \ldots, \beta_n)$ the equations

$$\alpha_1 x_1 + \alpha_2 x_2 + \cdots + \alpha_n x_n = \delta_1 \tag{11.2}$$

and

$$\beta_1 x_1 + \beta_2 x_2 + \cdots + \beta_n x_n = \delta_2 \tag{11.3}$$

are satisfied by p^{n-2} treatment combinations. That this is so follows from the theory of linear equations: Because the partitions $\boldsymbol{\alpha}$ and $\boldsymbol{\beta}$ are distinct, we can find $\alpha_i, \alpha_j, \beta_i, \beta_j$ such that

$$\begin{vmatrix} \alpha_i & \alpha_j \\ \beta_i & \beta_j \end{vmatrix} \neq 0$$

We can then choose all $x_k (k \neq i, j)$ freely, say $x_k = x_k^*$. Then (11.2) and (11.3) reduce to

$$\alpha_i x_i + \alpha_j x_j = \delta_1 - \sum_{k \neq i,j} \alpha_k x_k^* = \delta_1^*$$

$$\beta_i x_i + \beta_j x_j = \delta_2 - \sum_{k \neq i,j} \beta_k x_k^* = \delta_2^*$$

These equations have a unique solution in x_i and x_j, and since each x_k can take on p different values x_k^*, we have p^{n-2} different solutions to (11.2) and (11.3). This is true for all $\delta_2 = 0, 1, \ldots, p-1$ and δ_1 fixed, which implies that the p^{n-1} treatment combinations satisfying (11.2) can be divided into p distinct sets of p^{n-2} treatment combinations each satisfying one of the p equations (11.3) with $\delta_2 = 0, 1, \ldots, p-1$. Any contrast belonging to $E^{\boldsymbol{\alpha}}$ is therefore orthogonal to any contrast belonging to $E^{\boldsymbol{\beta}}$. It is in this sense then that the partitions $\boldsymbol{\alpha}$ and $\boldsymbol{\beta}$, and hence the interactions $E^{\boldsymbol{\alpha}}$ and $E^{\boldsymbol{\beta}}$, are orthogonal.

11.3 GENERALIZED INTERACTIONS

Any treatment combination $\boldsymbol{x}$ that satisfies the equations (11.2) and (11.3) also satisfies the equation

$$(\alpha_1 + \beta_1)x_1 + (\alpha_2 + \beta_2)x_2 + \cdots + (\alpha_n + \beta_n)x_n = \delta_1 + \delta_2 \tag{11.4}$$

As δ_1 and δ_2 take on all values in GF(p), $\delta_1 + \delta_2$ takes on each value in GF(p) exactly p times. Combining all $\boldsymbol{x}$ that satisfy (11.4) for a particular value of $\delta_1 + \delta_2$, we have partitioned the treatment combinations according to

$$(\alpha_1 + \beta_1)x_1 + (\alpha_2 + \beta_2)x_2 + \cdots + (\alpha_n + \beta_n)x_n = \delta$$

with $\delta \in$ GF(p). This is, of course, the partition $\boldsymbol{\alpha} + \boldsymbol{\beta}$, and the corresponding interaction $E^{\boldsymbol{\alpha}+\boldsymbol{\beta}}$ is, in conformity with previous usage, a GI of $E^{\boldsymbol{\alpha}}$ and $E^{\boldsymbol{\beta}}$.

We can make the same statements more generally about

$$(\alpha_1 + \lambda\beta_1)x_1 + (\alpha_2 + \lambda\beta_2)x_2 + \cdots + (\alpha_n + \lambda\beta_n)x_n = \delta_1 + \lambda\delta_2$$

with $\lambda \in \mathrm{GF}(p)$, $\lambda \neq 0$. This then leads to partitions $\boldsymbol{\alpha} + \lambda\boldsymbol{\beta}$ and hence to the totality of $p - 1$ GIs $E^{\alpha+\lambda\beta}$ of E^{α} and E^{β}. These GIs are defined by comparisons among the sets of treatment combinations satisfying

$$(\alpha_1 + \lambda\beta_1)x_1 + (\alpha_2 + \lambda\beta_2)x_2 + \cdots + (\alpha_n + \lambda\beta_n)x_n = \delta \qquad (11.5)$$

with $\delta \in \mathrm{GF}(p)$ and each $\lambda \in \mathrm{GF}(p)$, $\lambda \neq 0$.

Formally the GIs are obtained by multiplying the corresponding letters raised to certain powers into each other, reducing the powers mod p and modifying the powers (if necessary) such that the first letter included appears with power unity by multiplying every power by the same appropriate value in GF(p). For example, in a 5^3 factorial the GIs of AB and AC^2 are

$$\begin{aligned}
(AB) \times (AC^2) &= A^2BC^2 = (A^2BC^2)^3 = AB^3C \\
(AB) \times (AC^2)^2 &= A^3BC = (A^3BC)^2 = AB^2C^2 \\
(AB) \times (AC^2)^3 &= A^4BC = (A^4BC)^4 = AB^4C^4 \\
(AB) \times (AC^2)^4 &= BC^3
\end{aligned}$$

More generally we have the following theorem.

Theorem 11.1 The total number of GIs among q interactions $E^{\alpha_1}, E^{\alpha_2}, \ldots, E^{\alpha_q}$ is given by

$$\phi(q, p) = \sum_{j=2}^{q} \binom{q}{j} (p-1)^{j-1} \qquad (11.6)$$

Proof The q interactions are defined by the q sets of equations

$$\alpha_{j1}x_1 + \alpha_{j2}x_2 + \cdots + \alpha_{jn}x_n = \delta_j$$

with $\delta_j \in \mathrm{GF}(p)$, $j = 1, 2, \ldots, q$, or for short

$$\boldsymbol{\alpha}_j'\mathbf{x} = \delta_j \qquad (j = 1, 2, \ldots, q) \qquad (11.7)$$

Any treatment combination $\boldsymbol{x}$ that satisfies (11.7) for a given set $\delta_1, \delta_2, \ldots, \delta_q$ also satisfies any of the equations

$$\sum_{j=1}^{q} \lambda_j \boldsymbol{\alpha}_j'\boldsymbol{x} = \sum_{j=1}^{q} \lambda_j \delta_j \qquad (11.8)$$

with $\lambda_j \in \text{GF}(p)$ except $(\lambda_1, \lambda_2, \ldots, \lambda_q) = (0, 0, \ldots, 0)$. We also exclude any $\boldsymbol{\lambda}' = (\lambda_1, \lambda_2, \ldots, \lambda_q)$ with only one nonzero λ_j since such equations define the q interactions themselves. Given then a particular $\boldsymbol{\lambda}$, if we let the δ_j associated with the nonzero components of $\boldsymbol{\lambda}$ take on all values in GF(p), we obtain the partition $\boldsymbol{\lambda}'\boldsymbol{\alpha}$ that defines the interaction $E^{\boldsymbol{\lambda}'\boldsymbol{\alpha}}$, which by definition is a GI of $E^{\boldsymbol{\alpha}_1}, E^{\boldsymbol{\alpha}_2}, \ldots, E^{\boldsymbol{\alpha}_q}$. To have a unique enumeration of all possible $\boldsymbol{\lambda}$ and hence all possible GIs, we restrict the first nonzero component in $\boldsymbol{\lambda}$ to unity. All other components then take on all possible values in GF(p). Hence the possible number of $\boldsymbol{\lambda}$ vectors is

$$\sum_{j=2}^{q} \binom{q}{j} (p-1)^{j-1}$$

which is obtained by considering all admissible $\boldsymbol{\lambda}$ with $2, 3, \ldots, q$ nonzero components, respectively. □

11.4 SYSTEMS OF CONFOUNDING

With a p^n factorial and blocks of equal size the only block sizes are $k = p^\ell (\ell \leq n)$. If $\ell < n$, we need to confound certain interactions with blocks. More precisely, for a p^n factorial in blocks of size p^ℓ we have $p^{n-\ell}$ blocks and hence we must confound $(p^{n-\ell} - 1)/(p-1)$ interactions with blocks. To find such a system of confounding, we first state the following theorem.

Theorem 11.2 If in a p^n factorial with equal block sizes $p^\ell (\leq p^{n-2})$ two interactions $E^{\boldsymbol{\alpha}}$ and $E^{\boldsymbol{\beta}}$ are confounded with blocks, then so are their GIs $E^{\boldsymbol{\alpha}+\lambda\boldsymbol{\beta}}$, $\lambda \in \text{GF}(p)$, $\lambda \neq 0$.

Proof Consider the equations associated with $E^{\boldsymbol{\alpha}}$ and $E^{\boldsymbol{\beta}}$:

$$\begin{aligned} \alpha_1 x_1 + \alpha_2 x_2 + \cdots + \alpha_n x_n &= \delta_1 \\ \beta_1 x_1 + \beta_2 x_2 + \cdots + \beta_n x_n &= \delta_2 \end{aligned} \qquad (11.9)$$

For any pair (δ_1, δ_2) the equations (11.9) are satisfied by a set of p^{n-2} treatment combinations $\boldsymbol{x}' = (x_1, x_2, \ldots, x_n)$ denoted by $S(\boldsymbol{x}; \boldsymbol{\alpha}, \delta_1; \boldsymbol{\beta}, \delta_2)$ say. Each set $S(\boldsymbol{x}; \boldsymbol{\alpha}, \delta_1; \boldsymbol{\beta}, \delta_2)$ makes up one or several blocks. A contrast belonging to $E^{\boldsymbol{\alpha}}$ is given by

$$\sum_{\delta_1} \sum_{\delta_2} c_{\delta_1} \; S(\boldsymbol{x}; \boldsymbol{\alpha}, \delta_1; \boldsymbol{\beta}, \delta_2) \qquad (11.10)$$

and a contrast belonging to $E^{\boldsymbol{\beta}}$ by

$$\sum_{\delta_2}\sum_{\delta_1} d_{\delta_2}\ S(\boldsymbol{x};\boldsymbol{\alpha},\delta_1;\boldsymbol{\beta},\delta_2) \tag{11.11}$$

with $\sum_{\delta_1} c_{\delta_1} = \sum_{\delta_2} d_{\delta_2} = 0$. The contrasts (11.10) and (11.11), when evaluated in terms of treatment and block effects, are linear functions not only of treatment effects but also of block effects as each set $S(\boldsymbol{x};\ \boldsymbol{\alpha},\delta_1;\ \boldsymbol{\beta},\delta_2)$ contains a sum of block effects multiplied by a common constant, c_{δ_1} or c_{δ_2}. Now, for some $\lambda \in \mathrm{GF}(p)$,

$$S(\boldsymbol{x};\boldsymbol{\alpha}+\lambda\boldsymbol{\beta},\delta) = \sum_{\substack{\delta_1,\delta_2\\ \delta_1+\lambda\delta_2=\delta}} S(\boldsymbol{x};\boldsymbol{\alpha},\delta_1;\boldsymbol{\beta},\delta_2) \tag{11.12}$$

Hence a contrast belonging to $E^{\boldsymbol{\alpha}+\lambda\boldsymbol{\beta}}$ can be expressed as

$$\sum_{\delta} a_\delta\ S(\boldsymbol{x};\boldsymbol{\alpha}+\lambda\boldsymbol{\beta},\delta) = \sum_{\delta} a_\delta \left[\sum_{\substack{\delta_1,\delta_2\\ \delta_1+\lambda\delta_2=\delta}} S(\boldsymbol{x};\boldsymbol{\alpha},\delta_1;\boldsymbol{\beta},\delta_2)\right] \tag{11.13}$$

with $\sum_\delta a_\delta = 0$. Since (11.13) is of the same form as (11.10) and (11.11), that is, a linear combination of the same $S(\boldsymbol{x};\boldsymbol{\alpha},\delta_1;\boldsymbol{\beta},\delta_2)$, it is also a function of block effects, and hence $E^{\boldsymbol{\alpha}+\lambda\boldsymbol{\beta}}$ is also confounded with blocks. This holds for all $\lambda \in \mathrm{GF}(p)$, hence for all GIs of $E^{\boldsymbol{\alpha}}$ and $E^{\boldsymbol{\beta}}$. □

We now consider specifically a system of confounding of a p^n factorial in blocks of size p^ℓ.

Theorem 11.3 A system of confounding for a p^n factorial in blocks of size p^ℓ is completely determined by $q = n - \ell$ independent interactions $E^{\boldsymbol{\alpha}_1}, E^{\boldsymbol{\alpha}_2}, \ldots, E^{\boldsymbol{\alpha}_q}$.

Proof Since with any two interactions their GIs are also confounded with blocks, it follows immediately that with q interactions all their GIs are also confounded with blocks. By Theorem 11.1, the total number of GIs is $\phi(q, p)$ as given in (11.6). Hence together with the q independent interactions we confound

$$q + \phi(q,p) = \sum_{j=1}^{q} \binom{q}{j} (p-1)^{j-1}$$

interactions. Each interaction carries $p - 1$ d.f. The total number of degrees of freedom confounded with blocks is then

$$\sum_{j=1}^{q} \binom{q}{j} (p-1)^j = p^q - 1 = p^{n-\ell} - 1$$

This is the number of degrees of freedom among the $p^{n-\ell}$ blocks of size p^ℓ. □

Once the $n - \ell$ independent interactions have been chosen, it remains to obtain the composition of the blocks. One way to do this, obviously, is to consider the $q = n - \ell$ equations associated with $E^{\alpha_1}, E^{\alpha_2}, \ldots, E^{\alpha_q}$:

$$\begin{aligned} \alpha_{11}x_1 + \alpha_{12}x_2 + \cdots + \alpha_{1n}x_n &= \delta_1 \\ \alpha_{21}x_1 + \alpha_{22}x_2 + \cdots + \alpha_{2n}x_n &= \delta_2 \\ &\vdots \\ \alpha_{q1}x_1 + \alpha_{q2}x_2 + \cdots + \alpha_{qn}x_n &= \delta_q \end{aligned} \tag{11.14}$$

for all possible $p^{n-\ell}$ right-hand sides. For fixed $(\delta_1, \delta_2, \ldots, \delta_q)$ the equations (11.14) are satisfied by p^ℓ treatment combinations. Hence for all possible choices of $(\delta_1, \delta_2, \ldots, \delta_q)$ we obtain $p^q = p^{n-\ell}$ distinct sets of treatment combinations that then form the blocks.

11.5 INTRABLOCK SUBGROUP

Another way we have already used with the 2^n and 3^n factorial is to first obtain the IBSG and then use it to generate the remaining blocks. Denoting the equations (11.14) with $\delta_i = 0\,(i = 1, 2, \ldots, q)$ by (11.14′), we first prove the following theorem.

Theorem 11.4 The treatment combinations $\boldsymbol{x}' = (x_1, x_2, \ldots, x_n)$ satisfying the equations (11.14′) form a group, the IBSG, of size p^ℓ with componentwise addition as the operation among elements.

Proof (i) The solutions to (11.14′) contain the additive unit element $(0, 0, \ldots, 0)$; (ii) if $(x_1, x_2, \ldots, x_n)$ satisfies (11.14′), then $(\lambda x_1, \lambda x_2, \ldots, \lambda x_n)$ with $\lambda \in \mathrm{GF}(p)$ also satisfies (11.14′); (iii) if $(x_1, x_2, \ldots, x_n)$ and $(y_1, y_2, \ldots, y_n)$ satisfy (11.14′), then $(x_1 + y_1, x_2 + y_2, \ldots, x_n + y_n)$ satisfies (11.14′). □

Suppose now that $\boldsymbol{x}'$ is in the IBSG and that $\boldsymbol{z}' = (z_1, z_2, \ldots, z_n)$ is a treatment combination not belonging to the IBSG. Then the set $\{\boldsymbol{x}' + \boldsymbol{z}' : \boldsymbol{x}' \in \mathrm{IBSG}\}$ is a set of p^ℓ treatment combinations satisfying Eq. (11.14) for some vector

$(\delta_1, \delta_2, \ldots, \delta_q)$ other than $(0, 0, \ldots, 0)$. Hence these treatment combinations form a block. We shall refer to z' as the generator of this block. This process can be continued by choosing successively altogether $p^{n-\ell}$ such generators. These generators can obviously be found simply by inspection after another block has been constructed in this way. In certain situations it is possible, however, to give a complete characterization of the generators $z_1, z_2, \ldots, z_w$ $(w = p^{n-\ell} - 1)$, say, where $z_i' = (z_{i1}, z_{i2}, \ldots, z_{in})$.

Consider a particular system of confounding for a p^n factorial in blocks of size p^ℓ. Let $\boldsymbol{\alpha}_1, \boldsymbol{\alpha}_2, \ldots, \boldsymbol{\alpha}_q, \boldsymbol{\alpha}_{q+1}, \ldots, \boldsymbol{\alpha}_v$ be the partitions corresponding to the confounded interactions, where $v = (p^{n-\ell} - 1)/(p - 1)$ and $\boldsymbol{\alpha}_1, \boldsymbol{\alpha}_2, \ldots, \boldsymbol{\alpha}_q (q = n - \ell)$ represent the independent interactions. We then consider the set

$$\{\lambda \boldsymbol{\alpha}_i \mod p\colon i = 1, 2, \ldots, v;\ \lambda \in \mathrm{GF}(p),\ \lambda \neq 0\}$$

which we denote by

$$G\colon \{\boldsymbol{\alpha}_j;\ j = 1, \ldots, p^{n-\ell} - 1\}$$

We note that if we adjoin to G the element $\boldsymbol{\alpha}_0' = (0, 0, \ldots, 0)$ we have a group of $p^{n-\ell}$ elements of order $n - \ell$ with $\boldsymbol{\alpha}_1, \boldsymbol{\alpha}_2, \ldots, \boldsymbol{\alpha}_q$ as the generators. Denote this group by G^*.

We now state the following theorem.

Theorem 11.5 For a given system of confounding for a p^n factorial in blocks of size p^ℓ the composition of the blocks is obtained by using the IBSG together with the generators $z_j = \boldsymbol{\alpha}_j$ with $\boldsymbol{\alpha}_j \in G$ provided the $\boldsymbol{\alpha}_j$ are not contained in the IBSG.

Proof We have to show that each $z_j = \boldsymbol{\alpha}_j$ satisfies a different set of equations (11.14), that is, with different right-hand sides.

Suppose that $\boldsymbol{\alpha}_j$ and $\boldsymbol{\alpha}_{j'} (j \neq j')$ satisfy the same equations. It follows then from (11.14) that

$$\boldsymbol{\alpha}_1'(\boldsymbol{\alpha}_j - \boldsymbol{\alpha}_{j'}) = 0$$

$$\boldsymbol{\alpha}_2'(\boldsymbol{\alpha}_j - \boldsymbol{\alpha}_{j'}) = 0$$

$$\vdots$$

$$\boldsymbol{\alpha}_q'(\boldsymbol{\alpha}_j - \boldsymbol{\alpha}_{j'}) = 0$$

that is, $\boldsymbol{\alpha}_j - \boldsymbol{\alpha}_{j'}$ satisfies (11.14)$'$. Now $\boldsymbol{\alpha}_j - \boldsymbol{\alpha}_{j'} \in G^*$ and, by assumption, the only element in G^* that satisfies (11.14)$'$ is $\boldsymbol{\alpha}_0$. Hence $\boldsymbol{\alpha}_j = \boldsymbol{\alpha}_{j'}$. □

Example 11.1 Consider a 3^5 experiment in blocks of size 3^3. Suppose $ABCD$, $BCDE^2$, and hence $AB^2C^2D^2E^2$ and AE are confounded with blocks. The IBSG is determined by

$$x_1 + x_2 + x_3 + x_4 = 0$$

$$x_2 + x_3 + x_4 + 2x_5 = 0$$

and it can be seen that $\boldsymbol{\alpha}_1' = (1, 1, 1, 1, 0)$, $\boldsymbol{\alpha}_2' = (0, 1, 1, 1, 2)$, $\boldsymbol{\alpha}_3' = (1, 2, 2, 2, 2)$, $\boldsymbol{\alpha}_4' = (1, 0, 0, 0, 1)$ do not satisfy these equations. Hence the generators of the eight blocks besides the IBSG are $\boldsymbol{\alpha}_1', \boldsymbol{\alpha}_2', \boldsymbol{\alpha}_3', \boldsymbol{\alpha}_4'$, $\boldsymbol{\alpha}_5' = 2\boldsymbol{\alpha}_1' = (2, 2, 2, 2, 0)$, $\boldsymbol{\alpha}_6' = 2\boldsymbol{\alpha}_2' = (0, 2, 2, 2, 1,)$, $\boldsymbol{\alpha}_7' = 2\boldsymbol{\alpha}_3' = (2, 1, 1, 1, 1)$, $\boldsymbol{\alpha}_8' = 2\boldsymbol{\alpha}_4' = (2, 0, 0, 0, 2)$. □

11.6 ENUMERATING SYSTEMS OF CONFOUNDING

Concerning the number of possible systems of confounding, we state the following theorem.

Theorem 11.6 The total number of systems of confounding of a p^n factorial in blocks of size p^ℓ is

$$\frac{(p^n - 1)(p^n - p)\cdots(p^n - p^{n-\ell-1})}{(p^{n-\ell} - 1)(p^{n-\ell} - p)\cdots(p^{n-\ell} - p^{n-\ell-1})} \tag{11.15}$$

Proof It follows from Theorem 11.3 that the total number of systems of confounding is the same as the number of distinct sets of $n - \ell$ independent interactions. This number can be obtained as follows: The first interaction can be chosen in $(p^n - 1)/(p - 1)$ ways; the second in $(p^n - p)/(p - 1)$ ways; the third in

$$\frac{p^n - p}{p - 1} - 1 - \phi(2, p) = \frac{p^n - p^2}{p - 1}$$

ways, with $\phi(q, p)$ as defined in (10.6); the fourth in

$$\frac{p^n - p^2}{p - 1} - 1 - \phi(3, p) + \phi(2, p) = \frac{p^n - p^3}{p - 1}$$

ways; ...; the $(k + 1) - th(k \geq 3)$ in

$$\frac{p^n - p^{k-1}}{p - 1} - 1 - \phi(k, p) + \phi(k - 1, p) \tag{11.16}$$

ways, and since

$$\phi(k, p) - \phi(k-1, p) = p^{k-1} - 1$$

(11.16) becomes $(p^n - p^k)/(p-1)$. Hence the total number of sets generated in this way is

$$\left(\frac{p^n - 1}{p-1}\right)\left(\frac{p^n - p}{p-1}\right)\left(\frac{p^n - p^2}{p-1}\right)\cdots\left(\frac{p^n - p^{n-\ell-1}}{p-1}\right) \tag{11.17}$$

Any one set will, however, have been enumerated in

$$\left(\frac{p^{n-\ell} - 1}{p-1}\right)\left(\frac{p^{n-\ell} - p}{p-1}\right)\cdots\left(\frac{p^{n-\ell} - p^{n-\ell-1}}{p-1}\right) \tag{11.18}$$

different ways, which can be derived in the same way as (11.17). Dividing (11.17) by (11.18) gives the stated result (11.15). □

11.7 FISHER PLANS

In many cases (11.15) may be too large to actually write out all possible systems of confounding. A question of practical interest then is whether there exists a certain type of confounding and how one can find it if it exists. In particular, one may be interested to know whether there exists a system of confounding such that main effects and low-factor interactions remain unconfounded.

11.7.1 Existence and Construction

A first and important result in this direction is the following theorem.

Theorem 11.7 (Fisher, 1942, 1945) The maximum number of factors with p levels that can be accommodated in a design with blocks of size $p^\ell (\ell \geq 2)$ such that main effects and two-factor interactions remain unconfounded is

$$n = \frac{p^\ell - 1}{p - 1} \tag{11.19}$$

Proof The proof is based on the following correspondence between the $n = (p^\ell - 1)/(p-1)$ factors $F_1, F_2, \ldots, F_n$ each at p levels and the $(p^\ell - 1)/$

$(p-1)$ effects and interactions of a p^ℓ factorial with factors $A_1, A_2, \ldots, A_\ell$:

$$\begin{aligned} F_1 &= A_1 \\ F_2 &= A_2 \\ F_3 &= A_1 A_2 \\ F_4 &= A_1 A_2^2 \\ &\vdots \\ F_n &= A_1 A_2^{p-1} \cdots A_\ell^{p-1} \end{aligned}$$

or, if we label the factors F by $\boldsymbol{\alpha}$, where $\boldsymbol{\alpha}' = (\alpha_1, \alpha_2, \ldots, \alpha_\ell)$ designates a partition in the p^ℓ system,

$$F_{\boldsymbol{\alpha}} = A_1^{\alpha_1} A_2^{\alpha_2} \cdots A_\ell^{\alpha_\ell}$$

with $\alpha_i \in \mathrm{GF}(p) (i = 1, 2, \ldots, \ell)$, not all $\alpha_i = 0$ and the first nonzero α_i equal to unity. There are, of course, n such $\boldsymbol{\alpha}$'s and we denote the set of these $\boldsymbol{\alpha}$'s by $\mathcal{A}$. We then create the following two-way table: The rows are labeled by the p^ℓ factorial combinations $\boldsymbol{x}' = (x_1, x_2, \ldots, x_\ell)$ and the columns by the $(p^\ell - 1)/(p-1)$ interactions $A_1^{\alpha_1} A_2^{\alpha_2} \cdots A_\ell^{\alpha_\ell}$, that is, by the $F_{\boldsymbol{\alpha}}$. The table entries thus are the inner products $\boldsymbol{\alpha}'\boldsymbol{x}$ mod p [we note that these are the subscripts in the representation of $\tau(\boldsymbol{x})$ in terms of main effects and interactions]. Each row in this table then represents a treatment combination in the n factors $F_1, F_2, \ldots, F_n$, the level of $F_{\boldsymbol{\alpha}}$ in the treatment combination $z(\boldsymbol{x})$ corresponding to row $\boldsymbol{x}$ being $\boldsymbol{\alpha}'\boldsymbol{x}$. We thus have p^ℓ treatment combinations for n factors. These treatment combinations constitute the IBSG for the system of confounding to be considered. That this is so can be seen as follows: (i) The operation in this set is componentwise addition mod p; (ii) the set contains the addition identity; (iii) if the set contains

$$z(\boldsymbol{x}) = \left(x_1, x_2, x_1 + x_2, \ldots, x_1 + (p-1) \sum_{i=2}^{\ell} x_i \right)$$

then it also contains $\lambda \boldsymbol{x}$, $\lambda \in \mathrm{GF}(p)$, since with the p^ℓ treatment combination $\boldsymbol{x}$ we also have the p^ℓ treatment combination $\lambda \boldsymbol{x}$ because the set of p^ℓ treatment combinations forms a group; and (iv) if the set contains

$$z(\boldsymbol{x}) = \left(x_1, x_2, x_1 + x_2, \ldots, x_1 + (p-1) \sum_{i=2}^{\ell} x_i \right)$$

and

$$z(\boldsymbol{y}) = \left(y_1, y_2, y_1 + y_2, \ldots, y_1 + (p-1) \sum_{i=2}^{\ell} y_i \right)$$

then it also contains $z(x) + z(y)$ because with x and y the p^ℓ treatment combinations also contain $x + y$.

To show that this IBSG leads to a system of confounding that does not confound any of the main effects F_i $(i = 1, 2, \ldots, n)$ and any of the 2-factor interactions $F_i F_j^\gamma$, $\gamma \in \text{FG}(p)$, $\gamma \neq 0, i, j = 1, 2, \ldots, n, i \neq j$, we have to show that in the IBSG (i) each level of each factor F_i occurs the same number of times and (ii) each level combination for any two factors F_i, F_j occurs the same number of times: (1) The level $z \in \text{GF}(p)$ of factor $F_{\boldsymbol{\alpha}}$ occurs in the IBSG whenever $\boldsymbol{\alpha}' \boldsymbol{x} = z$. Since at least one $\alpha_i \neq 0$ in $\boldsymbol{\alpha}$, this equation has $p^{\ell-1}$ solutions for every $z \in \text{GF}(p)$ and every $\boldsymbol{\alpha} \in \mathcal{A}$. (2) The level combination (z_1, z_2) of factors $F_{\boldsymbol{\alpha}}$ and $F_{\boldsymbol{\beta}}$ occurs whenever $\boldsymbol{\alpha}' \boldsymbol{x} = z_1$ and $\boldsymbol{\beta}' z = z_2$. As shown previously (Section 11.2), there exist exactly $p^{\ell-2}$ such p^ℓ treatment combinations $\boldsymbol{x}$ that satisfy these equations and that hold for every $\boldsymbol{\alpha}, \boldsymbol{\beta} \in \mathcal{A}$ and every $z_1, z_2\, \text{GF}(p)$. □

We illustrate this theorem by the following example.

Example 11.2 Let $p = 3$ and $\ell = 2$, that is, $n = 4$. We set up the table, letting $A_1 = A$, $A_2 = B$,

	F_1	F_2	F_3	F_4
	A	B	AB	AB^2
x'	$\alpha' = (1, 0)$	$(0, 1)$	$(1, 1)$	$(1, 2)$
0, 0	0	0	0	0
0, 1	0	1	1	2
0, 2	0	2	2	1
1, 0	1	0	1	1
1, 1	1	1	2	0
1, 2	1	2	0	2
2, 0	2	0	2	2
2, 1	2	1	0	1
2, 2	2	2	1	0

Each row represents a treatment combination (z_1, z_2, z_3, z_4) for the factors F_1, F_2, F_3, F_4. The nine treatment combinations represent the IBSG for a system of confounding for a 3^4 factorial in blocks of size 3^2. □

If we write the IBSG as a $n \times p^\ell$ array, $\boldsymbol{W}$ say, that is, each column representing a treatment combination, the proof of Theorem 11.7 leads immediately to the following corollary.

Corollary 11.1 The IBSG for a p^n factorial in blocks of size p^ℓ, written as a matrix $\boldsymbol{W}$, is an orthogonal array of size p^ℓ, with $n = (p^\ell - 1)/(p - 1)$

constraints, p symbols, of strength 2 and index $p^{\ell-2}$, that is,

$$\mathrm{OA}\left[p^\ell, \frac{p^\ell - 1}{p - 1}, p, 2; p^{\ell-2}\right]$$

Such an orthogonal array was called by Rao (1946a) a hypercube of strength 2. More generally this leads to the following definition.

Definition 11.1 Let there be n factors each at p levels. A subset consisting of p^ℓ of the possible p^n treatment combinations is said to be a hypercube of strength d if all combinations of any d of the n factors occur $p^{\ell-d}$ times. Such an arrangement is denoted by $[n, p, \ell, d]$.

Still another way of expressing the result of Theorem 11.7 is the following corollary. □

Corollary 11.2 For a p^n factorial experiment there exists a system of confounding with main effects and 2-factor interactions unconfounded if the block size k is such that

$$k > n(p - 1) \tag{11.20}$$

For $p = 2, 3$, (11.20) is, of course, the result alluded to in Sections 8.2.3 and 10.6, respectively.

11.7.2 Identifying System of Confounding

Having constructed a system of confounding using the IBSG of Theorem 11.7, the question may then be asked: Which of the interactions in the p^n factorial are confounded? Let $z' = (z_1, z_2, \ldots, z_n)$ denote a treatment combination of the p^n factorial with the factors $F_1, F_2, \ldots, F_n$. We then know that the interaction $F_1^{\gamma_1} F_2^{\gamma_2} \cdots F_n^{\gamma_n}$ [$\gamma_i \in \mathrm{GF}(p), i = 1, 2, \ldots, n$, not all $\gamma_i = 0$, the first nonzero γ_i equals unity] is confounded with blocks if and only if

$$\gamma_1 z_1 + \gamma_2 z_2 + \cdots + \gamma_n z_n = 0 \qquad \bmod p \tag{11.21}$$

for every z' in the IBSG. The IBSG contains p^ℓ elements that form an additive group of order ℓ, that is, the group is generated by ℓ treatment combinations $z'_1, z'_2, \ldots, z'_\ell$, say with $z'_i = (z_{i1}, z_{i2}, \ldots, z_{in}), i = 1, 2, \ldots, \ell$. The system (11.21) then reduces to

$$\begin{aligned}
\gamma_1 z_{11} + \gamma_2 z_{12} + \cdots + \gamma_n z_{1n} &= 0 \\
\gamma_1 z_{21} + \gamma_2 z_{22} + \cdots + \gamma_n z_{2n} &= 0 \\
&\vdots \\
\gamma_1 z_{\ell 1} + \gamma_2 z_{\ell 2} + \cdots + \gamma_n z_{\ell n} &= 0
\end{aligned} \tag{11.22}$$

Equations (11.22) in the unknowns $\gamma_1, \gamma_2, \ldots, \gamma_n$ have $p^{n-\ell}$ solutions. One of these solutions is $(0, 0, \ldots, 0)$ so that $p^{n-\ell} - 1$ nontrivial solutions remain. However, if $\boldsymbol{\gamma}' = (\gamma_1, \gamma_2, \ldots, \gamma_n)$ is a solution, then $\lambda\boldsymbol{\gamma}'$ with $\lambda \in \mathrm{GF}(p)$ is a solution also. Hence there are $c = (p^{n-\ell} - 1)/(p - 1)$ distinct solutions $\boldsymbol{\gamma}_1, \boldsymbol{\gamma}_2, \ldots, \boldsymbol{\gamma}_c$, say, and these then correspond to the confounded interactions $E^{\boldsymbol{\gamma}_1}, E^{\boldsymbol{\gamma}_2}, \ldots, E^{\boldsymbol{\gamma}_c}$ in the $F_1, F_2, \ldots, F_n$ system. Of course, of the c interactions only $n - \ell$ are independent interactions.

To summarize: The general idea is to (i) obtain a set of generators $z_1', z_2', \ldots, z_\ell'$ of the IBSG, (ii) obtain all solutions to (11.22), and (iii) eliminate all redundancies from these solutions to obtain c distinct solutions $\boldsymbol{\gamma}_1, \boldsymbol{\gamma}_2, \ldots, \boldsymbol{\gamma}_c$.

We note that this confounding detection method works, of course, for any system of confounding. This is just the formal statement of the procedure mentioned in Section 8.4. The method of constructing a confounded design according to Theorem 11.7 is just one case in point where the confounded interactions are not known a priori.

11.7.3 Application to Fisher Plans

Returning now to the case of Theorem 11.7, what can we say specifically about $\boldsymbol{\gamma}_1, \boldsymbol{\gamma}_2, \ldots, \boldsymbol{\gamma}_c$? An explicit and complete characterization of the confounded interactions can be given as follows.

Let us return to Eqs. (11.22). We note that the treatment combinations $z_1', z_2', \ldots, z_\ell'$ are generated by the corresponding treatment combinations $\boldsymbol{x}_j' = (x_{j1}, x_{j2}, \ldots, x_{j\ell})$ $(j = 1, 2, \ldots, \ell)$. Rewriting (11.22) as

$$
\begin{aligned}
\gamma_1\boldsymbol{\alpha}_1'\boldsymbol{x}_1 + \gamma_2\boldsymbol{\alpha}_2'\boldsymbol{x}_1 + \cdots + \gamma_n\boldsymbol{\alpha}_n'\boldsymbol{x}_1 &= 0 \\
\gamma_1\boldsymbol{\alpha}_1'\boldsymbol{x}_2 + \gamma_2\boldsymbol{\alpha}_2'\boldsymbol{x}_2 + \cdots + \gamma_n\boldsymbol{\alpha}_n'\boldsymbol{x}_2 &= 0 \\
&\vdots \\
\gamma_1\boldsymbol{\alpha}_1'\boldsymbol{x}_\ell + \gamma_2\boldsymbol{\alpha}_2'\boldsymbol{x}_\ell + \cdots + \gamma_n\boldsymbol{\alpha}_n'\boldsymbol{x}_\ell &= 0
\end{aligned}
\tag{11.23}
$$

and taking $(\boldsymbol{x}_1, \boldsymbol{x}_2, \ldots, \boldsymbol{x}_\ell) = \boldsymbol{I}_\ell$, we obtain from (11.23) the equations

$$
\begin{aligned}
\alpha_{11}\gamma_1 + \alpha_{21}\gamma_2 + \cdots + \alpha_{n1}\gamma_n &= 0 \\
\alpha_{12}\gamma_1 + \alpha_{22}\gamma_2 + \cdots + \alpha_{n2}\gamma_n &= 0 \\
&\vdots \\
\alpha_{1\ell}\gamma_1 + \alpha_{2\ell}\gamma_2 + \cdots + \alpha_{n\ell}\gamma_n &= 0
\end{aligned}
\tag{11.24}
$$

which is a set of ℓ equations in the n unknowns $\gamma_1, \gamma_2, \ldots, \gamma_n$, where α_{ji} is the ith component of $\boldsymbol{\alpha}_j (j = 1, 2, \ldots, n; i = 1, 2, \ldots, \ell)$.

To characterize the solutions to (11.24), we write the $\boldsymbol{\alpha}_j \in \mathcal{A}$ in what we shall refer to as standard order in that $\boldsymbol{\alpha}'_1, \boldsymbol{\alpha}'_2, \ldots, \boldsymbol{\alpha}'_\ell$ represent the ℓ main effects $A_1, A_2, \ldots, A_\ell$, respectively, that is, $\boldsymbol{\alpha}'_\nu$ is the νth row in the identity matrix $\boldsymbol{I}_\ell (\nu = 1, 2, \ldots, \ell)$. From (11.24) we then obtain the equations

$$\begin{bmatrix} \gamma_1 \\ \gamma_2 \\ \vdots \\ \gamma_\ell \end{bmatrix} = - \begin{bmatrix} \alpha_{\ell+1,1} & \alpha_{\ell+2,1} & \cdots & \alpha_{n1} \\ \alpha_{\ell+1,2} & \alpha_{\ell+2,2} & \cdots & \alpha_{n2} \\ & & \vdots & \\ \alpha_{\ell+1,\ell} & \alpha_{\ell+2,\ell} & \cdots & \alpha_{n\ell} \end{bmatrix} \cdot \begin{bmatrix} \gamma_{\ell+1} \\ \gamma_{\ell+2} \\ \vdots \\ \gamma_{\ell+n} \end{bmatrix} \tag{11.25}$$

Since $\gamma_{\ell+1}, \gamma_{\ell+2}, \ldots, \gamma_n$ can be chosen freely, we can obtain $n - \ell$ independent solutions to (11.25) and hence $n - \ell$ independent partitions $\boldsymbol{\Gamma} = \left\{\boldsymbol{\gamma}_j = (\gamma_{j1}, \gamma_{j2}, \ldots, \gamma_{jn});\ j = 1, 2, \ldots, n - \ell\right\}$ by taking successively for $(\gamma_{\ell+1}, \gamma_{\ell+2}, \ldots, \gamma_n)$ the rows of $\boldsymbol{I}_{n-\ell}$. If we denote the jth row of $\boldsymbol{I}_{n-\ell}$ by $\boldsymbol{e}'_j$, then (11.25) yields successively

$$\boldsymbol{\gamma}'_j = (-\boldsymbol{\alpha}'_{\ell+j}, \boldsymbol{e}'_j) \tag{11.26}$$

$j = 1, 2, \ldots, n - \ell$, or if we convert $-\boldsymbol{\alpha}'_{\ell+j}$ to positive components mod p, (11.26) becomes

$$\boldsymbol{\gamma}'_j = (\boldsymbol{\alpha}'_{\ell+j}, (p-1)\,\boldsymbol{e}'_j) \tag{11.27}$$

This is so since in each $\boldsymbol{\alpha}'_{\ell+j}$ the first nonzero component is unity and hence adding p and multiplying by $p - 1$ yield (11.27). From (11.27) we then have that the $n - \ell$ interactions

$$E^{\boldsymbol{\gamma}_j} = F_1^{\alpha_{j+\ell,1}}\, F_2^{\alpha_{j+\ell,2}} \cdots F_\ell^{\alpha_{j+\ell,\ell}}\, F_{\ell+j}^{p-1}$$

$(j = 1, 2, \ldots, n - \ell)$ are confounded with blocks and so are their GIs, accounting for all $p^{n-\ell} - 1$ d.f. among blocks.

As an illustration consider the following example.

Example 11.3 Suppose $p = 2, \ell = 3$, and hence $n = 7$. The $\boldsymbol{\alpha}'$ in standard order are $\boldsymbol{\alpha}'_1 = (100)$, $\boldsymbol{\alpha}'_2 = (010)$, $\boldsymbol{\alpha}'_3 = (001)$, $\boldsymbol{\alpha}'_4 = (110)$, $\boldsymbol{\alpha}'_5 = (101)$, $\boldsymbol{\alpha}'_6 = (011)$, $\boldsymbol{\alpha}'_7 = (111)$. Then from (11.27) $\boldsymbol{\gamma}'_1 = (\boldsymbol{\alpha}'_4, \boldsymbol{e}'_1) = (1101000)$, $\boldsymbol{\gamma}'_2 = (\boldsymbol{\alpha}'_5, \boldsymbol{e}'_2) = (1010100)$, $\boldsymbol{\gamma}'_3 = (\boldsymbol{\alpha}'_6, \boldsymbol{e}'_3) = (0110010)$, $\boldsymbol{\gamma}'_4 = (\boldsymbol{\alpha}'_7, \boldsymbol{e}_4) = (1110001)$. Hence $F_1F_2F_4$, $F_1F_3F_5$, $F_2F_3F_6$, $F_1F_2F_3F_7$ and all their GIs are confounded. □

Example 11.4 Suppose $p = 3, \ell = 2$, and hence $n = 4$. We have (see Example 11.2) $\boldsymbol{\alpha}'_1 = (10)$, $\boldsymbol{\alpha}'_2 = (01)$, $\boldsymbol{\alpha}'_3 = (11)$, $\boldsymbol{\alpha}'_4 = (12)$. From (11.27) we then obtain $\boldsymbol{\gamma}'_1 = (1120)$, $\boldsymbol{\gamma}'_2 = (1202)$. Hence $F_1F_2F_3^2$, $F_1F_3^2F_4^2$ and their GIs $F_1F_3F_4$ and $F_2F_3F_4^2$ are confounded. The reader should verify that this is in agreement with the content of the IBSG given in Example 11.2. □

11.8 SYMMETRICAL FACTORIALS AND FINITE GEOMETRIES

It is of some interest to relate the concepts of a p^n factorial experiment and its systems of confounding to the notion of finite geometries, for example, Euclidean and projective geometries. This line of thought was developed initially by Bose (1947b) while deriving some known results in a different way and this then led to further extensions.

The p levels of the factors in a p^n factorial experiment may be identified with the elements of the GF(p). The treatment combinations $(x_1, x_2, \ldots, x_n)$ can then be thought of as the points of the Euclidean geometry EG(n, p) (see Appendix B). Any ($n - 1$)-space in EG(n, p) satisfies the equation

$$\alpha_0 + \alpha_1 x_1 + \alpha_2 x_2 + \cdots + \alpha_n x_n = 0 \tag{11.28}$$

Hence the ($n - 1$)-space consists of p^{n-1} points of the EG(n, p). Keeping $(\alpha_1, \alpha_2, \ldots, \alpha_n)$ fixed and varying $\alpha_0 \in$ GF(p) yield p distinct ($n - 1$)-spaces that together form, of course, the EG(n, p). More generally, a ($n - q$)-space of the EG(n, p) is determined by q linearly independent equations

$$\begin{aligned} \alpha_{10} + \alpha_{11} x_1 + \alpha_{12} x_2 + \cdots + \alpha_{1n} x_n &= 0 \\ \alpha_{20} + \alpha_{21} x_1 + \alpha_{22} x_2 + \cdots + \alpha_{2n} x_n &= 0 \\ &\vdots \\ \alpha_{q0} + \alpha_{q1} x_1 + \alpha_{q2} x_2 + \cdots + \alpha_{qn} x_n &= 0 \end{aligned} \tag{11.29}$$

which are satisfied by p^{n-q} points.

It is clear that the partitioning of the EG(n, p) into ($n - 1$)-spaces as given by (11.28) corresponds to the partitioning we have used earlier to define main effects and interactions. We represented such partitions by $(\alpha_1, \alpha_2, \ldots, \alpha_n)$, and two such partitions $(\alpha_1, \alpha_2, \ldots, \alpha_n)$ and $(\beta_1, \beta_2, \ldots, \beta_n)$, say, are identical, that is, lead to the same set of p distinct ($n - 1$)-spaces of EG(n, p), if $(\alpha_1, \alpha_2, \ldots, \alpha_n) = \lambda$ $(\beta_1, \beta_2, \ldots, \beta_n)$ for some $\lambda \in$ GF(p). We know that the number of distinct partitions is $(p^n - 1)/(p - 1)$, which is also the number of points in the projective geometry PG($n - 1$, p) (see Appendix B). It is therefore convenient to identify $(\alpha_1, \alpha_2, \ldots, \alpha_n)$ as a point in PG($n - 1$, p).

With two points $\boldsymbol{\alpha}' = (\alpha_1, \alpha_2, \ldots, \alpha_n)$ and $\boldsymbol{\beta}' = (\beta_1, \beta_2, \ldots, \beta_n) \in$ PG ($n - 1$, p) the points $\boldsymbol{\alpha}' + \lambda\boldsymbol{\beta}'$, $\lambda \in$ GF(p), define a line, that is, a 1-space, in PG($n - 1$, p). In terms of interactions this means that the points on the line represent the GIs of $E^{\boldsymbol{\alpha}}$ and $E^{\boldsymbol{\beta}}$. This implies that for a p^n factorial in blocks of size p^{n-2} any two points $\boldsymbol{\alpha}'$, $\boldsymbol{\beta}'$ in PG($n - 1$, p) determine such a system of confounding and the GIs, which are also confounded, are represented by the points on the line through $\boldsymbol{\alpha}'$ and $\boldsymbol{\beta}'$.

This can be generalized easily, and we then have the following: A system of confounding for a p^n factorial in blocks of size p^k is determined by $n - k$ points in PG$(n-1, p)$ such that the points do not lie on a subspace of PG$(n-1, p)$ of dimension $n-k-2$ or less, that is, the points are not conjoint.

11.9 PARAMETERIZATION OF TREATMENT RESPONSES

We have shown in Sections 7.4.2 and 10.3 how the true response of a treatment combination for 2^n and 3^n factorials can be expressed in terms of main-effect and interaction components. We shall now present such a parameterization for the p^n factorial.

In Section 11.2 we have partitioned the p^n treatment combinations into p mutually exclusive sets of p^{n-1} treatment combinations each through Eqs. (11.1), that is,

$$\alpha_1 x_1 + \alpha_2 x_2 + \cdots + \alpha_n x_n = \delta \tag{11.30}$$

with $\delta \in \mathrm{GF}(p)$ and a fixed $\boldsymbol{\alpha}' = (\alpha_1, \alpha_2, \ldots, \alpha_n)$ with $\alpha_i \in \mathrm{GF}(p)(i = 1, 2, \ldots, n)\boldsymbol{\alpha}' \neq (0, 0, \ldots, 0)$. The set of treatment combinations $\boldsymbol{x}' = (x_1, x_2, \ldots, x_n)$ that satisfy (11.30) for a given δ we denoted earlier by $S(\boldsymbol{x}; \boldsymbol{\alpha}, \delta)$.

Let $a(\boldsymbol{x})$ denote the true response of the treatment combination $\boldsymbol{x}$. We then define

$$E_\delta^{\boldsymbol{\alpha}} = \frac{1}{p^{n-1}} \sum_{\boldsymbol{x}^* \in S(\boldsymbol{x};\boldsymbol{\alpha},\delta)} a(\boldsymbol{x}^*) - \frac{1}{p^n} \sum_{\delta \in \mathrm{GF}(p)} \sum_{\boldsymbol{x}^* \in S(\boldsymbol{x};\boldsymbol{\alpha},\delta)} a(\boldsymbol{x}^*) \tag{11.31}$$

or, if we write for convenience

$$\frac{1}{p^n} \sum_{\delta \in \mathrm{GF}(p)} \sum_{\boldsymbol{x}^* \in S(\boldsymbol{x};\boldsymbol{\alpha},\delta)} a(\boldsymbol{x}^*) = M$$

the overall mean, then (11.31) becomes

$$E_\delta^{\boldsymbol{\alpha}} = \frac{1}{p^{n-1}} \sum_{\boldsymbol{x}^* \in S(\boldsymbol{x};\boldsymbol{\alpha},\delta)} a(\boldsymbol{x}^*) - M \tag{11.32}$$

Since $\delta = \boldsymbol{\alpha}'\boldsymbol{x}$, we can write (11.32) also as $E_{\boldsymbol{\alpha}'\boldsymbol{x}}^{\boldsymbol{\alpha}}$, and we shall use both notations interchangeably. We thus have p numbers $E_0^{\boldsymbol{\alpha}}, E_1^{\boldsymbol{\alpha}}, \ldots, E_{p-1}^{\boldsymbol{\alpha}}$ with $E_0^{\boldsymbol{\alpha}} + E_1^{\boldsymbol{\alpha}} + \cdots + E_{p-1}^{\boldsymbol{\alpha}} = 0$. Any linear contrast among these, that is,

$$c_0 E_0^{\boldsymbol{\alpha}} + c_1 E_1^{\boldsymbol{\alpha}} + \cdots + c_{p-1} E_{p-1}^{\boldsymbol{\alpha}}$$

with $c_0 + c_1 + \cdots + c_{p-1} = 0$ belongs to the interaction $E^{\boldsymbol{\alpha}} = A_1^{\alpha_1} A_2^{\alpha_2} \cdots A_n^{\alpha_n}$. We can do this for each admissible partition $\boldsymbol{\alpha}$, of which there are $(p^n - 1)/(p-1)$. We have denoted this set by $\mathcal{A}$.

We then have the following theorem.

Theorem 11.8 The true response $a(\boldsymbol{x})$ of a treatment combination $\boldsymbol{x}$ in a p^n factorial can be parameterized as

$$a(\boldsymbol{x}) = M + \sum_{\boldsymbol{\alpha} \in \mathcal{A}} E^{\boldsymbol{\alpha}}_{\boldsymbol{\alpha}'\boldsymbol{x}} \tag{11.33}$$

Proof We examine the right-hand side (RHS) of (11.33) and show that it equals $a(\boldsymbol{x})$. We have

$$\text{RHS} = M + \frac{1}{p_{n-1}} \sum_{\boldsymbol{\alpha} \in \mathcal{A}} \sum_{\boldsymbol{x}^* \in S(\boldsymbol{x};\boldsymbol{\alpha},\delta)} a(\boldsymbol{x}^*) - \frac{p^n - 1}{p - 1} M \tag{11.34}$$

and

$$\sum_{\boldsymbol{\alpha} \in \mathcal{A}} \sum_{\boldsymbol{x}^* \in S(\boldsymbol{x};\boldsymbol{\alpha},\delta)} a(\boldsymbol{x}^*) = \frac{p^n - 1}{p - 1} \cdot a(\boldsymbol{x}) + \sum_{\boldsymbol{\alpha} \in \mathcal{A}} \sum_{\substack{\boldsymbol{x}^* \in S(\boldsymbol{x};\boldsymbol{\alpha},\delta) \\ \boldsymbol{x}^* \neq \boldsymbol{x}}} a(\boldsymbol{x}^*) \tag{11.35}$$

Now consider a specific $\boldsymbol{x}^* \neq \boldsymbol{x}$. In how many terms of the second expression on the RHS of (11.35), that is, for how many $\boldsymbol{\alpha} \in \mathcal{A}$, does $a(\boldsymbol{x}^*)$ occur? It occurs if and only if

$$\boldsymbol{\alpha}'\boldsymbol{x} = \boldsymbol{\alpha}'\boldsymbol{x}^*$$

or if

$$\boldsymbol{\alpha}'(\boldsymbol{x} - \boldsymbol{x}^*) = 0$$

or if

$$\boldsymbol{\alpha}'\boldsymbol{u} = 0 \tag{11.36}$$

for $\boldsymbol{u} = \boldsymbol{x}^* - \boldsymbol{x}$. Now at least one of the components in $\boldsymbol{u}$ is not zero. Hence (11.36) as an equation in the unknown $\boldsymbol{\alpha}$ has p^{n-1} solutions. However, we have to exclude $\boldsymbol{\alpha}' = (0, 0, \ldots, 0)$, and since with $\boldsymbol{\alpha}$ also $\lambda\boldsymbol{\alpha}$ is a solution for $\lambda \in \text{GF}(p)$, $\lambda \neq 0$, the number of distinct solutions to (11.36) is $(p^{n-1} - 1)/(p - 1)$. Using this result together with (11.35) in (11.34), we obtain

$$\text{RHS} = M\left(1 - \frac{p^n - 1}{p - 1}\right) + \frac{p^n - 1}{p - 1} \frac{1}{p^n - 1} a(\boldsymbol{x})$$

$$+ \frac{p^{n-1} - 1}{p - 1} \frac{1}{p^n - 1} \sum_{\substack{\boldsymbol{x}^* \\ \boldsymbol{x}^* \neq \boldsymbol{x}}} a(\boldsymbol{x}^*)$$

$$= M\left(1 - \frac{p^n - 1}{p - 1}\right) + \frac{p^n - 1}{p - 1}\,\frac{1}{p^{n-1}}\,a(\boldsymbol{x})$$

$$+ \frac{p^{n-1} - 1}{p - 1}\,\frac{1}{p^{n-1}}\left[p^n M - a(\boldsymbol{x})\right]$$

$$= M\left(1 - \frac{p^n - 1}{p - 1} + \frac{(p^{n-1} - 1)p}{p - 1}\right) + \frac{(p^n - 1) - (p^{n-1} - 1)}{(p - 1)p^{n-1}}\,a(\boldsymbol{x})$$

$$= a(\boldsymbol{x}) \qquad \square$$

For an illustration of this theorem we refer to Example 10.2.

11.10 ANALYSIS OF p^n FACTORIAL EXPERIMENTS

We consider a particular system of confounding in s types of replicates. These s replicates form the basic pattern, which is then repeated q times. The usual model for data from such an experiment is

$$\boldsymbol{y} = \mu\mathfrak{I} + \boldsymbol{X}_\rho\boldsymbol{\rho} + \boldsymbol{X}_{\beta^*}\boldsymbol{\beta}^* + \boldsymbol{X}_\tau\boldsymbol{\tau} + \boldsymbol{e} \qquad (11.37)$$

where the various terms on the RHS of (11.37) refer to the overall mean, replicate effects, block-within-replicate effects, treatment effects, and error, respectively.

We recall that the blocks within each replicate are determined by the system of confounding for that replicate and that the basic pattern of replicates is determined by the various types of systems of confounding employed. We shall therefore divide the $(p^n - 1)/(p - 1)$ interactions $E^{\boldsymbol{\alpha}}$ into three mutually exclusive sets based on the amount of confounding:

$$\mathcal{E}_1 = \left\{E^{\boldsymbol{\alpha}_k},\ k = 1, 2, \ldots, n_1\right\}\text{: completely confounded}$$

$$\mathcal{E}_2 = \left\{E^{\boldsymbol{\gamma}_\ell},\ \ell = 1, 2, \ldots, n_2\right\}\text{: partially confounded}$$

$$\mathcal{E}_3 = \left\{E^{\boldsymbol{\delta}_m},\ m = 1, 2, \ldots, n_3\right\}\text{: not confounded}$$

and $E^{\boldsymbol{\gamma}_\ell} \in \mathcal{E}_2$ is confounded in $c(\boldsymbol{\gamma}_\ell)$ replicates and not confounded in $u(\boldsymbol{\gamma}_\ell) = s - c(\boldsymbol{\gamma}_\ell)$ replicates of the basic pattern ($\ell = 1, 2, \ldots, n_2$). Of course, $n_1 + n_2 + n_3 = (p^n - 1)/(p - 1)$.

Just as the interactions in $\mathcal{E}_1$ and $\mathcal{E}_2$ determine the composition of the blocks within replicates, the interaction components $E_0^{\boldsymbol{\alpha}}, E_1^{\boldsymbol{\alpha}}, \ldots, E_{p-1}^{\boldsymbol{\alpha}}$ are the parametric functions of the treatment effects $\boldsymbol{\tau}(\boldsymbol{x})$, which are of main interest from the analysis point of view. To estimate these quantities, we shall utilize the fact that the design under consideration is resolved in that each treatment combination occurs once in each replicate (we call a design *resolved* if it is resolvable and actually arranged in replicates). Hence, if we arrange the observations $\boldsymbol{y}$ in (11.37) according to replicates and if in each replicate we arrange the observations in the same order according to the p^n treatment combinations, we can write the design model matrix in (11.37), that is, the matrix

$$\boldsymbol{X} = (\mathfrak{I}, \boldsymbol{X}_\rho, \boldsymbol{X}_{\beta^*}, \boldsymbol{X}_\tau)$$

as

$$\left[\begin{array}{ccccccccc} \mathfrak{I} & \mathfrak{I} & & & \boldsymbol{X}_{\beta*(1)} & & & & \boldsymbol{I} \\ \mathfrak{I} & & \mathfrak{I} & & & \boldsymbol{X}_{\beta^*(2)} & & & \boldsymbol{I} \\ \vdots & & & \ddots & & & \ddots & & \vdots \\ \mathfrak{I} & & & \mathfrak{I} & & & & \boldsymbol{X}_{\beta^*(r)} & \boldsymbol{I} \end{array}\right] \tag{11.38}$$

$$\underbrace{\qquad\qquad}_{\boldsymbol{X}_\rho} \qquad \underbrace{\qquad\qquad\qquad\qquad}_{\boldsymbol{X}_{\beta^*}} \qquad \underbrace{\quad}_{\boldsymbol{X}_\tau}$$

where each matrix in (11.38) has p^n rows and $\boldsymbol{X}_{\beta^*(j)}$ represents the $\boldsymbol{X}_{\beta^*}$ matrix for the jth replicate ($j = 1, 2, \ldots, r = sq$). We now utilize this form of (11.38) and the structure of the $\boldsymbol{X}_{\beta^*(j)}$ to proceed with the analysis.

11.10.1 Intrablock Analysis

Consider the RNE (1.7) for an incomplete block design

$$\left(r\boldsymbol{I} - \frac{1}{k}\,\boldsymbol{N}\boldsymbol{N}'\right)\widehat{\boldsymbol{\tau}} = \boldsymbol{T} - \frac{1}{k}\,\boldsymbol{N}\boldsymbol{B} \tag{11.39}$$

where $\boldsymbol{N} = \boldsymbol{X}_\tau'\boldsymbol{X}_\beta$. In our case, because of the form of (11.38), we obtain

$$\boldsymbol{N} = \left(\boldsymbol{X}_{\beta^*(1)}, \boldsymbol{X}_{\beta^*(2)}, \ldots, \boldsymbol{X}_{\beta^*(r)}\right)$$

and hence (11.39) becomes

$$\left(r\mathbf{I} - \frac{1}{k}\sum_{j=1}^{r} \boldsymbol{X}_{\beta^*(j)}\,\boldsymbol{X}'_{\beta^*(j)}\right)\widehat{\boldsymbol{\tau}} = \boldsymbol{T} - \frac{1}{k}\,\boldsymbol{N}\boldsymbol{B} \tag{11.40}$$

Let us consider now the right-hand side of (11.40). If we write $\boldsymbol{y}'$ as $\left(\boldsymbol{y}'_{(1)}, \boldsymbol{y}'_{(2)}, \ldots, \boldsymbol{y}'_{(r)}\right)$, where $\boldsymbol{y}'_{(j)}$ represents the vector of observations in replicate j, then

$$\boldsymbol{B} = \begin{bmatrix} \boldsymbol{X}'_{\beta^*(1)}\boldsymbol{y}_{(1)} \\ \boldsymbol{X}'_{\beta^*(2)}\boldsymbol{y}_{(2)} \\ \vdots \\ \boldsymbol{X}'_{\beta^*(r)}\boldsymbol{y}_{(r)} \end{bmatrix}$$

and hence

$$\boldsymbol{N}\boldsymbol{B} = \sum_{j=1}^{r} \boldsymbol{X}_{\beta^*(j)}\boldsymbol{X}'_{\beta^*(j)}\boldsymbol{y}_{(j)}$$

Furthermore, because of the arrangement of the observations in the same order in each replicate, we have

$$\boldsymbol{T} = \sum_{j=1}^{r} \boldsymbol{y}_{(j)}$$

We then write (11.40) as

$$\sum_{j=1}^{r} \left(\boldsymbol{I} - \frac{1}{k}\,\boldsymbol{X}_{\beta^*(j)}\boldsymbol{X}'_{\beta^*(j)}\right)\widehat{\boldsymbol{\tau}} = \sum_{j=1}^{r} \left(\boldsymbol{I} - \frac{1}{k}\,\boldsymbol{X}_{\beta^*(j)}\,\boldsymbol{X}'_{\beta^*(j)}\right)\boldsymbol{y}_{(j)} \qquad (11.41)$$

with $k = p^{\ell}$. It follows from (11.41) that, in order to analyze the complete data set, we can analyze the data in each replicate separately, for example, obtain estimates of the main effects and interaction components, and then average the information over all contributing replicates.

Let us then consider the RNE for the jth replicate:

$$\left(\boldsymbol{I} - \frac{1}{k}\boldsymbol{X}_{\beta^*(j)}\,\boldsymbol{X}'_{\beta^*(j)}\right)\widehat{\boldsymbol{\tau}} = \left(\boldsymbol{I} - \frac{1}{k}\,\boldsymbol{X}_{\beta^*(j)}\,\boldsymbol{X}'_{\beta^*(j)}\right)\boldsymbol{y}_{(j)} \qquad (11.42)$$

We shall write the equation for treatment combination $\boldsymbol{x}$ as

$$\widehat{\tau}(\boldsymbol{x}) - \frac{1}{k}\sum_{\boldsymbol{x}^*}\lambda_j(\boldsymbol{x}, \boldsymbol{x}^*)\widehat{\tau}(\boldsymbol{x}^*) = y_{(j)}(\boldsymbol{x}) - \frac{1}{k}\sum_{\boldsymbol{x}^*}\lambda_j(\boldsymbol{x}, \boldsymbol{x}^*)\,y_{(j)}(\boldsymbol{x}^*) \quad (11.43)$$

where $\tau(\boldsymbol{x})$ and $y_{(j)}(\boldsymbol{x})$ represent the effect and response of treatment combination $\boldsymbol{x}$, respectively, and

$$\lambda_j(\boldsymbol{x}, \boldsymbol{x}^*) = \begin{cases} 1 & \text{if } \boldsymbol{x} \text{ and } \boldsymbol{x}^* \text{ are in the} \\ & \text{same block in replicate } j \\ 0 & \text{otherwise} \end{cases}$$

Suppose we want to estimate $E_\delta^{\boldsymbol{\alpha}}$, where [see (11.31)]

$$E_\delta^{\boldsymbol{\alpha}} = \frac{1}{p^{n-1}} \sum_{\boldsymbol{x}^* \in S(\boldsymbol{x};\boldsymbol{\alpha},\delta)} \tau(\boldsymbol{x}^*) - \frac{1}{p^n} \sum_{\boldsymbol{x} \in \mathcal{X}} \tau(\boldsymbol{x}) \tag{11.44}$$

with $S(\boldsymbol{x}; \boldsymbol{\alpha}, \delta)$ representing the set of all treatment combinations $\boldsymbol{x}$ satisfying the equation $\boldsymbol{\alpha}'\boldsymbol{x} = \delta$ and $\mathcal{X}$ being the set of all treatment combinations. We distinguish between two cases:

(i) Suppose $E^{\boldsymbol{\alpha}}$ is not confounded in replicate j, that is, $E^{\boldsymbol{\alpha}} \in \mathcal{E}_2$ or $\mathcal{E}_3$. Equations (11.44) and (11.42) then suggest that we consider

$$\sum_{\boldsymbol{x}}{}' \left[\widehat{\tau}(\boldsymbol{x}) - \frac{1}{k} \sum_{\boldsymbol{x}^*} \lambda_j(\boldsymbol{x}, \boldsymbol{x}^*) \widehat{\tau}(\boldsymbol{x}^*) \right] = \sum_{\boldsymbol{x}}{}' \left[y_{(j)}(\boldsymbol{x}) - \frac{1}{k} \sum_{\boldsymbol{x}^*} \lambda_j(\boldsymbol{x}, \boldsymbol{x}^*) y_{(j)}(\boldsymbol{x}^*) \right]$$

where

$$\sum_{\boldsymbol{x}}{}' = \sum_{\boldsymbol{x} \in S(\boldsymbol{x};\boldsymbol{\alpha},\delta)}$$

or

$$\begin{aligned} \sum_{\boldsymbol{x}}{}' \widehat{\tau}(\boldsymbol{x}) &- \frac{1}{k} \sum_{\boldsymbol{x}^*} \left[\sum_{\boldsymbol{x}}{}' \lambda_j(\boldsymbol{x}, \boldsymbol{x}^*) \right] \widehat{\tau}(\boldsymbol{x}^*) \\ &= \sum_{\boldsymbol{x}}{}' y_{(j)}(\boldsymbol{x}) - \frac{1}{k} \sum_{\boldsymbol{x}^*} \left[\sum_{\boldsymbol{x}}{}' \lambda_j(\boldsymbol{x}, \boldsymbol{x}^*) \right] y_{(j)}(\boldsymbol{x}^*) \end{aligned} \tag{11.45}$$

Recall now that each block contains p^ℓ treatment combinations, and since $E^{\boldsymbol{\alpha}}$ is not confounded in replicate j, the p^{n-1} treatment combinations satisfying $\boldsymbol{\alpha}'\boldsymbol{x} = \delta$ are equally distributed over the $p^{n-\ell}$ blocks. It follows then that $\sum_{\boldsymbol{x}}' \lambda_j(\boldsymbol{x}, \boldsymbol{x}^*) = p^{\ell-1}$ for each $\boldsymbol{x}^* \in \mathcal{X}$. Hence, using $k = p^\ell$, (11.45) becomes

$$\sum_{\boldsymbol{x}}{}' \widehat{\tau}(\boldsymbol{x}) - \frac{1}{p} \sum_{\boldsymbol{x}^*} \widehat{\tau}(\boldsymbol{x}^*) = \sum_{\boldsymbol{x}}{}' y_{(j)}(\boldsymbol{x}) - \frac{1}{p} \sum_{\boldsymbol{x}^*} y_{(j)}(\boldsymbol{x}^*) \tag{11.46}$$

Comparing (11.46) to (11.44) yields

$$p^{n-1}\widehat{E}^{\boldsymbol{\alpha}}_{\delta(j)} = \sum_{\boldsymbol{x}}{}' y_{(j)}(\boldsymbol{x}) - \frac{1}{p}\sum_{\boldsymbol{x}^*} y_{(j)}(\boldsymbol{x}^*) \tag{11.47}$$

or the best linear unbiased estimate (BLUE) of $E^{\boldsymbol{\alpha}}_{\delta}$ from the jth replicate is

$$\widehat{E}^{\boldsymbol{\alpha}}_{\delta(j)} = \frac{1}{p^{n-1}}\sum_{\boldsymbol{x}}{}' y_{(j)}(\boldsymbol{x}) - \overline{y}_{(j)} \tag{11.48}$$

with $\overline{y}_{(j)}$ being the mean of all the observations in replicate j.

(ii) Suppose $E^{\boldsymbol{\alpha}}$ is confounded in replicate j, that is, $E^{\boldsymbol{\alpha}} \in \mathcal{E}_1$ or $\mathcal{E}_2$. We consider again (11.45). Since $E^{\boldsymbol{\alpha}}$ is confounded, the p^{n-1} treatment combinations satisfying $\boldsymbol{\alpha}'\boldsymbol{x} = \delta$, that is, the treatment combinations in $S(\boldsymbol{x}; \boldsymbol{\alpha}, \delta)$, are distributed over $p^{n-\ell-1}$ of the $p^{n-\ell}$ blocks each containing p^{ℓ} of the $\boldsymbol{x}^* \in S(\boldsymbol{x}; \boldsymbol{\alpha}, \delta)$. Hence

$$\sum_{\boldsymbol{x}}{}' \lambda_j(\boldsymbol{x}, \boldsymbol{x}^*) = \begin{cases} p^{\ell} & \text{for } \boldsymbol{x}^* \in S(\boldsymbol{x}; \boldsymbol{\alpha}, \delta) \\ 0 & \text{otherwise} \end{cases}$$

It then follows that both sides of (11.45) are identically zero, and hence replicate j does not contribute to the estimation of $E^{\boldsymbol{\alpha}}_{\delta}$, that is, does not contribute to information about $E^{\boldsymbol{\alpha}}$.

The derivation above holds, of course, for any interaction $E^{\boldsymbol{\alpha}}$ ($\boldsymbol{\alpha} \in \mathcal{A}$) and any replicate j ($j = 1, 2, \ldots, r$). Specifically for $E^{\boldsymbol{\alpha}_k} \in \mathcal{E}_1$ it follows then that there is no intrablock information. For $E^{\boldsymbol{\gamma}_\ell} \in \mathcal{E}_2$ we obtain information from $qu(\boldsymbol{\gamma}_\ell)$ replicates. Let

$$\eta(\boldsymbol{\gamma}_\ell, j) = \begin{cases} 1 & \text{if } E^{\boldsymbol{\gamma}_\ell} \text{ is not confounded in replicate } j \\ 0 & \text{otherwise} \end{cases}$$

We then obtain from (11.48)

$$p^{n-1}\sum_{j=1}^{r} \eta(\boldsymbol{\gamma}_\ell, j)\, \widehat{E}^{\boldsymbol{\gamma}_\ell}_{\delta(j)} = \sum_{j=1}^{r} \eta(\boldsymbol{\gamma}_\ell, j)\left[\sum_{\boldsymbol{x}^* \in S(\boldsymbol{x}; \boldsymbol{\gamma}_\ell, \delta)} y_{(j)}(\boldsymbol{x}^*) - p^{n-1}\overline{y}_{(j)}\right]$$

or, if we let

$$\widehat{E}^{\boldsymbol{\gamma}_\ell}_{\delta} = \frac{1}{qu(\boldsymbol{\gamma}_\ell)}\sum_{j} \eta(\boldsymbol{\gamma}_\ell, j)\widehat{E}^{\boldsymbol{\gamma}_\ell}_{\delta(j)}$$

then

$$\widehat{E}_{\delta}^{\gamma_\ell} = \frac{1}{p^{n-1}qu(\boldsymbol{\gamma}_\ell)} \sum_{j=1}^{r} \eta(\boldsymbol{\gamma}_\ell, j) \left[\sum_{\boldsymbol{x}^* \in S(\boldsymbol{x}; \boldsymbol{\gamma}_\ell, \delta)} y_{(j)}(\boldsymbol{x}^*) - p^{n-1}\overline{y}_{(j)} \right] \tag{11.49}$$

Finally, information for $E^{\delta_m} \in \mathcal{E}_3$ is obtained from all replicates, so that

$$\widehat{E}_{\delta}^{\delta_m} = \frac{1}{p^{n-1}r} \sum_{j=1}^{r} \sum_{\boldsymbol{x}^* \in S(\boldsymbol{x}; \boldsymbol{\delta}_m, \delta)} y_{(j)}(\boldsymbol{x}^*) - \overline{y} \tag{11.50}$$

where $\overline{y}$ is the overall mean.

11.10.2 Disconnected Resolved Incomplete Block Designs

The result just derived is a special case of a more general result applying to disconnected resolved incomplete block designs, which we shall give now. Suppose we have r replicates with each treatment combination occurring exactly once in each replicate. According to (11.38), we write the model for the jth replicate as

$$\boldsymbol{y}_{(j)} \doteq \mu \mathfrak{I} + \rho_j \mathfrak{I} + \boldsymbol{X}_{\beta^*(j)} \boldsymbol{\beta}_j^* + \boldsymbol{I}\boldsymbol{\tau} \tag{11.51}$$

[the symbol $\doteq$, used in I.4.2 as a shorthand for "approximately described by," is used here to denote the expected value, $E(\cdot)$, of the expression on the left-hand side]. Consider now a linear function $\boldsymbol{\xi}'\boldsymbol{\tau}$ of the treatment effects, with $\boldsymbol{\xi}'\mathfrak{I} = 0$. Such a function may or may not be estimable, that is, it may or may not be confounded with blocks. We state the following definition.

Definition 11.2 A linear function $\boldsymbol{\xi}'\boldsymbol{\tau}$ with $\boldsymbol{\xi}'\mathfrak{I} = 0$ is *completely confounded* with blocks in the jth replicate of a resolved incomplete block design if $\boldsymbol{\xi}'\boldsymbol{y}_{(j)}$ is a linear function of block totals. □

To obtain a characterization of such confounded functions, we multiply (11.51) on both sides by $\boldsymbol{\xi}'$. This gives

$$\boldsymbol{\xi}'\boldsymbol{y}_{(j)} \doteq \boldsymbol{\xi}'\boldsymbol{X}_{\beta^*(j)}\ \boldsymbol{\beta}_j^* + \boldsymbol{\xi}'\boldsymbol{\tau} \tag{11.52}$$

the left-hand side of which must equal $\boldsymbol{\eta}'\boldsymbol{X}'_{\beta^*(j)}\boldsymbol{y}_{(j)}$ for some $\boldsymbol{\eta}$. This implies that

$$\boldsymbol{\xi}' = \boldsymbol{\eta}'\boldsymbol{X}'_{\beta^*(j)} \tag{11.53}$$

On the other hand, if (11.53) holds, it follows from (11.52) that

$$\boldsymbol{\eta}'\boldsymbol{X}_{\beta^*(j)}\boldsymbol{y}_{(j)} \doteq \boldsymbol{\eta}'\ \text{diag}\{k_{j1}, k_{j2}, \ldots, k_{jb}\}\boldsymbol{\beta}_j^* + \boldsymbol{\xi}'\boldsymbol{\tau}$$

since

$$\boldsymbol{X}'_{\beta^*(j)}\boldsymbol{X}_{\beta^*(j)} = \text{diag}\{k_{j1}, k_{j2}, \ldots, k_{jb}\}$$

where k_{ji} is the size of the ith block in the jth replicate ($j = 1, 2, \ldots, r$; $i = 1, 2, \ldots, b$), or with equal block size k, $\boldsymbol{X}_{\beta^*(j)}\boldsymbol{X}_{\beta^*(j)} = k\boldsymbol{I}$. Thus we have the following theorem.

Theorem 11.9 A linear function $\boldsymbol{\xi}'\boldsymbol{\tau}$ with $\boldsymbol{\xi}'\mathfrak{I} = 0$ is completely confounded with blocks of the jth replicate of a resolved incomplete block design if and only if $\boldsymbol{\xi} = \boldsymbol{X}_{\beta^*(j)}\boldsymbol{\eta}$ for some $\boldsymbol{\eta}$.

Using this theorem, we can now derive from basic principles of linear model theory that, if a linear function $\boldsymbol{\xi}'\boldsymbol{\tau}$ is completely confounded in c of the r replicates of a design, $\boldsymbol{\xi}'\boldsymbol{\tau}$ is estimated in the usual way from the $r - c$ replicates in which it is not confounded, thus corroborating our earlier results concerning the estimation of E^{α}_{δ}'s. Suppose then that $\boldsymbol{\xi}'\boldsymbol{\tau}$ is completely confounded in replicates $1, 2, \ldots, c$; that is, we have

$$\boldsymbol{\xi} = \boldsymbol{X}_{\beta^*(1)}\boldsymbol{\eta}_1 = \boldsymbol{X}_{\beta^*(2)}\boldsymbol{\eta}_2 = \cdots = \boldsymbol{X}_{\beta^*(c)}\boldsymbol{\eta}_c \tag{11.54}$$

and let

$$\widehat{\boldsymbol{\xi}'\boldsymbol{\tau}} = \sum_{j=1}^{r} \boldsymbol{a}'_j \boldsymbol{y}_{(j)}$$

Assuming a GMLM (the justification for this is given in I.6.3.5), we have

$$\text{var}\left(\widehat{\boldsymbol{\xi}'\boldsymbol{\tau}}\right) = \sum_{j=1}^{r} \boldsymbol{a}'_j \boldsymbol{a}_j \sigma_e^2$$

Hence, to find the BLUE for $\widehat{\boldsymbol{\xi}'\boldsymbol{\tau}}$, we minimize $\sum_{j=1}^{r} \boldsymbol{a}'_j \boldsymbol{a}_j$ subject to the unbiasedness conditions

$$\sum_{j=1}^{r} \boldsymbol{a}_j = \boldsymbol{\xi} \tag{11.55}$$

and

$$\boldsymbol{\xi}'\boldsymbol{a}_j = 0 \qquad \text{for } j = 1, 2, \ldots, c. \tag{11.56}$$

The condition (11.56) derives from the unbiasedness conditions $\boldsymbol{a}'_j \boldsymbol{X}_{\beta^*(j)} = \boldsymbol{\phi}$ together with (11.54) which implies

$$\boldsymbol{\xi}'\boldsymbol{a}_j = \boldsymbol{\eta}'_j \boldsymbol{X}'_{\beta^*(j)}\boldsymbol{a}_j = 0 \qquad \text{for } j = 1, 2, \ldots, c$$

The Lagrangian to be minimized is then

$$\mathcal{L} = \sum_{j=1}^{r} \boldsymbol{a}_j'\boldsymbol{a}_j - 2\boldsymbol{\nu}'\left(\sum_{j=1}^{r} \boldsymbol{a}_j - \boldsymbol{\xi}\right) + \sum_{j=1}^{c} \lambda_j \boldsymbol{\xi}'\boldsymbol{a}_j \tag{11.57}$$

where $\boldsymbol{\nu}$ and λ_j $(j = 1, 2, \ldots, c)$ are Lagrangian multipliers. The derivatives of (11.57) yield the equations

$$\boldsymbol{a}_j - \boldsymbol{\nu} + \lambda_j \boldsymbol{\xi} = \boldsymbol{\phi} \qquad (j = 1, 2, \ldots, c) \tag{11.58}$$

$$\boldsymbol{a}_j - \boldsymbol{\nu} = \boldsymbol{\phi} \qquad (j = c + 1, 2, \ldots, r) \tag{11.59}$$

$$\sum_{j=1}^{r} \boldsymbol{a}_j = \boldsymbol{\xi}$$

$$\boldsymbol{\xi}'\boldsymbol{a}_j = 0 \qquad (j = 1, 2, \ldots, c)$$

From (11.58) we obtain

$$\boldsymbol{\xi}'\boldsymbol{a}_j + \lambda_j \boldsymbol{\xi}'\boldsymbol{\xi} = \boldsymbol{\xi}'\boldsymbol{\nu} \qquad (j = 1, 2, \ldots, c)$$

which, using (11.56), reduces to

$$\lambda_j \boldsymbol{\xi}'\boldsymbol{\xi} = \boldsymbol{\xi}'\boldsymbol{\nu} \qquad (j = 1, 2, \ldots, c) \tag{11.60}$$

implying $\lambda_1 = \lambda_2 = \cdots = \lambda_c = \lambda$, say. Now it follows from (11.55), (11.56), and (11.59) that

$$\boldsymbol{\xi}'\boldsymbol{\xi} = \sum_{j=c+1}^{r} \boldsymbol{\xi}'\boldsymbol{a}_j = (r - c)\boldsymbol{\xi}'\boldsymbol{\nu} \tag{11.61}$$

Hence, from (11.60) and (11.61)

$$\lambda = \frac{1}{r - c} \tag{11.62}$$

and

$$\boldsymbol{\nu} = \frac{1}{r - c}\, \boldsymbol{\xi} \tag{11.63}$$

It follows then from (11.58), (11.59), (11.62), and (11.63) that

$$\boldsymbol{a}_j = \boldsymbol{\phi} \qquad \text{for } j = 1, 2, \ldots, c$$

and

$$\boldsymbol{a}_j = \frac{1}{r-c}\boldsymbol{\xi} \qquad \text{for } j = c+1, \ldots, r$$

Hence

$$\widehat{\boldsymbol{\xi}'\boldsymbol{\tau}} = \frac{\sum_{j=c+1}^{r} \boldsymbol{\xi}'\boldsymbol{y}_{(j)}}{r-c}$$

that is, the estimate of linear function $\boldsymbol{\xi}'\boldsymbol{\tau}$ that is completely confounded in c replicates is the average of the same linear function of the corresponding observations from the replicates in which $\boldsymbol{\xi}'\boldsymbol{\tau}$ is not confounded. If in particular $c = r$, then $\boldsymbol{\xi}'\boldsymbol{\tau}$ is, of course, not estimable.

11.10.3 Analysis of Variance Tables

We now return to the analysis of p^n factorial experiments in terms of main effects and interactions. Since we have used the reparameterization (11.33), a complete set of estimable functions of the $\tau(\boldsymbol{x})$ is given by

$$\left\{E_0^{\boldsymbol{\gamma}_\ell}, E_1^{\boldsymbol{\gamma}_\ell}, \ldots, E_{p-2}^{\boldsymbol{\gamma}_\ell}\colon E^{\boldsymbol{\gamma}_\ell} \in \mathcal{E}_2\right\}$$

and

$$\left\{E_0^{\boldsymbol{\delta}_m}, E_1^{\boldsymbol{\delta}_m}, \ldots, E_{p-2}^{\boldsymbol{\delta}_m}\colon E^{\boldsymbol{\delta}_m} \in \mathcal{E}_3\right\}$$

accounting for $(n_2 + n_3)(p-1)$ d.f. among treatments. Hence we have the T|B-ANOVA as given in Table 11.1a. It follows from the derivation of $\widehat{E}^{\boldsymbol{\gamma}_\ell}$ and $\widehat{E}^{\boldsymbol{\delta}_m}$, for example, (11.49) and (11.50), and the general principle (see Section 1.3.7)

$$SS(\boldsymbol{X}_\tau \mid \mathfrak{I}, \boldsymbol{X}_\beta) = \widehat{\boldsymbol{\tau}}'\boldsymbol{Q}$$

that

$$SS\left(E^{\boldsymbol{\gamma}_\ell}\right) = qu(\boldsymbol{\gamma}_\ell)p^{n-1}\sum_{\delta=0}^{p-1}\left[\widehat{E}_\delta^{\boldsymbol{\gamma}_\ell}\right]^2 \tag{11.64}$$

and

$$SS\left(E^{\boldsymbol{\delta}_m}\right) = rp^{n-1}\sum_{\delta=0}^{p-1}\left[\widehat{E}_\delta^{\boldsymbol{\delta}_m}\right]^2 \tag{11.65}$$

It is easy to verify that

$$E\left[SS\left(E^{\boldsymbol{\gamma}_\ell}\right)\right] = (p-1)\sigma_e^2 + qu(\boldsymbol{\gamma}_\ell)p^{n-1}\sum_{\delta=0}^{p-1}\left[E_\delta^{\boldsymbol{\gamma}_\ell}\right]^2$$

Table 11.1 T|B-ANOVA for p^n Factorial Experiment

Source	d.f.	SS
	a. Basic Partitioning	
$\boldsymbol{X}_\rho \mid \mathfrak{I}$	$sq-1$	Usual
$\boldsymbol{X}_{\beta^*} \mid \mathfrak{I}, \boldsymbol{X}_\rho$	$sq(p^{n-\ell}-1)$	Usual
$\boldsymbol{X}_\tau \mid \mathfrak{I}, \boldsymbol{X}_\rho, \boldsymbol{X}_{\beta^*}$	$(n_2+n_3)(p-1)$	$\sum_\ell \text{SS}(E^{\gamma_\ell}) + \sum_m \text{SS}(E^{\delta_m})$
$\boldsymbol{I} \mid \mathfrak{I}, \boldsymbol{X}_\rho, \boldsymbol{X}_{\beta^*}, \boldsymbol{X}_\tau$	ν_R	Difference
Total	sqp^n-1	Usual
	b. Partitioning of Block Sum of Squares	
$\boldsymbol{X}_{\beta^*} \mid \mathfrak{I}, \boldsymbol{X}_\rho$	$sq(p^{n-\ell}-1)$	Usual
$\{E^{\boldsymbol{\alpha}_k}\}$	$n_1(p-1)$	$\sum_k \text{SS}(E^{\boldsymbol{\alpha}_k})_c$
$\{E^{\boldsymbol{\gamma}_\ell}\}$	$n_2(p-1)$	$\sum_\ell \text{SS}(E^{\boldsymbol{\gamma}_\ell})_c$
Remainder	Difference $(=\nu)$	Difference

and

$$E\left[\text{SS}\left(E^{\delta_m}\right)\right] = (p-1)\sigma_e^2 + rp^{n-1}\sum_{\delta=0}^{p-1}\left[E_\delta^{\delta_\ell}\right]^2$$

Since all the comparisons belonging to $E^{\boldsymbol{\alpha}_k} \in \mathcal{E}_1$ and those comparisons belonging to $E^{\boldsymbol{\gamma}_\ell} \in \mathcal{E}_2$ from replicates in which $E^{\boldsymbol{\gamma}_\ell}$ is confounded do not contribute to $\text{SS}(\boldsymbol{X}_\tau \mid \mathfrak{I}, \boldsymbol{X}_\rho, \boldsymbol{X}_{\beta^*})$, they must contribute to $\text{SS}(\boldsymbol{X}_{\beta^*} \mid \mathfrak{I}, \boldsymbol{X}_\rho)$. This then leads to the partitioning of $\text{SS}(\boldsymbol{X}_{\beta^*} \mid \mathfrak{I}, \boldsymbol{X}_\rho)$ as given in Table 11.1b, where $\text{SS}(E^{\boldsymbol{\alpha}_k})_c$ is obtained from all replicates in the usual way [comparable to (11.65)] and $\text{SS}(E^{\boldsymbol{\gamma}_\ell})_c$ is obtained from all replicates in which $E^{\boldsymbol{\gamma}_\ell}$ is confounded [comparable to (11.64) with $c(\boldsymbol{\gamma}_\ell)$ substituted for $u(\boldsymbol{\gamma}_\ell)$]. The components $(E_\delta^{\boldsymbol{\alpha}_k})_c, (E_\delta^{\boldsymbol{\gamma}_k})_c$ for $E^{\boldsymbol{\alpha}_k} \in \mathcal{E}_1, E^{\boldsymbol{\gamma}_\ell} \in \mathcal{E}_2$ and $\delta = 0, 1, \ldots, p-2$ form a complete set of linearly independent functions involving both block effects and treatment effects, accounting for $(n_1+n_2)(p-1)$ d.f. of $\text{SS}(\boldsymbol{X}_{\beta^*} \mid \mathfrak{I}, \boldsymbol{X}_\rho)$ and constituting the treatment component of this sum of squares. The remaining comparisons belonging to $\text{SS}(\boldsymbol{X}_{\beta^*} \mid \mathfrak{I}, \boldsymbol{X}_\rho)$ are then pure block comparisons accounting for

$$\nu = sq(p^{n-\ell}-1) - (n_1+n_2)(p-1) \text{ d.f.} \tag{11.66}$$

11.11 INTERBLOCK ANALYSIS

We have shown in the previous section that intrablock information from several replicates can be combined additively to obtain the BLUE for an estimable function. We now turn our attention to interblock information for interaction components, which means we consider the situation where $\boldsymbol{\beta}^*$ is a vector of i.i.d. random variables with $E(\boldsymbol{\beta}^*) = \boldsymbol{\phi}$ and $E(\boldsymbol{\beta}^*\boldsymbol{\beta}^{*\prime}) = \boldsymbol{I}\sigma_\beta^2$. In particular, we shall be concerned about interblock estimates of $E^{\alpha_k} \in \mathcal{E}_1$ and $E^{\gamma_\ell} \in \mathcal{E}_2$ from those replicates in which E^{γ_ℓ} is confounded. First we shall establish, however, that similar to the case of intrablock information the estimates from several replicates can be combined additively to yield BLUEs, a result that does not hold generally.

11.11.1 Combining Interblock Information

We consider the model equation (11.51) for the observations in the jth replicate, assuming without loss of generality $\mu = \rho_j = 0$, so that we have

$$\boldsymbol{y}_{(j)} = \boldsymbol{I}\boldsymbol{\tau} + \boldsymbol{X}_{\beta^*(j)}\boldsymbol{\beta}_j^* + \boldsymbol{e}_{(j)} \tag{11.67}$$

We can rewrite (11.67) as

$$\boldsymbol{y}_{(j)} = \boldsymbol{I}\boldsymbol{\tau} + \boldsymbol{f}_{(j)} \tag{11.68}$$

with $E(\boldsymbol{f}_{(j)}) = \boldsymbol{\phi}$ and

$$\begin{aligned} E\left(\boldsymbol{f}_{(j)}\boldsymbol{f}'_{(j)}\right) &= \boldsymbol{I}\sigma_e^2 + \boldsymbol{X}_{\beta^*(j)}\boldsymbol{X}'_{\beta^*(j)}\sigma_\beta^2 \\ &= \left(\boldsymbol{I} + \rho^*\boldsymbol{X}_{\beta^*(j)}\boldsymbol{X}'_{\beta^*(j)}\right)\sigma_e^2 \\ &= \boldsymbol{V}_j\,\sigma_e^2 \end{aligned} \tag{11.69}$$

where $\rho^* = \sigma_\beta^2/\sigma_e^2$ and $\boldsymbol{V}_j = \boldsymbol{I} + \rho^*\boldsymbol{X}_{\beta^*(j)}\boldsymbol{X}'_{\beta^*(j)}$ $(j = 1, 2, \ldots, r)$. The model equation for the combined data from r replicates is thus

$$\begin{bmatrix} \boldsymbol{y}_{(1)} \\ \boldsymbol{y}_{(2)} \\ \vdots \\ \boldsymbol{y}_{(r)} \end{bmatrix} = \begin{bmatrix} \boldsymbol{I} \\ \boldsymbol{I} \\ \vdots \\ \boldsymbol{I} \end{bmatrix} \boldsymbol{\tau} + \begin{bmatrix} \boldsymbol{f}_{(1)} \\ \boldsymbol{f}_{(2)} \\ \vdots \\ \boldsymbol{f}_{(r)} \end{bmatrix} \tag{11.70}$$

with variance–covariance matrix

$$\boldsymbol{V}\sigma_e^2 = \text{diag}\{\boldsymbol{V}_1, \boldsymbol{V}_2, \ldots, \boldsymbol{V}_r\}\sigma_e^2 \tag{11.71}$$

The Aitken equation (see Section I.4.16.2) then is

$$(\boldsymbol{I}\ \boldsymbol{I} \cdots \boldsymbol{I})\, \text{diag}\{\boldsymbol{V}_1^{-1}, \boldsymbol{V}_2^{-1}, \ldots, \boldsymbol{V}_r^{1}\} \begin{pmatrix} \boldsymbol{I} \\ \boldsymbol{I} \\ \vdots \\ \boldsymbol{I} \end{pmatrix} \widetilde{\boldsymbol{\tau}}$$

$$= (\boldsymbol{I}\ \boldsymbol{I} \cdots \boldsymbol{I})\, \text{diag}\{\boldsymbol{V}_1^{-1}, \boldsymbol{V}_2^{-1}, \ldots, \boldsymbol{V}_r^{-1}\} \begin{pmatrix} \boldsymbol{y}_{(1)} \\ \boldsymbol{y}_{(2)} \\ \vdots \\ \boldsymbol{y}_{(r)} \end{pmatrix}$$

or

$$\left(\sum_{j=1}^{r} \boldsymbol{V}_j^{-1} \right) \widetilde{\boldsymbol{\tau}} = \sum_j \boldsymbol{V}_j^{-1} \mathbf{y}_{(j)} \tag{11.72}$$

and hence

$$\widetilde{\boldsymbol{\tau}} = \left(\sum_j \boldsymbol{V}_j^{-1} \right)^{-1} \left(\sum_j \boldsymbol{V}_j^{-1} \mathbf{y}_{(j)} \right)$$

If $\boldsymbol{\xi}'\boldsymbol{\tau}$ is an estimable function, then the BLUE is given by

$$\widetilde{\boldsymbol{\xi}'\boldsymbol{\tau}} = \boldsymbol{\xi}' \left(\sum_j \boldsymbol{V}_j^{-1} \right)^{-1} \left(\sum_j \boldsymbol{V}_j^{-1} \boldsymbol{y}_{(j)} \right) \tag{11.73}$$

Suppose now we consider the estimation of $\boldsymbol{\xi}'\boldsymbol{\tau}$ separately for each replicate. It follows from (11.67) that

$$\widetilde{\boldsymbol{\tau}} = \boldsymbol{y}_{(j)}$$

so that the BLUE of $\boldsymbol{\xi}'\boldsymbol{\tau}$ from replicate j, $\left(\widetilde{\boldsymbol{\xi}'\boldsymbol{\tau}}\right)_j$ say, is

$$\left(\widetilde{\boldsymbol{\xi}'\boldsymbol{\tau}}\right)_j = \boldsymbol{\xi}'\boldsymbol{y}_{(j)} \tag{11.74}$$

with

$$\text{var}\left(\widetilde{\boldsymbol{\xi}'\boldsymbol{\tau}}\right)_j = \boldsymbol{\xi}'\boldsymbol{V}_j\ \boldsymbol{\xi}\,\sigma_e^2 \tag{11.75}$$

Combining the estimates (11.74) with weights equal to the inverse of the variance (11.75), we obtain

$$\widetilde{\xi'\tau} = \frac{\sum_{j=1}^{r}(1/\xi' V_j\ \xi)\left(\widetilde{\xi'\tau}\right)_j}{\sum_{j=1}^{r}(1/\xi' V_j\ \xi)} \tag{11.76}$$

We know that (11.73) is BLUE for $\xi'\tau$. One can see easily that if $V_j = V$ for all $j = 1, 2, \ldots, r$, then $\widetilde{\xi'\tau}$ of (11.73) equals $\xi'\tau$ of (11.76), but in general these estimators will be different. We now give a condition under which (11.73) and (11.76) are identical, a condition we shall show to be met for the purpose of estimating interaction components.

Theorem 11.10 In a resolved design the interblock information for an estimable function $\xi'\tau$ obtained from each replicate can be combined additively as in (11.76) to yield the BLUE if, for each j,

$$V_j\ \xi = \nu_j \xi \tag{11.77}$$

that is, if ξ is an eigenvector of each variance–covariance matrix V_j ($j = 1, 2, \ldots, r$), with $\xi'\xi = 1$.

Proof (i) The estimator in (11.76) is, of course, unbiased.
(ii) We know that (11.76) is BLUE. Therefore, for (11.76) to be BLUE, we must have, using (11.73),

$$\xi'\left(\sum_i V_i^{-1}\right)^{-1}\left(\sum_j V_j^{-1} y_{(j)}\right) = \frac{\sum_j (1/\xi' V_j\ \xi)\,\xi' y_{(j)}}{\sum_i (1/\xi' V_i\ \xi)} \tag{11.78}$$

This must hold for all $y_{(j)}$. Hence from (11.78) we must have

$$\xi'\left(\sum_i V_i^{-1}\right)^{-1} V_j^{-1} = \frac{(1/\xi' V_j\ \xi)\xi'}{\sum_i (1/\xi' V_i\ \xi)} \tag{11.79}$$

Now, if (11.77) holds, it follows that

$$V_j^{-1}\xi = \frac{1}{\nu_j}\,\xi \tag{11.80}$$

and hence

$$\left(\sum_j V_j^{-1}\right)\xi = \left(\sum_j \frac{1}{\nu_j}\right)\xi$$

and

$$\left(\sum_j \boldsymbol{V}_j^{-1}\right)^{-1} \boldsymbol{\xi} = \left(\sum_j \frac{1}{\nu_j}\right)^{-1} \boldsymbol{\xi} \tag{11.81}$$

Using (11.77), (11.80), and (11.81), on both sides of (11.79) we obtain

$$\left(\sum_i \frac{1}{\nu_i}\right)^{-1} \frac{1}{\nu_j} \boldsymbol{\xi}' = \frac{(1/\nu_j)\,\boldsymbol{\xi}'}{\sum_i (1/\nu_i)}$$

which completes the proof. □

11.11.2 Estimating Confounded Interactions

We now use Theorem 11.10 to obtain interblock information for $E^{\boldsymbol{\alpha}_k} \in \mathcal{E}_1$ and $E^{\boldsymbol{\gamma}_\ell} \in \mathcal{E}_2$. Consider an $E^{\boldsymbol{\alpha}} \in \mathcal{E}_1$ or $\mathcal{E}_2$ and suppose that $E^{\boldsymbol{\alpha}}$ is confounded in replicate j. We then consider $\boldsymbol{\xi}'\boldsymbol{\tau} = E_\delta^{\boldsymbol{\alpha}}$ for some $\delta \in \mathrm{GF}(p)$. Obviously, we have $\boldsymbol{\xi}'\mathfrak{I} = 0$, and it follows from Theorem 11.9 that

$$\boldsymbol{\xi} = \boldsymbol{X}_{\beta^*(j)}\boldsymbol{\eta} \tag{11.82}$$

Using (11.82) in (11.69), we obtain

$$\begin{aligned}
\boldsymbol{V}_j\,\boldsymbol{\xi} &= (\boldsymbol{I} + \rho^* \boldsymbol{X}_{\beta^*(j)} \boldsymbol{X}'_{\beta^*(j)})\,\boldsymbol{\xi} \\
&= \boldsymbol{\xi} + \rho^* \boldsymbol{X}_{\beta^*(j)}) \boldsymbol{X}'_{\beta^*(j)}) \boldsymbol{X}_{\beta^*(j)} \boldsymbol{\eta} \\
&= \boldsymbol{\xi} + \rho^* k \boldsymbol{X}_{\beta^*(j)} \boldsymbol{\eta} \\
&= (1 + \rho^* k)\boldsymbol{\xi}
\end{aligned} \tag{11.83}$$

since $\boldsymbol{X}'_{\beta^*(j)}\boldsymbol{X}_{\beta^*(j)} = k\boldsymbol{I}$, where k is the block size. Hence the condition of Theorem 11.10 is satisfied, which implies that $E_\delta^{\boldsymbol{\alpha}_k}$ can be estimated by combining additively the estimates from each replicate. Denoting the overall estimate by $\widetilde{E}_\delta^{\boldsymbol{\alpha}_k}$, $\delta \in \mathrm{GF}(p)$, we then have, using (11.83) with $k = p^\ell$,

$$\mathrm{var}\left(\widetilde{E}_\delta^{\boldsymbol{\alpha}_k} - \widetilde{E}_{\delta'}^{\boldsymbol{\alpha}_k}\right) = \frac{2}{qsp^{n-1}}\left(\sigma_e^2 + p^\ell \sigma_\beta^2\right) \tag{11.84}$$

Similarly, we obtain $\widetilde{E}_\delta^{\boldsymbol{\gamma}_\ell}$ for $E^{\boldsymbol{\gamma}_\ell} \in \mathcal{E}_2$, $\delta \in \mathrm{GF}(p)$ from all replicates in which $E^{\boldsymbol{\gamma}_\ell}$ is confounded. Hence

$$\mathrm{var}\left(\widetilde{E}_\delta^{\boldsymbol{\gamma}_\ell} - \widetilde{E}_{\delta'}^{\boldsymbol{\gamma}_\ell}\right) = \frac{2}{qc(\boldsymbol{\gamma}_\ell)p^{n-1}}\left(\sigma_e^2 + p^\ell \sigma_\beta^2\right) \tag{11.85}$$

The $\widetilde{E}_\delta^{\alpha_k}$ and $\widetilde{E}_\delta^{\gamma_\ell}$ so obtained are, of course, used to yield

$$\text{SS}(E^{\alpha_k})_c = qsp^{n-1} \sum_{\delta=0}^{p-1} \left[\widetilde{E}_\delta^{\alpha_k}\right]^2$$

and

$$\text{SS}\left(E^{\gamma_\ell}\right)_c = qc(\boldsymbol{\gamma}_\ell)p^{n-1} \sum_{\delta=0}^{p-1} \left[\widetilde{E}_\delta^{\gamma_\ell}\right]^2$$

of Table 11.1b. Under the assumptions made earlier in this section it follows that

$$E[\text{SS}(E^{\alpha_k})_c] = (p-1)(\sigma_e^2 + p^\ell \sigma_\beta^2) + qsp^{n-1} \sum_{\delta=0}^{p-1} \left[E_\delta^{\alpha_k}\right]^2$$

and

$$E[\text{SS}(E^{\gamma_\ell})_c] = (p-1)(\sigma_e^2 + p^\ell \sigma_\beta^2) + qc(\boldsymbol{\gamma}_\ell)p^{n-1} \sum_{\delta=0}^{p-1} \left[E_\delta^{\gamma_\ell}\right]^2$$

Since, in Table 11.1b,

$$E[\text{MS(Remainder)}] = \sigma_e^2 + p^\ell \sigma_\beta^2 \tag{11.86}$$

we have a test for

$$H_0\colon E_0^{\alpha_k} = E_1^{\alpha_k} = \cdots = E_{p-1}^{\alpha_k} = 0$$

with $E^{\alpha_k} \in \mathcal{E}_1$, which was not available before from the intrablock analysis.

11.12 COMBINED INTRA- AND INTERBLOCK INFORMATION

To summarize the results so far, we have the following information available for

$E^{\alpha_k} \in \mathcal{E}_1$:	interblock information only
$E^{\gamma_\ell} \in \mathcal{E}_2$:	intrablock information from $qc(\boldsymbol{\gamma}_\ell)$ replicates and interblock information from $qc(\boldsymbol{\gamma}_\ell)$ replicates
$E^{\delta_m} \in \mathcal{E}_3$:	intrablock information only

We have discussed how to get these types of information in Sections 11.10.1 and 11.11.2. It remains then to combine the two types of information for $E^{\gamma_\ell} \in \mathcal{E}_2$.

11.12.1 Combined Estimators

We have seen that for $\boldsymbol{\xi}'\boldsymbol{\tau} = E_{\delta}^{\boldsymbol{\gamma}_\ell}$ the condition of Theorem 11.10 is satisfied for those replicates in which $E^{\boldsymbol{\gamma}_\ell}$ is confounded. For those replicates in which $E^{\boldsymbol{\gamma}_\ell}$ is not confounded, we have $\boldsymbol{\xi}'\boldsymbol{X}_{\beta^*(j)} = \boldsymbol{\phi}$ because of unbiasedness, and hence

$$V_j\,\xi = \left(\boldsymbol{I} + \rho^*\boldsymbol{X}_{\beta^*(j)}\boldsymbol{X}'_{\beta^*(J)}\right)\xi = \xi$$

so that the condition of Theorem 11.10 is trivially satisfied. It follows then that we can combine the estimates from all replicates additively by weighting inversely with the respective variances. Thus, letting $w = 1/\sigma_e^2$ and $w' = 1/(\sigma_e^2 + p^\ell\sigma_\beta^2)$, the combined estimator for $E_\delta^{\boldsymbol{\gamma}_\ell} - E_{\delta'}^{\boldsymbol{\gamma}_\ell}$, say $\widehat{\widehat{E}}_\delta^{\boldsymbol{\gamma}_\ell} - \widehat{\widehat{E}}_{\delta'}^{\boldsymbol{\gamma}_\ell}$, is given by

$$\widehat{\widehat{E}}_\delta^{\boldsymbol{\gamma}_\ell} - \widehat{\widehat{E}}_{\delta'}^{\boldsymbol{\gamma}_\ell} = \frac{u(\boldsymbol{\gamma}_\ell)w\left(\widehat{E}_\delta^{\boldsymbol{\gamma}_\ell} - \widehat{E}_{\delta'}^{\boldsymbol{\gamma}_\ell}\right) + c(\boldsymbol{\gamma}_\ell)w'\left(\widetilde{E}_\delta^{\boldsymbol{\gamma}_\ell} - \widetilde{E}_{\delta'}^{\boldsymbol{\gamma}_\ell}\right)}{u(\boldsymbol{\gamma}_\ell)w + c(\boldsymbol{\gamma}_\ell)w'} \tag{11.87}$$

and

$$\operatorname{var}\left(\widehat{\widehat{E}}_\delta^{\boldsymbol{\gamma}_\ell} - \widehat{\widehat{E}}_{\delta'}^{\boldsymbol{\gamma}_\ell}\right) = \frac{2\sigma_e^2}{u(\boldsymbol{\gamma}_\ell) + c(\boldsymbol{\gamma}_\ell)\rho^{-1}qp^{n-1}} \tag{11.88}$$

where $\rho = w/w'$. To estimate w and w' (or ρ), we use the B|T-ANOVA of Table 11.2. This yields

$$\widehat{\sigma}_e^2 = \frac{1}{\widehat{w}'} = \text{MS}(\boldsymbol{I}|\mathfrak{I}, \boldsymbol{X}_\rho, \boldsymbol{X}_{\beta^*}, \boldsymbol{X}_\tau) \tag{11.89}$$

and one obvious estimator for $\sigma_e^2 + p^\ell\ \sigma_\beta^2 = 1/w'$ is

$$\left(\widehat{\sigma_e^2 + p^\ell\sigma_\beta^2}\right)_1 = \frac{1}{(\widehat{w}')} = \text{MS(Remainder)} \tag{11.90}$$

Another estimator for $\sigma_e^2 + p^\ell\ \sigma_\beta^2$ can be obtained by evaluating $E[\text{MS}(\boldsymbol{X}_{\beta^*}|\mathfrak{I}, \boldsymbol{X}_\rho, \boldsymbol{X}_{\boldsymbol{\tau}})]$ in Table 11.2. In order to do that, we first obtain SS $\left(\widehat{E}^{\boldsymbol{\gamma}_\ell}\text{ vs. }\widetilde{E}^{\boldsymbol{\gamma}_\ell}\right)$ in Table 11.2b.

Let

$$\widehat{E}_\delta^{\boldsymbol{\gamma}_\ell} - \widetilde{E}_\delta^{\boldsymbol{\gamma}_\ell} = X_\delta^{\boldsymbol{\gamma}_\ell}$$

for $E^{\boldsymbol{\gamma}_\ell} \in \mathcal{E}_2$ and $\delta = 0, 1, \ldots, p-1$. Any $X_\delta^{\boldsymbol{\gamma}_\ell}$ is free of treatment effects but is a function of block effects and replicate effects. Hence comparisons among

Table 11.2 B|T-ANOVA for p^n Factorial Experiment

Source	d.f.	SS
	a. Basic Partitioning	
$\boldsymbol{X}_\rho \mid \mathfrak{I}$	$sq - 1$	Usual
$\boldsymbol{X}_\tau \mid \mathfrak{I}, \boldsymbol{X}_\rho$	$p^n - 1$	Usual
$\boldsymbol{X}_{\beta^*} \mid \mathfrak{I}, \boldsymbol{X}_\rho, \boldsymbol{X}_\tau$	$sq(p^{n-\ell} - 1) - n_1(p-1)$	Difference
$\boldsymbol{I} \mid \mathfrak{I}, \boldsymbol{X}_\rho, \boldsymbol{X}_\tau, \boldsymbol{X}_{\beta^*}$	ν_R	From Table 11.1
Total	$sqp^n - 1$	Usual
	b. Partitioning of Block Sum of Squares	
$\boldsymbol{X}_{\beta^*} \mid \mathfrak{I}, \boldsymbol{X}_\rho, \boldsymbol{X}_\tau$	$sq(p^{n-\ell} - 1) - n_1\ (p-1)$	
$\{\widehat{E}^{\boldsymbol{\gamma}_\ell} \text{ vs. } \widetilde{E}^{\boldsymbol{\gamma}_\ell}\}$	$n_2(p-1)$	See (11.93)
Remainder	ν	From Table 11.1

the $X_\delta^{\boldsymbol{\gamma}_\ell}$ for fixed $\boldsymbol{\gamma}_\ell$ and $\delta = 0, 1, \ldots, p-1$ contribute to $\text{SS}(\boldsymbol{X}_{\beta^*}|\mathfrak{I}, \boldsymbol{X}_\rho, \boldsymbol{X}^\tau)$. More specifically we have

$$\text{SS}\left(\widehat{E}^{\boldsymbol{\gamma}_\ell} \text{ vs. } \widetilde{E}^{\boldsymbol{\gamma}_\ell}\right) = \begin{bmatrix} X_0^{\boldsymbol{\gamma}_\ell} - X_{p-1}^{\boldsymbol{\gamma}_\ell} \\ X_1^{\boldsymbol{\gamma}_\ell} - X_{p-1}^{\boldsymbol{\gamma}_\ell} \\ \vdots \\ X_{p-2}^{\boldsymbol{\gamma}_\ell} - X_{p-1}^{\boldsymbol{\gamma}_\ell} \end{bmatrix}' \left[\frac{\boldsymbol{V}_{\boldsymbol{\gamma}_\ell}}{\sigma_e^2}\right]^{-1} \begin{bmatrix} X_0^{\boldsymbol{\gamma}_\ell} - X_{p-1}^{\boldsymbol{\gamma}_\ell} \\ X_1^{\boldsymbol{\gamma}_\ell} - X_{p-1}^{\boldsymbol{\gamma}_\ell} \\ \vdots \\ X_{p-2}^{\boldsymbol{\gamma}_\ell} - X_{p-1}^{\boldsymbol{\gamma}_\ell} \end{bmatrix} \tag{11.91}$$

where $\boldsymbol{V}_{\boldsymbol{\gamma}_\ell}$ is the variance–covariance matrix of the vector of comparisons $X_i^{\boldsymbol{\gamma}_\ell} - X_{p-1}^{\boldsymbol{\gamma}_\ell}$, $i = 0, 1, \ldots, p-2$. From the form of these comparisons it follows that

$$\boldsymbol{V}_{\boldsymbol{\gamma}_\ell} = \frac{s}{qu(\boldsymbol{\gamma}_\ell)\ c(\boldsymbol{\gamma}_\ell)\ p^{n-1}} (\boldsymbol{I} + \mathfrak{I}\mathfrak{I}')\sigma_e^2$$

and hence

$$\left[\frac{\boldsymbol{V}_{\boldsymbol{\gamma}_\ell}}{\sigma_e^2}\right]^{-1} = \frac{qu(\boldsymbol{\gamma}_\ell)\ c(\boldsymbol{\gamma}_\ell)\ p^{n-2}}{s} (p\boldsymbol{I} - \mathfrak{I}\mathfrak{I}')$$

We then obtain, from (11.91),

$$\text{SS}\left(\widehat{E}^{\boldsymbol{\gamma}_\ell} \text{ vs. } \widetilde{E}^{\boldsymbol{\gamma}_\ell}\right) = \frac{qu(\boldsymbol{\gamma}_\ell)\ c(\boldsymbol{\gamma}_\ell)\ p^{n-1}}{s} \sum_{\delta=0}^{p-1} [X_\delta^{\boldsymbol{\gamma}_\ell}]^2 \tag{11.92}$$

and summing (11.92) over all $E^{\gamma_\ell} \in \mathcal{E}_2$

$$\text{SS}\left(\widehat{E}^{\gamma_\ell} \text{ vs. } \widetilde{E}^{\gamma_\ell}\right) = \frac{qp^{n-1}}{s} \sum_\ell u(\boldsymbol{\gamma}_\ell)\, c(\boldsymbol{\gamma}_\ell) \sum_\delta \left[X_\delta^{\gamma_\ell}\right]^2 \tag{11.93}$$

Using the definition of $X_\delta^{\boldsymbol{\gamma}_\ell}$ and

$$\begin{aligned} E\left[X_\delta^{\boldsymbol{\gamma}_\ell}\right]^2 &= \text{var}\left(X_\delta^{\boldsymbol{\gamma}_\ell}\right) = \text{var}\left(\widehat{E}_\delta^{\boldsymbol{\gamma}_\ell}\right) + \text{var}\left(\widetilde{E}_\delta^{\boldsymbol{\gamma}_\ell}\right) \\ &= \frac{p-1}{qu(\boldsymbol{\gamma}_\ell)\, p^n}\, \sigma_e^2 + \frac{p-1}{qc(\boldsymbol{\gamma}_\ell)\, p^n} \left(\sigma_e^2 + p^\ell \sigma_\beta^2\right) \end{aligned}$$

it follows that

$$E\left[\text{SS}\left(\widehat{E}^{\gamma_\ell} \text{ vs. } \widetilde{E}^{\gamma_\ell}\right)\right] = (p-1)\left(\sigma_e^2 + \frac{u(\boldsymbol{\gamma}_\ell)}{s} p^\ell \sigma_\beta^2\right)$$

Hence

$$E\left[\text{SS}\left(\widehat{E}^{\gamma_\ell} \text{ vs. } \widetilde{E}^{\gamma_\ell}\right)\right] = (p-1)\left[n_2\sigma_e^2 + \frac{p^\ell}{s} \sum_\ell u(\boldsymbol{\gamma}_\ell)\, \sigma_\beta^2\right] \tag{11.94}$$

From (11.86) and (11.94) it follows then that

$$E\left[\text{MS}\left(\boldsymbol{X}_{\beta^*} \mid \mathfrak{I}, \boldsymbol{X}_\rho, \boldsymbol{X}_\tau\right)\right] = \sigma_e^2 + \Delta p^\ell \sigma_\beta^2 \tag{11.95}$$

where

$$\Delta = \frac{[(p-1)/s] \sum_\ell u(\boldsymbol{\gamma}_\ell) + \nu}{sq(p^{n-\ell}-1) - n_1(p-1)}$$

and ν as given in (11.66). A second estimator for w' is then obtained from (11.89) and (11.95) as

$$\begin{aligned} \left(\sigma_e^2 \,\widehat{+}\, p^\ell \sigma_\beta^2\right)_2 &= \frac{1}{\widetilde{w}'} \\ &= \left(1 - \frac{1}{\Delta}\right) \text{MS}(\boldsymbol{I} \mid \mathfrak{I}, \boldsymbol{X}_\rho, \boldsymbol{X}_{\beta^*}, \boldsymbol{X}_\tau) \\ &\quad + \frac{1}{\Delta} \text{MS}(\boldsymbol{X}_{\beta^*} \mid \mathfrak{I}, \boldsymbol{X}_\rho, \boldsymbol{X}_\tau) \end{aligned} \tag{11.96}$$

Using (11.89) and (11.96) in (11.87) yields the combined estimator for $E_\delta^{\boldsymbol{\gamma}_\ell} - E_{\delta'}^{\boldsymbol{\gamma}_\ell}$, $\delta, \delta' = 0, 1, \ldots, p-1$, $\delta \neq \delta'$.

11.12.2 Variance of Treatment Comparisons

Recalling the parameterization of the true response of a treatment combination as given in (11.33), we can estimate the difference of the effect of two treatment combinations $\boldsymbol{x}$ and $\boldsymbol{z}$, say, as

$$\widehat{\widehat{\tau}}(\boldsymbol{x}) - \widehat{\widehat{\tau}}(\boldsymbol{z}) = \sum_{\boldsymbol{\alpha} \in \mathcal{A}} \left(\widehat{\widehat{E}}^{\boldsymbol{\alpha}}_{\boldsymbol{\alpha}'\boldsymbol{x}} - \widehat{\widehat{E}}^{\boldsymbol{\alpha}}_{\boldsymbol{\alpha}'\boldsymbol{z}} \right) \tag{11.97}$$

with

$$\operatorname{var}\left(\widehat{\widehat{E}}^{\boldsymbol{\alpha}}_{\boldsymbol{\alpha}'\boldsymbol{x}} - \widehat{\widehat{E}}^{\boldsymbol{\alpha}}_{\boldsymbol{\alpha}'\boldsymbol{z}} \right) = \frac{2\sigma_e^2}{u(\boldsymbol{\alpha}) + c(\boldsymbol{\alpha})\rho^{-1}qp^{n-1}} \tag{11.98}$$

using the notation of (11.88). In (11.98) $u(\boldsymbol{\alpha}), c(\boldsymbol{\alpha}) = 0, 1, \ldots, s$ with $u(\boldsymbol{\alpha}) + c(\boldsymbol{\alpha}) = s$ for each $\boldsymbol{\alpha} \in \mathcal{A}$, with s being the totality of partitions for the p^n factorial. Since $\boldsymbol{\alpha}'\boldsymbol{x} = \boldsymbol{\alpha}'\boldsymbol{z}$ for $(p^{n-1} - 1)/(p - 1)$ different $\boldsymbol{\alpha} \in \mathcal{A}$, the right-hand side of (11.97) contains p^{n-1} terms corresponding to those $\boldsymbol{\alpha} \in \mathcal{A}^*$, say, for which $\boldsymbol{\alpha}'\boldsymbol{x} \neq \boldsymbol{\alpha}'\boldsymbol{z}$. It follows then from (11.97) that

$$\operatorname{var}\left(\widehat{\widehat{\tau}}(\boldsymbol{x}) - \widehat{\widehat{\tau}}(\boldsymbol{z})\right) = \sum_{\boldsymbol{\alpha} \in \mathcal{A}^*} \operatorname{var}\left(\widehat{\widehat{E}}^{\boldsymbol{\alpha}}_{\boldsymbol{\alpha}'\boldsymbol{x}} - \widehat{\widehat{E}}^{\boldsymbol{\alpha}}_{\boldsymbol{\alpha}'\boldsymbol{z}} \right) \tag{11.99}$$

with the appropriate expressions from (11.98) substituted in (11.99). Obviously $\mathcal{A}^*$ and hence (11.99) depend on $\boldsymbol{x}$ and $\boldsymbol{z}$ and the particular system of confounding that has been used.

Rather than compute the variance for each treatment comparison separately, it may in some situations be appropriate to consider an average variance. To obtain the average variance, we consider, for a fixed treatment combination $\boldsymbol{x}$, $\widehat{\widehat{\tau}}(\boldsymbol{x}) - \widehat{\widehat{\tau}}(\boldsymbol{z})$ for all $\boldsymbol{z} \neq \boldsymbol{x}$. For a fixed $\boldsymbol{\alpha} \in \mathcal{A}$ there exist p^{n-1} treatment combinations $\boldsymbol{z}$ that satisfy the equation

$$\alpha_1(x_1 - z_1) + \alpha_2(x_2 - z_2) + \cdots + \alpha_n(x_n - z_n) = \delta$$

for $\delta \in \mathrm{GF}(p)$. Since there are $p - 1$ nonzero $\delta \in \mathrm{GF}(p)$, it follows that for each $\boldsymbol{\alpha} \in \mathcal{A}$ there are $(p - 1)p^{n-1}$ different $\boldsymbol{z}$'s with $\boldsymbol{\alpha}'\boldsymbol{x} \neq \boldsymbol{\alpha}'\boldsymbol{z}$. Hence, using (11.98),

$$\begin{aligned} \text{av. var} &= \frac{1}{p^n - 1} \sum_{\boldsymbol{x}} \operatorname{var}\left(\widehat{\widehat{\tau}}(\boldsymbol{x}) - \widehat{\widehat{\tau}}(\boldsymbol{z})\right) \\ &= 2\frac{p-1}{p^n - 1}\sigma_e^2 \sum_{\boldsymbol{\alpha} \in \mathcal{A}} \frac{1}{(u(\boldsymbol{\alpha}) + c(\boldsymbol{\alpha})\rho^{-1})q} \end{aligned} \tag{11.100}$$

11.13 THE s^n FACTORIAL

To complete the discussion of the general symmetrical factorial experiment, we shall now indicate how the methodology developed for the p^n factorial can be extended to the s^n factorial, where $s = p^m$ and p is a prime. One way in which this is achieved is by replacing arithmetic in GF(p) for the p^n by arithmetic in GF(p^m) for the s^n factorial. Another way is to treat the $(p^m)^n$ factorial as a p^{mn} factorial, that is, mn factors with p levels each, these factors being referred to as pseudofactors.

11.13.1 Method of Galois Field Theory

Suppose we have factors $A_1, A_2, \ldots, A_n$. We denote a treatment combination as usual by $\boldsymbol{x}' = (x_1, x_2, \ldots, x_n)$ where $x_i (i = 1, 2, \ldots, n)$ takes on the values $0, 1, \ldots, s-1$. Also, we denote the elements of GF(p^m) by $u_0, u_1, \ldots, u_{s-1}$ (see Appendix A). The partitioning of the s^n treatment combinations into s mutually exclusive sets of s^{n-1} treatment combinations each associated with the interaction $E^\alpha = A_{A_1}^{\alpha_1} A_{A_2}^{\alpha_2} \ldots A_{A_n}^{\alpha_n}$ is obtained by considering the equations

$$u_{\alpha_1} u_{x_1} + u_{\alpha_2} u_{x_2} + \cdots + u_{\alpha_n} u_{x_n} = u_\delta \tag{11.101}$$

where $u_\delta \in \text{GF}(p^m)$. Each of the s equations (11.101) is satisfied by s^{n-1} treatment combinations as at least one $u_{\alpha_i} \neq u_0 = 0$. Altogether there are $(s^n - 1)/(s-1)$ such partitions and hence $(s^n - 1)/(s-1)$ effects and interactions, each accounting for $s-1$ d.f.

We illustrate this procedure in the following example.

Example 11.5 Take $p = 2$, $m = 2$, $n = 2$, that is, the 4^2 factorial. The elements of GF(2^2) are $u_0 = 0$, $u_1 = 1$, $u_2 = x$, $u_3 = x + 1$ with $P(x) = x^2 + x + 1$ (see Example A3 in Appendix A). The addition and multiplication tables for elements in GF(2^2) are

Addition

	u_0	u_1	u_2	u_3
u_0	u_0	u_1	u_2	u_3
u_1		u_0	u_3	u_2
u_2			u_0	u_1
u_3				u_0

Multiplication

	u_0	u_1	u_2	u_3
u_0	u_0	u_0	u_0	u_0
u_1		u_1	u_2	u_3
u_2			u_3	u_2
u_3				u_2

Denoting the factors by A and B, we have the following partitions and associated interactions:

Partition	Equations	Interaction
(1, 0)	$u_{x_1} = u_\delta$	A
(0, 1)	$u_{x_2} = u_\delta$	B
(1, 1)	$u_{x_1} + u_{x_2} = u_\delta$	AB
(1, 2)	$u_{x_1} + u_2 u_{x_2} = u_\delta$	AB^2
(1, 3)	$u_{x_1} + u_3 u_{x_2} = u_\delta$	AB^3

with $u_\delta \in \text{GF}(2^2)$. Using the addition and multiplication tables, we can determine easily for each treatment combination (x_1, x_2) which equation it satisfies for a particular interaction. The results of these computations are given in Table 11.3. It follows then, for example, that AB^2 is defined by comparisons among the four sets of treatment combinations:

$$
\begin{array}{lllll}
u_0: & (0,0), & (2,1), & (3,2), & (1,3) \\
u_1: & (1,0), & (3,1), & (2,2), & (0,3) \\
u_2: & (2,0), & (0,1), & (1,2), & (3,3) \\
u_3: & (3,0), & (1,1), & (0,2), & (2,3)
\end{array}
$$

□

Table 11.3 Calculation of Partitioning for Effects and Interactions for 4^2 Factorial

Treatment Combination		Effects		Interactions		
x_1	x_2	A	B	AB	AB^2	AB^3
0	0	u_0	u_0	u_0	u_0	u_0
1	0	u_1	u_0	u_1	u_1	u_1
2	0	u_2	u_0	u_2	u_2	u_2
3	0	u_3	u_0	u_3	u_3	u_3
0	1	u_0	u_1	u_1	u_2	u_3
1	1	u_1	u_1	u_0	u_3	u_2
2	1	u_2	u_1	u_3	u_0	u_1
3	1	u_3	u_1	u_2	u_1	u_0
0	2	u_0	u_2	u_2	u_3	u_1
1	2	u_1	u_2	u_3	u_2	u_0
2	2	u_2	u_2	u_0	u_1	u_3
3	2	u_3	u_2	u_1	u_0	u_2
0	3	u_0	u_3	u_3	u_1	u_2
1	3	u_1	u_3	u_2	u_0	u_3
2	3	u_2	u_3	u_1	u_3	u_0
3	3	u_3	u_3	u_0	u_2	u_1

11.13.2 Systems of Confounding

The method of constructing systems of confounding can now also be extended quite easily. Suppose we consider a s^n factorial in blocks of size s^ℓ. We choose $n-\ell$ independent interactions, $E^{\boldsymbol{\alpha}_1}, E^{\boldsymbol{\alpha}_2}, \ldots, E^{\boldsymbol{\alpha}_{n-\ell}}$ say, to be confounded with blocks. The $s^{n-\ell}$ blocks are then obtained from the equations

$$\begin{aligned}
u_{\alpha_{11}}u_{x_1} &+ u_{\alpha_{12}}u_{x_2} + \cdots + u_{\alpha_{1n}}u_{x_n} = u_{\delta_1} \\
u_{\alpha_{21}}u_{x_1} &+ u_{\alpha_{22}}u_{x_2} + \cdots + u_{\alpha_{2n}}u_{x_n} = u_{\delta_2} \\
&\vdots \\
u_{\alpha_{q1}}u_{x_1} &+ u_{\alpha_{q2}}u_{x_2} + \cdots + u_{\alpha_{qn}}u_{x_n} = u_{\delta_q}
\end{aligned} \tag{11.102}$$

where $q = n - \ell$, $\boldsymbol{\alpha}_i' = (\alpha_{i1}, \alpha_{i2}, \ldots, \alpha_{in})$, $u_{\delta_i} \in \text{GF}(s)$, $i = 1, 2, \ldots, q$. For a fixed set $(u_{\delta_1}, u_{\delta_2}, \ldots, u_{\delta_q})$ Eqs. (11.102) are satisfied by s^ℓ treatment combinations, which then form a block.

Further, any treatment combination that satisfies, say, the equations

$$u_{\alpha_1}u_{x_1} + u_{\alpha_2}u_{x_2} + \cdots + u_{\alpha_n}u_{x_n} = u_{\delta_1}$$

and

$$u_{\beta_1}u_{x_1} + u_{\beta_2}u_{x_2} + \cdots + u_{\beta_n}u_{x_n} = u_{\delta_2}$$

also satisfies the equation

$$\begin{aligned}
&\left(u_{\alpha_1} + u_\lambda u_{\beta_1}\right)u_{x_1} + \left(u_{\alpha_2} + u_\lambda u_{\beta_2}\right)u_{x_2} + \cdots \\
&\quad + \left(u_{\alpha_n} + u_\lambda u_{\beta_n}\right)u_{x_n} = u_{\delta_1} + u_\lambda u_{\delta_2}
\end{aligned} \tag{11.103}$$

for any $u_\lambda \in \text{GF}(p^m)$. If $u_{\delta_1}, u_{\delta_2}$ now take on all the values in $\text{GF}(p^m)$, then the right-hand side of (11.103) takes on all values of $\text{GF}(p^m)$ exactly $s = p^m$ times for fixed u_λ. Hence (11.103) provides a partition of the s^n treatment combinations. The interaction associated with it is, by analogy to the p^n case, referred to as a GI of $A_1^{\alpha_1}A_2^{\alpha_2}\ldots A_n^{\alpha_n}$ and $A_1^{\beta_1}A_2^{\beta_2}\ldots A_n^{\beta_n}$, denoted by $A_1^{\gamma_1}A_2^{\gamma_2}\ldots A_n^{\gamma_n}$ with $(\gamma_1, \gamma_2, \ldots, \gamma_n)$ to be determined from

$$\begin{aligned}
u_{\alpha_1} &+ u_\lambda u_{\beta_1} = u_{\gamma_1} \\
u_{\alpha_2} &+ u_\lambda u_{\beta_2} = u_{\gamma_2} \\
&\vdots \\
u_{\alpha_n} &+ u_\lambda u_{\beta_n} = u_{\gamma_n}
\end{aligned} \tag{11.104}$$

Since $u_\lambda \in \mathrm{GF}(p^m)$, $u_\lambda \neq u_0$, it follows that $A_1^{\alpha_1} A_2^{\alpha_2} \ldots A_n^{\alpha_n}$ and $A_1^{\beta_1} A_2^{\beta_2} \ldots A_n^{\beta_n}$ have $s - 1$ GIs among them. Extending this argument (see Section 11.4) implies then that with $E^{\alpha_1}, E^{\alpha_2}, \ldots, E^{\alpha_q}$ all their GIs are also confounded with blocks, accounting for the $s^{n-\ell}$ degrees of freedom among blocks.

As in the p^n case, the composition of the blocks can be obtained by varying the right-hand side of (11.102). Alternatively, we can obtain the IBSG by considering (11.102) with $u_{\delta_i} = u_0 (i = 1, 2, \ldots, q)$ and then generate the remaining blocks in the familiar way by adding treatment combinations to those of the IBSG using $\mathrm{GF}(p^m)$ arithmetic. As an illustration, consider the following example.

Example 11.6 Suppose we have a 4^3 factorial in blocks of size 4, that is, $p = 2, m = 2, n = 3, \ell = 1$. We have 16 blocks, that is, we confound five interactions with blocks. Suppose we confound AB and BC^2, that is, $\boldsymbol{\alpha}_1' = (1, 1, 0)$, $\boldsymbol{\alpha}_2' = (0, 1, 2)$. From (11.104) with $u_\lambda = u_1, u_2, u_3$ it follows, using the addition and multiplication tables from Example 11.5, that

$$\begin{array}{lll} u_1 + u_1 u_0 = u_1 & u_1 + u_2 u_0 = u_1 & u_1 + u_3 u_0 = u_1 \\ u_1 + u_1 u_1 = u_0 & u_1 + u_2 u_1 = u_3 & u_1 + u_3 u_1 = u_2 \\ u_0 + u_1 u_2 = u_2 & u_0 + u_2 u_2 = u_3 & u_0 + u_3 u_2 = u_1 \end{array}$$

and hence AC^2, AB^3C^3, and AB^2C are also confounded with blocks. As a result, 9 d.f. belonging to 2-factor interactions and 6 d.f. belonging to 3-factor interactions are confounded with blocks. The IBSG is obtained by considering the equations

$$\begin{aligned} u_{x_1} + u_{x_2} \qquad\qquad &= u_0 \\ u_{x_2} + u_2 u_{x_3} &= u_0 \end{aligned}$$

They are satisfied by the treatment combinations

$$\begin{aligned} (u_0, u_0, u_0) &\sim (0, 0, 0) \\ (u_1, u_1, u_3) &\sim (1, 1, 3) \\ (u_2, u_2, u_1) &\sim (2, 2, 1) \\ (u_3, u_3, u_2) &\sim (3, 3, 2) \end{aligned}$$

where the expressions on the right are the (x_1, x_2, x_3). Another block can be constructed by noting that, for example, the treatment combination (u_1, u_2, u_3) is not in the IBSG. Hence, adding this treatment combination to each in the

IBSG, we obtain the block consisting of

$$
\begin{aligned}
(u_0 + u_1, u_0 + u_2, u_0 + u_3) &= (u_1, u_2, u_3) \sim (1, 2, 3) \\
(u_1 + u_1, u_1 + u_2, u_3 + u_3) &= (u_0, u_3, u_0) \sim (0, 3, 0) \\
(u_2 + u_1, u_2 + u_2, u_1 + u_3) &= (u_3, u_0, u_2) \sim (3, 0, 2) \\
(u_3 + u_1, u_3 + u_2, u_2 + u_3) &= (u_2, u_1, u_1) \sim (2, 1, 1)
\end{aligned}
$$

This process is then continued in the obvious and familiar way. □

An equivalent system of confounding using SAS PROC FACTEX is given in Table 11.4. It, too, confounds 9 d.f. belonging to 2-factor interactions (as indicated by the Aliasing Structure) and 6 d.f. belonging to the 3-factor interaction. We repeat that SAS uses a different parameterization than the one we have used.

Finally, we mention that Theorem 11.7 holds for the s^n factorial also. All we need to do is replace p by $s = p^m$ and generate the IBSG by using GF(p^m) arithmetic. For example, it is possible to accommodate $n = 5$ factors with four levels each in blocks of size 16 without confounding main effects and 2-factor interactions.

11.13.3 Method of Pseudofactors

Let $A_1, A_2, \ldots, A_n$ again denote the factors in the $(p^m)^n = s^n$ system. We then associate with factor A_i the m pseudofactors $X_{i1}, X_{i2}, \ldots, X_{im}$ $(i = 1, 2, \ldots, n)$ each at p levels. This association implies the following: (i) If a_i is the level of factor A_i in a treatment combination and x_{ij} is the level of factor X_{ij} in a treatment combination belonging to the p^m system $\{X_{i1}, X_{i2}, \ldots, X_{im}\}$, we then set up the correspondence

$$
\begin{aligned}
(a_1, a_2, \ldots, a_n) = (&x_{11}, x_{12}, \ldots, x_{1m}; x_{21}, x_{22}, \ldots, x_{2m}; \\
&\ldots; x_{n1}, x_{n2}, \ldots, x_{nm})
\end{aligned}
$$

that is, each level a_i of the factor A_i corresponds to a level combination $(x_{i1}, x_{i2}, \ldots, x_{im})$ of the pseudofactors $X_{i1}, X_{i2}, \ldots, X_{im} (i = 1, 2, \ldots, n)$. (ii) The main effects and interactions in the s^n system correspond to sets of main effects and interactions within and among the n p^m systems. More formally we express this as follows: Let $\{E^{\boldsymbol{\alpha}_i}, \boldsymbol{\alpha}_i \in \mathcal{A}_i\}$ denote the set of main effects and interactions associated with the p^m system $\{X_{i1}, X_{i2}, \ldots, X_{im}\}$ and let $E^{\boldsymbol{\alpha}_i} \times E^{\boldsymbol{\alpha}_j}$, $(\boldsymbol{\alpha}_i, \boldsymbol{\alpha}_j) \in \mathcal{A}_i \times \mathcal{A}_j$, represent an interaction among factors in the p^m systems $\{X_{i1}, X_{i2}, \ldots, X_{im}\}$ and $\{X_{j1}, X_{j2}, \ldots, X_{jm}\}$ with $\mathcal{A}_i \times \mathcal{A}_j$ denoting the totality of the corresponding partitions $(\boldsymbol{\alpha}_i, \boldsymbol{\alpha}_j)$. Continuing in this fashion,

Table 11.4 4^3 Factorial in Blocks of Size 4 (Confounding 3 d.f. From Each 2-Factor Interaction and 6 d.f. From the 3-Factor Interaction)

```
options nodate pageno=1;
proc factex;
factors A B C/nlev=4;
blocks size=4;
model est=(A B C);
examine design confounding aliasing;
output out=design blockname=block nvals=(1 2 3 4 5 6 7 8 9
                                  10 11 12 13 14 15 16);
title1 'TABLE 11.4';
title2 '4**3 FACTORIAL IN BLOCKS OF SIZE 4';
title3 'CONFOUNDING 3 D.F. FROM EACH 2-FACTOR INTERACTION';
title4 'AND 6 D.F. FROM THE 3-FACTOR INTERACTION';
run;

proc print data=design;
run;
```

The FACTEX Procedure

Design Points

Experiment Number	A	B	C	Block
1	0	0	0	1
2	0	0	1	16
3	0	0	2	6
4	0	0	3	11
5	0	1	0	12
6	0	1	1	5
7	0	1	2	15
8	0	1	3	2
9	0	2	0	14
10	0	2	1	3
11	0	2	2	9
12	0	2	3	8
13	0	3	0	7
14	0	3	1	10
15	0	3	2	4
16	0	3	3	13
17	1	0	0	8
18	1	0	1	9
19	1	0	2	3
20	1	0	3	14
21	1	1	0	13
22	1	1	1	4
23	1	1	2	10

Table 11.4 (*Continued*)

24	1	1	3	7
25	1	2	0	11
26	1	2	1	6
27	1	2	2	16
28	1	2	3	1
29	1	3	0	2
30	1	3	1	15
31	1	3	2	5
32	1	3	3	12
33	2	0	0	10
34	2	0	1	7
35	2	0	2	13
36	2	0	3	4
37	2	1	0	3
38	2	1	1	14
39	2	1	2	8
40	2	1	3	9
41	2	2	0	5
42	2	2	1	12
43	2	2	2	2
44	2	2	3	15
45	2	3	0	16
46	2	3	1	1
47	2	3	2	11
48	2	3	3	6
49	3	0	0	15
50	3	0	1	2
51	3	0	2	12
52	3	0	3	5
53	3	1	0	6
54	3	1	1	11
55	3	1	2	1
56	3	1	3	16
57	3	2	0	4
58	3	2	1	13
59	3	2	2	7
60	3	2	3	10
61	3	3	0	9
62	3	3	1	8
63	3	3	2	14
64	3	3	3	3

Block Pseudofactor Confounding Rules

```
[B1] = (3*A)+(3*B)+(3*C)
[B2] = A+(2*B)+(3*C)
```

Table 11.4 *(Continued)*

```
Aliasing Structure

            A
        (2*A)
        (3*A)
            B
        (2*B)
        (3*B)
            C
        (2*C)
        (3*C)
            A +     B
[B]  =  (2*A) +     B
        (3*A) +     B
            A + (2*B)
        (2*A) + (2*B)
[B]  =  (3*A) + (2*B)
[B]  =      A + (3*B)
        (2*A) + (3*B)
        (3*A) + (3*B)
            A +     C
        (2*A) +     C
[B]  =  (3*A) +     C
[B]  =      A + (2*C)
        (2*A) + (2*C)
        (3*A) + (2*C)
            A + (3*C)
[B]  =  (2*A) + (3*C)
        (3*A) + (3*C)
            B +     C
[B]  =  (2*B) +     C
        (3*B) +     C
            B + (2*C)
        (2*B) + (2*C)
[B]  =  (3*B) + (2*C)
[B]  =      B + (3*C)
        (2*B) + (3*C)
        (3*B) + (3*C)
```

Obs	block	A	B	C
1	1	0	0	0
2	1	1	2	3
3	1	2	3	1
4	1	3	1	2
5	2	0	1	3
6	2	1	3	0
7	2	2	2	2

Table 11.4 (*Continued*)

8	2	3	0	1
9	3	0	2	1
10	3	1	0	2
11	3	2	1	0
12	3	3	3	3
13	4	0	3	2
14	4	1	1	1
15	4	2	0	3
16	4	3	2	0
17	5	0	1	1
18	5	1	3	2
19	5	2	2	0
20	5	3	0	3
21	6	0	0	2
22	6	1	2	1
23	6	2	3	3
24	6	3	1	0
25	7	0	3	0
26	7	1	1	3
27	7	2	0	1
28	7	3	2	2
29	8	0	2	3
30	8	1	0	0
31	8	2	1	2
32	8	3	3	1
33	9	0	2	2
34	9	1	0	1
35	9	2	1	3
36	9	3	3	0
37	10	0	3	1
38	10	1	1	2
39	10	2	0	0
40	10	3	2	3
41	11	0	0	3
42	11	1	2	0
43	11	2	3	2
44	11	3	1	1
45	12	0	1	0
46	12	1	3	3
47	12	2	2	1
48	12	3	0	2
49	13	0	3	3
50	13	1	1	0
51	13	2	0	2
52	13	3	2	1
53	14	0	2	0
54	14	1	0	3
55	14	2	1	1

Table 11.4 (*Continued*)

56	14	3	3	2
57	15	0	1	2
58	15	1	3	1
59	15	2	2	3
60	15	3	0	0
61	16	0	0	1
62	16	1	2	2
63	16	2	3	0
64	16	3	1	3

we then have the correspondence

$$A_i \cong \{E^{\alpha_i}: \boldsymbol{\alpha}_i \in \mathcal{A}_i\}$$

$$A_i \times A_j \cong \{E^{\alpha_i} \times E^{\alpha_j}: (\boldsymbol{\alpha}_i, \boldsymbol{\alpha}_j) \in \mathcal{A}_i \times \mathcal{A}_j\}$$

$$A_i \times A_j \times A_k \cong \{E^{\alpha_i} \times E^{\alpha_j} \times E^{\alpha_k}: (\boldsymbol{\alpha}_i, \boldsymbol{\alpha}_j, \boldsymbol{\alpha}_k) \in \mathcal{A}_i \times \mathcal{A}_j \times \mathcal{A}_k\}$$

and so on, up to

$$A_1 \times A_2 \times \cdots \times A_n \cong \{\Pi \times E^{\alpha_i}: (\boldsymbol{\alpha}_i, \boldsymbol{\alpha}_2, \ldots, \boldsymbol{\alpha}_n) \in \Pi \times \mathcal{A}_i\}$$

This implies, of course, that the s^n system can be handled entirely by GF(p) arithmetic rather than by GF(p^m) arithmetic. As an illustration, consider the following example.

Example 11.7 Suppose we have a $(2^2)^3$ factorial. For ease of notation we use the following correspondence for the factors $(A_i; X_{i1}, X_{i2})$:

4^3 System	2^6 System
A	X_1, X_2
B	Y_1, Y_2
C	Z_1, Z_2

and for the levels $(a_i; x_{i1}, x_{i2})$:

4^3 System	2^6 System
0	0, 0
1	1, 0
2	0, 1
3	1, 1

Table 11.5 Correspondence of Main Effects and Interactions for 4^3 and 2^6 System

4^3 System	2^6 System
A	X_1, X_2, X_1X_2
B	Y_1, Y_2, Y_1Y_2
C	Z_1, Z_2, Z_1Z_2
$A \times B$	$X_1Y_1, X_1Y_2, X_1Y_1Y_2$
	$X_2Y_1, X_2Y_2, X_2Y_1Y_2$
	$X_1X_2Y_1, X_1X_2Y_2, X_1X_2Y_1Y_2$
$A \times C$	$X_1Z_1, X_1Z_2, X_1Z_1Z_2$
	$X_2Z_1, X_2Z_2, X_2Z_1Z_2$
	$X_1X_2Z_1, X_1X_2Z_2, X_1X_2Z_1Z_2$
$B \times C$	$Y_1Z_1, Y_1Z_2, Y_1Z_1Z_2$
	$Y_2Z_1, Y_2Z_2, Y_2Z_1Z_2$
	$Y_1Y_2Z_1, Y_1Y_2Z_2, Y_1Y_2Z_1Z_2$
$A \times B \times C$	$X_1Y_1Z_1, X_1Y_1Z_2, X_1Y_1Z_1Z_2$
	$X_1Y_2Z_1, X_1Y_2Z_2, X_1Y_2Z_1Z_2$
	$X_1Y_1Y_2Z_1, X_1Y_1Y_2Z_2, X_1Y_1Y_2Z_1Z_2$
	$\vdots$
	$X_1X_2Y_1Y_2Z_1, X_1X_2Y_1Y_2Z_2, X_1X_2Y_1Y_2Z_1Z_2$

The correspondence of effects and interactions in both system is then as given in Table 11.5. This representation implies that all main effects and interactions in the 4^3 system are expressed as sets of single degrees of freedom contrasts; for example, the 3 d.f. for A are represented by the three orthogonal contrasts

$$\begin{aligned} x_1 = 0 &\quad \text{vs.} \quad & x_1 = 1 \\ x_2 = 0 &\quad \text{vs.} \quad & x_2 = 1 \\ x_1 + x_2 = 0 &\quad \text{vs.} \quad & x_1 + x_2 = 1 \end{aligned}$$ □

The value of the method of pseudofactors just described apart from the reduction to GF(p) arithmetic lies in its flexibility of constructing systems of confounding in blocks of size p^ℓ rather than $(p^m)^\ell$, although it requires a certain amount of care in order not to confound contrasts belonging to main effects or low-order interactions if that can be avoided.

Example 11.8 For the 4^3 factorial we may want to consider blocks of size 8. According to Theorem 11.7, we can find a system of confounding for the

2^6 factorial in blocks of size 2^3 without confounding main effects and 2-factor interactions. This assures us that no contrasts belonging to A, B, or C in the 4^3 system have to be confounded. A suitable system of confounding would be to confound $X_1X_2Y_2$, $X_1Y_1Z_1$, $X_2Y_1Z_2$ and their GIs $X_2Y_1Y_2Z_1$, $X_1Y_1Y_2Z_2$, $X_1X_2Z_1Z_2$, $Y_2Z_1Z_2$. Referring to Table 11.5, this means that we confound four contrasts belonging to $A \times B \times C$, namely ($X_1Y_1Z_1$, $X_2Y_1Z_2$, $X_2Y_1Y_2Z_1$, $X_1Y_1Y_2Z_2$), and one contrast each belonging to $A \times B$, namely ($X_1X_2Y_2$), $A \times C$, namely ($X_1X_2Z_1Z_2$), and $B \times C$, namely ($Y_2Z_1Z_2$). We mention that confounding $Y_2Z_1Z_2$ is not necessarily more desirable than confounding, for example, Y_2Z_1 since both belong to the $B \times C$ interaction. However, confounding X_1X_2, Y_1Y_2, Z_1Z_2 is undesirable.

It should be obvious how this method of pseudofactors is then used to analyze data from such a design. For Example 11.8, with q replicates, we have the T|B-ANOVA as given in Table 11.6, where

$$\sum{}' SS(X_1^{\alpha_1} X_2^{\alpha_2} Y_1^{\gamma_1} Y_1^{\gamma_2}) = SS(X_1Y_1) + SS(X_1Y_2) + SS(X_1Y_1Y_2) + SS(X_2Y_1) + SS(X_2Y_2) + SS(X_2Y_1Y_2) + SS(X_1X_2Y_1) + SS(X_1X_2Y_1Y_2)$$

as follows from Table 11.5 and the fact that $X_1X_2Y_2$ is confounded with blocks. The other sums of squares in Table 11.6 are obtained similarly. □

Table 11.6 T|B-ANOVA for 4^3 Factorial in Blocks of Size 8

Source	d.f.	SS
$\boldsymbol{X}_\rho \mid \mathfrak{I}$	$q-1$	Usual
$\boldsymbol{X}_{\beta^*} \mid \mathfrak{I}, \boldsymbol{X}_\rho$	$7q$	Usual
$\boldsymbol{X}_\tau \mid \mathfrak{I}, \boldsymbol{X}_\rho, \boldsymbol{X}_{\beta^*}$	56	
A	3	$SS(X_1) + SS(X_2) + SS(X_1X_2)$
B	3	$SS(Y_1) + SS(Y_2) + SS(Y_1Y_2)$
$A \times B$	8	$\sum{}' SS(X_1^{\alpha_1} X_2^{\alpha_2} Y_1^{\gamma_1} Y_2^{\gamma_2})$
C	3	$SS(Z_1) + SS(Z_2) + SS(Z_1Z_2)$
$A \times C$	8	$\sum{}' SS(X_1^{\alpha_1} X_2^{\alpha_2} Z_1^{\delta_1} Z_2^{\delta_2})$
$B \times C$	8	$\sum{}' SS(Y_1^{\gamma_1} Y_2^{\gamma_2} Z_1^{\delta_1} Z_2^{\delta_2})$
$A \times B \times C$	23	$\sum{}' SS(X_1^{\alpha_1} X_2^{\alpha_2} Y_1^{\gamma_1} Y_2^{\gamma_2} Z_1^{\delta_1} Z_2^{\delta_2})$
$\boldsymbol{I} \mid \mathfrak{I}, \boldsymbol{X}_\rho, \boldsymbol{X}_{\beta^*}, \boldsymbol{X}_\tau$	$56(q-1)$	Difference
Total	$64q-1$	Usual

11.13.4 The $(p_1 \times p_2 \times \cdots \times p_m)^n$ Factorial

The method of pseudofactors can also be used when the number of levels, s, is a product of prime numbers, say $s = p_1 p_2 \ldots p_m$, or even more generally, a product of prime powers, say $s = p_1^{n_1} p_2^{n_2} \ldots p_m^{n_m}$. Using pseudofactors then means that the $(p_1^{n_1} p_2^{n_2} \ldots p_m^{n_m})^n$ factorial can be treated by combining $\left(p_1^{n_1}\right)^n, \left(p_2^{n_2}\right)^n, \ldots, \left(p_m^{n_m}\right)^n$ factorials. Rather than discuss the general case, we shall describe briefly how to use this method for the 6^n factorial, that is, $s = 6$, $p_1 = 2$, $p_2 = 3$.

Denote the factors for the 6^n factorial by $A_1, A_2, \ldots, A_n$. For A_i introduce the pseudofactors X_{i1}, X_{i2}, where X_{i1} has $p_1 = 2$ levels and X_{i2} has $p_2 = 3$ levels $(i = 1, 2, \ldots, n)$. From our discussion in Section 11.13.3 it is clear then how we establish a correspondence between the levels of the six-level factor and those of the pair of two- and three-level factors:

6^n System	$2^n \cdot 3^n$ System
0	0, 0
1	1, 0
2	0, 1
3	1, 1
4	0, 2
5	1, 2

Similarly, the correspondence between main effects and interactions in the two systems is exemplified in Table 11.7.

This correspondence also suggests various possibilities for constructing systems of confounding. For example, for the 6^2 factorial we may consider blocks of size 18, 12, 9, 6. The allocation of the treatment combinations to the blocks can be achieved easily by using the methods for constructing systems of confounding for the 2^2 and/or 3^2 factorials discussed in Chapters 8 and 10. The basic ideas are summarized in Table 11.8. Notice that for blocks of size 18 and 9 we use confounding in the 2^2 system only, for blocks of size 12 we use the 3^2 system only, but for blocks of size 6 we use both the 2^2 and the 3^2 system.

The idea is, for example for blocks of size 12, to first construct blocks of size 3 in the 3^2 system by confounding X_{21} and then "augment" those blocks by "adding on" to each treatment combination (in the 3^2 system) in a given block all the treatment combinations for the 2^2 system (actually, this method can be more formally described as the Kronecker product design of a system of confounding for the 3^2 factorial with the full 2^2 factorial). If we denote a treatment combination by $(x_{11}, x_{21}, x_{12}, x_{22})$ where x_{11} and x_{21} refer to the 2^2 system and x_{12} and x_{22} to the 3^2 system, then confounding X_{21} leads to the three "intermediate" blocks given in Table 11.9a. These are augmented to give the blocks in Table 11.9b and, using the correspondence set up for the 6^n and $2^n 3^n$ systems, the final blocks in the 6^2 system in Table 11.9c. A closer look at Table 11.9c reveals the obvious, namely that not all contrasts defining the main

Table 11.7 Correspondence Between Main Effects and Interactions in 6^n and 2^n3^n Systems

6^n System	d.f.	2^n3^n System	d.f.
A_i	5	X_{i1}	1
		X_{i2}	2
		$X_{i1}X_{i2}$	2
A_iA_j	25	$X_{i1}X_{j1}$	1
		$X_{i1}X_{j2}$	2
		$X_{i2}X_{j1}$	2
		$X_{i2}X_{j2}$	2
		$X_{i2}X_{j2}^2$	2
		$X_{i1}X_{j1}X_{i2}$	2
		$X_{i1}X_{j1}X_{j2}$	2
		$X_{i1}X_{i2}X_{j2}$	2
		$X_{i1}X_{i2}X_{j2}^2$	2
		$X_{j1}X_{i2}X_{j2}$	2
		$X_{j1}X_{i2}X_{j2}^2$	2
		$X_{i1}X_{j1}X_{i2}X_{j2}$	2
		$X_{i1}X_{j1}X_{i2}X_{j2}^2$	2

Table 11.8 Systems of Confounding for 6^2 Factorial

	Confound	
Block Size	In 2^23^2 System	In 6^2 System[a]
18	$X_{11}X_{21}$	$A_1A_2(1)$
12	X_{12}	$A_1(2)$
9	X_{11}, X_{21}, $X_{11}X_{21}$	$A_1(1)$, $A_2(1)$, $A_1A_2(1)$
6	$X_{11}X_{21}$, $X_{12}X_{22}$, $X_{11}X_{12}X_{21}X_{22}$	$A_1A_2(5)$

[a]Number in parentheses indicates the number of degrees of freedom confounded from the indicated interaction.

effect A are estimable because they are confounded with blocks. This is also indicated in the third column of Table 11.8, where, for example, $A_1(2)$ means that 2 d.f. for main effect A_1 are confounded with blocks.

It is for this reason that this particular system of confounding by itself may not be the most useful. A system of partial confounding with X_{22} confounded in a second replicate solves this problem to some extent. A better method may be, however, to confound 2 d.f. from the A_1A_2 interaction, for example, to construct

Table 11.9 6^2 Factorial in Blocks of Size 12

Block		
1	2	3
a. Intermediate Blocks in 3^2 System		
(0 0)	(1 0)	(2 0)
(0 1)	(1 1)	(2 1)
(0 2)	(1 2)	(2 2)
b. Blocks in the $2^2 3^2$ System		
0 0 0 0	0 0 1 0	0 0 2 0
1 0 0 0	1 0 1 0	1 0 2 0
0 1 0 0	0 1 1 0	0 1 2 0
1 1 0 0	1 1 1 0	1 1 2 0
0 0 0 1	0 0 1 1	0 0 2 1
1 0 0 1	1 0 1 1	1 0 2 1
0 1 0 1	0 1 1 1	0 1 2 1
1 1 0 1	1 1 1 1	1 1 2 1
0 0 0 2	0 0 1 2	0 0 2 2
1 0 0 2	1 0 1 2	1 0 2 2
0 1 0 2	0 1 1 2	0 1 2 2
1 1 0 2	1 1 1 2	1 1 2 2
c. Blocks in the 6^2 System		
0 0	2 0	4 0
1 0	3 0	5 0
0 1	2 1	4 1
1 1	3 1	5 1
0 2	2 2	4 2
1 2	3 2	5 2
0 3	2 3	4 3
1 3	3 3	5 3
0 4	2 4	4 4
1 4	3 4	5 4
0 5	2 5	4 5
1 5	3 5	5 5

the intermediate blocks by confounding $X_{11}X_{22}$ (see Table 11.7). Such a method will be discussed in Section 11.14.

The design displayed in Table 11.9 can also be produced using SAS PROC FACTEX as given in Table 11.10

Finally, a method of combining systems of confounding for different component systems, as for the 6^2 factorial in blocks of size 6 (see Table 11.8), will be described in Section 12.2.

Table 11.10 6^2 Factorial in Blocks of Size 12 (Using Pseudofactors)

```
options nodate pageno=1;
proc factex;
factors C D/nlev=3;
blocks size=3;
model est=(D C*D);
output out=cddesign;
run;

factors A B;
output out=abdesign
designrep=cddesign;
run;

data abdesign (drop=A B C D);
 set abdesign;
 if A=-1 and C=-1 then F1=0;
 if A=1 and C=-1 then F1=1;
 if A=-1 and C=0 then F1=2;
 if A=1 and C=0 then F1=3;
 if A=-1 and C=1 then F1=4;
 if A=1 and C=1 then F1=5;
 if B=-1 and D=-1 then F2=0;
 if B=1 and D=-1 then F2=1;
 if B=-1 and D=0 then F2=2;
 if B=1 and D=0 then F2=3;
 if B=-1 and D=1 then F2=4;
 if B=1 and D=1 then F2=5;
run;

proc print data=abdesign;
title1 'TABLE 11.10';
title2 '6**2 FACTORIAL IN BLOCKS OF SIZE 12';
title3 '(USING PSEUDO-FACTORS)';
run;
```

Obs	BLOCK	F1	F2
1	1	0	0
2	1	0	1
3	1	1	0
4	1	1	1
5	1	0	2
6	1	0	3
7	1	1	2
8	1	1	3
9	1	0	4
10	1	0	5

Table 11.10 (*Continued*)

11	1	1	4
12	1	1	5
13	2	4	0
14	2	4	1
15	2	5	0
16	2	5	1
17	2	4	2
18	2	4	3
19	2	5	2
20	2	5	3
21	2	4	4
22	2	4	5
23	2	5	4
24	2	5	5
25	3	2	0
26	3	2	1
27	3	3	0
28	3	3	1
29	3	2	2
30	3	2	3
31	3	3	2
32	3	3	3
33	3	2	4
34	3	2	5
35	3	3	4
36	3	3	5

11.14 GENERAL METHOD OF CONFOUNDING FOR THE SYMMETRICAL FACTORIAL EXPERIMENT

We have so far considered the s^n factorial where s is (i) a prime, (ii) a prime power, or (iii) a product of primes. For (i) and (ii) our discussion has centered on the fact that the $s^n - 1$ d.f. can be partitioned orthogonally into $(s^n - 1)/(s - 1)$ components with $s - 1$ d.f. each. We saw that this provides a simple mechanism for constructing blocks of size $s^\ell (\ell < n)$ by confounding some of these components with blocks. For (iii) we made use of the same principle and indicated how, for example for $s = p_1 p_2$, blocks of size $p_1^{\ell_1} p_2^{\ell_2}$ $(\ell_2, \ell_2 \leq n)$ can be constructed. In all three cases this procedure may not always be entirely satisfactory for two reasons: (i) the block sizes are too restricted or (ii) information on important effects may have to be sacrificed (due to confounding). The second problem can be overcome to some extent by using a system of partial confounding at the cost of additional replications.

The problem of restricted block sizes can often also be overcome by using the factorial treatment design together with existing incomplete block designs, specifically BIB and PBIB designs. The drawback with this approach, however, is that we are giving up complete control over what information to sacrifice. In other words, we may have to give up partial information about many effects, including perhaps main effects and low-order interactions. This is expressed in reduced efficiency, that is, increased variance, rather than in terms of complete loss of certain degrees of freedom for some effects. We shall present an alternative method that uses generalized cyclic incomplete block designs (see, e.g., John, 1987) the construction of which attempts to take into account that we may want to retain information about main effects but may be willing to sacrifice some information about 2-factor interactions, for example, if that is possible at all. Before describing the method, however, it is useful to present some ideas about a calculus for factorial arrangements as developed by Kurkjian and Zelen (1962, 1963).

11.14.1 Factorial Calculus

We consider the s^n factorial, where s is a positive integer. Denote a treatment combination by $\boldsymbol{x} = (x_1, x_2, \ldots, x_n)'$ where x_i represents the level of the ith factor A_i with $0 \leq x_i \leq s-1$ $(i = 1, 2, \ldots, n)$. Let $\boldsymbol{\theta} = (0, 1, \ldots, s-1)'$ denote the vector of the s levels. Then the *symbolic direct product* (SDP) $\boldsymbol{\theta} \otimes \boldsymbol{\theta}$ is defined as (see Kurkjian and Zelen, 1962)

$$\begin{aligned}\boldsymbol{\theta} \otimes \boldsymbol{\theta} = [&(0,0), (0,1), \ldots, (0, s-1), (1,0),\\ &(1,1), \ldots, (1, s-1), \ldots, (s-1, 0),\\ &(s-1, 1), \ldots, (s-1, s-1)] \qquad (11.105)\end{aligned}$$

and $\boldsymbol{\theta} \otimes \boldsymbol{\theta} \otimes \theta$ as

$$\begin{aligned}\boldsymbol{\theta} \otimes \boldsymbol{\theta} \otimes \boldsymbol{\theta} = [&(0,0,0), (0,0,1), (0,0,2), \ldots, (0,0,s-1), (0,1,0),\\ &(0,1,1), \ldots, (0,1,s-1), \ldots, (s-1, s-1, s-1)]\end{aligned}$$

The array of all s^n treatment combinations can then be represented by the n-fold SDP $\boldsymbol{\theta}^{\otimes n} = \boldsymbol{\theta} \otimes \boldsymbol{\theta} \otimes \cdots \otimes \boldsymbol{\theta}$. This will be referred to as the standard arrangement (note that this is different from the standard order given in Chapter 7 for the 2^n factorial). Further, let $\tau(\boldsymbol{x})$ denote the true effect of the treatment combination $\boldsymbol{x}$ and let $\boldsymbol{\tau}$ be the vector of all treatment effects arranged in standard order. A useful reparameterization of $\boldsymbol{\tau}$ is given by expressing each $\tau(\boldsymbol{x})$ in terms of main effects, 2-factor interaction effects, and so on. We write

$$\begin{aligned}\tau(\boldsymbol{x}) = \sum_{i=1}^{n} a_i(x_i) + \sum_{\substack{i=1 \\ 1 \leq i < j \leq n}}^{n} \sum_{j=1}^{n} a_{ij}(x_i, x_j) + \cdots \\ + a_{12\ldots n}(x_1, x_2, \ldots, x_n) \qquad (11.106)\end{aligned}$$

where $a_i(x_i)$ represents the main effect of factor A_i at level x_i, $a_{ij}(x_i, x_j)$ represents the 2-factor interaction effect between factors A_i and A_j at levels x_i and x_j, respectively, and so on. These effects are defined such that

$$\sum_{x_i=0}^{s-1} a_i(x_i) = 0 \qquad \text{for every } i$$

$$\sum_{x_i=0}^{s-1} a_{ij}(x_i, x_j) = 0 = \sum_{x_j=0}^{s-1} a_{ij}(x_i, x_j) \qquad \text{for every pair } (i \neq j), \text{ etc.}$$

that is, every effect sums to zero over the levels of any factor. Note again that the reparameterization (11.106) is different from that given in (11.33). Each of the effects is expressible as a contrast in the $\tau(\boldsymbol{x})$, that is,

$$\Phi = \sum_{x_1} \sum_{x_2} \cdots \sum_{x_n} \phi(x_1, x_2, \ldots, x_n)\tau(x_1, x_2, \ldots, x_n) \tag{11.107}$$

with $\sum_{x_1} \sum_{x_2} \cdots \sum_{x_n} \phi(x_1, x_2, \ldots, x_n) = 0$. Following Bose (1947b) a contrast of the form (11.107) belongs to the q-factor interaction $A_{i_1} A_{i_2} \ldots A_{i_q}$ if the $\phi(x_1, x_2, \ldots, x_n)$ depend only on $x_{i_1}, x_{i_2}, \ldots, x_{i_q} (1 \leq i_1 < i_2 < \cdots < i_q \leq n; 1 \leq q \leq n)$.

Now define $z = (z_1, z_2, \ldots, z_n)$ with $z_i = 0, 1$ and consider

$$\boldsymbol{a}^z = a_1^{z_1} \otimes a_2^{z_2} \otimes \cdots \otimes a_n^{z_n} \tag{11.108}$$

The $a_i^{z_i}$ by themselves are abstract quantities [Kurkjian and Zelen (1962) refer to them as primitive elements]. It is only in the context of (11.108) that they have a quantitative meaning in the following sense.

In this expression we use the convention that $a_i^{z_i}$ is omitted if $z_i = 0$. Then $\boldsymbol{a}^z$ represents the vector of parameters of the interaction corresponding to the factors A_i for which $z_i = 1$, that is, all the terms corresponding to one of the parameters in (11.106). Specifically, if $z_{i_1} = z_{i_2} = \cdots = z_{i_q} = 1$, then $\boldsymbol{a}^z$ represents all the terms $a_{i_1 i_2} \ldots {}_{i_q}(x_{i_1}, x_{i_2}, \ldots, x_{i_q})$ of the q-factor interaction $A_{i_1} A_{i_2} \ldots A_{i_q}$. To express these terms as contrasts in the treatment effects, we define the contrast matrix

$$\boldsymbol{M} = s\boldsymbol{I} - \mathfrak{I}\mathfrak{I}' \tag{11.109}$$

where $\boldsymbol{I}$ is the $s \times s$ identity matrix and $\mathfrak{I}$ is an $s \times 1$ vector of unity elements. With

$$\boldsymbol{M}^{z_i} = \begin{cases} \mathfrak{I}' & \text{if } z_i = 0 \\ \boldsymbol{M} & \text{if } z_i = 1 \end{cases} \tag{11.110}$$

and with $\times$ denoting the Kronecker product,

$$\boldsymbol{M}^{z} = \boldsymbol{M}^{z_1} \times \boldsymbol{M}^{z_2} \times \cdots \times \boldsymbol{M}^{z_n} \tag{11.111}$$

we have

$$\boldsymbol{a}^{z} = \frac{1}{s^n}\boldsymbol{M}^{z}\boldsymbol{\tau} \tag{11.112}$$

To illustrate (11.112), consider, for example, the main effects for factor A_1, that is, $z_1 = 1$, $z_2 = z_3 = \cdots = z_n = 0$. For the sake of simplicity we shall take $s = 3$ and $n = 3$. Then (with subscripts indicating the dimensions of a matrix or a vector)

$$\begin{aligned}
\boldsymbol{M}^{(100)}_{3\times 27} &= \boldsymbol{M}^{1} \times \boldsymbol{M}^{0} \times \boldsymbol{M}^{0} \\
&= (3\,\boldsymbol{I} - \mathfrak{I}\mathfrak{I}') \times \mathfrak{I}' \times \mathfrak{I}' \\
&= 3\,\boldsymbol{I}_3 \times \mathfrak{I}'_9 - \mathfrak{I}_3\mathfrak{I}'_{27} \\
&= \begin{pmatrix} 2\mathfrak{I}'_9 & -\mathfrak{I}'_9 & -\mathfrak{I}'_9 \\ -\mathfrak{I}'_9 & 2\mathfrak{I}'_9 & -\mathfrak{I}'_9 \\ -\mathfrak{I}'_9 & -\mathfrak{I}'_9 & 2\mathfrak{I}'_9 \end{pmatrix}
\end{aligned}$$

and

$$\boldsymbol{a}^{(100)} = \begin{bmatrix} \frac{1}{9}\sum\limits_{jk}\tau(1,j,k) - \frac{1}{27}\sum\limits_{ijk}\tau(i,j,k) \\ \frac{1}{9}\sum\limits_{jk}\tau(2,j,k) - \frac{1}{27}\sum\limits_{ijk}\tau(i,j,k) \\ \frac{1}{9}\sum\limits_{jk}\tau(3,j,k) - \frac{1}{27}\sum\limits_{ijk}\tau(i,j,k) \end{bmatrix} = \begin{bmatrix} \overline{\tau}(1,\cdot,\cdot) - \overline{\tau}(\cdot,\cdot,\cdot) \\ \overline{\tau}(2,\cdot,\cdot) - \overline{\tau}(\cdot,\cdot,\cdot) \\ \overline{\tau}(3,\cdot,\cdot) - \overline{\tau}(\cdot,\cdot,\cdot) \end{bmatrix}$$

using familiar dot notation. Similarly for the main effects for factor A_2 we have

$$\begin{aligned}
\boldsymbol{M}^{(010)}_{3\times 27} &= \boldsymbol{M}^{0} \times \boldsymbol{M}^{1} \times \boldsymbol{M}^{0} \\
&= \mathfrak{I}' \times (3\,\boldsymbol{I} - \mathfrak{I}\mathfrak{I}') \times \mathfrak{I}' \\
&= 3\,\mathfrak{I}' \times \boldsymbol{I} \times \mathfrak{I}' - \mathfrak{I}' \times \mathfrak{I}\mathfrak{I}' \times \mathfrak{I}' \\
&= 3(\boldsymbol{I}_3 \times \mathfrak{I}'_3 \vdots \boldsymbol{I}_3 \times \mathfrak{I}'_3 \vdots \boldsymbol{I}_3 \times \mathfrak{I}'_3) - \mathfrak{I}_3\mathfrak{I}'_{27}
\end{aligned}$$

and

$$a^{(010)} = \begin{bmatrix} \overline{\tau}(\cdot, 1, \cdot) - \overline{\tau}(\cdot, \cdot, \cdot) \\ \overline{\tau}(\cdot, 2, \cdot) - \overline{\tau}(\cdot, \cdot, \cdot) \\ \overline{\tau}(\cdot, 3, \cdot) - \overline{\tau}(\cdot, \cdot, \cdot) \end{bmatrix}$$

For the interaction between factors A_1 and A_2 we obtain

$$\begin{aligned} \boldsymbol{M}^{(110)}_{9\times 27} &= \boldsymbol{M}^1 \times \boldsymbol{M}^1 \times \boldsymbol{M}^0 \\ &= (3\,\boldsymbol{I} - \mathfrak{I}\mathfrak{I}') \times (3\,\boldsymbol{I} - \mathfrak{I}\mathfrak{I}') \times \mathfrak{I}' \\ &= 9\,\boldsymbol{I}_3 \times \boldsymbol{I}_3 \times \mathfrak{I}'_3 - 3\,\boldsymbol{I}_3 \times \mathfrak{I}_3\mathfrak{I}'_3 \times \mathfrak{I}'_3 - 3\,\mathfrak{I}_3\mathfrak{I}'_3 \times \boldsymbol{I}_3 \times \mathfrak{I}'_3 \\ &\quad + \mathfrak{I}_3\mathfrak{I}'_3 \times \mathfrak{I}_3\mathfrak{I}'_3 \times \mathfrak{I}'_3 \\ &= 9\,\boldsymbol{I}_9 \times \mathfrak{I}'_3 - 3\,\boldsymbol{I}_3 \times \mathfrak{I}_3\mathfrak{I}'_9 - 3\,\mathfrak{I}_3\mathfrak{I}'_3 \times \boldsymbol{I}_3 \times \mathfrak{I}'_3 + \mathfrak{I}_9\mathfrak{I}'_{27} \end{aligned}$$

and

$$a^{(110)} = \begin{bmatrix} \overline{\tau}(1, 1, \cdot) - \overline{\tau}(1, \cdot, \cdot) - \overline{\tau}(\cdot, 1, \cdot) + \overline{\tau}(\cdot, \cdot, \cdot) \\ \overline{\tau}(1, 2, \cdot) - \overline{\tau}(1, \cdot, \cdot) - \overline{\tau}(\cdot, 2, \cdot) + \overline{\tau}(\cdot, \cdot, \cdot) \\ \overline{\tau}(1, 3, \cdot) - \overline{\tau}(1, \cdot, \cdot) - \overline{\tau}(\cdot, 3, \cdot) + \overline{\tau}(\cdot, \cdot, \cdot) \\ \overline{\tau}(2, 1, \cdot) - \overline{\tau}(2, \cdot, \cdot) - \overline{\tau}(\cdot, 1, \cdot) + \overline{\tau}(\cdot, \cdot, \cdot) \\ \overline{\tau}(2, 2, \cdot) - \overline{\tau}(2, \cdot, \cdot) - \overline{\tau}(\cdot, 2, \cdot) + \overline{\tau}(\cdot, \cdot, \cdot) \\ \overline{\tau}(2, 3, \cdot) - \overline{\tau}(2, \cdot, \cdot) - \overline{\tau}(\cdot, 3, \cdot) + \overline{\tau}(\cdot, \cdot, \cdot) \\ \overline{\tau}(3, 1, \cdot) - \overline{\tau}(3, \cdot, \cdot) - \overline{\tau}(\cdot, 1, \cdot) + \overline{\tau}(\cdot, \cdot, \cdot) \\ \overline{\tau}(3, 2, \cdot) - \overline{\tau}(3, \cdot, \cdot) - \overline{\tau}(\cdot, 2, \cdot) + \overline{\tau}(\cdot, \cdot, \cdot) \\ \overline{\tau}(3, 3, \cdot) - \overline{\tau}(3, \cdot, \cdot) - \overline{\tau}(\cdot, 3, \cdot) + \overline{\tau}(\cdot, \cdot, \cdot) \end{bmatrix}$$

It is easy to verify that, indeed,

$$\overline{\tau}(1, 1, \cdot) - \overline{\tau}(1, \cdot, \cdot) - \overline{\tau}(\cdot, 1, \cdot) + \overline{\tau}(\cdot, \cdot, \cdot) = a_{12}(1, 1)$$

for example. Hence, any contrast belonging to the 2-factor interaction $A_1 \times A_2$ can be expressed as

$$\Phi = \boldsymbol{\ell}'_{110}\boldsymbol{M}^{(110)}\boldsymbol{\tau}$$

where the elements of $\boldsymbol{\ell}_{110}$ add to zero. In general, the number of linearly independent contrasts belonging to the effect represented by $\boldsymbol{a}^z$ for a fixed z is given by

$$\nu(z) = \text{rank}(\boldsymbol{M}^{z}) = \prod_{i=1}^{n} \text{rank}(\boldsymbol{M}^{z_i}) = \prod_{i=1}^{n} (s-1)^{z_i}$$

11.14.2 Orthogonal Factorial Structure (OFS)

Our aim is to construct suitable systems of confounding for the general s^n factorial. In order to do so and to identify such systems that have certain desirable properties, it is useful to look ahead at the analysis and deduce from it structural properties that must hold for the error-control design and the treatment design.

Recall (see Chapter 1) that the reduced normal equations for an incomplete block design are given by

$$\boldsymbol{C}\widehat{\boldsymbol{\tau}} = \boldsymbol{Q} \tag{11.113}$$

where, for an equireplicate proper design,

$$\boldsymbol{C} = r\,\boldsymbol{I} - \frac{1}{k}\boldsymbol{N}\boldsymbol{N}' \tag{11.114}$$

and

$$\boldsymbol{Q} = \boldsymbol{T} - \frac{1}{k}\boldsymbol{N}'\boldsymbol{B}$$

The error-control design is determined by the incidence matrix $\boldsymbol{N}$ and the structure and properties of the design can best be characterized through the form of the concordance matrix $\boldsymbol{N}\boldsymbol{N}'$. All the systems of confounding derived in this and previous chapters possess the following two properties:

1. Each contrast belonging to a certain interaction is estimated orthogonally to any other contrast belonging to a different interaction. That is, if for the p^n factorial $\sum_{i=0}^{p-1} c_i\, E_i^{\alpha_1}$ and $\sum_{i=0}^{p-1} d_i\, E_i^{\alpha_2}$ are two contrasts belonging to E^{α_1} and E^{α_2}, respectively, then

$$\text{cov}\left(\sum_i c_i\, \widehat{E}_i^{\alpha_1}, \sum_i d_i\, \widehat{E}_i^{\alpha_2}\right) = 0$$

2. The treatment sum of squares $\text{SS}(\boldsymbol{X}_\tau|\mathfrak{I}, \boldsymbol{X}_\beta)$ is partitioned orthogonally into the individual interaction sums of squares:

$$\text{SS}(\boldsymbol{X}_\tau|\mathfrak{I}, \boldsymbol{X}_\beta) = \sum_{\alpha} \text{SS}(E^{\alpha}) \tag{11.115}$$

where $\text{SS}(E^\alpha)$ is the sum of squares due to E^α and $\sum_\alpha$ is the sum over all $\boldsymbol{\alpha}$ such that E^α is not completely confounded (i.e., belonging to $\mathcal{E}_2$ and $\mathcal{E}_3$ as defined in Section 11.6).

Property 2 follows, of course, from property 1. Factorial designs having these properties are said to have *orthogonal factorial structure* (OFS). The importance of OFS is that it simplifies greatly the interpretation of the analysis of a factorial experiment. For example, based on significance tests in the ANOVA certain interaction terms may be deleted from the model without affecting the estimates of other terms remaining in the model.

We know from earlier discussion (see Chapter 1) that

$$\mathrm{SS}(\boldsymbol{X}_\tau|\mathfrak{I}, \boldsymbol{X}_\beta) = \widehat{\boldsymbol{\tau}}'\boldsymbol{Q} = \widehat{\boldsymbol{\tau}}'\boldsymbol{C}\widehat{\boldsymbol{\tau}} = \boldsymbol{Q}'\boldsymbol{C}^-\boldsymbol{C}\boldsymbol{C}^-\boldsymbol{Q} \tag{11.116}$$

It is clear from (11.115) and (11.114) that $\mathrm{SS}(\boldsymbol{X}_\tau|\mathfrak{I}, \boldsymbol{X}_\beta)$ is a function of $\boldsymbol{N}\boldsymbol{N}'$. Now the structure of $\boldsymbol{C}$ is the same as the structure of $\boldsymbol{N}\boldsymbol{N}'$. We have seen earlier (see, e.g., Chapters 2 and 4) that for block designs the structure of $\boldsymbol{C}^-$ is the same as that of $\boldsymbol{C}$. We may then ask the questions: (i) For what structure of these matrices do we have an OFS and (ii) how can one exploit that structure to generate appropriate systems of confounding? We shall now give answers to these questions.

11.14.3 Systems of Confounding with OFS

Consider $\boldsymbol{a}^z$ as given in (11.112). It follows then from (11.113) that

$$\widehat{\boldsymbol{a}}^z = \frac{1}{s^n}\boldsymbol{M}^z\boldsymbol{C}^-\boldsymbol{Q}$$

with

$$\mathrm{var}\,(\widehat{\boldsymbol{a}}^z) = \frac{1}{s^{2n}}\boldsymbol{M}^z\boldsymbol{C}^-(\boldsymbol{M}^z)'\sigma_e^2$$

and

$$\mathrm{cov}\,(\widehat{\boldsymbol{a}}^z, \widehat{\boldsymbol{a}}^w) = \frac{1}{s^{2n}}\boldsymbol{M}^z\boldsymbol{C}^-(\boldsymbol{M}^w)'\sigma_e^2$$

for $\boldsymbol{z} \neq \boldsymbol{w}$. More generally, for two contrasts belonging to $\boldsymbol{a}^z$ and $\boldsymbol{a}^w$, respectively, we have

$$\begin{aligned}\mathrm{cov}[\boldsymbol{\ell}_z'\,\widehat{\boldsymbol{a}}^z, \boldsymbol{\ell}_w'\,\widehat{\boldsymbol{a}}^w] &= \frac{1}{s^{2n}}\,\mathrm{cov}[\boldsymbol{\ell}_z'\,\boldsymbol{M}^z\widehat{\boldsymbol{\tau}}, \boldsymbol{\ell}_w'\,\boldsymbol{M}^w\widehat{\boldsymbol{\tau}}] \\ &= \frac{1}{s^{2n}}\,\boldsymbol{\ell}_z'\,\boldsymbol{M}^z\boldsymbol{C}^-(\boldsymbol{M}^w)'\,\boldsymbol{\ell}_w' \end{aligned} \tag{11.117}$$

Furthermore, the sum of squares due to testing the significance of $\boldsymbol{a}^z$ is given by

$$\mathrm{SS}(\boldsymbol{a}^z) = (\boldsymbol{M}^z\widehat{\boldsymbol{\tau}})'\left[\boldsymbol{M}^z\boldsymbol{C}^-(\boldsymbol{M}^z)'\right]^-(\boldsymbol{M}^z\widehat{\boldsymbol{\tau}}) \tag{11.118}$$

With the reparameterization (11.106) OFS is achieved if expression (11.117) is zero and hence analogous to (11.115),

$$\mathrm{SS}(\boldsymbol{X}_\tau|\mathfrak{I}, \boldsymbol{X}_\beta) = \sum_z \mathrm{SS}(\boldsymbol{a}^z) \tag{11.119}$$

where $\sum_z$ is the sum over all binary n vectors except $(0, 0, \ldots, 0)$. We denote the set of all such vectors by Z. Since (11.117) must be equal to zero for all choices of $\boldsymbol{\ell}_z$ and $\boldsymbol{\ell}_w$, the following theorem holds.

Theorem 11.11 For the s^n factorial with parameterization (11.106) a block design will have OFS if

$$\boldsymbol{M}^z \boldsymbol{C}^- (\boldsymbol{M}^w)' = \boldsymbol{0} \tag{11.120}$$

for all $z, w \in Z (z \neq w)$, where $\boldsymbol{M}^z$ is defined in (11.111).

The following theorem, which will be instrumental in constructing block designs with OFS, is a special case of a more general theorem due to Cotter, John, and Smith (1973). We need to introduce some notation first. For simplicity of notation we shall denote the matrix $\boldsymbol{C}^-$, a generalized inverse of $\boldsymbol{C}$ in (11.114), by $\boldsymbol{D}$. We shall refer to a *cyclic partition* of a matrix $\boldsymbol{H}$ of order $m_1 m_2 \times m_1 m_2$ if $\boldsymbol{H}$ can be written as

$$\boldsymbol{H} = \begin{bmatrix} \boldsymbol{H}_1 & \boldsymbol{H}_2 & \boldsymbol{H}_3 & \cdots & \boldsymbol{H}_{m_1} \\ \boldsymbol{H}_{m_1} & \boldsymbol{H}_1 & \boldsymbol{H}_2 & \cdots & \boldsymbol{H}_{m_1-1} \\ \vdots & \vdots & \ddots & \vdots & \vdots \\ \boldsymbol{H}_2 & \boldsymbol{H}_3 & \cdots & \boldsymbol{H}_{m_1} & \boldsymbol{H}_1 \end{bmatrix} \tag{11.121}$$

where $\boldsymbol{H}_i$ $(i = 1, 2, \ldots, m_1)$ is of order $m_2 \times m_2$. We write (11.121) for short as

$$\boldsymbol{H} = \left\{ \boldsymbol{H}_1, \boldsymbol{H}_2, \ldots, \boldsymbol{H}_{m_1} \right\} = \{(\boldsymbol{H}_i)\} \tag{11.122}$$

We now state the following theorem.

Theorem 11.12 A block design $\boldsymbol{N}$ for an s^n factorial has OFS if the matrix $\boldsymbol{C}^- = \boldsymbol{D}$ has the following cyclic partition:

$$\begin{aligned} \boldsymbol{D} &= \left\{ \left(\boldsymbol{D}_{i_1} \right) \right\} \\ \boldsymbol{D}_{i_1} &= \left\{ \left(\boldsymbol{D}_{i_1 i_2} \right) \right\} \\ \boldsymbol{D}_{i_1 i_2} &= \left\{ \left(\boldsymbol{D}_{i_1 i_1 i_3} \right) \right\} \\ &\vdots \\ \boldsymbol{D}_{i_1 i_2 \ldots i_{n-2}} &= \left\{ \left(\boldsymbol{D}_{i_1 i_2 \ldots i_{n-1}} \right) \right\} \end{aligned}$$

for $i_j = 1, 2, \ldots, s$, $j = 1, 2, \ldots, n-1$, and if the $s \times s$ matrix $\boldsymbol{D}_{i_1 i_2 \ldots i_{n-1}}$ has row and column sums all equal.

Proof The proof is patterned after the results of Cotter, John, and Smith (1973) and John and Smith (1972). We show that (11.120) is true by using induction. First we show that the theorem is true for $n = 2$.

For $n = 2$ the set Z consists of three vectors: (1,0), (0,1), (1,1). Correspondingly we have, using (11.111),

$$\begin{aligned} \boldsymbol{M}^{(10)} &= (s\boldsymbol{I}_s - \mathfrak{I}_s\mathfrak{I}'_s) \times \mathfrak{I}'_s \\ &= s\boldsymbol{I}_s \times \mathfrak{I}'_s - \mathfrak{I}_s\mathfrak{I}'_s \times \mathfrak{I}'_s \\ &= s\boldsymbol{P}_1 - \mathfrak{I}_s\mathfrak{I}'_{s^2} \end{aligned}$$

where, using the notation of (11.122),

$$\boldsymbol{P}_1 = \left\{\mathfrak{I}'_s, \boldsymbol{0}'_s, \ldots, \boldsymbol{0}'_s\right\}$$

Also

$$\boldsymbol{M}^{(01)} = \mathfrak{I}'_s \times \left(s\boldsymbol{I}_s - \mathfrak{I}_s\mathfrak{I}'_s\right) = s\boldsymbol{P}_2 - \mathfrak{I}_s\mathfrak{I}'_{s^2}$$

with

$$\boldsymbol{P}_2 = (\boldsymbol{I}_s, \boldsymbol{I}_s, \ldots, \boldsymbol{I}_s)$$

and finally

$$\begin{aligned} \boldsymbol{M}^{(11)} &= \left(s\boldsymbol{I}_s - \mathfrak{I}_s\mathfrak{I}'_s\right) \times \left(s\boldsymbol{I}_s - \mathfrak{I}_s\mathfrak{I}'_s\right) \\ &= s^2\left\{\boldsymbol{I}_s, \boldsymbol{0}, \ldots, \boldsymbol{0}\right\} - s\left\{\mathfrak{I}_s, \mathfrak{I}'_s, \boldsymbol{0}, \ldots, \boldsymbol{0}\right\} \\ &\quad - s\left\{\boldsymbol{I}_s, \boldsymbol{I}_s, \ldots, \boldsymbol{I}_s\right\} + \left\{\mathfrak{I}_s\mathfrak{I}'_s, \mathfrak{I}_s\mathfrak{I}'_s, \ldots, \mathfrak{I}_s\mathfrak{I}'_s\right\} \end{aligned}$$

Now consider

$$\begin{aligned} \boldsymbol{M}^{(10)}\boldsymbol{D} &= s\,\boldsymbol{P}_1\boldsymbol{D} - \mathfrak{I}_s\mathfrak{I}'_s\boldsymbol{D} \\ &= s\left\{d_1\,\mathfrak{I}'_s, d_2\,\mathfrak{I}'_s, \ldots, d_s\,\mathfrak{I}'_s\right\} - \left(\sum_i d_i\right)\mathfrak{I}_s\mathfrak{I}'_s \end{aligned}$$

where d_i is the row and column sum of $\boldsymbol{D}_i$ in $\boldsymbol{D} = \{\boldsymbol{D}_1, \boldsymbol{D}_2, \ldots, \boldsymbol{D}_s\}$. Since $\mathfrak{I}\mathfrak{I}'\boldsymbol{M}^z = \boldsymbol{0}$ for $z \in Z$ it follows that

$$\begin{aligned} \boldsymbol{M}^{(10)}\boldsymbol{D}\left(\boldsymbol{M}^{(01)}\right)' &= s\left\{d_1\mathfrak{I}'_s, d_2\mathfrak{I}'_s, \ldots, d_s\mathfrak{I}'_s\right\}(s\boldsymbol{P}_2 - \mathfrak{I}_s\mathfrak{I}'_{s^2})' \\ &= s^2\left(\sum_i d_i\right)\mathfrak{I}_s\mathfrak{I}'_s - s^2\left(\sum_i d_i\right)\mathfrak{I}_s\mathfrak{I}'_s \\ &= \boldsymbol{0} \end{aligned}$$

Similarly,

$$\begin{aligned}
\boldsymbol{M}^{(10)}\boldsymbol{D}\boldsymbol{M}^{(11)'} &= s\left\{d_1\mathfrak{J}_s', d_2\mathfrak{J}_s', \ldots, d_s\mathfrak{J}_s'\right\}\ \boldsymbol{M}^{(11)} \\
&= s^3\left\{d_1\mathfrak{J}_s', d_2\mathfrak{J}_s', \ldots, d_s\mathfrak{J}_s'\right\} - s^3\left\{d_1\mathfrak{J}_s', d_2\mathfrak{J}_s', \ldots, d_s\mathfrak{J}_s'\right\} \\
&\quad - s^2\left(\sum d_i\right)\mathfrak{J}_s\mathfrak{J}_{s^2}' + s^2\left(\sum d_i\right)\mathfrak{J}_s\mathfrak{J}_{s^2}' = \boldsymbol{0}
\end{aligned}$$

Finally,

$$\begin{aligned}
\boldsymbol{M}^{(01)}\boldsymbol{D} &= s\boldsymbol{P}_2\boldsymbol{D} - \mathfrak{J}_s\mathfrak{J}_{s^2}'\boldsymbol{D} \\
&= s^2\left(\sum \boldsymbol{D}_i, \sum \boldsymbol{D}_i, \ldots, \sum \boldsymbol{D}_i,\right) - \left(\sum \boldsymbol{D}_i\right)\mathfrak{J}_s\mathfrak{J}_{s^2}'
\end{aligned}$$

and hence

$$\begin{aligned}
\boldsymbol{M}^{(01)}\boldsymbol{D}\boldsymbol{M}^{(11)'} &= s^3\left(\sum \boldsymbol{D}_i, \sum \boldsymbol{D}_i, \ldots, \sum \boldsymbol{D}_i,\right) \\
&\quad - s^2\left(\left(\sum d_i\right)\mathfrak{J}_s\mathfrak{J}_s', \left(\sum d_i\right)\mathfrak{J}_s\mathfrak{J}_s', \ldots, \left(\sum d_i\right)\mathfrak{J}_s\mathfrak{J}_s'\right) \\
&\quad - s^3\left(\sum \boldsymbol{D}_i, \sum \boldsymbol{D}_i, \ldots, \sum \boldsymbol{D}_i,\right) \\
&\quad + s^2\left(\left(\sum d_i\right)\mathfrak{J}_s\mathfrak{J}_s', \left(\sum d_i\right)\mathfrak{J}_s\mathfrak{J}_s', \ldots, \left(\sum d_i\right)\mathfrak{J}_s\mathfrak{J}_s'\right) \\
&= \boldsymbol{0}
\end{aligned}$$

Now assume that the theorem holds for the s^{n-1} factorial, say with factors $A_2, A_3, \ldots, A_n$. If we write the z vector for $n-1$ factors as $z_{*} = (z_2, z_3, \ldots, z_n)$, we then have for $\boldsymbol{z}_* \neq \boldsymbol{w}_* \neq (0, 0, \ldots, 0)$

$$\boldsymbol{M}^{\boldsymbol{z}_*}\boldsymbol{D}_i(\boldsymbol{M}^{\boldsymbol{w}_*})' = \boldsymbol{0} \tag{11.123}$$

for $i = 1, 2, \ldots, s$ with the row and column sum of $\boldsymbol{D}_i$ equal to d_i, say. Using (11.123), we now show that

$$\boldsymbol{M}^{\boldsymbol{z}}\boldsymbol{D}(\boldsymbol{M}^{\boldsymbol{w}})' = \boldsymbol{0}$$

for $\boldsymbol{z} \neq \boldsymbol{w}$. Now $\boldsymbol{M}^{\boldsymbol{z}}\boldsymbol{D}\boldsymbol{M}^{\boldsymbol{w}'}$ can be written as

$$\left(\boldsymbol{M}^{z_1} \times \boldsymbol{M}^{\boldsymbol{z}_*}\right)\{\boldsymbol{D}_1, \boldsymbol{D}_2, \ldots, \boldsymbol{D}_s\}\left(\boldsymbol{M}^{w_1} \times \boldsymbol{M}^{\boldsymbol{w}_*}\right)' \tag{11.124}$$

To show that (11.124) equals zero, it is sufficient to show that $\boldsymbol{M}^{\boldsymbol{z}_*}\boldsymbol{D}_i(\boldsymbol{M}^{\boldsymbol{w}_*})' = \boldsymbol{0}$ for every possible $\boldsymbol{z}_*$, $\boldsymbol{w}_*$ in $\boldsymbol{z} = (z_1, \boldsymbol{z}_*)$ and $\boldsymbol{w} = (w_1, \boldsymbol{w}_*)$ with $\boldsymbol{z} \neq \boldsymbol{w} \neq$

$(0, 0, \ldots, 0)$. This can be seen by simply multiplying **(11.124)** out. We distinguish between three cases:

1. $z_* = \mathbf{0}_* = (0, 0, \ldots, 0)$: We then have $\boldsymbol{w}_* \neq (0, 0, \ldots, 0)$ since otherwise $z = \boldsymbol{w}$. Now, by definition, $\boldsymbol{M}^{\mathbf{0}_*} = \mathfrak{I}_{s^{n-1}}$. Hence $\boldsymbol{M}^{\mathbf{0}_*}\boldsymbol{D}_i = d_i\mathfrak{I}'$, and since $\boldsymbol{M}^{\boldsymbol{w}_*}$ is a contrast matrix, it follows that $\boldsymbol{M}^{z_*}\boldsymbol{D}_i\boldsymbol{M}^{\boldsymbol{w}_*'} = \mathbf{0}$.
2. $z_* = \boldsymbol{w}_* \neq \mathbf{0}_*$: We then have $z_1 \neq w_1$ and without loss of generality assume $z_1 = 0$. Then

$$\boldsymbol{M}^{z} = \mathfrak{I}_s' \times \boldsymbol{M}^{z_*} = (\boldsymbol{M}^{z_*}, \boldsymbol{M}^{z_*}, \ldots, \boldsymbol{M}^{z_*})$$

and consequently

$$\begin{aligned}\boldsymbol{M}^{z}\boldsymbol{D} &= \left(\boldsymbol{M}^{z_*}\sum \boldsymbol{D}_i, \boldsymbol{M}^{z_*}\sum \boldsymbol{D}_i, \ldots, \boldsymbol{M}^{z_*}\sum \boldsymbol{D}_i\right)\\ &= \mathfrak{I}_s' \times \sum \boldsymbol{M}^{z_*}\boldsymbol{D}_i\end{aligned}$$

It follows then that

$$\begin{aligned}\boldsymbol{M}^{z}\boldsymbol{D}(\boldsymbol{M}^{\boldsymbol{w}})' &= \left(\mathfrak{I}_s' \times \sum \boldsymbol{M}^{z_*}\boldsymbol{D}_i\right)\left(\boldsymbol{M}^{w_1} \times \boldsymbol{M}^{\boldsymbol{w}_*}\right)'\\ &= \mathfrak{I}_s'(\boldsymbol{M}^{w_1})' \times \left(\sum \boldsymbol{M}^{z_*}\boldsymbol{D}_i(\boldsymbol{M}^{\boldsymbol{w}_*})'\right) = \mathbf{0}\end{aligned}$$

3. $z_* \neq \boldsymbol{w}_*$: By assumption the result is true.

We have therefore shown that if the theorem holds for the s^{n-1} factorial it also holds for the s^n factorial, and we have shown that it holds for the s^2 factorial. This completes the proof. □

11.14.4 Constructing Systems of Confounding

We shall now make use of Theorem 11.12 in our attempt to construct useful systems of confounding for the s^n factorial. Obviously, OFS is a desirable property of such a system of confounding. The idea then is to construct an incomplete block design for the s^n factorial such that the matrix $\boldsymbol{C}^- = \boldsymbol{D}$ has the structure given in Theorem 11.12. In some sense such an approach may seem to be backward since an incomplete block design is characterized first of all by its incidence matrix $\boldsymbol{N}$ and hence, equivalently, by $\boldsymbol{N}\boldsymbol{N}'$ or $\boldsymbol{C}$. The structure of $\boldsymbol{C}$ then determines the structure of $\boldsymbol{C}^-$. Now we know that for PBIB designs the structures of $\boldsymbol{C}$ and $\boldsymbol{C}^-$ are the same (see Chapter 4). We therefore want to find a PBIB design in which to accommodate the s^n factorial such that the $\boldsymbol{C}$ matrix has the form specified in Theorem 11.12 for $\boldsymbol{C}^-$, that is, hierarchical cyclic partitions. Such a method was proposed by John and Dean (1975) and can be described as follows.

As before, we denote a treatment combination by $\boldsymbol{x}' = (x_1, x_2, \ldots, x_n)$ where $0 \leq x_i \leq s-1 (i = 1, 2, \ldots, n)$. The totality of the treatment combinations, $\mathcal{X}$, is arranged in lexicographic order, that is, if $\boldsymbol{\theta} = (0, 1, \ldots, s-1)'$, then

$$\mathcal{X} = \boldsymbol{\theta} \otimes \boldsymbol{\theta} \otimes \cdots \otimes \boldsymbol{\theta}$$

(see Section 11.14.1). Following John (1973) we consider the use of generalized cyclic PBIB designs, GC/n-PBIB, (see Chapter 5) with blocks of size k. Recall that such a design is obtained from an initial block of k treatments. The jth block is obtained by adding the jth treatment combination to the treatments in the initial block. Here addition of two treatment combinations $\boldsymbol{x}' = (x_1, x_2, \ldots, x_n)$ and $\boldsymbol{y}' = (y_1, y_2, \ldots, y_n)$ is defined as

$$\boldsymbol{x}' + \boldsymbol{y}' = (x_1, +y_1, x_2, +y_2, \ldots, x_n + y_n) \bmod s \tag{11.125}$$

Let the highest common factor (HCF) of s and the nonzero elements of $\boldsymbol{x}$ be denoted by HCF$(s, \boldsymbol{x}) = h$ and let $q = s/h$. Then the treatment combinations $\boldsymbol{x}, 2\boldsymbol{x}, \ldots, (q-1)\,\boldsymbol{x}, q\boldsymbol{x} = \boldsymbol{0} \bmod s$ form a cyclic Abelian group G with addition as defined in (11.125). Here, G contains q distinct elements. The implication of this with respect to GC/n-PBIB designs is as follows. If $\boldsymbol{0}, \boldsymbol{x}, 2\boldsymbol{x}, \ldots, (q-1)\boldsymbol{x}$ represent the elements of the initial block of size $k = q$ and if additional blocks are obtained by adding in turn each treatment of the set $\mathcal{X}$ to those in the initial block, then the block obtained by adding $u\boldsymbol{x} (u < q)$ contains the same treatments as the initial block. Hence the initial block generates a fractional GC/n design with s^n/q distinct blocks such that each treatment combination occurs exactly once. We shall also say that $\boldsymbol{x}$ generates the fractional GC/n design.

More generally, we can consider g generators $\boldsymbol{x}_1, \boldsymbol{x}_2, \ldots, \boldsymbol{x}_g$ with corresponding Abelian groups $G_1, G_2, \ldots, G_g$. For the application to be described, the generators will have to be chosen such that the G_i have only the $\boldsymbol{0}$ element in common. Let HCF $(s, \boldsymbol{x}_i) = h_i, q_i = s/h_i$ and $q = \prod_{i=1}^{g} q_i$. We consider then the group G formed by taking the direct sum of $G_1, G_2, \ldots, G_g$, the general element being given by

$$u_1\boldsymbol{x}_1 + u_2\boldsymbol{x}_2 + \cdots + u_g\boldsymbol{x}_g \tag{11.126}$$

with $u_i = 0, 1, \ldots, q_i - 1$, $i = 1, 2, \ldots, g$. The q elements of G, taken as the initial block, then generate a GC/n design with s^n/q distinct blocks and every treatment combination in $\mathcal{X}$ replicated once. To illustrate this concept, consider the following example.

Example 11.9 Suppose we have $s = 6$, $n = 2$, that is, the 6^2 factorial. Let

$$\boldsymbol{x}_1' = (0, 3), \;\; \boldsymbol{x}_2' = (3, 0), \;\; \boldsymbol{x}_3' = (2, 2)$$

It follows then that

$$
\begin{aligned}
h_1 &= \text{HCF}\,[6, (0, 3)] = 3, \qquad & q_1 &= 2 \\
h_2 &= \text{HCF}\,[6, (3, 0)] = 3, \qquad & q_2 &= 2 \\
h_3 &= \text{HCF}\,[6, (2, 2)] = 2, \qquad & q_3 &= 3
\end{aligned}
$$

and $q = 12$. The elements of G are obtained from (11.126) with $u_1 = 0, 1$, $u_2 = 0, 1$, $u_3 = 0, 1, 2$ as

$$
\begin{array}{cccccc}
(0,0) & (0,3) & (3,0) & (3,3) & (2,2) & (2,5) \\
(5,2) & (5,5) & (4,4) & (4,1) & (1,4) & (1,1)
\end{array}
$$

This is the initial block, B_1 say. The remaining blocks are obtained by adding a treatment combination not in B_1 to each element in B_1. To obtain B_2, we add (1, 0), which yields

$$
\begin{array}{cccccc}
(1,0) & (1,3) & (4,0) & (4,3) & (3,2) & (3,5) \\
(0,2) & (0,5) & (5,4) & (5,1) & (2,4) & (2,1)
\end{array}
$$

and for B_3 we add (0, 1), which yields

$$
\begin{array}{cccccc}
(0,1) & (0,4) & (3,1) & (3,4) & (2,3) & (2,0) \\
(5,3) & (5,0) & (4,5) & (4,2) & (1,5) & (1,2)
\end{array}
$$

The design is summarized in Table 11.11.

We note here that the initial block, B_1 or G, acts as the IBSG discussed earlier. □

11.14.5 Verifying Orthogonal Factorial Structure

We shall now show that systems of confounding based on GC/n-PBIB designs have OFS. We do this by showing that the $\boldsymbol{C}$ matrix and hence the $\boldsymbol{C}^- = \boldsymbol{D}$ matrix have the structure of Theorem 11.12.

Let $\boldsymbol{R}_s^i$ denote an $s \times s$ circulant matrix with a unity element in the ith position of the first row and zero elements elsewhere. A circulant matrix

$$
\boldsymbol{A} = \begin{pmatrix}
a_1 & a_2 & a_3 & \ldots & a_s \\
a_s & a_1 & a_2 & \ldots & a_{s-1} \\
\ldots & \ldots & \ldots & \ldots & \ldots \\
a_2 & a_3 & a_4 & \ldots & a_1
\end{pmatrix}
$$

Table 11.11 6^2 in Blocks of Size 12 (Using GC/n Method)

Block 1	Block 2	Block 3
0 0	1 0	0 1
0 3	1 3	0 4
3 0	4 0	3 1
3 3	4 3	3 4
2 2	3 2	2 3
2 5	3 5	2 0
5 2	0 2	5 3
5 5	0 5	5 0
4 4	5 4	4 5
4 1	5 1	4 2
1 4	2 4	1 5
1 1	2 1	1 2

can then alternatively be written as

$$\boldsymbol{A} = \sum_{i=1}^{s} (\boldsymbol{R}_s^i \times a_i) = \sum_{i=1}^{s} a_i \, \boldsymbol{R}_s^i$$

As mentioned in Section 5.3, the matrix $\boldsymbol{NN}'$ and hence $\boldsymbol{C}$ for a cyclic PBIB design is a cyclic matrix, that is, $\boldsymbol{C}$ can be written as

$$\boldsymbol{C} = \sum_{i=1}^{t} c_i \, \boldsymbol{R}_t^i$$

For the situation considered here where a treatment consists of an n-tuple $(x_1, x_2, \ldots, x_n)$ this expression generalizes to

$$\boldsymbol{C} = \sum_{i_1=1}^{s} \sum_{i_2=1}^{s} \cdots \sum_{i_n=1}^{s} c_{i\, i_2 \ldots i_n} \left(\prod_{j=1}^{n} \times \boldsymbol{R}_s^{i_j} \right) \tag{11.127}$$

where

$$\prod_{j=1}^{n} \times \boldsymbol{R}_s^{i_j} = \boldsymbol{R}_s^{i_1} \times \boldsymbol{R}_s^{i_2} \times \cdots \times \boldsymbol{R}_s^{i_n}$$

and

$$c_{i_1 i_2 \ldots i_n} \begin{cases} = 1 - \dfrac{1}{q} & \text{for } i_1 = i_2 = \cdots = i_n = 1 \\ = \quad -\dfrac{1}{q} & \text{for } \boldsymbol{x}' = (i_1 - 1, i_2 - 1, \ldots i_n - 1) \text{ in } B_1 \\ & \text{but } \boldsymbol{x}' \neq (0, 0, \ldots, 0) \\ = 0 & \text{otherwise} \end{cases}$$

The reasons for the specific values of $c_{i_1 i_2 \ldots i_n}$ are, of course, (i) we have a single-replicate design, that is, $r = 1$; (ii) the block size k is equal to $q = \prod_{i=1}^{g} q_i$; and (iii) we have a disconnected design, that is, any two treatments occur together in the same block once or not at all. Alternatively, $\boldsymbol{C}$ can be written as

$$\boldsymbol{C} = \boldsymbol{I} - \frac{1}{q} \sum_{i_1} \cdots \sum_{i_n} c^*_{i_1 i_2 \ldots i_n} \prod_{j=1}^{n} \times \boldsymbol{R}_s^{i_j}$$

where

$$c^*_{i_1 i_2 \ldots i_n} \begin{cases} = 1 & \text{if } \boldsymbol{x}' = (i_1 - 1, i_2 - 1, \ldots, i_n - 1) \text{ in } B_1 \\ = 0 & \text{otherwise} \end{cases} \tag{11.128}$$

Now $\boldsymbol{C}$ in (11.127) can be written also (see John, 1973) as

$$\boldsymbol{C} = \sum_{i_1=1}^{s} \sum_{i_2=1}^{s} \cdots \sum_{i_\nu=1}^{s} \left\{ \left(\prod_{j=1}^{\nu} \times \boldsymbol{R}_s^{i_j} \right) \times \left(\sum_{i_n=1}^{s} c_{i_1 i_2 \ldots i_n} \boldsymbol{R}_s^{i_n} \right) \right\} \tag{11.129}$$

where $\nu = n - 1$. If we define

$$\boldsymbol{C}_{i_1 i_2 \ldots i_\nu} = \sum_{i_n=1}^{s} c_{i_1 i_2 \ldots i_n} \boldsymbol{R}_s^{i_n} \tag{11.130}$$

then (11.129) and hence (11.127) can be rewritten as

$$\boldsymbol{C} = \sum_{i_1} \cdots \sum_{i_\nu} \left\{ \left(\prod_{j=1}^{\nu} \times \boldsymbol{R}_s^{i_j} \right) \times \boldsymbol{C}_{i_1 i_2 \ldots i_\nu} \right\} \tag{11.131}$$

This, however, is just another way of writing the hierarchical cyclic partition required in Theorem 11.12. From the definition of $\boldsymbol{C}_{i_1 i_2 \ldots i_\nu}$ in (11.130) it follows immediately that the row and column totals for $\boldsymbol{C}_{i_1 i_2 \ldots i_\nu}$ are all equal, namely $\sum_{i_n} c_{i_1 i_2 \ldots i_n}$. It follows then from Theorem 11.12 that the GC/n-generated system of confounding has OFS.

11.14.6 Identifying Confounded Interactions

The question remains of which effects $\boldsymbol{a}^z$ are confounded with blocks or, more precisely, to which effects $\boldsymbol{a}^z$ do the $b - 1 = s^n/q - 1$ block degrees of freedom belong? Since we have a disconnected design with $b = s^n/q$ blocks it is clear that only $s^n - 1 - (b - 1)$ degrees of freedom remain for linearly independent estimable functions belonging to main effects and interactions. We know from our earlier discussion that without confounding the degrees of freedom for $\boldsymbol{a}^z$ are

$$\nu(z) = \prod_{i=1}^{n} (s - 1)^{z_i} \tag{11.132}$$

We shall denote the corresponding degrees of freedom for a given system of confounding by $\nu_c(z)$. We then show the following theorem.

Theorem 11.13 (John and Dean, 1975) For a GC/n-generated system of confounding the degrees of freedom for the effect $\boldsymbol{a}^z$ are given by

$$\nu_c(z) = \text{rank}(\boldsymbol{M}^z) - \frac{1}{q} \sum_{i_1} \cdots \sum_{i_n} c^*_{i_1 i_2 \ldots i_n} \left(\prod_{j=1}^{n} \rho_{i_j}^{z_j} \right) \tag{11.133}$$

where

$$\rho_{i_j}^{z_j} = \begin{cases} s - 1 & \text{if } i_j = 1 \text{ and } z_j = 1 \\ -1 & \text{if } i_j \neq 1 \text{ and } z_j = 1 \\ 1 & \text{if } z_j = 0 \end{cases}$$

and the $c^*_{i_1 i_2 \ldots i_n}$ are defined in (11.128).

Proof Recall from (11.118) that the adjusted sum of squares for $\boldsymbol{a}^z$ is given by

$$\text{SS}(\boldsymbol{a}^z) = (\boldsymbol{M}^z \widehat{\boldsymbol{\tau}})' (\boldsymbol{M}^z \boldsymbol{C}^- \boldsymbol{M}^{z\prime})^- (\boldsymbol{M}^z \widehat{\boldsymbol{\tau}})$$

John and Dean (1975) show that the degrees of freedom for $\text{SS}(\boldsymbol{a}^z)$ are given by

$$\nu_c(z) = \frac{1}{s^n} \left(\prod_{j=1}^{n} s^{z_j} \right)^{-1} \text{tr}\left(\boldsymbol{M}^z \boldsymbol{C} \boldsymbol{M}^{z\prime} \right) \tag{11.134}$$

where tr( ) denotes the trace of a matrix. Now

$$\text{tr}\left(\boldsymbol{M}^z \boldsymbol{C} \boldsymbol{M}^{z\prime} \right) = \text{tr}\left(\boldsymbol{M}^z \boldsymbol{M}^{z\prime} \right) - \frac{1}{q} \sum_{i_1} \cdots \sum_{i_n} c^*_{i_1 i_2 \ldots i_n} \prod_{j=1}^{n} \text{tr}\left\{ \boldsymbol{M}^{z_j} \boldsymbol{R}_s^{i_j} \boldsymbol{M}^{z_j\prime} \right\}$$

with

$$\operatorname{tr}\left(\boldsymbol{M}^{z}\boldsymbol{M}^{z\prime}\right) = s^{n}\left(\prod_{j} s^{z_j}\right)\left[\prod_{j}(s-1)^{z_j}\right] \tag{11.135}$$

and

$$\operatorname{tr}\left(\boldsymbol{M}^{z_j}\boldsymbol{R}_s^{i_j}\boldsymbol{M}^{z_j\prime}\right) = \begin{cases} s^2\rho_{i_j}^1 & \text{for } z_j = 1 \\ s & \text{for } z_j = 0 \end{cases} \tag{11.136}$$

Substituting (11.135) and (11.136) into (11.134) and using the fact that $\operatorname{rank}(\boldsymbol{M}^{z}) = \prod_j (s-1)^{z_j}$ prove the theorem. □

As an immediate result from Theorem 11.12, we state the following corollary.

Corollary 11.3 The number of contrasts belonging to $\boldsymbol{a}^{z}$ that are confounded with blocks is equal to

$$\nu^*(z) = \frac{1}{q}\sum_{i_1}\cdots\sum_{i_n} c^*_{i_1\ldots i_n}\left(\prod_{j=1}^{n}\rho_{i_j}^{z_j}\right) \tag{11.137}$$

It follows from (11.137) that (i) $\nu^*(z)$ is determined entirely by the treatment combinations in the initial block B_1; (ii) if $\nu^*(z) = 0$, then $\boldsymbol{a}^{z}$ is unconfounded; and (iii) if $\nu^*(z) = \prod_j (s-1)^{z_j}$, then $\boldsymbol{a}^{z}$ is completely confounded.

We shall illustrate Corollary 11.3 in terms of Example 11.9.

Example 11.9 (Continued) We need to find $\sum_{i_1=1}^{6}\sum_{i_2=1}^{6} c^*_{i_1 i_2}\rho_{i_1}^{z_1}\rho_{i_2}^{z_2}$. Using the definition of $c^*_{i_1 i_2}$ and B_1 given in Table 11.11, the values for $\rho_{i_1}^{z_1}\rho_{i_2}^{z_2}$ with nonzero coefficients and for $z = (1, 0), (0, 1), (1, 1)$ are given in Table 11.12. It follows then that $\nu^* = (1, 0) = \nu^*(0, 1) = 0$ and $\nu^*(1, 1) = \frac{24}{12} = 2$, that is, the main effects are unconfounded and 2 d.f. from the 2-factor interaction are confounded with blocks and as a consequence $\nu_c(1, 1) = 23$. □

11.15 CHOICE OF INITIAL BLOCK

It is worthwhile to compare the designs given in Tables 11.8 and 11.11, which were constructed by two different methods. The method of pseudofactors is easy to use but results in the confounding of degrees of freedom belonging to the main effect A_1. The advantage of the GC/2 design is that it leads to confounding of degrees of freedom belonging to the interaction $A_1 A_2$. This will in general be more desirable but not as easy to achieve since the design depends crucially on the choice of the initial block.

Table 11.12 Determination of $\rho_{i_1}^{z_1}\rho_{i_2}^{z_2}$ for Main Effects and Two-Factor Interaction for 6^2 Factorial in Blocks of Size 12

i_1, i_2	$z_1 = 1, z_2 = 0$	$z_1 = 0, z_2 = 1$	$z_1 = 1, z_2 = 1$
1, 1	$6 - 1$	$6 - 1$	$(6 - 1)^2$
1, 4	$6 - 1$	-1	$-(6 - 1)$
4, 1	-1	$6 - 1$	$-(6 - 1)$
4, 4	-1	-1	1
3, 3	-1	-1	1
3, 6	-1	-1	1
6, 3	-1	-1	1
6, 6	-1	-1	1
5, 5	-1	-1	1
5, 2	-1	-1	1
2, 5	-1	-1	1
2, 2	-1	-1	1
Sum	0	0	24

How then should one choose the initial block? Two questions need to be answered: (i) How many generators does one need? (ii) What should these generators be? First of all, only block sizes of the form $k = q = \prod_{i-1}^{g} q_i$ are permissible, where the q_i's are divisors (not necessarily different) of s (including s itself). For example, for the 6^2 in blocks of size $k = 12$, we could have $g = 3$ with $q_1 = q_2 = 2$, $q_3 = 3$ (see Example 11.9) or $g = 2$ with $q_1 = 6$, $q_2 = 2$. To answer (ii), one needs to choose the generators $\boldsymbol{x}_1, \boldsymbol{x}_2, \ldots, \boldsymbol{x}_g$ such that the groups $G_1, G_2, \ldots, G_g$ only have the zero element in common. The choice depends further on which effects one is willing to confound to the extent that one has a choice. One may have to do this by trial and error, computing for each choice of generators $\nu_c(z)$ for each z, and then decide what is the best choice for the situation at hand. If one wants to avoid confounding degrees of freedom belonging to main effects, one can proceed in the following way (this follows from the discussion by John and Dean, 1975). If we write the generators as $\boldsymbol{x}_i' = (x_{i1}, x_{i2}, \ldots, x_{in})(i = 1, 2, \ldots, g)$, we can consider the vector of the jth components $(x_{1j}, x_{2j}, \ldots, x_{gj})(j = 1, 2, \ldots, n)$. If for each j the nonzero x_{ij} are relative primes, then no degrees of freedom belonging to main effects are confounded. As an example consider for the 6^2 in blocks of size 12 the following two sets of generators: (1) $\boldsymbol{x}_1 = (1, 5)$, $\boldsymbol{x}_2 = (3, 0)$ and (2) $\boldsymbol{x}_1 = (3, 5)$, $\boldsymbol{x}_2 = (3, 0)$, giving, respectively, the initial blocks B_1:

00, 30, 15, 45, 24, 54, 33, 03, 42, 12, 51, 21

and

00, 30, 35, 05, 04, 34, 33, 03, 02, 32, 31, 01

Using Corollary 11.3 or, alternatively, simple inspection shows that, for system 1, 2 d.f. from A_1A_2 are confounded whereas, for system 2, 2 d.f. from A_1 are confounded.

We close this section by mentioning that the GC/n designs are most valuable for the s^n factorial where $s = p_1^{n_1} p_2^{n_2} \cdots p_m^{n_m}$ as the method provides some flexibility with block sizes and allows the constructing of systems of confounding such that main effects remain unconfounded (if that is possible at all). This will play a role when we discuss systems of confounding for asymmetrical factorial experiments in Chapter 12. Finally, for $s = p$ and p prime the GC/n designs are of the same type as those discussed in Section 11.3.

CHAPTER 12

Confounding in Asymmetrical Factorial Designs

12.1 INTRODUCTION

In previous chapters we have considered systems of confounding for symmetrical factorial experiments, that is, the s^n experiment in $s^{n-\ell}$ blocks of size s^ℓ. In many practical situations, however, we may not encounter such a pure system, but rather find that different factors have different numbers of levels. We call this an *asymmetrical* or *mixed factorial experiment* of the form $s_1^{n_1} \times s_2^{n_2} \times \cdots \times s_q^{n_q}$, where the ith set of factors contains n_i factors each at s_i levels ($i = 1, 2, \ldots, q$). A typical example is the 2×3^2 experiment, that is, one factor having two levels and two factors having three levels each. For such an experiment we may want to use blocks of size $s_1^{\ell_1} \times s_2^{\ell_2} \times \cdots \times s_q^{\ell_q}$, for example, for the 2×3^2 experiment we may be interested in finding a suitable arrangement of the treatment combinations in blocks of size $2 \times 3 = 6$.

Various methods have been proposed to deal with this situation. Generally speaking, they can be divided into the following broad categories:

1. By looking at each component pure system separately and using the procedures discussed in previous chapters, it may be possible to generate an appropriate design. For example, for the $2^2 \times 3^2$ factorial experiment, we can examine the 2^2 system in blocks of size 2 and the 3^2 system in blocks of size 3. Suitable arrangements then lead to blocks of size 6 or 12 for the asymmetrical factorial experiment.
2. One can generalize the formal method for pure systems by defining how to combine elements from different finite fields, for example, elements from residue fields mod 2 and mod 3.
3. The construction of suitable PBIB designs can lead to useful systems of confounding. We mention here in particular the EGD-PBIB design

Design and Analysis of Experiments. Volume 2: Advanced Experimental Design
By Klaus Hinkelmann and Oscar Kempthorne
ISBN 0-471-55177-5

(Section 4.6.9) as its association scheme is based upon the fact that the number of treatments t can be factored in various ways, for example, for the 2×3^2 factorial $t = 2 \cdot 3 \cdot 3$.

4. The factorial treatments can be embedded in a suitable PBIB design, where the form of the association scheme suggests a correspondence between the treatments in the usual sense and the factorial treatments.

5. By introducing pseudofactors, we may be able to reduce the asymmetrical factorial experiment at least formally to a pure (symmetrical) factorial experiment and then apply the familiar procedures. For example, a $2^3 \times 4$ factorial can be interpreted as a 2^5 factorial. However, care must be taken in that the interaction between the last two factors in the 2^5 system is really part of the main effect of the factor with four levels (see Section 11.13.3).

We shall discuss some of these methods in more detail and give suitable designs for some that we consider to be the most useful and most commonly occurring cases.

12.2 COMBINING SYMMETRICAL SYSTEMS OF CONFOUNDING

12.2.1 Construction of Blocks

To present the basic ideas, let us consider the situation involving only two sets of factors, the first set consisting of n_1 factors $A_1, A_2, \ldots, A_{n_1}$ each at s_1 levels and the second set consisting of n_2 factors $B_1, B_2, \ldots, B_{n_2}$ each at s_2 levels. For the sake of simplicity we shall take $s_1 = p_1$ and $s_2 = p_2$, where p_1 and p_2 are different primes, but the results can be generalized easily to the case where s_1 and/or s_2 are prime powers by using the methods described in Section 11.7. We denote a treatment combination by the $n_1 + n_2$ vector $(x_1, x_2, \ldots, x_{n_1}, y_1, y_2, \ldots, y_{n_2})$, where $x_i = 0, 1, 2, \ldots, p_1 - 1$ $(i = 1, 2, \ldots, n_1)$, $y_j = 0, 1, 2, \ldots, p_2 - 1$ $(j = 1, 2, \ldots, n_2)$.

Suppose then we want to arrange the $p_1^{n_1} p_2^{n_2}$ treatment combinations in blocks of size $p_1^{\ell_1} p_2^{\ell_2} (\ell_1 \le n_1, \ell_2 \le n_2$ with $\ell_1 + \ell_2 < n_1 + n_2)$. To this end we consider separately the two symmetrical factorial systems, the $p_1^{n_1}$ factorial in blocks of size $p_1^{\ell_1}$, and the $p_2^{n_2}$ factorial in blocks of size $p_2^{\ell_2}$. Using the methods described in Chapter 11, we choose $n_1 - \ell_1$ independent interactions to be confounded (together with their GIs) in the $p_1^{n_1}$ system. Suppose these effects are $E^{\alpha_1}, E^{\alpha_2}, \ldots, E^{\alpha_\nu} (\nu = n_1 - \ell_1)$. The arrangement of the $p_1^{n_1}$ treatment combinations in blocks of size $p_1^{\ell_1}$ is then obtained by solving the systems of equations

$$\begin{aligned}
\alpha_{11}x_1 + \alpha_{12}x_2 + \cdots + \alpha_{1n_1}x_{n_1} &= \delta_1 \\
\alpha_{21}x_1 + \alpha_{22}x_2 + \cdots + \alpha_{2n_1}x_{n_1} &= \delta_2 \\
&\vdots \\
\alpha_{\nu 1}x_1 + \alpha_{\nu 2}x_2 + \cdots + \alpha_{\nu n_1}x_{n_1} &= \delta_\nu
\end{aligned} \tag{12.1}$$

for the $p_1^\nu = p_1^{n_1-\ell_1}$ possible choices of $(\delta_1, \delta_2, \ldots, \delta_\nu)$, one for each block. Similarly, suppose that the $n_2 - \ell_2$ independent interactions to be confounded for the $p_2^{n_2}$ factorial in blocks of size $p_2^{\ell_2}$ are $E^{\boldsymbol{\beta}_1}, E^{\boldsymbol{\beta}_2}, \ldots, E^{\boldsymbol{\beta}_\mu} (\mu = n_2 - \ell_2)$. The corresponding systems of equations are then

$$\begin{aligned}
\beta_{11} y_1 + \beta_{12} y_2 + \cdots + \beta_{1n_2} y_{n_2} &= \gamma_1 \\
\beta_{21} y_1 + \beta_{22} y_2 + \cdots + \beta_{2n_2} y_{n_2} &= \gamma_2 \\
&\vdots \\
\beta_{\mu 1} y_1 + \beta_{\mu 2} y_2 + \cdots + \beta_{\mu n_2} y_{n_2} &= \gamma_\mu
\end{aligned} \tag{12.2}$$

We thus obtain p_1^ν blocks for the $p_1^{n_1}$ factorial, say $B_{11}, B_{12}, \ldots, B_{1p_1^\nu}$, and p_2^μ blocks for the $p_2^{n_2}$ factorial system, say $B_{21}, B_{22}, \ldots, B_{2p_2^\mu}$. To obtain the blocks for the $p_1^{n_1} \times p_2^{n_2}$ system, denoted by B_{ij}^* $(i = 1, 2, \ldots, p_1^\nu;\ j = 1, 2, \ldots, p_2^\mu)$, we combine the blocks from the individual systems as follows. To obtain the treatment combinations in block B_{ij}^*, we "adjoin" to each treatment combination $(x_1, x_2, \ldots, x_{n_1})$ in B_{1i} every treatment combination $(y_1, y_2, \ldots, y_{n_2})$ in B_{2j} to yield $p_1^{\ell_1} p_2^{\ell_2}$ treatment combinations of the form $(x_1, x_2, \ldots, x_{n_1}, y_1, y_2, \ldots, y_{n_2})$. Symbolically this can be expressed as

$$\boldsymbol{B}_{ij}^* = \boldsymbol{B}_{1i} \otimes \boldsymbol{B}_{2j} \tag{12.3}$$

which has been called the *symbolic direct product* (SDP) by Kurkjian and Zelen (1962).

Example 12.1 To illustrate the procedure, we consider the $2^3 \times 3^2$ factorial in blocks of size $2 \times 3 = 6$, that is, $p_1 = 2$, $p_2 = 3$, $n_1 = 3$, $n_2 = 2$, $\ell_1 = 1$, $\ell_2 = 1$. Suppose we choose to confound A_1A_2 and A_1A_3 (and hence A_2A_3) and B_1B_2, respectively, in the two symmetrical systems. The block compositions are determined by the equations

$$x_1 + x_2 = \delta_1 \quad \text{mod } 2$$

$$x_1 + x_3 = \delta_2 \quad \text{mod } 2$$

and

$$y_1 + y_2 = \gamma_1 \quad \text{mod } 3$$

respectively. This yields the blocks

$\boldsymbol{B}_{11}$	$\boldsymbol{B}_{12}$	$\boldsymbol{B}_{13}$	$\boldsymbol{B}_{14}$
000	100	010	001
111	011	101	110

and

B_{21}	B_{22}	B_{23}
00	10	20
12	01	02
21	22	11

The blocks for the asymmetrical factorial then are given in Table 12.1. The same design using SAS PROC FACTEX is given in Table 12.2. □

There is another way of representing the result given above. Let $\boldsymbol{x}$ represent the column vector of the treatment combinations for the $p_1^{n_1}$ system and $\boldsymbol{y}$ that for the $p_2^{n_2}$ system. Further, let $\boldsymbol{N}_1$ and $\boldsymbol{N}_2$ denote the incidence matrices for the two systems of confounding in blocks of size $p_1^{\ell_1}$ and $p_2^{\ell_2}$, respectively. The final design can then be expressed as the Kronecker product

$$\boldsymbol{N} = \boldsymbol{N}_1 \times \boldsymbol{N}_2 \tag{12.4}$$

where the vector of treatment combinations is given by the SDP $\boldsymbol{x} \otimes \boldsymbol{y}$ and the blocks are labeled as in (12.3).

12.2.2 Properties of Kronecker Product Method

We shall comment briefly on some of the properties of the procedure described above. For any interaction confounded in the $p_1^{n_1}$ system, say E^{α}, and any interaction confounded in the $p_2^{n_2}$ system, say E^{β}, also the interaction $E^{\alpha}E^{\beta}$ is

Table 12.1 Block Arrangement for the $2^3 \times 3^2$ Factorial in Blocks of Size 6

B_{11}^*	B_{12}^*	B_{13}^*	B_{21}^*	B_{22}^*	B_{23}^*
00000	00010	00020	10000	10010	10020
00012	00001	00002	10012	10001	10002
00021	00022	00011	10021	10022	10011
11100	11110	11120	01100	01110	01120
11112	11101	11102	01112	01101	01102
11121	11122	11111	01121	01122	01111
B_{31}^*	B_{32}^*	B_{33}^*	B_{41}^*	B_{42}^*	B_{43}^*
01000	01010	01020	00100	00110	00120
01012	01001	01002	00112	00101	00102
01021	01022	01011	00121	00122	00111
10100	10110	10120	11000	11010	11020
10112	10101	10102	11012	11001	11002
10121	10122	10111	11021	11022	11011

Table 12.2 $2^3 \times 3^2$ Factorial in Blocks of Size 6

```
options nodate pageno=1;
proc factex;
factors A B C;
blocks size=2;
model est=(A B C A*B*C);
output out=abcdesn;
run;


factors D E/nlev=3;
blocks size=3;
model est=(D E);
output out=dedesn
designrep=abcdesn;
run;


data dedesn(drop=block1 block2);
  set dedesn;
  block1=block;
  if block1=1 and block2=1 then block=1;
  if block1=1 and block2=2 then block=2;
  if block1=1 and block2=3 then block=3;
  if block1=2 and block2=1 then block=4;
  if block1=2 and block2=2 then block=5;
  if block1=2 and block2=3 then block=6;
  if block1=3 and block2=1 then block=7;
  if block1=3 and block2=2 then block=8;
  if block1=3 and block2=3 then block=9;
  if block1=4 and block2=1 then block=10;
  if block1=4 and block2=2 then block=11;
  if block1=4 and block2=3 then block=12;
run;


proc sort data=dedesn;
by block;
run;


proc print data=dedesn;
title1 'TABLE 12.2';
title2 '2**3x3**2 FACTORIAL IN BLOCKS OF SIZE 6';
run;
```

Table 12.2 *(Continued)*

Obs	BLOCK	A	B	C	D	E
1	1	-1	-1	1	-1	-1
2	1	-1	-1	1	0	1
3	1	-1	-1	1	1	0
4	1	1	1	-1	-1	-1
5	1	1	1	-1	0	1
6	1	1	1	-1	1	0
7	2	-1	-1	1	-1	1
8	2	-1	-1	1	0	0
9	2	-1	-1	1	1	-1
10	2	1	1	-1	-1	1
11	2	1	1	-1	0	0
12	2	1	1	-1	1	-1
13	3	-1	-1	1	-1	0
14	3	-1	-1	1	0	-1
15	3	-1	-1	1	1	1
16	3	1	1	-1	-1	0
17	3	1	1	-1	0	-1
18	3	1	1	-1	1	1
19	4	-1	1	1	-1	-1
20	4	-1	1	1	0	1
21	4	-1	1	1	1	0
22	4	1	-1	-1	-1	-1
23	4	1	-1	-1	0	1
24	4	1	-1	-1	1	0
25	5	-1	1	1	-1	1
26	5	-1	1	1	0	0
27	5	-1	1	1	1	-1
28	5	1	-1	-1	-1	1
29	5	1	-1	-1	0	0
30	5	1	-1	-1	1	-1
31	6	-1	1	1	-1	0
32	6	-1	1	1	0	-1
33	6	-1	1	1	1	1
34	6	1	-1	-1	-1	0
35	6	1	-1	-1	0	-1
36	6	1	-1	-1	1	1
37	7	-1	1	-1	-1	-1
38	7	-1	1	-1	0	1
39	7	-1	1	-1	1	0
40	7	1	-1	1	-1	-1
41	7	1	-1	1	0	1
42	7	1	-1	1	1	0

Table 12.2 *(Continued)*

43	8	-1	1	-1	-1	1
44	8	-1	1	-1	0	0
45	8	-1	1	-1	1	-1
46	8	1	-1	1	-1	1
47	8	1	-1	1	0	0
48	8	1	-1	1	1	-1
49	9	-1	1	-1	-1	0
50	9	-1	1	-1	0	-1
51	9	-1	1	-1	1	1
52	9	1	-1	1	-1	0
53	9	1	-1	1	0	-1
54	9	1	-1	1	1	1
55	10	-1	-1	-1	-1	-1
56	10	-1	-1	-1	0	1
57	10	-1	-1	-1	1	0
58	10	1	1	1	-1	-1
59	10	1	1	1	0	1
60	10	1	1	1	1	0
61	11	-1	-1	-1	-1	1
62	11	-1	-1	-1	0	0
63	11	-1	-1	-1	1	-1
64	11	1	1	1	-1	1
65	11	1	1	1	0	0
66	11	1	1	1	1	-1
67	12	-1	-1	-1	-1	0
68	12	-1	-1	-1	0	-1
69	12	-1	-1	-1	1	1
70	12	1	1	1	-1	0
71	12	1	1	1	0	-1
72	12	1	1	1	1	1

confounded. This can be seen easily by referring to Table 12.2. Recall that the $p_1 - 1$ d.f. for E^α arise from $p_1 - 1$ linearly independent contrasts among the components $E_0^\alpha, E_1^\alpha, \ldots, E_{p_1-1}^\alpha$ and the $p_2 - 1$ d.f. for E^β arise from contrasts among $E_0^\beta, E_1^\beta, \ldots, E_{p_2-1}^\beta$. The $\left(E_i^\alpha, E_j^\beta\right)$ entry in Table 12.2 is essentially the average of the responses for all treatment combinations in the $p_1^{n_1} p_2^{n_2}$ system that satisfy the equations

$$\alpha_1 x_1 + \alpha_2 x_2 + \cdots + \alpha_{n_1} x_{n_1} = i \mod p_1$$

and

$$\beta_1 y_1 + \beta_2 y_2 + \cdots + \beta_{n_2} y_{n_2} = j \mod p_2$$

and the $(p_1 - 1)(p_2 - 1)$ d.f. for $E^\alpha E^\beta$ arise entirely from contrasts among the $p_1 p_2$ entries $\left(E_i^\alpha, E_j^\beta\right)$ in Table 12.3 with $i = 0, 1, \ldots, p_1 - 1$ and $j = 0,$

Table 12.3 Representation of Interaction Contrasts for $p_1^{n_1} p_2^{n_2}$ Factorial

	E_0^β	E_1^β	E_2^β	$\cdots$	$E_{p_2-1}^\beta$
E_0^α					
E_1^α					
E_2^α					
$\vdots$					
$E_{p_1-1}^\alpha$					

1, …, $p_2 - 1$. This implies then that with this procedure we cannot confound interactions between A factors and B factors without confounding at the same time also the corresponding A factor interaction and the B factor interaction. In terms of Example 12.1 we confound with A_1A_2, A_1A_3, A_2A_3, and B_1B_2 also $A_1A_2B_1B_2$, $A_1A_3B_1B_2$, and $A_2A_3B_1B_2$, each accounting for 2 d.f. The possibilities for useful systems of confounding are therefore somewhat limited in that one may sometimes be forced to confound low-order interactions or even main effects. Partial confounding, if feasible, may resolve this problem to some extent.

From the construction of the designs it is clear that these systems of confounding have OFS as defined in Chapter 11. Expressions for sums of squares associated with effects in the pure systems can be written out easily using the methods discussed in earlier chapters. The same is true for interaction sums of squares involving factors from different pure systems, for example, $\text{SS}(A_1 \times B_1)$ in Example 12.1, as long as the interaction is not confounded. There do not, however, seem to be simple expressions for interaction sums of squares when the interactions involve factors from different pure systems and are partially confounded, for example, $\text{SS}(A_1 \times A_2 \times B_1 \times B_2)$ in Example 12.1. Because $A_1A_2B_1B_2$ is confounded with blocks, we can write formally

$$\text{SS}(A_1 \times A_2 \times B_1 \times B_2) = \text{SS}(A_1A_2B_1B_2^2)$$

with 2 d.f. The difficulty in writing out an expression for $\text{SS}(A_1A_2B_1B_2^2)$ using familiar methods is the fact that this would involve arithmetic with elements from different fields. An attempt has been made by White and Hultquist (1965) and a more useful formulation will be given in Section 12.4. A general way is, of course, to use the least-squares principle and obtain, using the notation of Chapter I.4, directly the partial sum of squares $\text{SS}(\boldsymbol{X}_{A_1A_2B_1B_2} | \mathfrak{I}, \boldsymbol{X}_\beta$, all other factorial effects), where $\boldsymbol{X}_\beta$ refers to the observation–block incidence matrix, which in this case is the treatment–block incidence matrix.

12.2.3 Use of Pseudofactors

The method of constructing systems of confounding described in Section 12.2.1 can be extended naturally to include pseudofactors (see Section 11.8). For example, this allows us to deal with asymmetrical factorials of the form $2^{n_1} \times 3^{n_2} \times 4^{n_3} \times 5^{n_4} \times 6^{n_5} (n_i \geq 0,\ i = 1,\ 2,\ 3,\ 4,\ 5)$, which includes most practical situations. All we need to do is to rewrite this as $2^{n_1+2n_3+n_5} \times 3^{n_2+n_5} \times 5^{n_4}$ and remember that some interactions in the symmetrical components of this modified asymmetrical factorial are possibly main effects in the original asymmetrical factorial. To illustrate this, we consider the following example.

Example 12.2 Suppose we want to construct a system of confounding for a $2 \times 4 \times 6$ factorial in blocks of size 12 and avoid confounding main effects. Denote the three factors by A_1, A_2, A_3, respectively. Then we replace A_2 by its pseudofactors X_1, X_2 each at two levels and A_3 by its pseudofactors Y_1 with two levels and Y_2 with three levels. We then have four factors at two levels each and one factor at three levels, that is, a $2^4 \times 3$ factorial. In order to construct blocks of size 12, we need to confound three contrasts with blocks. To avoid confounding main effect contrasts, we need to consider interactions among factors from different pure systems. For example, we may confound A_1X_1, A_1Y_1 and hence X_1Y_1, which correspond to interaction contrasts belonging to A_1A_2, A_1A_3, and A_2A_3. The initial block in terms of the pseudofactors and the original factors is then as given below:

A_1	X_1	X_2	Y_1	Y_2		A_1	A_2	A_3
0	0	0	0	0		0	0	0
0	0	0	0	1		0	0	1
0	0	0	0	2		0	0	2
0	0	1	0	0		0	1	0
0	0	1	0	1		0	1	1
0	0	1	0	2		0	1	2
1	1	0	1	0		1	2	3
1	1	0	1	1		1	2	4
1	1	0	1	2		1	2	5
1	1	1	1	0		1	3	3
1	1	1	1	1		1	3	4
1	1	1	1	2		1	3	5

The same design (apart from different block numbering) as produced by SAS PROC FACTEX is given in Table 12.4. □

Table 12.4 2 x 4 x 6 Factorial in Blocks of Size 12 (Using the Method of Pseudofactors)

```
options ate pageno=1;
proc factex;
factors A1 X1 X2 Y1;
blocks size=4;
model est=(A1 X1 X2 A1*X2 X1*X2 A1*X1*X2 Y1 Y1*X2 Y1*A1*X1
 Y1*A1*X2 Y1*X1*X2 Y1*A1*X1*X2);
examine design confounding;
output out=twodesn;
run;

factors Y2/nlev=3;
output out=threeds
  designrep=twodesn;
run;

data threeds (drop=X1 X2 Y1 Y2);
  set threeds;
 if A1=-1 then A1=0;
 if X1=-1 and X2=-1 then A2=0;
 if X1=-1 and X2=1 then A2=1;
 if X1=1 and X2=-1 then A2=2;
 if X1=1 and X2=1 then A2=3;
 if Y1=-1 and Y2=-1 then A3=0;
 if Y1=-1 and Y2=0 then A3=1;
 if Y1=-1 and Y2=1 then A3=2;
 if Y1=1 and Y2=-1 then A3=3;
 if Y1=1 and Y2=0 then A3=4;
 if Y1=1 and Y2=1 then A3=5;
run;

proc print data=threeds;
title1 'TABLE 12.4';
title2 '2x4x6 FACTORIAL IN BLOCKS OF SIZE 12';
title3 '(USING THE METHOD OF PSEUDO-FACTORS)';
run;
```

```
                     The FACTEX Procedure

                        Design Points

      Experiment
          Number     A1      X1      X2      Y1      Block
      -----------------------------------------------------
               1     -1      -1      -1      -1          4
               2     -1      -1      -1       1          1
```

Table 12.4 (*Continued*)

3	-1	-1	1	-1	4
4	-1	-1	1	1	1
5	-1	1	-1	-1	3
6	-1	1	-1	1	2
7	-1	1	1	-1	3
8	-1	1	1	1	2
9	1	-1	-1	-1	2
10	1	-1	-1	1	3
11	1	-1	1	-1	2
12	1	-1	1	1	3
13	1	1	-1	-1	1
14	1	1	-1	1	4
15	1	1	1	-1	1
16	1	1	1	1	4

Block Pseudofactor Confounding Rules

[B1] = X1*Y1
[B2] = A1*Y1

Design Points

Experiment Number	Y2
1	-1
2	0
3	1

Obs	BLOCK	A1	A2	A3
1	1	0	0	3
2	1	0	0	4
3	1	0	0	5
4	1	0	1	3
5	1	0	1	4
6	1	0	1	5
7	1	1	2	0
8	1	1	2	1
9	1	1	2	2
10	1	1	3	0
11	1	1	3	1

Table 12.4 (*Continued*)

12	1	1	3	2
13	2	0	2	3
14	2	0	2	4
15	2	0	2	5
16	2	0	3	3
17	2	0	3	4
18	2	0	3	5
19	2	1	0	0
20	2	1	0	1
21	2	1	0	2
22	2	1	1	0
23	2	1	1	1
24	2	1	1	2
25	3	0	2	0
26	3	0	2	1
27	3	0	2	2
28	3	0	3	0
29	3	0	3	1
30	3	0	3	2
31	3	1	0	3
32	3	1	0	4
33	3	1	0	5
34	3	1	1	3
35	3	1	1	4
36	3	1	1	5
37	4	0	0	0
38	4	0	0	1
39	4	0	0	2
40	4	0	1	0
41	4	0	1	1
42	4	0	1	2
43	4	1	2	3
44	4	1	2	4
45	4	1	2	5
46	4	1	3	3
47	4	1	3	4
48	4	1	3	5

12.3 THE GC/*n* METHOD

The method of using generalized cyclic designs, as described in Section 11.9 for symmetrical factorial experiments, was extended to asymmetrical factorials by Dean and John (1975). The general ideas are the same, and the statements of theorems and proofs have to be modified in an obvious manner. We shall not go into the details but rather give a brief description of the method as it applies to factorials of the form $s_1 \times s_2 \times \cdots \times s_n$, that is, we have n factors A_1, A_2, ..., A_n with A_i having s_i levels (the s_i are not necessarily all distinct). We denote the total number of treatment combinations by $t = \prod_{i=1}^{n} s_i$.

12.3.1 Description of the Method

Let μ be the lowest common multiple (LCM) of s_1, s_2, ..., s_n, that is,

$$\mu = \text{LCM}(s_1, s_2, \ldots, s_n) \tag{12.5}$$

and for a given treatment combination $\boldsymbol{x} = (x_1, x_2, \ldots, x_n)$ with $0 \le x_i \le s_i - 1 (i = 1, 2, \ldots, n)$ let

$$h = \text{HCF}\left(\mu, \frac{\mu x_1}{s_1}, \frac{\mu x_2}{s_2}, \ldots, \frac{\mu x_n}{s_n}\right) \tag{12.6}$$

be the *highest common factor* of the quantities in parentheses. The treatment combination $u\boldsymbol{x} = (ux_1, ux_2, \ldots, ux_n)$ is obtained by reducing ux_i mod $s_i (i = 1$, $2, \ldots, n)$. Since $\mu\boldsymbol{x}/h \equiv 0$, a cyclic Abelian group G of size μ/h is given by the treatment combinations 0, $\boldsymbol{x}$, $2\boldsymbol{x}$, ..., $(\mu/h - 1)\boldsymbol{x}$, and $\boldsymbol{x}$ is called the *generator* of G. In the context of factorials the group G constitutes the initial block, B_1, of size $k = \mu/h$. The remaining blocks are then obtained by adding each treatment combination to those in B_1 and retaining only the set of $b = t \cdot h/\mu$ different blocks. Each treatment combination will occur in only one block.

In general, we may have g generators $\boldsymbol{x}_1, \boldsymbol{x}_2, \ldots, \boldsymbol{x}_g$ with corresponding Abelian groups $G_1, G_2, \ldots, G_g$ having only the zero element in common. Let

$$h_i = \text{HCF}\left(\mu, \frac{\mu x_{i1}}{s_1}, \frac{\mu x_{i2}}{s_2}, \ldots, \frac{\mu x_{in}}{s_n}\right)$$

where x_{ij} is the level of A_j in $\boldsymbol{x}_i (i = 1, 2, \ldots, g;\ j = 1, 2, \ldots, n)$. Further, let $q_i = \mu/h_i$ and $q = \prod_{i=1}^{g} q_i$. The group G is then formed by the elements

$$u_1\boldsymbol{x}_1 + u_2\boldsymbol{x}_2 + \cdots + u_g\boldsymbol{x}_g (u_i = 0, 1, \ldots, q_i - 1; i = 1, 2, \ldots, g)$$

These q treatment combinations form the initial block B_1. Cyclic development of B_1 as described above yields then a single replicate GC/n design with blocks of size $k = q$ and number of blocks $b = t/q$.

Extensive tables of GC/n designs and their generators for $n \le 5$, $t \le 56$, $k \le 30$, $s_i \le 7 (i = 1, 2, 3, 4, 5)$ were given by Dean and John (1975) and additional designs for $n \leq 7$, $t \leq 200$, $k \leq 100$, $s_i \leq 7 (i = 1, 2, \ldots, 7)$ by Lewis (1982). In all these designs main effects are unconfounded. We should point out that all the designs listed can be easily constructed by the methods described in Section 12.2.

12.3.2 Choice of Generators

To illustrate the GC/*n* method and discuss the possible choice of generators, we consider the following example.

Example 12.3 Suppose we wish to construct a $2 \times 5 \times 6$ factorial in blocks of size 10. Now LCM $(2, 5, 6) = 30$. We have $k = q = 10 = 2 \cdot 5$, suggesting that we should have two generators with $q_1 = 2$ and $q_2 = 5$. Now $q_1 = 30/h_1$, which implies

$$h_1 = \text{HCF}\left(30, \frac{30x_{11}}{2}, \frac{30x_{12}}{5}, \frac{30x_{13}}{6}\right) = 15 \tag{12.7}$$

Since $0 \leq x_{11} \leq 1$, $0 \leq x_{12} \leq 4$, $0 \leq x_{13} \leq 5$, it follows that (12.7) will be satisfied by $\boldsymbol{x}_1 = (1, 0, 3)$. Similarly, $q_2 = 30/h_2$ implies

$$h_2 = \text{HCF}\left(30, \frac{30x_{21}}{2}, \frac{30x_{22}}{5}, \frac{30x_{23}}{6}\right) = 6$$

which leads to $\boldsymbol{x}_2 = (0, 1, 0)$ as a possible generator. The initial block B_1 and the remaining five blocks are given in Table 12.5. □

12.3.3 Loss of Degrees of Freedom

To obtain the loss of degrees of freedom due to confounding for the various main effects and interactions $\boldsymbol{a}^{\boldsymbol{z}}$ with $\boldsymbol{z} = (z_1, z_2, \ldots, z_n)$, the expression (11.137) in Corollary 11.3,

$$\nu^*(\boldsymbol{z}) = \frac{1}{q} \sum_{i_1} \cdots \sum_{i_n} c^*_{i_1 \ldots i_n} \left(\prod_{j=1}^{n} \rho_{i_j}^{z_j} \right) \tag{12.8}$$

has to be modified by defining

$$\rho_{i_j}^{z_j} = \begin{cases} s_i - 1 & \text{if } i_j = 1 \text{ and } z_j = 1 \\ -1 & \text{if } i_j \neq 1 \text{ and } z_j = 1 \\ 1 & \text{if } z_j = 0 \end{cases}$$

Table 12.5 GC/3-Design for 2 x 5 x 6 Factorial in Blocks of Size 10

Block					
1	2	3	4	5	6
000	100	011	111	002	102
103	003	114	014	105	005
010	110	021	121	012	112
113	013	124	024	115	015
020	120	031	131	022	122
123	023	134	034	125	025
030	130	041	141	032	132
133	033	144	044	135	035
040	140	001	101	042	142
143	043	104	004	145	045

(see Dean and John, 1975). Recall that $c^*_{i_1 i_2 \ldots i_n} = 1$ for all treatment combinations $\boldsymbol{x} = (i_1 - 1, i_2 - 1, \ldots, i_n - 1)$ in B_1 and zero otherwise. Thus (12.8) depends only on the treatment combinations in B_1 and hence can be evaluated easily for each $z \in Z$.

Example 12.3 (Continued) The $\nu^*(z)$ follow easily from inspection of the design given in Table 12.5. Every level of factor A_1 occurs the same number of times in each block; hence $\nu^*(100) = 0$. The same holds true for A_2; hence $\nu^*(010) = 0$. For factor A_3, however, only two levels occur equally often in each block, 0 and 3 in B_1 and B_2, 1 and 4 in B_3 and B_4, and 2 and 5 in B_5 and B_6. Hence only three contrasts belonging to main effect A_3 can be estimated; hence $\nu^*(001) = 2 = 5 - 3$. For every contrast belonging to a 2-factor interaction each level combination has to occur the same number of times. It follows then that $\nu^*(110) = \nu^*(011) = 0$ but $\nu^*(101) > 0$. To illustrate the method according to (12.8), however, we develop the $\nu^*(z)$ in Table 12.6. We obtain $\nu^*(001) = \frac{20}{10} = 2$, $\nu^*(101) = \frac{30}{10} = 3$. These 5 d.f. account then for the 5 d.f. among blocks.

We note here that the design given in Example 12.3 is slightly better than what could have been obtained with the Kronecker product method of Section 12.2. Both result in confounding main effects (2 d.f. from A_3 vs. 1 d.f. from A_1 and 2 d.f. from A_3), but in this situation that seems to be unavoidable. This shows that the GC/n method leads to useful designs not available otherwise.

For a detailed discussion of other properties of GC/n designs we refer the reader to Gupta and Mukerjee (1989). □

12.4 METHOD OF FINITE RINGS

The methods of constructing systems of confounding for symmetrical factorials as discussed in Chapters 8, 9, 10, and 11 (Sections 11.3, 11.7, and 11.8) are based on the arithmetic in Galois fields. To transfer these procedures to asymmetrical

Table 12.6 Determination of $\rho_{i_1}^{z_1}\rho_{i_2}^{z_2}\rho_{i_3}^{z_3}$ for the Design of Table 12.5

			z						
i_1	i_2	i_3	100	010	110	001	101	011	111
1	1	1	1	4	4	5	5	20	20
2	1	4	−1	4	−4	−1	1	−4	4
1	2	1	1	−1	−1	5	5	−5	−5
2	2	4	−1	−1	1	−1	1	1	−1
1	3	1	1	−1	−1	5	5	−5	−5
2	3	4	−1	−1	1	−1	1	1	−1
1	4	1	1	−1	−1	5	5	−5	−5
2	4	4	−1	−1	1	−1	1	1	−1
1	5	1	1	−1	−1	5	5	−5	−5
2	5	4	−1	−1	1	−1	1	1	−1
	Sum		0	0	0	20	30	0	0

factorials involves combining elements from different finite fields. Using results from abstract algebra (e.g., van der Waerden, 1966), White and Hultquist (1965) and Raktoe (1969, 1970) provided the foundation for generalizing the methods for symmetrical factorials to asymmetrical factorials (Hinkelmann, 1997). We shall give a brief description of the method (following Raktoe, 1969) without going into all mathematical details and proofs.

12.4.1 Mathematics of Ideals and Rings

First, we need the following mathematical results. Suppose we have m distinct primes $p_1, p_2, \ldots, p_m$. Let GF(p_i) be the Galois field associated with $p_i (i = 1, 2, \ldots, m)$, the elements of which are the residues mod p_i. Let $q = \prod_{i=1}^{m} p_i$ and $R(q)$ be the ring of residue classes mod q. We denote by $I(w)$ an ideal generated by an arbitrary element w of $R(q)$. We then state the following results (for proofs see Raktoe, 1969):

1. The element

$$a_j = \prod_{\substack{i=1 \\ i \neq j}}^{m} p_i - p_j = c_j - p_j \qquad \text{with } c_j = \prod_{i \neq j} p_i$$

   in $R(q)$ is prime to q and hence a_j^{-1} exists. The a_j's belong to the multiplicative group of nonzero divisors in $R(q)$.
2. The element $b_j = c_j a_j^{-1}$ generates the ideal $I(b_j)$ in $R(q)$.
3. The element b_j is the multiplicative identity in $I(b_j)$.
4. The multiplicative identity element 1 in $R(q)$ is the sum of the multiplicative identities in $I(b_j)$, that is, $1 = \sum_{j=1}^{m} b_j$.

5. The ring $R(q)$ is the direct sum of ideals $I(b_j)$, that is, $R(q) = \sum_{j=1}^{m} \otimes I(b_j)$.

6. The field GF(p_j) is isomorphic to the ideal $I(b_j)$; for $x \in$ GF(p_j) and $y \in I(b_j)$ the mapping σ: GF$(p_j) \to I(b_j)$ is defined by $\sigma(x) = b_j x = y$.

7. Addition and multiplication of elements from different Galois fields $x \in$ GF(p_j), $x^* \in GF(p_i)$ are defined by

$$x + x^* = \sigma(x) + \sigma(x^*) \bmod q$$

$$xx^* = \sigma(x)\sigma(x^*) \bmod q$$

8. Addition and multiplication of an element $r \in R(q)$ and an element $x \in GF(p_j)$ are defined by

$$r + x = r + \sigma(x) \bmod q$$

$$rx = r\sigma(x) \bmod q$$

9. The ring $R(q)$ is the direct sum of the GF(p_j), that is, $R(q) = \sum_{j=1}^{m} \otimes GF(p_j)$.

To illustrate these results, we consider the following example.

Example 12.4 Suppose we have $p_1 = 2$, $p_2 = 3$. Then $p = 6$ and the generators of the ideals are

$$b_1 = 3(3-2)^{-1} = 3 \cdot 1 = 3 \bmod 6$$

and

$$b_2 = 2(2-3)^{-1} = 2(5)^{-1} = 2 \cdot 5 = 4 \bmod 6$$

$[(5)^{-1} = 5$ since $5 \cdot 5 = 25 = 1 \bmod 6]$. We thus have the two mappings

$$\text{GF}(2) \to I(3) \qquad \text{and} \qquad \text{GF}(3) \to I(4)$$

$$\left.\begin{matrix}0\\1\end{matrix}\right\} \xrightarrow{\sigma} \left\{\begin{matrix}0\\3\end{matrix}\right. \qquad\qquad \left.\begin{matrix}0\\1\\2\end{matrix}\right\} \xrightarrow{\tau} \left\{\begin{matrix}0\\4\\2\end{matrix}\right.$$

and the direct sums

GF(2) ⊗ GF(3)	=	***I*(3) ⊗ *I*(4)**	=	***R*(6)**
$0+0$	=	$\sigma(0)+\tau(0)$	=	0
$1+0$	=	$\sigma(1)+\tau(0)$	=	3
$0+1$	=	$\sigma(0)+\tau(1)$	=	4
$1+1$	=	$\sigma(1)+\tau(1)$	=	1
$0+2$	=	$\sigma(0)+\tau(2)$	=	2
$1+2$	=	$\sigma(1)+\tau(2)$	=	5

□

12.4.2 Treatment Representations

We now apply these concepts and results to asymmetrical factorials of the form $p_1^{n_1} \times p_2^{n_2} \times \cdots \times p_m^{n_m}$. For ease of notation we shall consider specifically the $2^2 \times 3^2$ factorial. Generalizations will then become obvious.

Ordinarily, treatment combinations would be represented in the form of quadruples, say $\boldsymbol{x} = (x_{11}, x_{12}, x_{21}, x_{22})$, where x_{11}, x_{12} refer to the levels of the two-level factors, A_1, A_2 say, and x_{21}, x_{22} refer to the levels of the three-level factors, B_1, B_2, say, that is, $x_{1i} = 0, 1$, $x_{2i} = 0, 1, 2$. We shall refer to this as the *elementary representation of treatment combinations* and denote it by $T(\boldsymbol{x})$, remembering that the components in $\boldsymbol{x}$ are elements from the respective Galois fields. In order to perform arithmetic with treatment combinations, we now replace $T(\boldsymbol{x})$ by a new representation $T(\boldsymbol{v})$, where the components in $\boldsymbol{v} = (v_{11}, v_{12}, v_{21}, v_{22})$ represent the corresponding elements from the respective ideals, that is, $v_{11}, v_{12} \in I(3)$, $v_{21}, v_{22} \in I(4)$. The correspondence is as follows:

$T(\boldsymbol{x})$				$T(\boldsymbol{v})$			
x_{11}	x_{12}	x_{21}	x_{22}	v_{11}	v_{12}	v_{21}	v_{22}
0	0	0	0	0	0	0	0
0	1	0	0	0	3	0	0
1	0	0	0	3	0	0	0
1	1	0	0	3	3	0	0
0	0	0	1	0	0	0	4
0	0	0	2	0	0	0	2
0	0	1	0	0	0	4	0
	⋮				⋮		
1	1	2	2	3	3	2	2

12.4.3 Representation of Main Effects and Interactions

Concerning the representation of main effects and interactions, we recall that for a p^n factorial an effect E^{α} is represented by contrasts among the "levels" $E_0^{\alpha}, E_1^{\alpha}, \ldots, E_{p-1}^{\alpha}$ of E^{α}. Here the components in $\boldsymbol{\alpha}' = (\alpha_1, \alpha_2, \ldots, \alpha_n)$ also take on the values $0, 1, \ldots, p-1$ with the understanding that the first nonzero α_i equals 1. We note also that for the p^n factorial we can write an interaction in general as $A_1^{z_1} \times A_2^{z_2} \times \cdots \times A_n^{z_n}$, where the z_i are either 0 or 1 with the understanding that if $z_i = 0$ then the letter A_i is dropped from the interaction. An interaction $A_{i_1} \times A_{i_2} \times \cdots \times A_{i_\ell}$ consists then of $(p-1)^{\ell-1}$ components $A_{i_1} A_{i_2}^{\alpha_2} \cdots A_{i_\ell}^{\alpha_\ell}$ where the α_i take on all possible values between 1 and $p-1$. Each such component represents $p-1$ d.f.

To generalize this representation for asymmetrical factorials, we now write an interaction for the $2^2 \times 3^2$ factorial as $A_1^{z_{11}^*} \times A_2^{z_{12}^*} \times B_1^{z_{21}^*} \times B_2^{z_{22}^*}$, where z_{11}^* and

z_{12}^* are either 0 or 3, that is, elements in $I(3)$, and z_{21}^* and z_{22}^* are either 0 or 4, that is, elements in $I(4)$, with the understanding that if $z_{ij}^* = 0$ then that factor is dropped from the interaction. We may have, for example, $A_1^3 \times B_1^4 \times B_2^4$, which refers to the interaction among the factors A_1, B_1, and B_2. The components of this interaction are, in analogy to the symmetrical component systems, given by $A_1^3 \times B_1^4 \times B_2^4 = \{A_1^3 B_1^4 B_2^4, A_1^3 B_1^4 B_2^2\}$. The number of "levels," L say, for these interaction components is determined by the fact that they involve factors with two and three levels. The L is equal to the number of residue classes for the direct sum of the ideals corresponding to the factors involved. In our case this is the number of residue classes for $I(3) \otimes I(4) = R(6)$, that is, $L = 6$. More generally, if the interaction involves only ℓ factors out of m, say $F_{i_1}, F_{i_2}, \ldots, F_{i_\ell} (\ell < m)$, then L is equal to the number of residue classes in $\boldsymbol{I}(b_{i_1}) \otimes \boldsymbol{I}(b_{i_2}) \otimes \cdots \otimes \boldsymbol{I}(b_{i_\ell})$ that is generated by the greatest common divisor of $b_{i_1}, b_{i_2}, \ldots, b_{i_\ell}$. Trivially, this includes the main effects, that is, for main effect F_i with p_i levels, L equals the number of residue classes for $I(b_i)$, which equals of course p_i.

The reason why we mention the levels of an interaction is that the total number of treatment combinations can be partitioned into L equal-sized sets (of treatment combinations) that can be identified with blocks and a recognizable system of confounding. Let us consider the $2^2 \times 3^2$ factorial. For it we have for the various main effects and interactions the following L's:

$$L(A_1^3) = L(A_2^3) = L(A_1^3 A_2^3) = 2$$

$$L(B_1^4) = L(B_2^4) = L(B_1^4 B_2^4) = L(B_1^4 B_2^2) = 3$$

$$L(A_1^3 B_1^4) = L(A_1^3 B_2^4) = \cdots = L(A_1^{\varphi_1} A_1^{\varphi_2} B_1^{\eta_1} B_2^{\eta_2}) = \cdots = L(A_1^3 A_2^3 B_1^4 B_2^2) = 6$$

where $\varphi_1, \varphi_2, \in I(3)$, $\eta_1, \eta_2 \in I(4)$. For $A_1^3 A_2^3 B_1^4 B_2^2$, for example, this leads to the partitioning of the 36 treatment combinations $(v_{11}, v_{12}, v_{21}, v_{22})$ into six sets according to the equations.

$$3v_{11} + 3v_{12} + 4v_{21} + 2v_{22} = \delta \tag{12.9}$$

where $\delta = 0, 1, 2, 3, 4, 5$, that is, $\delta \in R(6)$. The sets are given in Table 12.7. Each set represents a block of size 6. We note that the set $\delta = 0$ represents the IBSG and all other sets can be obtained easily by adding to each treatment combination in the IBSG a treatment not in the IBSG and not in any other already constructed set. As a consequence of constructing blocks according to (12.9), it follows that the interaction associated with (12.9), that is, $A_1^3 A_2^3 B_1^4 B_2^2$, is confounded with blocks. This statement, however, is to be understood only in the sense that together with $A_1^3 A_2^3 B_1^4 B_2^2$ also $A_1^3 A_2^3$ and $B_1^4 B_2^2$, that is, the components from the symmetrical systems, are confounded with blocks. This is so because to satisfy (12.9) for a given δ we must have

$$3v_{11} + 3v_{12} = \delta_1 \bmod 6 \tag{12.10}$$

Table 12.7 Sets of Treatment Combinations Satisfying $3v_{11} + 3v_{12} + 4v_{21} + 2v_{22} = \delta$

$\delta = 0$	$\delta = 1$	$\delta = 2$
0 0 0 0	3 0 4 0	0 0 2 0
3 3 0 0	0 3 4 0	3 3 2 0
0 0 4 4	3 0 2 4	0 0 0 4
3 3 4 4	0 3 2 4	3 3 0 4
0 0 2 2	3 0 0 2	0 0 4 2
3 3 2 2	0 3 0 2	3 3 4 2
$\delta = 3$	$\delta = 4$	$\delta = 5$
3 0 0 0	0 0 4 0	3 0 2 0
0 3 0 0	3 3 4 0	0 3 2 0
3 0 4 4	0 0 2 4	3 0 0 4
0 3 4 4	3 3 2 4	0 3 0 4
3 0 2 2	0 0 0 2	3 0 4 2
0 3 2 2	3 3 0 2	0 3 4 2

and

$$4v_{21} + 2v_{22} = \delta_2 \bmod 6 \tag{12.11}$$

with $\delta_1 + \delta_2 = \delta$. Thus we see that the system of confounding generated in this manner is the same as that obtained by using the Kronecker product method of Section 12.2 by considering confounding in the 2^2 and 3^2 systems separately and then combining them. This is true in general (see Voss, 1986) and thus may detract somewhat from the value of this method, but we shall discuss below that it provides for an extension of a useful result for symmetrical factorials that other methods do not do.

12.4.4 Parameterization of Treatment Responses

Recall from Chapter 11 that we expressed the true response $a(\boldsymbol{x})$ of a treatment combination $\boldsymbol{x}$ as a linear function of interaction components $E^{\boldsymbol{\alpha}}$ as follows:

$$a(\boldsymbol{x}) = M + \sum_{\boldsymbol{\alpha}} E^{\boldsymbol{\alpha}}_{\boldsymbol{\alpha}'\boldsymbol{x}} \tag{12.12}$$

where $\sum_{\boldsymbol{\alpha}}$ is the summation over all partitions $\boldsymbol{\alpha}$. To write down the counterpart to (12.12) for asymmetrical factorials, we consider again the $2^2 \times 3^2$ factorial. The partitions $\boldsymbol{\alpha}$ are determined by the various main effects and interactions and are given in Table 12.8. Associated with each partition is an effect/interaction component and each component has a number, L, of levels as mentioned earlier.

Table 12.8 Partitions for the $2^2 \times 3^2$ Factorial

Interaction $\mathbf{z}^{*\prime} = (z_{11}^*, z_{12}^*, z_{21}^*, z_{22}^*)$	Partition $\boldsymbol{\alpha}' = (\alpha_{11}, \alpha_{12}, \alpha_{21}, \alpha_{22})$	Interaction/Modified Interaction Component
3 0 0 0	3 0 0 0	$(A_1^3)_\delta (\delta = 0, 3)$
0 3 0 0	0 3 0 0	$(A_2^3)_\delta (\delta = 0, 3)$
3 3 0 0	3 3 0 0	$(A_1^3 A_2^3)_\delta (\delta = 0, 3)$
0 0 4 0	0 0 4 0	$(B_1^4)_\delta (\delta = 0, 4, 2)$
0 0 0 4	0 0 0 4	$(B_2^4)_\delta (\delta = 0, 4, 2)$
0 0 4 4	0 0 4 4	$(B_1^4 B_2^4)_\delta (\delta = 0, 4, 2)$
	0 0 4 2	$(B_1^4 B_2^2)_\delta (\delta = 0, 4, 2)$
3 0 4 0	3 0 4 0	$(A_1^3 B_1^4)_\delta^* = (A_1^3 B_1^4)_\delta - (A_1^3)_{\delta_1} - (B_1^4)_{\delta_2}$ $(\delta = 0, 1, 2, 3, 4, 5;\ \delta_1 = 0, 3;\ \delta_2 = 0, 4, 2)$
3 0 0 4	3 0 0 4	$(A_1^3 B_2^4)_\delta^* = (A_1^3 B_2^4)_\delta - (A_1^3)_{\delta_1} - (B_2^4)_{\delta_2}$
0 3 4 0	0 3 4 0	$(A_2^3 B_1^4)_\delta^* = (A_2^3 B_1^4)_\delta - (A_2^3)_{\delta_1} - (B_1^4)_{\delta_2}$
0 3 0 4	0 3 0 4	$(A_2^3 B_2^4)_\delta^* = (A_2^3 B_2^4)_\delta - (A_2^3)_{\delta_1} - (B_2^4)_{\delta_2}$
3 3 4 0	3 3 4 0	$(A_1^3 A_2^3 B_1^4)_\delta^* = (A_1^3 A_2^3 B_1^4)_\delta - (A_1^3 A_2^3)_{\delta_1} - (B_1^4)_{\delta_2}$
3 3 0 4	3 3 0 4	$(A_1^3 A_2^3 B_2^4)_\delta^* = (A_1^3 A_2^3 B_2^4)_\delta - (A_1^3 A_2^3)_{\delta_1} - (B_2^4)_{\delta_2}$
3 0 4 4	3 0 4 4	$(A_1^3 B_1^4 B_2^4)_\delta^* = (A_1^3 B_1^4 B_2^4)_\delta - (A_1^3)_{\delta_1} - (B_1^4 B_2^4)_{\delta_2}$
	3 0 4 2	$(A_1^3 B_1^4 B_2^2)_\delta^* = (A_1^3 B_1^4 B_2^2)_\delta - (A_1^3)_{\delta_1} - (B_1^4 B_2^2)_{\delta_2}$

Table 12.8 ***(Continued)***

Interaction $z^{*\prime} = (z_{11}^*, z_{12}^*, z_{21}^*, z_{22}^*)$	Partition $\boldsymbol{\alpha}' = (\alpha_{11}, \alpha_{12}, \alpha_{21}, \alpha_{22})$	Interaction/Modified Interaction Component
0 3 4 4	0 3 4 4	$(A_2^3 B_1^4 B_2^4)_\delta^* = (A_2^3 B_1^4 B_2^4)_\delta - (A_2^3)_{\delta_1} - (B_1^4 B_2^4)_{\delta_2}$
	0 3 4 2	$(A_2^3 B_1^4 B_2^2)_\delta^* = (A_2^3 B_1^4 B_2^2)_\delta - (A_2^3)_{\delta_1} - (B_1^4 B_2^2)_{\delta_2}$
3 3 4 4	3 3 4 4	$(A_1^3 A_2^3 B_1^4 B_2^4)_\delta^* = (A_1^3 A_2^3 B_1^4 B_2^4)_\delta - (A_1^3 A_2^3)_{\delta_1} - (B_1^4 B_2^4)_{\delta_2}$
	3 3 4 2	$(A_1^3 A_2^3 B_1^4 B_2^2)_\delta^* = (A_1^3 A_2^3 B_1^4 B_2^2)_\delta - (A_1^3 A_2^3)_{\delta_1} - (B_1^4 B_2^2)_{\delta_2}$

The levels are the elements δ in $I(3)$, $I(4)$, $R(6)$, respectively, as indicated in Table 12.8. We define the components as

$$\begin{aligned} E_\delta^{\boldsymbol{\alpha}} &= (A_1^{\alpha_{11}} A_2^{\alpha_{12}} B_1^{\alpha_{21}} B_2^{\alpha_{22}})_\delta \\ &= \{\text{average of all treatment responses } a(v_{11}, v_{12}, v_{21}, v_{22}) \\ &\quad \text{with } \alpha_{11}v_{11} + \alpha_{12}v_{12} + \alpha_{21}v_{21} + \alpha_{22}v_{22} = \delta (\text{mod } q)\} \\ &\quad -\{\text{average of all treatment responses}\} \end{aligned} \tag{12.13}$$

where in our case $q = 6$. For interactions involving factors with different numbers of levels we now introduce what we shall call *modified interaction components* defined as

$$\begin{aligned} (E^{\boldsymbol{\alpha}})_\delta^* &= (A_1^{\alpha_{11}} A_2^{\alpha_{12}} B_1^{\alpha_{21}} B_2^{\alpha_{22}})_\delta^* \\ &= (A_1^{\alpha_{11}} A_2^{\alpha_{12}} B_1^{\alpha_{21}} B_2^{\alpha_{22}})_\delta \\ &\quad -(A_1^{\alpha_{11}} A_2^{\alpha_{12}})_{\delta_1} - (B_1^{\alpha_{21}} B_2^{\alpha_{22}})_{\delta_2} \end{aligned} \tag{12.14}$$

We shall refer to the partitions $\boldsymbol{\alpha}$ involving only factors with the same number of levels as *pure partitions* and those that involve factors not all having the same number of levels as *mixed partitions*. Let $\mathcal{A}_1$ denote the set of all pure partitions and $\mathcal{A}_2$ the set of all mixed partitions. We can then write

$$a(\boldsymbol{v}) = \boldsymbol{M} + \sum_{\boldsymbol{\alpha} \in \mathcal{A}_1} E^{\boldsymbol{\alpha}}_{\boldsymbol{\alpha}'\boldsymbol{v}} + \sum_{\boldsymbol{\alpha} \in \mathcal{A}_2} (E^{\boldsymbol{\alpha}})^*_{\boldsymbol{\alpha}'\boldsymbol{v}} \tag{12.15}$$

where $\boldsymbol{M}$ represents the overall mean. Expression (12.15) is the extension of (12.12) for the asymmetrical factorial and is used in the same way as (12.12) is used for the symmetrical factorial.

12.4.5 Characterization and Properties of the Parameterization

The proposed parameterization (12.15) is a major result in that it unifies the theory of symmetrical and asymmetrical factorials, both with respect to the construction of design, that is, systems of confounding, and the analysis of such designs. Although we have only illustrated the idea in terms of a simple example, the result is true in general. The same holds for the following comments about further characterizations and properties of the modified components [defined in (12.14)] as part of the parameterization (12.15).

1. Expressing each $a(\boldsymbol{v})$ as (12.15) represents a reparameterization of the treatment effects. This reparameterization is singular since for each $E^{\boldsymbol{\alpha}}$ we have $\sum_\delta E^{\boldsymbol{\alpha}}_\delta = 0$.
2. For the modified components the values for δ_1 and δ_2 are determined uniquely from the addition table, for example,

		δ_2		
		0	4	2
δ_1	0	0	4	2
	3	3	1	5

   where the entries in the table are $\delta = \delta_1 + \delta_2$.
3. For the modified components we have $\sum_{\delta_1} (E^{\boldsymbol{\alpha}})^*_{\delta_1+\delta_2} = 0$ for each δ_2 and $\sum_{\delta_2} (E^{\boldsymbol{\alpha}})^*_{\delta_1+\delta_2} = 0$ for each δ_1.
4. The degrees of freedom $\nu(\boldsymbol{\alpha})$ associated with the component $E^{\boldsymbol{\alpha}}$ are given by $\nu(\boldsymbol{\alpha}) = L(E^{\boldsymbol{\alpha}}) - 1$; for example, for $A_1^3 A_2^3 B_1^4 B_2^2$ we have $L(A_1^3 A_2^3 B_1^4 B_2^2) = 6$ and hence $\nu(3, 3, 4, 2) = 5$.
5. The degrees of freedom $\nu^*(\boldsymbol{\alpha})$ associated with the modified component $(A_1^{\alpha_{11}} A_2^{\alpha_{12}} B_1^{\alpha_{21}} B_2^{\alpha_{22}})^*$ are given by

$$\begin{aligned} \nu^*(\alpha_{11}\alpha_{12}\alpha_{21}\alpha_{22}) &= \nu(\alpha_{11}\alpha_{12}\alpha_{21}\alpha_{22}) - \nu(\alpha_{11}\alpha_{12}00) - \nu(00\alpha_{21}\alpha_{22}) \\ &= \nu(\alpha_{11}\alpha_{12}00) \cdot \nu(00\alpha_{21}\alpha_{22}) \end{aligned}$$

6. If a block design is constructed by confounding $A_1^{\alpha_{11}} A_2^{\alpha_{12}} B_1^{\alpha_{21}} B_2^{\alpha_{22}}$ and if $A_1^{\alpha_{11}} A_2^{\alpha_{12}} B_1^{\alpha_{21}} B_2^{\alpha_{22}}$ is a mixed component, it follows from comment 3 and the definition of the modified component that $A_1^{\alpha_{11}} A_2^{\alpha_{12}}$ and $B_1^{\alpha_{21}} B_2^{\alpha_{22}}$ are also confounded (as pointed out earlier).

7. The component $E_\delta^{\boldsymbol{\alpha}}$ is estimated by

$$\widehat{E}_\delta^{\boldsymbol{\alpha}} = \{\text{mean of observed responses for treatments } \boldsymbol{v} \text{ satisfying } \boldsymbol{\alpha}'\boldsymbol{v} = \delta\} - \{\text{overall observed mean}\}$$

8. If the confounded interactions, that is, $A_1^{\alpha_{11}} A_2^{\alpha_{12}} B_1^{\alpha_{21}} B_2^{\alpha_{22}}$, $A_1^{\alpha_{11}} A_2^{\alpha_{12}}$, $B_1^{\alpha_{21}} B_2^{\alpha_{22}}$, are assumed to be negligible, then differences between two treatment responses (more generally, linear contrasts among treatment responses), say $a(\boldsymbol{v}) - a(\boldsymbol{w})$, can be estimated by

$$\widehat{a}(\boldsymbol{v}) - \widehat{(\boldsymbol{w})} = \sum_{\boldsymbol{\alpha} \in \mathcal{A}_1}{}' \left[\widehat{E}^{\boldsymbol{\alpha}}_{\boldsymbol{\alpha}'\boldsymbol{v}} - \widehat{E}^{\boldsymbol{\alpha}}_{\boldsymbol{\alpha}'\boldsymbol{w}}\right] + \sum_{\boldsymbol{\alpha} \in \mathcal{A}_2}{}' \left[(\widehat{E}^{\boldsymbol{\alpha}})^*_{\boldsymbol{\alpha}'\boldsymbol{v}} - (\widehat{E}^{\boldsymbol{\alpha}})^*_{\boldsymbol{\alpha}'\boldsymbol{w}}\right] \tag{12.16}$$

where $\sum'$ refers to summation over all $\boldsymbol{\alpha}$ except those belonging to confounded interactions. Using (12.14), we find, for example,

$$\begin{aligned}
a(3040) - a(0000) = {} & \left[(A_1^3)_3 - (A_1^3)_0\right] + \left[(A_1^3 A_2^3)_3 - (A_1^3 A_2^3)_0\right] \\
& + \left[(B_1^4)_4 - (B_1^4)_0\right] + \left[(B_1^4 B_2^4)_4 - (B_1^4 B_2^4)_0\right] \\
& + \left[(A_1^3 B_1^4)_1^* - (A_1^3 B_1^4)_0^*\right] + \left[(A_1^3 B_2^4)_3^* - (A_1^3 B_2^4)_0^*\right] \\
& + \left[(A_2^3 B_1^4)_4^* - (A_2^3 B_1^4)_0^*\right] \\
& + \left[(A_1^3 A_2^3 B_1^4)_1^* - (A_1^3 A_2^3 B_1^4)_0^*\right] \\
& + \left[(A_1^3 A_2^3 B_2^4)_3^* - (A_1^3 A_2^3 B_2^4)_0^*\right] \\
& + \left[(A_1^3 B_1^4 B_2^4)_1^* - (A_1^3 B_1^4 B_2^4)_0^*\right] \\
& + \left[(A_1^3 B_1^4 B_2^2)_1^* - (A_1^3 B_1^4 B_2^2)_0^*\right] \\
& + \left[(A_2^3 B_1^4 B_2^4)_4^* - (A_2^3 B_1^4 B_2^4)_0^*\right]
\end{aligned}$$

$$+\left[(A_2^3B_1^4B_2^2)_4^* - (A_2^3B_1^4B_2^2)_0^*\right]$$

$$+\left[(A_1^3A_2^3B_1^4B_2^4)_1^* - (A_1^3A_2^3B_1^4B_2^4)_0^*\right] \tag{12.17}$$

Observed means (based on different numbers of observations) are then substituted for the true means to obtain $\widehat{a}(3040) - \widehat{a}(0000)$.

9. It is obvious that variances for differences of the form $\left[(\widehat{A_1^3})_3 - (\widehat{A_1^3})_0\right]$ can be written out easily in terms of the number of observations satisfying the equation $3v_{11} = \delta$ for a given δ. If N is the total number of treatment combinations, then the number of treatment combinations satisfying $3v_{11} = \delta$ is given by $N/L(A_1^3)$. Hence

$$\text{var}\left[(\widehat{A_1^3})_3 - (\widehat{A_1^3})_0\right] = 2\frac{L(A_1^3)}{N}\sigma_e^2$$

which equals $\frac{1}{9}\sigma_e^2$ for the $2^2 \times 3^2$ factorial.

For differences of the form $\left[(\widehat{A_1^3B_2^4})_3^* - (\widehat{A_1^3B_2^4})_0^*\right]$, that is, involving modified components, it can be worked out that

$$\text{var}\left[(\widehat{A_1^3B_2^4})_3^* - (\widehat{A_1^3B_2^4})_0^*\right] = \frac{2}{N}\left[L(A_1^3B_2^4) - L(A_1^3) - L(B_2^4)\right]\sigma_e^2$$

which for the $2^2 \times 3^2$ factorial equals

$$\frac{2}{36}[6 - 2 - 3]\sigma_e^2 = \frac{1}{18}\sigma_e^2$$

Finally, it can be shown that the covariances between differences of components are zero, for example,

$$\text{cov}\left\{\left[(\widehat{A_1^3})_3 - (\widehat{A_1^3})_0\right], \left[(\widehat{B_1^4B_2^2})_4 - (\widehat{B_1^4B_2^2})_0\right]\right\} = 0$$

$$\text{cov}\left\{\left[(\widehat{A_1^3})_3 - (\widehat{A_1^3})_0\right], \left[(\widehat{A_1^3B_1^4})_1^* - (\widehat{A_1^3B_1^4})_0^*\right]\right\} = 0$$

$$\text{cov}\left\{\left[(\widehat{A_1^3B_1^4})_1^* - (\widehat{A_1^3B_1^4})_0^*\right], \left[(\widehat{A_1^3A_2^3B_1^4})_1^* - (\widehat{A_1^3A_2^3B_1^4})_0^*\right]\right\} = 0$$

10. Using (12.16) and the results in comment 9, one can obtain easily the variances of the estimates of treatment contrasts, that is,

$$\begin{aligned}\text{var}[\widehat{a}(\boldsymbol{v}) - \widehat{a}(\boldsymbol{w})] &= \text{var}[\widehat{\tau}(\boldsymbol{v}) - \widehat{\tau}(\boldsymbol{w})] \\ &= (\text{sum of the variances of all estimated} \\ &\qquad \text{component differences})\end{aligned}$$

for example, for (12.17) we find

$$\operatorname{var}[\widehat{a}(3040) - \widehat{a}(0000)] = \left[\frac{2}{18} + \frac{2}{18} + \frac{2}{12} + \frac{2}{12} + 10\left(\frac{2}{6} - \frac{2}{18} - \frac{2}{12}\right)\right]\sigma_e^2 = \frac{20}{18}\sigma_e^2$$

11. The sums of squares for unconfounded effects can be written out similarly to those for symmetrical factorials:

$$\mathrm{SS}(A_1^3) = \frac{N}{L(A_1)} \sum_{\delta \in I(3)} \left[(\widehat{A_1^3})_\delta\right]^2$$

$$\mathrm{SS}(B_1^4 B_2^2) = \frac{N}{L(B_1^4 B_2^2)} \sum_{\delta \in I(4)} \left[(\widehat{B_1^4 B_2^2})_\delta\right]^2$$

$$\mathrm{SS}(A_1^3 B_1^4) = \frac{N}{L(A_1^3 B_1^4)} \sum_{\delta \in R(6)} \left[(\widehat{A_1^3 B_1^4})_\delta^*\right]^2$$

and because of comment 9 these sums of squares are orthogonal to each other. This makes testing of hypotheses concerning main effects and interactions easy.

12.4.6 Other Methods for Constructing Systems of Confounding

In addition to the method discussed in this section, alternative but essentially equivalent methods of constructing systems of confounding for asymmetrical factorials have been proposed. A method based on finite groups and rings is discussed by Banerjee (1970), Worthley and Banerjee (1974) (see also Raktoe, Rayner, and Chalton, 1978), and Sihota and Banerjee (1981), who extended the method to include any number of levels, thus avoiding the use of pseudofactors. A different technique based on the Chinese remainder theorem was introduced by Lin (1986) and extended by Huang (1989).

12.5 BALANCED FACTORIAL DESIGNS (BFD)

The incomplete block designs for asymmetrical factorials we have discussed so far have resulted in the confounding of certain contrasts belonging to some main effects and/or interactions. As a consequence the number of degrees of freedom for such effects has been reduced (possibly to zero). The lost degrees of freedom are said to be confounded with blocks. As we have seen in the previous sections, the confounding of interactions may sometimes also lead to the confounding of main effects, for example, with the methods of Sections 12.2 and 12.4. If we confound, for example, a two-factor interaction involving factors with different numbers of levels, then the corresponding main effects are confounded also. This is generally an undesirable feature. If it cannot be avoided, the method of partial confounding, that is, using different systems of confounding in different replicates, provides some relief. This, however, may not always be feasible from a purely practical point of view because of the usually large number of experimental units required.

In this section we shall discuss a method that results in a different kind of partial confounding. Rather than confound completely certain degrees of freedom in some replicates and not at all in others, the method to be described results in some loss of information for some degrees of freedom over the whole design. We shall illustrate this with a very simple example.

Example 12.5 Consider the $2^2 \times 3$ factorial in blocks of size 6. Using the Kronecker product method of Section 12.2, a reasonable design may be obtained by using three replicates of two blocks each. If we denote the factors by A_1, A_2, and B with two, two, and three levels, respectively, we may construct the design by confounding A_1 in replicate I, A_2 in replicate II, and A_1A_2 in replicate III. The allocation of the treatment combinations is given in Table 12.9a. Compared to a RCBD with $b = 3$ blocks (replicates), the relative information for main effects A_1, A_2 and the interaction A_1A_2 is $\frac{2}{3}$ and 1 for all other effects.

In Table 12.9b an arrangement due to Yates (1937b) is given that is the only arrangement not resulting in the confounding of main effects. Kempthorne (1952)

Table 12.9 $2^2 \times 3$ Factorial in Blocks of Size 6

	Replicate					
	I		II		III	
Block	1	2	3	4	5	6
	a. Kronecker Product Method					
	000	100	000	010	000	100
	001	101	001	011	001	101
	002	102	002	012	002	102
	010	110	100	110	110	010
	011	111	101	111	111	011
	012	112	102	112	112	012
	b. Yates Method					
	000	100	100	000	100	000
	110	010	010	110	010	110
	101	001	001	101	101	001
	011	111	111	011	011	111
	102	002	102	002	002	102
	012	112	012	112	112	012

shows, from first principles, using the least-squares method, that the relative information for A_1A_2 is $\frac{8}{9}$, for A_1A_2B is $\frac{5}{9}$, and for all other effects is 1.

If we define the *total relative loss in information*, TRL say, as the weighted sum of relative losses for the various effects, with the weights being the degrees of freedom for each effect, we find that TRL is the same for both arrangements, namely

$$\text{TRL(a)} = 1 \cdot \tfrac{1}{3} + 1 \cdot \tfrac{1}{3} + 1 \cdot \tfrac{1}{3} = 1$$

$$\text{TRL(b)} = 1 \cdot \tfrac{1}{9} + 2 \cdot \tfrac{4}{9} = 1$$

Even so, we would generally prefer arrangement (b) over arrangement (a) since its loss of information is mainly associated with the three-factor interaction. □

Both designs of Example 12.5 belong to the class of *balanced factorial designs* (BFD), a notion defined by Shah (1958) for the general factorial experiment. Although Shah (1958, 1960) pointed out the connection between BFDs and PBIB designs, Kurkjian and Zelen (1963) and Kshirsagar (1966) established the one-to-one correspondence between BFDs and EGD-PBIB designs as defined by Hinkelmann and Kempthorne (1963) and Hinkelmann (1964) (see Section 4.6.9). We shall now give some definitions of and results for BFDs and then give some methods of construction.

12.5.1 Definitions and Properties of BFDs

Suppose we have n factors $A_1, A_2, \ldots, A_n$ with $s_1, s_2, \ldots, s_n$ levels, respectively, where the s_i $(i = 1, 2, \ldots, n)$ do not necessarily have to be all distinct. Following Kurkjian and Zelen (1963) and extending the notation of Section 11.9, we define the $s_i \times s_i$ matrix

$$\boldsymbol{M}_i = s_i \boldsymbol{I} - \mathfrak{I}\mathfrak{I}' \tag{12.18}$$

and

$$\boldsymbol{M}_i^{z_i} = \begin{cases} \mathfrak{I}'_{s_i} & \text{if } z_i = 0 \\ \boldsymbol{M}_i & \text{if } z_i = 1 \end{cases}$$

$$\boldsymbol{M}^{z} = \boldsymbol{M}_1^{z_1} \times \boldsymbol{M}_2^{z_2} \times \cdots \times \boldsymbol{M}_n^{z_n} \tag{12.19}$$

for $z = (z_1, z_2, \ldots, z_n)$ and $z_i = 0, 1 (i = 1, 2, \ldots, n)$. Let $\boldsymbol{Z} = \{z\}$ be the collection of all possible vectors z except $z = (0, 0, \ldots, 0)$. Further, let $t = \prod_{i=1}^{n} s_i$ denote the number of treatment combinations. Then for any $z \in \boldsymbol{Z}$ the interaction components in the n-factor factorial can be expressed in terms of the treatment effects $\tau(x_1, x_2, \ldots, x_n)(0 \leq x_i \leq s_i - 1; i = 1, 2, \ldots, n)$ arranged in standard order (see Section 11.9) in the vector $\boldsymbol{\tau}$ as

$$\boldsymbol{a}^{z} = \frac{1}{t} \boldsymbol{M}^{z} \boldsymbol{\tau} \tag{12.20}$$

Let

$$\phi(z) = \boldsymbol{\ell}_z' \boldsymbol{a}^z = \boldsymbol{\varphi}'(z)\boldsymbol{\tau} \tag{12.21}$$

denote a normalized contrast belonging to $\boldsymbol{a}^z$ (note that we are using $\boldsymbol{a}^z$ for two different things but it should always be clear from the context what is meant). With

$$\widehat{\boldsymbol{\tau}} = \boldsymbol{C}^- \boldsymbol{Q}$$

as a solution to the reduced normal equations we obtain the BLUE of $\phi(z)$ in (12.21) as

$$\widehat{\phi}(z) = \boldsymbol{\ell}_z' \widehat{\boldsymbol{a}}^z = \frac{1}{t} \boldsymbol{\ell}_z' \boldsymbol{M}^z \widehat{\boldsymbol{\tau}} \tag{12.22}$$

Following Gupta and Mukerjee (1989), we then give the following definition.

Definition 12.1 In a factorial design an interaction $\mathbf{a}^z (z \in Z)$ is said to be *balanced* if either (a) all contrasts belonging to $\boldsymbol{a}^z$ are estimable and the BLUEs of all normalized contrasts belonging to $\boldsymbol{a}^z$ have the same variance or (b) no contrast belonging to $\boldsymbol{a}^z$ is estimable. □

If we adopt the convention that if a contrast is nonestimable its BLUE has variance ∞, then parts (a) and (b) in Definition 12.1 can be combined, and it is in this sense that we shall use the definition. For an overall characterization of a factorial design, we then have the following definition.

Definition 12.2 A factorial design is said to be a BFD if the interaction $\boldsymbol{a}^z$ is balanced for every $z \in \boldsymbol{Z}$.

It is of historical interest to note that Li (1944, p. 457) already refers to BFD (although not by that name) when he says, "An interaction which has more than 1 degree of freedom should be confounded as evenly as possible." □

Gupta and Mukerjee (1989) show that an interaction $\boldsymbol{a}^z$ is balanced in the sense of Definition 12.1 if and only if the BLUEs of any two orthogonal contrasts belonging to $\boldsymbol{a}^z$ are uncorrelated. Recall that a similar result holds for factorial designs with OFS except that in that case BLUEs of contrasts belonging to different interactions are uncorrelated. Hence in a BFD with OFS we have between-interaction and within-interaction orthogonality (Gupta and Mukerjee, 1989). As we have pointed out earlier (e.g., I.7), orthogonality is a useful property as it leads to ease of analysis and ease of interpretation of the results from an experiment. We have seen in Chapter 11 that OFS requires a certain structure of $\boldsymbol{C}^-$(see Theorem 11.12) and hence of $\boldsymbol{C}$ and ultimately $\boldsymbol{NN}'$. The results of Shah (1958, 1960), Kurkjian and Zelen (1963), Kshirsagar (1966), and Gupta and Mukerjee (1989) have led to similar structures for $\boldsymbol{NN}'$ (and hence $\boldsymbol{C}$ and

$\boldsymbol{C}^-$) for BFDs. We shall give a brief description of the major results, deferring details to the above references.

The basic result for giving an algebraic characterization of a BFD with OFS is what Kurkjian and Zelen (1963) have called property A of the matrix $\boldsymbol{NN}'$. Let $\boldsymbol{D}_i^{\delta_i}$ be an $s_i \times s_i$ matrix defined as

$$\boldsymbol{D}_i^{\delta_i} = \begin{cases} \boldsymbol{I}_{s_i} & \text{if } \delta_i = 0 \\ \mathfrak{I}_{s_i}\mathfrak{I}'_{s_i} & \text{if } \delta_i = 1 \end{cases}$$

for $i = 1, 2, \ldots, n$, and

$$\boldsymbol{D}^{\boldsymbol{\delta}} = \prod_{i=1}^{n} \times \boldsymbol{D}_i^{\delta_i}$$

with $\boldsymbol{\delta} = (\delta_1, \delta_2, \ldots, \delta_n)$. Denote the set of all $\boldsymbol{\delta}$ (with $\delta_i = 0, 1$) by $\boldsymbol{Z}^* = \boldsymbol{Z} \cup \{(0, 0, \ldots, 0)\}$. We then state formally:

Definition 12.3 (Gupta and Mukerjee, 1989) A $t \times t$ matrix $\boldsymbol{G}$ with $t = \prod_{i=1}^n s_i$ is said to have *property* A if it is of the form

$$\boldsymbol{G} = \sum_{\boldsymbol{\delta} \in \boldsymbol{Z}^*} h(\boldsymbol{\delta}) \boldsymbol{D}^{\boldsymbol{\delta}} \tag{12.23}$$

where the $h(\boldsymbol{\delta})$ are constants depending on $\boldsymbol{\delta}$.

Now suppose the matrix $\boldsymbol{C}$ has property A, that is,

$$\boldsymbol{C} = r\boldsymbol{I} - \frac{1}{k}\boldsymbol{NN}' = \sum_{\boldsymbol{\delta} \in \boldsymbol{Z}^*} h(\boldsymbol{\delta}) \boldsymbol{D}^{\boldsymbol{\delta}} \tag{12.24}$$

We want to show that if $\boldsymbol{C}$ is of the form (12.24) then all normalized contrasts belonging to $\boldsymbol{a}^{\boldsymbol{z}}$ are estimated with the same variance and the estimators of any two orthogonal contrasts belonging to $\boldsymbol{a}^{\boldsymbol{z}}$ are uncorrelated. Let $\boldsymbol{H}'_i$ be an $(s_i - 1) \times s_i$ matrix of normalized orthogonal contrasts, that is, $\boldsymbol{H}'_i \mathfrak{I}_{s_i} = \boldsymbol{0}$ and $\boldsymbol{H}'_i \boldsymbol{H}_i = \boldsymbol{I}_{s_i - 1}$, and define

$$\boldsymbol{H}_i^{z_i} = \begin{cases} \boldsymbol{H}'_i & \text{if } z_i = 1 \\ 1 & \text{if } z_i = 0 \end{cases}$$

and

$$\boldsymbol{H}^{\boldsymbol{z}} = \prod_{i=1}^{n} \times \boldsymbol{H}_i^{z_i}$$

A complete set of normalized orthogonal contrasts belonging to $\boldsymbol{a}^z$ is then given by $\boldsymbol{H}^z\boldsymbol{a}^z$. More specifically, the contrasts belonging to the p-factor interaction $A_{i_1} \times A_{i_2} \times \cdots \times A_{i_p} (1 \leq p \leq n)$, that is, to the $s_{i_1}s_{i_2}\cdots s_{i_p}$ elements of $a_{i_1} \otimes a_{i_2} \otimes \cdots \otimes a_{i_p}$, are given by

$$\left(\boldsymbol{H}'_{i_1} \times \boldsymbol{H}'_{i_2} \times \cdots \times \boldsymbol{H}'_{i_p}\right)\left(a_{i_1} \otimes a_{i_2} \otimes \cdots \otimes a_{i_p}\right)$$

and there are $(s_{i_1} - 1)(s_{i_2} - 1)\cdots(s_{i_p} - 1)$ such contrasts. We can then prove the following theorem. □

Theorem 12.1 If for an $s_1 \times s_2 \times \cdots \times s_n$ factorial design the $\boldsymbol{C}$ matrix satisfies property A, then the design is a BFD and the variance–covariance matrix for $\boldsymbol{H}^z\widehat{\boldsymbol{a}}^z$ is given by

$$\text{var}(\boldsymbol{H}^z\widehat{\boldsymbol{a}}^z) = \frac{1}{\prod_i s_i^{1-z_i} r E(z)} \boldsymbol{I}\sigma_e^2 \tag{12.25}$$

where r is the number of replications for each treatment combination and $E(z)$, the efficiency factor for $\boldsymbol{a}^z$, is a constant depending only on z.

Proof Let

$$\boldsymbol{C} = \sum_{\delta \epsilon Z^*} h(\delta)\boldsymbol{D}^\delta$$

Then

$$\boldsymbol{M}^z\boldsymbol{C} = \sum_{\delta \epsilon \boldsymbol{Z}^*} h(\delta)\boldsymbol{M}^z\boldsymbol{D}^\delta$$

$$= \sum_{\delta \epsilon \boldsymbol{Z}^*} h(\delta) \prod_i \times \boldsymbol{M}_i^{z_i}\boldsymbol{D}_i^{\delta_i} \tag{12.26}$$

Now

$$\boldsymbol{M}_i^{z_i}\boldsymbol{D}_i^{\delta_i} = (1 - z_i\delta_i)s_i^{(1-z_i)\delta_i}\boldsymbol{M}_i^{z_i}$$

and

$$\prod_i \times \boldsymbol{M}_i^{z_i}\boldsymbol{D}_i^1 = \boldsymbol{0}$$

so that (12.26) can be written as

$$\boldsymbol{M}^z\boldsymbol{C} = \sum_{\delta \in \boldsymbol{Z}^{**}} h(\delta) \prod_i \times \left[(1 - z_i\delta_i)s_i^{(1-z)\delta_i}\boldsymbol{M}_i^{z_i}\right]$$

$$= rE(z)\boldsymbol{M}^z \tag{12.27}$$

where, by definition,

$$rE(z) = \sum_{\delta \in Z^{**}} h(\delta) \prod_i (1 - z_i \delta_i) s_i^{(1-z_i)\delta_i} \tag{12.28}$$

and $Z^{**} = Z^* - \{(1, 1, \ldots, 1)\}$. It follows now from

$$C\widehat{\tau} = Q$$

and (12.27) that

$$M^z C\widehat{\tau} = M^z Q$$

can be written as

$$rE(z) M^z \widehat{\tau} = M^z Q$$

and, using (12.20),

$$rtE(z)\widehat{a}^z = M^z Q$$

so that

$$H^z \widehat{a}^z = \frac{1}{rtE(z)} H^z M^z Q$$

Then

$$\text{var}(H^z \widehat{a}^z) = \frac{1}{[rtE(z)]^2} H^z M^z C (M^z)'(H^z)'\sigma_e^2$$

and using (12.27) again

$$\text{var}(H^z \widehat{a}^z) = \frac{1}{t^2 rE(z)} H^z M^z (M^z)'(H^z)'\sigma_e^2 \tag{12.29}$$

To simplify (12.29), we write

$$\begin{aligned} H^z M^z &= \left(H_1^{z_1} \times H_2^{z_2} \times \cdots \times H_n^{z_n}\right)\left(M_1^{z_1} \times M_2^{z_2} \times \cdots \times M_n^{z_n}\right) \\ &= \left(H_1^{z_1} M_1^{z_1}\right) \times \left(H_2^{z_2} M_2^{z_2}\right) \times \cdots \times \left(H_n^{z_n} M_n^{z_n}\right) \end{aligned}$$

From the definitions of $H_i^{z_i}$ and $M_i^{z_i}$ it follows that

$$H_1^{z_i} M_1^{z_i} = \begin{cases} \mathfrak{J}'_{s_i} & \text{if } z_i = 0 \\ s_i H'_i & \text{if } z_i = 1 \end{cases}$$

and

$$H^z M^z (M^z)'(H^z)' = (M^z H^z)(M^z H^z)' = \prod_i s_i^{1-z_i} \prod_i [s_i^{z_i}]^2 I \quad (12.30)$$

Substituting (12.30) into (12.29) yields

$$\begin{aligned} \text{var}(H^z \widehat{a}^z) &= \frac{1}{t^2 r E(z)} \prod_i s_i^{1+z_i} I \sigma_e^2 \\ &= \frac{1}{tr E(z)} \prod_i s_i^{z_i} I \sigma_e^2 \\ &= \frac{1}{\prod_i s_i^{1-z_i} r E(z)} I \sigma_e^2 \end{aligned}$$

using the fact that $t = \prod_i s_i$. This is the desired result. □

We comment briefly on the efficiency factors $E(z) = (z_1, z_2, \ldots, z_n)$ associated with the interaction a^z. Recall that the efficiency factor of an IBD is defined as the ratio of the average variance for simple treatment comparisons using an RCBD over the corresponding variance for the IBD. We saw in Chapter 4 that if the IBD is a PBIB(m) design then we may have up to m different efficiencies E_1, $E_2, \ldots, E_m$, where E_i refers to the efficiency factor for comparing two treatments that are ith associates ($i = 1, 2, \ldots, m$). In the context of the BFD we have up to $2^n - 1$ different efficiency factors $E(z)$ associated with treatment comparisons belonging to the various interactions (which in our terminology also include the main effects). Let $\ell'_{z,i} a^z (i = 1, 2, \ldots, \nu(z))$ be a set of linearly independent normalized contrasts belonging to a^z [e.g., the rows of H^z constitute such a set of $\ell'_{z,i}$, where $\nu(z)$ represents the number of degrees of freedom for a^z]. We know that with an RCBD each such contrast is estimated in terms of treatment means, averaging over the levels of those factors not involved in the interaction, that is, the factors for which $z_i = 0$ in a^z. With r replications (blocks) we then have

$$\text{var}_{\text{RCBD}}(\ell'_{z,i} a^z) = \frac{1}{r \prod_i s_i^{1-z_i}} \sigma_e^2 \quad (12.31)$$

From (12.25) we have

$$\text{var}_{\text{IBD}}(\ell'_{z,i} a^z) = \frac{1}{r \prod_i s_i^{1-z_i} E(z)} \sigma_e^2 \quad (12.32)$$

The ratio of (12.31) to (12.32) is then equal to $E(z)$. We shall return to evaluating $E(z)$ shortly.

We have shown in Theorem 12.1 that if for a factorial design the $\boldsymbol{C}$ matrix has a certain structure, as defined by property A, then the design is a BFD. Kshirsagar (1966) proved the converse so that we can now state the following theorem.

Theorem 12.2 For an $s_1 \times s_2 \times \cdots \times s_n$ factorial design to be a BFD with OFS, it is necessary and sufficient that the $\boldsymbol{C}$ matrix of the design satisfies property A.

We proved the sufficiency part in Theorem 12.1 and for the necessity part we refer the reader to Kshirsagar (1966) (see also Gupta and Mukerjee, 1989).

12.5.2 EGD-PBIBs and BFDs

To provide further insight into BFDs, we recall that the association matrices $\boldsymbol{B}^{\boldsymbol{\gamma}}$ for an EGD-PBIB design can be expressed as Kronecker products of (the trivial) association matrices for a BIBD, $\boldsymbol{B}_0 = \boldsymbol{I}$ and $\boldsymbol{B}_1 = \mathfrak{I}\mathfrak{I}' - \boldsymbol{I}$ (see Section 5.4). More specifically, using the notation of this section, we define

$$\boldsymbol{B}_i^{\gamma_i} = \begin{cases} \boldsymbol{I}_{s_i} & \text{for } \gamma_i = 0 \\ \mathfrak{I}_{s_i}\mathfrak{I}_{s_i}' - \boldsymbol{I}_{s_i} & \text{for } \gamma_i = 1 \end{cases}$$

Then the association matrix, $\boldsymbol{B}^{\boldsymbol{\gamma}}$, for $\boldsymbol{\gamma}$th associates [with $\boldsymbol{\gamma} = (\gamma_1, \gamma_2, \ldots, \gamma_n)$ and $\gamma_i = 0, 1$ for $i = 1, 2, \ldots, n$] can be written as

$$\boldsymbol{B}^{\boldsymbol{\gamma}} = \boldsymbol{B}_1^{\gamma_1} \times \boldsymbol{B}_2^{\gamma_2} \times \cdots \times \boldsymbol{B}_n^{\gamma_n}$$

with

$$\boldsymbol{N}\boldsymbol{N}' = \sum_{\boldsymbol{\gamma} \in \boldsymbol{Z}^*} \lambda(\boldsymbol{\gamma})\boldsymbol{B}^{\boldsymbol{\gamma}}$$

or

$$\boldsymbol{C} = \sum_{\boldsymbol{\gamma} \in \boldsymbol{Z}^*} c(\boldsymbol{\gamma})\boldsymbol{B}^{\boldsymbol{\gamma}} \tag{12.33}$$

where

$$c(0, 0, \ldots, 0) = \lambda(0, 0, \ldots, 0)\frac{k-1}{k}$$

$$c(\boldsymbol{\gamma}) = -\frac{1}{k}\lambda(\boldsymbol{\gamma}) \qquad \text{for all } \boldsymbol{\gamma} \neq (0, 0, \ldots, 0)$$

It is not difficult to speculate that there exists some relationship between BFDs and EGD-PBIB designs. Indeed, Kshirsagar (1966) proved the following (for an alternative proof see Gupta and Mukerjee, 1989).

Theorem 12.3 Every n-factor BFD is an EGD-PBIB $(2^n - 1)$ design.

This result is useful for several reasons, two of which we shall discuss in some detail:

i. For any given EGD-PBIB it is easy to obtain the efficiency factors;
ii. Existing methods for constructing EGD-PBIBs can be used to obtain BFDs.

Let us consider first i by returning to Example 12.5.

Example 12.5 (Continued) For arrangement b in Table 12.9 it is easy to verify, using the association scheme for EGD-PBIBs, that

$$\begin{array}{ll} \lambda(000) = 3 & \lambda(001) = 1 \\ \lambda(100) = 0 & \lambda(101) = 2 \\ \lambda(010) = 0 & \lambda(011) = 2 \\ \lambda(110) = 3 & \lambda(111) = 1 \end{array}$$

Further, we can express each $\boldsymbol{B}^{\gamma}$ easily in terms of the $\boldsymbol{D}_i^{\delta_i}$ as used in defining property A [see (12.23)]. We note that

$$\boldsymbol{B}_1^0 = \boldsymbol{B}_2^0 = \boldsymbol{I}_2 = \boldsymbol{D}_1^0 = \boldsymbol{D}_2^0$$

$$\boldsymbol{B}_1^1 = \boldsymbol{B}_2^1 = \mathfrak{I}_2\mathfrak{I}_2' - \boldsymbol{I}_2 = \boldsymbol{D}_1^1 - \boldsymbol{D}_1^0 = \boldsymbol{D}_2^1 - \boldsymbol{D}_2^0$$

and

$$\boldsymbol{B}_3^0 = \boldsymbol{I}_3 = \boldsymbol{D}_3^0,\ \boldsymbol{B}_3^1 = \mathfrak{I}_3\mathfrak{I}_3' - \boldsymbol{I}_3 = \boldsymbol{D}_3^1 - \boldsymbol{D}_3^0$$

so that, for example,

$$\boldsymbol{B}^{(000)} = \boldsymbol{D}_1^0 \times \boldsymbol{D}_2^0 \times \boldsymbol{D}_3^0$$

$$\boldsymbol{B}^{(100)} = \boldsymbol{D}_1^1 \times \boldsymbol{D}_2^0 \times \boldsymbol{D}_3^0 - \boldsymbol{D}_1^0 \times \boldsymbol{D}_2^0 \times \boldsymbol{D}_3^0$$

$$\begin{aligned} \boldsymbol{B}^{(110)} &= \boldsymbol{D}_1^1 \times \boldsymbol{D}_2^1 \times \boldsymbol{D}_3^0 - \boldsymbol{D}_1^1 \times \boldsymbol{D}_2^0 \times \boldsymbol{D}_3^0 \\ &\quad - \boldsymbol{D}_1^0 \times \boldsymbol{D}_2^1 \times \boldsymbol{D}_3^0 + \boldsymbol{D}_1^0 \times \boldsymbol{D}_2^0 \times \boldsymbol{D}_3^0 \end{aligned}$$

and so on. Collecting terms, we can rewrite (12.33) as

$$\boldsymbol{C} = \left(3 - \tfrac{8}{6}\right) \boldsymbol{D}^{(000)} - \tfrac{4}{6}\boldsymbol{D}^{(100)} - \tfrac{4}{6}\boldsymbol{D}^{(010)} - \tfrac{2}{6}\boldsymbol{D}^{(110)}$$

$$+ \tfrac{2}{6}\boldsymbol{D}^{(001)} - \tfrac{1}{6}\boldsymbol{D}^{(101)} - \tfrac{1}{6}\boldsymbol{D}^{(011)} - \tfrac{1}{6}\boldsymbol{D}^{(111)}$$

that is, in the form (12.23) with

$$h(000) = \tfrac{10}{6} \quad h(100) = -\tfrac{4}{6} \quad h(010) = -\tfrac{4}{6} \quad h(110) = -\tfrac{2}{6}$$

$$h(001) = \tfrac{2}{6} \quad h(101) = -\tfrac{1}{6} \quad h(011) = -\tfrac{1}{6} \quad h(111) = -\tfrac{1}{6}$$

Using (12.28), we can now obtain the efficiency factors $E(z)$. Let us write

$$W_i(z_i, \delta_i) = (1 - z_i\delta_i)s_i^{(1-z_i)\delta_i}$$

with

$$W_i(z_i, \delta_i) = \begin{cases} 1 & \text{for } z_i = 0, \quad \delta_i = 0 \\ 1 & \text{for } z_i = 1, \quad \delta_i = 0 \\ s_i & \text{for } z_i = 0, \quad \delta_i = 1 \\ 0 & \text{for } z_i = 1, \quad \delta_i = 1 \end{cases}$$

Then (12.28) can be rewritten as

$$rE(z_1, z_2, \ldots, z_n) = \sum_{\delta \in Z^{**}} h(z_1, z_2, \ldots, z_n) \prod_{i=1}^{n} W_i(z_i, \delta_i) \qquad (12.34)$$

So, for example,

$$\begin{aligned} 6 \cdot 3E(110) &= 10W_1(1, 0)\ W_2(1, 0)\ W_3(0,\ 0) \\ &\quad -4W_1(1, 1)\ W_2(1, 0)\ W_3(0, 0) \\ &\quad -4W_1(1, 0)\ W_2(1, 1)\ W_3(0, 0) \\ &\quad -2W_1(1, 1)\ W_2(1, 1)\ W_3(0, 0) \\ &\quad +2W_1(1, 0)\ W_2(1, 0)\ W_3(0, 1) \\ &\quad -W_1(1, 1)\ W_2(1, 0)\ W_3(0, 1) \\ &\quad -W_1(1, 0)\ W_2(1, 1)\ W_3(0, 1) \\ &= 10 + 2 \cdot 3 = 16 \end{aligned}$$

Hence

$$E(110) = \tfrac{8}{9}$$

the same value we obtained earlier and referred to as relative information for the two-factor interaction $A_1 \times A_2$.

If we define the efficiency factor for a BFD as

$$E = \sum_{z \in Z} \prod_i \frac{(s_i - 1)^{z_i} E(z)}{t - 1} = 1 - \frac{TRL}{t - 1} \qquad (12.35)$$

with TRL being the total relative loss defined earlier, we obtain, for both designs of Table 12.9, $E = \frac{10}{11}$. To choose among competing designs, it is therefore in general advisable to consider not only E for each design but also the individual $E(z)$ and then choose the design that is most appropriate for the experiment under consideration. □

12.5.3 Construction of BFDs

We shall now turn to the construction of BFDs or, alternatively, to the construction of EGD-PBIB designs. For some of these methods we shall refer to Chapter 5 and provide here just an illustration, for other methods we shall give a brief description of their special features in the context of factorial experiments.

12.5.3.1 Kronecker Product Designs

The method is described in Section 5.5, and we know from Chapter 4 that the Kronecker product of BIBDs is an EGD-PBIB design. As an illustration let us consider Example 5.5 to construct a BFD for a 4 × 3 factorial in blocks of size 6. Using the SDP notation of Section 11.9, the treatment combinations are written in the order (0, 0), (0, 1), (0, 2), (1, 0), (1, 1), (1, 2), (2, 0), (2, 1), (2, 2), (3, 0), (3, 1), (3, 2). The 12 blocks are then as given in Table 12.10. Following our earlier discussion, it is easy to work out that $E(1,0)=\frac{31}{36}$, $E(0,1)=\frac{29}{36}$, and $E(1,1)=\frac{35}{36}$, and the overall efficiency factor is $E=1-\frac{35}{396}$.

12.5.3.2 GD-PBIB(2) Designs

If in an EGD(3)-PBIB design we have $\lambda(10)=\lambda(11)$, then the design reduces to a GD-PBIB(2) design with parameters $n_1=n(01)$, $n_2=n(10)+n(11)$ and $\lambda_1=\lambda(01)$, $\lambda_2=\lambda(10)=\lambda(11)$. It is therefore possible to obtain a BFD for a $t=s_1\times s_2$ factorial if there exists a GD-PBIB design for $t=s_1s_2$ treatments.

Table 12.10 4 x 3 Factorial in Blocks of Size 6 Using a Kronecker Product Design

Block					
1	2	3	4	5	6
00	00	01	00	00	01
01	02	02	01	02	02
10	10	11	10	10	11
11	12	12	11	12	12
20	20	21	30	30	31
21	22	22	31	32	32
Block					
7	8	9	10	11	12
00	00	01	10	10	11
01	02	02	11	12	12
20	20	21	20	20	21
21	22	22	21	22	22
30	30	31	30	30	31
31	32	32	31	32	32

All we need to do is to arrange the factorial treatment combinations in the rectangular array

$$\begin{array}{cccc}
(0,0) & (0,1) & \cdots & (0, s_2-1) \\
(1,0) & (1,1) & \cdots & (1, s_2-1) \\
\vdots & \vdots & \ddots & \vdots \\
(s_1-1,0) & (s_1-1,1) & \cdots & (s_1-1, s_2-1)
\end{array}$$

and use the association scheme for a GD-PBIB design: Any two treatments in the same row are first associates and any two treatments in different rows are second associates. We can then use any suitable GD-PBIB design to obtain a BFD. This method was proposed by Kramer and Bradley (1957) and Zelen (1958). We shall illustrate this method for the 4×3 factorial in blocks of size 6 using design S 27 of Clatworthy (1973).

Using the following correspondence between ordinary treatments and factorial treatments,

$$\begin{array}{ccc} 1 & 5 & 9 \\ 2 & 6 & 10 \\ 3 & 7 & 11 \\ 4 & 8 & 12 \end{array} \longrightarrow \begin{array}{ccc} 00 & 01 & 02 \\ 10 & 11 & 12 \\ 20 & 21 & 22 \\ 30 & 31 & 32 \end{array}$$

the BFD is given in Table 12.11 [we note here that this labeling does not agree with that obtained from the SDP of $(0\ 1\ 2\ 3)'$ and $(0\ 1\ 2)'$ but rather it agrees with the labelling for GD-PBIB designs as given by Clatworthy (1973) and that for EGD-PBIB designs]. With $\lambda_0 = r = 3$, $\lambda_1 = 3$, $\lambda_2 = 1$ for the GD-PBIB and hence $\lambda(00) = r = 3$, $\lambda(01) = 3$, $\lambda(10) = \lambda(11) = 1$ for the EGD-PBIB we can obtain the efficiency factors in the usual way except for one small change. From

Table 12.11 4 x 3 Factorial in Blocks of Size 6 Using a GD-PBIB

Block					
1	2	3	4	5	6
00	20	02	12	01	11
10	30	22	32	31	21
01	21	00	10	02	12
11	31	20	30	32	22
02	22	01	11	00	10
12	32	21	31	30	20

the structure of the GD-PBIB design we know that we can write

$$
\begin{aligned}
\boldsymbol{B}_0 &= \boldsymbol{I}_t \\
\boldsymbol{B}_1 &= \left(\mathfrak{I}_{s_2}\mathfrak{I}'_{s_2} - \boldsymbol{I}_{s_2}\right) \times \boldsymbol{I}_{s_1} \\
&= \mathfrak{I}_{s_2}\mathfrak{I}'_{s_2} \times \boldsymbol{I}_{s_1} - \boldsymbol{I}_t \\
\boldsymbol{B}_2 &= \mathfrak{I}_{s_2}\mathfrak{I}'_{s_2} \times \left(\mathfrak{I}_{s_1}\mathfrak{I}'_{s_1} - \boldsymbol{I}_{s_1}\right) \\
&= \mathfrak{I}_t\mathfrak{I}'_t - \mathfrak{I}_{s_2}\mathfrak{I}'_{s_2} \times \boldsymbol{I}_{s_1}
\end{aligned}
$$

and

$$
\begin{aligned}
\boldsymbol{B}_0 &= \boldsymbol{B}^{(00)} = \boldsymbol{D}_2^0 \times \boldsymbol{D}_1^0 \\
\boldsymbol{B}_1 &= \boldsymbol{B}^{(01)} = \boldsymbol{D}_2^1 \times \boldsymbol{D}_1^0 - \boldsymbol{D}_2^0 \times \boldsymbol{D}_1^0 \\
\boldsymbol{B}_2 &= \boldsymbol{B}^{(10)} + \boldsymbol{B}^{(11)} = \boldsymbol{D}_2^1 \times \boldsymbol{D}_1^1 - \boldsymbol{D}_2^1 \times \boldsymbol{D}_1^0
\end{aligned}
$$

Notice that because of the labeling the order of the $\boldsymbol{D}$ matrices is reversed (this is, of course, essential only for $\boldsymbol{D}_2^1 \times \boldsymbol{D}_1^0$). Hence, in general,

$$
\boldsymbol{C} = \left(r - \frac{r - \lambda_1}{k}\right) \boldsymbol{D}_2^0 \times \boldsymbol{D}_1^0 - \frac{\lambda_1 - \lambda_2}{k} \boldsymbol{D}_2^1 \times \boldsymbol{D}_1^0 - \frac{\lambda_2}{k} \boldsymbol{D}_2^1 \times \boldsymbol{D}_1^1
$$

so that, using (12.34) and the properties of the W_i,

$$
rE(1,0) = \left(r - \frac{r - \lambda_1}{k}\right) W_1(1,0)W_2(0,0) - \frac{\lambda_1 - \lambda_2}{k} W_1(1,0)W_2(0,1)
$$

$$
rE(0,1) = \left(r - \frac{r - \lambda_1}{k}\right) W_1(0,0)W_2(1,0)
$$

and

$$
rE(1,1) = \left(r - \frac{\lambda_1}{k}\right) W_2(1,0)W_1(1,0)
$$

Then, for the specific example of Table 12.11 we find

$$
E(1,0) = \tfrac{2}{3} \qquad E(0,1) = 1 \qquad E(1,1) = 1
$$

that is, we have full information on the main effect A_2 and the interaction $A_1 \times A_2$ and two-thirds information on the main effect A_1.

The GD-PBIB design for $t = t_1 t_2$ treatments can, of course, also be used for factorials with more than two factors, that is, for an $s_1 \times s_2 \times \cdots \times s_n$ factorial, as long as $s_{i_1} s_{i_2} \cdots s_{i_m} = t_1$ for any m factors ($m < n$) and $s_{j_1} s_{j_2} \cdots s_{j_\ell} = t_2$ for the remaining $\ell = n - m$ factors. Thus the design in Table 12.11 can also be used for the $2 \times 2 \times 3$ factorial in blocks of size 6 with $t_1 = 4 = s_1 s_2 = 2 \cdot 2$. All we need to do is relabel appropriately the levels of A_1. It follows then immediately that

$$E(100) = E(010) = E(110) = \tfrac{2}{3}$$

and all other $E(z) = 1$.

12.5.3.3 Other Methods

The methods of constructing EGD-PBIB designs given by Aggarwal (1974) and Chang and Hinkelmann (1987) (see Section 5.5) can be used to obtain different systems of confounding for the $s_1 \times s_2 \times \cdots \times s_n$ factorial. If the factors are labeled such that $s_1 \leq s_2 \leq \cdots \leq s_n$, then Aggarwal's method leads to BFDs with block size $k = s_1$. The method by Chang and Hinkelmann is somewhat more flexible in that it can be used to obtain BFDs with block sizes $k = s_i$ $(i = 1, 2, \ldots, n-1)$. It leads, however, to disconnected designs, which means that some effects are completely confounded with blocks. Both methods can be modified easily to generate a larger class of BFDs by combining several factors into "superfactors" and thereby changing the order of the levels, that is, some $s_i \neq s_1$ becomes the smallest number of levels, or creating new levels, that is, s_n may no longer be the largest number of levels. The first modification is useful for the Aggarwal method, the second for the Chang and Hinkelmann method.

There exist other ad hoc methods used by Yates (1937b) and Li (1944) for specific asymmetrical factorials. For example, the design for the $2^2 \times 3$ factorial given in Table 12.9b can be obtained by combining the levels of factor A_3 with treatment combinations for factors A_1 and A_2 determined by the equations

$$x_1 + x_2 = 0 : (0, 0), (1, 1)$$

and

$$x_1 + x_2 = 1 : (1, 0), (0, 1)$$

If we denote $\{(0, 0), (1, 1)\}$ by α and $\{(1, 0), (0, 1)\}$ by β, then the design can be written in compact form as follows (Kempthorne, 1952):

Block

1	**2**	**3**	**4**	**5**	**6**
$\alpha 0$	$\beta 0$	$\beta 0$	$\alpha 0$	$\beta 0$	$\alpha 0$
$\beta 1$	$\alpha 1$	$\alpha 1$	$\beta 1$	$\beta 1$	$\alpha 1$
$\beta 2$	$\alpha 2$	$\beta 2$	$\alpha 2$	$\alpha 2$	$\beta 2$

Rather than describe the ad hoc methods in detail, we reproduce some of the more useful designs given by Li (1944) for up to four factors with two, three, and four levels. They are given in Appendix D together with some relevant parameters.

Other methods of construction have been proposed by Kishen and Srivastava (1959) and Muller (1966). They consider $s \times s_1 \times \cdots \times s_{n-1}$ factorials, where s is a prime or prime power and various block sizes.

It is quite obvious that BFDs require in general several replications that may lead to more blocks than is feasible from a practical point of view. If one is willing to give up the property of balancedness, then useful designs can be obtained by a suitable reduction of the number of replications. Such possibilities are mentioned for some situations in Appendix D. Even though these designs are no longer BFDs, they yield full or partial information on all degrees of freedom. This may make these designs more useful than comparable designs obtained by methods discussed in Sections 11.2 and 11.3.

CHAPTER 13

Fractional Factorial Designs

13.1 INTRODUCTION

So far, we have considered possible arrangements of the complete factorial system; that is, we have considered only arrangements in which each individual treatment combination of the factorial set is represented the same number of times. For example, with 4 factors each at 2 levels, we have discussed arrangements of the 16 treatment combinations in blocks of size 2, blocks of size 4, 8×8 squares, and so on, but always with the restriction that each of the 16 combinations is to be tested an equal number of times. Such a restriction is practicable with a small number of factors, but, when we consider the case of, say, 10 factors each at 2 levels, it would result in the necessity of testing 1024 combinations or a multiple of these. The main reason for imposing the restriction is that it results in the estimates of effects and interactions having maximum precision and being uncorrelated. Thus, to take a simple example, suppose we are evaluating 2 factors, A and B, each at 2 levels, and test the treatment combination (1), a, b, ab with r_1, r_2, r_3, r_4 replications, respectively. Then using the definitions of Section 7.2.1, each effect and interaction is estimated with variance

$$\frac{\sigma_e^2}{4}\left(\frac{1}{r_1}+\frac{1}{r_2}+\frac{1}{r_3}+\frac{1}{r_4}\right)$$

but the estimates are correlated thus:

$$\text{cov}\left(\widehat{A}, \widehat{B}\right) = \frac{\sigma_e^2}{4}\left[\frac{1}{r_1}-\frac{1}{r_2}-\frac{1}{r_3}+\frac{1}{r_4}\right]$$

$$\text{cov}\left(\widehat{A}, \widehat{AB}\right) = \frac{\sigma_e^2}{4}\left[-\frac{1}{r_1}-\frac{1}{r_2}+\frac{1}{r_3}+\frac{1}{r_4}\right]$$

Design and Analysis of Experiments. Volume 2: Advanced Experimental Design
By Klaus Hinkelmann and Oscar Kempthorne
ISBN 0-471-55177-5

and

$$\operatorname{cov}\left(\widehat{B}, \widehat{AB}\right) = \frac{\sigma_e^2}{4}\left[-\frac{1}{r_1} + \frac{1}{r_2} - \frac{1}{r_3} + \frac{1}{r_4}\right]$$

These covariances will be zero if $r_1 = r_2 = r_3 = r_4$. Furthermore, if we fix $(r_1 + r_2 + r_3 + r_4)$, the total number of observations, we obtain minimum variance for each of the effects and interaction with $r_1 = r_2 = r_3 = r_4$.

As we have seen throughout, the general problem of the design of experiments is that we are given a model and we wish to determine parametric functions of such a model with as low a variance as possible. There are two methods by which this may be accomplished: namely, by choosing the pattern of observations, for example, equal r's above, or by reducing the error variance σ_e^2 by choice of the error control design.

The question we ask in this chapter is whether it is always necessary to test all the factorial combinations equally frequently or whether we can omit some of them. The question is of considerable relevance, for factorial experiments are most appropriate for exploratory research, and in such research the number of possible factors that should be tested is by no means so small as 2 or 3. For example, in research into the possible importance of the various vitamins in the nutrition of an organism, the number of factors that might be used is at least of the order of 10. It is, furthermore, important to use a factorial system because we cannot assume all interactions to be negligible. The testing of 1024 treatment combinations may be virtually impossible from the practical viewpoint. If the whole 1024 combinations were tested, the subdivision of the 1023 treatment comparisons would be

Main effects	10
2-factor interactions	45
3-factor interactions	120
4-factor interactions	210
5-factor interactions	252
6-factor interactions	210
7-factor interactions	120
8-factor interactions	45
9-factor interactions	10
10-factor interactions	1
Total	1023

It may well be reasonable to assume that high-order interactions are negligible, and, if, for instance, all interactions involving 3 or more factors could be ignored, the testing of the 1024 combinations would give an estimate of error based on about 950 degrees of freedom. The accuracy that would result for the estimation of main effects and interactions (a variance of $\sigma_e^2/256$ where σ_e^2 is the experimental error variance) might well be unnecessarily high. We may then ask what information can be obtained from the testing of a lesser number of combinations. We shall, in fact, find that information on main effects and interactions for all the 10 factors can be obtained under some mild assumptions from the testing of 512, 256, or perhaps even 128 of the total of 1024 combinations. The general process by which this is accomplished is known as *fractional replication*. The resulting design is called a *fractional factorial design*, which was introduced first by Finney (1945).

For an annotated bibliography of applications papers using some of the concepts developed in the following chapters, see Prvan and Street (2002).

13.2 SIMPLE EXAMPLE OF FRACTIONAL REPLICATION

We shall illustrate the concept of a fractional factorial design in terms of a simple, though somewhat unrealistic, example. Suppose we are testing three factors A, B, and C, each of two levels, and for some reason we are able to observe only four of the eight possible treatment combinations. Let these four treatment combinations be a, b, c, and abc. The question then is: What information can we obtain from these four combinations? This question can be answered easily by referring to Table 13.1, which shows the relationship between the true yields of these four treatment combinations and the effects and interactions for the ordinary 2^3 design. It shows that, for these four treatment combinations, M and $\frac{1}{2}ABC$ have always the same sign, and so have $\frac{1}{2}A$ and $\frac{1}{2}BC$, $\frac{1}{2}B$ and $\frac{1}{2}AC$, $\frac{1}{2}C$, and $\frac{1}{2}AB$. It will therefore be impossible to separate the mean from the ABC interaction, the A effect from the BC interaction, the B effect from the AC interaction, and the C effect from the AB interaction. In fact, we have the following four estimating relations:

$$\tfrac{1}{4}(a + b + c + abc) \text{ estimates } M + \tfrac{1}{2}ABC$$

$$\tfrac{1}{2}(a - b - c + abc) \text{ estimates } A + BC$$

$$\tfrac{1}{2}(-a + b - c + abc) \text{ estimates } B + AB$$

$$\tfrac{1}{2}(-a - b + c + abc) \text{ estimates } C + AB$$

where a, b, c, abc are now the observed yields. We may say that, with these four observations only, A is completely *confounded* with BC because there is

Table 13.1 Relationship Between True Yields and Effects and Interactions for 2^3 Design

True Yield	M	$\frac{1}{2}A$	$\frac{1}{2}B$	$\frac{1}{2}AB$	$\frac{1}{2}C$	$\frac{1}{2}AC$	$\frac{1}{2}BC$	$\frac{1}{2}ABC$
a	+	+	−	−	−	−	+	+
b	+	−	+	−	−	+	−	+
c	+	−	−	+	+	−	−	+
abc	+	+	+	+	+	+	+	+

no possibility of estimating A alone and BC alone but only their sum. We also express this by saying that A is an *alias* of BC (and vice versa) or that A is *aliased* with BC. Similarly, B is completely confounded with AC, C with AB, and M with ABC. This confounding, however, need not cause the experimenter any worry if he is prepared to assume that the interactions are negligible: Then the second, third, and fourth comparisons above may be used to estimate the three main effects, and the estimates are uncorrelated. In this simple example, there is not, of course, any possibility of estimating error variance, and we would not, generally, use this design for any practical purpose. It serves, however, to bring out the main idea of fractional replication, namely that a suitably chosen subset of the full factorial set can provide worthwhile information. The above example utilizes a $\frac{1}{2}$ replication of the 2^3 system because we have tested only four of the eight combinations.

We could equally well have used the treatment combinations (1), ab, ac, and bc, and it may be verified that the estimating equations will be

$$M - \tfrac{1}{2}ABC = \tfrac{1}{4}[(1) + ab + ac + bc]$$

$$A - BC = \tfrac{1}{2}[-(1) + ab + ac - bc]$$

$$B - AC = \tfrac{1}{2}[-(1) + ab - ac + bc]$$

$$C - AB = \tfrac{1}{2}[-(1) - ab + ac - bc]$$

The confounding among the mean, main effects, and interactions is the same as before, only the functions that can be estimated are different.

The dominant feature of this example is the choice of the interaction ABC and the selection of either the treatments entering positively in this interaction or those entering negatively, or alternatively, either those that satisfy the equation

$$x_1 + x_2 + x_3 = 1 \bmod 2$$

or those satisfying the equation

$$x_1 + x_2 + x_3 = 0 \bmod 2, \text{ respectively}$$

For most purposes we may disregard which half of the treatment combinations we choose because we shall always choose an interaction in such a way that those interactions that are confounded with the effects or interactions we wish to estimate are assumed on the basis of prior knowledge to be negligible.

If we use the equality sign to denote "confounded with," we may write the confounding relations above in the form:

$$\begin{aligned} A &= BC \\ B &= AC \\ C &= AB \end{aligned} \tag{13.1}$$

and, if by convention we denote M by I, also

$$I = ABC \tag{13.2}$$

We note the equalities (13.1) are obtained from (13.2) by regarding the relationship (13.2) as an algebraic identity and I as unity and using multiplication with the rule that the square of any letter is to be replaced by unity. Thus, (13.2) multiplied by A gives

$$A = A^2BC = BC$$

when multiplied by B gives

$$B = AB^2C = AC$$

and when multiplied by C gives

$$C = ABC^2 = AB$$

In Table 13.2 we show how this design and the alias structure can be obtained using SAS PROC FACTEX. We comment briefly on the input statements and the output:

1. The number of treatment combinations to be used can be specified either by "size design = # of treatment combinations" or "size fraction = denominator of fraction."
2. The model can either be specified by listing the effects we wish to estimate or by giving the resolution of the design (see Section 13.3.2).
3. The factor confounding rule is equivalent to the defining relation in the sense that $C = A * B$ is formally identical to $C^2 = A * B * C$ and $C^2 = I$; operationally $C = A * B$ indicates how from a complete 2^2 factorial the levels of the factor C are obtained, namely by multiplying the levels of factors A and B into each other when the levels are denoted by -1 and $+1$.

If, instead of a, b, c, abc, we had chosen the treatment combinations that enter positively into AB, namely (1), ab, c, and abc, we can obtain the confounding relationships formally by multiplying each effect or interaction into

$$I = AB \tag{13.3}$$

Thus

$$\begin{aligned} A &= A^2B = B \\ C &= ABC \\ AC &= A^2BC = BC \end{aligned} \tag{13.4}$$

Table 13.2 1/2 Fraction of 2^3 Factorial With Defining Contrast I = ABC

```
 options nodate pageno=1;
proc factex;
factors A B C;
size design=4;
model est=(A B C);
examine design aliasing confounding;
title1 'TABLE 13.2';
title2 '1/2 FRACTION OF 2**3 FACTORIAL';
title3 'WITH DEFINING CONTRAST I=ABC';
run;

proc factex;
factors A B C;
size fraction=2;
model res=3;
examine design aliasing confounding;
run;
```

```
                      The FACTEX Procedure

                         Design Points

            Experiment
                Number      A      B      C
            ----------------------------------
                     1     -1     -1      1
                     2     -1      1     -1
                     3      1     -1     -1
                     4      1      1      1

                   Factor Confounding Rules

                           C = A*B

Aliasing Structure

A = B*C
B = A*C
C = A*B
```

The equations (13.4) we interpret then by saying that A and B are confounded, C and ABC are confounded, and AC and BC are confounded. Of course, (13.3) means that M and AB are confounded. Alternatively, we can say that with the four observations above we can only estimate $M + \frac{1}{2}AB$, $A + B$, $C + ABC$, and

$AC + BC$. This can be verified easily by using an argument similar to the one we used at the beginning of this section.

In each of these two examples we have used a $\frac{1}{2}$ replicate of the full set of factorial combinations. The information obtained from these two designs is quite different: If we can assume that the three factors do not interact, the first replicate is a satisfactory design (apart from the fact that no estimate of error can be made) because we can estimate each of the three main effects A, B, and C. In the second $\frac{1}{2}$ replicate, however, the effect A is confounded with the effect B, and this is generally undesirable.

The crucial part of the specification of the design is clearly the choice of the relationship $I = ABC$ or $I = AB$. This relation is known as the *defining relation* or *identity relationship*. Once this relationship is chosen, we can specify the functions of the parameters that can be estimated, and the choice of an identity relationship is based on this fact. Thus, we would typically not use a relationship that results in confounding of main effects with each other.

In passing we note that the defining relation can be written in such a way that it identifies uniquely the two complimentary $\frac{1}{2}$ replicates, for example,

$$I = +ABC, \qquad \text{and} \qquad I = -ABC$$

the first denoting the $\frac{1}{2}$ replicate based on the treatment combinations entering positively into ABC, and the second based on those entering negatively into ABC. The specific form of the defining relation also tells us which functions of the parameters can be estimated; for example, $A + BC$ from the first and $A - BC$ from the second $\frac{1}{2}$ replicate. In either case, A and BC are confounded, and so forth. The totality of all confounding relationships is also referred to as the *alias structure* of the fractional factorial design.

13.3 FRACTIONAL REPLICATES FOR 2^n FACTORIAL DESIGNS

It is clear from the discussion in Section 13.2 that for a $\frac{1}{2}$ replicate of a 2^n factorial design will generally be determined by the identity relationship of the form

$$I = A_1 A_2 \cdots A_n \tag{13.5}$$

so that main effects and low-order interactions are aliased with interactions of as high an order as possible. The actual choice of the specific $\frac{1}{2}$ replicate based on (13.5) is often made so that it contains the treatment combination $(0, 0, \ldots, 0)$, which is often the control.

13.3.1 The $\frac{1}{2^\ell}$ Fraction

In a practical situation it may be that a $\frac{1}{2}$ replicate is still too large a fraction to be accommodated in a reasonable experimental design. Thus smaller fractions may have to be considered, such as a $\frac{1}{2^\ell}$ fraction of a 2^n factorial design. We denote such a design by $2^{n-\ell}$. The treatment combinations of a $2^{n-\ell}$ design are obtained by specifying ℓ equations:

$$\begin{aligned}\alpha_{11}x_1 + \alpha_{12}x_2 + \cdots + \alpha_{1n}x_n &= \delta_1\\ \alpha_{21}x_1 + \alpha_{22}x_2 + \cdots + \alpha_{2n}x_n &= \delta_2\\ &\vdots\\ \alpha_{\ell 1}x_1 + \alpha_{\ell 2}x_2 + \cdots + \alpha_{\ell n}x_n &= \delta_\ell\end{aligned} \tag{13.6}$$

where the $\boldsymbol{\alpha}_i' = (\alpha_{i1}, \alpha_{i2}, \ldots, \alpha_{in})$ represent independent interactions $E^{\alpha_i}(i = 1, 2, \ldots, \ell)$ (see Section 8.2) and the $\delta_i (i = 1, 2, \ldots, \ell)$ are either 0 or 1 mod 2. It follows then that there are exactly $2^{n-\ell}$ treatment combinations that satisfy (13.6) and these constitute the specified $2^{n-\ell}$ fractional factorial. We note the following:

1. All treatment combinations satisfying (13.6) enter with the same sign into $E^{\alpha_i} (i = 1, 2, \ldots, \ell)$.
2. Any treatment combination that satisfies (13.6) also satisfies any linear combination of the equations in (13.6).
3. Because of conditions 1 and 2 all treatment combinations satisfying (13.6) also enter with the same sign into any GI of the $E^{\alpha_i} (i = 1, 2, \ldots, \ell)$.

It then follows that the E^{α_i} and all their GIs are confounded with (or inseparable from) the mean M, so that the identity relationship is then

$$\begin{aligned}I &= E^{\alpha_1} = E^{\alpha_2} = E^{\alpha_1+\alpha_2} = E^{\alpha_3} = E^{\alpha_1+\alpha_3} = E^{\alpha_2+\alpha_3}\\ &= E^{\alpha_1+\alpha_2+\alpha_3} = \cdots = E^{\alpha_1+\alpha_2+\cdots+\alpha_\ell}\end{aligned} \tag{13.7}$$

that is, (13.7) contains $2^\ell - 1$ interactions. The alias structure is obtained by taking an E^β not represented in (13.7) and multiplying it into each term in (13.7), that is, E^β is confounded with

$$E^{\alpha_1+\beta},\ E^{\alpha_2+\beta},\ E^{\alpha_1+\alpha_2+\beta},\ E^{\alpha_3+\beta},\ \ldots,\ E^{\alpha_1+\alpha_2+\ldots+\alpha_\ell+\beta} \tag{13.8}$$

There are $s = 2^{2-\ell} - 1$ such interactions, say $E^{\beta_j} (j = 1, 2, \ldots, s)$ and each is aliased with $s^\ell - 1$ other interactions according to (13.8). We thus have the complete alias structure for the $2^{n-\ell}$ design based on (13.7).

It is, of course, clear from Eqs. (13.6) that there exist 2^ℓ different fractions $2^{n-\ell}$ depending on the choice of the $\delta_i (i = 1, 2, \ldots, \ell)$. To specify which fraction we are choosing we may attach a plus or minus sign to the independent interactions $E^{\alpha_i} (i = 1, 2, \ldots, \ell)$ so that $+E^{\alpha_i}(-E^{\alpha_i})$ means that we take all treatment combinations that enter positively (negatively) into E^{α_i}. Regardless of what fraction we choose, as long as it is based on one of the possible sets of Eqs. (13.6), the alias structure remains the same, only the estimable linear functions of effects and interactions change.

To illustrate these concepts we consider the following example.

Example 13.1 Let $n = 5$ and $\ell = 2$, that is, we consider a $\frac{1}{4}$ replicate of a 2^5 design or a 2^{5-2} design. We may specify (13.6) as

$$\begin{aligned} x_1 + x_2 + x_3 \qquad\qquad &= 0 \\ x_1 \qquad\qquad + x_4 + x_5 &= 0 \end{aligned}$$

which are satisfied by the treatment combinations

$$(1), bc, abd, acd, abe, ace, de, bcde$$

These treatment combinations also satisfy the equation

$$x_2 + x_3 + x_4 + x_5 = 0$$

and, moreover, these treatment combinations enter negatively into ABC and ADE and hence positively into their GI $BCDE$. The identity relationship therefore is

$$I = -ABC = -ADE = +BCDE \tag{13.9}$$

From (13.9) we then obtain the alias structure given in Table 13.3. It, together with (13.9), also shows which parametric functions are estimable, for example, $A - BC - DE + ABCDE$, which is defined as $\frac{1}{4}[abd + acd + abe + ace - (1) - bc - de - bcde]$. Had we selected the eight treatment combinations according to the equations

$$\begin{aligned} x_1 + x_2 + x_3 \qquad\qquad &= 1 \\ x_1 \qquad\qquad + x_4 + x_5 &= 0 \\ x_2 + x_3 + x_4 + x_5 &= 1 \end{aligned}$$

Table 13.3 Alias Structure for 2^{5-2} Design

$A = BC = DE = BCDE$
$B = AC = ABDE = CDE$
$C = AB = ACDE = BDE$
$D = ABCD = AE = BCE$
$E = ABCE = AD = BCD$
$BD = ACD = ABE = CE$
$BE = ACE = ABD = CD$

that is,

$$b, c, ad, ae, bde, cde, abce, abcd$$

the identity relationship would have been

$$I = ABC = -ADE = -BCDE \tag{13.10}$$

From the alias structure given in Table 13.3 and from (13.10) we see that now, for example, $A + BC - DE - ABCDE$ is estimable. □

Rather than specifying the identity relationship (13.7) by attaching plus or minus signs to the E^{α_i}, we can indicate the specific fractional replicate by writing $E_{\delta_i}^{\alpha_i}$, where δ_i is the ith right-hand side in (13.6), so that according to the definition of $E_{\delta_i}^{\alpha_i}$ given in Section 7.4 this implies that we select all treatment combinations that satisfy the equation

$$\alpha_{i1}x_1 + \alpha_{i2}x_2 + \cdots + \alpha_{in}x_n = \delta_i$$

($i = 1, 2, \ldots, \ell$). We shall see that this notation is useful when we consider the $p^{n-\ell}$ system (see Section 13.5).

13.3.2 Resolution of Fractional Factorials

As has been mentioned earlier the value of fractional replication lies in the fact that often higher-order interactions are negligible. One would therefore consider fractions that result in the confounding of main effects and low-order interactions with higher-order interactions. The extent to which this is possible is expressed clearly by the form of the identity relationship. Basically, the minimum number of factors involved in an interaction included in the identity relationship determines the *resolution* of the design, that is, the extent to which main effects and low-order interactions are estimable under the assumption that certain higher-order interactions are negligible. Following the terminology introduced by Box and Hunter (1961a, 1961b), we distinguish between the following types of fractional replicates.

Resolution III Designs No main effect is confounded with another main effect, but main effects are confounded with 2-factor interactions.

Resolution IV Designs No main effect is confounded with another main effect or a 2-factor interaction, but 2-factor interactions are confounded with one another.

Resolution V Designs No main effect is confounded with another main effect or 2-factor interaction, and 2-factor interactions are confounded with 3-factor interactions.

These definitions apply, strictly speaking, only to so-called regular fractional factorial designs for symmetrical factorials. A regular fraction is, for example, a $\frac{1}{2^\ell}$ of a 2^n factorial as compared to an irregular fraction such as a $\frac{3}{2^\ell}$ of a 2^n factorial (see, e.g., Addelman, 1961; Webb, 1971). A more general definition was given by Webb (1964).

Definition 13.1 A fractional factorial design is of resolution $(2R + 1)$ if it permits the estimation of all effects up to R-factor interactions assuming that all interactions involving more than R factors are negligible ($R = 1, 2, \ldots$). A design is of resolution $2R$ if it permits the estimation of all effects up to $(R - 1)$-factor interactions assuming that all interactions involving $(R + 1)$ factors are negligible ($R = 2, 3, \ldots$). □

As a consequence of this definition, a resolution III design (i.e., $R = 1$) permits the estimation of main effects assuming that all interactions are negligible, and a resolution IV design (i.e., $R = 2$) permits the estimation of main effects assuming that all 3-factor and higher-order interactions are negligible. The design given in Example 13.1 is a resolution III design, which we indicate by calling it a 2^{5-2}_{III} design. Note also that the lowest order interaction in the identity relationship involves three factors or, using different terminology, the smallest word (i.e., interaction) contains three letters (i.e., factors). A $2^{n-\ell}_{\text{IV}}$ is given in the following example.

Example 13.2 Let $n = 6$, $\ell = 2$, and

$$I = ABCD = CDEF = ABEF \tag{13.11}$$

The 16 treatment combinations are obtained by solving the equations

$$x_1 + x_2 + x_3 + x_4 = 0$$

$$x_3 + x_4 + x_5 + x_6 = 0$$

It is clear from (13.11) that main effects are confounded with 3-factor interactions.

We can use SAS PROC FACTEX to generate this design. It is given in Table 13.4

The output shows that the defining relationship

$$I = BCDE = ACDF = ABEF$$

was used, which is of the same form as (13.11). As a consequence seven of the fifteen 2-factor interactions are confounded with one other 2-factor interaction and one 2-factor interaction is confounded with two other 2-factor interactions. □

13.3.3 Word Length Pattern

In discussing the resolution of a fractional factorial design, we have referred to the interactions in the defining relationship as words, and the number of factors involved in an interaction is called the *word length*. Typically, a defining relationship contains several words of possibly different length. This is referred to as the *word length pattern* (WLP) of the defining relationship. A $2^{n-\ell}$ fraction may be constructed by using different defining relationships with different WLPs. More precisely, a defining relationship can be characterized with respect to its general properties by giving the distribution of words of different length. Denoting the number of words with in letters, that is, an m-factor interaction, by L_m, we denote the WLP of an identity relationship by

$$\text{WLP}(L_3, L_4, \ldots, L_n)$$

We ignore designs with words of length 1 and 2 since they are obviously of no practical interest as the smallest $L_m \neq 0$ determines the resolution of the design. For a $2^{n-\ell}$ design we then have

$$\sum_{m=3}^{n} L_m = 2^\ell - 1$$

For the designs in Example 13.2 and in Table 13.4 we have WLP(0, 3, 0, 0), and it is easy to see that this is the only possible WLP for a 2^{6-2} design.

13.3.4 Criteria for Design Selection

In general, however, there may be different WLPs for a $2_R^{n-\ell}$ design, that is, different identity relationships. In such situations the particular WLP can be used to distinguish among competing designs and select the one that is "best." And best in our context means the largest number of estimable, that is, unconfounded, effects or interactions. For example, for a resolution III we would like the number of main effects that are confounded with 2-factor interactions as small as possible, or expressed differently, the number of main effects not confounded with 2-factor interactions as large as possible. For a resolution IV design we would like to have as few 2-factor interactions confounded with each other. Wu and Chen (1992) called a main effect or 2-factor interaction *clear* if none of its aliases are main effects or 2-factor interactions, and *strongly clear* if none of its aliases are main effects, 2-factor interactions or 3-factor interactions (see also Wu and Hamada, 2000).

Another criterion to distinguish among competing designs with maximum possible resolution, $R_{\max}$ say, was introduced by Fries and Hunter (1980).

Table 13.4 1/4 Replicate of 2^6 Factorial of Resolution IV

```
options nodate pageno=1;
proc factex;
factors A B C D E F;
size fraction=4;
model res=4;
examine design aliasing confounding;
title1 'TABLE 13.4';
title2 '1/4 REPLICATE OF 2**6 FACTORIAL';
title3 'OF RESOLUTION IV';
run;
```

```
                    The FACTEX Procedure

                       Design Points

     Experiment
         Number      A      B      C      D      E      F
     ----------------------------------------------------
              1     -1     -1     -1     -1     -1     -1
              2     -1     -1     -1      1      1      1
              3     -1     -1      1     -1      1      1
              4     -1     -1      1      1     -1     -1
              5     -1      1     -1     -1      1     -1
              6     -1      1     -1      1     -1      1
              7     -1      1      1     -1     -1      1
              8     -1      1      1      1      1     -1
              9      1     -1     -1     -1     -1      1
             10      1     -1     -1      1      1     -1
             11      1     -1      1     -1      1     -1
             12      1     -1      1      1     -1      1
             13      1      1     -1     -1      1      1
             14      1      1     -1      1     -1     -1
             15      1      1      1     -1     -1     -1
             16      1      1      1      1      1      1

                  Factor Confounding Rules

                        E = B*C*D
                        F = A*C*D

Aliasing Structure

A
B
C
D
E
F
```

Table 13.4 (*Continued*)

```
A*B = E*F
A*C = D*F
A*D = C*F
A*E = B*F
A*F = B*E = C*D
B*C = D*E
B*D = C*E
```

Suppose we have two competing $2^{n-\ell}$ designs, D_1 and D_2 say, of maximum resolution and suppose that the WLP are given as

$$D_1\colon \ \text{WLP}(L_3, L_4, \ldots, L_n)$$

and

$$D_2\colon \ \text{WLP}(L'_3, L'_4, \ldots, L'_n)$$

Consider then the first i for which $L_i \neq L'_i$ ($R_{\max} \leq i \leq n$). If $L_i < L'_i$, then D_1 is said to have less aberration than D_2, otherwise D_2 has less aberration. Among all competing designs the one with the smallest L_i is said to be the *minimum aberration design.*

As an illustration of these concepts, we consider the following example.

Example 13.3 Let $n = 8$, $\ell = 3$, and $R =$ IV; that is, we are interested in a 2^{8-3}_{IV} design consisting of 32 runs.

Design 1

$$\begin{aligned} I &= ABCDEF = CDEG = ABFG \\ &= BDEH = ACFH = BCGH = ADEFGH \end{aligned}$$

with WLP(0, 5, 0, 2, 0, 0,). This design, as constructed by SAS PROC FACTEX, is given in Table 13.5. Inspection of the alias structure shows that all main effects are clear, but not strongly clear. Also, only 4 of the 28 two-factor interactions are clear (actually, strongly clear), all other 2-factor interactions are confounded with other 2-factor interactions.

Design 2

$$\begin{aligned} I &= CDEF = ABDEG = ABCFG \\ &= ABCEH = ABDFH = CDGH = EFGH \end{aligned}$$

with WLP(0, 3, 4, 0, 0, 0). This design, given in Table 13.6, was constructed by SAS PROC FACTEX using the option "minabs" to produce a minimum

Table 13.5 1/8th Fraction of the 2^8 Factorial of Resolution IV

```
options nodate pageno=1;
proc factex;
factors A B C D E F G H;
size design=32;
model res=4;
examine design aliasing confounding;
title1 'TABLE 13.5';
title2 '1/8TH FRACTION OF THE 2**8 FACTORIAL';
title3 'OF RESOLUTION IV';
run;
```

The FACTEX Procedure

Design Points

Experiment Number	A	B	C	D	E	F	G	H
1	-1	-1	-1	-1	-1	-1	-1	-1
2	-1	-1	-1	-1	1	1	1	1
3	-1	-1	-1	1	-1	1	1	1
4	-1	-1	-1	1	1	-1	-1	-1
5	-1	-1	1	-1	-1	1	1	-1
6	-1	-1	1	-1	1	-1	-1	1
7	-1	-1	1	1	-1	-1	-1	1
8	-1	-1	1	1	1	1	1	-1
9	-1	1	-1	-1	-1	1	-1	1
10	-1	1	-1	-1	1	-1	1	-1
11	-1	1	-1	1	-1	-1	1	-1
12	-1	1	-1	1	1	1	-1	1
13	-1	1	1	-1	-1	-1	1	1
14	-1	1	1	-1	1	1	-1	-1
15	-1	1	1	1	-1	1	-1	-1
16	-1	1	1	1	1	-1	1	1
17	1	-1	-1	-1	-1	1	-1	-1
18	1	-1	-1	-1	1	-1	1	1
19	1	-1	-1	1	-1	-1	1	1
20	1	-1	-1	1	1	1	-1	-1
21	1	-1	1	-1	-1	-1	1	-1
22	1	-1	1	-1	1	1	-1	1
23	1	-1	1	1	-1	1	-1	1
24	1	-1	1	1	1	-1	1	-1
25	1	1	-1	-1	-1	-1	-1	1
26	1	1	-1	-1	1	1	1	-1
27	1	1	-1	1	-1	1	1	-1
28	1	1	-1	1	1	-1	-1	1
29	1	1	1	-1	-1	1	1	1
30	1	1	1	-1	1	-1	-1	-1
31	1	1	1	1	-1	-1	-1	-1
32	1	1	1	1	1	1	1	1

Table 13.5 ***(Continued)***

```
                         Factor Confounding Rules

                              F = A*B*C*D*E
                              G = C*D*E
                              H = B*D*E

Aliasing Structure

A
B
C
D
E
F
G
H
A*B = F*G
A*C = F*H
A*D
A*E
A*F = B*G = C*H
A*G = B*F
A*H = C*F
B*C = G*H
B*D = E*H
B*E = D*H
B*H = C*G = D*E
C*D = E*G
C*E = D*G
D*F
E*F
```

aberration design ($L_4 = 3$ in D_2 vs. $L_4 = 5$ in D_1). All main effects are clear, of course, but only A and B are strongly clear. Now thirteen 2-factor interactions are clear, but none of them are strongly clear. □

Intuitively, minimum aberration designs should be the best designs with respect to the estimation of 2-factor interactions. However, Wu and Hamada (2000) in their listing of useful $2^{n-\ell}$ designs give a few examples of resolution IV designs, where a judicious choice of the identity relationship leads to a design with more clear 2-factor interactions than the corresponding minimum aberration design (see also Wu and Wu, 2002). Hence, both concepts should be used when selecting a design for a particular application.

Still other criteria have been introduced to further enable the search for and study of the properties of "desirable" designs, such as estimation capacity (Cheng, Steinberg, and Sun, 1999) and estimation index (Chen and Cheng, 2004). In

Table 13.6 1/8th Fraction of the 2^8 Factorial of Resolution IV and Minimum Aberration

```
options nodate pageno=1;
proc factex;
factors A B C D E F G H;
size design=32;
model res=4/minabs;
examine design aliasing confounding;
title1 'TABLE 13.6';
title2 '1/8TH FRACTION OF THE 2**8 FACTORIAL';
title3 'OF RESOLUTION IV AND MINIMUM ABERRATION';
run;
```

The FACTEX Procedure

Design Points

Experiment Number	A	B	C	D	E	F	G	H
1	-1	-1	-1	-1	-1	-1	1	1
2	-1	-1	-1	-1	1	1	-1	-1
3	-1	-1	-1	1	-1	1	-1	1
4	-1	-1	-1	1	1	-1	1	-1
5	-1	-1	1	-1	-1	1	1	-1
6	-1	-1	1	-1	1	-1	-1	1
7	-1	-1	1	1	-1	-1	-1	-1
8	-1	-1	1	1	1	1	1	1
9	-1	1	-1	-1	-1	-1	-1	-1
10	-1	1	-1	-1	1	1	1	1
11	-1	1	-1	1	-1	1	1	-1
12	-1	1	-1	1	1	-1	-1	1
13	-1	1	1	-1	-1	1	-1	1
14	-1	1	1	-1	1	-1	1	-1
15	-1	1	1	1	-1	-1	1	1
16	-1	1	1	1	1	1	-1	-1
17	1	-1	-1	-1	-1	-1	-1	-1
18	1	-1	-1	-1	1	1	1	1
19	1	-1	-1	1	-1	1	1	-1
20	1	-1	-1	1	1	-1	-1	1
21	1	-1	1	-1	-1	1	-1	1
22	1	-1	1	-1	1	-1	1	-1
23	1	-1	1	1	-1	-1	1	1
24	1	-1	1	1	1	1	-1	-1
25	1	1	-1	-1	-1	-1	1	1
26	1	1	-1	-1	1	1	-1	-1
27	1	1	-1	1	-1	1	-1	1
28	1	1	-1	1	1	-1	1	-1
29	1	1	1	-1	-1	1	1	-1
30	1	1	1	-1	1	-1	-1	1
31	1	1	1	1	-1	-1	-1	-1
32	1	1	1	1	1	1	1	1

Table 13.6 *(Continued)*

```
                        Factor Confounding Rules

                          F = C*D*E
                          G = A*B*D*E
                          H = A*B*C*E

Aliasing Structure

A
B
C
D
E
F
G
H
A*B
A*C
A*D
A*E
A*F
A*G
A*H
B*C
B*D
B*E
B*F
B*G
B*H
C*D = E*F = G*H
C*E = D*F
C*F = D*E
C*G = D*H
C*H = D*G
E*G = F*H
E*H = F*G
```

many cases application of these criteria leads to the same design. For a detailed discussion of the relationship among these criteria, we refer the reader to Chen and Cheng (2004).

13.4 FRACTIONAL REPLICATES FOR 3^n FACTORIAL DESIGNS

The basic ideas for the $2^{n-\ell}$ design can be extended easily to the $3^{n-\ell}$ design, that is, the $\frac{1}{3^\ell}$ fraction of the 3^n factorial. To obtain the treatment combinations for such a fraction we have to specify an identity relationship containing ℓ independent interactions. If we denote these interactions by $E^{\alpha_1}, E^{\alpha_2}, \ldots, E^{\alpha_\ell}$, we can then write

$$\boldsymbol{I} = E_{\delta_1}^{\alpha_1} = E_{\delta_2}^{\alpha_2} = \cdots = E_{\delta_\ell}^{\alpha_\ell} \tag{13.12}$$

The treatment combinations are obtained by solving the equations (mod 3)

$$\begin{aligned} \alpha_{11}x_1 &+ \alpha_{12}x_2 + \cdots + \alpha_{1n}x_n = \delta_1 \\ \alpha_{21}x_1 &+ \alpha_{22}x_2 + \cdots + \alpha_{2n}x_n = \delta_2 \\ &\vdots \\ \alpha_{\ell 1}x_1 &+ \alpha_{\ell 2}x_2 + \cdots + \alpha_{\ell n}x_n = \delta_\ell \end{aligned} \tag{13.13}$$

For any choice of $\delta_1, \delta_2, \ldots, \delta_\ell$ there will be exactly $3^{n-\ell}$ treatment combinations satisfying (13.13). Each such set, and there will be 3^ℓ possible sets, constitutes a $3^{n-\ell}$ fractional factorial. If, for some reason, we want the fraction to contain the "control," that is, the treatment combination $(0, 0, \ldots, 0)$, then we choose $\delta_1 = \delta_2 = \cdots = \delta_\ell = 0$.

We know, of course, that any treatment combination $\boldsymbol{x} = (x_1, x_2, \ldots, x_n)'$ satisfying Eqs. (13.13), that is, $\boldsymbol{\alpha}_i'\boldsymbol{x} = \delta_i\,(i = 1,\ 2, \ldots, \ell)$, also satisfies the equations

$$\left(\sum_{i=1}^{\ell} \lambda_i\,\boldsymbol{\alpha}_i\right)' \boldsymbol{x} = \sum_{i=1}^{\ell} \lambda_i\,\delta_i \tag{13.14}$$

where $\lambda_i = 0, 1, 2 (i = 1, 2, \ldots, \ell)$. For a given set of $\lambda_1, \lambda_2, \ldots, \lambda_\ell$ (not all zero) let us denote $\sum \lambda_i \boldsymbol{\alpha}_i$ by $\boldsymbol{\alpha}^*$ and $\sum \lambda_i \delta_i$ by δ^*. Thus (13.14) becomes

$$(\boldsymbol{\alpha}^*)'\boldsymbol{x} = \delta^* \tag{13.15}$$

Each $\boldsymbol{\alpha}^*$ corresponds to an effect $E^{\boldsymbol{\alpha}^*}$ and the treatment combinations satisfying (13.15) belong to the component $E_{\delta^*}^{\boldsymbol{\alpha}^*}$ with $\delta^* = 0, 1, 2$. It is not difficult to see that, using our convention that the first nonzero element in $\boldsymbol{\alpha}^*$ is equal to one, there are altogether $(3^\ell - 1)/2$ such $\boldsymbol{\alpha}^*$ including $\boldsymbol{\alpha}_1, \boldsymbol{\alpha}_2, \ldots, \boldsymbol{\alpha}_\ell$. And it is obvious that the $\boldsymbol{\alpha}^*$ other than $\boldsymbol{\alpha}_1,\ \boldsymbol{\alpha}_2, \ldots, \boldsymbol{\alpha}_\ell$ correspond to all possible GIs among $E^{\alpha_1}, E^{\alpha_2}, \ldots, E^{\alpha_\ell}$. Hence the complete identity relationship is obtained by adding to (13.12) all the GIs, that is,

$$\begin{aligned} \boldsymbol{I} = E_{\delta_1}^{\alpha_1} &= E_{\delta_2}^{\alpha_2} = \cdots = E_{\delta_\ell}^{\alpha_\ell} = E_{\delta_1+\delta_2}^{\alpha_1+\alpha_2} \\ &= E_{\delta_1+2\delta_2}^{\alpha_1+2\alpha_2} = \cdots = E_{\delta_1+2\delta_2+\ldots+2\delta_\ell}^{\alpha_1+2\alpha_2+\ldots+2\alpha_\ell} \end{aligned} \tag{13.16}$$

In order to obtain the treatment combinations for the $3^{n-\ell}$ design, we need to know only (13.12) and solve Eqs. (13.13), but in order to obtain the alias structure we need to use the entire set of interactions in the identity relationship as given in (13.16). Suppose $E^{\boldsymbol{\beta}}$ is an effect not contained in (13.16). We know that because of orthogonality (see Chapter 10) exactly $\frac{1}{3}$ of the treatment combinations satisfying (13.13) also satisfy the equation

$$\beta_1 x_1 + \beta_2 x_2 + \cdots + \beta_n x_n = \delta \tag{13.17}$$

for $\delta = 0, 1, 2$ and each treatment combination satisfies only one of the three equations. Then $E^{\boldsymbol{\beta}}$ is defined, as with the full set of 3^n treatment combinations, by contrasts of the form

$$c_0 E_0^{\boldsymbol{\beta}} + c_1 E_1^{\boldsymbol{\beta}} + c_2 E_2^{\boldsymbol{\beta}} \tag{13.18}$$

with $c_0 + c_1 + c_2 = 0$, where $E_\delta^{\boldsymbol{\beta}}$ is obtained (apart from the overall mean) from the $3^{n-\ell-1}$ treatment combinations satisfying (13.17) for $\delta = 0, 1, 2$. Let us take now any of the interactions in (13.16), say $E_{\delta_i}^{\boldsymbol{\alpha}_i}$, and consider the GI between $E^{\boldsymbol{\beta}}$ and $E^{\boldsymbol{\alpha}_i}$, denoted by $E^{\boldsymbol{\beta}+\boldsymbol{\alpha}_i}$. A contrast belonging to this effect is given by

$$c_0 \boldsymbol{E}_0^{\boldsymbol{\beta}+\boldsymbol{\alpha}_i} + c_1 \boldsymbol{E}_1^{\boldsymbol{\beta}+\boldsymbol{\alpha}_i} + c_2 \boldsymbol{E}_2^{\boldsymbol{\beta}+\boldsymbol{\alpha}_i}$$

where the leading mean in $E_\delta^{\boldsymbol{\beta}+\boldsymbol{\alpha}_i}$ is obtained from the treatment combinations satisfying

$$(\beta_1 + \alpha_{i1})\, x_1 + (\beta_2 + \alpha_{i2})\, x_2 + \cdots + (\beta_n + \alpha_{in})\, x_n = \delta \tag{13.19}$$

We know that

$$\alpha_{i1} x_1 + \alpha_{i2} x_2 + \cdots + \alpha_{in} x_n = \delta_i \tag{13.20}$$

Substituting (13.20) in (13.19) yields

$$\beta_1 x_1 + \beta_2 x_2 + \cdots + \beta_n x_n = \delta - \delta_i$$

Now $\delta - \delta_i$ takes on the values 0, 1, 2 (mod 3), and it is therefore obvious that a contrast belonging to $E^{\boldsymbol{\beta}}$ is also a contrast belonging to $E^{\boldsymbol{\beta}+\boldsymbol{\alpha}_i}$, that is, $E^{\boldsymbol{\beta}}$ and $E^{\boldsymbol{\alpha}_i}$ are confounded. The same argument holds for the other GI between $E^{\boldsymbol{\beta}}$ and $E^{\boldsymbol{\alpha}_i}$, namely $E^{\boldsymbol{\beta}+2\boldsymbol{\alpha}_i}$. This then explains how we use (13.16) to obtain the alias structure for the $3^{n-\ell}$ design: Any effect $E^{\boldsymbol{\beta}}$ not in (13.16) is confounded with all GIs obtained by formally multiplying $E^{\boldsymbol{\beta}}$ into $E^{\boldsymbol{\gamma}}$ and $E^{2\boldsymbol{\gamma}}$ where $E^{\boldsymbol{\gamma}}$ is any effect listed in (13.16), that is, $E^{\boldsymbol{\beta}}$ is confounded with $E^{\boldsymbol{\beta}+\boldsymbol{\gamma}}$ and $E^{\boldsymbol{\beta}+2\boldsymbol{\gamma}}$. We illustrate this below.

Example 13.4 Consider the 3^{4-2} design with the identity relationship given by

$$I = ABC_0^2 = ACD_0 = AB^2D_0^2 = BCD_0^2 \qquad (13.21)$$

where ABC^2 and ACD are the two independent interactions of (13.12) and AB^2D^2, BCD^2 are their GIs. The treatment combinations are obtained from the equations

$$\begin{aligned} x_1 + x_2 + 2x_3 \qquad &= 0 \\ x_1 \qquad + x_3 + x_4 &= 0 \end{aligned}$$

They are

0 0 0 0	0 1 1 2	0 2 2 1
1 0 1 1	1 1 2 0	1 2 0 2
2 0 2 2	2 1 0 1	2 2 1 0

The alias structure is given by multiplying each effect not contained in (13.21) into each interaction in (13.21) and its square (ignoring the subscripts). We obtain (using reduction mod 3):

$$\begin{aligned} A &= A(ABC^2) = A(ABC^2)^2 = A(ACD) = A(ACD)^2 \\ &= A(AB^2D^2) = A(AB^2D^2)^2 = A(BCD^2) = A(BCD^2)^2 \\ &= AB^2C = BC^2 = AC^2D^2 = CD = ABD = BD = ABCD^2 = AB^2C^2D \end{aligned}$$

Similarly

$$\begin{aligned} B &= AB^2C^2 = AC^2 = ABCD = AB^2CD = AD^2 = ABD^2 \\ &= BC^2D = CD^2 \end{aligned}$$

$$\begin{aligned} C &= AB = ABC = AC^2D = AD = AB^2CD^2 = AB^2C^2D^2 \\ &= BC^2D^2 = BD^2 \end{aligned}$$

$$\begin{aligned} D &= ABC^2D = ABC^2D^2 = ACD^2 = AC = AB^2 = AB^2D \\ &= BC = BCD \end{aligned}$$

Together with the effects in (13.21) these account for all the effects of a 3^4 factorial. We recognize, of course, that this design is a resolution III design, which allows the estimation of all main effects only if all interactions are negligible.

An equivalent design, using SAS PROC FACTEX, is given in Table 13.7.

The factor confounding rule states that the levels of C are obtained as twice the levels of A plus twice the levels of B, and that the levels of D are obtained as the levels of A plus twice the levels of B and reduction mod 3. The output shows

Table 13.7 1/9 Fraction of the 3^4 Factorial of Resolution III

```
options nodate pageno=1;
proc factex;
factors A B C D/nlev=3;
size fraction=9;
model res=3;
examine design aliasing confounding;
title1 'TABLE 13.7';
title2 '1/9 FRACTION OF THE 3**4 FACTORIAL';
title3 'OF RESOLUTION III';
run;
```

```
                    The FACTEX Procedure

                       Design Points

          Experiment
              Number       A       B       C       D
          ------------------------------------------
                   1      -1      -1      -1      -1
                   2      -1       0       1       1
                   3      -1       1       0       0
                   4       0      -1       1       0
                   5       0       0       0      -1
                   6       0       1      -1       1
                   7       1      -1       0       1
                   8       1       0      -1       0
                   9       1       1       1      -1

                  Factor Confounding Rules

                     C = (2*A) + (2*B)
                     D = A + (2*B)

Aliasing Structure

    A   = (2*B) + (2*C) = B + D = C + (2*D)
 (2*A)  = B + C = (2*B) + (2*D) = (2*C) + D
    B   = (2*A) + (2*C) = A + (2*D) = C + D
 (2*B)  = A + C = (2*A) + D = (2*C) + (2*D)
    C   = (2*A) + (2*B) = A + D = B + (2*D)
 (2*C)  = A + B = (2*A) + (2*D) = (2*B) + D
    D   = A + (2*B) = (2*A) + C = B + (2*C)
 (2*D)  = (2*A) + B = A + (2*C) = (2*B) + C
```

that the levels of C are indeed computed in this fashion, but the levels of D are obtained by subtracting 1 from the results of the rule given above (presumably this is done to include the control in the fraction). The alias structure shows that, just as in Example 13.4, the main effect A is confounded with 2 d.f. of each of the 2-factor interactions $B \times C$, $B \times D$, and $C \times D$ (and higher-order interactions).

The notions of clear effects and minimum aberration design can be carried over in an obvious fashion to $3^{n-\ell}$ designs. A listing of useful designs is given by Wu and Hamada (2000). □

13.5 GENERAL CASE OF FRACTIONAL REPLICATION

13.5.1 Symmetrical Factorials

From our discussion above it is quite obvious how the concept of fractional replication can be extended to the general case, that is, to the $1/s^\ell$ fraction of the s^n factorial where s is a prime or prime power. All we need to do is specify ℓ independent interactions, say $E^{\alpha_1}, E^{\alpha_2}, \ldots, E^{\alpha_\ell}$ and the $s^{n-\ell}$ treatment combinations are then obtained by solving the ℓ equations

$$\alpha_{i1}x_1 + \alpha_{i2}x_2 + \cdots + \alpha_{in}x_n = \delta_i (i = 1, 2, \ldots, \ell)$$

where α_{ij}, x_j, δ_i are elements in GF(s) ($i = 1, 2, \ldots, \ell; j = 1, 2, \ldots, n$). The identity relationship is given by

$$I = E^{\alpha_1} = E^{\alpha_2} = \cdots = E^{\alpha_\ell} = E^{\alpha_1+\alpha_2} = E^{\alpha_1+u_2\alpha_2}$$

$$= \cdots = E^{\alpha_1+u_{s-1}\alpha_2} = \cdots = E^{\alpha_1+u_{s-1}\alpha_2+\cdots+u_{s-1}\alpha_\ell} \tag{13.22}$$

[$u_j \in$ GF(s)] that is, (13.22) contains the ℓ independent interactions and all their GIs. It follows then that an effect E^α is aliased with all the GIs between E^α and the interactions in (13.22).

13.5.2 Asymmetrical Factorials

The basic ideas of fractional replication for symmetrical factorial experiments as presented above can be extended and applied also to asymmetrical factorials. Suppose we want to choose a fraction of an $s_1^{n_1} \times s_2^{n_2} \times \cdots \times s_1^{n_q}$ factorial. An obvious and easy method would be to consider a $1 - in - s_1^{\ell_1} s_2^{\ell_2}, \ldots, s_q^{\ell_q}$ replicate consisting of $s_1^{n_1-\ell_1}, s_2^{n_2-\ell_2}, \ldots, s_q^{n_q-\ell_q}$ treatment combinations. The idea, of course, is to consider each of the q symmetrical component systems separately, obtain a $1/s_i^{\ell_i}$ fraction of the $s_i^{n_i}$ factorial ($i = 1, 2, \ldots, q$) and then combine those fractions as a Kronecker product design. More specifically, let $\boldsymbol{x}_i = (\boldsymbol{x}_{i1}, \boldsymbol{x}_{i2}, \ldots, \boldsymbol{x}_{i\nu_i})'$ denote the $\nu_i = s_i^{n-\ell_i}$ treatment combinations for the fractional factorial design of the $s_i^{n_i}$ factorial, abbreviated FFD(s_i, n_i, ℓ_i), for $i = 1, 2, \ldots, q$, with $\boldsymbol{x}_{ij} = (x_{ij1}, x_{ij2}, \ldots, x_{ijn_i})'$ and x_{ijk} the level of the kth factor in the jth treatment combination of FFD(s_i, n_i, ℓ_i). Then the treatment combinations for the fractional replicate of the asymmetrical factorial $s_1^{n_1} \times s_2^{n_2} \times \cdots \times s_q^{n_q}$ are given by the symbolic direct product (see Section 11.4.1) $\boldsymbol{x}_1 \otimes \boldsymbol{x}_2 \otimes \cdots \otimes \boldsymbol{x}_q$. Symbolically, we may write this also as $\prod_{i=1}^q \times$ FFD(s_i, n_i, ℓ_i).

Now each FFD (s_i, n_i, ℓ_i) is based on an identity relationship, I_i say, containing ℓ_i independent interactions and all their GI's. Thus I_i determines the alias structure for FFD(s_i, n_i, ℓ_i) involving the n_i factors in the $s_i^{n_i}$ factorial. We simply apply the rules given above. In addition, we have confounding among mixed interactions, that is, interactions involving factors from different symmetrical factorials. We note here that the alias structure for mixed interactions may be more advantageous than those for interactions in the symmetrical component systems, advantageous in the sense of being confounded with higher-order interactions that may be assumed negligible.

We shall illustrate these ideas with the following example.

Example 13.5 Consider a $\frac{1}{6}$ replicate of the $2^3 \times 3^3$ factorial; that is, $q = 2$, $s_1 = 2$, $n_1 = 3$, $\ell_1 = 1$, $s_2 = 3$, $n_2 = 3$, $\ell_2 = 1$. Let A_1, A_2, A_3 and B_1, B_2, B_3 denote the factors for the two symmetrical component systems, respectively. To obtain FFD(2, 3, 1) we use

$$I_1 = (A_1\,A_2\,A_3)_0$$

and for FFD(3, 3, 1) we use

$$I_2 = (B_1\,B_2\,B_3^2)_0$$

It follows that

$$\boldsymbol{x}_1 = \begin{pmatrix} 0 & 0 & 0 \\ 1 & 1 & 0 \\ 1 & 0 & 1 \\ 0 & 1 & 1 \end{pmatrix}$$

and

$$\boldsymbol{x}_2 = \begin{pmatrix} 0 & 0 & 0 \\ 1 & 2 & 0 \\ 2 & 1 & 0 \\ 0 & 1 & 1 \\ 1 & 0 & 1 \\ 2 & 2 & 1 \\ 0 & 2 & 2 \\ 1 & 1 & 2 \\ 2 & 0 & 2 \end{pmatrix}$$

The 36 treatment combinations are obtained by combining every treatment combination in $\boldsymbol{x}_1$ with every treatment combination in $\boldsymbol{x}_2$. The alias structure for FFD(2, 3, 1) is given by

$$A_1 = A_2 A_3$$

$$A_2 = A_1 A_3$$

$$A_3 = A_1 A_2$$

and for FFD(3, 3, 1) by

$$B_1 = B_1B_2^2B_3 = B_2B_3^2$$
$$B_2 = B_1B_2^2B_3^2 = B_1B_3^2$$
$$B_3 = B_1B_2 = B_1B_2B_3$$
$$B_1B_2^2 = B_1B_3 = B_2B_3$$

The alias structure for mixed interactions is illustrated by

$$\begin{aligned} A_1 \times B_1 = A_1 \times B_1B_2^2B_3 &= A_1 \times B_2B_3^2 = A_2A_3 \times B_1 \\ &= A_2A_3 \times B_1B_2^2B_3 = A_2A_3 \times B_2B_3^2 \end{aligned}$$

This shows that the 2-factor interaction $A_1 \times B_1$ is confounded with 3-factor or higher-order interactions.

An isomorphic design using SAS PROC FACTEX is given in Table 13.8. □

13.5.3 General Considerations

From a practical point of view highly fractionated designs may be the only feasible designs, requiring only a relatively small number of treatment combinations. This has, of course, fairly severe consequences as each effect will be confounded with a "large" number of other effects. For such a design to be useful the identity relationship must be chosen carefully in order to avoid having effects of interest confounded with each other. For example, if main effects and some or all 2-factor interactions are considered to be important, then certainly main effects should not be confounded with other main effects and 2-factor interactions. If only some 2-factor interactions are important, then confounding of certain 2-factor interactions is permissible, but if all 2-factor interactions are important, then they must only be confounded with 3-factor or higher-order interactions. In other words, resolution IV or V designs are called for. It is obvious that the more effects we want to estimate the more treatment combinations we need. It is typical that for experiments having to use fractional factorial designs the two requirements of a limited number of treatment combinations and of information about a specified number of effects are often in conflict with each other and compromises will have to be made. It is therefore important to be aware of methods of constructing fractional factorial designs of minimal size to achieve certain objectives such as those mentioned above. This leads also to consideration of fractional replicates other than the regular fractions discussed so far. A general discussion of the structures and properties of fractional factorial designs is given by Raktoe, Hedayat, and Federer (1981), and a description of construction methods for certain fractional factorial designs is provided by Dey (1985). We shall discuss some of these issues in Section 13.6.

Table 13.8 1/6 Fraction of the $2^3 \times 3^3$ Factorial Using a Kronecker Product Design

```
options nodate pageno=1;
proc factex;
factors A1 A2 A3;
size design=4;
model res=3;
examine design confounding aliasing;
output out=two;
title1 'TABLE 13.8';
title2 '1/6 FRACTION OF THE 2**3x3**3 FACTORIAL';
title3 'USING A KRONECKER PRODUCT DESIGN';
run;
factors B1 B2 B3/nlev=3;
size design=9;
model res=3;
examine design confounding aliasing;
output out=three
designrep=two;
run;

proc print data=three;
run;
```

```
                         The FACTEX Procedure

                            Design Points

                  Experiment
                      Number   A1       A2       A3
                  ---------------------------------
                           1   -1       -1        1
                           2   -1        1       -1
                           3    1       -1       -1
                           4    1        1        1

                      Factor Confounding Rules

                            A3 = A1*A2


Aliasing Structure

  A1 = A2*A3
  A2 = A1*A3
  A3 = A1*A2
```

Table 13.8 (*Continued*)

Design Points

Experiment Number	B1	B2	B3
1	-1	-1	-1
2	-1	0	1
3	-1	1	0
4	0	-1	1
5	0	0	0
6	0	1	-1
7	1	-1	0
8	1	0	-1
9	1	1	1

Factor Confounding Rules

B3 = (2*B1) + (2*B2)

Aliasing Structure

```
   B1    = (2*B2) + (2*B3)
 (2*B1)  = B2 + B3
   B2    = (2*B1) + (2*B3)
 (2*B2)  = B1 + B3
   B3    = (2*B1) + (2*B2)
 (2*B3)  = B1 + B2
 (2*B1)+     B2   = B1 + (2*B3)  = (2*B2) + B3
   B1 + (2*B2)  = (2*B1) + B3  = B2 + (2*B3)
```

Obs	A1	A2	A3	B1	B2	B3
1	-1	-1	1	-1	-1	-1
2	-1	-1	1	-1	0	1
3	-1	-1	1	-1	1	0
4	-1	-1	1	0	-1	1
5	-1	-1	1	0	0	0
6	-1	-1	1	0	1	-1
7	-1	-1	1	1	-1	0
8	-1	-1	1	1	0	-1
9	-1	-1	1	1	1	1
10	-1	1	-1	-1	-1	-1
11	-1	1	-1	-1	0	1
12	-1	1	-1	-1	1	0
13	-1	1	-1	0	-1	1
14	-1	1	-1	0	0	0
15	-1	1	-1	0	1	-1
16	-1	1	-1	1	-1	0
17	-1	1	-1	1	0	-1
18	-1	1	-1	1	1	1
19	1	-1	-1	-1	-1	-1

Table 13.8 (*Continued*)

20	1	-1	-1	-1	0	1
21	1	-1	-1	-1	1	0
22	1	-1	-1	0	-1	1
23	1	-1	-1	0	0	0
24	1	-1	-1	0	1	-1
25	1	-1	-1	1	-1	0
26	1	-1	-1	1	0	-1
27	1	-1	-1	1	1	1
28	1	1	1	-1	-1	-1
29	1	1	1	-1	0	1
30	1	1	1	-1	1	0
31	1	1	1	0	-1	1
32	1	1	1	0	0	0
33	1	1	1	0	1	-1
34	1	1	1	1	-1	0
35	1	1	1	1	0	-1
36	1	1	1	1	1	1

13.5.4 Maximum Resolution Design

The discussion in Section 13.6 will allow us to make statements about the minimum size of a design for a specified resolution. An equally important question arises if we are given a fixed number, say N, of runs and we want to know what is the maximum resolution that can be achieved under this constraint.

It is clear, of course, that a $1/s$ fraction of the s^n factorial has resolution $R = n$. For the general case of the $s^{n-\ell}$ fractional factorial an upper bound of the resolution was given by Fujii (1976). We state his result in the following theorem without proof.

Theorem 13.1 (Fujii, 1976) Let $M = (s^\ell - 1)/(s - 1)$ and $n = qM + m$, with $q \geq 0$ integer. Then, for $\ell \geq 2$, the maximum resolution $R_\ell(n, s)$ of the $s^{n-\ell}$ design (with s prime or prime power) is given by

$$
\begin{aligned}
R_\ell(n, s) &= s^{\ell-1}q \quad \text{if } m = 0, 1 \\
&\leq s^{\ell-1}q + \left[s^{\ell-2}(s-1)(m-1)/(s^{\ell-1}-1)\right] \\
&\qquad\qquad \text{if } m = 2, 3, \ldots, s^{\ell-1} - 1 \\
&\leq s^{\ell-1}q + [(s-1)m/s] \\
&\qquad\qquad \text{if } m = s^{\ell-1}, s^{\ell-1} + 1, \ldots, M - 1
\end{aligned}
$$

where $[x]$ is the greatest integer not exceeding x.

In Table 13.9 we present $R_{\max}$ values for $s = 2, 3, 4, 5$ and $4 \leq n \leq 10$, $2 \leq \ell \leq 7$ with the number of runs ≤ 256. An asterisk indicates that $R_{\max} < R_\ell(n, s)$

Table 13.9 Maximum Resolution $s^{n-\ell}$ Designs

s	n	ℓ	$R_{\max}$	Number of Runs
2	5	2	3	8
	6	2	4	16
	6	3	3	8
	7	2	4	32
	7	3	4	16
	7	4	3	8
	8	2	5	64
	8	3	4	32
	8	4	4	16
	9	3	4	64
	9	4	4	32
	9	5	3*	16
	10	3	5	128
	10	4	4*	64
	10	5	4	32
	10	6	3*	16
3	4	2	3	9
	5	2	3	27
	6	2	4	81
	6	3	3	27
	7	2	5	243
	7	3	4	81
	7	4	3*	27
	8	3	5	243
	8	4	4	81
	8	5	3*	27
	9	4	5	243
	9	5	3*	81
	9	6	3*	27
	10	5	5*	243
	10	6	3*	81
	10	7	3*	27
4	4	2	3	16
	5	2	4	64
	5	3	3	16
	6	2	4	256
	6	3	4	64
	7	3	4	256
	7	4	3*	64
	8	4	4*	256
	8	5	3*	64
	9	5	4*	256
	9	6	3*	64
	10	6	4*	256
	10	7	3*	64

Table 13.9 *(Continued)*

s	n	ℓ	$R_{\max}$	Number of Runs
5	4	2	3	25
	5	2	4	125
	5	3	3	25
	6	3	4	125
	6	4	3*	25
	7	4	3*	125
	8	5	3*	125
	9	6	3*	125
	10	7	3*	125

as given above [it shows that $R_\ell(n, s)$ is not particularly good for larger values of ℓ, a fact also observed by Fries and Hunter (1980) who provided a different bound for $s = 2$].

13.6 CHARACTERIZATION OF FRACTIONAL FACTORIAL DESIGNS OF RESOLUTION III, IV, AND V

13.6.1 General Formulation

Recall from (12.15) that each true treatment response, $a(\boldsymbol{v})$, for a $s_1^{n_1} \times s_2^{n_2} \times \cdots \times s_q^{n_q}$ factorial with $s_1 < s_2 < \cdots < s_q$ can be expressed in terms of the mean response, a component from each pure partition and a component from each mixed partition. Of interest here are only all components from pure partitions belonging to main effects and 2-factor interactions and all components from mixed partitions belonging to 2-factor interactions. As an illustration we shall consider the $2^2 \times 3^2$ factorial.

Let A_1, A_2 denote the two factors with two levels each, and B_1, B_2 the two factors with three levels. Recall from Example 12.4 that all arithmetic has to be done in $R(6)$ and that the levels of A_1, A_2 are denoted by 0, 3 and those of B_1, B_2 by 0, 4, 2. To simplify the notation we shall write the treatment combination $\boldsymbol{v} = (v_{11}, v_{12}, v_{21}, v_{22})'$ as $\boldsymbol{v} = (\boldsymbol{v}_1', \boldsymbol{v}_2')'$ where $\boldsymbol{v}_1 = (v_{11}, v_{12})'$, $\boldsymbol{v}_2 = (v_{21}, v_{22})'$. We then have

$$\begin{aligned} a(\boldsymbol{v}) = a(\boldsymbol{v}_1, \boldsymbol{v}_2) = &+ (A_1^3)_{(30)\boldsymbol{v}_1} + (A_2^3)_{(03)\boldsymbol{v}_1} + (A_1^3 A_2^3)_{(33)\boldsymbol{v}_1} \\ &+ (B_1^4)_{(40)\boldsymbol{v}_2} + (B_2^4)_{(04)\boldsymbol{v}_2} + (B_1^4 B_2^4)_{(44)\boldsymbol{v}_2} \\ &+ (B_1^4 B_2^2)_{(42)\boldsymbol{v}_2} + (A_1^3 B_1^4)^*_{(3040)\boldsymbol{v}} + (A_1^3 B_2^4)^*_{(3004)\boldsymbol{v}} \\ &+ (A_2^3 B_1^4)^*_{(0340)\boldsymbol{v}} + (A_2^3 B_2^4)^*_{(0304)\boldsymbol{v}} \end{aligned} \quad (13.23)$$

Alternatively, and equivalently, we can express $a(\boldsymbol{v})$ in terms of the traditional model for factorial structures as

$$\begin{aligned} a(\boldsymbol{v}) = \mu + \alpha_i + \beta_j + (\alpha\beta)_{ij} + \gamma_k + \delta_\ell \\ +(\gamma\delta)_{k\ell} + (\alpha\gamma)_{ik} + (\beta\gamma)_{jk} \\ +(\alpha\delta)_{i\ell} + (\beta\delta)_{j\ell} \end{aligned} \tag{13.24}$$

with $i, j = 0, 1, k, \ell = 0, 1, 2$. Models of the type (13.24) are used in SAS, for example.

We ignore 3-factor and higher-order interactions since those are assumed to be negligible for resolution III, IV, and V designs. In addition, for resolution III designs 2-factor interactions are also negligible, and we are interested only in estimating the mean and main effects. For resolution IV designs we are mainly interested in estimating main effects, but the mean and 2-factor interactions are assumed to be unknown (and not necessarily zero). For resolution V designs we wish, of course, to estimate the mean, main effects, and 2-factor interactions. All three situations can be expressed in very general terms in the form of a three-part linear model, the first part referring to the mean, the second part referring to the main effects, and the third part to the 2-factor interactions. We formulate this as follows (see Raktoe, Hedayat, and Federer, 1981).

Let D denote an FFD consisting of N treatment combinations (not necessarily all distinct) and let $\boldsymbol{a}_D$ denote the $N \times 1$ vector of the true treatment responses associated with D. Further, let M denote the mean, $\boldsymbol{E}_1$ the $q_1 \times 1$ vector of main effect components and $\boldsymbol{E}_2$ the $q_2 \times 1$ vector of 2-factor interaction components [using either (13.23) or (13.24)], with $\boldsymbol{E}_1$ and $\boldsymbol{E}_2$ conformable to D and the factorial from which D was obtained. We then write, analogously to (12.15),

$$\boldsymbol{a}_D = (\mathfrak{I}, \boldsymbol{X}_{1D}, \boldsymbol{X}_{2D}) \begin{pmatrix} M \\ \boldsymbol{E}_1 \\ \boldsymbol{E}_2 \end{pmatrix} \tag{13.25}$$

where $\boldsymbol{X}_{1D}$ and $\boldsymbol{X}_{2D}$ are $N \times q_1$ and $N \times q_2$ incidence matrices, respectively. If $\boldsymbol{y}_D$ represents the $N \times 1$ vector of observations corresponding to $\boldsymbol{a}_D$, then we write (13.25) alternatively and in a more familiar form as

$$E[\boldsymbol{y}_D] = (\mathfrak{I}, \boldsymbol{X}_{1D}, \boldsymbol{X}_{2D}) \begin{pmatrix} M \\ \boldsymbol{E}_1 \\ \boldsymbol{E}_2 \end{pmatrix} \tag{13.26}$$

In order to determine whether a given FFD is of resolution III, IV, or V it is sufficient to consider, instead of D, the unreplicated design D_0, which contains the same treatment combinations as D except that each treatment combination is represented only once. Suppose D_0 contains N_0 treatment combinations. We then replace (13.26), using obvious notation, by

$$E[\boldsymbol{y}_{D_0}] = (\mathfrak{I}, \boldsymbol{X}_{1D_0}, \boldsymbol{X}_{2D_0}) \begin{pmatrix} M \\ \boldsymbol{E}_1 \\ \boldsymbol{E}_2 \end{pmatrix} \tag{13.27}$$

We shall now consider (13.27) and the resulting NE for each of the three situations.

13.6.2 Resolution III Designs

For a resolution III design we have, of course, $\boldsymbol{E}_2 = \boldsymbol{0}$ and (13.27) reduces to the two-part model (dropping the subscript D_0 for convenience)

$$E(\boldsymbol{y}) = (\mathfrak{I}, \boldsymbol{X}_1) \begin{pmatrix} M \\ \boldsymbol{E}_1 \end{pmatrix} \tag{13.28}$$

and the NE are of the form

$$\begin{pmatrix} \mathfrak{I}'\mathfrak{I} & \mathfrak{I}'\boldsymbol{X}_1 \\ \boldsymbol{X}_1'\mathfrak{I} & \boldsymbol{X}_1'\boldsymbol{X}_1 \end{pmatrix} \begin{pmatrix} M \\ \boldsymbol{E}_1 \end{pmatrix} = \begin{pmatrix} \mathfrak{I}'\boldsymbol{y} \\ \boldsymbol{X}_1'\boldsymbol{y} \end{pmatrix} \tag{13.29}$$

Suppose we have n factors $A_1, A_2, \ldots, A_n$ and factor A_i has s_i levels ($i = 1, 2, \ldots, n$), where the s_i need not be all different. Then $\boldsymbol{E}_1$ has $q_1 = \sum_i s_i$ components, but for factor A_i the s_i components add to zero ($i = 1, 2, \ldots, n$). Hence $\boldsymbol{E}_1$ contains only $\sum_i (s_i - 1) = q_1 - n$ independent components. It follows then also that the coefficient matrix in (13.29) is not of full rank. A solution to (13.29) can be written as

$$\begin{pmatrix} \widehat{M} \\ \widehat{\boldsymbol{E}}_1 \end{pmatrix} = \begin{pmatrix} \mathfrak{I}'\mathfrak{I} & \mathfrak{I}'\boldsymbol{X}_1 \\ \boldsymbol{X}_1'\mathfrak{I} & \boldsymbol{X}_1'\boldsymbol{X}_1 \end{pmatrix}^{-} \begin{pmatrix} \mathfrak{I}'\boldsymbol{y} \\ \boldsymbol{X}_1'\boldsymbol{y} \end{pmatrix} \tag{13.30}$$

or, if we write $\boldsymbol{X}_1^* = (\mathfrak{I}, \boldsymbol{X}_1)$,

$$\begin{pmatrix} \widehat{M} \\ \widehat{\boldsymbol{E}}_1 \end{pmatrix} = \left(\boldsymbol{X}_1^{*\prime}\boldsymbol{X}_1^*\right)^{-} \boldsymbol{X}_1^{*\prime}\boldsymbol{y} \tag{13.31}$$

It is obvious then that for the mean and all main effects to be estimable we must have

$$\operatorname{rank}\left(\boldsymbol{X}_1^{*\prime}\boldsymbol{X}_1^*\right) = 1 + q_1 - n \tag{13.32}$$

Condition (13.32) can in certain situations be achieved with $N_0 = 1 + q_1 - n$ treatment combinations, and we shall give some examples later (see Chapter 14).

For such a design no d.f. are available for estimating the variance σ_e^2 in the context of the ANOVA. The design D_0 is then referred to as a ü *saturated design*. If we use a design D with N runs where $N > 1 + q_1 - n$ and (13.32) is satisfied, then σ_e^2 can, of course, be estimated with $N - (1 + q_1 - n)$ d.f. Using $(\boldsymbol{X}_{1D}^{*\prime}\boldsymbol{X}_{1D}^{*})^{-}$ as the variance–covariance matrix we can then, within the factor space of the experiment, compare different treatments, for example, $a(\boldsymbol{v}) - a(\boldsymbol{w})$ for some treatment combinations $\boldsymbol{v}$ and $\boldsymbol{w}$ using the main effects model (13.28). For the analysis of a satu:ated design D_0 see Section 13.9.

13.6.3 Resolution IV Designs

For this case it is convenient to rewrite (13.27) (dropping again the subscript D_0) as

$$E(\boldsymbol{y}) = (\boldsymbol{X}_1, \boldsymbol{X}_2^*)\begin{pmatrix}\boldsymbol{E}_1\\ \boldsymbol{E}_2^*\end{pmatrix} \tag{13.33}$$

with

$$\boldsymbol{X}_2^* = (\mathfrak{I}, \boldsymbol{X}_2) \qquad \boldsymbol{E}_2^* = \begin{pmatrix}\boldsymbol{M}\\ \boldsymbol{E}_2\end{pmatrix}.$$

The reason to include M in $\boldsymbol{E}_2^*$, the vector of nuisance parameters, is mostly one of mathematical convenience. It simply means that we are not interested in estimating the mean just as we are not interested in estimating 2-factor interactions. A plausible argument for proceeding in this fashion was given by Raktoe, Hedayat, and Federer (1981) who point out that because of the possible presence of 2-factor interactions we cannot use (13.27) in the same way as we used (13.28) to make comparisons among treatments. Our only aim here can be to estimate main effects and draw conclusions from that concerning the importance of the various factors. The important difference between estimating main effects from a resolution IV design as compared to a resolution III design is, of course, that in a resolution IV design 3-factor interactions are assumed to be negligible, that is, main effects are clear, whereas in a resolution III design 2-factor interactions are also assumed to be negligible.

Returning now to model (13.33) the NE are

$$\begin{pmatrix}\boldsymbol{X}_1'\boldsymbol{X}_1 & \boldsymbol{X}_1'\boldsymbol{X}_2^*\\ \boldsymbol{X}_2^{*\prime}\boldsymbol{X}_1 & \boldsymbol{X}_2^{*\prime}\boldsymbol{X}_2^*\end{pmatrix}\begin{pmatrix}\boldsymbol{E}_1\\ \boldsymbol{E}_2^*\end{pmatrix} = \begin{pmatrix}\boldsymbol{X}_1'\boldsymbol{y}\\ \boldsymbol{X}_2^{*\prime}\boldsymbol{y}\end{pmatrix} \tag{13.34}$$

Using results of Chapter I.4 the RNE for $\boldsymbol{E}_1$ are given by

$$\begin{aligned}&\left(\boldsymbol{X}_1'\boldsymbol{X}_1 - \boldsymbol{X}_1'\boldsymbol{X}_2^*\left(\boldsymbol{X}_2^{*\prime}\boldsymbol{X}_2^*\right)^{-}\boldsymbol{X}_2^{*\prime}\boldsymbol{X}_1\right)\boldsymbol{E}_1\\ &\quad = \left(\boldsymbol{X}_1' - \boldsymbol{X}_1'\boldsymbol{X}_2^*\left(\boldsymbol{X}_2^{*\prime}\boldsymbol{X}_2^*\right)^{-}\boldsymbol{X}_2^{*\prime}\right)\boldsymbol{y}\end{aligned} \tag{13.35}$$

or

$$\boldsymbol{C}\boldsymbol{E}_1 = \boldsymbol{Q} \tag{13.36}$$

where $\boldsymbol{C}$ and $\boldsymbol{Q}$ are defined in an obvious way. For all main effects to be estimable we must have

$$\operatorname{rank}\left(\boldsymbol{X}_1'\boldsymbol{X}_1 - \boldsymbol{X}_1'\boldsymbol{X}_2^*\left(\boldsymbol{X}_2^{*\prime}\boldsymbol{X}_2^*\right)^- \boldsymbol{X}_2^{*\prime}\boldsymbol{X}_1\right) = \sum_i (s_i - 1) = q_1 - n \tag{13.37}$$

To illustrate these ideas we consider the following example.

Example 13.6 Suppose we have four factors, A, B, C, F, each at two levels and we have a $\frac{1}{2}$ replicate $D_0 = \{(0\ 0\ 0\ 0), (0\ 1\ 1\ 0), (1\ 0\ 1\ 0), (1\ 1\ 0\ 0), (0\ 0\ 1\ 1), (0\ 1\ 0\ 1), (1\ 0\ 0\ 1), (1\ 1\ 1\ 1)\}$ with eight treatment combinations. Notice that D_0 is based on the identity relationship $I = ABCF$. So we already know that D_0 is a resolution IV design. We now have

$$\boldsymbol{X}_1 = \begin{bmatrix} A_0 & A_1 & B_0 & B_1 & C_0 & C_1 & F_0 & F_1 \\ 1 & 0 & 1 & 0 & 1 & 0 & 1 & 0 \\ 1 & 0 & 0 & 1 & 0 & 1 & 1 & 0 \\ 0 & 1 & 1 & 0 & 0 & 1 & 1 & 0 \\ 0 & 1 & 0 & 1 & 1 & 0 & 1 & 0 \\ 1 & 0 & 1 & 0 & 0 & 1 & 0 & 1 \\ 1 & 0 & 0 & 1 & 1 & 0 & 0 & 1 \\ 0 & 1 & 1 & 0 & 1 & 0 & 0 & 1 \\ 0 & 1 & 0 & 1 & 0 & 1 & 0 & 1 \end{bmatrix}$$

$$\boldsymbol{X}_2^* = \begin{bmatrix} M & AB_0 & AB_1 & AC_0 & AC_1 & AF_0 & AF_1 & BC_0 & BC_1 & BF_0 & BF_1 & CF_0 & CF_1 \\ 1 & 1 & 0 & 1 & 0 & 1 & 0 & 1 & 0 & 1 & 0 & 1 & 0 \\ 1 & 0 & 1 & 0 & 1 & 1 & 0 & 1 & 0 & 0 & 1 & 0 & 1 \\ 1 & 0 & 1 & 1 & 0 & 0 & 1 & 0 & 1 & 1 & 0 & 0 & 1 \\ 1 & 1 & 0 & 0 & 1 & 0 & 1 & 0 & 1 & 0 & 1 & 1 & 0 \\ 1 & 1 & 0 & 0 & 1 & 0 & 1 & 0 & 1 & 0 & 1 & 1 & 0 \\ 1 & 0 & 1 & 1 & 0 & 0 & 1 & 0 & 1 & 1 & 0 & 0 & 1 \\ 1 & 0 & 1 & 0 & 1 & 1 & 0 & 1 & 0 & 0 & 1 & 0 & 1 \\ 1 & 1 & 0 & 1 & 0 & 1 & 0 & 1 & 0 & 1 & 0 & 1 & 0 \end{bmatrix}$$

$$\boldsymbol{X}_1'\boldsymbol{X}_1 = \begin{bmatrix} 4 & 0 & 2 & 2 & 2 & 2 & 2 & 2 \\ 0 & 4 & 2 & 2 & 2 & 2 & 2 & 2 \\ 2 & 2 & 4 & 0 & 2 & 2 & 2 & 2 \\ 2 & 2 & 0 & 4 & 2 & 2 & 2 & 2 \\ 2 & 2 & 2 & 2 & 4 & 0 & 2 & 2 \\ 2 & 2 & 2 & 2 & 0 & 4 & 2 & 2 \\ 2 & 2 & 2 & 2 & 2 & 2 & 4 & 0 \\ 2 & 2 & 2 & 2 & 2 & 2 & 0 & 4 \end{bmatrix}$$

$$= 4\boldsymbol{I}_4 \times \boldsymbol{I}_2 + 2(\mathfrak{I}_4\mathfrak{I}_4' - \boldsymbol{I}_4) \times \mathfrak{I}_2\mathfrak{I}_2'$$

$$X_2^{*\prime} X_2^* = \begin{bmatrix} 8 & 4\,\mathfrak{I}_{12}' \\ 4\mathfrak{I}_{12} & 4\,I_6 \times I_2 + 2(\mathfrak{I}_6\mathfrak{I}_6' - I_6) \times \mathfrak{I}_2\mathfrak{I}_2' \end{bmatrix}$$

$$(X_2^{*\prime} X_2^*)^- = \begin{bmatrix} \frac{1}{8} & \mathbf{0}_{12}' \\ -\frac{1}{8}\,\mathfrak{I}_{12} & \frac{1}{4}\,I_{12} \end{bmatrix}$$

$$X_1' X_2^* (X_2^{*\prime} X_2^*)^- X_2^{*\prime} X_1 = 2\,\mathfrak{I}_8 \mathfrak{I}_8'$$

and

$$C = X_1' X_1 - X_1' X_2^* (X_2^{*\prime} X_2^*)^- X_2^{*\prime} X_1 = 2\,I_4 \times \begin{pmatrix} 1 & -1 \\ -1 & 1 \end{pmatrix}.$$

It is clear then that

$$\text{rank } C = 4$$

and hence (13.37) is satisfied since $q_1 = 8$, $n = 4$. All main effects are estimable (which we knew already, of course). □

Let us now return to (13.36). Since C is not of full rank, we write a solution to **(13.36)** as

$$\widehat{E}_1 = C^- Q$$

where C^- is a generalized inverse of C. We also have for any estimable function $c' E_1$

$$\text{var}(c' \widehat{E}_1) = c' C^- c \sigma_e^2$$

The estimable functions here are, of course, those belonging to the various main effects. To illustrate this we consider again Example 13.6.

Example 13.6 (Continued) A generalized inverse for C is

$$C^- = \begin{bmatrix} \frac{1}{2} & & & & & & & \\ & 0 & & & & & & 0 \\ & & \frac{1}{2} & & & & & \\ & & & 0 & & & & \\ & & & & \frac{1}{2} & & & \\ & 0 & & & & 0 & & \\ & & & & & & \frac{1}{2} & \\ & & & & & & & 0 \end{bmatrix}$$

It is verified easily that

$$X_1' X_2^* (X_2^{*\prime} X_2^*)^- X_2^{*\prime} = \tfrac{1}{2}\, \mathfrak{I}_8 \mathfrak{I}_8'$$

and hence

$$Q = \left[X_1' - \tfrac{1}{2}\, \mathfrak{I}_8\mathfrak{I}_8'\right] y = \frac{1}{2}\begin{bmatrix} 1 & 1 & -1 & -1 & 1 & 1 & -1 & -1 \\ -1 & -1 & 1 & 1 & -1 & -1 & 1 & 1 \\ 1 & -1 & 1 & -1 & 1 & -1 & 1 & -1 \\ -1 & 1 & -1 & 1 & -1 & 1 & -1 & 1 \\ 1 & -1 & -1 & 1 & -1 & 1 & 1 & -1 \\ -1 & 1 & 1 & -1 & 1 & -1 & -1 & 1 \\ 1 & 1 & 1 & 1 & -1 & -1 & -1 & -1 \\ -1 & -1 & -1 & -1 & 1 & 1 & 1 & 1 \end{bmatrix} y$$

Finally, we obtain

$$\widehat{E}_1 = C^- Q = \frac{1}{4}\begin{bmatrix} 1 & 1 & -1 & -1 & 1 & 1 & -1 & -1 \\ 0 & 0 & 0 & 0 & 0 & 0 & 0 & 0 \\ 1 & -1 & 1 & -1 & 1 & -1 & 1 & -1 \\ 0 & 0 & 0 & 0 & 0 & 0 & 0 & 0 \\ 1 & -1 & -1 & 1 & -1 & 1 & 1 & -1 \\ 0 & 0 & 0 & 0 & 0 & 0 & 0 & 0 \\ 1 & 1 & 1 & 1 & -1 & -1 & -1 & -1 \\ 0 & 0 & 0 & 0 & 0 & 0 & 0 & 0 \end{bmatrix}\begin{bmatrix} y(0000) \\ y(0110) \\ y(1010) \\ y(1100) \\ y(0011) \\ y(0101) \\ y(1001) \\ y(1111) \end{bmatrix}$$

We thus have

$$\begin{bmatrix} \widehat{A}_0 \\ \widehat{A}_1 \\ \widehat{B}_0 \\ \widehat{B}_1 \\ \widehat{C}_0 \\ \widehat{C}_1 \\ \widehat{F}_0 \\ \widehat{F}_1 \end{bmatrix} = \begin{bmatrix} \overline{y}(0\ldots) - \overline{y}(1\ldots) \\ 0 \\ \overline{y}(.\,0\,.\,.) - \overline{y}(.\,1\,.\,.) \\ 0 \\ \overline{y}(.\,.\,0\,.) - \overline{y}(.\,.\,1\,.) \\ 0 \\ \overline{y}(\ldots 0) - \overline{y}(\ldots 1) \\ 0 \end{bmatrix}$$

and hence for the four estimable functions $A = A_1 - A_0$, $B = B_1 - B_0$, $C = C_1 - C_0$, $F = F_1 - F_0$, we obtain the estimators

$$\begin{aligned} \widehat{A} = \widehat{A}_1 - \widehat{A}_0 &= \overline{y}(1\ldots) - \overline{y}(0\ldots) \\ \widehat{B} = \widehat{B}_1 - \widehat{B}_0 &= \overline{y}(.\,1\,.\,.) - \overline{y}(.\,0\,.\,.) \\ \widehat{C} = \widehat{C}_1 - \widehat{C}_0 &= \overline{y}(.\,.\,1\,.) - \overline{y}(.\,.\,0\,.) \\ \widehat{F} = \widehat{F}_1 - \widehat{F}_0 &= \overline{y}(\ldots 1) - \overline{y}(\ldots 0) \end{aligned}$$

with

$$\text{var}(\widehat{A}) = \text{var}(\widehat{B}) = \text{var}(\widehat{C}) = \text{var}(\widehat{F}) = \tfrac{1}{2}\sigma_e^2$$

This agrees, of course, with our earlier results. □

We noted that for resolution III designs the minimum number of distinct treatment combinations, that is, the number of treatment combinations in D_0, is $1 + q_1 - n$, and designs with the number of runs can be constructed. No such general statement, simply based on the rank condition (13.37), can be made for resolution IV designs because of the presence of nuisance parameters, for example, 2-factor interactions.

It is possible, however, to give a lower bound (LB) of the number of runs needed for a resolution IV design. For a $s_1^{n_1} \times s_2^{n_2} \times \cdots \times s_q^{n_q}$ factorial with $s_1 < s_2 < \cdots < s_q$ such a bound was derived by Margolin (1969) as

$$\text{LB} = s_q \left\{ \left(\sum_{i=1}^{q} (s_i - 1) n_i \right) - (s_q - 2) \right\} \tag{13.38}$$

(see also Seely and Birkes (1984) for a comment on its validity). Specifically for symmetrical factorials, that is, s^n, we have the following lower bounds:

s	**LB**
2	$2n$
3	$3(2n-1)$
4	$4(3n-2)$
5	$5(4n-3)$

We can see that the design in Example 13.6 achieves this lower bound. But a quick look at Table 13.9 shows that only few of those designs achieve the lower bound, namely for $s = 2$, $n = 8$, $\ell = 4$ with 16 runs and $s = 4$, $n = 6$, $\ell = 3$ with 64 runs.

13.6.4 Foldover Designs

Experiments are often initiated with resolution III designs because they are relatively small in size and provide some preliminary information about possibly active factors. However, the fact that main effects are not clear may make it sometimes difficult to interpret the results from such an experiment. One way to solve or partially solve this problem is to augment the original design with another fractional replicate such that the resulting design is a resolution IV design. For the $2_{\text{III}}^{n-\ell}$ fraction this can be achieved by using the foldover principle or producing a *foldover design* (Box and Wilson, 1951).

The idea behind a foldover design simply is to replace in each treatment combination of the original fraction all 0 levels by 1 levels and all 1 levels by 0 levels. We shall illustrate this principle in the following example.

Example 13.7 Consider the 2_{III}^{5-2} fraction given by

$$I = ABCD = CDE = ABE \tag{13.39}$$

and with the design obtained by solving the equations

$$\begin{aligned} x_1 + x_2 + x_3 + x_4 \quad\quad &= 0 \\ x_3 + x_4 + x_5 &= 0 \\ x_1 + x_2 \quad\quad\quad\quad + x_5 &= 0 \end{aligned} \tag{13.40}$$

This yields the design points

(00000)	(01101)
(11110)	(10011)
(00110)	(01011)
(11000)	(10101)

The foldover design then is

(11111)	(10010)
(00001)	(01100)
(11001)	(10100)
(00111)	(01010)

It is easy to see that the treatment combinations of the foldover design satisfy the equations

$$\begin{aligned} x_1 + x_2 + x_3 + x_4 \quad\quad &= 0 \\ x_3 + x_4 + x_5 &= 1 \\ x_1 + x_2 \quad\quad\quad\quad + x_5 &= 1 \end{aligned}$$

Alternatively, we could have written (13.39) more precisely as

$$I = ABCD = -CDE = -ABE$$

and then the identity relationship for the foldover fraction becomes

$$I = ABCD = CDE = ABE$$

so that the identity relationship for the combined design is simply

$$I = ABCD$$

that is, we have obtained a 2_{IV}^{5-1} fraction. □

Within the practical context the augmentation of the original fraction by the foldover fraction results in a sequential design in that the two fractions can be thought of as occurring in two different blocks. Formally this can be expressed by adding another factor to the original design and having that factor occur at the same level in the original fraction. Applying the foldover principle to the new design yields then the same results as described above. We illustrate this briefly with a continuation of Example 13.7

Example 13.7 (Continued) Denote the new (blocking) factor by F. We then have a 2^{6-3} fraction defined by

$$I = ABCD = -CDE = -ABE = F = ABCDF = -CDEF = -ABEF$$

The foldover design is thus defined by

$$I = ABCD = CDE = ABE = -F = -ABCDF = -CDEF = -ABEF$$

The two fractions combined are then defined by

$$I = ABCD = -CDEF = -ABEF \qquad (13.41)$$

hence yielding a 2_{IV}^{6-2} fraction. Assuming no block treatment interaction, we see from (13.41) that with respect to the original factors $A, \ldots, E$ we obtain again a 2_{IV}^{5-1} design as before. □

From Example 13.7 we can easily deduce the general idea associated with the foldover principle: Starting with a resolution III design, we know that the identity relationship contains one or more words of length 3 and other words of length > 3. Or, alternatively, the defining equations [such as (13.40)] contain three or more unknowns. Using the foldover principle changes the right-hand sides of all equations with an odd number of unknowns, but does not change the right-hand sides of equations with an even number of unknowns. This implies that the identity relationship for the final design contains only interactions with an even word length, the smallest of which will be 4. The foldover principle is thus an easy way to achieve dealiasing of main effects and 2-factor interactions.

Such dealiasing can also be achieved by constructing what we might call a partial foldover design. Rather than changing the level of each factor in each treatment combination, we may choose to change only the levels of a judiciously chosen subset of factors, in particular only one factor. This may have some merit from the practical point of view.

It is, of course, clear that the foldover principle leads to a design with twice the number of design points as the original fraction. Especially for larger designs, that is, a large number of factors, this may be a disadvantage if not all main effects may need to be dealiased. Other techniques, such as optimality based techniques (see Wu and Hamada, 2000), may then be used instead by focusing on a particular model as suggested by the data from the original experiment.

13.6.5 Resolution V Designs

For resolution V designs we wish to estimate the mean, main effects, and 2-factor interactions. Hence models (13.26) and (13.27) are appropriate. For purposes of our discussion we write the model for short as

$$E[\boldsymbol{y}] = \boldsymbol{X} \cdot \boldsymbol{E} \tag{13.42}$$

where

$$\boldsymbol{X} = (\mathfrak{I}, \boldsymbol{X}_1, \boldsymbol{X}_2) \tag{13.43}$$

and

$$\boldsymbol{E} = (\boldsymbol{M}, \boldsymbol{E}_1', \boldsymbol{E}_2')'$$

with $q = 1 + q_1 + q_2$ parameters. The NE for model (13.42) is

$$(\boldsymbol{X}'\boldsymbol{X})\boldsymbol{E} = \boldsymbol{X}'\boldsymbol{y}$$

and we know that because of the parameterization used

$$\operatorname{rank}(\boldsymbol{X}'\boldsymbol{X}) < q$$

We also know that the number of degrees of freedom associated with the mean and all main effects and 2-factor interactions included in $\boldsymbol{E}$ is

$$1 + \sum_{i=1}^{n}(s_i - 1) + \sum_{i<j}(s_i - 1)(s_j - 1) = \nu(\boldsymbol{E})$$

say. For a design D to give rise to $\nu(\boldsymbol{E})$ degrees of freedom for SS($\boldsymbol{X}$), the sum of squares due to fitting the model (13.42), and hence to the estimation of the mean, all main effects and 2-factor interactions, we must have

$$\operatorname{rank}(\boldsymbol{X}'\boldsymbol{X}) = \nu(\boldsymbol{E})$$

A design D with $\nu(\boldsymbol{E})$ runs can always be achieved, a result similar to that for resolution III designs. Such a design is obviously a saturated design.

We illustrate this with the following example.

Example 13.8 Consider $n = 4$ and $s_1 = s_2 = s_3 = s_4 = 2$. Then $\nu(\boldsymbol{E}) = 11$, and it can be shown that the following (nonorthogonal) FFD allows the estimation of all main effects and 2-factor interactions:

(0000)	(1010)	(1110)
(0011)	(1001)	(1101)
(1100)	(0110)	(1011)
(1111)	(0101)	

□

13.7 FRACTIONAL FACTORIALS AND COMBINATORIAL ARRAYS

The treatment combinations of a factorial design can conveniently be represented in the form of an array or a matrix with n rows representing the factors $A_1, A_2, \ldots, A_n$ and N columns representing the treatment combinations (runs). The elements of this array represent the levels of the factors, that is, if factor A_i has s_i levels, then the elements in the ith row take on values from the set $S_i = \{0, 1, 2, \ldots, s_i - 1\}$ depending on which treatment combinations are part of the design. In this section we shall give an overview of several types of arrays and their relationship to certain types of fractional factorials.

13.7.1 Orthogonal Arrays

Let us consider an orthogonal array of size N, with n constraints, s levels, strength d, and index λ, denoted by OA$[N, n, s, d; \lambda]$ (see Appendix C). From our discussion above it follows that the rows represent the factors, the columns represent the runs, and each factor has s levels. We are considering here specifically the case $N = s^{n-\ell} = \lambda s^d$.

Rao (1947b, 1950) has shown that for $d = p + q - 1$ we can estimate from such an array main effects and interactions involving up to p factors if interactions involving $q (q > p)$ or more factors are assumed to be negligible. Of particular interest are the following cases:

d	*p*	*q*	**Resolution**
2	1	2	III
3	1	3	IV
4	2	3	V

An extension of this result to asymmetrical factorial designs was given by Chakravarti (1956). Suppose we consider the $s_1^{n_1} \times s_2^{n_2} \times \cdots \times s_g^{n_g}$ factorial setting and we wish to construct a subset of size N from the complete asymmetrical factorial such that certain effects are estimable assuming that other effects (interactions) are negligible. The solution given by Chakravarti (1956) can be stated in the following theorem.

Theorem 13.2 If orthogonal arrays OA$[N_i, n_i, s_i, d_i; \lambda_i]$ $(i = 1, 2, \ldots, g)$ exist, then the Kronecker product array of the g orthogonal arrays above yields a fractional factorial designs in

$$N = \prod_{i=1}^{g} N_i = \prod_{i=1}^{g} \lambda_i \, s_i^{d_i}$$

runs. If the ith orthogonal array is of strength $d_i = p_i + q_i - 1$ (with p_i, q_i as defined above), then in this Kronecker product array all main effects and interactions involving up to $r = \sum_{i=1}^{g} r_i$ $(0 \le r_i \le p_i;\ 0 < r \le \sum p_i)$ factors are estimable, where r_i factors are chosen from the ith group of factors at s_i levels $(i = 1, 2, \ldots, g)$.

As an illustration of this procedure we consider the following example.

Example 13.9 (Chakravarti, 1956) Suppose we have three factors A_1, A_2, A_3 with two levels each and four factors B_1, B_2, B_3, B_4 with three levels each; that is, we have the $2^3 \times 3^4$ asymmetrical factorial. There exist resolution III designs for the individual systems with 4 and 9 runs, respectively, given by OA$[2^2, 3, 2, 2;\ 1]$

	1	2	3	4
A_1	0	1	1	0
A_2	0	1	0	1
A_3	0	0	1	1

and OA$(3^2, 4, 3, 2;\ 1)$

	1	2	3	4	5	6	7	8	9
B_1	0	0	0	1	1	1	2	2	2
B_2	0	1	2	1	2	0	2	0	1
B_3	0	2	1	1	0	2	2	1	0
B_4	0	1	2	0	1	2	0	1	2

which correspond to the identity relationships

$$I = A_1 A_2 A_3$$

and

$$I = B_1 B_2 B_3 \;=\; B_2 B_3^2 B_4 \;=\; B_1 B_2^2 B_4 \;=\; B_1 B_3^2 B_4^2$$

respectively. The Kronecker product design is then obtained by combining every column of OA$[2^2, 3, 2, 2;\ 1]$ with every column of OA$[3^2, 4, 3, 2;\ 1]$ resulting in an array of $n_1 + n_2 = 7$ rows and $N_1 N_2 = 4 \cdot 9 = 36$ columns. Since $p_1 = p_2 = 1$ and $q_1 = q_2 = 2$, it follows that in addition to the main effects only 2-factor interactions involving one factor $A_i (i = 1, 2, 3)$ and one factor $B_j (j = 1, 2, 3, 4)$, each with 2 d.f., are estimable assuming that all other interactions are negligible. □

This method can produce a large number of asymmetrical fractional factorials, but, as is the case with many Kronecker product designs, the number of runs becomes quite large. For this reason we mention only a few other special cases that lead to useful designs with the same properties as the design in Example 13.9:

1. For the $2^n \cdot s$ complete factorial $(3 < n \leq 7)$ combine each level of the factor with s levels with each column of the array OA$[8, n, 2, 2;\ 2]$ to obtain a design with $8s$ runs.
2. For the $2^n \cdot s$ complete factorial $(7 < n \leq 11)$ combine each level of the factor with s levels with each column of the array OA$[12, n, 2, 2;\ 3]$ to obtain a design with $12s$ runs.

3. For the $2^{n_1} \cdot 3^{n_2}$ complete factorial with $3 < n_1 \leq 7$ and $1 < n_2 \leq 4$ obtain the Kronecker product design of OA[8, n_1, 2, 2; 2] and OA[3^2, n_2, 3, 2; 1] with 72 runs.

For a more general characterization of designs in the above classes as well as other factor combinations, see Chakravarti (1956).

13.7.2 Balanced Arrays

Just as orthogonal arrays are related to fractional factorial designs, so are balanced arrays (see Appendix C). B-arrays of strength 2, 3, 4 represent resolution III, IV, V designs, respectively. As an illustration consider the following example.

Example 13.10 BA(5, 4, 2, 2; (1, 1, 2)) is given by

$$\boldsymbol{B} = \begin{pmatrix} 0 & 0 & 1 & 1 & 1 \\ 0 & 1 & 0 & 1 & 1 \\ 0 & 1 & 1 & 0 & 1 \\ 0 & 1 & 1 & 1 & 0 \end{pmatrix}$$

This represents a main effect plan for four factors in five runs. □

However, contrary to fractional factorial designs in the form of orthogonal arrays, the designs from B-arrays are nonorthogonal, but as shown by Chakravarti (1956) and Srivastava (1965) the variance–covariance matrix for estimates of the parameters is balanced. For Example 13.10 this means that var($\widehat{A}_i$) is the same for each $i = 1, 2, 3, 4$, cov($\widehat{\mu}$, $\widehat{A}_i$) is the same for each i, and cov($\widehat{A}_i$, $\widehat{A}_j$) is the same for each pair (i, j) with $i \neq j$.

In a series of studies Srivastava and Chopra (1971) and Chopra and Srivastava (1974, 1975) have provided methods of constructing B-arrays of strength 4 and a list of resolution V designs for $4 \leq n \leq 8$ factors and a wide range of N runs.

13.8 BLOCKING IN FRACTIONAL FACTORIALS

13.8.1 General Idea

As we have discussed, the purpose of FFDs is to reduce the number of experimental runs consonant with the objectives of the experiment. That may still leave us with a relatively large number of treatment combinations such that in the practical setting some form of blocking may be required. To achieve such blocked designs we simply need to combine the ideas of systems of confounding and fractional factorials. The concept can be described in general terms as follows.

Let us consider an unreplicated $s^{n-\ell}$ fractional factorial to be used in s^m blocks. We then know the following:

1. We have $s^{n-\ell}$ treatment combinations.
2. The available treatment combinations are determined by ℓ independent interaction components in the identity relationship.

3. We can estimate $(s^{n-\ell} - 1)/(s - 1)$ functions of main effects and interaction components.
4. The estimable functions are determined by the identity relationship in the form of the alias structure.
5. We now have s^m blocks of size $2^{n-\ell-m}$.
6. We have $s^m - 1$ d.f. among blocks.
7. We need to confound $(s^m - 1)/(s - 1)$ interaction components with blocks.
8. From the available estimable functions (as given by the alias structure) we need to select m functions to confound with blocks.
9. Then all their generalized interactions (and aliases) are also confounded with blocks.

We shall illustrate these steps and some other ideas in the following sections with particular reference to the 2^n and 3^n systems.

13.8.2 Blocking in $2^{n-\ell}$ Designs

Consider a $2^{n-\ell}$ FFD defined by the identity relationship (see Section 13.3.1)

$$I = E^{\alpha_1} = E^{\alpha_2} = \cdots = E^{\alpha_\ell} = E^{\alpha_1+\alpha_2} = \cdots = E^{\alpha_1+\alpha_2+\ldots+\alpha_\ell} \tag{13.44}$$

where $E^{\alpha_1}, E^{\alpha_2}, \ldots, E^{\alpha_\ell}$ are the independent interactions. We shall denote the $2^{n-\ell} - 1 = q$ estimable effects by $E^{\beta_1}, E^{\beta_2}, \ldots, E^{\beta_q}$. The alias structure is obtained by multiplying each $E^{\beta_i} (i = 1, 2, \ldots, q)$ into (13.44).

We now consider placing the treatment combinations specified by (13.44) into 2^m blocks of size $2^{n-\ell-m}$. To accomplish this we need to select m effects from among the $E^{\beta_i} (i = 1, 2, \ldots, q)$ to confound with blocks, say, without loss of generality, $E^{\beta_1}, E^{\beta_2}, \ldots, E^{\beta_m}$. We know that then all their generalized interactions are also confounded with blocks, giving rise to the required 2^m blocks.

To generate the intrablock subgroup (IBS) we combine the ℓ equations for generating the fraction specified by (13.44) and the m equations for the specified system of confounding, that is,

$$\begin{aligned}
\alpha_{11}x_1 + \alpha_{12}x_2 + \cdots + \alpha_{1n}x_n &= 0 \\
\alpha_{21}x_1 + \alpha_{22}x_2 + \cdots + \alpha_{2n}x_n &= 0 \\
&\vdots \\
\alpha_{\ell 1}x_1 + \alpha_{\ell 2}x_2 + \cdots + \alpha_{\ell n}x_n &= 0 \\
\beta_{11}x_1 + \beta_{12}x_2 + \cdots + \beta_{1n}x_n &= 0 \\
\beta_{21}x_1 + \beta_{22}x_2 + \cdots + \beta_{2n}x_n &= 0 \\
&\vdots \\
\beta_{m1}x_1 + \beta_{m2}x_2 + \cdots + \beta_{mn}x_n &= 0
\end{aligned} \tag{13.45}$$

Equations (13.45) show that the IBS constitutes a $2^{n-(\ell+m)}$ fraction, with the remaining blocks representing the other fractions obtained by changing successively the right-hand sides of the second set of equations in (13.45).

In the practical context the interactions $E^{\boldsymbol{\beta}_i} (i = 1, 2, \ldots, m)$ would have to be chosen carefully so as to avoid the confounding of "important" effects with blocks. We shall illustrate the preceding procedure with the following example.

Example 13.11 Consider the 2^{7-3} design in four blocks. We start with the resolution IV design defined by the identity relationship

$$I = \underline{ABCD} = \underline{CDEF} = ABEF = \underline{BCEG} = ADEG = BDFG = ACFG$$

where the three independent interactions are underlined. The 15 estimable functions (d.f.) are accounted for by seven main effects (+ 3-factor and higher-order interactions), seven 2-factor interactions (+ higher-order interactions), and one 3-factor interaction (+ higher-order interactions). In order to set up blocks of size 4 we need to confound three 2-factor interactions (and their aliases), choosing, say, AB and AC as the two independent interactions. Thus, to generate the design we chose in (13.45)

$$\boldsymbol{\alpha}_1' = (1\,1\,1\,1\,0\,0\,0) \qquad \boldsymbol{\alpha}_2' = (0\,0\,1\,1\,1\,1\,0) \qquad \boldsymbol{\alpha}_3' = (0\,1\,1\,0\,1\,0\,1)$$

$$\boldsymbol{\beta}_1' = (1\,1\,0\,0\,0\,0\,0) \qquad \boldsymbol{\beta}_2' = (1\,0\,1\,0\,0\,0\,0)$$

An isomorphic design using SAS PROC FACTEX is given in Table 13.10. □

13.8.3 Optimal Blocking

As we have mentioned earlier, there may be various ways to set up the blocks for a particular fractional factorial, for example, by choosing different $\boldsymbol{\beta}_j$ to use in (13.45). Such choices for the $2^{n-\ell}$ fraction in 2^m blocks may be made by first inspecting the alias structure for the $2^{n-\ell}$ and then selecting m appropriate interactions to generate the system of confounding. We obviously need to be careful in this selection so that some of the resulting $2^m - m - 1$ generalized interactions do not end up being desirable effects.

A more formal approach to this problem makes use of the fact that blocking in a $2^{n-\ell}$ can be considered as a special case of fractionation (Kempthorne, 1952; Lorenzen and Wincek, 1992) by introducing additional factors as blocking factors. More specifically, a $2^{n-\ell}$ fractional factorial in 2^m blocks of size $2^{n-\ell-m} (m < n - \ell)$ can be considered formally as a $2^{(n+m)-(\ell+m)}$ (Chen and Cheng, 1999). The factors are divided into n treatment factors $A_1, A_2, \ldots, A_n$ and m blocking factors $b_1, b_2, \ldots, b_m$. The 2^m "level" combinations of the blocking factors are used to divide the $2^{n-\ell}$ treatment combinations into 2^m blocks. By treating this as a special case of a fractional factorial, the identity relationship will contain two types of words: treatment-defining words are those that contain only treatment

Table 13.10 2^{7-3} Design in 4 Blocks

```
options nodate pageno=1;
proc factex;
factors A B C D E F G;
size design=16;
model res=4;
blocks size=4;
examine design aliasing(3) confounding;
output out=design
    blockname=block nvals=(1 2 3 4);
title1 'TABLE 13.10';
title2 '2**(7-3) DESIGN IN 4 BLOCKS';
run;

proc print data=design;
run;
```

The FACTEX Procedure

Design Points

Experiment Number	A	B	C	D	E	F	G	Block
1	-1	-1	-1	-1	-1	-1	-1	4
2	-1	-1	-1	1	1	1	1	1
3	-1	-1	1	-1	1	1	-1	1
4	-1	-1	1	1	-1	-1	1	4
5	-1	1	-1	-1	1	-1	1	3
6	-1	1	-1	1	-1	1	-1	2
7	-1	1	1	-1	-1	1	1	2
8	-1	1	1	1	1	-1	-1	3
9	1	-1	-1	-1	-1	1	1	3
10	1	-1	-1	1	1	-1	-1	2
11	1	-1	1	-1	1	-1	1	2
12	1	-1	1	1	-1	1	-1	3
13	1	1	-1	-1	1	1	-1	4
14	1	1	-1	1	-1	-1	1	1
15	1	1	1	-1	-1	-1	-1	1
16	1	1	1	1	1	1	1	4

Factor Confounding Rules

E = B*C*D
F = A*C*D
G = A*B*D

Table 13.10 (*Continued*)

```
               Block Pseudofactor Confounding Rules

                         [B1]  = A*B*C*D
                         [B2]  = C*D

Aliasing Structure

      A = B*D*G = B*E*F = C*D*F = C*E*G
      B = A*D*G = A*E*F = C*D*E = C*F*G
      C = A*D*F = A*E*G = B*D*E = B*F*G
      D = A*B*G = A*C*F = B*C*E = E*F*G
      E = A*B*F = A*C*G = B*C*D = D*F*G
      F = A*B*E = A*C*D = B*C*G = D*E*G
      G = A*B*D = A*C*E = B*C*F = D*E*F
[B] = A*B = D*G = E*F
      A*C = D*F = E*G
      A*D = B*G = C*F
[B] = A*E = B*F = C*G
[B] = A*F = B*E = C*D
      A*G = B*D = C*E
      B*C = D*E = F*G
      A*B*C = A*D*E = A*F*G = B*D*F = B*E*G = C*D*G = C*E*F
```

Obs	block	A	B	C	D	E	F	G
1	1	-1	-1	-1	1	1	1	1
2	1	-1	-1	1	-1	1	1	-1
3	1	1	1	-1	1	-1	-1	1
4	1	1	1	1	-1	-1	-1	-1
5	2	-1	1	-1	1	-1	1	-1
6	2	-1	1	1	-1	-1	1	1
7	2	1	-1	-1	1	1	-1	-1
8	2	1	-1	1	-1	1	-1	1
9	3	-1	1	-1	-1	1	-1	1
10	3	-1	1	1	1	1	-1	-1
11	3	1	-1	-1	-1	-1	1	1
12	3	1	-1	1	1	-1	1	-1
13	4	-1	-1	-1	-1	-1	-1	-1
14	4	-1	-1	1	1	-1	-1	1
15	4	1	1	-1	-1	1	1	-1
16	4	1	1	1	1	1	1	1

factors, and block-defining words are those that contain at least one blocking factor (Chen and Cheng, 1999). We shall illustrate this in the following example.

Example 13.12 Consider the 2^{7-3} in 2^2 blocks as a $2^{(7+2)-(3+2)}$ fraction. We need three independent treatment-defining words, say $A_2A_3A_4A_5$, $A_1A_3A_4A_6$, and $A_1A_2A_4A_7$, and two independent block-defining words, say $A_1A_2A_3A_4b_1$ and $A_3A_4b_2$. To obtain the desired design we thus need to solve the following equations:

$$
\begin{aligned}
x_2 + x_3 + x_4 + x_5 \qquad\qquad\qquad &= 0 \\
x_1 \qquad + x_3 + x_4 \qquad + x_6 \qquad\qquad &= 0 \\
x_1 + x_2 \qquad + x_4 \qquad\qquad + x_7 \qquad &= 0 \\
x_1 + x_2 + x_3 + x_4 \qquad\qquad\qquad + y_1 &= 0 \\
x_3 + x_4 \qquad\qquad\qquad\qquad + y_2 &= 0
\end{aligned}
\tag{13.46}
$$

where in (13.46) $x_1, x_2, \ldots, x_7$ refer to the treatment factors and y_1, y_2 refer to the block factors, with $x_i = 0, 1, (i = 1, 2, \ldots, 7)$, $y_j = 0, 1 (j = 1, 2)$. The solution of Eqs. (13.46) leads to the design given in Table 13.10 if we use the following association between block numbers and block factor level combinations:

Block No.	y_1	y_2
1	1	1
2	0	1
3	1	0
4	0	0

□

Inspection of the identity relationship in Example 13.12 illustrates some of the difficulties encountered in extending the notions of resolution and minimum aberration to situations involving block-defining words. The presence of a three-letter word, for example, $A_3A_4b_2$, would suggest that the design is a resolution III design, when in fact it is clearly a resolution IV design with regard to the treatment factors. Similar problems have been addressed and solutions proposed by, for example, Bisgaard (1994), Sitter, Chen, and Feder (1997), and Sun, Wu, and Chen (1997), but as Chen and Cheng (1999) point out, none of these solutions are satisfactory with regard to properly characterizing blocked designs and identifying optimal blocking schemes.

The problem is that there are two different word length patterns, one for treatment-defining words and one for block-defining words. Suppose we consider only designs where no main effects are confounded either with other main effects or block effects. Following Sun, Wu, and Chen (1997), let $L_{i,0}$ be the number of treatment-defining words containing i treatment letters, and let $L_{i,1}$ be the number of block-defining words also containing i treatment letters and one or more block letters. We then have, for a given design, two types of word length patterns, one for treatment-defining words

$$\text{WLP}_t = (L_{3,0}, L_{4,0}, \ldots) \tag{13.47}$$

and one for block-defining words

$$\text{WLP}_{bt} = (L_{2,1}, L_{3,1}, \ldots) \tag{13.48}$$

Clearly, (13.47) and (13.48) have different meanings and are of different importance. An intuitive method to proceed would be to search for a design that has minimum aberration (see Section 13.3.4) with respect to both treatment and block factors. Zhang and Park (2000), however, have shown that no such design exists. Instead, in order to extend the concept of minimum aberration we somehow may need to combine (13.47) and (13.48). To this end Chen and Cheng (1999) propose the following ordering, based on the hierarchical assumption that lower-order effects are more important than higher-order effects and that effects of the same order are of equal importance:

$$(L_{3,0}, L_{2,1}), L_{4,0}, (L_{5,0}, L_{3,1}), L_{6,0}, (L_{7,0}, L_{4,1})\, L_{8,0}, \ldots \tag{13.49}$$

This means, for example, that if $L_{3,0} > 0$ some 2-factor interactions are aliased with main effects, and if $L_{2,1} > 0$ then some 2-factor interactions are confounded with blocks. In either case those 2-factor interactions cannot be estimated. Further, designs with $L_{3,0} > 0$ and/or $L_{2,1} > 0$ are clearly less desirable than designs with $L_{3,0} = L_{2,1} = 0$ and $L_{4,0} > 0$.

Based on such considerations, the criterion of minimum aberration for blocked fractional factorial designs might be defined by sequentially minimizing $L_{3,0}, L_{2,1}, L_{4,0}, L_{5,0}, L_{3,1}, L_{6,0}, L_{7,0}, L_{4,1}, L_{8,0}, \ldots$. But Chen and Cheng (1999) show that even that is not quite satisfactory. Based on the notion of estimation capacity [for details see Cheng, Steinberg, and Sun, (1999)], they propose to minimize sequentially what they call the blocking word length pattern:

$$\text{WLP}_b = \left(L_3^b, L_4^b, \ldots, L_{n+[n/2]}^b\right) \tag{13.50}$$

where

$$L_j^b = \begin{cases} L_{j,0} & \text{for even } j \le n \\ \binom{j}{(j+1)/2} L_{j,0} + L_{(j+1)/2,1} & \text{for odd } j \le n \\ L_{j-[n/2],1} & \text{for } n+1 \le j \le n+[n/2] \end{cases} \tag{13.51}$$

and $[n/2]$ is the largest integer less than or equal to $n/2$. As can be seen from (13.51) the word length for odd $j \le n$ is a linear combination of the two associated word lengths enclosed in parentheses in (13.49). For example, $L_3^b = 3L_{3,0} + L_{2,1}$ and $L_5^b = 10\, L_{5,0} + L_{3,1}$.

For detailed methods of identifying minimum aberration blocked fractional factorial designs we refer the reader to Chen and Cheng (1999). In Tables 13.11, 13.12, and 13.13 we give, in a different form, some of their results for 8-, 16-, 32-run minimum aberration $2^{n-\ell}$ designs in 2^m blocks for $n = 4, 5, \ldots, 10$;

Table 13.11 8-Run Minimum Aberration $2^{n-\ell}$ Designs

			Independent Defining Words		
n	ℓ	m	Treatments	Blocks	WLP_b[a]
4	1	1	$ABCD$	ABb_1	2 1 0
4	1	2	$ABCD$	ABb_1, ACb_2	6 1 0
5	2	1	ABD, ACE	BCb_1	8 1 2
6	3	1	ABD, ACE, BCF	$ABCb_1$	15 3 4

[a]Giving only L_3^b, L_4^b, L_5^b.

Table 13.12 16-Run Minimum Aberration $2^{n-\ell}$ Designs

			Independent Defining Words		
n	ℓ	m	Treatments	Blocks	WLP_b[a]
5	1	1	$ABCE$	$ABDb_1$	0 1 2
5	1	2	$ABCE$	$ABb_1, ACDb_2$	2 1 4
5	1	3	$ABCE$	ABb_1, ACb_2, ADb_3	10 1 0
6	2	1	$ABCE, ABDF$	$ACDb_1$	0 3 4
6	2	2	$ABCE, ABDF$	$ABb_1, ACDb_2$	3 3 8
6	2	3	$ABCE, ABDF$	ABb_1, ACb_2, ADb_3	15 3 0
7	3	1	$ABCE, ABDF, ACDG$	$BCDb_1$	0 7 7
7	3	2	$ABCE, ABDF, ACDG$	ABb_1, ACb_2	9 7 0
7	3	3	$ABCE, ABDF, ACDG$	ABb_1, ACb_2, ADb_3	21 7 0
8	4	1	$ABCE, ABDF, ACDG, BCDH$	ABb_1	4 14 0
8	4	2	$ABCE, ABDF, ACDG, BCDH$	ABb_1, ACb_2	12 14 0
8	4	3	$ABCE, ABDF, ACDG, BCDH$	ABb_1, ACb_2, ADb_3	28 14 0
9	5	1	$ABE, ACF, ADG, BCDH, ABCDJ$	BCb_1,	16 14 84
9	5	2	$ABE, ACF, ADG, BCDH, ABCDJ$	BCb_1, BDb_2	24 14 92
10	6	1	$ABE, ACF, BCG, ADH, BCDJ, ABCDK$	BDb_1	28 18 96
10	6	2	$ABE, ACF, BCG, ADH, BCDJ, ABCDK$	$ABCb_1, BDb_2$	37 18 184

[a]Giving only L_3^b, L_4^b, L_5^b.

Table 13.13 32-Run Minimum Aberration $2^{n-\ell}$ Designs

			Independent Defining Words		
n	ℓ	m	Treatments	Blocks	WLP_b[a]
6	1	1	*ABCDEF*	$ABCb_1$	0 0 2
6	1	2	*ABCF*	$ABDb_1, ACEb_2$	0 1 4
6	1	3	*ABCDEF*	$ABb_1, CDb_2, ACEb_3$	3 0 8
6	1	4	*ABCDEF*	$ABb_1, ACb_2,$ ADb_3, AEb_4	15 0 0
7	2	1	*ABCF*, *ABDEG*	$ACDb_1$	0 1 22
7	2	2	*ABCF*, *ABDG*	$ACDb_1, ABEb_2$	0 3 7
7	2	3	*ABCF*, *ABDEG*	$ACb_1, ABDb_2,$ $ABEb_3$	5 1 32
7	2	4	*ABCF*, *ADEG*	ABb_1, ACb_2 ADb_3, AEb_4	21 2 0
8	3	1	*ABCF*, *ABDG*, *ACDEH*	$ABEb_1$	0 3 43
8	3	2	*ABCF*, *ABDG*, *ACDEH*	$ABEb_1, BCDEb_2$	1 3 50
8	3	3	*ABCF*, *ABDG*, *ACEH*	$ABb_1, ACDb_2,$ AEb_3	7 5 18
8	3	4	*ABCF*, *ABDG*, *ACEH*	$ABb_1, ACb_2,$ ADb_3, AEb_4	28 5 0
9	4	1	*ABCF*, *ABDG*, *ABEH*, *ACDEJ*	$BCDEb_1$	0 6 84
9	4	2	*ABCF*, *ABDG*, *ACEH*, *ADEJ*	$BCb_1, BDEb_2$	2 9 14
9	4	3	*ABCF*, *ABDG*, *ACEH*, *ADEJ*	$ABb_1, ACDb_2,$ AEb_3	9 9 27
9	4	4	*ABCF*, *ABDG*, *ACEH*, *ADEJ*	$ABb_1, ACb_2,$ DEb_3, AEb_4	36 9 0
10	5	1	*ABCF*, *ABDG*, *ACEH*, *ADEJ*, *ABCDEK*	$ACDb_1$	0 15 10
10	5	2	*ABCF*, *ABDG*, *ACEH*, *ADEJ*, *ABCDEK*	$ABb_1, ACDb_2$	3 15 20
10	5	3	*ABCF*, *ABDG*, *ACDH*, *ABEJ*, *ACEK*	$ABb_1, ACb_2,$ $ADEb_3$	12 16 36
10	5	4	*ABCF*, *ABDG*, *ACEH*, *ADEJ*, *ABCDEK*	ABb_1, ACb_2 ADb_3, AEb_4	45 15 0

[a]Giving only L_3^b, L_4^b, L_5^b.

$\ell = 1, 2 \ldots, 6$; $m = 1, 2, 3, 4$. For possible combinations of n, ℓ, and m we give the independent words for the identity relationship from which the design can be obtained by solving a set of equations as illustrated in Example 13.12. We also list the first three components, L_3^b, L_4^b, and L_5^b, of each associated WLP$_b$. In Tables 13.11, 13.12, and 13.13 we denote the treatment factors by A, B, C, D, E, F, G, H, J, K and the blocking factors by b_1, b_2, b_3, b_4.

13.9 ANALYSIS OF UNREPLICATED FACTORIALS

One reason for using fractional factorial designs is to reduce the number of runs in the experiments. But even a highly fractional design for a large number of factors may involve quite a large number of runs. In that case we may be able to afford only a few or no replications of the treatment combinations included in the design. As we know replications are important for estimating the variance of the estimable effects and interactions.

For unreplicated 2^n factorials we have described how negligible interactions can be used to estimate σ_e^2 (see Section 7.6.6). This method is, of course, applicable also for the general s^n factorial and mixed factorials. The problem becomes more complicated for fractional factorials where often only main effects and low-order interactions are estimable. In this case we have to fall back on the assumption of effect sparsity, which says that typically only a few effects and/or interactions are real. We shall mention here briefly the essential ideas of two methods that are helpful with the analysis of data and the interpretation of the results from unreplicated fractional factorials.

13.9.1 Half-Normal Plots

Let us consider a $2^{n-\ell}$ fraction and suppose that the effects and interactions E^{α_1}, $E^{\alpha_2}, \ldots, E^{\alpha_m} (m = 2^{n-\ell} - 1)$ are estimable (these are, of course, aliased with other interactions which we assume to be negligible). Noting that $\mid \widehat{E^{\alpha_i}} \mid = \mid \widehat{E_1^{\alpha_i}} - \widehat{E_0^{\alpha_i}} \mid$ can be viewed as the range of a pair of (approximately) normally distributed random variables suggests that the $\mid \widehat{E^{\alpha_i}} \mid$ are themselves half-normally distributed, that is, follow the distribution function

$$\begin{aligned} f(x) &= [2/(\pi\,\sigma_e^2)]^{1/2} \exp[-x^2/(2\,\sigma_e^2)] \qquad &\text{for } x \geq 0 \\ &= 0 &\text{for } x < 0 \end{aligned}$$

Its cdf can be linearized by using the upper half of normal probability paper, relabeling the P axis for $P \geq .5$ as $P' = 2P - 1$. Based on this idea Daniel (1959) proposed a procedure that has come to be known as *half-normal plot*.

It consists of plotting the absolute values of the m estimates, approximating their P' values by $P' = (i - \frac{1}{2})/m$ for $i = 1, 2, \ldots, m$. If, as assumed, most of the estimates are, in fact, estimating zero, then they should fall near a straight line through the origin. The 68% point (obtained from $P' = 2 \times .84 - 1 = .68$)

would provide an approximate value of the standard error, say SE_0, of the effects plotted, that is, an estimate of σ_e^2 since $SE(\widehat{E}^{\alpha_i}) = \widehat{\sigma}_e/(2^{n-\ell-2})^{1/2}$. Furthermore, to the extent that some of the largest estimates deviate from the straight line is an indication that the corresponding effects are different from zero.

The above describes in essence the procedure, but we need to elaborate on two points: (1) How do we judge an estimate to be significantly different from zero? and (2) should we reestimate σ_e^2 in light of some significant estimates?

1. The significance of effects is judged by whether they fall outside certain "guardrails" (Daniel, 1959). The α-level guardrails are constructed by drawing a line through the α-level critical values for the largest standardized absolute effects, say $t_h (h = m, m-1, \ldots, m-K)$, standardized by the initial estimate of σ_e. Zahn (1975) provided the critical values for $m = 15$, 31, 63, 127 and $\alpha = .05, .20, .40$. In Table 13.14 we give the critical values for $m = 15, 31$ and $\alpha = .05, .20, .40$ for some t_h.
2. Based upon the test performed by using the procedure in point 1 above, we exclude not only the significant effects at $\alpha = .05$ but also "borderline" effects at $\alpha = .20$ or even $\alpha = .40$, say there are K_α such effects altogether, and replot the $m - K_\alpha$ remaining effects. The value at $P' = .68$ provides then the final value for $SE(\widehat{E}^{\alpha_i})$, say SE_f, and from it we obtain the final estimate of σ_e [for a more refined explanation see Zahn (1975)].

The half-normal plot procedure is illustrated in the following example.

Example 13.13 The following data are taken from Example 10.4 in Davies (1956), which describes a 2^{5-1} factorial investigating the quality of a basic dyestuff. The factors are A (temperature), B (quality of starting material), C

Table 13.14 Guardrail Critical Values

$m = 15$:	h	$\alpha = .05$	$\alpha = .20$	$\alpha = .40$
	15	3.230	2.470	2.066
	14	2.840	2.177	1.827
	13	2.427	1.866	1.574
	12	2.065	1.533	1.298
$m = 31$:	h	$\alpha = .05$	$\alpha = .20$	$\alpha = .40$
	31	3.351	2.730	2.372
	30	3.173	2.586	2.247
	29	2.992	2.439	2.121
	28	2.807	2.288	1.991
	27	2.615	2.133	1.857

(reduction pressure), D (pressure), and E (vacuum leak). The $\frac{1}{2}$-fraction is based on $I = -ABCDE$. The treatment combinations and observations Y are given in Table 13.15.

Pertinent computed values to draw the half-normal plot are given in Table 13.16, and the half-normal plot is given in Figure 13.1.

From Table 13.16 we see that $\text{SE}_0 = \widehat{BC} = 8.31$, the 11th-order statistic, is closest to the .68 percentile and hence is the initial estimate of $\text{SE}(\widehat{E}^{\alpha_i})$. The standardized effects are then obtained as $t_i = | \widehat{E}^{\alpha_i} | /8.31$. The guardrails for $\alpha = .05$, .20, and .40 are constructed by drawing lines through the corresponding

Table 13.15 2^{5-1} Fractional Factorial

Treatment	Y	Treatment	Y
00000	201.5	00011	255.5
10001	178.0	10010	240.5
01001	183.5	01010	208.5
11000	176.0	11011	244.0
00101	188.5	00110	274.0
10100	178.5	10111	257.5
01100	174.5	01111	256.0
11101	196.5	11110	274.5

Table 13.16 Calculations for Half-Normal Plot of Figure 13.1

Effect	$\widehat{E}^{\alpha_i}$	Order(i)	$(i - .5)/15$	t_i
DE	− .0625	1	.0333	.00752
A	.4375	2	.1000	.05263
AE	−2.3125	3	.1667	.27820
AC	3.0625	4	.2333	.36842
BD	−3.5625	5	.3000	.42857
E	3.9375	6	.3667	.47368
CE	−4.6875	7	.4333	.56391
AD	5.1875	8	.5000	.62406
B	−7.5625	9	.5667	.90977
BE	7.6875	10	.6333	.92481
BC	8.3125	11	.7000	1.00000
C	14.0625	12	.7667	1.69173
CD	14.3125	13	.8333	1.72180
AB	16.6875	14	.9000	2.00752
D	66.6875	15	.9667	8.02256

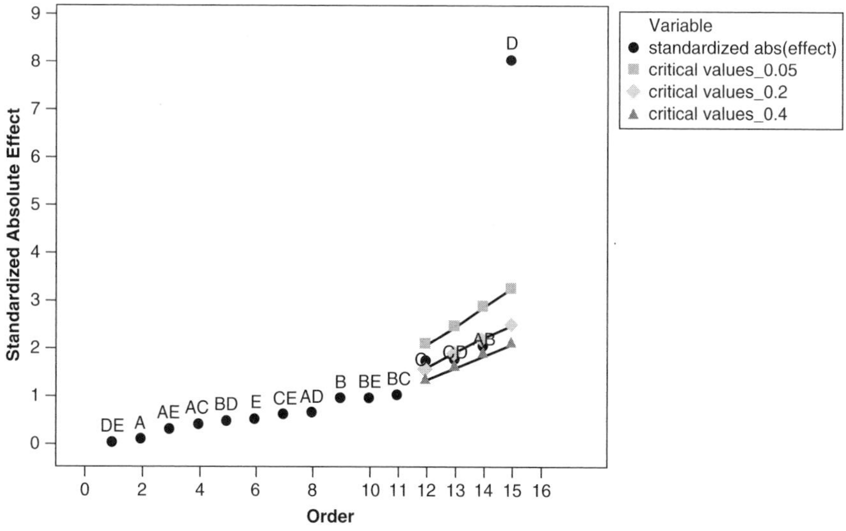

Figure 13.1 Half-normal plot with guardrails.

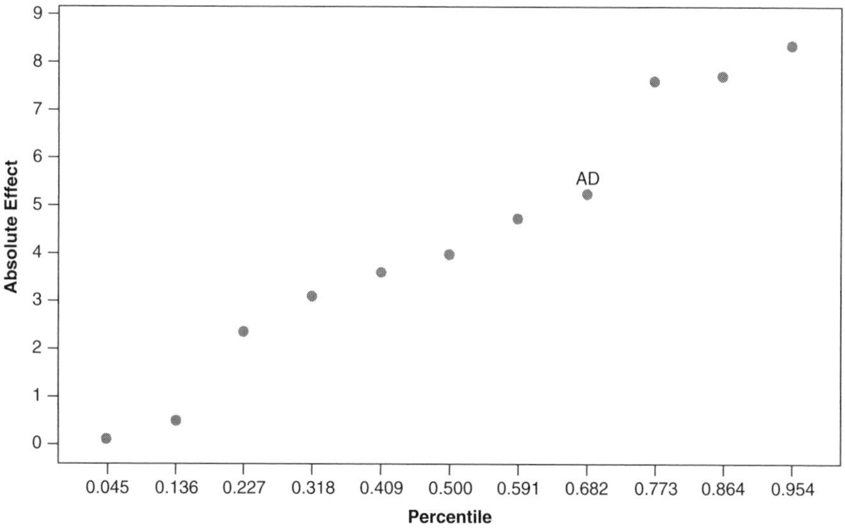

Figure 13.2 Half-normal plot for nonsignificant effects.

critical values for t_{15}, t_{14}, t_{13}, and t_{12}. From Figure 13.1 we see that for $\alpha = .05$ only D is significant, for $\alpha = .20$ D and C are significant, and for $\alpha = .40$ D, C, CD, and AB are significant. Using the latter result, we obtain the final estimate of σ_e from the .68 percentile of the half-normal plot for the 11 smallest effects in Figure 13.2. From it we see that $\widehat{AD} = 5.188 = \text{SE}(\widehat{E}^{\alpha_i}) = \widehat{\sigma}_e/(2^{5-1-2})^{1/2} = \text{SE}_f$, and hence $\widehat{\sigma}_e = 2 \times 5.188 = 10.376$. □

13.9.2 Bar Chart

This procedure, proposed by Lenth (1989), is based on a simple formula for the standard error of the estimated effects. The effects are plotted and the usual t procedures are used to judge significance and interpret the results.

Let

$$s_0 = 1.5 \times \text{median} \ | \ \widehat{E}^{\alpha_i} \ |$$

and define the *pseudostandard error* (PSE) of the estimated effect by

$$\text{PSE} = 1.5 \times r\text{-median} \ | \ \widehat{E}^{\alpha_i} \ | \tag{13.52}$$

where r-median is the restricted median over all $| \ \widehat{E}^{\alpha_i} \ | < 2.5 s_0$. PSE (corresponding to SE_f in Section 13.9.1) in (13.52) is an estimate of $\sigma_e / (2^{n-\ell-2})^{1/2}$. It can be used to define two useful quantities that in turn can be used to assess the significance of effects.

Let

$$\text{ME} = t_{.975,d} \times \text{PSE} \tag{13.53}$$

denote the *margin of error* for $\widehat{E}^{\alpha_i}$, where $t_{.975,d}$ is the .975th quantile of the t distribution with $d = m/3$ d.f. ($m = 2^{n-\ell} - 1$). An approximate 95% confidence interval for E^{α_i} is then given by $\widehat{E}^{\alpha_i} \pm \text{ME}$. A *simultaneous margin of error* (SME) is defined:

$$\text{SME} = t_{\gamma,d} \times \text{PSE} \tag{13.54}$$

where $\gamma = (1 + .95^{1/m})/2$. For some values of m we have

m	$t_{.975,d}$	$t_{\gamma,d}$
7	3.76	9.01
15	2.57	5.22
31	2.22	4.22
63	2.08	3.91

The procedure proposed by Lenth (1989) then is as follows:

1. Construct a bar chart of the signed effects.
2. Add reference lines at $\pm$ ME [from (13.53)] and $\pm$ SME [from (13.54)].
3. Judge an effect as clearly active if $\widehat{E}^{\alpha_i}$ extends beyond the SME lines, inactive if $\widehat{E}^{\alpha_i}$ does not extend beyond the ME lines, and of debatable significance if $\widehat{E}^{\alpha_i}$ falls between the ME and SME lines.

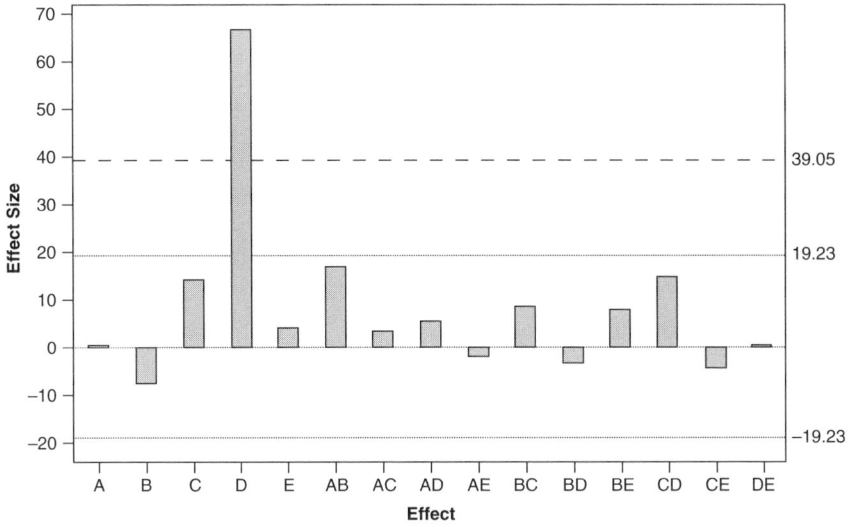

Figure 13.3 Bar chart for Example 13.14.

We illustrate this procedure in the following example.

Example 13.14 Consider the data from Example 13.13. We obtain $s_0 = 1.5 \times 5.1875 = 7.78$, PSE $= 1.5 \times 4.9875 = 7.48$, ME $= 19.23$, SME $= 39.05$. The bar chart is given in Figure 13.3. From it we see that only D is significant at $\alpha = .05$. This agrees with the result in Example 13.13. □

13.9.3 Extension to Nonorthogonal Design

The previous analysis procedures are applicable for orthogonal designs. There are, however, situations where we may encounter nonorthogonal designs, either by construction (see Section 14.4) or as a consequence of some mishap in the experiment. For example, we may have some replicates for one or a few treatment combinations. Or we may have missing values. Both cases lead to nonorthogonal designs, complicating the analysis and interpretation of results.

A data-driven method for nonorthogonal designs was proposed by Kunert (1997), extending the method of Daniel (1959) as modified by Zahn (1975) and the method of Lenth (1989). For details we refer the reader to the original article by Kunert (1997).

CHAPTER 14

Main Effect Plans

14.1 INTRODUCTION

Main effect plans (MEP) or resolution III designs are commonly used in exploratory studies when a large number of factors need to be considered or screened. The idea is to try to detect factors that exhibit "large" main effects and discard factors with no noticeable effects from further study. The crucial assumption here is that all interactions are negligible, including 2-factor interactions. This may not always be realistic, but as a first approximation this is nevertheless a very valuable method. In some cases it may be helpful to know with which 2-factor interactions each main effect is aliased. We shall make some remarks about that later.

14.2 ORTHOGONAL RESOLUTION III DESIGNS FOR SYMMETRICAL FACTORIALS

Among the resolution III designs those that admit uncorrelated estimates of main effects are of major importance. Not only are they easy to construct and analyze, but the results from such experiments are also easy to interpret. Because of their special feature, these designs are referred to as orthogonal resolution III designs or orthogonal main effect plans (OMEP). We shall now describe some methods of constructing OMEPs.

14.2.1 Fisher Plans

In discussing systems of confounding (see Section 11.4) for an s^n factorial in blocks of size s^ℓ with $n = (s^\ell - 1)/(s - 1)$ we have already provided a method of constructing a $1/s^{n-\ell}$ fraction of an s^n factorial. Such a FFD is a saturated

Design and Analysis of Experiments. Volume 2: Advanced Experimental Design
By Klaus Hinkelmann and Oscar Kempthorne
ISBN 0-471-55177-5

OMEP, designated by $s_{\text{III}}^{n-(n-\ell)}$, indicating that this is a $1/s^{n-\ell}$ fraction of an s^n factorial resulting in $s^{n-(n-\ell)} = s^\ell$ treatment combinations (runs, trials) and is of resolution III. Here $s = p^m$ and p is a prime. The method of construction is based on results given by Fisher (1942, 1945) on how to obtain the IBSG for an s^n factorial in blocks of size s^ℓ. The IBSG is then the OMEP. The method of construction guarantees that each level of each factor occurs the same number of times, namely $s^{\ell-1}$, and each level combination for any two factors occurs the same number of times, namely $s^{\ell-2}$ (see Section 11.3). The latter is a sufficient condition for the estimates of the main effects to be uncorrelated.

To see this, consider model (13.28) with $\boldsymbol{X}_1$ having the following properties:

1. Each column has $s^{\ell-1}$ unity elements, the remaining $s^{\ell-1}(s-1)$ elements are zero.
2. The inner product of any two columns belonging to the same factor [i.e., columns belonging to $(A_i)_\delta$ and $(A_i)_\gamma$, $\delta,\ \gamma = 0, 1, \ldots, s-1$; $\delta \neq \gamma$] is equal to zero.
3. The inner product of any two columns belonging to different factors [i.e., columns belonging to $(A_i)_\delta$ and $(A_j)_\gamma$, $i,\ j = 1, 2, \ldots, n$; $i \neq j$] is equal to $s^{\ell-2}$.

Considering (13.28) as a two-part linear model and using the result of Section I.4.8 we can write the RNE for $\boldsymbol{E}_1$ as

$$\boldsymbol{X}_1'\left(\boldsymbol{I} - \frac{1}{s^\ell}\mathfrak{I}\mathfrak{I}'\right)\boldsymbol{X}_1\boldsymbol{E}_1 = \boldsymbol{X}_1'\left(\boldsymbol{I} - \frac{1}{s^\ell}\mathfrak{I}\mathfrak{I}'\right)\boldsymbol{y} \tag{14.1}$$

The coefficient matrix in (14.1) can be written alternatively as

$$\left[\boldsymbol{X}_1'\left(\boldsymbol{I} - \frac{1}{s^\ell}\mathfrak{I}\mathfrak{I}'\right)\right]\left[\left(\boldsymbol{I} - \frac{1}{s^\ell}\mathfrak{I}\mathfrak{I}'\right)\boldsymbol{X}_1\right] = \widetilde{\boldsymbol{X}}_1'\widetilde{\boldsymbol{X}}_1 \tag{14.2}$$

with

$$\widetilde{\boldsymbol{X}}_1 = \left(\boldsymbol{I} - \frac{1}{s^\ell}\mathfrak{I}\mathfrak{I}'\right)\boldsymbol{X}_1$$

that is, $\widetilde{\boldsymbol{X}}_1$ is obtained from $\boldsymbol{X}_1$ by subtracting from each element the corresponding column mean. Because of property 1 each column mean is $1/s$. Because of properties 2 and 3, it follows then that

$$\widetilde{\boldsymbol{X}}_1'\widetilde{\boldsymbol{X}}_1 = \text{diag}\,\underbrace{[\boldsymbol{M}, \boldsymbol{M}, \ldots, \boldsymbol{M}]}_{n}$$

with $\boldsymbol{M}$ an $s \times s$ matrix of the form

$$\boldsymbol{M} = s^{\ell-2}(s\boldsymbol{I} - \mathfrak{J}\mathfrak{J}')$$

Using the result from Theorem 1.1 we then have

$$\left(\widetilde{\boldsymbol{X}}_1'\widetilde{\boldsymbol{X}}_1\right)^- = \text{diag}(\boldsymbol{M}^-, \boldsymbol{M}^-, \ldots, \boldsymbol{M}^-) \tag{14.3}$$

with

$$\boldsymbol{M}^- = \frac{1}{s^{\ell-1}}\boldsymbol{I}$$

Since $\left(\widetilde{\boldsymbol{X}}_1'\widetilde{\boldsymbol{X}}_1\right)^-$ is the variance–covariance matrix for estimable functions associated with the various main effects, it follows then from (14.3) that estimable functions associated with different main effects, such as $\sum_\delta c_\delta(\widehat{A_i})_\delta$ and $\sum_\delta d_\delta(\widehat{A_j})_\delta$ for the two factors A_i and A_j, are uncorrelated. The same, of course, is true for the estimates of orthogonal contrasts associated with the main effects of the same factor, such as $\sum_\delta c_\delta(\widehat{A_i})_\delta$ and $\sum_\delta d_\delta(\widehat{A_i})_\delta$ with $\sum_\delta c_\delta d_\delta = 0$. The estimates of the estimable functions are obtained from (14.1) as

$$\widehat{\boldsymbol{E}}_1 = \left(\widetilde{\boldsymbol{X}}_1'\widetilde{\boldsymbol{X}}_1\right)^- \widetilde{\boldsymbol{X}}_1'\boldsymbol{y}$$

Examples of OMEPs that can be obtained by this method are given in Table 14.1. In this table n represents the maximum number of factors that can be accommodated in $N = s^\ell$ runs resulting in a saturated OMEP. It should be clear how this method can also be used to construct OMEPs for $n^* < n = (s^\ell - 1)/(s - 1)$ factors in N runs. All we need to do is to construct the plan for n factors and then delete $n - n^*$ factors from that plan. The resulting design is obviously an OMEP, but it is no longer a saturated OMEP. Instead it provides $(n - n^*)(s - 1)$ d.f. for error.

Table 14.1 Orthogonal Main Effect Plans (Fisher's Method)

s^n	$N = s^\ell$
2^7	8
2^{15}	16
2^{31}	32
3^4	9
3^{13}	27
3^{40}	81
4^5	16
4^{21}	64
5^6	25

14.2.2 Collapsing Factor Levels

The condition of equal frequencies for all level combinations for any pair of factors in an OMEP can be relaxed to that of proportional frequencies. This was shown first by Plackett (1946) and later by Addelman and Kempthorne (1961) for asymmetrical factorials (see Theorem 14.1). This result leads to the construction of additional OMEPs by using what is called *collapsing of factor levels*. For example, we may collapse the levels of a three-level factor to those of a two-level factor by setting up the correspondence

Three-Level Factor		Two-Level Factor
0	$\longrightarrow$	0
1	$\longrightarrow$	1
2	$\longrightarrow$	0

Using this correspondence we can, for example, use the 3_{III}^{4-2} and obtain an OMEP for the 2^4 factorial with nine runs.

14.2.3 Alias Structure

Even though for resolution III designs we make the assumption that all interactions are negligible, it may in certain situations be useful to know with which 2-factor interactions the main effects are confounded. The reason, of course, is that among interactions the 2-factor interactions are least likely to be negligible and hence to know the aliasing between main effects and 2-factor interactions may be helpful for diagnostic purposes (see also Section 13.6.4). We shall illustrate this later.

The general method of investigating the alias structure is, of course, to make use of models (13.28) and (13.33). That is, we estimate the main effects from a given design using (13.31) and then consider the expected value of the estimates under model (13.33). To illustrate this we have from (13.31), with $\boldsymbol{E}_1^{*\prime} = (M, \boldsymbol{E}_1')$,

$$\widehat{\boldsymbol{E}}_1^* = (\boldsymbol{X}_1^{*\prime}\boldsymbol{X}_1^*)^- \boldsymbol{X}_1^{*\prime}\boldsymbol{y}$$

and, using (13.33)

$$\begin{aligned} E[\boldsymbol{X}_1^*\widehat{\boldsymbol{E}}_1^*] &= \boldsymbol{X}_1^*(\boldsymbol{X}_1^{*\prime}\boldsymbol{X}_1^*)^- \boldsymbol{X}_1^{*\prime}(\boldsymbol{X}_1^*\boldsymbol{E}_1^* + \boldsymbol{X}_2\boldsymbol{E}_2) \\ &= \boldsymbol{X}_1^*\boldsymbol{E}_1^* + \boldsymbol{X}_1^*(\boldsymbol{X}_1^{*\prime}\boldsymbol{X}_1^*)^- \boldsymbol{X}_1^{*\prime}\boldsymbol{X}_2\boldsymbol{E}_2 \end{aligned}$$

or, for an estimable function $\boldsymbol{c}'\boldsymbol{E}_1^*$,

$$E[\boldsymbol{c}'\widehat{\boldsymbol{E}}_1^*] = \boldsymbol{c}'\boldsymbol{E}_1^* + \boldsymbol{c}'(\boldsymbol{X}_1^{*\prime}\boldsymbol{X}_1^*)^- \boldsymbol{X}_1^{*\prime}\boldsymbol{X}_2\boldsymbol{E}_2 \tag{14.4}$$

Expressions for this alias structure can be stated explicitly (see below) for certain OMEPs. In other cases computer programs may be useful to obtain expressions for (14.4).

Since the OMEP $s^{n-(n-\ell)}$ with $n = (s^\ell - 1)/(s-1)$ is equivalent to the IBSG of an s^n experiment in blocks of size s^ℓ, it is obvious that the interactions that are confounded with blocks are also the interactions included in the identity relationship. Altogether there are $(s^{n-\ell} - 1)/(s-1)$ interactions in the identity relationship, $n - \ell$ independent interactions plus all their GIs. For our purpose here we are concerned only with knowing the 3-factor interactions contained in the identity relationship. Some 3-factor interactions are directly identified from (11.27), namely those $E^{\boldsymbol{\gamma}_j}$ for which the partition $\boldsymbol{\alpha}_{j+\ell}$ corresponds to a 2-factor interaction in the s^ℓ system. Others may be obtained as GIs among the $n - \ell$ independent interactions (11.27). Now if, for example, $F_{i_1} F_{i_2}^{\delta} F_{i_3}^{\gamma}$ is a 3-factor interaction in the identity relationship for some i_1, i_2, i_3 with $1 \le i_1, i_2, i_3 \le n$ and i_1, i_2, i_3 different and some δ, γ with $1 \le \delta, \gamma \le s-1$, then F_{i_1} will be confounded with $F_{i_2}^{\delta} F_{i_3}^{\gamma}$ (or an appropriate power of $F_{i_2}^{\delta} F_{i_3}^{\gamma}$ to bring it into standard form), and so forth. An alternative method, using existing computer programs, of obtaining the relevant alias structure is given below. We now illustrate our discussion with the following example.

Example 14.1 Suppose $s = 3$, $\ell = 2$, and hence $n = 4$. The 3_{III}^{4-2} was given as the IBSG in Example 11.2. From Example 11.4 it follows that the identity relationship is given by

$$I = F_1 F_2 F_3^2 = F_1 F_2^2 F_4^2 = F_1 F_3 F_4 = F_2 F_3 F_4^2 \tag{14.5}$$

or, more precisely,

$$I = (F_1 F_2 F_3^2)_0 = (F_1 F_2^2 F_4^2)_0 = (F_1 F_3 F_4)_0 = (F_2 F_3 F_4^2)_0 \tag{14.6}$$

where $F_1 F_2 F_3^2$ and $F_1 F_2^2 F_4^2$ are the independent interactions obtained from (11.27) and $F_1 F_3 F_4$, $F_2 F_3 F_4^2$ are their GIs. It follows then from (14.5) that the relevant aliasing for F_1, for example, is given by

$$F_1 = F_2 F_3^2 = F_2 F_4 = F_3 F_4$$

If the 2-factor interactions were indeed not negligible, then instead of estimating, for example, $(F_1)_2 - (F_1)_0$ we would actually estimate [see also (14.4)]

$$\begin{aligned}
&[(F_1)_2 - (F_1)_0] + [(F_2 F_3^2)_1 - (F_2 F_3^2)_0] \\
&\qquad + [(F_2 F_4)_2 - (F_2 F_4)_0] + [(F_3 F_4)_1 - (F_3 F_4)_0]
\end{aligned} \tag{14.7}$$

making use of (14.6). Clearly, if all 2-factor interactions were nonnegligible using the OMEP would be useless. But suppose the investigator is fairly sure that all 2-factor interactions except possibly $F_2 \times F_3$ are negligible. Then we would have to be concerned about the confounding of F_1 and $F_2F_3^2$ and F_4 and F_2F_3. If then an estimated contrast belonging to F_1 were "large," we may suspect that this is partly due to a corresponding contrast belonging to $F_2F_3^2$, and so forth. Of course, we have no way of knowing, but this may suggest at least further investigation, for example, by augmenting the original design to achieve the desired dealiasing. It is in this sense that knowledge of the alias structure can be helpful as a diagnostic tool.

The form of the estimable function (14.7) can also be obtained by using the E-option in SAS PROC GLM. The (fictitious) data and the ensuing analysis are given in Table 14.2.

Concerning the SAS input and output in Table 14.2 we make the following comments:

1. The factors are now labeled A, B, C, D.
2. The model includes all 2-factor interactions similar to the model for a resolution IV design.
3. The option $e1$ lists all type I estimable functions (see comment 4) and produces the type I sums of squares (see I.4.11.1).
4. Using a model similar to (13.23) (see also Section 7.2.1), we see that by putting $L2 = 1$ and all other $Li = 0 (i = 3, 5, 6, 8, 9, 11, 12)$ the following function is estimable:

$$\begin{aligned}
&\alpha_0 - \alpha_2 + \tfrac{1}{3}[(\alpha\beta)_{00} + (\alpha\beta)_{01} + (\alpha\beta)_{02} - (\alpha\beta)_{20} - (\alpha\beta)_{21} - (\alpha\beta)_{22}] \\
&\quad + \tfrac{1}{3}[(\alpha\gamma)_{00} + (\alpha\gamma)_{01} + (\alpha\gamma)_{02} - (\alpha\gamma)_{20} - (\alpha\gamma)_{21} - (\alpha\gamma)_{22}] \\
&\quad + \tfrac{1}{3}[(\alpha\delta)_{00} + (\alpha\delta)_{01} + (\alpha\delta)_{02} - (\alpha\delta)_{20} - (\alpha\delta)_{21} - (\alpha\delta)_{22}] \\
&\quad + \tfrac{1}{3}[(\beta\gamma)_{00} - (\beta\gamma)_{02} - (\beta\gamma)_{10} + (\beta\gamma)_{11} - (\beta\gamma)_{21} + (\beta\gamma)_{22}] \\
&\quad + \tfrac{1}{3}[(\beta\delta)_{00} - (\beta\delta)_{02} - (\beta\delta)_{11} + (\beta\delta)_{12} - (\beta\delta)_{20} + (\beta\delta)_{21}] \\
&\quad + \tfrac{1}{3}[(\gamma\delta)_{00} - (\gamma\delta)_{01} - (\gamma\delta)_{10} + (\gamma\delta)_{12} + (\gamma\delta)_{21} - (\gamma\delta)_{22}] \qquad (14.8)
\end{aligned}$$

Defining the interaction terms such that, for example,

$$\sum_{j=0}^{2} (\alpha\beta)_{ij} = 0 \quad \text{for} \quad i = 0, 1, 2$$

then in (14.8) all terms in $(\alpha\beta)_{ij}$, $(\alpha\gamma)_{ik}$, $(\alpha\delta)_{i\ell}$ drop out. Concerning the remaining interaction terms, we make use of the two equivalent parameterizations, for example,

$$B \times C = BC + BC^2$$

Table 14.2 Orthogonal Main Effect Plan 3^{4-2}

```
options nodate pageno=1;
data OMEP;
input A B C D Y @@;
datalines;
0 0 0 0 5   1 1 2 0 7   2 2 1 0 6
0 1 1 2 8   1 0 1 1 4   0 2 2 1 5
1 2 0 2 8   2 0 2 2 9   2 1 0 1 7
;
run;

proc print data=OMEP;
title1 'TABLE 14.2';
title2 'ORTHOGONAL MAIN EFFECT PLAN 3**(4-2)';
run;

proc glm data=OMEP;
class A B C D;
model y=A B C D A*B A*C A*D B*C B*D C*D/e1;
title3 'ANALYSIS AND ESTIMABLE FUNCTIONS';
run;
```

Obs	A	B	C	D	Y
1	0	0	0	0	5
2	1	1	2	0	7
3	2	2	1	0	6
4	0	1	1	2	8
5	1	0	1	1	4
6	0	2	2	1	5
7	1	2	0	2	8
8	2	0	2	2	9
9	2	1	0	1	7

The GLM Procedure

ANALYSIS AND ESTIMABLE FUNCTIONS

Class Level Information

Class	Levels	Values
A	3	0 1 2
B	3	0 1 2
C	3	0 1 2
D	3	0 1 2

Table 14.2 (*Continued*)

Number of observations 9

Type I Estimable Functions

Effect		Coefficients A	Coefficients B	Coefficients C
Intercept		0	0	0
A	0	L2	0	0
A	1	L3	0	0
A	2	-L2-L3	0	0
B	0	0	L5	0
B	1	0	L6	0
B	2	0	-L5-L6	0
C	0	0	0	L8
C	1	0	0	L9
C	2	0	0	-L8-L9
D	0	0	0	0
D	1	0	0	0
D	2	0	0	0
A*B	0 0	0.3333*L2	0.3333*L5	0.3333*L8
A*B	0 1	0.3333*L2	0.3333*L6	0.3333*L9
A*B	0 2	0.3333*L2	-0.3333*L5-0.3333*L6	-0.3333*L8-0.3333*L9
A*B	1 0	0.3333*L3	0.3333*L5	0.3333*L9
A*B	1 1	0.3333*L3	0.3333*L6	-0.3333*L8-0.3333*L9
A*B	1 2	0.3333*L3	-0.3333*L5-0.3333*L6	0.3333*L8
A*B	2 0	-0.3333*L2-0.3333*L3	0.3333*L5	-0.3333*L8-0.3333*L9
A*B	2 1	-0.3333*L2-0.3333*L3	0.3333*L6	0.3333*L8
A*B	2 2	-0.3333*L2-0.3333*L3	-0.3333*L5-0.3333*L6	0.3333*L9
A*C	0 0	0.3333*L2	0.3333*L5	0.3333*L8
A*C	0 1	0.3333*L2	0.3333*L6	0.3333*L9
A*C	0 2	0.3333*L2	-0.3333*L5-0.3333*L6	-0.3333*L8-0.3333*L9
A*C	1 0	0.3333*L3	-0.3333*L5-0.3333*L6	0.3333*L8
A*C	1 1	0.3333*L3	0.3333*L5	0.3333*L9
A*C	1 2	0.3333*L3	0.3333*L6	-0.3333*L8-0.3333*L9
A*C	2 0	-0.3333*L2-0.3333*L3	0.3333*L6	0.3333*L8
A*C	2 1	-0.3333*L2-0.3333*L3	-0.3333*L5-0.3333*L6	0.3333*L9
A*C	2 2	-0.3333*L2-0.3333*L3	0.3333*L5	-0.3333*L8-0.3333*L9
A*D	0 0	0.3333*L2	0.3333*L5	0.3333*L8
A*D	0 1	0.3333*L2	-0.3333*L5-0.3333*L6	-0.3333*L8-0.3333*L9
A*D	0 2	0.3333*L2	0.3333*L6	0.3333*L9
A*D	1 0	0.3333*L3	0.3333*L6	-0.3333*L8-0.3333*L9
A*D	1 1	0.3333*L3	0.3333*L5	0.3333*L9

Table 14.2 (*Continued*)

Type I Estimable Functions

Effect		Coefficients D	A*B	A*C	A*D	B*C	B*D	C*D
Intercept		0	0	0	0	0	0	0
A	0	0	0	0	0	0	0	0
A	1	0	0	0	0	0	0	0
A	2	0	0	0	0	0	0	0
B	0	0	0	0	0	0	0	0
B	1	0	0	0	0	0	0	0
B	2	0	0	0	0	0	0	0
C	0	0	0	0	0	0	0	0
C	1	0	0	0	0	0	0	0
C	2	0	0	0	0	0	0	0
D	0	L11	0	0	0	0	0	0
D	1	L12	0	0	0	0	0	0
D	2	-L11-L12	0	0	0	0	0	0
A*B	0 0	0.3333*L11	0	0	0	0	0	0
A*B	0 1	-0.3333*L11-0.3333*L12	0	0	0	0	0	0
A*B	0 2	0.3333*L12	0	0	0	0	0	0
A*B	1 0	0.3333*L12	0	0	0	0	0	0
A*B	1 1	0.3333*L11	0	0	0	0	0	0
A*B	1 2	-0.3333*L11-0.3333*L12	0	0	0	0	0	0
A*B	2 0	-0.3333*L11-0.3333*L12	0	0	0	0	0	0
A*B	2 1	0.3333*L12	0	0	0	0	0	0
A*B	2 2	0.3333*L11	0	0	0	0	0	0
A*C	0 0	0.3333*L11	0	0	0	0	0	0
A*C	0 1	-0.3333*L11-0.3333*L12	0	0	0	0	0	0
A*C	0 2	0.3333*L12	0	0	0	0	0	0
A*C	1 0	-0.3333*L11-0.3333*L12	0	0	0	0	0	0
A*C	1 1	0.3333*L12	0	0	0	0	0	0
A*C	1 2	0.3333*L11	0	0	0	0	0	0
A*C	2 0	0.3333*L12	0	0	0	0	0	0
A*C	2 1	0.3333*L11	0	0	0	0	0	0
A*C	2 2	-0.3333*L11-0.3333*L12	0	0	0	0	0	0
A*D	0 0	0.3333*L11	0	0	0	0	0	0
A*D	0 1	0.3333*L12	0	0	0	0	0	0
A*D	0 2	-0.3333*L11-0.3333*L12	0	0	0	0	0	0
A*D	1 0	0.3333*L11	0	0	0	0	0	0
A*D	1 1	0.3333*L12	0	0	0	0	0	0

Table 14.2 *(Continued)*

Type I Estimable Functions

Effect		------------------	Coefficients	------------------
		A	B	C
A*D	1 2	0.3333*L3	-0.3333*L5-0.3333*L6	0.3333*L8
A*D	2 0	-0.3333*L2-0.3333*L3	-0.3333*L5-0.3333*L6	0.3333*L9
A*D	2 1	-0.3333*L2-0.3333*L3	0.3333*L6	0.3333*L8
A*D	2 2	-0.3333*L2-0.3333*L3	0.3333*L5	-0.3333*L8-0.3333*L9
B*C	0 0	0.3333*L2	0.3333*L5	0.3333*L8
B*C	0 1	0.3333*L3	0.3333*L5	0.3333*L9
B*C	0 2	-0.3333*L2-0.3333*L3	0.3333*L5	-0.3333*L8-0.3333*L9
B*C	1 0	-0.3333*L2-0.3333*L3	0.3333*L6	0.3333*L8
B*C	1 1	0.3333*L2	0.3333*L6	0.3333*L9
B*C	1 2	0.3333*L3	0.3333*L6	-0.3333*L8-0.3333*L9
B*C	2 0	0.3333*L3	-0.3333*L5-0.3333*L6	0.3333*L8
B*C	2 1	-0.3333*L2-0.3333*L3	-0.3333*L5-0.3333*L6	0.3333*L9
B*C	2 2	0.3333*L2	-0.3333*L5-0.3333*L6	-0.3333*L8-0.3333*L9
B*D	0 0	0.3333*L2	0.3333*L5	0.3333*L8
B*D	0 1	0.3333*L3	0.3333*L5	0.3333*L9
B*D	0 2	-0.3333*L2-0.3333*L3	0.3333*L5	-0.3333*L8-0.3333*L9
B*D	1 0	0.3333*L3	0.3333*L6	-0.3333*L8-0.3333*L9
B*D	1 1	-0.3333*L2-0.3333*L3	0.3333*L6	0.3333*L8
B*D	1 2	0.3333*L2	0.3333*L6	0.3333*L9
B*D	2 0	-0.3333*L2-0.3333*L3	-0.3333*L5-0.3333*L6	0.3333*L9
B*D	2 1	0.3333*L2	-0.3333*L5-0.3333*L6	-0.3333*L8-0.3333*L9
B*D	2 2	0.3333*L3	-0.3333*L5-0.3333*L6	0.3333*L8
C*D	0 0	0.3333*L2	0.3333*L5	0.3333*L8
C*D	0 1	-0.3333*L2-0.3333*L3	0.3333*L6	0.3333*L8
C*D	0 2	0.3333*L3	-0.3333*L5-0.3333*L6	0.3333*L8
C*D	1 0	-0.3333*L2-0.3333*L3	-0.3333*L5-0.3333*L6	0.3333*L9
C*D	1 1	0.3333*L3	0.3333*L5	0.3333*L9
C*D	1 2	0.3333*L2	0.3333*L6	0.3333*L9
C*D	2 0	0.3333*L3	0.3333*L6	-0.3333*L8-0.3333*L9
C*D	2 1	0.3333*L2	-0.3333*L5-0.3333*L6	-0.3333*L8-0.3333*L9
C*D	2 2	-0.3333*L2-0.3333*L3	0.3333*L5	-0.3333*L8-0.3333*L9

Type I Estimable Functions

Effect		------------------	Coefficients	------------------				
		D	A*B	A*C	A*D	B*C	B*D	C*D
A*D	1 2	-0.3333*L11-0.3333*L12	0	0	0	0	0	0
A*D	2 0	0.3333*L11	0	0	0	0	0	0
A*D	2 1	0.3333*L12	0	0	0	0	0	0
A*D	2 2	-0.3333*L11-0.3333*L12	0	0	0	0	0	0

Table 14.2 (*Continued*)

B*C	0 0	0.3333*L11	0	0	0	0	0	0
B*C	0 1	0.3333*L12	0	0	0	0	0	0
B*C	0 2	-0.3333*L11-0.3333*L12	0	0	0	0	0	0
B*C	1 0	0.3333*L12	0	0	0	0	0	0
B*C	1 1	-0.3333*L11-0.3333*L12	0	0	0	0	0	0
B*C	1 2	0.3333*L11	0	0	0	0	0	0
B*C	2 0	-0.3333*L11-0.3333*L12	0	0	0	0	0	0
B*C	2 1	0.3333*L11	0	0	0	0	0	0
B*C	2 2	0.3333*L12	0	0	0	0	0	0
B*D	0 0	0.3333*L11	0	0	0	0	0	0
B*D	0 1	0.3333*L12	0	0	0	0	0	0
B*D	0 2	-0.3333*L11-0.3333*L12	0	0	0	0	0	0
B*D	1 0	0.3333*L11	0	0	0	0	0	0
B*D	1 1	0.3333*L12	0	0	0	0	0	0
B*D	1 2	-0.3333*L11-0.3333*L12	0	0	0	0	0	0
B*D	2 0	0.3333*L11	0	0	0	0	0	0
B*D	2 1	0.3333*L12	0	0	0	0	0	0
B*D	2 2	-0.3333*L11-0.3333*L12	0	0	0	0	0	0
C*D	0 0	0.3333*L11	0	0	0	0	0	0
C*D	0 1	0.3333*L12	0	0	0	0	0	0
C*D	0 2	-0.3333*L11-0.3333*L12	0	0	0	0	0	0
C*D	1 0	0.3333*L11	0	0	0	0	0	0
C*D	1 1	0.3333*L12	0	0	0	0	0	0
C*D	1 2	-0.3333*L11-0.3333*L12	0	0	0	0	0	0
C*D	2 0	0.3333*L11	0	0	0	0	0	0
C*D	2 1	0.3333*L12	0	0	0	0	0	0
C*D	2 2	-0.3333*L11-0.3333*L12	0	0	0	0	0	0

Dependent Variable: Y

Source	DF	Sum of Squares	Mean Square	F Value	Pr > F
Model	8	22.22222222	2.77777778	.	.
Error	0	0.00000000	.		
Corrected Total	8	22.22222222			

R-Square	Coeff Var	Root MSE	Y Mean
1.000000	.	.	6.555556

Source	DF	Type I SS	Mean Square	F Value	Pr > F
A	2	2.88888889	1.44444444	.	.
B	2	2.88888889	1.44444444	.	.

Table 14.2 *(Continued)*

C	2	1.55555556	0.77777778	.	.
D	2	14.88888889	7.44444444	.	.
A*B	0	0.00000000	.	.	.
A*C	0	0.00000000	.	.	.
A*D	0	0.00000000	.	.	.
B*C	0	0.00000000	.	.	.
B*D	0	0.00000000	.	.	.
C*D	0	0.00000000	.	.	.

(see Section 10.2) where the 2 d.f. of BC consist of comparisons among BC_0, BC_1, and BC_2, with

$$BC_0\colon\ (\beta\gamma)_{00} + (\beta\gamma)_{12} + (\beta\gamma)_{21}$$

$$BC_1\colon\ (\beta\gamma)_{10} + (\beta\gamma)_{01} + (\beta\gamma)_{22}$$

$$BC_2\colon\ (\beta\gamma)_{20} + (\beta\gamma)_{02} + (\beta\gamma)_{11}$$

and similarly

$$BC_0^2\colon\ (\beta\gamma)_{00} + (\beta\gamma)_{11} + (\beta\gamma)_{22}$$

$$BC_1^2\colon\ (\beta\gamma)_{10} + (\beta\gamma)_{02} + (\beta\gamma)_{21}$$

$$BC_2^2\colon\ (\beta\gamma)_{20} + (\beta\gamma)_{01} + (\beta\gamma)_{12}$$

We can write out similar expressions involving the $(\beta\delta)_{jk}$ and $(\gamma\delta)_{k\ell}$ terms. Utilizing these equivalencies in (14.8) leads to the same result as given in (14.7). □

14.2.4 Plackett–Burman Designs

For 2^n factorials a special method of constructing a large number of OMEPs was provided by Plackett and Burman (1946). The method is based on the existence of Hadamard matrices.

Definition 14.1 A square matrix $\boldsymbol{H}$ of order N with entries -1, 1 is called a *Hadamard matrix* of order N if

$$\boldsymbol{H}\boldsymbol{H}' = \boldsymbol{H}'\boldsymbol{H} = N\boldsymbol{I}_N \tag{14.9}$$

□

Condition (14.9) is equivalent to saying that for a Hadamard matrix $\boldsymbol{H}$ the rows are pairwise orthogonal and the columns are pairwise orthogonal. Given a Hadamard matrix $\boldsymbol{H}$, an equivalent matrix with all elements in the first column equal to $+1$ can be obtained by multiplying by -1 each element in every row of $\boldsymbol{H}$ whose first element is -1. Such a matrix will be called *seminormalized.* Through a similar operation one can transform a Hadamard matrix $\boldsymbol{H}$ such that all elements in the first row are equal to $+1$. If the first column and first row contain only $+1$ elements, then the matrix is said to be *normalized.* Except for the trivial Hadamard matrices

$$\boldsymbol{H}_1 = (1), \boldsymbol{H}_2 = \begin{pmatrix} 1 & 1 \\ 1 & -1 \end{pmatrix}$$

it is known that a necessary condition of the existence of a Hadamard matrix of order N is that $N \equiv 0 \bmod 4$ (Paley, 1933). Hadamard matrices for $N \leq 200$ have been constructed (see, e.g., Hedayat and Wallis, 1978; Hedayat, Sloane and Stufken, 1999). For $N = 2^\ell$, for example, $\boldsymbol{H}$ can be constructed as the ℓ-fold Kronecker product of $\boldsymbol{H}_2$ with itself. Hadamard matrices for $N = 4t$ and $t = 1$, 2, ..., 8 are given by Hedayat and Wallis (1978) and for $t = 1, 2, \ldots, 12$ by Dey (1985).

Consider now a Hadamard matrix $\boldsymbol{H}$ of order N in normalized (or seminormalized) form and delete from it the first column. Denote the resulting matrix by $\widetilde{\boldsymbol{H}}$ and identify the columns with factors $F_1, F_2, \ldots, F_{N-1}$. We then take

$$\boldsymbol{H}^* = \tfrac{1}{2}\left(\widetilde{\boldsymbol{H}} + \mathfrak{I}\mathfrak{I}'\right)$$

and consider each row of $\boldsymbol{H}^*$ as a treatment combination from the 2^{N-1} factorial. These N runs constitute an OMEP for $N-1$ factors, with $N = 4t(t = 1, 2, \ldots)$. The fact that this FFD is an OMEP follows immediately from the properties of a Hadamard matrix in seminormalized (and hence normalized) form, namely for any column (factor) we have the same number of $+1$ and -1 elements in $\widetilde{\boldsymbol{H}}$ (and hence the same number of $+1$ and 0 elements in $\boldsymbol{H}^*$, and for any pair of columns we have the same number of $(+1, +1)$, $(+1, -1)$, $(-1, +1)$, $(-1, -1)$ combinations in $\widetilde{\boldsymbol{H}}$). FFDs constructed in this manner are called *Plackett–Burman designs.* Actually, Plackett and Burman (1946) list designs for $t = 1, 2, \ldots, 25$ (except 23 which was given by Baumert, Golomb, and Hall, 1962) using $\boldsymbol{H}^* = \frac{1}{2}(-\widetilde{\boldsymbol{H}} + \mathfrak{I}\mathfrak{I}')$. In this way the OMEP includes the control, that is, the treatment combination $(0, 0, \ldots, 0)$. We note here that for $N = 2^\ell$ the Plackett–Burman designs are the same as the Fisher plans for $n = N - 1$ factors discussed earlier. Also, the Plackett–Burman designs are relevant for the problem of weighing $N - 1$ light objects in N weighings, a problem first discussed by Yates (1935) for a spring balance and by Hotelling (1944) for a chemical balance. Hotelling (1944) showed that the optimum *weighing design* was obtained by using Hadamard matrices. Kempthorne (1948) discussed the problem from the viewpoint of factorial experimentation and showed how fractional factorials could be used. For a detailed discussion of weighing designs the reader is referred to Banerjee (1975).

In contrast to the Fisher plans the alias structure for the Plackett–Burman designs is rather complicated and difficult to determine (Margolin, 1968; Lin and

Draper, 1993). In any case, each main effect is aliased with a long string of 2-factor interactions and no expressions like (14.7) can be written out easily and studied in case the estimate of a main effect turns out to be "large."

Of course, folding-over the Plackett–Burman design leads to a resolution IV design, thereby dealiasing the main effects. In many cases less drastic measures may, however, achieve the goal of dealiasing some main effects, and that may be all that is needed since main effect plans are often used for screening purposes and only some of the main effects may be important; that is, only some of the factors may be active. This leads to investigating the projections of a Plackett–Burman design into smaller, say k, dimensions, where $k < n$. Here projection simply means to ignore the remaining $n - k$ factors after k factors have been identified, at least in a rough and preliminary sense, as being active. For example, Lin and Draper (1992) have shown that for the 2^{11} Plackett–Burman design in 12 runs the projection to $k = 4$ dimensions shows that one needs to add only one more run to identify a 2^{4-1}_{IV} design among the 13 runs. Or for $k = 5$ one needs to add 8 more runs to obtain a 2^{5-1}_{IV} design plus additional runs. Similar results have been obtained for other Plackett–Burman designs (Lin and Draper, 1993).

We shall use the 2^{11} Plackett–Burman design in 12 runs, as given in Table 14.3, to make some additional comments about projections and the estimation of main effects and 2-factor interactions:

1. $k = 2$: Since the Plackett–Burman designs are orthogonal arrays of strength 2 (see Section 14.2.5), it follows that each combination of the levels of two factors occurs equally often. Hence, for the two specified factors we can estimate the main effects and the 2-factor interaction.
2. $k = 3$: Inspection of the design in Table 14.3 shows that it contains a complete 2^3 factorial, allowing the estimation of the three main effects and three 2-factor interactions. The 12 runs no longer constitute an orthogonal design and hence the estimators will be correlated.
3. $k = 4$: It turns out that the available 11 d.f. allow the estimation of the four main effects and their six 2-factor interactions. This is shown in Table 14.3 using the E-option in SAS PROC GLM (keeping in mind that, e.g., $EF_{01} + EJ_{01} + EL_{01} = 0$). The ANOVA in Table 14.3 also shows that the design is no longer orthogonal as indicated by different numerical values for the type I and type III sums of squares.

14.2.5 Other Methods

We mention here briefly two other methods of constructing OMEPs. Addelman and Kempthorne (1961) describe a procedure for constructing an OMEP for an s^n factorial in $2s^\ell$ runs, where $n = [2(s^\ell - 1)/(s - 1) - 1]$. This procedure is an extension of the method used for the Fisher plans, using, however, a different correspondence between the factors and factor combinations from the 2^ℓ factorial and the n factors for the FFD. This method leads to OMEPs such as $N = 18 = 2 \cdot 3^2$ runs for the 3^7, $N = 54 = 2 \cdot 3^3$ runs for the 3^{25}, or $N = 32 = 2 \cdot 4^2$ runs for the 4^9 factorial.

Table 14.3 Plackett–Burman Design for 11 Factors in 12 Runs

```
options pageno=1 nodate;
data PBdesign;
input A B C D E F G H J K L Y;
datalines;
1 1 0 1 1 1 0 0 0 1 0 10
0 1 1 0 1 1 1 0 0 0 1 15
1 0 1 1 0 1 1 1 0 0 0 20
0 1 0 1 1 0 1 1 1 0 0 25
0 0 1 0 1 1 0 1 1 1 0 23
0 0 0 1 0 1 1 0 1 1 1 32
1 0 0 0 1 0 1 1 0 1 1 29
1 1 0 0 0 1 0 1 1 0 1 33
1 1 1 0 0 0 1 0 1 1 0 30
0 1 1 1 0 0 0 1 0 1 1 31
1 0 1 1 1 0 0 0 1 0 1 28
0 0 0 0 0 0 0 0 0 0 0 30
;
run;

proc print data=PBdesign;
title1 'TABLE 14.3';
title2 'PLACKETT-BURMAN DESIGN';
title3 'FOR 11 FACTORS IN 12 RUNS';
run;

proc glm data=PBdesign;
class A B C D E F G H J K L;
model Y=A B C D E F G H J K L;
title4 'ANALYSIS OF VARIANCE';
run;

proc glm data=PBdesign;
class E F J L;
model Y=E|F|J|L@2/e;
title3 'PROJECTION ON FACTORS E,F,J,L';
title4 'ANALYSIS WITH MAIN EFFECTS AND TWO-FACTOR INTERACTIONS';
run;
```

Obs	A	B	C	D	E	F	G	H	J	K	L	Y
1	1	1	0	1	1	1	0	0	0	1	0	10
2	0	1	1	0	1	1	1	0	0	0	1	15
3	1	0	1	1	0	1	1	1	0	0	0	20
4	0	1	0	1	1	0	1	1	1	0	0	25
5	0	0	1	0	1	1	0	1	1	1	0	23
6	0	0	0	1	0	1	1	0	1	1	1	32
7	1	0	0	0	1	0	1	1	0	1	1	29
8	1	1	0	0	0	1	0	1	1	0	1	33
9	1	1	1	0	0	0	1	0	1	1	0	30
10	0	1	1	1	0	0	0	1	0	1	1	31
11	1	0	1	1	1	0	0	0	1	0	1	28
12	0	0	0	0	0	0	0	0	0	0	0	30

Table 14.3 (*Continued*)

Class Level Information

Class	Levels	Values
A	2	0 1
B	2	0 1
C	2	0 1
D	2	0 1
E	2	0 1
F	2	0 1
G	2	0 1
H	2	0 1
J	2	0 1
K	2	0 1
L	2	0 1

Number of observations 12

Dependent Variable: Y

Source	DF	Sum of Squares	Mean Square	F Value	Pr > F
Model	11	575.0000000	52.2727273	.	.
Error	0	0.0000000	.		
Corrected Total	11	575.0000000			

R-Square	Coeff Var	Root MSE	Y Mean
1.000000	.	.	25.50000

Source	DF	Type I SS	Mean Square	F Value	Pr > F
A	1	3.0000000	3.0000000	.	.
B	1	27.0000000	27.0000000	.	.
C	1	12.0000000	12.0000000	.	.
D	1	16.3333333	16.3333333	.	.
E	1	176.3333333	176.3333333	.	.
F	1	133.3333333	133.3333333	.	.
G	1	1.3333333	1.3333333	.	.
H	1	21.3333333	21.3333333	.	.
J	1	108.0000000	108.0000000	.	.
K	1	1.3333333	1.3333333	.	.
L	1	75.0000000	75.0000000	.	.

Table 14.3 (*Continued*)

Source	DF	Type III SS	Mean Square	F Value	Pr > F
A	1	3.0000000	3.0000000	.	.
B	1	27.0000000	27.0000000	.	.
C	1	12.0000000	12.0000000	.	.
D	1	16.3333333	16.3333333	.	.
E	1	176.3333333	176.3333333	.	.
F	1	133.3333333	133.3333333	.	.
G	1	1.3333333	1.3333333	.	.
H	1	21.3333333	21.3333333	.	.
J	1	108.0000000	108.0000000	.	.
K	1	1.3333333	1.3333333	.	.
L	1	75.0000000	75.0000000	.	.

PROJECTION ON FACTORS E, F, J, L
ANALYSIS WITH MAIN EFFECTS AND TWO-FACTOR INTERACTIONS

Class Level Information

Class	Levels	Values
E	2	0 1
F	2	0 1
J	2	0 1
L	2	0 1

Number of observations 12

General Form of Estimable Functions

Effect		Coefficients
Intercept		L1
E	0	L2
E	1	L1-L2
F	0	L4
F	1	L1-L4
E*F	0 0	L6
E*F	0 1	L2-L6
E*F	1 0	L4-L6
E*F	1 1	L1-L2-L4+L6
J	0	L10
J	1	L1-L10

Table 14.3 *(Continued)*

```
                    E*J          0 0       L12
                    E*J          0 1       L2-L12
                    E*J          1 0       L10-L12
                    E*J          1 1       L1-L2-L10+L12

                    F*J          0 0       L16
                    F*J          0 1       L4-L16
                    F*J          1 0       L10-L16
                    F*J          1 1       L1-L4-L10+L16

                    L            0         L20
                    L            1         L1-L20

                    E*L          0 0       L22
                    E*L          0 1       L2-L22
                    E*L          1 0       L20-L22
                    E*L          1 1       L1-L2-L20+L22

                    F*L          0 0       L26
                    F*L          0 1       L4-L26
                    F*L          1 0       L20-L26
                    F*L          1 1       L1-L4-L20+L26

                    J*L          0 0       L30
                    J*L          0 1       L10-L30
                    J*L          1 0       L20-L30
                    J*L          1 1       L1-L10-L20+L30

Dependent Variable: Y

                                   Sum of
Source                DF          Squares    Mean Square    F Value    Pr > F

Model                 10      574.5000000     57.4500000     114.90    0.0725

Error                  1        0.5000000      0.5000000

Corrected Total       11      575.0000000


        R-Square      Coeff Var      Root MSE        Y Mean

        0.999130       2.772968      0.707107      25.50000


Source        DF        Type I SS    Mean Square     F Value     Pr > F

E              1      176.3333333    176.3333333      352.67     0.0339
F              1      133.3333333    133.3333333      266.67     0.0389
```

Table 14.3 (*Continued*)

E*F	1	65.3333333	65.3333333	130.67	0.0556
J	1	66.6666667	66.6666667	133.33	0.0550
E*J	1	2.6666667	2.6666667	5.33	0.2601
F*J	1	112.6666667	112.6666667	225.33	0.0423
L	1	10.8000000	10.8000000	21.60	0.1349
E*L	1	5.4857143	5.4857143	10.97	0.1867
F*L	1	0.1758242	0.1758242	0.35	0.6592
J*L	1	1.0384615	1.0384615	2.08	0.3862

Source	DF	Type III SS	Mean Square	F Value	Pr > F
E	1	61.03846154	61.03846154	122.08	0.0575
F	1	61.03846154	61.03846154	122.08	0.0575
E*F	1	9.34615385	9.34615385	18.69	0.1447
J	1	61.03846154	61.03846154	122.08	0.0575
E*J	1	0.11538462	0.11538462	0.23	0.7149
F*J	1	72.11538462	72.11538462	144.23	0.0529
L	1	9.34615385	9.34615385	18.69	0.1447
E*L	1	5.65384615	5.65384615	11.31	0.1840
F*L	1	0.11538462	0.11538462	0.23	0.7149
J*L	1	1.03846154	1.03846154	2.08	0.3862

Another method is derived from the fact that each OMEP is an orthogonal array of strength 2 (Rao, 1946a). More precisely, if n denotes the number of factors, N the number of runs, s the number of levels, then an OMEP can be denoted by OA[N, n, s, 2]. Bose and Bush (1952) show that an OA[λs^2, λs, s, 2] can be constructed if λ and s are both powers of the same prime p. Suppose $\lambda = p^u$, $s = p^v$. Then some examples of OMEPs obtained with this method are given below:

Factorial	**Number of Runs (N)**
2^4	8
2^8	16
3^9	27
3^{27}	81

Obviously, these will not be saturated OMEPs.

14.3 ORTHOGONAL RESOLUTION III DESIGNS FOR ASYMMETRICAL FACTORIALS

In many practical situations it is the case that we shall use factors with different numbers of levels. This is dictated by the nature of the experiment and the type of factors included in the experiment as well as the suspected nature of their response at different levels. And again, if we want to investigate a large number of factors, concentrating on estimating main effects may be a prudent thing to do.

14.3.1 Kronecker Product Design

An easy method to construct an OMEP for an $s_1^{n_1} \times s_2^{n_2} \times \cdots \times s_q^{n_q}$ asymmetrical factorial is to consider each component symmetrical factorial separately, construct an OMEP for it, and then take the Kronecker product of these designs. For example, if we have a $2^7 \times 3^4$ we can construct a 2_{III}^{7-4} and a 3_{III}^{4-2} and the Kronecker product of these two designs will yield a FFD, which is an OMEP, with $N = 8 \cdot 9 = 72$ runs. Obviously, this design contains far too many treatment combinations to account for the $7 + 4 \cdot 2 = 15$ d.f. for the main effects. It is for this reason that other methods of constructing OMEPs for asymmetrical factorials have been developed.

14.3.2 Orthogonality Condition

A method that yields a very large number of OMEPs for many different situations was developed by Addelman and Kempthorne (1961) (see also Addelman, 1962a). These designs are based on OMEPs for symmetrical factorials with subsequent collapsing and replacement of levels for certain factors. This procedure will be described more precisely in Section 14.3.3. For now we mention only that as a consequence of this procedure the levels of one factor no longer occur together with the same frequency with each of the levels of any other factor, as was the case for the OMEPs for symmetrical factorials. They do, however, occur together with proportional frequencies and that, as we shall show, is sufficient to guarantee orthogonal estimates of main effects.

To reduce the amount of writing let us consider the case of two factors, A and B say, where A has a levels and B has b levels. Suppose that in a given FFD the ith level of A occurs together with the jth level of B for N_{ij} times. Let $\sum_j N_{ij} = N_{i.}$, $\sum_i N_{ij} = N_{.j}$, $\sum_i N_{i.} = \sum_j N_{.j} = N$, the total number of runs in the FFD. If the level combinations occur with proportional frequencies, we have $N_{ij} = N_{i.}N_{.j}/N$ for every i and j ($i = 0, 1, \ldots, a-1$; $j = 0\ 1, \ldots, b-1$). We then state the following theorem.

Theorem 14.1 A sufficient condition for a main effect plan for two factors A and B with a and b levels, respectively, to be an OMEP is that each level combination for the two factors occurs with proportional frequency.

Proof Consider model (13.28) and rewrite it as

$$E(\boldsymbol{y}) = (\mathfrak{I}, \boldsymbol{X}_{11}, \boldsymbol{X}_{12}) \begin{pmatrix} \mu \\ \boldsymbol{A} \\ \boldsymbol{B} \end{pmatrix}$$

where $\boldsymbol{A}' = (A_0, A_1, \ldots, A_{a-1})$ and $\boldsymbol{B}' = (B_0, B_1, \ldots, B_{b-1})$ represent the main effect components for the two factors A and B, respectively. We then rewrite the N E (13.29) as

$$\begin{bmatrix} \mathfrak{I}'\mathfrak{I} & \mathfrak{I}'\boldsymbol{X}_{11} & \mathfrak{I}'\boldsymbol{X}_{12} \\ \boldsymbol{X}'_{11}\mathfrak{I} & \boldsymbol{X}'_{11}\boldsymbol{X}_{11} & \boldsymbol{X}'_{11}\boldsymbol{X}_{12} \\ \boldsymbol{X}'_{12}\mathfrak{I} & \boldsymbol{X}'_{12}\boldsymbol{X}_{11} & \boldsymbol{X}'_{12}\boldsymbol{X}_{12} \end{bmatrix} \begin{bmatrix} \mu \\ \boldsymbol{A} \\ \boldsymbol{B} \end{bmatrix} = \begin{bmatrix} \mathfrak{I}'\boldsymbol{y} \\ \boldsymbol{X}'_{11}\boldsymbol{y} \\ \boldsymbol{X}'_{12}\boldsymbol{y} \end{bmatrix} \tag{14.10}$$

Denote the coefficient matrix in (14.10) by $\boldsymbol{K}$. Then we have

$$\boldsymbol{K} = \begin{bmatrix} N & N_{0.} & N_{1.} & \ldots & N_{a-1,.} & N_{.0} & N_{.1} & \ldots & N_{.,b-1} \\ N_{0.} & N_{0.} & 0 & \ldots & 0 & N_{00} & N_{01} & \ldots & N_{0,b-1} \\ N_{1.} & 0 & N_{1.} & \ldots & 0 & N_{10} & N_{11} & \ldots & N_{1,b-1} \\ \vdots & & & & & & & & \vdots \\ N_{a-1,.} & 0 & 0 & \ldots & N_{a-1,.} & N_{a-1,0} & N_{a-1,1} & \ldots & N_{a-1,b-1} \\ N_{.0} & N_{00} & N_{10} & \ldots & N_{a-1,0} & N_{.0} & 0 & \ldots & 0 \\ N_{.1} & N_{01} & N_{11} & \ldots & N_{a-1,1} & 0 & N_{.1} & \ldots & 0 \\ \vdots & & & & & & & & \vdots \\ N_{.,b-1} & N_{0,b-1} & N_{1,b-1} & \ldots & N_{a-1,b-1} & 0 & 0 & \ldots & N_{.,b-1} \end{bmatrix}$$

To obtain a solution to (14.10) we need to find a generalized inverse for $\boldsymbol{K}$. If $N_{ij} = N_{i.}N_{.j}/N$ for each i and j, then such a g inverse is given by

$$\boldsymbol{K}^- = \begin{bmatrix} \frac{1}{N} & 0 & \ldots & & 0 & 0 & \ldots & & 0 \\ -\frac{1}{N} & \frac{1}{N_{0.}} & & & & & & & \\ -\frac{1}{N} & & \frac{1}{N_{1.}} & & & & \mathbf{0} & & \\ \vdots & & & \ddots & & & & & \\ -\frac{1}{N} & & & & \frac{1}{N_{a-1,.}} & & & & \\ -\frac{1}{N} & & & & & \frac{1}{N_{.0}} & & & \\ -\frac{1}{N} & & & & & & \frac{1}{N_{.1}} & & \\ \vdots & & \mathbf{0} & & & & & \ddots & \\ -\frac{1}{N} & & & & & & & & \frac{1}{N_{.,b-1}} \end{bmatrix} \tag{14.11}$$

This follows easily by verifying that $\boldsymbol{K}\boldsymbol{K}^-\boldsymbol{K} = \boldsymbol{K}$ using the proportionality condition. The reader may recognize that the solution and hence $\boldsymbol{K}^-$ can be obtained by imposing the conditions $\sum_i N_{i.}\widehat{A}_i = 0 = \sum_j N_{.j}\widehat{B}_j$. Consider now the contrasts $\sum c_i A_i$ and $\sum d_j B_j$ (with $\sum c_i = 0$, $\sum d_j = 0$) belonging to the main effects A and B, respectively. It follows then immediately that

$$\operatorname{cov}\left(\sum c_i \widehat{A}_i, \sum d_j \widehat{B}_j\right) = (0, \boldsymbol{c}', \boldsymbol{0})\boldsymbol{K}^- \begin{pmatrix} 0 \\ \boldsymbol{0} \\ \boldsymbol{d} \end{pmatrix} = 0 \tag{14.12}$$

where $\boldsymbol{c}' = (c_0, c_1, \ldots, c_{a-1})$, $\boldsymbol{d}' = (d_0, d_1, \ldots, d_{b-1})$, that is, estimates of contrasts for the two main effects are uncorrelated. Hence we have an OMEP. □

Obviously the proof can be extended in a straightforward manner to any number of factors with the proportionality condition holding for any pair of factors. We mention also that Addelman and Kempthorne (1961) proved that proportional frequency is actually a necessary *and* sufficient condition for a main effect plan to be an OMEP (see also Dey, 1985).

The methods of constructing main effect plans as described in the following section make use of the proportionality condition of Theorem 14.1. Before proceeding, however, we need to clarify an argument put forth by Srivastava and Ghosh (1996). They argue that main effect plans with proportional (and not equal) frequencies are not OMEPs. Clearly, Eq. (14.12) shows that the estimators of contrasts of main effects belonging to two different factors are uncorrelated. The concern of Srivastava and Ghosh (1996) has to do with estimators of contrasts belonging to the same factor, for example,

$$\sum_i c_i \widehat{A}_i \qquad \sum_i d_i \widehat{A}_i \tag{14.13}$$

with $\sum_i c_i = 0 = \sum_i d_i$ and $\sum_i c_i d_i = 0$. Using $\boldsymbol{K}^-$ from (14.11), it follows immediately that the contrasts in (14.13) are correlated (see also Section I.7.3). In order to produce uncorrelated contrasts Addelman (1962a) considers contrasts of the form

$$\sum_i N_{i.} c_i^* A_i \qquad \sum_i N_{i.} d_i^* A_i \tag{14.14}$$

with $\sum_i N_{i.} c_i^* = 0 = \sum_i N_{i.} d_i^*$ and $\sum_i N_{i.} c_i^* d_i^* = 0$.

We emphasize that contrasts of the form (14.13) or (14.14) are considered only for purposes of posthoc analyses (see Chapter I.7). For qualitative factors such contrasts are prespecified, for quantitative factors they represent various forms of trends, such as linear, quadratic, and so forth. In either case the contrasts must be chosen such that they are meaningful in the context of the experiment, whether they are orthogonal or not. And that does not affect the properties of the design. Whether a main effect plan is an OMEP is determined only by whether the condition of Theorem 14.1 is satisfied.

14.3.3 Addelman–Kempthorne Methods

We now describe three methods for constructing OMEPs for asymmetrical factorials as developed by Addelman and Kempthorne (1961).

Method 1

We consider the $s_1^{n_1} \times s_2^{n_2} \times \cdots \times s_q^{n_q}$ factorial and assume that $s_1 > s_2 > \cdots > s_q$.

Collapsing Factors

We can construct an OMEP with $N = s_1^\ell$ runs, where s_1 is a prime or prime power and

$$\sum_{i=1}^{q} n_i \leq \frac{s_1^\ell - 1}{s_1 - 1} \tag{14.15}$$

in the following way. We first construct an OMEP for the s_1^n factorial in s_1^ℓ trials where ℓ is chosen so that (14.15) is satisfied and where $n = (s_1^\ell - 1)/(s_1 - 1)$ using the method described in Sections 14.2 and 11.7. We write the plan as an $N \times n$ array, D say, such that the columns denote the factors and each row represents a treatment combination (run) for the n factors with s_1 levels each. We shall refer to D also as the auxiliary OMEP. In the context of the $s_1^{n_1} \times s_2^{n_2} \times \cdots \times s_q^{n_q}$ factorial the first n_1 columns in D correspond to the first set of factors, each having s_1 levels. For the following factors we use what Addelman and Kempthorne (1961) refer to as *collapsing factors* with s_1 levels to factors occurring at s_i levels ($i = 2, 3, \ldots, q$) by setting up a many–one correspondence between the set of s_1 levels and the set of s_i levels. This means that different levels of the s_1-level factor are mapped into the same level of the s_i-level factor. As an example, suppose we have $s_1 = 5$, $s_2 = 3$. We then might have the following mapping:

Five-Level Factor		**Three-Level Factor**
0	$\longrightarrow$	0
1	$\longrightarrow$	1
2	$\longrightarrow$	2
3	$\longrightarrow$	0
4	$\longrightarrow$	1

Proceeding in this way we replace the next n_2 factors by s_2-level factors, the following n_3 factors by s_3-level factors, and so on, until all factors have been accommodated. Since $\sum_{i=1}^{q} n_i \leq n$ it may be necessary to delete some factors from the basic OMEP if the inequality holds. Since the collapsing procedure leads automatically to proportional frequencies (for any collapsing scheme) the final design is an OMEP.

We illustrate this procedure with the following example.

Example 14.2 Consider the $5^3 \times 3 \times 2^2$ factorial. Since for $\ell = 2$, $(5^2 - 1)/(5-1) = 6$ we can construct an OMEP with 25 runs from an OMEP for the 5^6 factorial in 25 runs, using the method of Section 11.7.1. The construction of this plan is illustrated in Table 14.4a. Using the collapsing schemes

$$\begin{array}{ccccccc} 0 & \longrightarrow & 0 & & 0 & \longrightarrow & 0 \\ 1 & \longrightarrow & 1 & & 1 & \longrightarrow & 1 \\ 2 & \longrightarrow & 2 & \text{and} & 2 & \longrightarrow & 0 \\ 3 & \longrightarrow & 0 & & 3 & \longrightarrow & 1 \\ 4 & \longrightarrow & 1 & & 4 & \longrightarrow & 0 \end{array}$$

we then obtain the final design given in Table 14.4b. It can easily be verified that the proportionality conditions are met. For example, for factors F_4 and F_5 we have $N_{00} = 6$, $N_{0.} = 10$, $N_{.0} = 15$, and with $N = 25$ we have $N_{00} = 10 \cdot 15/25 = 6$. □

Expanding Factors

Rather than collapsing the s_1-level factor to an s_i-level factor we may, in certain situations, expand the s_1-level factor into a set of s_i-level factors if $s_1 = s_i^m$ for some i. We know that there exists an OMEP for $\nu = (s_i^m - 1)/(s_i - 1)$ factors at s_i levels with s_i^m runs, that is, an $s_{i\text{III}}^{\nu-(\nu-m)}$. We then construct the $s_{1\text{III}}^{n-(n-\ell)}$ for a suitable n as discussed before and ℓ determined from $n \le (s_1^\ell - 1)/(s_1 - 1)$. Next, we replace each of the s_1 levels for one factor by a treatment combination from the $s_{i\text{III}}^{\nu-(\nu-m)}$. We thus replace an s_1-level factor by $\nu = (s_i^m - 1)/(s_i - 1)$ s_i-level factors. This method will be most useful for $s_1 = 4 = 2^2$ and perhaps for $s_1 = 8 = 2^3$, $s_1 = 9 = 3^2$. We now illustrate this method in the following example.

Example 14.3 Consider the $4^3 \times 3 \times 2^3$ factorial. Instead of using the collapsing factors method with $n_1 = 3$, $n_2 = 1$, $n_3 = 3$ which would lead to an OMEP with 64 runs, we use the fact that a 4-level factor can be replaced by $\nu = 3$ two-level factors. We thus consider an OMEP for the 4^5 factorial that can be obtained with 16 runs and then use replacing and collapsing as given in Table 14.5. For details of the construction of the auxiliary OMEP using GF(2^2) arithmetic, we refer to Sections 11.7 and 11.13.1 (Example 11.5). In the final plan (Table 14.5b), the elements in GF(2^2) are replaced by 0, 1, 2, and 3, respectively. We use the following collapsing scheme for F_4:

Four-Level Factor (F_4)		Three-Level Factor (F_4)
0	⟶	0
1	⟶	1
2	⟶	2
3	⟶	0

Table 14.4 Orthogonal Main Effect Plan for $5^3 \times 3 \times 2^2$ Factorial

x \ α		F_1 (1 0)	F_2 (0 1)	F_3 (1 1)	F_4 (1 2)	F_5 (1 3)	F_6 (1 4)
		a. Auxiliary Plan (5^6 in 25 Runs)					
0	0	0	0	0	0	0	0
0	1	0	1	1	2	3	4
0	2	0	2	2	4	1	3
0	3	0	3	3	1	4	2
0	4	0	4	4	3	2	1
1	0	1	0	1	1	1	1
1	1	1	1	2	3	4	0
1	2	1	2	3	0	2	4
1	3	1	3	4	2	0	3
1	4	1	4	0	4	3	2
2	0	2	0	2	2	2	2
2	1	2	1	3	4	0	1
2	2	2	2	4	1	3	0
2	3	2	3	0	3	1	4
2	4	2	4	1	0	4	3
3	0	3	0	3	3	3	3
3	1	3	1	4	0	1	2
3	2	3	2	0	2	4	1
3	3	3	3	1	4	2	0
3	4	3	4	2	1	0	4
4	0	4	0	4	4	4	4
4	1	4	1	0	1	2	3
4	2	4	2	1	3	0	2
4	3	4	3	2	0	3	1
4	4	4	4	3	2	1	0

b. Final Plan

F_1	F_2	F_3	F_4	F_5	F_6
0	0	0	0	0	0
0	1	1	2	1	0
0	2	2	1	1	1
0	3	3	1	0	0
0	4	4	0	0	1
1	0	1	1	1	1
1	1	2	0	0	0
1	2	3	0	0	0
1	3	4	2	0	1
1	4	0	1	1	0
2	0	2	2	0	0

Table 14.4 (*Continued*)

F_1	F_2	F_3	F_4	F_5	F_6
2	1	3	1	0	1
2	4	4	1	1	0
2	3	0	0	1	0
2	4	1	0	0	1
3	0	3	0	1	1
3	1	4	0	1	0
3	2	0	2	0	1
3	3	1	1	0	0
3	4	2	1	0	0
4	0	4	1	0	0
4	1	0	1	0	1
4	2	1	0	0	0
4	3	2	0	1	1
4	4	3	2	1	0

Table 14.5 Orthogonal Main Effect Plan for $4^3 \times 3 \times 2^3$ Factorial

x \ α	F_1 (u_1, u_0)	F_2 (u_0, u_1)	F_3 (u_1, u_1)	F_4 (u_1, u_2)	F_5 (u_1, u_3)
			a. Auxiliary OMEP		
$u_0,\ u_0$	u_0	u_0	u_0	u_0	u_0
$u_0,\ u_1$	u_0	u_1	u_1	u_2	u_3
$u_0,\ u_2$	u_0	u_2	u_2	u_3	u_1
$u_0,\ u_3$	u_0	u_3	u_3	u_1	u_2
$u_1,\ u_0$	u_1	u_0	u_1	u_1	u_1
$u_1,\ u_1$	u_1	u_1	u_0	u_3	u_2
$u_1,\ u_2$	u_1	u_2	u_3	u_2	u_0
$u_1,\ u_3$	u_1	u_3	u_2	u_0	u_3
$u_2,\ u_0$	u_2	u_0	u_2	u_2	u_2
$u_2,\ u_1$	u_2	u_1	u_3	u_0	u_1
$u_2,\ u_2$	u_2	u_2	u_0	u_1	u_3
$u_2,\ u_3$	u_2	u_3	u_1	u_3	u_0
$u_3,\ u_0$	u_3	u_0	u_3	u_3	u_3
$u_3,\ u_1$	u_3	u_1	u_2	u_1	u_0
$u_3,\ u_2$	u_3	u_2	u_1	u_0	u_2
$u_3,\ u_3$	u_3	u_3	u_0	u_2	u_1

Table 14.5 ***(Continued)***

b. Final Plan

F_1	F_2	F_3	F_4	$F_{5,1}$	$F_{5,2}$	$F_{5,3}$
0	0	0	0	0	0	0
0	1	1	2	1	1	0
0	2	2	0	0	1	1
0	3	3	1	1	0	1
1	0	1	1	0	1	1
1	1	0	0	1	0	1
1	2	3	2	0	0	0
1	3	2	0	1	1	0
2	0	2	2	1	0	1
2	1	3	0	0	1	1
2	2	0	1	1	1	0
2	3	1	0	0	0	0
3	0	3	0	1	1	0
3	1	2	1	0	0	0
3	2	1	0	1	0	1
3	3	0	2	0	1	1

and the following replacement (expansion):

F_5		$F_{5,1}$	$F_{5,2}$	$F_{5,3}$
0	$\longrightarrow$	0	0	0
1	$\longrightarrow$	0	1	1
2	$\longrightarrow$	1	0	1
3	$\longrightarrow$	1	1	0

□

Method 2

For the $s \times s_1^{n_1} \times s_2^{n_2} \times \cdots \times s_q^{n_q}$ factorial we can construct an OMEP with $2s_1^\ell$ runs, where s_1 is a prime or prime power and

$$s_1 > s_2 > \cdots > s_1 \qquad s_1 < s \leq 2s_1$$

and

$$1 + \sum_{i=1}^{q} n_i \leq \frac{s_1^\ell - 1}{s_1 - 1} \tag{14.16}$$

We construct first the OMEP for the s_1^ν factorial where $\nu = (s_1^\ell - 1)/(s_1 - 1)$ and ℓ is chosen such that (14.16) is satisfied. We then duplicate the plan except that the levels 0, 1, 2, ..., $s_1 - 1$ for the first factor, F_1, are replaced by s_1, $s_1 + 1$, $s_1 + 2$, ..., $2s_1 - 1$, respectively. This will be an OMEP with $2s_1^\ell$ runs for the $2s_1 \times s_1^\nu$ factorial. If $s < 2s_1$, we obtain the final plan by collapsing F_1 and then proceed as in method 1a. We illustrate this with the following example.

Example 14.4 We consider the $5 \times 3^2 \times 2$ factorial. We have $1 + 2 + 1 = 4 \leq (3^\ell - 1)/(3 - 1)$. Hence $\ell = 2$, and we construct the 3_{III}^{4-2}, duplicate it (see Table 14.6a) and collapse factors F_1 and F_4 as follows:

Six-Level Factor (F_1)		Five-Level Factor (F_1)
0	$\longrightarrow$	0
1	$\longrightarrow$	1
2	$\longrightarrow$	2
3	$\longrightarrow$	3
4	$\longrightarrow$	4
5	$\longrightarrow$	0

Three-Level Factor (F_4)		Two-Level Factor (F_4)
0	$\longrightarrow$	0
1	$\longrightarrow$	1
2	$\longrightarrow$	0

The final design is given in Table 14.6b. □

Method 3

In some cases the methods described above lead to an unnecessarily large number of runs since the auxiliary plan is based on an OMEP involving factors with the largest number of levels. The method to be described is, by contrast, based on an appropriate OMEP for factors having the smallest number of levels. The basic idea is, as a converse to method 1b, to replace several factors with the smallest number of levels by a factor with a larger number of levels. More precisely, we can construct an OMEP for the $s_1^{n_1} \times s_2^{n_2} \times \cdots \times s_q^{n_q}$ factorial with s_1^ℓ runs, where s_1 is a prime or prime power, $s_1 < s_2, \ldots, < s_q$ and

$$\sum_{i=1}^{q} \lambda_i n_i \leq \frac{s_1^\ell - 1}{s_1 - 1} \tag{14.17}$$

with $1 = \lambda_1 < \lambda_2 \leq \lambda_3 \leq \cdots \leq \lambda_q$ and λ_i ($i = 1, 2, \ldots, q$) an integer. The basic idea of this procedure is based on the following theorem.

Theorem 14.2 In the $s_{\text{III}}^{n-(n-\ell)}$ with $n = (s^\ell - 1)/(s - 1)$ and s a prime or prime power, a factor with t levels, $s^{m-1} < t \leq s^m$, can be introduced as a replacement for a suitably chosen set of $\nu = (s^m - 1)/(s - 1)$ factors that will preserve the orthogonality of the main effect estimates.

Table 14.6 Orthogonal Main Effect Plan for 5 x 3^2 x 2 Factorial

a. Auxiliary Plan				b. Final Plan			
F_1	F_2	F_3	F_4	F_1	F_2	F_3	F_4
0	0	0	0	0	0	0	0
0	1	1	2	0	1	1	0
0	2	2	1	0	2	2	1
1	0	1	1	1	0	1	1
1	1	2	0	1	1	2	0
1	2	0	2	1	2	0	0
2	0	2	2	2	0	2	0
2	1	0	1	2	1	0	1
2	2	1	0	2	2	1	0
3	0	0	0	3	0	0	0
3	1	1	2	3	1	1	0
3	2	2	1	3	2	2	1
4	0	1	1	4	0	1	1
4	1	2	0	4	1	2	0
4	2	0	2	4	2	0	0
5	0	2	2	0	0	2	0
5	1	0	1	0	1	0	1
5	2	1	0	0	2	1	0

Proof Let $t = s^m$. There exists an $s_{\text{III}}^{\nu-(\nu-m)}$ with $\nu = (s^m - 1)/(s-1)$. Hence a factor with $t = s^m$ levels can replace ν factors with s levels each. If $s^{m-1} < t < s^m$, then a factor with s^m levels can be collapsed.

We illustrate this method with the following example. □

Example 14.5 We consider the $2^4 \times 4$ factorial, that is, $s_1 = 2$, $s_2 = 4$, $n_1 = 4$, $n_2 = 1$. Now $t = 4 = 2^2$, that is, $m = 2$ and $\nu = 3$ and a 4-level factor can replace three 2-level factors in the $2_{\text{III}}^{n-(n-\ell)}$ where n and ℓ are determined as follows so that (14.17) is satisfied: With $n_1 = 4$, $n_2 = 1$, and $\nu = 3$ we have for (14.17) with $\lambda_2 = \nu = 3$

$$4 + 3 \cdot 1 = n \leq (2^\ell - 1)/(2-1)$$

which yields $\ell = 3$. Hence we can obtain the OMEP with $N = 2^3 = 8$ runs from the 2_{III}^{7-4}. The final plan is given in Table 14.7 using the replacement according to the 2_{III}^{3-1} given below:

Two-Level Factors in 2_{III}^{3-1}		**Four-Level Factor**
0 0 0	$\longrightarrow$	0
0 1 1	$\longrightarrow$	1
1 0 1	$\longrightarrow$	2
1 1 0	$\longrightarrow$	3

Table 14.7 Orthogonal Main Effect Plan for $2^4 \times 4$ Factorial

x \ α	F_1 (1 0 0)	F_2 (0 1 0)	F_3 (1 1 0)	F_4 (0 0 1)	F_5 (1 0 1)	F_6 (0 1 1)	F_7 (1 1 1)
				a. Auxiliary Plan			
0 0 0	0	0	0	0	0	0	0
0 0 1	0	0	0	1	1	1	1
0 1 0	0	1	1	0	0	1	1
0 1 1	0	1	1	1	1	0	0
1 0 0	1	0	1	0	1	0	1
1 0 1	1	0	1	1	0	1	0
1 1 0	1	1	0	0	1	1	0
1 1 1	1	1	0	1	0	0	1

b. Final Plan

F_1'	F_4	F_5	F_6	F_7
0	0	0	0	0
0	1	1	1	1
1	0	0	1	1
1	1	1	0	0
2	0	1	0	1
2	1	0	1	0
3	0	1	1	0
3	1	0	0	1

It follows then that in the auxiliary plan (Table 14.5a) factors F_1, F_2, F_3 with two levels are replaced by factor F_1' with four levels. □

This procedure of constructing an OMEP is most useful if $s < t \leq s^2$. It follows then from Theorem 14.2 that the maximum number of t-level factors that can be introduced into the $s_{\text{III}}^{n-(n-\ell)}$ with $n = (s^\ell - 1)/(s - 1)$ is equal to $(s^\ell - 1)/(s - 1)$ if ℓ is even, and equal to the largest integer less than or equal to $(s^\ell - 1)/(s - 1) - 1$ if ℓ is odd.

The three methods described above lead to a very large number of OMEPs for asymmetrical factorials. An extensive catalog of such plans is provided by Addelman and Kempthorne (1961), Addelman (1962a, 1962b), and Dey (1985). Dey also lists some additional OMEPs not derivable by the Addelman–Kempthorne methods but instead derivable from Hadamard matrices or orthogonal arrays. For details the reader is referred to Dey (1985).

For the $2^n \times 3^m$ Addelman–Kempthorne designs Margolin (1968) has investigated the aliasing of main effects with 2-factor interactions. This is important for the reasons discussed in Section 14.2 for OMEPs for symmetrical factorials. We shall not discuss the rather complicated results here but refer to the method illustrated in Table 14.2.

14.4 NONORTHOGONAL RESOLUTION III DESIGNS

Orthogonal main effect plans are useful because they are easy to analyze and the results are easy to interpret. Sometimes there may, however, be practical reasons why an OMEP may not be the most suitable design. In that case nonorthogonal MEPs may be more appropriate. We shall give some examples.

As we have mentioned earlier, a saturated OMEP does not provide any degrees of freedom for the estimation of the error variance. If no information about the error variance is available, additional observations must be obtained. To replicate the OMEP may lead to too large an experiment. Hence only some runs may be replicated leading to nonproportional frequencies and as a consequence to correlated estimates of the main effects.

In contrast to the just described situation an OMEP may not be a saturated plan and may therefore have too many runs. It is always possible to obtain a saturated MEP. One such plan is the *one-at-a-time plan*, where if the runs are considered in sequence, only the level of one factor is changed from one run to the next. As an example consider the following MEP for a 2^8 in nine runs (apart from randomization):

Run	F_1	F_2	F_3	F_4	F_5	F_6	F_7	F_8
1	0	0	0	0	0	0	0	0
2	1	0	0	0	0	0	0	0
3	1	1	0	0	0	0	0	0
4	1	1	1	0	0	0	0	0
5	1	1	1	1	0	0	0	0
6	1	1	1	1	1	0	0	0
7	1	1	1	1	1	1	0	0
8	1	1	1	1	1	1	1	0
9	1	1	1	1	1	1	1	1

An alternative to the plan above is one where each run, other than the control, has only one factor at the high level and all other factors at the low level (apart from randomization):

Run	F_1	F_2	F_3	F_4	F_5	F_6	F_7	F_8
1	0	0	0	0	0	0	0	0
2	1	0	0	0	0	0	0	0
3	0	1	0	0	0	0	0	0
4	0	0	1	0	0	0	0	0
5	0	0	0	1	0	0	0	0
6	0	0	0	0	1	0	0	0
7	0	0	0	0	0	1	0	0
8	0	0	0	0	0	0	1	0
9	0	0	0	0	0	0	0	1

Obviously, this method can be adapted to both symmetrical and asymmetrical factorials with factors at any number of levels. It should be clear to the reader that if a saturated OMEP exists for a given factorial it will be more efficient than the comparable nonorthogonal MEP. But often statistical properties may have to be balanced against practical considerations.

In listing the designs above we have added for emphasis the cautionary note "apart from randomization." There may be settings, especially in industrial applications, where it would be most economical and feasible to execute the experiment using the run order given above. This may have serious consequences for the analysis and interpretation, since not only is there lack of randomization (see I.9.9) but this would also invoke a split-plot type of experiment if the levels of factors are not reset for each run (see I.13). For a discussion of such problems see Daniel (1973) and Webb (1968).

CHAPTER 15

Supersaturated Designs

15.1 INTRODUCTION AND RATIONALE

We have seen in the preceding chapters that very often practical considerations lead the experimenter to using fractional factorial designs. The aim, of course, is to reduce the size of the experiment without sacrificing important information. Even then, the suggested design may be quite large. This may be true even for the most highly fractionated design, the saturated main effect plan, if the number of factors under investigation is large. And it is not unusual that in many physical, chemical, or industrial experiments the number of factors that one would like to study is quite large, but consideration of cost calls for a small number of runs, that is, treatment combinations. Moreover, it is suspected that among the large number of factors only few are important or active. Such situations have led to the development of what are called *supersaturated designs* (for a definition see Section 15.3).

The idea is to somehow select N treatment combinations from the totality of all possible treatment combinations, say from the $s_1 \times s_2 \times \cdots \times s_n$ combinations of a factorial with factors $A_1, A_2, \ldots, A_n$ where factor A_i has p_i levels $(i = 1, 2, \ldots, n)$ and $n > N - 1$. Obviously, orthogonality as with fractional factorials can no longer be achieved. Rather, orthogonality is being replaced by some measure of near orthogonality. Also the analysis of data from such an experiment will have to be approached differently from the usual analysis of factorial experiments. Simultaneous estimation of effects will no longer be possible. Instead, based on the assumption of scarcity of active factors and absence of interactions, stepwise forward regression provides a plausible approach.

15.2 RANDOM BALANCE DESIGNS

One way to select the N treatment combinations is by using a random sampling process. This idea has led to the notion of *random balance designs* (Satterthwaite, 1959).

Design and Analysis of Experiments. Volume 2: Advanced Experimental Design
By Klaus Hinkelmann and Oscar Kempthorne
ISBN 0-471-55177-5

Here the notion of random balance is understood in the following sense (Raghavarao, 1986): Factor A_j is *randomly balanced* with respect to factor A_i if the random sampling process used to select the p_j levels of factor A_j is the same for each level of factor A_i. If every factor has random balance with respect to every other factor, then Satterthwaite (1959, p. 113) refers to this as a *pure random balance design*.

Satterthwaite (1959, p. 114) states that "pure random balance designs are almost always inefficient in some sense" (see Section 15.4.1), but at the same time he argues that "there are important and large classes of problems for which these inefficiencies are often unimportant or even trivial." These views were largely contradicted by Youden et al. (1959), who, among other things, called for a systematic, rather than random, construction of designs for the earlier stated purpose.

15.3 DEFINITION AND PROPERTIES OF SUPERSATURATED DESIGNS

Generalizing some results of Yamada and Lin (1999), we consider an experiment with n factors $A_1, A_2, \ldots, A_n$, each with p levels denoted by $1, 2, \ldots, p$. Let N be the number of runs (treatment combinations) with (i) $(p-1)n > N-1$ and (ii) $N = p \cdot q$ and q a positive integer. Further, let $\mathcal{C}(N)$ be the set of N-dimensional vectors each containing each of the p levels exactly q times.

A p-level supersaturated design D can be described as a selection of n vectors in $\mathcal{C}(N)$, say $\boldsymbol{c}_1, \boldsymbol{c}_2, \ldots, \boldsymbol{c}_n$, by some specified rule. The design D can then be represented by the $N \times n$ array

$$[\boldsymbol{c}_1, \boldsymbol{c}_2, \ldots, \boldsymbol{c}_n] \tag{15.1}$$

Obviously, the rule by which the selection leading to (15.1) is made is of major importance. As mentioned earlier, orthogonality of the vectors in (15.1) cannot be achieved, but we would like to stay as "close" as possible to this property, which is exhibited in the Fisher and Plackett–Burman main effect plans (see Sections 14.2.1 and 14.2.4). Following Yamada and Lin (1999) we shall adopt the following χ^2 statistic as a measure of deviation from orthogonality and choose the design D, which achieves the smallest amount of deviation. Such a design we declare to exhibit near-optimality.

Let $n^{ab}(\boldsymbol{c}_i, \boldsymbol{c}_j)$ denote the number of rows in the $N \times 2$ array $[\boldsymbol{c}_i, \boldsymbol{c}_j]$ whose values are $[a, b]$, where $\boldsymbol{c}_i$, $\boldsymbol{c}_j$, $(i < j)$ are two columns from (15.1). Obviously,

$$\sum_{a,b=1}^{p} n^{ab}(\boldsymbol{c}_i, \boldsymbol{c}_j) = N$$

for any pair $i,\ j (1 \le i < j \le n)$. The χ^2 statistic mentioned above is then defined as

$$\chi_{ij}^2 = \sum_{a,b=1}^{p} \frac{[n^{ab}(\boldsymbol{c}_i, \boldsymbol{c}_j) - N/p^2]^2}{N/p^2} \tag{15.2}$$

For orthogonality (or complete independence) of $\boldsymbol{c}_i$ and $\boldsymbol{c}_j$ we have $n^{ab}(\boldsymbol{c}_i, \boldsymbol{c}_j) = N/p^2$ for each pair $[a, b]$ $(1 \le a,\ b \le p)$, whereas for complete confounding of factors A_i, A_j (or complete dependence) we have, without loss of generality,

$$n^{ab}(\boldsymbol{c}_i, \boldsymbol{c}_j) = \begin{cases} N/p & \text{for } a = b \\ 0 & \text{otherwise} \end{cases}$$

It is then easy to see that the χ^2 value of (15.2) lies between 0 and $(p-1)N$.

For the overall evaluation of deviation from orthogonality for the design D, we then use either

$$\text{av}\ \chi^2 = \frac{\sum\limits_{\substack{i,j=1 \\ i<j}}^{n} \chi_{i,j}^2}{n(n-1)/2} \tag{15.3}$$

or

$$\max\ \chi^2 = \max\ \chi_{ij}^2 \qquad 1 \le i < j \le n \tag{15.4}$$

with χ_{ij}^2 given in (15.2). Using either (15.3) or (15.4) the design D with the smallest value will be chosen among competing arrangements.

For an extension of the χ^2 criterion to mixed-level supersaturated designs see Yamada and Matsui (2002).

15.4 CONSTRUCTION OF TWO-LEVEL SUPERSATURATED DESIGNS

15.4.1 Computer Search Designs

Most of the emphasis of devising rules for obtaining a suitable array (15.1) has been concentrated on 2-level supersaturated designs, that is, $p = 2$, denoting the levels by -1 and $+1$. Booth and Cox (1962) provided the first set of such designs for the following pairs of $(n,\ N)$ values: (16, 12), (20, 12), (24, 12), (24, 18), (30, 18), (36, 18), and (30, 24). These designs were obtained through computer search.

To compare their designs with random balance designs, Booth and Cox (1962) consider $s_{ij} = \boldsymbol{c}_i'\boldsymbol{c}_j$ $(i < j)$ as a measure of nonorthogonality of factors A_i and A_j. As an overall measure of the non-orthogonality for the design under consideration, they then define

$$E(s^2) = \frac{\sum\limits_{i<j} s_{ij}^2}{n(n-1)/2} \tag{15.5}$$

Yamada and Lin (1999) show that (15.5) is equivalent to (15.3) for $p = 2$. For a random balance design (15.5) is given by

$$E_{\text{RBD}}(s^2) = \frac{N^2}{N-1} \tag{15.6}$$

(Booth and Cox, 1962). It turns out that for all designs constructed by Booth and Cox (1962)

$$E(s^2) < E_{\text{RBD}}(s^2)$$

indicating a smaller degree of nonorthogonality, that is, less confounding among the factors, of their designs, a result that holds true in general for other supersaturated designs as well.

15.4.2 Hadamard-Type Designs

A more systematic way to construct a special class of 2-level supersaturated designs was provided by Lin (1993). Just as the Plackett–Burman designs (see Section 14.2.4) are based on Hadamard matrices of order M, so are the proposed supersaturated designs. Their construction can be described in general as follows: Consider a Hadamard matrix of order M (which is not constructed via foldover) in seminormalized form. Delete the first column of $+1$'s and label the remaining columns $1, 2, \ldots, M-1$. Choose one of these columns as the branching column, that is, the column that divides the Hadamard matrix into two half-fractions according to its $+1$'s and -1's. Delete the branching column and choose a sign, say $+$. Then the $M/2$ rows that corresponding to $+1$ in the branching column form a supersaturated design with $n = M - 2$ factors and $N = M/2$ runs, where each factor is, of course, represented $M/4$ times at the high and low level.

Letting $\boldsymbol{H}$ be an $M \times M$ Hadamard matrix, then the procedure described above can be formally represented as

$$\boldsymbol{H} = \begin{bmatrix} \mathfrak{I} & \mathfrak{I} & \boldsymbol{H}^h \\ \mathfrak{I} & -\mathfrak{I} & * \end{bmatrix} \tag{15.7}$$

Table 15.1 Supersaturated Design for $n = 10$ Factors in $N = 6$ Runs

			Factor										
Row			1	2	3	4	5	6	7	8	9	10	Run
1	+	+	+	−	+	+	+	−	−	−	+	−	1
2	+	+	−	+	+	+	−	−	−	+	−	+	2
3	+	+	+	+	−	−	−	+	−	+	+	−	3
4	+	+	+	−	−	−	+	−	+	+	−	+	4
5	+	+	−	−	−	+	−	+	+	−	+	+	5
6	+	+	−	+	+	−	+	+	+	−	−	−	6
7	+	−	+	+	−	+	+	+	−	−	−	+	
8	+	−	+	+	+	−	−	−	+	−	+	+	
9	+	−	−	−	+	−	+	+	−	+	+	+	
10	+	−	−	+	−	+	+	−	+	+	+	−	
11	+	−	+	−	+	+	−	+	+	+	−	−	
12	+	−	−	−	−	−	−	−	−	−	−	−	

(Bulutoglu and Cheng, 2003), where $\mathfrak{I}$ represents a vector of $N/2$ unity elements and $\boldsymbol{H}^h$ represents the supersaturated design, that is, the columns of $\boldsymbol{H}^h$ represent the columns $[\boldsymbol{c}_1, \boldsymbol{c}_2, \ldots, \boldsymbol{c}_n]$ of (15.1) with $n = M - 2$.

Example 15.1 We shall illustrate this method with the design for $M = 12$ given by Lin (1993). Table 15.1 contains the Hadamard matrix in the form indicated by (15.7), where $+$ represents $+1$ and $-$ represents -1.

It is easy to see that for this from (15.5) design all $| s_{ij} | = 2$ and hence $E(s^2) = 4$, compared to $E_{\text{RBD}}(s^2) = 7.2$ from (15.6). □

For $n < M - 2$ we can delete certain columns from the complete design. In practice it does not matter which column we choose as the branching column and which columns we delete subsequently (Lin, 1993).

Another Hadamard-type supersaturated design was proposed by Wu (1993). Starting with a seminormalized Hadamard matrix of order M, it makes use of all its columns except the first and augments it by interaction columns to generate a supersaturated design with $N = M$ runs for n factors, where $M \leq n \leq 2M - 3$. The procedure can be described as follows.

Let the generator Hadamard matrix $\boldsymbol{H}$ be given by $\boldsymbol{H} = (\mathfrak{I}, \boldsymbol{x}_1, \boldsymbol{x}_2, \ldots, \boldsymbol{x}_{M-1})$. Obtain $M - 2$ interaction columns $\boldsymbol{x}_{i_0 j} = \boldsymbol{x}_{i_0} \circ \boldsymbol{x}_j$ as the elementwise product of $\boldsymbol{x}_{i_0}$ and $\boldsymbol{x}_j$ $(j \neq i_0, i_0$ fixed). For cyclic Hadamard matrices we can choose $i_0 = 1$; for noncyclic Hadamard matrices we choose i_0 to minimize $\max | \boldsymbol{x}_i' \boldsymbol{x}_{i_0 j} |$ over i and j. Thus, (15.1) is given by

$$[\boldsymbol{x}_1, \boldsymbol{x}_2, \ldots, \boldsymbol{x}_{M-1}, \boldsymbol{x}_{i_0 j} (j = 1, 2, \ldots, M-1;\ j \neq i_0)]$$

Among all inner products of the columns only $\boldsymbol{x}_i' \boldsymbol{x}_{i_0 j}$ with i, i_0, j all different may be nonzero. Hence the choice of i_0 ensures that the confounding among factors is kept to a minimum. It can be shown (Wu, 1993) that for the above designs

$$E(s^2) = \frac{M^2(n - M + 1)}{n(n-1)/2} \tag{15.8}$$

If $n < 2M - 3$, we simply choose $n - M + 1$ columns from the $M - 2$ interaction columns, keeping $E(s^2)$ as small as possible.

A third method to construct Hadamard-type supersaturated designs was proposed by Tang and Wu (1997). The method essentially consists of adjoining k Hadamard matrices (after deleting their first column of ones) of order M. The resulting array yields a supersaturated design for $n = k(M - 1)$ factors in M runs. For $n = (k - 1)(M - 1) + j$ factors (with $1 \leq j \leq M - 2$) we delete $M - 1 - j$ from the full design, that is, the design for $k(M - 1)$ factors.

We shall conclude this section by making some remarks about the possible optimality with regard to the $E(s^2)$ criterion. Nguyen (1996) and Tang and Wu (1997) have shown that a lower bound for $E(s^2)$ is given by

$$E(s^2) \geq \frac{M^2(n - M + 1)}{(n-1)(M-1)} \tag{15.9}$$

Nguyen (1996) has shown that the half-fraction Hadamard designs of Lin (1993) achieve the lower bound (15.9). This is true also for the full designs given by Tang and Wu (1997). Comparing (15.8) with (15.9) shows that for $n = 2M - 3$ the designs of Wu (1993) have $E(s^2)$ values only slightly larger than the lower bound. A similar statement can be made about the designs of Tang and Wu (1997) when $n = (k - 1)(M - 1) + j$. Obviously, designs that achieve the lower bound are $E(s^2)$-optimal.

15.4.3 BIBD-Based Supersaturated Designs

Interesting relationships between certain BIB designs and supersaturated designs were established by Nguyen (1996), Cheng (1997), and Liu and Zhang (2000). We shall not go into all the details here, but just describe one method based on the cyclic development of initial blocks as described by Liu and Zhang (2000). They establish the equivalence of existence of a supersaturated design with $N = 2v$, $n = c(2v - 1)$ (with $v \geq 3$, $c \geq 2$ and when v is odd, c is even) and that of a BIB design with $t = 2v - 1$, $b = c(2v - 1)$, $r = c(v - 1)$, $k = v - 1$, $\lambda = c(v - 2)/2$, which is cyclically developed from c initial blocks.

Given a BIB design with parameters specified above and the treatments labeled as $0, 1, 2, \ldots, t-1 = 2v-2$, the procedure to generate a supersaturated design can be described as follows:

1. Develop the first initial block cyclically, giving rise to blocks $1, 2, \ldots, 2v-1$.
2. Consider an array of $2v-1$ columns, each column containing the elements $0, 1, 2, \ldots, 2v-1$ (in that order), which correspond to the $2v-1$ blocks obtained in step 1. In column j replace the $v-1$ elements corresponding to the treatments in block j by $+1$ and the remaining v elements by -1.
3. Add a $+1$ in the first position to each column, thus generating the first $2v-1$ columns of the array (15.1).
4. Repeat steps 1, 2, and 3 for each of the remaining $c-1$ initial blocks, thus generating altogether the $n = c(2v-1)$ columns of (15.1).

We shall illustrate this procedure in the next example.

Example 15.2 For $v = 4$, $c = 2$ the two initial blocks are given by Liu and Zhang (2000) as (0 1 3) and (0 1 5). The individual steps then are:

(1) Develop (0 1 3) and (0 1 5) mod 7:

$$\begin{array}{ccccccc} 0 & 1 & 2 & 3 & 4 & 5 & 6 \\ 1 & 2 & 3 & 4 & 5 & 6 & 0 \\ 3 & 4 & 5 & 6 & 0 & 1 & 2 \end{array}$$

$$\begin{array}{ccccccc} 0 & 1 & 2 & 3 & 4 & 5 & 6 \\ 1 & 2 & 3 & 4 & 5 & 6 & 0 \\ 5 & 6 & 0 & 1 & 2 & 3 & 4 \end{array}$$

(2) + (3) + (4) and writing + for +1 and − for −1:

+	+	+	+	+	+	+	+	+	+	+	+	+	+
+	−	−	−	+	−	+	+	−	+	−	−	−	+
+	+	−	−	−	+	−	+	+	−	+	−	−	−
−	+	+	−	−	−	+	−	+	+	−	+	−	−
+	−	+	+	−	−	−	−	−	+	+	−	+	−
−	+	−	+	+	−	−	−	−	−	+	+	−	+
−	−	+	−	+	+	−	+	−	−	−	+	+	−
−	−	−	+	−	+	+	−	+	−	−	−	+	+

The columns represent the supersaturated design (15.1) with $N = 8$, $n = 14$.

Liu and Zhang (2000) provide a list of initial blocks for constructing a large number of supersaturated designs with $N = 2v$ and $n = c(2v - 1)$ for $3 \leq v \leq 12$. All these designs are $E(s^2)$ optimal. □

15.5 THREE-LEVEL SUPERSATURATED DESIGNS

Clearly, from a practical point of view two-level supersaturated designs are the most important. But in some situations three-level designs may be called for if the few active factors are expected to exhibit linear and/or quadratic effects.

Whereas there exist several methods of constructing two-level supersaturated designs, there exist only very few for constructing three-level designs (see Yamada and Lin, 1999; Yamada et al., 1999; Fang, Lin, and Ma, 2000). We shall describe briefly one method due to Yamada and Lin (1999).

Let $\boldsymbol{C} = [\boldsymbol{c}_1, \boldsymbol{c}_2, \ldots, \boldsymbol{c}_{n*}]$ be a two-level supersaturated design or saturated main effect plan in N^* runs. Let $\phi^{ab}(\boldsymbol{C})$ be an operator that changes the elements in $\boldsymbol{C}$ from -1 (or $-$) to a and $+1$ (or $+$) to b. Then consider the matrix

$$\boldsymbol{D} = \begin{bmatrix} \phi^{12}(\boldsymbol{C}) & \phi^{12}(\boldsymbol{C}) & \phi^{13}(\boldsymbol{C}) & \phi^{23}(\boldsymbol{C}) \\ \phi^{23}(\boldsymbol{C}) & \phi^{13}(\boldsymbol{C}) & \phi^{23}(\boldsymbol{C}) & \phi^{12}(\boldsymbol{C}) \\ \phi^{31}(\boldsymbol{C}) & \phi^{23}(\boldsymbol{C}) & \phi^{12}(\boldsymbol{C}) & \phi^{13}(\boldsymbol{C}) \end{bmatrix} \tag{15.10}$$

$$= \begin{bmatrix} \boldsymbol{D}_1 & \boldsymbol{D}_2 & \boldsymbol{D}_3 & \boldsymbol{D}_4 \end{bmatrix} \tag{15.11}$$

Array (15.10) represents a three-level supersaturated design with $N = 3N^*$ runs and $n = 4n^*$ factors. Each level of each factor occurs N^* times, and $2n > N$.

For $n = 2n^*$ or $n = 3n^*$ a combination of subdesigns $\boldsymbol{D}_i$ $(i = 1, 2, 3, 4)$ from (15.11) can be used to form a suitable supersaturated design. We shall illustrate this with the following example.

Example 15.3 Suppose we want to examine 14 three-level factors accounting for 28 effects, and suppose further that we are allowed 24 runs. A suitable design can be constructed by choosing for $\boldsymbol{C}$ the saturated main effect plan with $N^* = 8$ and $n^* = 7$, given by the following Hadamard matrix

$$\boldsymbol{C} = \begin{bmatrix} + & + & + & - & + & - & - \\ - & + & + & + & - & + & - \\ - & - & + & + & + & - & + \\ + & - & - & + & + & + & - \\ - & + & - & - & + & + & + \\ + & - & + & - & - & + & + \\ + & + & - & + & - & - & + \\ - & - & - & - & - & - & - \end{bmatrix} \tag{15.12}$$

Using $\boldsymbol{D}_2$ and $\boldsymbol{D}_3$ from (15.11) and applying the transformations to (15.12) yields the supersaturated design D:

2	2	2	1	2	1	1	3	3	3	1	3	1	1
1	2	2	2	1	2	1	1	3	3	3	1	3	1
1	1	2	2	2	1	2	1	1	3	3	3	1	3
2	1	1	2	2	2	1	3	1	1	3	3	3	1
1	2	1	1	2	2	2	1	3	1	1	3	3	3
2	1	2	1	1	2	2	3	1	3	1	1	3	3
2	2	1	2	1	1	2	3	3	1	3	1	1	3
1	1	1	1	1	1	1	1	1	1	1	1	1	1
3	3	3	1	3	1	1	3	3	3	2	3	2	2
1	3	3	3	1	3	1	2	3	3	3	2	3	2
1	1	3	3	3	1	3	2	2	3	3	3	2	3
3	1	1	3	3	3	1	3	2	2	3	3	3	2
1	3	1	1	3	3	3	2	3	2	2	3	3	3
3	1	3	1	1	3	3	3	2	3	2	2	3	3
3	3	1	3	1	1	3	3	3	2	3	2	2	3
1	1	1	1	1	1	1	2	2	2	2	2	2	2
3	3	3	2	3	2	2	2	2	2	1	2	1	1
2	3	3	3	2	3	2	1	2	2	2	1	2	1
2	2	3	3	3	2	3	1	1	2	2	2	1	2
3	2	2	3	3	3	2	2	1	1	2	2	2	1
2	3	2	2	3	3	3	1	2	1	1	2	2	2
3	2	3	2	2	3	3	2	1	2	1	1	2	2
3	3	2	3	2	2	3	2	2	1	2	1	1	2
2	2	2	2	2	2	2	1	1	1	1	1	1	1

□

Using the max χ^2 criterion (15.4), Yamada and Lin (1999) show that if $\max |s_{ij}| = d$, say, over all i, j $(1 \leq i < j \leq n^*)$, then for the design D in (15.10) or a combination of subdesigns from (15.11) we have

$$\max \chi^2 = \max \left\{ \frac{(N + 9d)^2}{8N}, \frac{N}{2} \right\}$$

This result is useful because it allows us to compare competing designs with regard to the maximum dependency between two factors, the smaller this value, the better the design.

15.6 ANALYSIS OF SUPERSATURATED EXPERIMENTS

Because the number of runs, N, is smaller than the number of parameters to be estimated, $(p - 1)n$, we cannot apply the usual analysis to the data from a supersaturated experiment, that is, an experiment using a supersaturated design.

Therefore, the analysis of such data must be predicated on the principle (or assumption) of active factor sparsity (Box and Meyer, 1986a, 1986b) or, in the case of p-level designs, effect sparsity as each factor is associated with $p - 1$ effects. Furthermore, we assume that all interactions are negligible.

To put this in the context of a linear regression model, we express the observation vector $\boldsymbol{y}' = (y_1, y_2, \ldots, y_N)$ as

$$\boldsymbol{y} = \boldsymbol{X}\boldsymbol{\beta} + \boldsymbol{e} \tag{15.13}$$

where $\boldsymbol{X}$ is an $N \times [(p-1)n + 1]$ known matrix, which is determined by the design matrix, $\boldsymbol{\beta}$ is the $[(p-1)n+1]$ vector of unknown parameters, and $\boldsymbol{e}$ is the error vector with the variance–covariance structure for a completely randomized design (see I.6.3.4). Based on the assumption of effect sparsity we may rewrite (15.13) as

$$\boldsymbol{y} = (\boldsymbol{X}_1, \boldsymbol{X}_2)\begin{pmatrix}\boldsymbol{\beta}_1\\ \boldsymbol{\beta}_2\end{pmatrix} + \boldsymbol{e} \tag{15.14}$$

where $\boldsymbol{\beta}_1$ represents the vector of nonzero or large effects (including the intercept) and $\boldsymbol{\beta}_2$ represents the vector of zero or negligible effects. Our aim, obviously, is to identify $\boldsymbol{\beta}_1$, that is, the "true" model.

This necessitates applying variable selection procedures. An obvious choice is stepwise variable selection. This method may, however, not always be the best choice. If the number of factors is very large, the probability is high that some of the inert factors may be selected (type I error). Also, mainly because of the nonorthogonality of the $\boldsymbol{X}$ matrix, it may not always select active factors (type II error). A thorough discussion of these difficulties has been provided by Westfall, Young, and Lin (1998). To overcome in particular type I error problems, they recommend using adjusted P values at each step of the stepwise procedure, using familywise error rates. The familywise error rate should not be higher than .50 for inclusion of a variable in the model.

It is for these reasons that Abraham, Chipman, and Vijayan (1999) recommend the use of all-subsets regression. For obvious reasons, this method has to be implemented in such a way that only subsets up to a certain prespecified size ($< N$) will be considered. An advantage of this method is that it may produce different competing models, which allow the experimenter to make a choice based perhaps on subject matter considerations.

Using similar arguments against the use of stepwise variable selection, Li and Lin (2002) propose a penalized least-squares approach to the variable selection problem. Let us rewrite (15.13) as

$$y_i = \boldsymbol{x}_i' \boldsymbol{\beta} + e_i \qquad (i = 1, 2, \ldots, N) \tag{15.15}$$

where $\boldsymbol{x}_i$ is the vector of input variables. For example, for the two-level design (15.15) is

$$y_i = \beta_0 + x_{1i}\beta_1 + x_{2i}\beta_2 + \cdots + x_{ni}\beta_n + e_i$$

or for the three-level design we have

$$y_i = \beta_0 + x_{1i}\beta_1 + x_{1i}^2\beta_{11} + x_{2i}\beta_2 + x_{2i}^2\beta_{22} + \cdots + x_{ni}\beta_n + x_{ni}^2\beta_{nn} + e_i$$

A form of penalized least squares is defined as

$$Q(\boldsymbol{\beta}) = \frac{1}{2N}\sum_{i=1}^{N}(y_i - \boldsymbol{x}_i'\boldsymbol{\beta})^2 + \sum_{j=0}^{(p-1)n} p_{\lambda_N}(\mid\beta_j\mid) \tag{15.16}$$

(Fan and Li, 2001; Li and Lin, 2002), where $p_{\lambda_N}(\cdot)$ is a penalty function and λ_N is a tuning parameter, which is usually chosen by some data-driven approach.

Since this method cannot be applied to the full model (15.15), Li and Lin (2002) recommend to proceed in two steps: Apply stepwise variable selection (or subset selection, we might add) in step 1, using a low threshold, that is, a high α value, and then, in step 2, apply (15.16) to the model obtained in step 1. More formally, in (15.16) $\boldsymbol{\beta} = \boldsymbol{\beta}_1$ and $\boldsymbol{x}_i = \boldsymbol{x}_{1i}$, say, where $\boldsymbol{x}_{1i}$ conforms to $\boldsymbol{\beta}_1$.

An appropriate penalty function for supersaturated experiments is the smoothly clipped absolute deviation (SCAD) penalty, whose first-order derivative is

$$p_\lambda'(|\beta|) = \lambda\left\{I(|\beta| \le \lambda) + \frac{(a\lambda - |\beta|)_+}{(a-1)\lambda}I(|\beta| > \lambda)\right\} \tag{15.17}$$

for $|\beta| > 0$ with $a = 3.7$, and $p_\lambda(0) = 0$ (Fan and Li, 2001; Li and Lin, 2002).

Suppose $\boldsymbol{\beta}_1$ in (15.14) has been identified in step 1 above to consist of d terms, say $\beta_{(1)}, \beta_{(2)}, \ldots, \beta_{(d)}$. Let $\boldsymbol{\beta}_1^{(0)}$ be an initial value close to the true value $\boldsymbol{\beta}_1$, when $\beta_{(j)}$ is not very close to zero. The penalty function (15.17) can be locally approximated by a quadratic function as

$$\left[p_\lambda(|\beta_{(j)}|)\right]' \approx \left\{\frac{p_\lambda'\left(|\beta_{(j)}^{(0)}|\right)}{|\beta_{(j)}^{(0)}|}\right\}\beta_{(j)} \tag{15.18}$$

and $\widehat{\beta}_{(j)} = 0$ if $\beta_{(j)}^{(0)}$ is close to zero. With the local approximation (15.18), the solution to the penalized least square can be obtained iteratively as

$$\boldsymbol{\beta}_1^{(1)} = \left\{\boldsymbol{X}_1'\boldsymbol{X}_1 + N\sum\nolimits_\lambda\left(\boldsymbol{\beta}_1^{(0)}\right)\right\}^{-1}\boldsymbol{X}_1'\boldsymbol{y} \tag{15.19}$$

with

$$\sum\nolimits_\lambda\left(\boldsymbol{\beta}_1^{(0)}\right) = \text{diag}\left\{\frac{p_\lambda'\left(|\beta_{(j)}^{(0)}|\right)}{|\beta_{(j)}^{(0)}|}\right\}_{j=1,2,\ldots,d}$$

For more details about the ridge regression (15.19) and description of the properties of the penalized least-squares estimators we refer the reader to Li and

Lin (2002). See also Li and Lin (2003) and Holcomb, Montgomery, and Carlyle (2003) for a comparison of analysis procedures.

Regardless of which type of analysis is used, special care has to be exercised to interpret the results. We always have to remember that supersaturated designs are used for screening purposes, and the primary objective is to eliminate a large portion of inactive factors from further consideration. The nature of the designs do not guarantee that we always find the right answer. In any case, smaller follow-up experiments will be needed.

CHAPTER 16

Search Designs

16.1 INTRODUCTION AND RATIONALE

The usefulness of factorial experiments is based on the premise that high-order interactions are negligible. In previous chapters we have shown how this premise can be used to construct useful and efficient factorial designs through confounding and fractionation. Extreme examples of the latter category are, of course, main effect plans and supersaturated designs. They are based on the severe and often nonsustainable assumption that *all* interactions are negligible.

It is this point that leads to the question whether one can construct designs that (1) have a relatively small number of runs and (2) allow the exploration of a limited number of low-level interactions. The emphasis here is on "limited number," because even for a 2^n factorial there are $n(n-1)/2$ two-factor and $n(n-1)(n-2)/6$ three-factor interactions and to investigate them all would necessitate a resolution VII design. In most cases this may lead to a prohibitively large number of runs. But how can we limit the number of interactions to be investigated since we do not know which of the interactions are possibly nonnegligible?

Clearly, we need to make some assumptions to overcome this difficulty. Rather than specifying the particular interactions we would like to investigate, we specify the number of interactions to be investigated. In other words, we assume that a specified number, say k, 2- or 3-factor interactions are possibly nonnegligible. We would then like to construct a design that allows us to (1) identify these nonnegligible interactions and (2) estimate these effects in addition to the main effects. This idea was first proposed and investigated by Srivastava (1975) when he introduced the concepts of *search design* and *search linear model*. We shall describe in this chapter some of his results and subsequent developments on this topic.

16.2 DEFINITION OF SEARCH DESIGN

Let us consider here the case of n factors, each at two levels. We know from Section 7.4.2 that the true response for each treatment combination can be

Design and Analysis of Experiments. Volume 2: Advanced Experimental Design
By Klaus Hinkelmann and Oscar Kempthorne
ISBN 0-471-55177-5

expressed as a linear function of the mean, all main effects and all interactions. We can divide the $2^n - 1$ factorial effects into three categories:

$\mathcal{C}_1$: contains the effects we want to estimate in any case, such as the main effects;

$\mathcal{C}_2$: contains effects we are interested in, but we know that only a few of them are nonnegligible, such as 2-factor interactions, $\mathcal{C}_2(2)$, or 2- and 3-factor interactions, $\mathcal{C}_2(2, 3)$;

$\mathcal{C}_3$: contains the remaining effects (such as higher-order interactions), which we consider to be negligible.

Based on the above scenario we can then express a set of N observations, given by the $N \times 1$ vector $\boldsymbol{y}$, in terms of the following linear model:

$$\boldsymbol{y} = \boldsymbol{A}_1\boldsymbol{\xi}_1 + \boldsymbol{A}_2\boldsymbol{\xi}_2 + \boldsymbol{e} \tag{16.1}$$

where $\boldsymbol{A}_1$, $\boldsymbol{A}_2$ are known matrices of order $N \times \nu_1$ and $N \times \nu_2$, respectively, $\boldsymbol{\xi}_1$ is a $\nu_1 \times 1$ vector of unknown parameters consisting of the mean μ and the effects in $\mathcal{C}_1$, $\boldsymbol{\xi}_2$ is a $\nu_2 \times 1$ vector of unknown parameters pertaining to the effects in $\mathcal{C}_2$, and $\boldsymbol{e}$ is the $N \times 1$ vector of errors with

$$E(\boldsymbol{e}) = \boldsymbol{0} \qquad \text{var}(\boldsymbol{e}) = \boldsymbol{V} \tag{16.2}$$

where $\boldsymbol{V}$ represents the variance–covariance matrix for $\boldsymbol{e}$ under the error structure of the completely randomized design (see Section I.6.3.5).

We assume now that at most k of the effects in $\mathcal{C}_2$ are nonnegligible, where k is quite small compared to ν_2. Let D be the design that gave rise to the observations $\boldsymbol{y}$ in (16.1). If $\boldsymbol{y}$ and hence $\boldsymbol{A}_1$ and $\boldsymbol{A}_2$ are such that we can estimate $\boldsymbol{\xi}_1$ and the k nonnegligible effects of $\boldsymbol{\xi}_2$, then D is said to be a *search design* of resolving (or revealing) power $(\boldsymbol{\xi}_1; \boldsymbol{\xi}_2, k)$ and the model (16.1) is said to be a *search linear model* (Srivastava, 1975).

16.3 PROPERTIES OF SEARCH DESIGNS

16.3.1 General Case

In the words of Srivastava (1984), model (16.1) represents a supermodel from which we want to extract an individual model based on the above criteria. Obviously, a search design that accomplishes this task must have certain properties. More specifically, the search design must have a sufficiently large number of runs and it must include a particular subset of all possible runs.

Consider a design D with N runs and suppose that D is composed of two parts or subdesigns, D_1 with N_1 runs, and D_2 with N_2 runs. Let the observation vector $\boldsymbol{y}$ in (16.1) be partitioned conformably as $\boldsymbol{y}' = (\boldsymbol{y}_1', \boldsymbol{y}_2')$, where $\boldsymbol{y}_1$ is $N_1 \times 1$ and is associated with D_1, while $\boldsymbol{y}_2$ is $N_2 \times 1$ and is associated with D_2. Very often D_1 represents a main effect plan to which treatment combinations in the form of D_2 are added to generate a search design. In essence, this is a process of dealiasing main effects from certain interactions.

A fundamental result that provides a characterization of the properties that such a design must possess is due to Srivastava (1975) and can be stated as follows:

Theorem 16.1 (i) Given model (16.1), a necessary condition for a design D to be a search design of revealing power $(\boldsymbol{\xi}_1; \boldsymbol{\xi}_2, k)$ is that

$$\text{rank}(A_1, A_{20}) = \nu_1 + 2k \tag{16.3}$$

for every $N \times 2k$ submatrix A_{20} of A_2.

(ii) In the noiseless case, that is, for $\boldsymbol{e} \equiv \mathbf{0}$ in (16.1) and (16.2), the condition (16.3) is also sufficient.

We shall comment briefly on the noiseless case. It implies, of course, that there is no experimental error and no observational error (see Section I.6.3.4). Such a situation does not occur in practice. Even so, the noiseless case is important with regard to the construction of search designs, for if the structure of designs D_1 and D_2 and hence the structure of $\boldsymbol{A}_1$ and $\boldsymbol{A}_2$ in (16.1) is such that the search and estimation problem cannot be solved in this case, then it certainly cannot be solved when noise is present. Moreover, in the noiseless case the observations $\boldsymbol{y}$ in (16.1) are nonstochastic and hence the search problem is deterministic. This implies that the nonnegligible effects in $\boldsymbol{\xi}_2$ can be identified with probability one if D satisfies the condition (16.3) of Theorem 16.1 In the stochastic case such identification can be achieved only with a certain probability, usually less than one.

The following theorems provide not only a tool to check whether for a given design D condition (16.3) is satisfied but can also be used to obtain more explicit conditions that a search design has to satisfy.

Theorem 16.2 (Srivastava and Ghosh, 1977) Consider model (16.1) and suppose that $\text{rank}(\boldsymbol{A}_1) = \nu_1$. Then condition (16.3) holds if and only if for every $N \times 2k$ submatrix $\boldsymbol{A}_{20}$ of $\boldsymbol{A}_2$ we have

$$\text{rank}\,[\boldsymbol{A}_{20}'\boldsymbol{A}_{20} - \boldsymbol{A}_{20}'\boldsymbol{A}_1(\boldsymbol{A}_1'\boldsymbol{A}_1)^{-1}\boldsymbol{A}_1'\boldsymbol{A}_{20}] = 2k \tag{16.4}$$

A similar result is given in the following theorem.

Theorem 16.3 (Srivastava and Gupta, 1979) Consider model (16.1) and suppose that $\text{rank}(\boldsymbol{A}_1) = \nu_1$. Let $\boldsymbol{A}_{11}$ be a $\nu_1 \times \nu_1$ submatrix of $\boldsymbol{A}_1$ of rank ν_1. Rearrange the observation vector $\boldsymbol{y}$ and write (16.1) as

$$\begin{pmatrix} \boldsymbol{y}_1^* \\ \boldsymbol{y}_2^* \end{pmatrix} = \begin{pmatrix} \boldsymbol{A}_{11} \\ \boldsymbol{A}_{21} \end{pmatrix} \boldsymbol{\xi}_1 + \begin{pmatrix} \boldsymbol{A}_{12} \\ \boldsymbol{A}_{22} \end{pmatrix} \boldsymbol{\xi}_2 + \boldsymbol{e} \tag{16.5}$$

where $\boldsymbol{y}_1^*$, $\boldsymbol{A}_{11}$, and $\boldsymbol{A}_{12}$ have ν_1 rows and y_2^*, $\boldsymbol{A}_{21}$, and $\boldsymbol{A}_{22}$ have $N - \nu_1$ rows. A necessary and sufficient condition for (16.3) to be satisfied is that every set of

$2k$ columns of

$$Q = A_{22} - A_{21}A_{11}^{-1}A_{12} \tag{16.6}$$

are linearly independent.

Note that in Theorem 16.3 the vector y_1^* is not necessarily equal to y_1 obtained from D_1 and, hence, y_2^* is not necessarily equal to y_2 obtained from D_2 as described earlier. However, if D_1 is a saturated main effect plan with $N_1 = \nu_1 = n + 1$, then $y_1^* = y_1$ and, hence, $y_2^* = y_2$.

16.3.2 Main Effect Plus One Plans

As stated earlier, in the context of search designs we assume *a priory* that at most k effects in ξ_2 of (16.1) are nonnegligible, where k is generally small. The most important and tractable situation from both the theoretical and practical point of view is the case $k = 1$. In addition, the most obvious starting point is to consider the case where ξ_1 is the vector of the mean, μ, and the n main effects, say $F_1, F_2, \ldots, F_n$, of the n factors $F_1, F_2, \ldots, F_n$, each at two levels. Thus, D_1 might be a main effects plan. The resulting search design D is then referred to as a main effect plus one plan, or MEP.1 for short, with resolution III.1.

Let us denote a treatment combination by $x' = (x_1, x_2, \ldots, x_n)$ with $x_\ell = 0$ or $1 (\ell = 1, 2, \ldots, n)$. Then we denote the N_i treatment combinations of $D_i (i = 1, 2)$ by $x'_{ij} = (x_{ij1}, x_{ij2}, \ldots, x_{ijn})$ for $j = 1, 2, \ldots, N_i$. Now consider the following D_1 with $N_1 = n + 2$ treatment combinations and given by its design matrix

$$D_1 = \begin{pmatrix} \mathfrak{I}'_n \\ I_n \\ 0'_n \end{pmatrix} \tag{16.7}$$

Recall from Section 7.4.2 that the true response of treatment combination x can be expressed as

$$a(x) = \mu + \frac{1}{2}\sum_{\alpha}(-1)^{\sum_i \alpha_i(1-x_i)} E^{\alpha}$$

or, if we define $E^{*\alpha} = \frac{1}{2}E^{\alpha}$, as

$$a(x) = \mu + \sum_{\alpha}(-1)^{\sum_i \alpha_i(1-x_i)} E^{*\alpha} \tag{16.8}$$

Taking y_1^* to be the first $n + 1$ observations in D_1 it follows from (16.7) and (16.8) that A_{11} in Theorem 16.3 can be written as the patterned matrix

$$A_{11} = \begin{pmatrix} 1 & \mathfrak{I}'_n \\ \mathfrak{I}_n & 2I_n - \mathfrak{I}_n\mathfrak{I}'_n \end{pmatrix}$$

with $\boldsymbol{A}_{11}^{-1}$ having the same pattern. Following lengthy but elementary arguments we can then derive the general form of $\boldsymbol{Q}$ in (16.6) such that the conditions of Theorem 16.3 are satisfied (see Srivastava and Gupta, 1979), leading to the following theorem.

Theorem 16.4 (Srivastava and Gupta, 1979) Let $\boldsymbol{D} = \binom{\boldsymbol{D}_1}{\boldsymbol{D}_2}$ be the design matrix for the design D, where $\boldsymbol{D}_1$ is given by (16.7). Let $\boldsymbol{D}_{21}$ and $\boldsymbol{D}_{22}$ be any two submatrices of $\boldsymbol{D}_2$ of order $(N - \nu_1) \times n_1$ and $(N - \nu_1) \times n_2$, respectively, with $n_1, n_2 \geq 2$. Let n_{ij} be the number of ones in the ith row of $\boldsymbol{D}_{2j} (j = 1, 2)$ and let for an integer z

$$\phi(z) = \begin{cases} z & \text{if } z \text{ is even} \\ z - 1 & \text{if } z \text{ is odd} \end{cases}$$

Then a necessary and sufficient condition for D to be a search design, where at most one element of $\boldsymbol{\xi}_2$ is nonnegligible, is that for every pair of distinct submatrices $\boldsymbol{D}_{21}$ and $\boldsymbol{D}_{22}$ there exists a row in $\boldsymbol{D}_2$ (say, the ith row), such that

$$\frac{\phi(n_{i1})}{\phi(n_1)} \neq \frac{\phi(n_{i2})}{\phi(n_2)}$$

[where i may be different for different pairs $(\boldsymbol{D}_{21}, \boldsymbol{D}_{22})$].

This provides a convenient and simple check on $\boldsymbol{D}_2$ for D to be an MEP.1.

Whereas for the MEP.1 discussed above the vector $\boldsymbol{\xi}_2$ in (16.1) consists of all possible interactions in 2^n factorial, it seems more reasonable to restrict $\boldsymbol{\xi}_2$ to include only 2- and 3-factor interactions, such that $\nu_2 = \binom{n}{2} + \binom{n}{3}$. A search design for identifying at most one nonnegligible effect among those interactions has been referred to by Gupta (1990) as a weak MEP.1. The structure of such an MEP.1 has been considered by Ghosh (1981) and Ohnishi and Shirakura (1985), and we shall give a brief description of their results without going through all the derivations.

Let D_1 be given by its design matrix

$$\boldsymbol{D}_1 = \begin{pmatrix} \boldsymbol{\Im}_n' \\ \boldsymbol{0}_n' \\ \boldsymbol{\Im}_n \boldsymbol{\Im}_n' - \boldsymbol{I}_n \end{pmatrix} \tag{16.9}$$

where D_1 consists of $N_1 = n + 2$ treatment combinations and represents a main effects plan. Let $\boldsymbol{y}_1$ denote the observation vector obtained from D_1 and let

$$\boldsymbol{y}_1 = \boldsymbol{A}_1^* \boldsymbol{\xi}_1 + \boldsymbol{A}_2^* \boldsymbol{\xi}_2 + \boldsymbol{e}_1 \tag{16.10}$$

with $\boldsymbol{\xi}_1$ and $\boldsymbol{\xi}_2$ as defined above. Using the parameterization (16.8) it follows that

$$A_1^* = \begin{pmatrix} 1 & \mathfrak{I}_n' \\ 1 & -\mathfrak{I}_n' \\ \mathfrak{I}_n & \mathfrak{I}_n\mathfrak{I}_n' - 2\boldsymbol{I}_n \end{pmatrix} \tag{16.11}$$

Now the columns of $\boldsymbol{A}_2^*$ in (16.9) and of $\boldsymbol{A}_2$ in (16.1) are obtained by appropriate multiplication of the elements in the columns of $\boldsymbol{A}_1^*$ (and $\boldsymbol{A}_1$). For example, to obtain the column corresponding to $F_{i_1}F_{i_2}$ in $\boldsymbol{A}_2^*$ (and in $\boldsymbol{A}_2$) we simply multiply the corresponding elements of the columns belonging to F_{i_1} and F_{i_2}. Since D_1 is a main effects plan, the columns of $\boldsymbol{A}_2^*$ are linear combinations of the columns of $\boldsymbol{A}_1^*$. This fact can be exploited to obtain conditions for D_2 such that condition (16.3) of Theorem 16.1 is satisfied for $k = 1$. This is done by considering all possible types of pairs of 2- and/or 3-factor interactions such as $(F_{i_1i_2}, F_{i_1i_3})$, $(F_{i_1i_2}, F_{i_3i_4})$, $(F_{i_1i_2}, F_{i_1i_3i_4})$, and so forth with distinct indices i_1, i_2, i_3, i_4. This leads to the following theorem.

Theorem 16.5 (Ohnishi and Shirakura, 1985) Let $\omega(z')$ denote the weight of the vector z, which is defined as the number of nonzero elements in z. Further, let $\boldsymbol{D}_1$ be given by (16.9) with row vectors of weights 0, $n - 1$, and n. Then a necessary and sufficient condition on D_2 with N_2 treatment combinations, such that $D = D_1 + D_2$ is a search design, is that there exist treatment combinations $(x_1, x_2, \ldots, x_n)$ in D_2 satisfying the following conditions for any distinct integers $i_k (k = 1, 2, 3, 4, 5, 6)$ belonging to the set $\{1, 2, \ldots, n\}$:

(1)	$\omega(x_{i_1}, x_{i_2}) = 1$	$x_{i_3} = 0$	
(2)	$\omega(x_{i_1}, x_{i_2}) = 1$	$\omega(x_{i_3}, x_{i_4}) = 1$	
(3a)	$\omega(x_{i_1}, x_{i_2}) \geq 1$	$\omega(x_{i_3}, x_{i_4}) = 0$	or
(3b)	$\omega(x_{i_1}, x_{i_2}) = 0$	$\omega(x_{i_3}, x_{i_4}) \geq 1$	
(4a)	$x_{i_1} = 0, x_{i_2} = 1$	$\omega(x_{i_3}, x_{i_4}) = 1$	or
(4b)	$\omega(x_{i_1}, x_{i_2}) \geq 1$	$\omega(x_{i_3}, x_{i_4}) = 0$	or
(4c)	$\omega(x_{i_1}, x_{i_2}) = 0$	$\omega(x_{i_3}, x_{i_4}) = 2$	
(5a)	$\omega(x_{i_1}, x_{i_2}) = 2$	$\omega(x_{i_3}, x_{i_4}) = 0$	or
(5b)	$\omega(x_{i_1}, x_{i_2}) = 0$	$\omega(x_{i_3}, x_{i_4}) = 2$	or
(5c)	$\omega(x_{i_1}, x_{i_2}, x_{i_3}, x_{i_4}) = 1$	$x_{i_5} = 1$	or
(5d)	$\omega(x_{i_1}, x_{i_2}, x_{i_3}, x_{i_4}) = 3$	$x_{i_5} = 0$	

(6a) $\omega(x_{i_1}, x_{i_2}) = 0$ $\omega(x_{i_3}, x_{i_4}, x_{i_5}) \geq 2$ or

(6b) $\omega(x_{i_1}, x_{i_2}) \geq 1$ $\omega(x_{i_3}, x_{i_4}, x_{i_5}) \leq 1$

(7a) $\omega(x_{i_1}, x_{i_2}, x_{i_3}) \leq 1$ $\omega(x_{i_4}, x_{i_5}, x_{i_6}) \geq 2$ or

(7b) $\omega(x_{i_1}, x_{i_2}, x_{i_3}) \geq 2$ $\omega(x_{i_4}, x_{i_5}, x_{i_6}) \leq 1$

Even though the conditions in Theorem 16.5 are intended to provide a mechanism for checking condition (16.3) for all possible submatrices $\boldsymbol{A}_{20}$, they themselves are not very easy to check since the number of cases to be considered is quite large. The conditions prove, however, to be useful in a sequential construction of search designs (see Section 16.4).

16.3.3 Resolution V Plus One Plans

In the context of fractional factorial designs resolution V designs are of particular importance as long as the assumption of negligible interactions involving three or more factors is reasonable. If we suspect, however, that at most k of those interactions are nonnegligible, then a search design may be appropriate. Again, the case $k = 1$ is of interest. Such a search design is referred to as a resolution V.1 design.

For resolution V designs the parameter vector $\boldsymbol{\xi}_1$ in (16.1) contains the mean, the main effects, and the 2-factor interactions, with

$$\nu_1 = 1 + n + \binom{n}{2}$$

components, and the vector of negligible effects, $\boldsymbol{\xi}_2$, consists of the remaining interactions with $\nu_2 = 2^n - \nu_1$ terms. For purposes of a search design $\boldsymbol{\xi}_2$ may be (i) as described above or (ii) restricted to involve only the 3-factor interactions, in which case $\nu_2 = \binom{n}{3}$.

The case (i) above was considered by Srivastava and Ghosh (1976, 1977), whereas case (ii) was considered by Shirakura and Ohnishi (1985). In both cases the starting points are balanced resolution V plans (see Section 13.6.5). The reason for the restriction to these balanced arrays is the fact that it reduces greatly the number of cases one has to consider in order to verify whether condition (16.4) is indeed met. Exploiting the form of balanced arrays in light of the rank condition (16.4) leads then to the identification of those resolution V designs that are also resolution V.1 designs. We shall list a few examples in Section 16.4.2.

16.3.4 Other Search Designs

In the previous section we have considered search designs for 2^n factorials that allow us to search for one ($k = 1$) additional interaction among those interactions that are ordinarily assumed to be negligible. Extensions to $k = 2$ have been considered using Theorem 16.1 [for resolution III.2 plans see, e.g., Shirakura (1991) and Chatterjee, Deng, and Lin (2001)]. Another extension is that to factors with more than two levels. Ghosh and Burns (2001), for example, consider search designs for the 3^n factorial, whereas Chatterjee and Mukerjee (1986, 1993) consider the most general case for symmetrical and asymmetrical factorials for $k = 1$ and $k \geq 3$, respectively.

Another application of the principles of a search design is given by Ghosh and Burns (2002). They assume that $\boldsymbol{\xi}_1$ in (16.1) contains only the mean and $\boldsymbol{\xi}_2$ contains the main effects of the n factors. By assuming that at most $k\,(< n)$ factors have nonnegligible effects, they are able to reduce the number of treatment combinations necessary to identify and estimate those effects compared to the number of treatment combinations in an orthogonal main effects plan. Note that this idea is similar to that used in supersaturated designs (see Section 15.1) except that the number of factors here is "small."

16.4 LISTING OF SEARCH DESIGNS

There are not very many methods of constructing search designs. The existing methods fall essentially into three categories:

1. Start with a "minimal" design D_1 that allows estimating $\boldsymbol{\xi}_1$ in (16.1) assuming that $\boldsymbol{\xi}_2$ is negligible; then add treatment combinations in the form of D_2 such that the final design $D = D_1 + D_2$ satisfies the conditions of a search design.
2. Start with a "larger" design D_1 that allows estimating $\boldsymbol{\xi}_1$ and show that D_1 itself satisfies the conditions of a search design.
3. Start with an "intuitive" design D_1 and show that it satisfies the conditions of a search design, or modify it accordingly.

Methods 1 and 2 have been used for resolution III.1 and V.1, respectively, and we shall give some examples below. Method 3 is essentially a trial-and-error method and requires a great amount of intuition and insight.

16.4.1 Resolution III.1 Designs

Starting with the design D_1 given in (16.9), Ghosh (1981) suggested to construct the design D_2 iteratively. More precisely, let $D_{2,n}$ be the design D_2 for n factors with $N_{2,n}$ treatment combinations, then $D_{2,n+1}$ can be represented by the design matrix

$$\boldsymbol{D}_{2,n+1} = \begin{pmatrix} \boldsymbol{D}_{2,n} & z_{1,n+1} \\ \boldsymbol{D}^*_{2,n} & z_{2,n+1} \end{pmatrix} \tag{16.12}$$

where $\boldsymbol{D}_{2,n+1}$ is a $N_{2,n+1} \times (n+1)$ matrix, $\boldsymbol{D}^*_{2,n}$ is a $(N_{2,n+1} - N_{2,n}) \times n$ matrix, and $z_{n+1} = (z'_{1,n+1}, z'_{2,n+1})'$ is an $N_{2,n+1} \times 1$ vector. All the elements in $\boldsymbol{D}_{2,n+1}$ are, of course, either 0 or 1. We note here that it is possible that $N_{2,n+1} = N_{2,n}$.

To proceed iteratively has the advantage that it greatly reduces the number of cases that have to be considered to check the conditions of Theorem 16.5. Using this method, Ohnishi and Shirakura (1985) have provided designs $D_{2,n}$ for $3 \leq n \leq 8$, which are given in Table 16.1 with a summary of the various model and design parameters as given in Table 16.2. It illustrates the point that the total number N of treatment combinations required is relatively small but increases proportionately, of course, as n, the number of factors, increases.

16.4.2 Resolution V.1 Designs

In order to reduce the computational load of checking the rank condition (16.4), construction of resolution V.1 designs has been limited to balanced arrays (see Section 13.6.5). Srivastava and Ghosh (1977) and Shirakura and Ohnishi (1985) provide tables of such search designs for $n = 4, 5, 6, 7, 8$, and we present some examples of their results in Table 16.3.

Table 16.1 Design Matrices $D_{2,n}$ for $3 \leq n \leq 8$ Factors

		n							
		1	2	3	4	5	6	7	8
	1	1	0	0	1	0	0	1	0
	2	0	0	1	1	0	0	1	1
	3	1	0	0	0	1	0	1	1
	4	0	1	0	0	1	0	1	1
$N_{2,n}$	5	0	0	1	0	1	1	0	0
	6	1	0	0	1	0	1	0	1
	7	1	0	0	0	1	1	0	1
	8	1	0	0	1	1	0	0	0
	9	1	0	0	0	1	1	0	0
	10	0	1	0	0	1	0	1	0

Table 16.2 Number of Parameters and Runs for Search Design of Resolution III.1

n	ν_1	ν_2	N_1	N_2	N
3	4	4	5	2	7
4	5	10	6	5	11
5	6	20	7	5	12
6	7	35	8	8	16
7	8	56	9	8	17
8	9	84	10	10	20

Table 16.3 Resolution V.1 Designs for $4 \le n \le 8$ Factors

		Srivastava and Ghosh (1977)			Ohnishi and Shirakura (1985)		
n	ν_1	ν_2	N	λ'	ν_2	N	λ'
4	11	5	15	1 1 1 1 0			
5	16	16	21	1 1 1 0 1 0			
6	22	42	28	1 1 1 0 0 1 0	20	28	1 0 0 1 0 1 1
7	29	99	36	1 1 1 0 0 0 1 0	35	35	0 1 0 0 0 1 1 0
8	37	219	45	1 1 1 0 0 0 0 1 0	56	44	0 1 0 0 0 0 1 1 0

Let $S_{n,j}$ denote the set of all treatment combinations with exactly j factors at level 1 and $n - j$ factors at level 0, and let λ_j denote the number of times that each treatment combination in $S_{n,j}$ appears in the design (array) ($j = 0, 1, \ldots, n$). Of course, each vector (treatment combination) in $S_{n,j}$ has weight j and $\boldsymbol{\lambda}' = (\lambda_0, \lambda_1, \ldots, \lambda_n)$ is referred to as the index set of the array. The resolution V.1 designs, which are also referred to as BARE 5.1 plans by Srivastava and Ghosh (1976), are then completely described by their index sets, as given in Table 16.3.

The Srivastava–Ghosh designs contain

$$N_n = 1 + 2n + \binom{n}{2}$$

runs and can be represented easily as $D = S_{n,0} \cup S_{n,1} \cup S_{n,2} \cup \ldots \cup S_{n,n-1}$. Both types of search designs given in Table 16.3 are minimal designs, but designs with a larger number of runs are given in the two references. We also note that restricting $\boldsymbol{\xi}_2$ in (16.1) does not reduce the size of the design appreciably.

16.5 ANALYSIS OF SEARCH EXPERIMENTS

16.5.1 General Setup

We shall call an experiment that employs a search design a *search experiment*. The analysis of data from a search experiment are based, of course, on submodels of (16.1), which we write as

$$\boldsymbol{y} = \boldsymbol{A}_1\boldsymbol{\xi}_1 + \boldsymbol{A}_{21}(\boldsymbol{\zeta})\boldsymbol{\zeta} + \boldsymbol{e} \tag{16.13}$$

where $\boldsymbol{\zeta}$ refers to k possible components of $\boldsymbol{\xi}_2$ in (16.1) and $\boldsymbol{A}_{21}(\boldsymbol{\zeta})$ refers to the column vectors in $\boldsymbol{A}_2$ corresponding to $\boldsymbol{\zeta}$. The analysis will then be performed for each possible $\boldsymbol{\zeta}$ from $\boldsymbol{\xi}_2$.

For purposes of characterizing and comparing the outcomes from the various analyses we shall denote by $\boldsymbol{\zeta}_0$ the true (unknown) nonnegligible effects in $\boldsymbol{\xi}_2$. We shall then refer to model (16.13) with $\boldsymbol{\zeta} = \boldsymbol{\zeta}_0$ as $M0$, and with $\boldsymbol{\zeta} \neq \boldsymbol{\zeta}_0$ as $M1$, where $M1$ represents a series of competing models to $M0$.

16.5.2 Noiseless Case

Even though in practice there does not exist a situation where $\boldsymbol{e} \equiv \boldsymbol{0}$ in (16.3), we have pointed out earlier that this noiseless case is important for purposes of constructing search designs. The noiseless case is equally important and instructive for describing and comparing the various analyses based on $M0$ and $M1$.

We shall illustrate this for the case $k = 1$ by first constructing a data set with known parameters $\boldsymbol{\xi}_1$ and $\boldsymbol{\xi}_2$ and then using (16.13) with $\boldsymbol{e} \equiv \boldsymbol{0}$ to analyze those data.

Example 16.1 Let us consider the case of $n = 3$ factors and the search design with D_2 given in Table 16.1. Thus, in model (16.1),

$$\boldsymbol{\xi}_1 = \begin{pmatrix} \mu \\ A \\ B \\ C \end{pmatrix} \qquad \boldsymbol{\xi}_2 = \begin{pmatrix} AB \\ AC \\ BC \\ ABC \end{pmatrix}$$

Let $\mu = 10,\ A = 4,\ B = -2,\ C = 3,\ AB = 2,\ AC = 0,\ BC = 0,\ ABC = 0.$ Then using (16.8) [see also (7.30)] with $A^* = 2$, $B^* = -1$, $C^* = 1.5$, $(AB)^* = 1$ we obtain the true yield $y = a(\boldsymbol{x})$ for the seven treatment combinations of the design D as follows:

Treatment Combination	μ **10**	A^* **2**	B^* **−1**	C^* **1.5**	$(AB)^*$ **1**	y
1 1 1	1	−1	−1	−1	1	13.5
0 0 0	1	−1	−1	−1	1	8.5
0 1 1	1	−1	1	1	−1	7.5
1 0 1	1	1	−1	1	−1	13.5
1 1 0	1	1	1	−1	1	10.5
1 0 0	1	1	−1	−1	−1	10.5
0 0 1	1	−1	−1	1	−1	11.5

The analysis of these data, using SAS PROC GLM, is presented in Tables 16.4a–16.4c for the models with $\zeta = AB$, AC, and BC, respectively. We shall comment briefly on some aspects of the output:

1. Since $AB = \zeta_0$, the model used in Table 16.4 a provides a perfect fit to the data as indicated by the fact that SS(Error) = SSE($M0$) = 0, which is equivalent to the fact that the estimated effects are equal to the true effects, for example, $\widehat{A} = A$, and that the predicted values $\widehat{y}$ are equal to the observed values y. In other words, the true model is identified correctly.

Table 16.4 Data and Analysis for Search Experiment (Noiseless Case)

```
options nodate pageno=1;
data search;
input A B C y;
datalines;
1 1 1 13.5
0 0 0 8.5
0 1 1 7.5
1 0 1 13.5
1 1 0 10.5
1 0 0 10.5
0 0 1 11.5
;
run;

proc print data=search;
title1 'TABLE 16.4';
title2 'DATA FOR SEARCH EXPERIMENT';
title3 '(NOISELESS CASE)';
run;

proc glm data=search;
class A B C;
model y=A B C A*B/e;
estimate 'A' A -1 1;
estimate 'B' B -1 1;
estimate 'C' C -1 1 ;
estimate 'A*B' A*B 1 -1 -1 1/divisor=2;
output out=predicta p=yhat r=resid stdr=eresid;
title1 'TABLE 16.4a';
title2 'ANALYSIS WITH MODEL Y=A B C A*B';
title3 '(NOISELESS CASE)';
run;

proc print data=predicta;
run;

proc glm data=search;
class A B C;
model y=A B C A*C/e;
estimate 'A' A -1 1;
estimate 'B' B -1 1;
estimate 'C' C -1 1 ;
estimate 'A*C' A*C 1 -1 -1 1/divisor=2;
output out=predictb p=yhat r=resid stdr=eresid;
title1 'TABLE 16.4b';
```

Table 16.4 ***(Continued)***

```
title2 'ANALYSIS WITH MODEL A B C A*C';
title3 '(NOISELESS CASE)';
run;

proc print data=predictb;
run;

proc glm data=search;
class A B C;
model y=A B C B*C/e;
estimate 'A' A -1 1;
estimate 'B' B -1 1;
estimate 'C' C -1 1;
estimate 'B*C' B*C -1 1 1 -1/divisor=2;
output out=predictc p=yhat r=resid stdr=eresid;
title1 'TABLE 16.4c';
title2 'ANALYSIS WITH MODEL Y=A B C B*C';
title3 '(NOISELESS CASE)';
run;

proc print data=predictc;
run;
```

Obs	A	B	C	y
1	1	1	1	13.5
2	0	0	0	8.5
3	0	1	1	7.5
4	1	0	1	13.5
5	1	1	0	10.5
6	1	0	0	10.5
7	0	0	1	11.5

A. ANALYSIS WITH MODEL Y=A B C A*B
(NOISELESS CASE)

The GLM Procedure

Class Level Information

Class	Levels	Values
A	2	0 1
B	2	0 1
C	2	0 1

Table 16.4 (*Continued*)

```
Number of observations Read    7
Number of observations Used    7

General Form of Estimable Functions

   Effect            Coefficients

   Intercept         L1

   A         0       L2
   A         1       L1-L2

   B         0       L4
   B         1       L1-L4

   C         0       L6
   C         1       L1-L6

   A*B       0 0     L8
   A*B       0 1     L2-L8
   A*B       1 0     L4-L8
   A*B       1 1     L1-L2-L4+L8
```

Dependent Variable: y

Source	DF	Sum of Squares	Mean Square	F Value	Pr > F
Model	4	31.42857143	7.85714286	Infty	<.0001
Error	2	0.00000000	0.00000000		
Corrected Total	6	31.42857143			

R-Square	Coeff Var	Root MSE	y Mean
1.000000	0	0	10.78571

Source	DF	Type I SS	Mean Square	F Value	Pr > F
A	1	13.76190476	13.76190476	Infty	<.0001
B	1	1.66666667	1.66666667	Infty	<.0001
C	1	10.00000000	10.00000000	Infty	<.0001
A*B	1	6.00000000	6.00000000	Infty	<.0001

Table 16.4 ***(Continued)***

Source	DF	Type III SS	Mean Square	F Value	Pr > F
A	1	24.00000000	24.00000000	Infty	<.0001
B	1	6.00000000	6.00000000	Infty	<.0001
C	1	13.50000000	13.50000000	Infty	<.0001
A*B	1	6.00000000	6.00000000	Infty	<.0001

Parameter	Estimate	Standard Error	t Value	Pr > |t|
A	4.00000000	0	Infty	<.0001
B	-2.00000000	0	-Infty	<.0001
C	3.00000000	0	Infty	<.0001
A*B	2.00000000	0	Infty	<.0001

Obs	A	B	C	y	yhat	resid	eresid
1	1	1	1	13.5	13.5	0	.
2	0	0	0	8.5	8.5	0	.
3	0	1	1	7.5	7.5	-2.6645E-15	.
4	1	0	1	13.5	13.5	0	.
5	1	1	0	10.5	10.5	0	.
6	1	0	0	10.5	10.5	0	.
7	0	0	1	11.5	11.5	0	.

B. ANALYSIS WITH MODEL A B C A*C
(NOISELESS CASE)

The GLM Procedure

General Form of Estimable Functions

Effect		Coefficients
Intercept		L1
A	0	L2
A	1	L1-L2
B	0	L4
B	1	L1-L4
C	0	L6
C	1	L1-L6
A*C	0 0	L8
A*C	0 1	L2-L8
A*C	1 0	L6-L8

Table 16.4 (*Continued*)

Dependent Variable: y

Source	DF	Sum of Squares	Mean Square	F Value	Pr > F
Model	4	26.09523810	6.52380952	2.45	0.3106
Error	2	5.33333333	2.66666667		
Corrected Total	6	31.42857143			

R-Square	Coeff Var	Root MSE	y Mean
0.830303	15.14033	1.632993	10.78571

Source	DF	Type I SS	Mean Square	F Value	Pr > F
A	1	13.76190476	13.76190476	5.16	0.1511
B	1	1.66666667	1.66666667	0.62	0.5120
C	1	10.00000000	10.00000000	3.75	0.1924
A*C	1	0.66666667	0.66666667	0.25	0.6667

Source	DF	Type III SS	Mean Square	F Value	Pr > F
A	1	16.66666667	16.66666667	6.25	0.1296
B	1	2.66666667	2.66666667	1.00	0.4226
C	1	8.16666667	8.16666667	3.06	0.2222
A*C	1	0.66666667	0.66666667	0.25	0.6667

Parameter	Estimate	Standard Error	t Value	Pr > |t|
A	3.33333333	1.33333333	2.50	0.1296
B	-1.33333333	1.33333333	-1.00	0.4226
C	2.33333333	1.33333333	1.75	0.2222
A*C	0.66666667	1.33333333	0.50	0.6667

Obs	A	B	C	y	yhat	resid	eresid
1	1	1	1	13.5	12.8333	0.66667	0.94281
2	0	0	0	8.5	8.5000	0.00000	.
3	0	1	1	7.5	8.8333	-1.33333	0.94281
4	1	0	1	13.5	14.1667	-0.66667	0.94281
5	1	1	0	10.5	9.8333	0.66667	0.94281
6	1	0	0	10.5	11.1667	-0.66667	0.94281
7	0	0	1	11.5	10.1667	1.33333	0.94281

Table 16.4 (*Continued*)

```
            C. ANALYSIS WITH MODEL Y=A B C B*C
                     (NOISELESS CASE)

                    The GLM Procedure

           General Form of Estimable Functions

              Effect              Coefficients

              Intercept           L1

               A        0         L2
               A        1         L1-L2

               B        0         L4
               B        1         L1-L4

               C        0         L6
               C        1         L1-L6

               B*C      0 0       L8
               B*C      0 1       L4-L8
               B*C      1 0       L6-L8
               B*C      1 1       L1-L4-L6+L8

Dependent Variable: y
```

Source	DF	Sum of Squares	Mean Square	F Value	Pr > F
Model	4	26.09523810	6.52380952	2.45	0.3106
Error	2	5.33333333	2.66666667		
Corrected Total	6	31.42857143			

R-Square	Coeff Var	Root MSE	y Mean
0.830303	15.14033	1.632993	10.78571

Source	DF	Type I SS	Mean Square	F Value	Pr > F
A	1	13.76190476	13.76190476	5.16	0.1511
B	1	1.66666667	1.66666667	0.62	0.5120
C	1	10.00000000	10.00000000	3.75	0.1924
B*C	1	0.66666667	0.66666667	0.25	0.6667

Table 16.4 (*Continued*)

Source	DF	Type III SS	Mean Square	F Value	Pr > F
A	1	16.66666667	16.66666667	6.25	0.1296
B	1	2.66666667	2.66666667	1.00	0.4226
C	1	8.16666667	8.16666667	3.06	0.2222
B*C	1	0.66666667	0.66666667	0.25	0.6667

Parameter	Estimate	Standard Error	t Value	Pr > |t|
A	3.33333333	1.33333333	2.50	0.1296
B	-1.33333333	1.33333333	-1.00	0.4226
C	2.33333333	1.33333333	1.75	0.2222
B*C	0.66666667	1.33333333	0.50	0.6667

Obs	A	B	C	y	yhat	resid	eresid
1	1	1	1	13.5	12.1667	1.33333	0.94281
2	0	0	0	8.5	7.8333	0.66667	0.94281
3	0	1	1	7.5	8.8333	-1.33333	0.94281
4	1	0	1	13.5	14.1667	-0.66667	0.94281
5	1	1	0	10.5	10.5000	0.00000	.
6	1	0	0	10.5	11.1667	-0.66667	0.94281
7	0	0	1	11.5	10.8333	0.66667	0.94281

2. For the models with $\zeta = AC$ and BC, Tables 16.4 b and 16.4 c show that SS(Error)=SSE($M1$)>0, as they should because the wrong models are used.
3. In all three cases the expressions for estimable functions confirm the theoretical results that, individually, AB, AC, and BC are estimable.
4. It is, of course, clear that ABC cannot be estimated since its definition depends on all eight treatment combinations of the 2^3 factorial. □

16.5.3 Noisy Case

As we have seen in the previous section, in the noiseless case the true model is identified correctly. This may not necessarily happen in the noisy case, that is, when $\text{var}(\boldsymbol{e}) = \boldsymbol{V} \neq \boldsymbol{0}$. However, based on the results for the noiseless case, where SSE($M0$)=0, it seems plausible to select that model as the correct model for which SSE achieves the smallest value among competing models. This, in fact, was suggested by Srivastava (1975).

We illustrate this procedure in the following example.

Example 16.2 We consider the same search design as in Example 16.1. We also use the same parameter values as in Example 16.1. The observations are obtained by adding $N(0, 1)$ noise to each true value. The data are given in Table 16.5.

Table 16.5 Data and Analysis for Search Experiment (With N(0, 1) Noise)

Obs	A	B	C	y
1	1	1	1	13.792
2	0	0	0	9.013
3	0	1	1	7.445
4	1	0	1	15.187
5	1	1	0	10.176
6	1	0	0	10.606
7	0	0	1	11.638

A. ANALYSIS WITH MODEL Y=A B C A*B
(WITH N(0, 1) NOISE)

The GLM Procedure

Class Level Information

Class	Levels	Values
A	2	0 1
B	2	0 1
C	2	0 1

Number of observations 7

Dependent Variable: y

Source	DF	Sum of Squares	Mean Square	F Value	Pr > F
Model	4	42.09222138	10.52305535	22.00	0.0439
Error	2	0.95654033	0.47827017		
Corrected Total	6	43.04876171			

R-Square	Coeff Var	Root MSE	y Mean
0.977780	6.217804	0.691571	11.12243

Source	DF	Type I SS	Mean Square	F Value	Pr > F
A	1	16.20876430	16.20876430	33.89	0.0283
B	1	4.81496682	4.81496682	10.07	0.0866
C	1	15.73393923	15.73393923	32.90	0.0291
A*B	1	5.33455104	5.33455104	11.15	0.0792

Table 16.5 *(Continued)*

Source	DF	Type III SS	Mean Square	F Value	Pr > F
A	1	29.79504504	29.79504504	62.30	0.0157
B	1	11.74600417	11.74600417	24.56	0.0384
C	1	19.51928067	19.51928067	40.81	0.0236
A*B	1	5.33455104	5.33455104	11.15	0.0792

Parameter	Estimate	Standard Error	t Value	Pr > |t|
A	4.45683333	0.56466519	7.89	0.0157
B	-2.79833333	0.56466519	-4.96	0.0384
C	3.60733333	0.56466519	6.39	0.0236
A*B	1.88583333	0.56466519	3.34	0.0792

Obs	A	B	C	y	yhat	resid	eresid
1	1	1	1	13.792	13.7877	0.00433	0.39928
2	0	0	0	9.013	8.5218	0.49117	0.39928
3	0	1	1	7.445	7.4450	-0.00000	.
4	1	0	1	15.187	14.7002	0.48683	0.39928
5	1	1	0	10.176	10.1803	-0.00433	0.39928
6	1	0	0	10.606	11.0928	-0.48683	0.39928
7	0	0	1	11.638	12.1292	-0.49117	0.39928

- -

B. ANALYSIS WITH MODEL Y=A B C A*C
(WITH N(0, 1) NOISE)

The GLM Procedure

Class Level Information

Class	Levels	Values
A	2	0 1
B	2	0 1
C	2	0 1

Number of observations 7

Dependent Variable: y

Source	DF	Sum of Squares	Mean Square	F Value	Pr > F
Model	4	39.22872871	9.80718218	5.13	0.1696
Error	2	3.82003300	1.91001650		
Corrected Total	6	43.04876171			

Table 16.5 ***(Continued)***

R-Square	Coeff Var	Root MSE	y Mean
0.911263	12.42564	1.382033	11.12243

Source	DF	Type I SS	Mean Square	F Value	Pr > F
A	1	16.20876430	16.20876430	8.49	0.1004
B	1	4.81496682	4.81496682	2.52	0.2533
C	1	15.73393923	15.73393923	8.24	0.1030
A*C	1	2.47105837	2.47105837	1.29	0.3733

Source	DF	Type III SS	Mean Square	F Value	Pr > F
A	1	20.14284038	20.14284038	10.55	0.0832
B	1	6.03605400	6.03605400	3.16	0.2174
C	1	11.88633750	11.88633750	6.22	0.1301
A*C	1	2.47105837	2.47105837	1.29	0.3733

Parameter	Estimate	Standard Error	t Value	Pr > |t|
A	3.66450000	1.12842560	3.25	0.0832
B	-2.00600000	1.12842560	-1.78	0.2174
C	2.81500000	1.12842560	2.49	0.1301
A*C	1.28350000	1.12842560	1.14	0.3733

Obs	A	B	C	y	yhat	resid	eresid
1	1	1	1	13.792	13.4865	0.3055	0.79792
2	0	0	0	9.013	9.0130	0.0000	.
3	0	1	1	7.445	8.5385	-1.0935	0.79792
4	1	0	1	15.187	15.4925	-0.3055	0.79792
5	1	1	0	10.176	9.3880	0.7880	0.79792
6	1	0	0	10.606	11.3940	-0.7880	0.79792
7	0	0	1	11.638	10.5445	1.0935	0.79792

- -

C. ANALYSIS WITH MODEL Y=A B C B*C
(WITH N(0, 1) NOISE)

The GLM Procedure

Class Level Information

Class	Levels	Values
A	2	0 1
B	2	0 1
C	2	0 1

Table 16.5 (*Continued*)

Number of observations 7

Dependent Variable: y

Source	DF	Sum of Squares	Mean Square	F Value	Pr > F
Model	4	37.33955238	9.33488810	3.27	0.2477
Error	2	5.70920933	2.85460467		
Corrected Total	6	43.04876171			

R-Square	Coeff Var	Root MSE	y Mean
0.867378	15.19055	1.689558	11.12243

Source	DF	Type I SS	Mean Square	F Value	Pr > F
A	1	16.20876430	16.20876430	5.68	0.1400
B	1	4.81496682	4.81496682	1.69	0.3236
C	1	15.73393923	15.73393923	5.51	0.1434
B*C	1	0.58188204	0.58188204	0.20	0.6959

Source	DF	Type III SS	Mean Square	F Value	Pr > F
A	1	21.99952017	21.99952017	7.71	0.1090
B	1	7.07094704	7.07094704	2.48	0.2562
C	1	13.32209004	13.32209004	4.67	0.1633
B*C	1	0.58188204	0.58188204	0.20	0.6959

Parameter	Estimate	Standard Error	t Value	Pr > |t|
A	3.82966667	1.37951795	2.78	0.1090
B	-2.17116667	1.37951795	-1.57	0.2562
C	2.98016667	1.37951795	2.16	0.1633
B*C	0.62283333	1.37951795	0.45	0.6959

Obs	A	B	C	y	yhat	resid	eresid
1	1	1	1	13.792	12.5333	1.25867	0.97547
2	0	0	0	9.013	7.8947	1.11833	0.97547
3	0	1	1	7.445	8.7037	-1.25867	0.97547
4	1	0	1	15.187	15.3273	-0.14033	0.97547
5	1	1	0	10.176	10.1760	0.00000	.
6	1	0	0	10.606	11.7243	-1.11833	0.97547
7	0	0	1	11.638	11.4977	0.14033	0.97547

The individual analyses for different models are given in Tables 16.5a–16.5c. The summary below shows that in this case we would have selected the true model over the others:

	Model	SS(Error)
$M0$:	A, B, C, AB	0.478
$M1$:	A, B, C, AC	1.910
$M1$:	A, B, C, BC	2.855

□

16.6 SEARCH PROBABILITIES

The discussion of the preceding section can be expressed in terms of probabilities as follows: For the noiseless case we have

$$P[\mathrm{SSE}(M0) < \mathrm{SSE}(M1) \mid M0, M1, \boldsymbol{V} = \boldsymbol{0}] = 1$$

and for the noisy case we have

$$P[\mathrm{SSE}(M0) < \mathrm{SSE}(M1) \mid M0, M1, \boldsymbol{V} \neq \boldsymbol{0}] < 1 \tag{16.14}$$

Let us assume, for operational purposes, that $\mathrm{var}(\boldsymbol{e}) = \boldsymbol{V} = \sigma_e^2 \boldsymbol{I}$. If $\sigma_e^2 = \infty$, then $M0$ and $M1$ are indistinguishable, and hence we can rewrite (16.14) more precisely as

$$P[\mathrm{SSE}(M0) < \mathrm{SSE}(M1) \mid M0, M1, \sigma_e^2 = \infty] = \tfrac{1}{2}$$

The probability $P[\mathrm{SSE}(M0) < \mathrm{SSE}M1) \mid M0, M1, \sigma_e^2]$ is called a *search probability*, which assumes values between $\frac{1}{2}$ and 1 (Ghosh and Teschmacher, 2002).

The expression in (16.14) can be written out more explicitly as follows. To simplify the notation, let us write

$$\boldsymbol{\theta} = \begin{pmatrix} \xi_1 \\ \zeta \end{pmatrix} \qquad \boldsymbol{A}(\zeta) = [\boldsymbol{A}_1, \boldsymbol{A}_{21}(\zeta)] \qquad \boldsymbol{M}(\zeta) = \boldsymbol{A}(\zeta)'\boldsymbol{A}(\zeta)$$

Then the error sum of squares, SSE, for model (16.13), which we shall denote more precisely by $s(\zeta)^2$, can be written as

$$s(\zeta)^2 = \boldsymbol{y}'[\boldsymbol{I} - \boldsymbol{Q}(\zeta)]\boldsymbol{y} \tag{16.15}$$

where

$$\boldsymbol{Q}(\zeta) = \boldsymbol{A}(\zeta)\,[\boldsymbol{M}(\zeta)]^{-}\boldsymbol{A}(\zeta)'$$

(see Section I.4.4.3). Rather than using (16.15) in (16.14) and minimizing $s(\zeta)^2$, Shirakura, Takahashi, and Srivastava (1996) propose maximizing

$$h(\zeta\,, y) = \boldsymbol{y}'[\boldsymbol{Q}(\zeta) - \boldsymbol{Q}]\boldsymbol{y}$$

where

$$Q = A_1(A_1'A_1)^{-1}A_1$$

and

$$s(\zeta)^2 = y'(I - Q)y - h(\zeta, y)$$

They then define

$$P = \min_{\zeta_0 \subset \xi_2} \min_{\zeta \in \mathcal{A}(\xi_2;\, \zeta)} P[h(\zeta_0, y) > h(\zeta, y)] \tag{16.16}$$

as the search probability for a given search design, where $\mathcal{A}(\xi_z;\ \zeta)$ denotes the set of all possible ζ of ξ_2, of which at least one effect is not of ζ_0, and ζ_0 represents the vector of the nonnegligible effects in ξ_2.

For the case $k = 1$ and assuming that $e \sim N(\mathbf{0}, I\sigma_e^2)$, Shirakura, Takahashi, and Srivastava (1996) derive an explicit expression for

$$P(\zeta_0, \zeta, \sigma_e^2) = P[h(\zeta_0, y) - h(\zeta, y)] \tag{16.17}$$

after rewriting (16.13) as

$$y = A_1\xi_1 + a(\zeta)\zeta + e \tag{16.18}$$

where ζ is one effect of ξ_2 and $a(\zeta)$ is the column vector in A_2 corresponding to ζ, and ζ_0 represents the nonnegligible effect in ξ_2. The expression for (16.17) is then given as follows (see Ghosh and Teschmacher, 2002):

$$P(\zeta_0, \zeta, \sigma_e^2) = .5 + 2[\phi(\rho c_1(\zeta_0, \zeta)) - .5]$$

$$\times [\phi(\rho\sqrt{r(\zeta_0, \zeta_0) - c_1^2(\zeta_0, \zeta)}) - .5] \tag{16.19}$$

where $\phi(\cdot)$ is the standard normal cdf, $r(\gamma, \eta) = a'(\gamma)[I - A_1(A_1'A_1)^{-1}A_1']a(\eta)$ for $(\gamma, \eta) = (\zeta_0, \zeta_0)$, (ζ_0, ζ), (ζ, ζ), $x(\zeta_0, \zeta) = r(\zeta_0, \zeta)/\sqrt{r(\zeta_0, \zeta_0)\,r(\zeta, \zeta)}$, $c_1^2(\zeta_0, \zeta) = [r(\zeta_0, \zeta_0)/2][1 - |\,x(\zeta_0, \zeta)\,|]$, and $\rho = |\,\zeta_0\,|/\sigma_e$.

It follows immediately from (16.19) that $.5 \le P(\zeta_0, \zeta, \sigma_e^2) \le 1$, as stated before, and that $P(\zeta_0, \zeta, \sigma_e^2)$ increases as ρ increases.

Using the search design and parameters in Example 16.2 we illustrate the use of (16.19) in the next example.

Example 16.3 From Example 16.1 we obtain for use in (16.18)

$$a(\zeta_0) = a'(AB) = \begin{pmatrix}1 & 1 & -1 & -1 & 1 & -1 & 1\end{pmatrix}$$
$$a(\zeta_1) = a'(AC) = \begin{pmatrix}1 & 1 & -1 & 1 & -1 & -1 & -1\end{pmatrix}$$
$$a(\zeta_2) = a'(BC) = \begin{pmatrix}1 & 1 & 1 & -1 & -1 & 1 & -1\end{pmatrix}$$

and

$$A_1 = \begin{pmatrix} 1 & 1 & 1 & 1 \\ 1 & -1 & -1 & -1 \\ 1 & -1 & 1 & 1 \\ 1 & 1 & -1 & 1 \\ 1 & 1 & 1 & -1 \\ 1 & 1 & -1 & -1 \\ 1 & -1 & -1 & 1 \end{pmatrix}$$

with $r(\zeta_0, \zeta_0) = 6$, $r(\zeta_0, \zeta_1) = 3$, $r(\zeta_0, \zeta_2) = -1$, $r(\zeta_1, \zeta_1) = 1.5$, $r\zeta_2, \zeta_2) = 5.5$, $x(\zeta_0, \zeta_1) = 1$, $x(\zeta_0, \zeta_2) = -0.1741$, $c_1^2(\zeta_0, \zeta_1) = 0$, $c_1^2(\zeta_0, \zeta_2) = 2.4777$, $\rho = 1$. We then obtain $P(\zeta_0, \zeta_1, 1) = .5$, $P(\zeta_0, \zeta_2, 1) = .9144$. Individually, the results show that there is a much higher probability of distinguishing between the true model (including AB) and the model including BC. If we adopt (16.16) as the overall measure for the given design and parameters, then the search probability P is only .5. □

Ghosh and Teschmacher (2002) show how search probabilities can be used to compare different search designs. They develop criteria so that such comparisons can be made without requiring a specific value of $\rho = |\,\zeta_0\,|/\sigma_e$.

CHAPTER 17

Robust-Design Experiments

As we have mentioned before, factorial designs in their various forms were originally developed and used in the context of agricultural experimentation. They have since found wide application in almost all fields of empirical research. Important contributions to factorial experimentation has come from applications in engineering and manufacturing, a development usually associated with the name Taguchi. The so-called Taguchi method and variations of this method are concerned with quality engineering and product improvement. According to Taguchi (1993, p. 4), "the objective of quality engineering is to discover the countermeasure which results in an overall reduction in the total loss to both the company and society. It is therefore important to evaluate the two kinds of losses—the loss due to function and one due to harmful effects." Thus, the basic philosophy here rests on the concept of societal loss due to imperfect production and "environmental" factors and the role that experimental design can play to reduce that loss. We shall discuss some of the underlying ideas in the following sections.

17.1 OFF-LINE QUALITY CONTROL

An important aspect of any production process is quality control. Typically this is performed during the manufacturing phase of the process. Various forms of inspection sampling can be applied to ensure that the product meets certain specification and that the process is in control (see, e.g., Montgomery, 1991). There exists, however, another form of quality control. It is conducted at the product or process design stage, thus preceding the manufacturing stage. This type of quality control is therefore referred to as *off-line quality control*. Its aim is to improve product manufacturability and reliability, and to reduce product development and lifetime costs (Kackar, 1985).

Design and Analysis of Experiments. Volume 2: Advanced Experimental Design
By Klaus Hinkelmann and Oscar Kempthorne
ISBN 0-471-55177-5

It is in this context that Taguchi (1986, 1987) introduced the application of what he called *robust parameter design* (parameter here refers to the characteristic of a product). According to Taguchi (1986, p. 98) its purpose "is to adjust the parameter levels so that the objective characteristic will not vary much even if the system and environmental parameters change. It is a search for the parameter levels at which the characteristic is stable." More explicitly, Montgomery (1999) reformulates this in terms of making (1) products and processes insensitive to environmental factors or other factors hard to control and to variation transmitted from components and (2) finding levels of the process variables that achieve a mean performance at a specified (target) value and at the same time reducing variability around that value.

17.2 DESIGN AND NOISE FACTORS

With the previous statements we have implicitly acknowledged the existence of two categories of factors that affect the characteristic of the product: Design factors and noise factors. This categorization is a new feature that we have not encountered before, at least not in this explicit form. It is, however, an important feature in conducting and analyzing robust-design experiments.

Design factors, sometimes also referred to as design parameters (Kackar, 1985) or control factors (Taguchi, 1993), are those factors that can be controlled easily during the manufacturing process, the values of which remain essentially fixed at "optimal" levels. Noise factors, on the other hand, are factors that are hard to control, such as environmental factors, and, as a consequence, their levels will vary within a certain range during the production process and/or during subsequent use of the product [for a detailed description of various types of noise factors see, e.g., Wu and Hamada (2000)].

As an example we mention an experiment performed in a semiconductor factory to determine the influence of five factors on the transistor gain for a particular device (Montgomery, 1999). The three factors *implant dose, drive-in time*, and *vacuum level* are easy to control and hence represent the design factors. The factors *oxide thickness* and *temperature* are hard to control during the manufacturing process and hence represent the noise factors.

The underlying idea of the Taguchi method is to incorporate the noise factors into the product design phase by varying their settings systematically over a range representative of the variation during typical manufacturing and/or use conditions. The aim then is to find settings for the design factors so that the influence of the noise factors on the product performance is minimized or, to express it differently, to make the product robust with respect to the noise factors. It is for this reason that this methodology is referred to as robust design or robust parameter design (or parameter design for short).

17.3 MEASURING LOSS

As mentioned earlier, the notion of loss to society and to the company forms the underpinning of the Taguchi method. To assess the loss we need to specify a loss function as a function of a value of the performance characteristic of the product, say Y, and the target value, say τ, for that performance characteristic. We can think of many loss functions, but in practice it is usually difficult to come up with the actual loss function. However, in many situations an appropriate loss function is given by

$$\ell(Y) = c(Y - \tau)^2 \tag{17.1}$$

where c is some constant, which can be determined if $\ell(Y)$ is known for some value of Y.

One way to determine c in (17.1) was described by Kackar (1985). Let $(\tau - \Delta, \tau + \Delta)$ be the consumer's tolerance interval in the sense that if Y falls outside this interval, then the product needs either to be repaired or discarded. Suppose the cost for that action is A dollars. It follows then from (17.1) that $A = c\Delta^2$ and hence $c = A/\Delta^2$.

The loss function (17.1) is symmetric around τ. This may not always be appropriate as the loss may depend on whether $Y < \tau$ or $Y > \tau$. In this case a simple variation of (17.1) may lead to the form

$$\ell(Y) = \begin{cases} c_1(Y - \tau)^2 & \text{if } Y < \tau \\ c_2(Y - \tau)^2 & \text{if } Y > \tau \end{cases} \tag{17.2}$$

where c_1 and c_2 are constants that can be determined using arguments similar to those given above.

A special case of (17.2) occurs if $\tau = 0$. Then (17.2) reduces to

$$\ell(Y) = c_2 Y^2 \qquad \text{if } Y > 0 \tag{17.3}$$

In either of the cases (17.1), (17.2), (17.3) we consider the expected loss

$$L = E[\ell(Y)] \tag{17.4}$$

as the criterion to assess the quality of a product or process. The expectation in (17.4) is taken with respect of the distribution of Y during the product's lifetime and across different users. The latter is represented to a great extent by some or all of the noise factors.

17.4 ROBUST-DESIGN EXPERIMENTS

In order to minimize the average loss (17.4) we perform an experiment, the robust-design experiment. The purpose of this experiment is to determine the optimal levels for the control factors in conjunction with reducing the product's performance around the target value. There are several ways in which such an experiment can be performed making use of some of the factorial designs we have discussed in previous chapters or certain types of response surface designs (see I.12.3 and I.12.4).

17.4.1 Kronecker Product Arrays

Distinguishing between the two sets of factors, design and noise factors, suggests the use of the following experimental design: Choose an appropriate fractional factorial design for the design factors and another one for the noise factors and then combine these two designs in the form of a Kronecker product design; that is, each design factor level combination occurs with each noise factor level combination. Taguchi (1986) referred to these two component designs as *inner array* and *outer array*, respectively. Moreover, he suggested the use of main effect plans in the form of orthogonal arrays.

As an illustration consider the following example.

Example 17.1 Suppose we have four design factors A_1, A_2, A_3, A_4, each at three levels, and three noise factors, B_1, B_2, B_3, each at two levels. The inner array is then given by OA(9, 4, 3, 2; 1) and the outer array by OA(4, 3, 2, 2; 1) (see Example 13.9). Then the Kronecker product design or the so-called *cross array* consists of $9 \times 4 = 36$ runs. □

An important aspect of the cross array is that the interactions between individual design and noise factors can be estimated (see Example 13.9). If such interactions exist, they can be exploited to choose the design factor settings such that the product variation over the noise factor settings will be reduced (we shall return to this point in Section 17.5).

On the other hand, the number of runs required for a cross array may be quite large, even when we use minimal designs for both the design and noise factors. This disadvantage of cross arrays will become even more pronounced if we allow at least some 2-factor interactions between design factors by using a resolution IV or resolution V design instead of a main effect plan.

17.4.2 Single Arrays

The last two comments above provide a strong argument to look for alternative experimental designs in connection with robust-design experiments. This can be achieved by combining the design and noise factors into one set of factors and then obtain an appropriate fractional factorial design that achieves the objectives of the experiment. Such experimental designs are referred to as *single array* or *common array*. We shall illustrate this approach with some examples. For a listing of useful single arrays see Wu and Hamada (2000).

Example 17.2 Suppose we have three design factors A_1, A_2, A_3 and two noise factors B_1, B_2, each at two levels. The usual Taguchi method would cross the 2^{3-1}_{III} array with the 2^2 array with $4 \times 4 = 16$ runs. Suppose, however, that we are interested also in estimating 2-factor interactions between design factors. In that case the crossed array would be of the form $2^3 \times 2^2$ with $8 \times 4 = 32$ runs. A single array can be obtained by considering the 2^{5-1}_{V} array based on the identity relationship $I = A_1A_2A_3\,B_1B_2$ with only 16 runs. □

Example 17.3 Suppose we have five design factors A_1, A_2, A_3, A_4, A_5 and two noise factors B_1, B_2, each at two levels. The usual cross array would be of the form $2^{5-2}_{\text{III}} \times 2^2$ with $8 \times 4 = 32$ runs. The single array 2^{7-2}_{IV}, also with 32 runs, based on the identity relationship

$$I = A_1A_2A_3A_5 = A_1A_2A_4B_1B_2 = A_3A_4A_5B_1B_2$$

has the additional property that it allows us to estimate at least some of the 2-factor interactions among design factors, assuming that only interactions involving three or more factors are negligible. □

Example 17.4 Suppose we have three design factors A_1, A_2, A_3, each at three levels, and two noise factors, B_1, B_2, each at two levels. The usual cross array would be of the form $3^{3-1}_{\text{III}} \times 2^2$ with $9 \times 4 = 36$ runs. Here the 3^{3-1}_{III}, referred to $L_9(3^3)$ by Taguchi, is based on $I = A_1A_2A_3^2$. An alternative single array in the form of a modified central composite design for five factors (see I.12.4.3) was proposed by Myers, Khuri, and Vining (1992). Their design consists of three parts: (1) The factorial part is a 2^{5-1}_{V} based on $I = A_1A_2A_3B_1B_2$ and using only the high and low levels of each factor; (2) the axial part consists of the axial points for the control factors, keeping the noise factors at their intermediate level (assuming that the noise factors are quantitative factors), and (3) n_0 center runs. Using the response surface methodology convention of denoting the low, intermediate and high levels by -1, 0, 1, respectively, the final design is given in Table 17.1.

This design, consisting of $16 + 6 + n_0 = 22 + n_0$ runs, allows us to estimate all main effects, all 2-factor interactions between design factors, and the 2-factor interactions between design and noise factors. □

The single arrays discussed above provide some insight into their general structure. It also shows that single arrays may be more economical than crossed arrays or that they provide more information. These are aspects that may determine which type of array we should choose for a given situation. However, to some extent this choice will also be influenced by the type of modeling we intend to use for data from such experiments, that is, how we intend to use the data in order to arrive at a robust product design. We shall discuss some of these issues in the following sections.

Table 17.1 Central Composite Design for Three Design Factors and Two Noise Factors

A_1	A_2	A_3	B_1	B_2	
−1	−1	−1	−1	−1	
−1	−1	1	1	1	
−1	1	−1	1	1	
−1	1	1	−1	1	
−1	1	1	1	−1	
1	−1	−1	1	1	
1	−1	1	−1	1	
1	−1	1	1	−1	
1	1	−1	−1	1	
1	1	−1	1	−1	
1	1	1	−1	−1	
−1	−1	−1	−1	1	
−1	−1	−1	1	−1	
−1	−1	1	−1	−1	
−1	1	−1	−1	−1	
1	−1	−1	−1	−1	
−1	0	0	0	0	
1	0	0	0	0	
0	−1	0	0	0	
0	1	0	0	0	
0	0	−1	0	0	
0	0	1	0	0	
0	**0**	**0**	**0**	**0**	n_0

17.5 MODELING OF DATA

Just as there are different approaches to deciding on a statistical design for a robust-design experiment, there are different ways of analyzing the data from such an experiment. The proposal by Taguchi (1986) to use various forms of signal-to-noise ratios (S/N ratios) to analyze and interpret such data has come under a certain amount of criticism (e.g., Box, 1988) and as a consequence other methods of looking at the data from these experiments have been proposed. We shall indicate some of these ideas in the following sections.

17.5.1 Location and Dispersion Modeling

As we have mentioned earlier, the novel idea associated with robust-design experiments is not only to identify the factors that affect the average response, but also to identify the factors that contribute to the variability of the response. In the practical setting of product design this means more specifically that we need to model both the location and the dispersion of the response or, as it is often referred to, the performance characteristic in order to find an appropriate level combination of the control factors that will achieve the desired value of or target for the performance characteristic.

Let us consider the cross array of Section 17.4.1. Suppose the inner or design array D_1 consists of N_1 treatment combinations and the outer or noise array D_2 consists of N_2 treatment combinations. The total number of runs in $D = D_1 \times D_2$ then is $N = N_1 N_2$. Let $\boldsymbol{A}$ denote the design factor settings, then, following Nair (1986), we can write a location-scale model for the (possibly transformed) observations as

$$Y = \mu(\boldsymbol{A}) + \sigma(\boldsymbol{A})e \tag{17.5}$$

where the random component e does not depend on $\boldsymbol{A}$ and is assumed to have mean 0 and variance 1. The terms $\mu(\boldsymbol{A})$ and $\sigma(\boldsymbol{A})$ represent the location and dispersion effects, respectively.

We should emphasize that using model (17.5) implies that the experiment has been carried out according to a completely randomized design. In practice this may not always be possible and some form of blocking may have to be used as discussed in Chapters 11 and 12. Another possibility is the use of a split-plot-type design (see I.13.4), where, for example, the treatment combinations of the design array represent the whole-plot treatments and the treatment combinations of the noise array represent the split-plot treatments. In either case it is easy to modify model (17.5) by either incorporating the block effects or the whole-plot error term, both of which are independent of the design factors.

Using the loss function (17.1) the expected loss (17.4) for the response (17.5) can be written as

$$\begin{aligned} L &= E[\ell(Y)] = c\,E(Y - \tau)^2 \\ &= c[\mu(\boldsymbol{A}) - \tau]^2 + c\,\sigma^2(\boldsymbol{A}) \end{aligned} \tag{17.6}$$

This form of L shows explicitly that in order to minimize the mean-squared error (17.6) we need to identify the setting of the design factors such that $\mu(\boldsymbol{A})$ achieves the target value τ and $\sigma^2(\boldsymbol{A})$ is minimized.

Now let $\mathcal{A} = (A_1, A_2, \ldots, A_n)$ be the set of design factors employed in the experiment. Then the aim of the subsequent analysis is to divide $\mathcal{A}$ into three sets, say, $\mathcal{A} = (\mathcal{A}_1, \mathcal{A}_2, \mathcal{A}_3)$ (Phadke, 1982), where

$\mathcal{A}_1$ represents the factors that affect dispersion (they are referred to as control factors);

$\mathcal{A}_2$ represents the factors that affect location but not dispersion (they are referred to as adjustment factors);

$\mathcal{A}_3$ represents the factors that neither affect location nor dispersion.

According to this division the factors included in $\mathcal{A}_1$ and $\mathcal{A}_2$ represent the actual design factors (Box, 1988). Such a division is possible only if we can get rid of dispersion effects that arise because of a dependence of σ on μ (Box and Meyer, 1986b; Box, 1988). This can often be achieved through an appropriate transformation of the original data or through the use of a performance measure independent of adjustment (PerMIA) as proposed by Leon, Shoemaker, and Kackar (1987). We assume that Y in (17.5) is of that form.

To achieve the division of $\mathcal{A}$ let us now consider the observations Y from the experiment based on the cross array D. More specifically, let y_{ij} be the (possibly transformed) observation for the ith treatment combination of D_1 and the jth treatment combination of D_2 ($i = 1, 2, \ldots, N_1$; $j = 1, 2, \ldots, N_2$). In some sense the N_2 observations, stemming from D_2, for each treatment combination of D_1 can be considered as replications for that treatment combination. It is, therefore, plausible to analyze

$$\overline{y}_{i.} = \frac{1}{N_2} \sum_{j=1}^{N_2} y_{ij} \quad \text{and} \quad s_i^2 = \frac{1}{N_2 - 1} \sum_{j=1}^{N_2} (y_{ij} - \overline{y}_{i.})^2$$

for $i = 1, 2, \ldots, N_1$, separately as measures of location and variability, respectively. The analysis of the $\overline{y}_{i.}$ conforms, of course, to the usual analysis of fractional factorial designs. The s_i^2 are obtained over the same design points of D_2 and, hence, any differences among them can be attributed to the design factors. This provides justification for analyzing the s_i^2, or more appropriately $\log s_i^2$, in the same way as the $\overline{y}_{i.}$ in order to identify the factors belonging to $\mathcal{A}_1$ and $\mathcal{A}_2$. For both variables we can use the tools of the analysis of variance.

We shall illustrate this approach with the data obtained from a cross-array experiment described by Engel (1992).

Example 17.5 (Engel, 1992) Seven design factors and three noise factors, each at two levels, were included in the experiment in an effort to identify the design factors that had an effect on the mean of and variation in the percentage of shrinkage of products made by injection moulding. The factors are (rather than using the labels $A_1, A_2, \ldots, A_7$ and B_1, B_2, B_3 we retain the labels used by Engel):

Design Factors	**Noise Factors**
A: cycle time	M: percentage regrind
B: mould temperature	N: moisture content
C: cavity thickness	O: ambient temperature
D: holding pressure	
E: injection speed	
F: holding time	
G: gate size	

The cross-array D is given by crossing $D_1 = 2^{7-4}_{\text{III}}$ with $D_2 = 2^{3-1}_{\text{III}}$, were D_1 has the identity relationship

$$I = ABC = DEFG = ABCDEFG = ADE = BCDE$$

$$= AFG = BCFG = BEG = ACEG = BDF = ACDF$$

and D_2 has the identity relationship

$$I = MNO$$

The 32 design points with their observations are listed in Table 17.2 together with the values for s_i^2 (note that as a matter of convenience each s_i^2 value is listed four times for the four replications of each design point in D_1). The results of the analysis of variance for the $\overline{y}_{i.}$ and the log s_i^2 are given in Table 17.2.

Concerning the location modeling aspect, we may conclude by looking informally and somewhat subjectively at the P values (or half-normal plots) for the estimated main effects, that the factors A, D, and possibly G constitute $\mathcal{A}_2$ and hence represent the adjustment factors. A separate regression analysis using A, D, and G yields the prediction equation

$$\widehat{Y} = 2.3375 + 0.85x_A - 0.5625x_D - 0.4625x_6$$

where x_A, x_D, x_G represent the coded levels of A, D, and G, respectively. Together with engineering and production considerations this equation can then be used to adjust the process so that a particular target, τ, can be achieved.

With regard to the dispersion modeling the results point clearly to factor F as having a pronounced effect on the variability and hence constituting the only control factor. A look at the data shows, indeed, that the s_i^2 for those treatment combinations in D_1 with $F = 0$ are much smaller than for those with $F = 1$.

For a different approach to dispersion modeling and analysis of the same data set we refer the reader to Engel (1992). □

17.5.2 Dual-Response Modeling

We have pointed out earlier that cross-array experiments allow for the estimation of 2-factor interactions between design and noise factors. Yet, in the preceding discussion we have assumed that such interactions either do not exist or have been eliminated through an appropriate transformation of the original observations. Also, the location and dispersion modeling approach as discussed in the previous section is obviously not appropriate for single-array experiments.

In order to deal with these situations we shall describe now a two-step modeling procedure that Myers, Khuri, and Vining (1992) and Vining and Myers (1990) referred to as a dual-response approach. The basic idea is to express an

Table 17.2 Robust-Design Experiment Cross Array Data Points, Data, and Analysis

```
options nodate pageno=1 LS=75;
data robust;
input A B C D E F G M N O Y S @@;
logS=log(S);
datalines;
0 0 0 0 0 0 0 0 0 0 2.2 0.0092    0 0 0 0 0 0 0 0 1 1 2.1 0.0092
0 0 0 0 0 0 0 1 0 1 2.3 0.0092    0 0 0 0 0 0 0 1 1 0 2.3 0.0092
0 0 0 1 1 1 1 0 0 0 0.3 1.7700    0 0 0 1 1 1 1 0 1 1 2.5 1.7700
0 0 0 1 1 1 1 1 0 1 2.7 1.7700    0 0 0 1 1 1 1 1 1 0 0.3 1.7700
0 1 1 0 0 1 1 0 0 0 0.5 2.1000    0 1 1 0 0 1 1 0 1 1 3.1 2.1000
0 1 1 0 0 1 1 1 0 1 0.4 2.1000    0 1 1 0 0 1 1 1 1 0 2.8 2.1000
0 1 1 1 1 0 0 0 0 0 2.0 0.0092    0 1 1 1 1 0 0 0 1 1 1.9 0.0092
0 1 1 1 1 0 0 1 0 1 1.8 0.0092    0 1 1 1 1 0 0 1 1 0 2.0 0.0092
1 0 1 0 1 0 1 0 0 0 3.0 0.0025    1 0 1 0 1 0 1 0 1 1 3.1 0.0025
1 0 1 0 1 0 1 1 0 1 3.0 0.0025    1 0 1 0 1 0 1 1 1 0 3.0 0.0025
1 0 1 1 0 1 0 0 0 0 2.1 1.8733    1 0 1 1 0 1 0 0 1 1 4.2 1.8733
1 0 1 1 0 1 0 1 0 1 1.0 1.8733    1 0 1 1 0 1 0 1 1 0 3.1 1.8733
1 1 0 0 1 1 0 0 0 0 4.0 1.7625    1 1 0 0 1 1 0 0 1 1 1.9 1.7625
1 1 0 0 1 1 0 1 0 1 4.6 1.7625    1 1 0 0 1 1 0 1 1 0 2.2 1.7625
1 1 0 1 0 0 1 0 0 0 2.0 0.0067    1 1 0 1 0 0 1 0 1 1 1.9 0.0067
1 1 0 1 0 0 1 1 0 1 1.9 0.0067    1 1 0 1 0 0 1 1 1 0 1.8 0.0067
;
run;

proc print data=robust;
title1 'TABLE 17.2';
title2 'ROBUST-DESIGN EXPERIMENT';
title3 'CROSS ARRAY DATA POINTS AND DATA';
run;

proc glm data=robust;
class A B C D E F G;
model Y logS = A B C D E F G;
estimate 'A' A -1 1/divisor=16;
estimate 'B' B -1 1/divisor=16;
estimate 'C' C -1 1/divisor=16;
estimate 'D' D -1 1/divisor=16;
estimate 'E' E -1 1/divisor=16;
estimate 'F' F -1 1/divisor=16;
estimate 'G' G -1 1/divisor=16;
title2 'IDENTIFICATION OF ADJUSTMENT AND DISPERSION FACTORS';
title3 'THROUGH ANALYSIS OF VARIANCE';
run;

proc reg data=robust;
model Y = A D G;
title2 'ESTIMATING PREDICTION EQUATION';
title3 'USING ADJUSTMENT FACTORS A, D, G';
run;
quit;
```

Table 17.2 (*Continued*)

Obs	A	B	C	D	E	F	G	M	N	O	Y	S	LogS
1	0	0	0	0	0	0	0	0	0	0	2.2	0.0092	-4.68855
2	0	0	0	0	0	0	0	0	1	1	2.1	0.0092	-4.68855
3	0	0	0	0	0	0	0	1	0	1	2.3	0.0092	-4.68855
4	0	0	0	0	0	0	0	1	1	0	2.3	0.0092	-4.68855
5	0	0	0	1	1	1	1	0	0	0	0.3	1.7700	0.57098
6	0	0	0	1	1	1	1	0	1	1	2.5	1.7700	0.57098
7	0	0	0	1	1	1	1	1	0	1	2.7	1.7700	0.57098
8	0	0	0	1	1	1	1	1	1	0	0.3	1.7700	0.57098
9	0	1	1	0	0	1	1	0	0	0	0.5	2.1000	0.74194
10	0	1	1	0	0	1	1	0	1	1	3.1	2.1000	0.74194
11	0	1	1	0	0	1	1	1	0	1	0.4	2.1000	0.74194
12	0	1	1	0	0	1	1	1	1	0	2.8	2.1000	0.74194
13	0	1	1	1	1	0	0	0	0	0	2.0	0.0092	-4.68855
14	0	1	1	1	1	0	0	0	1	1	1.9	0.0092	-4.68855
15	0	1	1	1	1	0	0	1	0	1	1.8	0.0092	-4.68855
16	0	1	1	1	1	0	0	1	1	0	2.0	0.0092	-4.68855
17	1	0	1	0	1	0	1	0	0	0	3.0	0.0025	-5.99146
18	1	0	1	0	1	0	1	0	1	1	3.1	0.0025	-5.99146
19	1	0	1	0	1	0	1	1	0	1	3.0	0.0025	-5.99146
20	1	0	1	0	1	0	1	1	1	0	3.0	0.0025	-5.99146
21	1	0	1	1	0	1	0	0	0	0	2.1	1.8733	0.62770
22	1	0	1	1	0	1	0	0	1	1	4.2	1.8733	0.62770
23	1	0	1	1	0	1	0	1	0	1	1.0	1.8733	0.62770
24	1	0	1	1	0	1	0	1	1	0	3.1	1.8733	0.62770
25	1	1	0	0	1	1	0	0	0	0	4.0	1.7625	0.56673
26	1	1	0	0	1	1	0	0	1	1	1.9	1.7625	0.56673
27	1	1	0	0	1	1	0	1	0	1	4.6	1.7625	0.56673
28	1	1	0	0	1	1	0	1	1	0	2.2	1.7625	0.56673
29	1	1	0	1	0	0	1	0	0	0	2.0	0.0067	-5.00565
30	1	1	0	1	0	0	1	0	1	1	1.9	0.0067	-5.00565
31	1	1	0	1	0	0	1	1	0	1	1.9	0.0067	-5.00565
32	1	1	0	1	0	0	1	1	1	0	1.8	0.0067	-5.00565

Table 17.2 *(Continued)*

IDENTIFICATION OF ADJUSTMENT AND DISPERSION FACTORS
THROUGH ANALYSIS OF VARIANCE

The GLM Procedure

Dependent Variable: Y

Source	DF	Sum of Squares	Mean Square	F Value	Pr > F
Model	7	11.00000000	1.57142857	1.67	0.1646
Error	24	22.60000000	0.94166667		
Corrected Total	31	33.60000000			

R-Square	Coeff Var	Root MSE	y Mean
0.327381	43.12867	0.970395	2.250000

Source	DF	Type I SS	Mean Square	F Value	Pr > F
A	1	5.78000000	5.78000000	6.14	0.0207
B	1	0.18000000	0.18000000	0.19	0.6659
C	1	0.12500000	0.12500000	0.13	0.7188
D	1	2.53125000	2.53125000	2.69	0.1141
E	1	0.66125000	0.66125000	0.70	0.4103
F	1	0.01125000	0.01125000	0.01	0.9139
G	1	1.71125000	1.71125000	1.82	0.1902

Source	DF	Type III SS	Mean Square	F Value	Pr > F
A	1	5.78000000	5.78000000	6.14	0.0207
B	1	0.18000000	0.18000000	0.19	0.6659
C	1	0.12500000	0.12500000	0.13	0.7188
D	1	2.53125000	2.53125000	2.69	0.1141
E	1	0.66125000	0.66125000	0.70	0.4103
F	1	0.01125000	0.01125000	0.01	0.9139
G	1	1.71125000	1.71125000	1.82	0.1902

Parameter	Estimate	Standard Error	t Value	Pr > |t|
A	0.05312500	0.02144291	2.48	0.0207
B	-0.00937500	0.02144291	-0.44	0.6659
C	0.00781250	0.02144291	0.36	0.7188
D	-0.03515625	0.02144291	-1.64	0.1141
E	0.01796875	0.02144291	0.84	0.4103
F	-0.00234375	0.02144291	-0.11	0.9139
G	-0.02890625	0.02144291	-1.35	0.1902

Table 17.2 *(Continued)*

IDENTIFICATION OF ADJUSTMENT AND DISPERSION FACTORS
THROUGH ANALYSIS OF VARIANCE

The GLM Procedure

Dependent Variable: LogS

Source	DF	Sum of Squares	Mean Square	F Value	Pr > F
Model	7	266.4310915	38.0615845	Infty	<.0001
Error	24	0.0000000	0.0000000		
Corrected Total	31	266.4310915			

R-Square	Coeff Var	Root MSE	LogS Mean
1.000000	0	0	-2.233358

Source	DF	Type I SS	Mean Square	F Value	Pr > F
A	1	1.5111751	1.5111751	Infty	<.0001
B	1	0.6003957	0.6003957	Infty	<.0001
C	1	0.2841756	0.2841756	Infty	<.0001
D	1	0.3835367	0.3835367	Infty	<.0001
E	1	0.7414489	0.7414489	Infty	<.0001
F	1	261.7830683	261.7830683	Infty	<.0001
G	1	1.1272911	1.1272911	Infty	<.0001

Source	DF	Type III SS	Mean Square	F Value	Pr > F
A	1	1.5111751	1.5111751	Infty	<.0001
B	1	0.6003957	0.6003957	Infty	<.0001
C	1	0.2841756	0.2841756	Infty	<.0001
D	1	0.3835367	0.3835367	Infty	<.0001
E	1	0.7414489	0.7414489	Infty	<.0001
F	1	261.7830683	261.7830683	Infty	<.0001
G	1	1.1272911	1.1272911	Infty	<.0001

Parameter	Estimate	Standard Error	t Value	Pr > |t|
A	-0.02716392	0	-Infty	<.0001
B	0.01712197	0	Infty	<.0001
C	-0.01177954	0	-Infty	<.0001
D	0.01368480	0	Infty	<.0001
E	-0.01902723	0	-Infty	<.0001
F	0.35752449	0	Infty	<.0001
G	-0.02346135	0	-Infty	<.0001

Table 17.2 *(Continued)*

ESTIMATING PREDICTION EQUATION
USING ADJUSTMENT FACTORS A, D, G

The REG Procedure
Model: MODEL1
Dependent Variable: Y

Analysis of Variance

Source	DF	Sum of Squares	Mean Square	F Value	Pr > F
Model	3	10.02250	3.34083	3.97	0.0178
Error	28	23.57750	0.84205		
Corrected Total	31	33.60000			

Root MSE	0.91763	R-Square	0.2983
Dependent Mean	2.25000	Adj R-Sq	0.2231
Coeff Var	40.78377		

Parameter Estimates

Variable	DF	Parameter Estimate	Standard Error	t Value	Pr > |t|
Intercept	1	2.33750	0.32443	7.20	<.0001
A	1	0.85000	0.32443	2.62	0.0140
D	1	-0.56250	0.32443	-1.73	0.0940
G	1	-0.46250	0.32443	-1.43	0.1650

observation as a function of design and noise factor effects as well as the 2-factor interactions between design and noise factors, or

$$Y = \mu(\boldsymbol{A}) + \nu(\boldsymbol{B}) + \lambda(\boldsymbol{A}, \boldsymbol{B}) + \sigma e \tag{17.7}$$

where $\boldsymbol{B}$ denotes the noise factor settings. Contrary to (17.5), in (17.7) σ is independent of the design factors. Even though the noise factors are considered to be random factors during the use of the product, we need to remember that for the purpose of the robust-design experiment their levels are selected by the experimenter. It is for this reason that we first consider (17.7) conditionally on $\boldsymbol{B}$ in order to identify important design and noise factors. Using those factors we then consider the predicted value $\widehat{Y}$ and consider $\text{var}(\widehat{Y})$ as a second response in an effort to reduce the variability.

More specifically, let us now write (17.7) as

$$\boldsymbol{Y} = \boldsymbol{X\beta} + \boldsymbol{Z\gamma} + \boldsymbol{W\delta} + \sigma\boldsymbol{e} \tag{17.8}$$

where $\boldsymbol{X}_{N\times p}$, $\boldsymbol{Z}_{N\times q}$, and $\boldsymbol{W}_{N\times s}$ are known design model matrices associated with the design factor effects $\boldsymbol{\beta}_{p\times 1}$, the noise factor effects $\boldsymbol{\gamma}_{q\times 1}$, and the design–noise factor interactions $\boldsymbol{\delta}_{s\times 1}$, respectively. It is convenient and useful here to think of model (17.8) as a regression or response surface model. Depending on the experimental design used, $\boldsymbol{\beta}$ may include linear and quadratic main effects and possibly some interaction effects between design factors, and $\boldsymbol{\delta}$ may include some or all 2-factor interactions between design and noise factors.

As indicated earlier, in step 1 we analyze $\boldsymbol{Y}$ conditional on $\boldsymbol{\gamma}$ and $\boldsymbol{\delta}$ using analysis of variance. Suppose we identify $\boldsymbol{\beta}_0 \subset \boldsymbol{\beta}$, $\boldsymbol{\gamma}_0 \subset \boldsymbol{\gamma}$, and $\boldsymbol{\delta}_0 \subset \boldsymbol{\delta}$ as important effects. We write a model for the predicted value then as

$$\widehat{y} = \widehat{\boldsymbol{\beta}}_0'\boldsymbol{x}_0 + \widehat{\boldsymbol{\gamma}}_0'z_0 + \widehat{\boldsymbol{\delta}}_0'\boldsymbol{w}_0 \tag{17.9}$$

where $\boldsymbol{x}_0$ represents the linear and quadratic terms for design factors retained in (17.9), z_0 represents linear terms for noise factors retained in (17.9), and $\boldsymbol{w}_0$ represents cross-product terms xz associated with design and noise factor interaction terms retained in (17.9). For example, (17.9) might be of the form

$$\widehat{y} = \widehat{\beta}_0 + \widehat{\beta}_1x_1 + \widehat{\beta}_{11}x_1^2 + \widehat{\beta}_2x_2 + \widehat{\gamma}_2z_2 + \widehat{\gamma}_3z_3 + \widehat{\delta}_{12}x_1z_2 + \widehat{\delta}_{23}x_2z_3 \tag{17.10}$$

in which case $\widehat{\boldsymbol{\beta}}_0' = (\widehat{\beta}_0, \widehat{\beta}_1, \widehat{\beta}_{11}, \widehat{\beta}_2)$, $\widehat{\boldsymbol{\gamma}}_0' = (\widehat{\gamma}_2, \widehat{\gamma}_3)$, $\widehat{\boldsymbol{\delta}}_0' = (\widehat{\delta}_{12}, \widehat{\delta}_{23})$ and $\boldsymbol{x}_0' = (1, x_1, x_1^2, x_2)$, $z_0' = (z_2, z_3)$, $\boldsymbol{w}_0' = (x_1z_2, x_2z_3)$, with x_i referring to the ith design factor and z_j to the jth noise factor.

Given $\widehat{\boldsymbol{\beta}_0}$, $\widehat{\boldsymbol{\gamma}_0}$, $\widehat{\boldsymbol{\delta}_0}$ and assuming now that the z and w terms are random variables, we model the location as

$$\widehat{E}(y \mid \widehat{\boldsymbol{\beta}}_0, \widehat{\boldsymbol{\gamma}}_0, \widehat{\boldsymbol{\delta}}_0) = \widehat{\boldsymbol{\beta}}_0'\boldsymbol{x}_0 \tag{17.11}$$

and the variability as

$$\begin{aligned}\widehat{\text{var}}(y|\widehat{\boldsymbol{\beta}}_0, \widehat{\boldsymbol{\gamma}}_0, \widehat{\boldsymbol{\delta}}_0) &= \text{var}(\widehat{\boldsymbol{\gamma}}_0'z_0) + \text{var}(\widehat{\boldsymbol{\gamma}}_0'\boldsymbol{w}_0)\\ &\quad + 2\,\text{cov}(\widehat{\boldsymbol{\gamma}}_0'z_0, \widehat{\boldsymbol{\gamma}}_0'\boldsymbol{w}_0) + \sigma^2\end{aligned} \tag{17.12}$$

To illustrate (17.11) and (17.12) we consider (17.10) and find

$$\widehat{E}(y) = \widehat{\beta}_0 + \widehat{\beta}_1x_1 + \widehat{\beta}_{11}x_1^2 + \widehat{\beta}_2x_2 \tag{17.13}$$

and

$$\begin{aligned}\operatorname{var}(y) &= (\widehat{\gamma}_2)^2\sigma_{z_2}^2 + (\widehat{\gamma}_3)^2\sigma_{z_3}^2 + (\widehat{\delta_{12}x_1})^2\sigma_{z_2}^2 \\ &\quad + (\widehat{\delta_{23}x_2})^2\sigma_{z_3}^2 + 2\widehat{\gamma}_2\widehat{\delta}_{12}x_1\,\sigma_{z_2}^2 \\ &\quad + 2\widehat{\gamma}_3\,\widehat{\delta}_{23}x_2\,\sigma_{z_3}^2 + \sigma^2 \\ &= [(\widehat{\gamma}_2)^2 + (\widehat{\delta_{12}x_1})^2 + 2\widehat{\gamma}_2\widehat{\delta}_{12}x_1]\,\sigma_{z_2}^2 \\ &\quad + [(\widehat{\gamma}_3)^2 + (\delta_{23}x_2)^2 + 2\widehat{\gamma}_3\widehat{\delta}_{23}x_2]\sigma_{z_3}^2 \\ &\quad + \sigma^2 \end{aligned} \tag{17.14}$$

Assuming that the levels for the noise factors B_2 and B_3 are chosen such that $\sigma_{z_2}^2 = \sigma_{z_3}^2 = 1$, then an estimator for (17.14) can be written as

$$\begin{aligned}\widehat{\operatorname{var}}(y|\widehat{\boldsymbol{\beta}}_0, \widehat{\boldsymbol{\gamma}}_0, \widehat{\boldsymbol{\delta}}_0) &= [(\widehat{\gamma}_2)^2 + (\widehat{\delta_{12}x_1})^2 + 2\widehat{\gamma}_2\widehat{\delta}_{12}x_{12}] \\ &\quad + [(\widehat{\gamma}_3)^2 + (\widehat{\delta_{23}x_2})^2 + 2\widehat{\gamma}_3\{\widehat{\delta}_{23}x_2] + \mathrm{MS}(E) \end{aligned} \tag{17.15}$$

where MS(E) is obtained from the ANOVA using model (17.8). By choosing the levels of the noise factors such that $\sigma_{z_j}^2 = 1$, we also assume that under field conditions the z values represent a random sample of levels within the same range.

The response surface (17.11) in general or (17.13) for the illustrated case will then be used to adjust the process to the target value. At the same time the response surface (17.12) in general or (17.15) for the illustrated case will serve to reduce the variance by using the identified adjustment factors. In general, these two objectives cannot be achieved independently of each other and hence require engineering as well as statistical considerations. For examples, see Myers, Khuri, and Vining (1992) and Vining and Myers (1990), who also point out how this may involve sequential experimentation.

Another way of making use of the design–noise factor interactions is to consider, for each interaction, the interaction plot and choose the design factor level that has the flattest response over the range of the particular noise factor (see Wu and Hamada, 2000).

Finally, an extension of model (17.8) and the analysis based on it has been proposed by Wolfinger and Tobias (1998). Their mixed linear model allows a more direct approach of modeling variability in addition to location. However, this approach requires the assumption of normality for the distribution of the dispersion and random effects in order to perform maximum likelihood or REML estimation.

CHAPTER 18

Lattice Designs

18.1 DEFINITION OF QUASI-FACTORIAL DESIGNS

We have previously discussed the desirability and usefulness of incomplete block designs (see Chapter 1) when the number of treatments is large and/or complete blocks are unavailable or inappropriate. In this chapter we describe yet other methods of constructing certain types of resolvable incomplete block designs, some of which are in fact BIB or PBIB designs, but others are not. These methods are based on a correspondence between the treatments to be compared and the treatment combinations of a factorial set such that the procedures for constructing systems of confounding (see Chapter 11) can be utilized. Because of this correspondence, these designs are referred to as *quasi-factorial* or *lattice designs*. We emphasize, however, that the actual treatments do not have a factorial structure.

Historically, lattice designs were developed for large-scale agricultural experiments (Yates, 1936b) in which large numbers of varieties were to be compared. The main application since then has been and continues to be in agriculture.

A special feature of lattice designs is that the number of treatments, t, is related to the block size, k, in the form $t = k^2$ or $t = k^3$ or $t = k(k + 1)$ (see Section 18.2). Even though this limits the number of possible designs (see Section 6.2), lattice designs represent an important class of designs nevertheless, in particular when one is dealing with a large number of treatments. In certain types of agronomic experiments the number of treatments can easily be 100 or more, for example, in breeding experiments.

18.1.1 An Example: The Design

We shall illustrate the basic idea of a lattice design with a simple example. Suppose we have $t = 4$ treatments t_1, t_2, t_3, t_4 and blocks of size 2. We set

Design and Analysis of Experiments. Volume 2: Advanced Experimental Design
By Klaus Hinkelmann and Oscar Kempthorne
ISBN 0-471-55177-5

up the following correspondence between the four treatments and the treatment combinations of a 2^2 factorial (using the notation of Section 7.2.4):

Treatment	t_1	t_2	t_3	t_4
Quasi-factorial combination	(1)	a	b	ab

To arrange the treatments in blocks of size 2 we make use of a system of partial confounding (see Chapter 9). We may, for example, use three different types of confounding in three replicates, that is,

Replicate	Confounding	Blocks
I	A	(1), b and a, ab
II	B	(1), a and b, ab
III	AB	(1), ab and a, b

In terms of the treatments t_1, t_2, t_3, t_4 the arrangement is then as follows:

Replicate I	**Replicate II**	**Replicate III**
t_1, t_3	t_1, t_2	t_1, t_4
t_2, t_4	t_3, t_4	t_2, t_3

Suppose that this basic pattern is repeated q times, so that there are $3q$ replicates altogether.

18.1.2 Analysis

Using the results of Chapter 9 we obtain the estimates $\widehat{A}_{\mathrm{II,III}}$, $\widehat{B}_{\mathrm{I,III}}$, $\widehat{AB}_{\mathrm{I,II}}$. In order to make comparisons among the treatments t_1, t_2, t_3, t_4 or their effects τ_1, τ_2, τ_3, τ_4, respectively, we apply the parameterization (7.49). In this way we obtain, for example,

$$\widehat{\tau}_1 - \widehat{\tau}_2 = -\widehat{A}_{\mathrm{II,III}} + \widehat{AB}_{\mathrm{I,II}}$$

$$\widehat{\tau}_1 - \widehat{\tau}_3 = -\widehat{B}_{\mathrm{I,III}} + \widehat{AB}_{\mathrm{I,II}}$$

$$\widehat{\tau}_1 - \widehat{\tau}_4 = -\widehat{A}_{\mathrm{II,III}} + \widehat{B}_{\mathrm{I,III}}$$

and other contrasts can be estimated similarly. Since $\widehat{A}_{\mathrm{II,III}}$, $\widehat{B}_{\mathrm{I,III}}$, and $\widehat{AB}_{\mathrm{I,III}}$ are independent with

$$\mathrm{var}(\widehat{A}_{\mathrm{II,III}}) = \mathrm{var}(\widehat{B}_{\mathrm{I,III}}) = \mathrm{var}(\widehat{AB}_{\mathrm{I,II}}) = \sigma_e^2/2q$$

Table 18.1 T | B-ANOVA for Lattice Design

Source	d.f.	SS
$\boldsymbol{X}_\rho \mid \mathfrak{I}$	$3q-1$	Usual
$\boldsymbol{X}_{\beta^*} \mid \mathfrak{I}, \boldsymbol{X}_\rho$	$3q$	Usual
$\boldsymbol{X}_\tau \mid \mathfrak{I}, \boldsymbol{X}_\rho, \boldsymbol{X}_{\beta^*}$	3	$2q\{[\widehat{A}_{\text{II,III}}]^2 + [\widehat{B}_{\text{I,III}}]^2 + [\widehat{AB}_{\text{I,II}}]^2\}$
$\boldsymbol{I} \mid \mathfrak{I}, \boldsymbol{X}_\rho, \boldsymbol{X}_{\beta^*}, \boldsymbol{X}_\tau$	$3(2q-1)$	Subtraction
Total	$12q-1$	Usual

it follows then that

$$\text{var}(\widehat{\tau}_i - \widehat{\tau}_j) = \sigma_e^2/2q$$

for $i, j = 1, 2, 3, 4, i \neq j$.

To estimate σ_e^2 we turn to the T|B-ANOVA given in Table 18.1, which follows from the familiar model [see (9.1)]

$$\mathbf{y} = \mu\mathfrak{I} + \boldsymbol{X}_\rho\boldsymbol{\rho} + \boldsymbol{X}_{\beta^*}\boldsymbol{\beta}^* + \boldsymbol{X}_\tau\boldsymbol{\tau} + \boldsymbol{e}$$

Hence MS($\boldsymbol{I}|\mathfrak{I}, \boldsymbol{X}_\rho, \boldsymbol{X}_{\beta^*}, \boldsymbol{X}_\tau$) is an estimate of σ_e^2.

Based upon the factorial correspondence it is also easy to obtain the combined estimator for treatment comparisons. Using the notation of Section 9.2.3, we combine the intrablock estimator $\widehat{A}_{\text{II,III}}$ and the interblock estimator $\widetilde{A}_{\text{I}}$ and obtain the combined estimator for A as

$$\begin{aligned}\widehat{\widehat{A}} &= \frac{2qw\widehat{A}_{\text{II,III}} + qw'\widetilde{A}_{\text{I}}}{2qw + qw'} \\ &= \frac{w\widehat{A}_{\text{II}} + w\widehat{A}_{\text{III}} + w'\widetilde{A}_{\text{I}}}{2w + w'}\end{aligned} \tag{18.1}$$

where we have used the fact that $\widehat{A}_{\text{II,III}} = \frac{1}{2}(\widehat{A}_{\text{II}} + \widehat{A}_{\text{III}})$ and $w = 1/\sigma_e^2$, $w' = 1/[\sigma_e^2 + 2\sigma_\beta^2]$. Similarly

$$\widehat{\widehat{B}} = \frac{w\widehat{B}_{\text{I}} + w\widehat{B}_{\text{III}} + w'\widetilde{B}_{\text{II}}}{2w + w'} \tag{18.2}$$

and

$$\widehat{\widehat{AB}} = \frac{w\widehat{AB}_{\text{I}} + w\widehat{AB}_{\text{II}} + w'\widetilde{AB}_{\text{III}}}{2w + w'} \tag{18.3}$$

Since

$$\text{var}\left(\widehat{\widehat{A}}\right) = \text{var}\left(\widehat{\widehat{B}}\right) = \text{var}\left(\widehat{\widehat{AB}}\right) = \frac{1}{q(2w + w')}$$

Table 18.2 B | T-ANOVA for Lattice Design

Source	d.f.	SS	E(MS)
$\boldsymbol{X}_\rho \mid \mathfrak{I}$	$3q-1$	Usual	
$\boldsymbol{X}_\tau \mid \mathfrak{I}, \boldsymbol{X}_\rho$	3	Usual	
$\boldsymbol{X}_{\beta^*} \mid \mathfrak{I}, \boldsymbol{X}_\rho, \boldsymbol{X}_\tau$	$3q$		$\sigma_e^2 + 2\left(1 - \frac{1}{3q}\right)\sigma_\beta^2$
$A \times$ reps I	$q-1$	$\sum_{i=1}^{q}\left[\widetilde{A}_{\mathrm{I},i}\right]^2 - q\left[\widetilde{A}_{\mathrm{I}}\right]^2$	$\sigma_e^2 + 2\sigma_\beta^2$
$B \times$ reps II	$q-1$	$\sum_{i=1}^{q}\left[\widetilde{B}_{\mathrm{II},i}\right]^2 - q\left[\widetilde{B}_{\mathrm{II}}\right]^2$	$\sigma_e^2 + 2\sigma_\beta^2$
$AB \times$ reps III	$q-1$	$\sum_{i=1}^{q}\left[\widetilde{AB}_{\mathrm{III},i}\right]^2 - q\left[\widetilde{AB}_{\mathrm{III}}\right]^2$	$\sigma_e^2 + 2\sigma_\beta^2$
$\widehat{A}_{\mathrm{II,III}}$ vs. $\widetilde{A}_{\mathrm{I}}$	1	$\frac{2q}{3}\left[\widehat{A}_{\mathrm{II,III}} - \widetilde{A}_{\mathrm{I}}\right]^2$	$\sigma_e^2 + \frac{4}{3}\sigma_\beta^2$
$\widehat{B}_{\mathrm{I,III}}$ vs. $\widetilde{B}_{\mathrm{II}}$	1	$\frac{2q}{3}\left[\widehat{B}_{\mathrm{I,III}} - \widetilde{B}_{\mathrm{II}}\right]^2$	$\sigma_e^2 + \frac{4}{3}\sigma_\beta^2$
$\widehat{AB}_{\mathrm{I,II}}$ vs. $\widetilde{AB}_{\mathrm{III}}$	1	$\frac{2q}{3}\left[\widehat{AB}_{\mathrm{I,II}} - \widetilde{AB}_{\mathrm{III}}\right]^2$	$\sigma_e^2 + \frac{4}{3}\sigma_\beta^2$
$\boldsymbol{I} \mid \mathfrak{I}, \boldsymbol{X}_{\beta^*}\boldsymbol{X}_\tau$	$3(2q-1)$	From Table 18.1	σ_e^2

it follows then that, for example,

$$\operatorname{var}\left(\widehat{\widehat{\tau}}_1 - \widehat{\widehat{\tau}}_2\right) = \operatorname{var}\left(-\widehat{\widehat{A}} + \widehat{\widehat{AB}}\right) = \frac{2}{q(2w + w')}$$

The variance of other contrasts can be worked out similarly. To estimate w and w' use is made of the T | B-ANOVA of Table 18.1 and the B | T-ANOVA of Table 18.2, which follows the format of Table 9.11. We thus obtain

$$\widehat{\sigma}_e^2 = \frac{1}{\widehat{w}} = \mathrm{MS}(\boldsymbol{I} \mid \mathfrak{I}, \boldsymbol{X}_\rho, \boldsymbol{X}_{\beta^*}, \boldsymbol{X}_\tau) \tag{18.4}$$

$$\widehat{\sigma}_e^2 + 2\widehat{\sigma}_\beta^2 = \frac{1}{\widehat{w}'} = \frac{3q}{3q-1}\mathrm{MS}(\boldsymbol{X}_{\beta^*} \mid \mathfrak{I}, \boldsymbol{X}_\rho, \boldsymbol{X}_\tau) - \frac{1}{3q-1}\mathrm{MS}(\boldsymbol{I} \mid \mathfrak{I}, \boldsymbol{X}_\rho, \boldsymbol{X}_{\beta^*}, \boldsymbol{X}_\tau) \tag{18.5}$$

Substituting (18.4) and (18.5) into (18.1), (18.2), and (18.3) then yields the combined estimator for A, B, and AB, respectively, and with that the combined estimators for $\tau_i - \tau_j$, for example,

$$\widetilde{\tau}_1 - \widetilde{\tau}_2 = -\frac{2\widehat{w}\widehat{A}_{\text{II,III}} + \widehat{w}'\widetilde{A}_{\text{I}}}{2\widehat{w} + \widehat{w}'} + \frac{2\widehat{w}\widehat{AB}_{\text{I,II}} + \widehat{w}'\widetilde{AB}_{\text{III}}}{2\widehat{w} + \widehat{w}'}$$

with estimated variance

$$\widehat{\text{var}}\left(\widetilde{\tau}_i - \widetilde{\tau}_j\right) = \frac{2}{2q\widehat{w} + q\widehat{w}'} \tag{18.6}$$

for all $i, j = 1, 2, 3, 4, i \neq j$.

The fact that each treatment comparison is estimated with the same variance, whether we use intrablock information only or combined intra- and interblock information, suggests that this lattice design is in fact a BIB design. This can be verified easily, and this design serves as an illustration of the construction procedure referred to in Section 3.4. Because of the feature mentioned above, this particular lattice design is called a *balanced lattice*. It is just one of several types of lattice designs as we shall describe in the next section.

18.1.3 General Definition

Even though we have only discussed a simple example, the general idea of a lattice design can be stated as follows: The treatments are identified with the treatment combinations of a symmetrical factorial. The implied factorial structure is then used to construct a resolvable design in incomplete blocks by using different systems of partial confounding of main effects and interactions. The same factorial (or better: quasi-factorial) structure can then be used to obtain various types of estimators, that is, intrablock, interblock, combined estimators, for contrasts of treatment effects, together with their variances. This latter point is no longer as important as it was at the time when these designs were developed. We can now simply refer to the procedures described in Sections 1.8 and 1.14. On the other hand, the derivations in terms of the factorial are useful for establishing general properties of lattice designs, as, for example, whether they are BIB or PBIB designs.

18.2 TYPES OF LATTICE DESIGNS

The existing lattice designs can be classified according to

1. Number of treatments, t
2. Block size, k
3. Number of different systems of confounding used, s
4. Number of restrictions imposed on randomization, c

Using these parameters a partial list consisting of the most commonly used lattice designs is given in Table 18.3. Although this table mentions only lattice designs with $c = 1$, 2, and $t = K^2$, K^3, $K(K-1)$, it should be clear how this classification can be extended to n-dimensional lattices, that is, $t = K^n (n = 4, 5, \ldots,)$ with c restrictions, $c = 1, 2, 3, \ldots$ (e.g., Kempthorne and Federer, 1948a, 1948b; Federer, 1955).

Balanced lattices have the same properties as BIB designs, in that each treatment occurs together once with every other treatment in the same block, which

Table 18.3 Types of Lattice Designs

Type	c	t	k	s
One-restrictional two-dimensional (square) lattices	1	K^2		
Simple (double) lattice			K	2
Triple lattice			K	3
Quadruple lattice			K $(\neq 6)$	4
Balanced lattice			K^a	$K+1$
One-restrictional three-dimensional (cubic) lattices	1	K^3		
Triple lattice			K	3
Quadruple lattice			K^a	4
Balanced lattice			K^a	K^2+K+1
Two-restrictional two-dimensional lattices	2	K^2		
Semibalanced lattice square			K^a	$(K+1)/2$
Balanced lattice square			K^a	$K+1$
Unbalanced lattice square			K^a	≥ 2
Unbalanced lattice square			6	≤ 3
Unbalanced lattice square			10	≤ 4
Unbalanced lattice square			12	≤ 7
Unbalanced lattice square			14	≤ 5
Unbalanced lattice square			15	≤ 6
One-restrictional two-dimensional rectangular lattices	1	$K(K-1)$		
Simple rectangular lattice			$K-1$	2
Triple rectangular lattice			$K-1$	3
Near balanced rectangular lattice			$K-1^a$	K

[a] K is prime or prime power.

implies that all treatment differences are estimated with the same variance. The number of different systems of confounding (s) needed to achieve such balance depends on the dimensionality (n) of the design, increasing roughly exponentially. Even for moderate values of n, it may therefore be difficult, from a practical point of view, to achieve balance.

Lattice designs other than balanced lattices have at least the properties of PBIB designs with two or more associate classes (see Section 18.7). The method of constructing lattice designs using different systems of confounding provides thus an easy and convenient way to construct PBIB designs of certain types.

With regard to rectangular lattices one can envision more general designs with $t = KL$ in blocks of size $k = L$ with $L \leq K - 1$. The most important case, however, and the one usually referred to as rectangular lattice is that of $L = K - 1$. Apart from using blocks of size $k = K - 1$ instead of K as for the other (square) lattice designs, the rectangular lattice of the above type is suitable for t values between $(K - 1)^2$ and K^2, thus providing additional incomplete block designs. Even though the rectangular lattices can be constructed easily (see Section 18.11), they are not based on systems of confounding, that is, they are not based on the method of pseudofactors. It has been shown (Nair, 1951) that rectangular lattices are PBIB designs or even more general than PBIB designs.

18.3 CONSTRUCTION OF ONE-RESTRICTIONAL LATTICE DESIGNS

18.3.1 Two-Dimensional Lattices

We consider first the case where K is prime or prime power. The K^2 treatments can then be represented by the treatment combinations of a K^2 factorial with pseudofactors A and B. A simple lattice with blocks of size K is obtained by confounding A in one replicate (sometimes referred to as the X arrangement) and B in the other replicate (sometimes referred to as the Y arrangement). For a triple lattice we may confound A, B, and AB, respectively, in the three replicates, and for a quadruple lattice A, B, AB, and AB^2, respectively. Continuing in this fashion, a balanced lattice is obtained by confounding $A, B, AB, AB^2, \ldots, AB^{K-1}$, respectively, in the $K + 1$ replicates.

This method of using systems of confounding to construct lattice designs is, of course, equivalent to arranging the K^2 treatments in a square array with the cell in the ith row and jth column containing the treatment combination $a_i b_j$ and using for a simple lattice the rows as blocks for one replicate and the columns as blocks for the other replicate. For a triple lattice the rows, columns, and Latin letters of a Latin square constitute the three replicates. Since a Latin square exists for any K, this implies that we can construct a triple lattice for every K. For a quadruple lattice the blocks of the four replicates are generated by the rows, columns, Latin letters, and Greek letters, respectively, of a Graeco-Latin square of order K. This is possible for all K except $K = 6$. The types of lattice designs that can be constructed for a given K depends then on how many mutually orthogonal Latin squares (MOLS) exist for that K. For example, for $K = 10$, there exist only 2 MOLS, hence only simple, triple and quadruple lattices can be constructed; for $K = 12$ there exist 5 MOLS so that lattices up to a septuple lattice can be constructed, similarly, for $K = 14, 15$ there exist 3, 4 MOLS, respectively, so that lattices up to a quintuple, sextuple lattice, respectively, can be constructed (see Abel et al., 1996). Hence, $K - 1$ MOLS and balanced lattices exist, however, only for K prime or prime power.

18.3.2 Three-Dimensional Lattices

The replicates of these designs consist of K^2 blocks of size K. Expressing the K^3 treatments as the treatment combinations of a K^3 factorial with factors A, B, C such replicates can be obtained for K a prime or prime power by confounding any two effects or interactions and their GIs. In order to avoid complete confounding of any effect or interaction, which would lead to a disconnected design, at least three replicates are needed giving a triple lattice. We may specify a set of three replicates as follows:

Rep I: Confound $A, B, AB, AB^2, \ldots, AB^{K-1}$
Rep II: Confound $A, C, AC, AC^2, \ldots, AC^{K-1}$
Rep III: Confound $B, C, BC, BC^2, \ldots, BC^{K-1}$

which are sometimes referred to as X, Y, Z arrangements, respectively.

Alternatively, we can arrange the treatments in a $K \times K \times K$ cube such that the treatment in the ith row, jth column, and ℓth layer is $a_i b_j c_\ell$. The blocks for the three replicates can then be specified as follows:

Rep I: $\{a_i b_j c_\ell, \ell = 1, 2, \ldots, K\}$ for each $i, j = 1, 2, \ldots, K$
Rep II: $\{a_i b_j c_\ell, j = 1, 2, \ldots, K\}$ for each $i, \ell = 1, 2, \ldots, K$
Rep III: $\{a_i b_j c_\ell, i = 1, 2, \ldots, K\}$ for each $j, \ell = 1, 2, \ldots, K$

Obviously, such an arrangement is possible for every K, so that for a triple lattice K does not have to be prime or prime power.

For a quadruple lattice with K prime or prime power we can specify in addition to the three replicates above

Rep IV: Confound AB, AC and all GIs

This process of adding further independent systems of confounding can be continued to generate further lattices. It follows from (11.15) that the total number of possible systems of confounding is

$$\frac{(K^3 - 1)(K^3 - K)}{(K^2 - 1)(K^2 - K)} = \frac{K^3 - 1}{K - 1} = K^2 + K + 1$$

Using all such possible systems of confounding yields a balanced lattice.

18.3.3 Higher-Dimensional Lattices

The methods described in the previous sections can be extended immediately to higher dimensions such as $t = K^4$ or $t = K^5$. For $t = K^4$ and blocks of size K, we need at least four replicates to obtain a connected design. This may be achieved as follows:

Rep I:	Confound A, B, C, and GIs
Rep II:	Confound A, B, D, and GIs
Rep III:	Confound A, C, D, and GIs
Rep IV:	Confound B, C, D, and GIs

which is equivalent to writing the blocks as

Rep I:	$\{a_i b_j c_\ell d_m, m = 1, 2, \ldots, K\}$ for $i, j, \ell = 1, 2, \ldots, K$
Rep II:	$\{a_i b_j c_\ell d_m, \ell = 1, 2, \ldots, K\}$ for $i, j, m = 1, 2, \ldots, K$
Rep III:	$\{a_i b_j c_\ell d_m, j = 1, 2, \ldots, K\}$ for $i, \ell, m = 1, 2, \ldots, K$
Rep IV:	$\{a_i b_j c_\ell d_m, i = 1, 2, \ldots, K\}$ for $j, \ell, m = 1, 2, \ldots, K$

This shows that a lattice design is possible for every K.

For a balanced lattice with $t = K^4$ treatments we need $(K^4 - 1)/(K - 1)$ different replicates. Already for $K = 3$ this number, 40, is prohibitively large for most practical purposes. An enumeration of such a set of systems of confounding is given by Kempthorne (1952).

For $t = K^5$ the same arguments can be made, except that in this case at least five replicates are needed to achieve a connected design.

Rather than considering only blocks of size K we may also envisage blocks of size K^2 or K^3. Using the factorial correspondence, it should be clear how such designs can be constructed for K prime or prime power. Care must be taken to avoid complete confounding of any main effect or interaction. For other values of K the arrangement of the treatments in a hypercube suggests immediately how replicates with blocks of size K^2 or K^3 can be found by simply extending the principle used above.

18.4 GENERAL METHOD OF ANALYSIS FOR ONE-RESTRICTIONAL LATTICE DESIGNS

Since lattice designs are incomplete block designs, their analysis follows along the general principles discussed in Chapter 1, leading to intrablock or combined intra- and interblock estimates of estimable functions of the treatment effects. However, since the construction of lattice designs for K prime or prime power is based on the methods of factorial experiments, the analysis can be carried out alternatively as that of confounded factorial experiments as presented in Chapter 11 making use of the parameterization of treatment effects in terms of main effects and interactions (cf. Section 11.9). We shall now reiterate the major steps for this type of analysis, making extensive references to Chapter 11.

The basic model for any one-restrictional lattice design is

$$y_{ij\ell} = \mu + \rho_i + \beta^*_{ij} + \tau_\ell + e_{ij\ell} \tag{18.7}$$

where ρ_i is a replicate effect ($i = 1, 2, \ldots, sq$, $q =$ number of repetitions of basic lattice design), β^*_{ij} is a block/replicate effect ($j = 1, 2, \ldots, b$ with b depending on the type of lattice), and τ_ℓ is a treatment effect ($\ell = 1, 2, \ldots, t$). Further, if treatment ℓ corresponds to the pseudotreatment combination $\boldsymbol{x}' = (x_1, x_2, \ldots, x_K)$, we have, using the notation of Section 11.9),

$$\tau_\ell \equiv \tau(\boldsymbol{x}) = a(\boldsymbol{x}) - M = \sum_{\boldsymbol{\alpha} \in \mathcal{A}} E^{\boldsymbol{\alpha}}_{\boldsymbol{\alpha}'\boldsymbol{x}} \tag{18.8}$$

We are interested in estimating treatment contrasts

$$\sum_{\ell=1}^{t} c_\ell \tau_\ell \equiv \sum_{\boldsymbol{x} \in \chi} c(\boldsymbol{x})\, \tau(\boldsymbol{x}) = \sum_{\boldsymbol{x} \in \chi} \sum_{\boldsymbol{\alpha} \in \mathcal{A}} c(\boldsymbol{x})\, E^{\boldsymbol{\alpha}}_{\boldsymbol{\alpha}'\boldsymbol{x}} \tag{18.9}$$

for

$$\sum_{\ell=1}^{t} c_\ell = \sum_{\boldsymbol{x} \in \chi} c(\boldsymbol{x}) = 0$$

using the right-hand side (RHS) of (18.9), that is, using estimates of the $E^{\boldsymbol{\alpha}}_{\boldsymbol{\alpha}'\boldsymbol{x}}$. From the construction of lattice designs it follows that the interaction components in (18.9) fall into only two of the three sets defined in Section 11.6, namely $E^{\boldsymbol{\gamma}_\ell} \in \mathcal{E}_2$ (partially confounded) and $E^{\boldsymbol{\delta}_m} \in \mathcal{E}_3$ (unconfounded). Thus, if we use intrablock information only, we obtain $\widehat{E}^{\boldsymbol{\gamma}_\ell}_{\boldsymbol{\gamma}'_\ell \boldsymbol{x}}$ and $\widehat{E}^{\boldsymbol{\delta}_\ell}_{\boldsymbol{\delta}'_\ell \boldsymbol{x}}$ in the usual way and then

$$\sum_{\ell=1}^{t} c_\ell \widehat{\tau}_\ell = \sum_{\boldsymbol{x} \in \chi} c(\boldsymbol{x})\, \widehat{\tau}(\boldsymbol{x}) = \sum_{\boldsymbol{x} \in \chi} c(\boldsymbol{x}) \left[\sum_{\mathcal{E}_2} \widehat{E}^{\boldsymbol{\gamma}_\ell}_{\boldsymbol{\gamma}'_\ell \boldsymbol{x}} + \sum_{\mathcal{E}_3} \widehat{E}^{\boldsymbol{\delta}_m}_{\boldsymbol{\delta}'_m \boldsymbol{x}} \right] \tag{18.10}$$

If we use intra- and interblock information, as is usually the case with lattice designs, we obtain

$$\sum_{\ell=1}^{t} c_\ell \widehat{\widehat{\tau}}_\ell = \sum_{\boldsymbol{x} \in \chi} c(\boldsymbol{x})\, \widehat{\widehat{\tau}}(\boldsymbol{x}) = \sum_{\boldsymbol{x} \in \chi} c(\boldsymbol{x}) \left[\sum_{\mathcal{E}_2} \widehat{\widehat{E}}^{\boldsymbol{\gamma}_\ell}_{\boldsymbol{\gamma}'_\ell \boldsymbol{x}} + \sum_{\mathcal{E}_3} \widehat{E}^{\boldsymbol{\delta}_m}_{\boldsymbol{\delta}'_m \boldsymbol{x}} \right] \tag{18.11}$$

where $\widehat{\widehat{E}}^{\boldsymbol{\gamma}_\ell}_{\boldsymbol{\gamma}'_\ell \boldsymbol{x}}$ is given in (11.87).

A specific form of (18.9) is $\tau_\ell - \tau_{\ell'} = \tau(\boldsymbol{x}) - \tau(\boldsymbol{z})$, say. Then (18.11) is given by

$$\widehat{\widehat{\tau}}(\boldsymbol{x}) - \widehat{\widehat{\tau}}(\boldsymbol{z}) = \sum_{\mathcal{E}_2} \left[\widehat{\widehat{E}}^{\boldsymbol{\gamma}_\ell}_{\boldsymbol{\gamma}'_\ell \boldsymbol{x}} - \widehat{\widehat{E}}^{\boldsymbol{\gamma}_\ell}_{\boldsymbol{\gamma}'_\ell \boldsymbol{z}} \right] + \sum_{\mathcal{E}_3} \left[\widehat{E}^{\boldsymbol{\delta}_m}_{\boldsymbol{\delta}'_m \boldsymbol{x}} - \widehat{E}^{\boldsymbol{\delta}_m}_{\boldsymbol{\delta}'_m \boldsymbol{z}} \right] \tag{18.12}$$

Recall that $\mathcal{E}_2$ contains n_2 interactions and $\mathcal{E}_3$ contains n_3 interactions with $n_2 + n_3 = (K^n - 1)/(K - 1)$. There are, however, $(K^{n-1} - 1)/(K - 1)$ different $E^{\boldsymbol{\gamma}_\ell} \in \mathcal{E}_2$ and $E^{\boldsymbol{\delta}_m} \in \mathcal{E}_3$ for which $\boldsymbol{\gamma}'_\ell \boldsymbol{x} = \boldsymbol{\gamma}'_\ell \boldsymbol{z}$ and $\boldsymbol{\delta}'_m \boldsymbol{x} = \boldsymbol{\delta}'_m \boldsymbol{z}$, for given $\boldsymbol{x}$ and $\boldsymbol{z}$, so that the right-hand side of (18.12) contains only K^{n-1} terms. Denoting the sets of interactions for which $\boldsymbol{\gamma}'_\ell \boldsymbol{x} \neq \boldsymbol{\gamma}'_\ell \boldsymbol{z}$ and $\boldsymbol{\delta}'_m \boldsymbol{x} \neq \boldsymbol{\delta}'_m \boldsymbol{z}$ by $\mathcal{E}^*_2(\boldsymbol{x}, \boldsymbol{z})$ and $\mathcal{E}^*_3(\boldsymbol{x}, \boldsymbol{z})$, respectively, and using (11.98) we then have for the estimator (18.12)

$$\begin{aligned} &\operatorname{var}\left[\widehat{\widehat{\tau}}(\boldsymbol{x}) - \widehat{\widehat{\tau}}(\boldsymbol{z})\right] \\ &= 2\sigma_e^2 \left\{ \sum_{\mathcal{E}^*_2(\boldsymbol{x},\boldsymbol{z})} \left[(u(\boldsymbol{\gamma}_\ell) + c(\boldsymbol{\gamma}_\ell)\rho^{-1}) K^{n-1} \right]^{-1} + \sum_{\mathcal{E}^*_3(\boldsymbol{x},\boldsymbol{z})} [s K^{n-1}]^{-1} \right\} \end{aligned} \tag{18.13}$$

Except for balanced lattices it will be the case that (18.13) will take on different values so that different treatment comparisons may be estimated with different variances. Rather than use the variance as given by (18.13) for a given comparison it is common practice in the analysis of lattice designs, in particular within the context of multiple comparison procedures, to use the average variance of all such comparisons. It follows from (11.100) that such average variance is given by

$$\text{av. var} = 2 \frac{K - 1}{K^n - 1} \sigma_e^2 \sum_{\boldsymbol{\alpha} \in \mathcal{A}} [u(\boldsymbol{\alpha}) + c(\boldsymbol{\alpha})\rho^{-1}]^{-1}$$

where $\mathcal{A}$ is the set of all partitions for a K^n factorial and $0 \leq u(\boldsymbol{\alpha}) \leq s$, $0 \leq c(\boldsymbol{\alpha}) \leq s - 1$ with $u(\boldsymbol{\alpha}) + c(\boldsymbol{\alpha}) = s$ as determined by the type of lattice. Alternatively, (18.14) can be written as

$$\text{av. var} = 2 \frac{K - 1}{K^n - 1} \sum_{i=1}^{s} \frac{m_i}{i + (s - i)\rho^{-1}} \sigma_e^2 \tag{18.14}$$

where m_i denotes the number of $E^{\boldsymbol{\alpha}}$ unconfounded in i replicates, that is, for which $u(\boldsymbol{\alpha}) = i$, and hence confounded in $(s - 1)$ replicates, that is, for which $c(\boldsymbol{\alpha}) = s - i$.

We illustrate this with the following example.

Example 18.1 Triple three-dimensional lattice with $K = 3$, that is, $t = 27$ treatments in blocks of size $k = 3$ with three replicates specifying three different systems of confounding. Let those systems of confounding be (see Section 18.3.2)

Table 18.4 Plan for Triple Three-Dimensional Lattice with 27 Treatments

Block	Rep I	Rep II	Rep III
1	000, 001, 002	000, 010, 020	000, 100, 200
2	100, 101, 102	100, 110, 120	010, 110, 210
3	200, 201, 202	200, 210, 220	020, 120, 220
4	110, 111, 112	001, 011, 021	001, 101, 201
5	220, 221, 222	002, 012, 022	002, 102, 202
6	010, 011, 012	101, 111, 121	011, 111, 211
7	020, 021, 022	102, 112, 122	012, 112, 212
8	120, 121, 122	201, 211, 221	021, 121, 221
9	210, 211, 212	202, 212, 222	022, 122, 222

Rep I: $A,\ B,\ AB,\ AB^2$
Rep II: $A,\ C,\ AC,\ AC^2$
Rep III: $B,\ C,\ BC,\ BC^2$

The experimental plan is then as given in Table 18.4. It is easy to check that there are three types of comparisons between treatments; that is, (18.13) leads to three distinct expressions for all possible comparisons. Denoting these three expressions by V_i $(i = 0,\ 1,\ 2)$, where V_i is the variance of a treatment comparison when both treatments have i levels of the three pseudofactors in common, we find from (18.13)

$$V_0 = \frac{2}{9}\left[\frac{3}{1+2\rho^{-1}} + \frac{3}{2+\rho^{-1}} + \frac{3}{3}\right]\sigma_e^2$$

$$V_1 = \frac{2}{9}\left[\frac{2}{1+2\rho^{-1}} + \frac{5}{2+\rho^{-1}} + \frac{2}{3}\right]\sigma_e^2$$

$$V_2 = \frac{2}{9}\left[\frac{1}{1+2\rho^{-1}} + \frac{4}{2+\rho^{-1}} + \frac{4}{3}\right]\sigma_e^2$$

and from (18.14)

$$\text{av. var} = \frac{2}{13}\left[\frac{3}{1+2\rho^{-1}} + \frac{6}{2+\rho^{-1}} + \frac{4}{3}\right]\sigma_e^2 \qquad \square$$

It is a property of all lattice designs discussed up to this point that (18.13) will result in only a few distinct expressions for all possible comparisons, similar to the findings in Example 18.4. This means that comparisons can be put into distinct sets such that all treatment comparisons in a given set are made with the same precision and that the precisions are different for the different sets. This is, of course, also the property of PBIB designs. We shall return to this topic in Section 18.7.

To estimate $\text{var}[\widehat{\widetilde{\tau}}(\boldsymbol{x}) - \widehat{\widetilde{\tau}}(\boldsymbol{z})]$ as given by (18.13), we need to estimate σ_e^2 and ρ, or alternatively w and w'. As with factorial experiments this is achieved via the T | B- and B | T-ANOVA tables. The results of Sections 11.10 and 11.12, in particular Tables 11.1a and 11.2 are applicable to lattice designs putting $p = K$, $n_1 = 0$, and $n_2 + n_3 = (K^n - 1)/(K - 1)$. In most cases we have only one repetition of the basic pattern; that is, $q = 1$. We note that in the terminology of lattice designs the sources *Remainder* and "$\{\widehat{E}^{\gamma_\ell}$ vs. $\widetilde{E}^{\gamma_\ell}\}$" of $(\boldsymbol{X}_{\beta^*} \mid \mathfrak{I}, \boldsymbol{X}_\rho, \boldsymbol{X}_\tau)$ in Table 11.2b are often referred to as *component* (a) and *component* (b), respectively.

From the discussion in Section 11.12 it should then be clear how w and w' are estimated for lattice designs. As an illustration we elaborate Example 18.1.

Example 18.1 (Continued) Using the model

$$y_{ij\ell} = \mu + \rho_i + \beta_{ij}^* + \tau_\ell + e_{ij\ell}$$

$(i = 1, 2, 3;\ j = 1, 2, \ldots, 9;\ \ell = 1, 2, \ldots, 27)$, the T | B-ANOVA is given in Table 18.5. Here SS(Treatments) =SS$(\boldsymbol{X}_\tau \mid \mathfrak{I}, \boldsymbol{X}_\rho, \boldsymbol{X}_{\beta^*})$, the component SSs of which are computed from those replicates in which the respective effects or interactions are unconfounded, for example, $\text{SS}(A \times B)_{\text{II,III}}$ is the usual sum of squares for the $A \times B$ interaction obtained from replicates II and III. The B | T-ANOVA is presented in Table 18.6. The partitioning of SS(Blocks/reps) = SS$(\boldsymbol{X}_{\beta^*} \mid \mathfrak{I}, \boldsymbol{X}_\rho, \boldsymbol{X}_\tau)$ follows the general procedure outlined in Section 11.12 with regard to computing sums of squares and determining the E(MS). We then obtain $\widehat{\sigma}_e^2 = 1/\widehat{w} = \text{MS(Residual)}$ and with $\Delta = \frac{2}{3}$,

$$\widehat{\sigma}_e^2 + 3\widehat{\sigma}_\beta^2 = \frac{1}{\widehat{w}'} = \tfrac{3}{2}\ \text{MS(Blocks/reps)} - \tfrac{1}{2}\ \text{MS(Residual)}$$

using (11.96). □

18.5 EFFECTS OF INACCURACIES IN THE WEIGHTS

We recall that an estimate of $\tau(\boldsymbol{x}) - \tau(\boldsymbol{z})$ is obtained by replacing in (18.12) w and w' by $\widehat{w}$ and $\widehat{w}'$, respectively. We thus obtain

$$\begin{aligned} \widetilde{\widetilde{\tau}}(\boldsymbol{x}) - \widetilde{\widetilde{\tau}}(\boldsymbol{z}) &= \sum_{\mathcal{E}_2} \left[\widetilde{\widetilde{E}}^{\boldsymbol{\gamma}_\ell}_{\boldsymbol{\gamma}'_\ell \boldsymbol{x}} - \widetilde{\widetilde{E}}^{\boldsymbol{\gamma}_\ell}_{\boldsymbol{\gamma}'_\ell \boldsymbol{z}} \right] \\ &\quad + \sum_{\mathcal{E}_3} \left[\widehat{E}^{\boldsymbol{\delta}_m}_{\boldsymbol{\delta}'_m \boldsymbol{x}} - \widehat{E}^{\boldsymbol{\delta}_m}_{\boldsymbol{\delta}'_m \boldsymbol{z}} \right] = \sum \left[\widetilde{\widetilde{E}}^{\boldsymbol{\alpha}}_{\boldsymbol{\alpha}' \boldsymbol{x}} - \widetilde{\widetilde{E}}^{\boldsymbol{\alpha}}_{\boldsymbol{\alpha}' \boldsymbol{z}} \right] \end{aligned} \tag{18.15}$$

Table 18.5 T | B-ANOVA for Triple Three-Dimensional Lattice with 27 Treatments

Source	d.f.	SS
Replicates	2	$\frac{1}{27}\sum_i y_{i..}^2 - \frac{1}{81} y_{...}^2$
Blocks/Reps	24	$\frac{1}{3}\sum_{ij} y_{ij.}^2 - \frac{1}{27}\sum_i y_{i..}^2$
Treatments	26	sum of SS's below
A	2	$SS(A)_{\text{III}}$
B	2	$SS(B)_{\text{II}}$
$A \times B$	4	$SS(A \times B)_{\text{II,III}}$
C	2	$SS(C)_{\text{I}}$
$A \times C$	4	$SS(A \times C)_{\text{I,III}}$
$B \times C$	4	$SS(B \times C)_{\text{I,II}}$
$A \times B \times C$	8	$SS(A \times B \times C)_{\text{I,II,III}}$
Residual	28	Difference = SS(Residual)
Total	80	$\sum_{ij\ell} y_{ij\ell}^2 - \frac{1}{81} y_{...}^2$

say, where

$$\widetilde{\widetilde{E}}^{\alpha}_{(\alpha' x)} - \widetilde{\widetilde{E}}^{\alpha}_{(\alpha' z)} = \frac{u(\boldsymbol{\alpha})\widehat{w}\left[\widehat{E}^{\alpha}_{\alpha' x} - \widehat{E}^{\alpha}_{\alpha' z}\right] + c(\boldsymbol{a})\widehat{w}'\left[\widetilde{E}^{\alpha}_{\alpha' x} - \widetilde{E}^{\alpha}_{\alpha' z}\right]}{u(\boldsymbol{\alpha})\widehat{w} + c(\boldsymbol{\alpha})\widehat{w}'} \tag{18.16}$$

using the notation of Section 11.12. It can be shown that

$$E\left[\widetilde{\widetilde{E}}^{\alpha}_{\alpha' x} - \widetilde{\widetilde{E}}^{\alpha}_{\alpha' z}\right] = E^{\alpha}_{\alpha' x} - E^{\alpha}_{\alpha' z}$$

and it follows then that

$$E\left[\widetilde{\widetilde{\tau}}(\boldsymbol{x}) - \widetilde{\widetilde{\tau}}(\boldsymbol{z})\right] = \tau(\boldsymbol{x}) - \tau(\boldsymbol{z})$$

Turning now to $\text{var}[\widetilde{\widetilde{\tau}}(\boldsymbol{x}) - \widetilde{\widetilde{\tau}}(\boldsymbol{z})]$ we note first that

$$\text{var}\left[\widehat{E}^{\alpha}_{\alpha' x} - \widehat{E}^{\alpha}_{\alpha' z}\right] = \frac{2}{K^{n-1}} \cdot \frac{1}{u(\boldsymbol{\alpha})w} \tag{18.17}$$

Table 18.6 B | T-ANOVA for Triple Three-Dimensional Lattice with 27 Treatments

Source	d.f.	SS	E(MS)
Replicates	2	$\frac{1}{27}\sum_i y_{i..}^2 - \frac{1}{81}y_{...}^2$	
Treatments	26	$\frac{1}{3}\sum_\ell y_{..\ell}^2 - \frac{1}{81}y_{...}^2$	
Blocks/Reps	24	$SS(a)+SS(b)$	$\sigma_e^2 + 2\sigma_\beta^2$
Component a	6	$SS(a)$ = sum of SSs below	$\sigma_e^2 + 3\sigma_\beta^2$
	2	$SS(A)_{I} + SS(A)_{II} - SS(A)_{I,II}$	$\sigma_e^2 + 3\sigma_\beta^2$
	2	$SS(B)_{I} + SS(B)_{III} - SS(B)_{I,III}$	$\sigma_e^2 + 3\sigma_\beta^2$
	2	$SS(C)_{II} + SS(C)_{III} - SS(C)_{II,III}$	$\sigma_e^2 + 3\sigma_\beta^2$
Component b	18	$SS(b)$ = sum of SSs below	$\sigma_e^2 + \frac{5}{3}\sigma_\beta^2$
	2	$SS(\widehat{A} \text{ vs. } \widetilde{A})$	$\sigma_e^2 + \sigma_\beta^2$
	2	$SS(\widehat{B} \text{ vs. } \widetilde{B})$	$\sigma_e^2 + \sigma_\beta^2$
	2	$SS(\widehat{AB} \text{ vs. } \widetilde{AB})$	$\sigma_e^2 + 2\sigma_\beta^2$
	2	$SS(\widehat{AB^2} \text{ vs. } \widetilde{AB^2})$	$\sigma_e^2 + 2\sigma_\beta^2$
	2	$SS(\widehat{C} \text{ vs. } \widetilde{C})$	$\sigma_e^2 + \sigma_\beta^2$
	2	$SS(\widehat{AC} \text{ vs. } \widetilde{AC})$	$\sigma_e^2 + 2\sigma_\beta^2$
	2	$SS(\widehat{AC^2} \text{ vs. } \widetilde{AC^2})$	$\sigma_e^2 + 2\sigma_\beta^2$
	2	$SS(\widehat{BC} \text{ vs. } \widetilde{BC})$	$\sigma_e^2 + 2\sigma_\beta^2$
	2	$SS(\widehat{BC^2} \text{ vs. } \widetilde{BC^2})$	$\sigma_e^2 + 2\sigma_\beta^2$
Residual	28	From Table 18.5	σ_e^2

and

$$\text{var}\left[\widetilde{E}^{\boldsymbol{\alpha}}_{\boldsymbol{\alpha}'x} - \widetilde{E}^{\boldsymbol{\alpha}}_{\boldsymbol{\alpha}'z}\right] = \frac{2}{K^{n-1}} \cdot \frac{1}{c(\boldsymbol{\alpha})w'} \tag{18.18}$$

for $\boldsymbol{\alpha}'\boldsymbol{x} \neq \boldsymbol{\alpha}'\boldsymbol{z}$. Using (18.17) and (18.18) in (18.16) we obtain

$$\frac{K^{n-1}}{2}\,\text{var}\left[\widetilde{\widetilde{E}}^{\boldsymbol{\alpha}}_{\boldsymbol{\alpha}'x} - \widetilde{\widetilde{E}}^{\boldsymbol{\alpha}}_{\boldsymbol{\alpha}'z} \mid \widehat{w}, \widehat{w}'\right] = \frac{u(\boldsymbol{\alpha})\frac{\widehat{w}^2}{w} + c(\boldsymbol{\alpha})\frac{\widehat{w}'^2}{w'}}{[u(\boldsymbol{\alpha})\widehat{w} + c(\boldsymbol{\alpha})\widehat{w}']^2} \tag{18.19}$$

Note that if $\widehat{w} = w$ and $\widehat{w}' = w'$, then (18.19) reduces to $1/[u(\boldsymbol{\alpha})w + c(\boldsymbol{\alpha})w']$. If we put

$$F = \frac{w'}{\widehat{w}'}\,\frac{\widehat{w}}{w}$$

then, following Kempthorne (1952), the right-hand side of (18.19) can be written as

$$\frac{1}{u(\boldsymbol{\alpha})w + c(\boldsymbol{\alpha})w'} + \frac{u(\boldsymbol{\alpha})c(\boldsymbol{\alpha})\,\frac{w}{w'}\,\widehat{w}'^2(F-1)^2}{[u(\boldsymbol{\alpha})\widehat{w} + c(\boldsymbol{\alpha})\widehat{w}']^2\,[u(\boldsymbol{\alpha})w + c(\boldsymbol{\alpha})w']}$$

or

$$\frac{1}{w}\left[\frac{1}{u(\boldsymbol{\alpha}) + c(\boldsymbol{\alpha})\,\rho^{-1}} + \frac{u(\boldsymbol{\alpha})c(\boldsymbol{\alpha})\rho^{-1}(F-1)^2}{\left[u(\boldsymbol{\alpha})F + c(\boldsymbol{\alpha})\rho^{-1}\right]^2\left[u(\boldsymbol{\alpha}) + c(\boldsymbol{\alpha})\rho^{-1}\right]}\right] \tag{18.20}$$

where $w/w' = \rho$. The proportional increase in the variance of $\widetilde{\widetilde{E}}^{\boldsymbol{\alpha}}_{(\boldsymbol{\alpha}'x)} - \widetilde{\widetilde{E}}^{\boldsymbol{\alpha}}_{(\boldsymbol{\alpha}'z)}$ due to inaccuracies in the weights, that is, due to the fact that we substitute $\widehat{w}$ and $\widehat{w}'$ for w and w', is then given by the second term in (18.20). We note that this increase depends on the weights only through $w'/w = \rho^{-1}$ and $\widehat{w}/\widehat{w}'$. We note that if $w'/w = \widehat{w}'/\widehat{w}$ then $F = 1$ and hence there is no increase in variance. Also, if $\widehat{w} < \widehat{w}'$ we take, following the Yates procedure, $\widehat{w}/\widehat{w}' = 1$. Then $F = \rho^{-1}$ and hence the proportional increase in (18.20) becomes

$$\frac{1}{w} \cdot \frac{u(\boldsymbol{\alpha})c(\boldsymbol{\alpha})(\rho^{-1} - 1)^2}{\rho^{-1}s\left[u(\boldsymbol{\alpha}) + c(\boldsymbol{\alpha})\rho^{-1}\right]} \tag{18.21}$$

where we have used the fact that $u(\boldsymbol{\alpha}) + c(\boldsymbol{\alpha}) = s$. Adding over all possible effects and interactions, we shall obtain the increase in variance for a treatment comparison due to inaccuracies in the weights, which is a constant if $F < \rho^{-1}$ and is a function of F if $F > \rho^{-1}$. The percentage loss in information due to the inaccuracies in the weights may then be obtained. We shall illustrate this by continuing Example 18.1.

Example 18.1 (Continued) Using the results of Section 18.4 and (18.20) and (18.21), we find the following increases in

$$V_0:\quad \frac{2}{9}\cdot\frac{1}{w}\left[\frac{3\cdot 1\cdot 2}{(F+2\rho^{-1})^2(1+2\rho^{-1})} + \frac{3\cdot 2\cdot 1}{(2F+\rho^{-1})^2(2+\rho^{-1})}\right]\cdot\rho^{-1}(F-1)^2$$

$$V_1:\quad \frac{2}{9}\cdot\frac{1}{w}\left[\frac{2\cdot 1\cdot 2}{(F+2\rho^{-1})^2(1+2\rho^{-1})} + \frac{5\cdot 2\cdot 1}{(2F+\rho^{-1})^2(2+\rho^{-1})}\right]\cdot\rho^{-1}(F-1)^2$$

$$V_2:\quad \frac{2}{9}\cdot\frac{1}{w}\left[\frac{1\cdot 1\cdot 2}{(F+2\rho^{-1})^2(1+2\rho^{-1})} + \frac{4\cdot 2\cdot 1}{(2F+\rho^{-1})^2(2+\rho^{-1})}\right]\cdot\rho^{-1}(F-1)^2$$

and

$$\text{av. var:}\quad \frac{2}{13}\cdot\frac{1}{w}\left[\frac{3\cdot 1\cdot 2}{(F+2\rho^{-1})^2(1+2\rho^{-1})} + \frac{6\cdot 2\cdot 1}{(2F+\rho^{-1})^2(2+\rho^{-1})}\right]\cdot\rho^{-1}(F-1)^2$$

for $F > \rho^{-1}$, and

$$V_0: \quad \frac{2}{9} \cdot \frac{1}{w}\left[\frac{3 \cdot 1 \cdot 2}{1+2\rho^{-1}} + \frac{3 \cdot 2 \cdot 1}{2+\rho^{-1}}\right] \frac{(\rho^{-1}-1)^2}{3\rho^{-1}}$$

$$V_1: \quad \frac{2}{9} \cdot \frac{1}{w}\left[\frac{2 \cdot 1 \cdot 2}{1+2\rho^{-1}} + \frac{5 \cdot 2 \cdot 1}{2+\rho^{-1}}\right] \frac{(\rho^{-1}-1)^2}{3\rho^{-1}}$$

$$V_2: \quad \frac{2}{9} \cdot \frac{1}{w}\left[\frac{1 \cdot 1 \cdot 2}{1+2\rho^{-1}} + \frac{4 \cdot 2 \cdot 1}{2+\rho^{-1}}\right] \frac{(\rho^{-1}-1)^2}{3\rho^{-1}}$$

and

$$\text{av. var:} \quad \frac{2}{13} \cdot \frac{1}{w}\left[\frac{3 \cdot 1 \cdot 2}{1+2\rho^{-1}} + \frac{6 \cdot 2 \cdot 1}{2+\rho^{-1}}\right] \frac{(\rho^{-1}-1)^2}{3\rho^{-1}}$$

for $F = \rho^{-1}$. □

If we assume a Gauss–Markov normal linear model (GMNLM) and if we use

$$\widehat{w} = 1/\text{MS(Residual)}$$

and

$$\widehat{w}' = 1/\text{MS(Remainder)} = 1/\text{MS(Component } a) \tag{18.22}$$

[see (11.90)], then

$$F = \frac{w'}{\widehat{w}'} \cdot \frac{\widehat{w}}{w}$$

follows an F distribution with ν and ν_R d.f. (see Table 11.2). This would allow us to determine, for a given ρ, the expected increase in variance due to inaccuracies in the weights. If instead of (18.22) we use

$$\widehat{w}' = 1/\text{MS}(\boldsymbol{X}_{\beta^*}|\mathfrak{I}, \boldsymbol{X}_\rho \boldsymbol{X}_\tau)$$

matters become more complicated because of the heterogeneity of MS $(\boldsymbol{X}_{\beta^*} \mid \mathfrak{I}, \boldsymbol{X}_\rho, \boldsymbol{X}_\tau)$, that is, because the different components of MS $(\boldsymbol{X}_{\beta^*} \mid \mathfrak{I}, \boldsymbol{X}_\rho, \boldsymbol{X}_\tau)$ have different expected values (see Table 18.6). For details we refer to Kempthorne (1952).

The results published by Yates (1939) and Kempthorne (1952) indicate that the effect of inaccuracies of the weights on the variance of treatment comparisons is of the order of 1, 2, or 3% and, therefore, trivial. The effect of inaccuracies of the weights will increase as the number of degrees of freedom for interblock error [ν or $s(K^{n-1} - 1)$] and intrablock error (ν_R) decrease. Usually the number of degrees of freedom for intrablock error is sufficiently large, and as a practical rule with lattice designs, it is probably safe to say that the effect of inaccuracies will be inappreciable if the interblock error is based on 10 or more degrees of freedom. With a smaller number of degrees of freedom, it is best to assume that $w' = 0$ and to utilize intrablock information only.

18.6 ANALYSIS OF LATTICE DESIGNS AS RANDOMIZED COMPLETE BLOCK DESIGNS

Yates (1939) has shown that certain types of lattice designs may be analyzed as randomized complete block designs (RCBD). His method of proof depends on the fact that, if, under a particular analysis, the expectation of the treatment mean square, in the absence of treatment effects, is equal to the expectation of the error mean square, this analysis gives an unbiased estimate of the error variance for treatment comparisons. This, however, is not necessarily true, for the error sum of squares must be homogeneous. In this section we shall discuss briefly to what extent such an analysis is valid.

Suppose we have an experiment on K^n treatments in a lattice design with blocks of size K and s replicates. For the analysis as an RCBD we use the model

$$\boldsymbol{y} = \mu \mathfrak{I} + \boldsymbol{X}_\rho \boldsymbol{\rho} + \boldsymbol{X}_\tau \boldsymbol{\tau} + \boldsymbol{\omega} \tag{18.23}$$

with

$$\boldsymbol{\omega} = \boldsymbol{X}_{\beta^*} \boldsymbol{\beta}^* + \boldsymbol{e} \tag{18.24}$$

and the analysis of variance as outlined in Table 18.7 (see Section I.9.2.4). Using the quasi-factorial approach as employed for lattice designs, we can partition $SS(\boldsymbol{X}_\tau \mid \mathfrak{I}, \boldsymbol{X}_\rho)$ into $(K^n - 1)/(K - 1)$ components $E^{\boldsymbol{\alpha}}(\boldsymbol{\alpha} \in \mathcal{A})$ each with $K - 1$ d.f. and similarly $SS(\boldsymbol{I} \mid \mathfrak{I}, \boldsymbol{X}_\rho, \boldsymbol{X}_\tau)$ into $(K^n - 1)/(K - 1)$ components $E^{\boldsymbol{\alpha}} \times$ reps $(\boldsymbol{\alpha} \in \mathcal{A})$ each with $(s - 1)(K - 1)$ d.f. We shall now consider these components and their expected mean squares.

Consider $E^{\boldsymbol{\alpha}}$ and let $E^{\boldsymbol{\alpha}}_{ij}$ denote the average response of all treatment combinations in replicate j satisfying the equation $\sum \alpha_\ell x_\ell = i$ expressed as deviation from the average response of replicate j. Note that if $E^{\boldsymbol{\alpha}}$ is not confounded in

Table 18.7 ANOVA Structure for Lattice Design as RCBD

Source	d.f.
$\boldsymbol{X}_\rho \mid \mathfrak{I}, \boldsymbol{X}_\tau$	$s - 1$
$\boldsymbol{X}_\tau \mid \mathfrak{I}, \boldsymbol{X}_\rho$	$K^n - 1$
$\boldsymbol{I} \mid \mathfrak{I}, \boldsymbol{X}_\rho, \boldsymbol{X}_\tau$	$(s - 1)(K^n - 1)$
Total	$sK^n - 1$

replicate j, then $\widetilde{E}^{\boldsymbol{\alpha}}_{ij} = \widehat{E}^{\boldsymbol{\alpha}}_{ij}$ $(i = 0, 1, \ldots, K-1)$. We know (see Section 11.10.3) that, under model (18.23), the sum of squares for $E^{\boldsymbol{\alpha}}$ is

$$\mathrm{SS}(E^{\boldsymbol{\alpha}}) = \frac{K^{n-1}}{s} \sum_{i=0}^{K-1} \Big[\sum_{j=1}^{s} \widetilde{E}^{\boldsymbol{\alpha}}_{ij} \Big]^2 \tag{18.25}$$

If $E^{\boldsymbol{\alpha}}$ is unconfounded in $u(\boldsymbol{\alpha})$ replicates and confounded in $c(\boldsymbol{\alpha})$ replicates, with $u(\boldsymbol{\alpha}) + c(\boldsymbol{\alpha}) = s$, then the expected value of $\big[\sum_j \widetilde{E}^{\boldsymbol{\alpha}}_{ij}\big]^2$ for each $i = 0, 1, \ldots, K-1$, using (18.23) with $\boldsymbol{\tau} = \mathbf{0}$ and (18.24) is

$$\begin{aligned}
&\left\{ c(\boldsymbol{\alpha}) \left[\frac{1}{K^{n-1}} \left(1 - \frac{1}{K}\right) \right]^2 K^{n-2} + \frac{1}{K^{2n}} \left(K^{n-1} - K^{n-2}\right) K^2 \right\} \sigma^2_\beta \\
&\quad + \left\{ s \left[\frac{1}{K^{n-1}} \left(1 - \frac{1}{K}\right) \right]^2 K^{n-1} + \frac{s}{K^{2n}} \left(K^{n-1} - K^{n-2}\right) K \right\} \sigma^2_\beta \\
&= \frac{c(\boldsymbol{\alpha})(K-1)}{K^{n-1}} \sigma^2_\beta + \frac{s(K-1)}{K^n} \sigma^2_e
\end{aligned}$$

Hence

$$E[\mathrm{SS}(E^{\boldsymbol{\alpha}})] = (K-1)\left[\sigma^2_e + \frac{c(\boldsymbol{\alpha})}{s} K\sigma^2_\beta\right] \tag{18.26}$$

We now consider $E^{\boldsymbol{\alpha}} \times$ reps and its associated sum of squares, SS($E^{\boldsymbol{\alpha}} \times$ reps) say, which is a component of $\mathrm{SS}(\boldsymbol{I} \mid \mathfrak{I}, \boldsymbol{X}_\rho, \boldsymbol{X}_\tau)$, accounting for $(s-1)(K-1)$ d.f. This sum of squares, in turn, can be partitioned into three components as presented in Table 18.8. It follows then that

$$\begin{aligned}
E[\mathrm{MS}(E^{\boldsymbol{\alpha}} \times \text{ reps})] &= \sigma^2_e + \frac{1}{s-1}\left[\frac{u(\boldsymbol{\alpha})}{s} + c(\boldsymbol{\alpha}) - 1\right] K\sigma^2_\beta \\
&= \sigma^2_e + \frac{c(\boldsymbol{\alpha})}{s} K\sigma^2_\beta
\end{aligned} \tag{18.27}$$

Table 18.8 Partitioning of $\mathrm{SS}(\boldsymbol{I} \mid \mathfrak{I}, \boldsymbol{X}_\rho, \boldsymbol{X}_\tau)$

Source	d.f.	E(MS)
$\widehat{E}^{\alpha}$ vs. $\widetilde{E}^{\alpha}$	$K-1$	$\sigma^2_e + \dfrac{u(\boldsymbol{\alpha})}{s} K\sigma^2_\beta$
$\widehat{E}^{\alpha} \times$ reps	$[u(\alpha) - 1](K-1)$	σ^2_e
$\widetilde{E}^{\alpha} \times$ reps	$[c(\alpha) - 1](K-1)$	$\sigma^2_e + K\sigma^2_\beta$

It follows further from (18.26) and (18.27) that, with $\boldsymbol{\tau} = \mathbf{0}$, MS($E^{\alpha}$) and MS($E^{\alpha}$ × reps) and hence

$$\mathrm{MS}(\boldsymbol{X}_{\tau} \mid \mathfrak{I}, \boldsymbol{X}_{\rho}) = \sum_{\alpha} \frac{\mathrm{SS}(E^{\alpha})}{K^n - 1}$$

and

$$\mathrm{MS}(\boldsymbol{I} \mid \mathfrak{I}, \boldsymbol{X}_{\rho}, \boldsymbol{X}_{\tau}) = \sum_{\alpha} \frac{\mathrm{SS}(E^{\alpha} \times \mathrm{reps})}{(s-1)(K^n - 1)}$$

have the same expected value.

An ordinary RCBD analysis of a one-restrictional lattice design is, therefore, valid insofar as the expectation of the treatment mean squares is the same as the expectation of the RCBD error mean square. It follows then, from consideration of the randomization test, that the RCBD analysis gives a reasonably valid test of the null hypothesis that there are no differences among treatment effects. A test may also be made, of course, with the intrablock treatment and error mean squares. Difficulties do arise, however, as soon as we wish to ascribe variances to treatment comparisons obtained from the ordinary treatment averages, because we have shown (see Table 18.8) that the error sum of squares, SS($\boldsymbol{I} \mid \mathfrak{I}, \boldsymbol{X}_{\rho}, \boldsymbol{X}_{\tau}$), with $(s-1)(K^n - 1)$ d.f. is not homogeneous if σ_{β}^2 is greater than zero.

To illustrate the problem we consider the comparison of two treatments. Based on the RCBD analysis the difference of the treatment effects is estimated by the corresponding treatment means, with estimated variance $(2/s)\,\mathrm{MS}(\boldsymbol{I} \mid \mathfrak{I}, \boldsymbol{X}_{\rho}, \boldsymbol{X}_{\tau})$. Now, from (18.24),

$$\begin{aligned}
\frac{2}{s} E[\mathrm{MS}(\boldsymbol{I} \mid \mathfrak{I}, \boldsymbol{X}_{\rho}, \boldsymbol{X}_{\tau})] &= \frac{2}{s} \frac{K-1}{K^n - 1} \sum_{\alpha} \left[\sigma_e^2 + \frac{c(\boldsymbol{\alpha})}{s} K\sigma_{\beta}^2\right] \\
&= \frac{2}{s}\left[\sigma_e^2 + \frac{K^{n-1} - 1}{K^n - 1} K\sigma_{\beta}^2\right] \\
&= \frac{2}{s}\left[\sigma_e^2 + \sigma_{\beta}^2\left(1 - \frac{K-1}{K^n - 1}\right)\right] \qquad (18.28)
\end{aligned}$$

since $\sum_{\alpha} c(\boldsymbol{\alpha}) = s(K^{n-1} - 1)/(K-1)$ as $(K^{n-1} - 1)/(K-1)$ sets of $K-1$ d.f. are confounded in each replicate. We now compare (18.28) with the true variance of the difference of two treatment means. Suppose that the two treatments occur together in a block in, say, q out of the s replicates. Then, using (18.23) and (18.24) the variance is given by

$$\frac{2}{s^2}\left[s\sigma_e^2 + (s-q)\sigma_{\beta}^2\right] = \frac{2}{s}\left[\sigma_e^2 + \sigma_{\beta}^2\left(1 - \frac{q}{s}\right)\right] \qquad (18.29)$$

Hence the difference between the expectation of the estimated variance (18.29) and the true variance (18.28) is

$$\frac{2}{s}\left[\frac{q}{s} - \frac{K-1}{K^n-1}\right]\sigma_\beta^2 \tag{18.30}$$

which can be either positive or negative: If (18.30) is positive, we overestimate the variance; if (18.30) is negative, we underestimate the variance. As an illustration we consider again Example 18.1.

Example 18.1 (Continued) It follows from Table 18.4 that there are now two types of comparisons with either $q = 0$ or $q = 1$. We find for $q = 0$: (18.30) equals $-\frac{2}{39}\sigma_\beta^2 < 0$, $q = 1$: (18.30) equals $\frac{2}{3}\left[\frac{1}{3} - \frac{1}{13}\right]\sigma_\beta^2 > 0$, so that one variance is underestimated and the other variance is overestimated. □

In practical situations the decision whether to analyze a lattice design as an incomplete block design or as an RCBD is often based on the efficiency of lattice designs relative to the RCBD. In light of what has been said above this, too, may be a questionable procedure, since the efficiency is based on (18.28). More specifically, the efficiency of a lattice design relative to the RCBD is given by the ratio of (18.14) over (18.28), that is,

$$E = \frac{\dfrac{K^{n-1}-1}{\rho^{-1}} + K^{n-1}(K-1)}{s(K-1)\displaystyle\sum_{i=1}^{s}\frac{m_i}{i+(s-i)\rho^{-1}}} \tag{18.31}$$

where the numerator is obtained by rewriting (18.28) in terms of ρ^{-1} and w. For different values of ρ and various lattice designs Kempthorne (1952) has tabulated E. As is to be expected, E increases quite rapidly with increasing ρ, which would indicate that in most situations an RCB analysis is not warranted. In a practical situation this question may be decided by using $\widehat{\rho}$ in (18.31) and computing $\widehat{E}$, say, and do a RCBD analysis if $1.00 \leq \widehat{E} \leq 1.05$.

18.7 LATTICE DESIGNS AS PARTIALLY BALANCED INCOMPLETE BLOCK DESIGNS

In the preceding sections we have discussed one-restrictional lattice designs with K^n treatments in blocks of size K as they relate to factorial experiments. This discussion presupposes that K is a prime or a prime power although we did not mention this explicitly. We know, however, from Table 18.3 that certain lattice designs exist even if K is not a prime or prime power. For these cases it is useful to know that lattice designs are PBIB designs and hence can be constructed and analyzed as such.

Let us consider the one-restrictional two-dimensional lattice designs. We can arrange the K^2 treatments in a $K \times K$ array. A simple lattice can then be constructed by taking the rows of this array as the blocks for one replicate and the columns of the array as the blocks for the second replicate. We can check easily that this results in a PBIB(2) design with a Latin square association scheme, that is, a L_2-PBIB(2) design (see Section 4.6.3). The relevant parameters of this design are $t = K^2$, $r = 2$, $k = K$, $b = 2K$, $\lambda_1 = 1$, $\lambda_2 = 0$. Accordingly, two treatments are compared with variance V_1, say, if they occur together in the same block, and with variance V_2, say, if they do not occur together in the same block. The reader should check that this agrees with the result obtained from (18.13).

Using the same $K \times K$ array we can obtain triple, quadruple, and so forth lattices by superimposing one, two, and so on mutually orthogonal Latin squares (if they exist), respectively. Blocks of size K are then formed by the rows, the columns, and the letters of each of the superimposed MOLS, respectively. Thus the triple lattice represents an L_3-PBIB(2) design with parameters $t = K^2, r = 3$, $k = K, b = 3K, \lambda_1 = 1, \lambda_2 = 0$; the quadruple lattice an L_4-PBIB(2) design with parameters $t = K^2, r = 4, k = K, b = 4K, \lambda_1 = 1, \lambda_2 = 0$; in general the i-tuple lattice an L_i-PBIB(2) design (see Section 4.6.3) with parameters $t = K^2$, $r = i$, $k = K$, $b = iK$, $\lambda_1 = 1$, $\lambda_2 = 0$.

The triple three-dimensional lattice with $t = K^3$ treatments in blocks of size K represents a PBIB(3) design with the cubic association scheme (Section 4.6.8). Following the geometric interpretation of that association scheme and denoting the three axes by x, y, and z, respectively, the arrangement in blocks is such that for the first replicate the treatments in a block lie on the same x axis, that is, are of the form $\{(x, y, z)\colon x = 0, 1, \ldots, K - 1\}$ for fixed y and z ($y, z = 0, 1, \ldots, K - 1$), for the second replicate the treatments in a block lie on the same y axis, that is, are of the form $\{(x, y, z)\colon y = 0, 1, \ldots, K - 1\}$ for fixed x and z ($x, z = 0, 1, \ldots, K - 1$), and for the third replicate the treatments in a block lie on the same z axis, that is, are of the form $\{(x, y, z)\colon z = 0, 1, \ldots, K - 1\}$ for fixed x and y ($x, y = 0, 1, \ldots, K - 1$). For an illustration we refer to Example 18.4 (with an appropriate renumbering of the replicates). It follows then immediately that this design has the parameters $t = K^2$, $r = 3$, $k = K$, $b = 3K^2$, $\lambda_1 = 1$, $\lambda_2 = 0$, $\lambda_3 = 0$.

18.8 LATTICE DESIGNS WITH BLOCKS OF SIZE K^ℓ

The most commonly used lattice designs are those for K^n treatments in blocks of size K. There are, however, situations where blocks of size K^ℓ ($1 < \ell < n$) are more appropriate. There are no difficulties extending the previously discussed methods to this case:

1. If n is a multiple of ℓ, say $n = m\ell$, then we can represent the experiment as one of m factors with K^ℓ levels and employ the methods described in

Sections 18.3 and 18.4 [using GF(K^ℓ) arithmetic for K prime as discussed in Appendix A or Section 18.7].

2. If n is not a multiple of ℓ, then we employ appropriate systems of confounding as presented in Chapter 11 for K prime. The practical problem is to choose out of all possible systems of confounding a system for each replicate such that the confounding is distributed as evenly as possible over all the effects and interactions and such that no effect or interaction is completely confounded over the entire design. The second requirement is necessary for the design to be connected with respect to the intrablock analysis. For more details and examples of practically useful designs the reader is referred to Kempthorne and Federer (1948b).

18.9 TWO-RESTRICTIONAL LATTICES

In Section 9.8 we considered the problem of double confounding in a 2^n factorial experiment as a device for eliminating heterogeneity in two directions using incomplete blocks. This principle can be extended to other factorial experiments and can be applied also to construct two-restrictional two-dimensional lattices, which are also referred to as *lattice squares* and *lattice rectangles* (Yates, 1940b). They represent special types of resolvable row–column designs (see Section 6.6).

We shall consider designs for K^2 treatments arranged in K rows and K columns. To construct suitable arrangements we distinguish two cases: (1) K prime (or prime power) and (2) K not prime.

18.9.1 Lattice Squares with K Prime

For K prime (or prime power) we can use the correspondence between the $t = K^2$ treatments and the treatment combinations of the K^2 factorial experiment to partially confound any or all of the $K + 1$ effects and interactions, E^α, with rows or columns in s squares of order K. If $s = K + 1$, we confound each E^α once with rows and once with columns yielding a balanced lattice square. If $s = (K + 1)/2$, we confound each E^α once with rows or columns giving a semibalanced lattice square. If $2 \leq s \leq (K + 1)/2$ we confound any E^α at most once with rows or columns, and if $(K + 1)/2 < s < K$, we try to confound the E^α as equally as possible with rows and columns.

The analysis of lattice squares is based on the linear model

$$\boldsymbol{y} = \mu \mathfrak{I} + \boldsymbol{X}_s \boldsymbol{s} + \boldsymbol{X}_r \boldsymbol{r} + \boldsymbol{X}_c \boldsymbol{c} + \boldsymbol{X}_\tau \boldsymbol{\tau} + \boldsymbol{e} \tag{18.32}$$

where $\boldsymbol{s}$, $\boldsymbol{r}$, and $\boldsymbol{c}$ represent the effects of squares (replicates), rows within squares, and columns within squares, respectively. For the combined analysis $\boldsymbol{r}$ and $\boldsymbol{c}$ are considered to be vectors of random variables with $E(\boldsymbol{r}) = \boldsymbol{0}$, $\text{var}(\boldsymbol{r}) = \boldsymbol{I}\sigma_r^2$, $E(\boldsymbol{c}) = \boldsymbol{0}$, $\text{var}(\boldsymbol{c}) = \boldsymbol{I}\sigma_c^2$.

Following the notation of Section 18.4, we can write out the intrablock and the combined estimators for a linear contrast, $\sum c_k \tau_k$, of treatment effects as in (18.10) and (18.11), respectively. In particular, the combined estimator $\widehat{\widehat{\tau}}_k - \widehat{\widehat{\tau}}_{k'}$ for a simple treatment comparison, $\tau_k - \tau_{k'}$, is given by (18.12). To find $\text{var}\big(\widehat{\widehat{\tau}}_k - \widehat{\widehat{\tau}}_{k'}\big)$ we denote by $c_r(\boldsymbol{\gamma}_\ell)$ and $c_c(\boldsymbol{\gamma}_\ell)$ the number of squares in which $E^{\boldsymbol{\gamma}_\ell} \in \mathcal{E}_2$ is confounded with rows and columns, respectively, with $c_r(\boldsymbol{\gamma}_\ell) + c_c(\boldsymbol{\gamma}_\ell) = c(\boldsymbol{\gamma}_\ell)$ being the number of squares in which $E^{\boldsymbol{\gamma}_\ell}$ is confounded in the design. Further, let

$$w_r = \frac{1}{\sigma_e^2 + K\sigma_r^2} \qquad w_c = \frac{1}{\sigma_e^2 + K\sigma_c^2} \tag{18.33}$$

denote the reciprocals of the interrow and intercolumn variances, and

$$\rho_r = \frac{w}{w_r} \qquad \rho_c = \frac{w}{w_c}$$

We can then extend (18.13) easily and write

$$\begin{aligned} \text{var}\left[\widehat{\widehat{\tau}}_k - \widehat{\widehat{\tau}}_{k'}\right] &= \text{var}\left[\widehat{\widehat{\tau}}(\boldsymbol{x}) - \widehat{\widehat{\tau}}(\boldsymbol{z})\right] \\ &= 2\sigma_e^2 \left\{ \sum_{\mathcal{E}_2^*(\boldsymbol{x},\boldsymbol{z})} \left[\left(u(\boldsymbol{\gamma}_\ell) + c_r(\boldsymbol{\gamma}_\ell)\rho_r^{-1} + c_c(\boldsymbol{\gamma}_\ell)\rho_c^{-1}\right) K \right]^{-1} \right. \\ &\qquad \left. + \sum_{\mathcal{E}_3^*(\boldsymbol{x},\boldsymbol{z})} [sK]^{-1} \right\} \end{aligned} \tag{18.34}$$

We shall consider (18.34) for the special case of a balanced lattice square. Recall (see Table 18.3) that for a balanced lattice square $s = K + 1$ such that each effect and interaction is confounded once with rows and once with columns, which implies that $\mathcal{E}_3 = \phi$, $\mathcal{E}_2 = \mathcal{A}$, $c_r(\boldsymbol{\gamma}_\ell) = 1 = c_c(\boldsymbol{\gamma}_\ell)$, $u(\boldsymbol{\gamma}_\ell) = K - 1$ for all $E^{\boldsymbol{\gamma}_\ell} \in \mathcal{E}_2$, $\mathcal{E}_2^*(\boldsymbol{x}, \boldsymbol{z})$ consists of K interactions for each $(\boldsymbol{x}, \boldsymbol{z})$. It follows then from (18.34) that

$$\text{var}\left[\widehat{\widehat{\tau}}(\boldsymbol{x}) - \widehat{\widehat{\tau}}(\boldsymbol{z})\right] = \frac{2_e^2}{K - 1 + \rho_r^{-1} + \rho_c^{-1}} \tag{18.35}$$

for any two treatments $(\boldsymbol{x}, \boldsymbol{z})$, as one would expect for a balanced design. Obviously, (18.35) is also the average variance of all such treatment comparisons, which is an extension of (18.14) for this special case.

In order to complete the actual analysis, we still have to estimate w, w_r, and w_c. This is easily accomplished with the help of ANOVA tables as given

Table 18.9 T | *R*, C-ANOVA for Lattice Square

Source	d.f.
$X_s \mid \mathfrak{I}$	$s-1$
$X_r \mid \mathfrak{I}, X_s$	$s(K-1)$
$X_c \mid \mathfrak{I}, X_s$	$s(K-1)$
$X_\tau \mid \mathfrak{I}, X_s, X_r, X_c$	K^2-1
$I \mid \mathfrak{I}, X_s, X_r, X_c, X_\tau$	$(K-1)[s(K-1)-(K+1)] = \nu_R$
Total	sK^2-1

Table 18.10 R | C, T- and C | R, T-ANOVA for Lattice Square

Source	d.f.	Source
$X_s \mid \mathfrak{I}$	$s-1$	$X_s \mid \mathfrak{I}$
$X_\tau \mid \mathfrak{I}, X_s$	$K^2-1)$	$X_\tau \mid \mathfrak{I}, X_s$
$X_c \mid \mathfrak{I}, X_s, X_\tau$	$s(K-1)$	$X_r \mid \mathfrak{I}, X_s, X_\tau$
$\left\{\left(\widetilde{E}^{\gamma_\ell}\right)_{cc} \text{ vs. } \left(\widetilde{E}^{\gamma_\ell}\right)_{uc}\right\}$		$\left\{\left(\widetilde{E}^{\gamma_\ell}\right)_{cr} \text{ vs. } \left(\widetilde{E}^{\gamma_\ell}\right)_{ur}\right\}$
$X_r \mid \mathfrak{I}, X_s, X_c, X_\tau$	$s(K-1)$	$X_c \mid \mathfrak{I}, X_s, X_r, X_\tau$
$\left\{\left(\widetilde{E}^{\gamma_\ell}\right)_{cr} \text{ vs. } \left(\widehat{E}^{\gamma_\ell}\right)\right\}$		$\left\{\left(\widetilde{E}^{\gamma_\ell}\right)_{cc} \text{ vs. } \left(\widehat{E}^{\gamma_\ell}\right)\right\}$
$I \mid \mathfrak{I}, X_s, X_r, X_c, X_\tau$	ν_R	$I \mid \mathfrak{I}, X_s, X_r, X_c, X_\tau$

in Tables 18.9 and 18.10. In Table 18.9, the treatment-after-rows-and-columns ANOVA or T | R, C-ANOVA, the sums of squares are obtained in the familiar fashion, where the components of

$$\text{SS}(X_\tau \mid \mathfrak{I}, X_x, X_r, X_c) = \text{SS}(A) + \text{SS}(B) + \text{SS}(AB) + \cdots + \text{SS}\left(AB^{K-1}\right)$$

are obtained from the intrablock analysis (see Section 11.10) and

$$\text{SS(Residual)} = \text{SS}(I \mid X_s, X_r, X_c, X_\tau)$$

is obtained by subtraction.

The row-after-treatments-and-columns or R | C, T-ANOVA and the column-after-treatments-and-rows or C | R, T-ANOVA, as outlined in Table 18.10, are used to estimate w_r and w_c, respectively. In order to do so it is useful to comment briefly on the form of some of the sums of squares in Table 18.10. For $E^{\boldsymbol{\gamma}_\ell} \in \mathcal{E}_2$ let $u_r(\boldsymbol{\gamma}_\ell) = s - c_r(\boldsymbol{\gamma}_\ell)$ and $u_c(\boldsymbol{\gamma}_\ell) = s - c_c(\boldsymbol{\gamma}_\ell)$ denote the number of squares in which E^{γ_ℓ} is unconfounded with rows and columns, respectively, and let $\left(\widetilde{E}_\delta^{\gamma_\ell}\right)_{cc}$, $\left(\widetilde{E}_\delta^{\gamma_\ell}\right)_{cr}$, $\left(\widetilde{E}_\delta^{\gamma_\ell}\right)_{uc}$, $\left(\widetilde{E}_\delta^{\gamma_\ell}\right)_{ur}$ be the estimate of $E_\delta^{\gamma_\ell}$ obtained from those squares in which E^{γ_ℓ} is confounded with columns, confounded with rows, unconfounded with columns, unconfounded with rows, respectively. It is then easy to verify that, in Table 18.10,

$$\text{SS}(\boldsymbol{X}_c \mid \mathfrak{I}, \boldsymbol{X}_s, \boldsymbol{X}_\tau) = \text{SS}\left[\left\{\left(\widetilde{E}^{\gamma_\ell}\right)_{cc} \text{ vs. } \left(\widetilde{E}^{\gamma_\ell}\right)_{uc}\right\}\right] \tag{18.36}$$

$$\text{SS}(\boldsymbol{X}_r \mid \mathfrak{I}, \boldsymbol{X}_s, \boldsymbol{X}_c\, \boldsymbol{X}_\tau) = \text{SS}\left[\left\{\left(\widetilde{E}^{\gamma_\ell}\right)_{cr} \text{ vs. } \left(\widehat{E}^{\gamma_\ell}\right)\right\}\right] \tag{18.37}$$

$$\text{SS}(\boldsymbol{X}_r \mid \mathfrak{I}, \boldsymbol{X}_s, \boldsymbol{X}_\tau) = \text{SS}\left[\left\{\left(\widetilde{E}^{\gamma_\ell}\right)_{cr} \text{ vs. } \left(\widetilde{E}^{\gamma_\ell}\right)_{ur}\right\}\right] \tag{18.38}$$

$$\text{SS}(\boldsymbol{X}_c \mid \mathfrak{I}, \boldsymbol{X}_s, \boldsymbol{X}_r\, \boldsymbol{X}_\tau) = \text{SS}\left[\left\{\left(\widetilde{E}^{\gamma_\ell}\right)_{cc} \text{ vs. } \left(\widehat{E}^{\gamma_\ell}\right)\right\}\right] \tag{18.39}$$

The RHS of (18.36)–(18.39) can be computed following (11.93) with the proper modifications, that is, computing the quantities $X_\delta^{\gamma_\ell}$ from the squares as indicated in each expression and then modifying $c(\boldsymbol{\gamma}_\ell)$ and $u(\boldsymbol{\gamma}_\ell)$ accordingly [this includes replacing s by $c_r(\boldsymbol{\gamma}_\ell) + u(\boldsymbol{\gamma}_\ell)$ in (18.38) and by $c_c(\boldsymbol{\gamma}_\ell) + u(\boldsymbol{\gamma}_\ell)$ in (18.39)]. With regard to estimating w_r and w_c, the sums of squares (18.38) and (18.39) and their expected values are needed. Again, these are easily obtained from (11.94) as

$$E\left[\text{MS}(\boldsymbol{X}_r \mid \mathfrak{I}, \boldsymbol{X}_s, \boldsymbol{X}_c, \boldsymbol{X}_\tau)\right] = \sigma_e^2 + \Delta_r\, K\sigma_r^2 \tag{18.40}$$

where

$$\Delta_r = \frac{1}{s}\sum_{\mathcal{E}_{2,r}} \frac{u(\boldsymbol{\gamma}_\ell)}{(u(\boldsymbol{\gamma}_\ell) + c_r(\boldsymbol{\gamma}_\ell))} \tag{18.41}$$

and $\sum_{\mathcal{E}_{2,r}}$ denotes summation over all $E^{\gamma_\ell} \in \mathcal{E}_2$, which are confounded with rows, and

$$E\left[\text{MS}(\boldsymbol{X}_c \mid \mathfrak{I}, \boldsymbol{X}_s, \boldsymbol{X}_r, \boldsymbol{X}_\tau)\right] = \sigma_e^2 + \Delta_c\, K\sigma_c^2 \tag{18.42}$$

where

$$\Delta_c = \frac{1}{s}\sum_{\mathcal{E}_{2,c}} \frac{u(\boldsymbol{\gamma}_\ell)}{(u(\boldsymbol{\gamma}_\ell) + c_c(\boldsymbol{\gamma}_\ell))} \tag{18.43}$$

and $\sum_{\mathcal{E}_{2,c}}$ denotes summation over all $E^{\gamma_\ell} \in \mathcal{E}_2$, which are confounded with columns. We then obtain

$$\widehat{\sigma}_e^2 = \frac{1}{\widehat{w}} = \text{MS}(\boldsymbol{I} \mid \mathfrak{I}, \boldsymbol{X}_s, \boldsymbol{X}_r, \boldsymbol{X}_c, \boldsymbol{X}_\tau) \tag{18.44}$$

from (18.40) and (18.44)

$$\widehat{\sigma}_e^2 + K\widehat{\sigma}_r^2 = \frac{1}{\widehat{w}_r} = \left(1 - \frac{1}{\Delta_r}\right) \text{MS}(\boldsymbol{I} \mid \mathfrak{I}, \boldsymbol{X}_s, \boldsymbol{X}_r, \boldsymbol{X}_c, \boldsymbol{X}_\tau)$$

$$+ \frac{1}{\Delta_r} \text{MS}(\boldsymbol{X}_r \mid \mathfrak{I}, \boldsymbol{X}_s, \boldsymbol{X}_c, \boldsymbol{X}_\tau)$$

and from (18.42) and (18.44)

$$\widehat{\sigma}_e^2 + K\widehat{\sigma}_c^2 = \frac{1}{\widehat{w}_c} = \left(1 - \frac{1}{\Delta_c}\right) \text{MS}(\boldsymbol{I} \mid \mathfrak{I}, \boldsymbol{X}_s, \boldsymbol{X}_r, \boldsymbol{X}_c, \boldsymbol{X}_\tau)$$

$$+ \frac{1}{\Delta_c} \text{MS}(\boldsymbol{X}_c \mid \mathfrak{I}, \boldsymbol{X}_s, \boldsymbol{X}_r, \boldsymbol{X}_\tau)$$

We illustrate the above by the following example.

Example 18.2 Suppose we have $t = 5^2$ treatments in an unbalanced lattice square with $s = 4$ squares. We can use the following system of confounding:

	Square			
	I	II	III	IV
Rows	A	B	AB^3	AB
Columns	AB	AB^2	AB^4	AB^3

The experimental plan (apart from randomization of rows and columns) is then as given in Table 18.11, and the relevant parts of the R | C, T-ANOVA and the C | R, T-ANOVA are given in Tables 18.12 and 18.13, respectively. Concerning treatment comparisons such as, for example, (0, 0) versus (1, 1), we have

$$\widehat{\widehat{\tau}}(0,0) - \widehat{\widehat{\tau}}(1,1) = \left(\widehat{\widehat{A}}_0 - \widehat{\widehat{A}}_1\right) + \left(\widehat{\widehat{B}}_0 - \widehat{\widehat{B}}_1\right)$$
$$+ \left(\widehat{\widehat{AB}}_0 - \widehat{\widehat{AB}}_2\right) + \left(\widehat{\widehat{AB}}{}^2_0 - \widehat{\widehat{AB}}{}^2_3\right) + \left(\widehat{\widehat{AB}}{}^3_0 - \widehat{\widehat{AB}}{}^3_4\right)$$

with [see (18.34)]

$$\text{var}\left(\widehat{\widehat{\tau}}(0,0) - \widehat{\widehat{\tau}}(1,1)\right) = \frac{4}{5}\sigma_e^2 \left[\frac{1}{3 + \rho_r^{-1}} + \frac{1}{2 + \rho_r^{-1} + \rho_c^{-1}} + \frac{1}{3 + \rho_c^{-1}}\right] \qquad \square$$

18.9.2 Lattice Squares for General K

We now turn to the case where K is not prime or prime power. Suppose there exist q MOLS of order K. Then the row and column classification and the

Table 18.11 Experimental Plan for Lattice Square with 25 Treatments in 4 Squares

		Square I		
(0, 0)	(0, 1)	(0, 2)	(0, 3)	(0, 4)
(1, 4)	(1, 0)	(1, 1)	(1, 2)	(1, 3)
(2, 3)	(2, 4)	(2, 0)	(2, 1)	(2, 2)
(3, 2)	(3, 3)	(3, 4)	(3, 0)	(3, 1)
(4, 1)	(4, 2)	(4, 3)	(4, 4)	(4, 0)
		Square II		
(0, 0)	(1, 0)	(2, 0)	(3, 0)	(4, 0)
(1, 2)	(2, 2)	(3, 2)	(4, 2)	(0, 2)
(2, 4)	(3, 4)	(4, 4)	(0, 4)	(1, 4)
(3, 1)	(4, 1)	(0, 1)	(1, 1)	(2, 1)
(4, 3)	(0, 3)	(1, 3)	(2, 3)	(3, 3)
		Square III		
(0, 0)	(1, 3)	(2, 1)	(3, 4)	(4, 2)
(1, 1)	(2, 4)	(3, 2)	(4, 0)	(0, 3)
(2, 2)	(3, 0)	(4, 3)	(0, 1)	(1, 4)
(3, 3)	(4, 1)	(0, 4)	(1, 2)	(2, 0)
(4, 4)	(0, 2)	(1, 0)	(2, 3)	(3, 1)
		Square IV		
(0, 0)	(1, 4)	(2, 3)	(3, 2)	(4, 1)
(1, 3)	(2, 2)	(3, 1)	(4, 0)	(0, 4)
(2, 1)	(3, 0)	(4, 4)	(0, 3)	(1, 2)
(3, 4)	(4, 3)	(0, 2)	(1, 1)	(2, 0)
(4, 2)	(0, 1)	(1, 0)	(2, 4)	(3, 3)

classifications with respect to the q languages, say $L_1, L_2, \ldots, L_q$, can be used in pairs to construct various numbers of squares, that is, if we denote the orthogonal classifications by $R = L_{q+1}, C = L_{q+2}, L_1, \ldots, L_q$, respectively, then any pair (L_i, L_j) with $i \neq j$ $(i, j = 1, 2, \ldots, q+2)$ defines a suitable square arrangement of the $t = K^2$ treatments, where the levels of L_i determine the rows and the levels of L_j determine the columns of the square.

Example 18.3 For $K = 6$ we have $q = 1$. A suitable arrangement would be as follows:

	Square		
	I	II	III
Rows	L_2	L_1	L_3
Columns	L_3	L_2	L_1

□

Table 18.12 R | C, T-ANOVA for Example 18.2

Source	d.f.	E(MS)
$\boldsymbol{X}_s \mid \mathfrak{J}$	3	
$\boldsymbol{X}_\tau \mid \mathfrak{J}, \boldsymbol{X}_s$	24	
$\boldsymbol{X}_c \mid \mathfrak{J}, \boldsymbol{X}_s, \boldsymbol{X}_\tau$	16	
$\widetilde{AB}_{\text{I}}$ vs. $\widetilde{AB}_{\text{II,III,IV}}$	4	
$\widetilde{AB}^2_{\text{II}}$ vs. $\widetilde{AB}^2_{\text{I,III,IV}}$	4	
$\widetilde{AB}^3_{\text{IV}}$ vs. $\widetilde{AB}^3_{\text{I,II,III}}$	4	
$\widetilde{AB}^4_{\text{III}}$ vs. $\widetilde{AB}^4_{\text{I,II,IV}}$	4	
$\boldsymbol{X}_r \mid \mathfrak{J}, \boldsymbol{X}_s, \boldsymbol{X}_c, \boldsymbol{X}_\tau$	16	$\sigma_e^2 + \frac{7}{12}\,5\sigma_r^2$
$\widetilde{A}_{\text{I}}$ vs. $\widehat{A}_{\text{II,III,IV}}$	4	$\sigma_e^2 + \frac{3}{4}\,5\sigma_r^2$
$\widetilde{B}_{\text{II}}$ vs. $\widehat{B}_{\text{I,III,IV}}$	4	$\sigma_e^2 + \frac{3}{4}\,5\sigma_r^2$
$\widetilde{AB}^3_{\text{III}}$ vs. $\widehat{AB}^3_{\text{I,II}}$	4	$\sigma_e^2 + \frac{2}{3}\,5\sigma_r^2$
$\widetilde{AB}_{\text{IV}}$ vs. $\widehat{AB}_{\text{II,III}}$	4	$\sigma_e^2 + \frac{2}{3}\,5\sigma_r^2$
$\boldsymbol{I} \mid \mathfrak{J}, \boldsymbol{X}_s, \boldsymbol{X}_r, \boldsymbol{X}_c, \boldsymbol{X}_\tau$	40	σ_e^2

Table 18.13 C | R, T-ANOVA for Example 18.3

Source	d.f.	E(MS)
$\boldsymbol{X}_s \mid \mathfrak{J}$	3	
$\boldsymbol{X}_\tau \mid \mathfrak{J}, \boldsymbol{X}_s$	24	
$\boldsymbol{X}_r \mid \mathfrak{J}, \boldsymbol{X}_s, \boldsymbol{X}_\tau$	16	
$\widetilde{A}_{\text{I}}$ vs. $\widetilde{A}_{\text{II,III,IV}}$	4	
$\widetilde{B}_{\text{II}}$ vs. $\widetilde{B}_{\text{I,III,IV}}$	4	
$\widetilde{AB}^3_{\text{III}}$ vs. $\widetilde{AB}^3_{\text{I,II,IV}}$	4	
$\widetilde{AB}_{\text{IV}}$ vs. $\widetilde{AB}_{\text{I,II,III}}$	4	
$\boldsymbol{X}_c \mid \mathfrak{J}, \boldsymbol{X}_s, \boldsymbol{X}_r, \boldsymbol{X}_\tau$	16	$\sigma_e^2 + \frac{7}{12}\,5\sigma_c^2$
$\widetilde{AB}_{\text{I}}$ vs. $\widehat{AB}_{\text{II,III}}$	4	$\sigma_e^2 + \frac{2}{3}\,5\sigma_c^2$
$\widetilde{AB}^2_{\text{II}}$ vs. $\widehat{AB}^2_{\text{I,III,IV}}$	4	$\sigma_e^2 + \frac{3}{4}\,5\sigma_c^2$
$\widetilde{AB}^4_{\text{III}}$ vs. $\widehat{AB}^4_{\text{I,II,IV}}$	4	$\sigma_e^2 + \frac{3}{4}\,5\sigma_c^2$
$\widetilde{AB}^3_{\text{IV}}$ vs. $\widehat{AB}^3_{\text{I,II}}$	4	$\sigma_e^2 + \frac{2}{3}\,5\sigma_c^2$
$\boldsymbol{I} \mid \mathfrak{J}, \boldsymbol{X}_s, \boldsymbol{X}_r, \boldsymbol{X}_c, \boldsymbol{X}_\tau$	40	σ_e^2

Example 18.4 For $K = 10$ we have $q = 2$. A suitable arrangement would be as follows:

	Square			
	I	II	III	IV
Rows	L_1	L_2	L_3	L_4
Columns	L_2	L_3	L_4	L_1

Other arrangements are, of course, possible. □

18.10 LATTICE RECTANGLES

The concept of the lattice square as a design that enables elimination of heterogeneity in two directions can be extended to situations where the rows and columns do not contain the same number of experimental units. The notion of double confounding presented in Section 9.8 may be used to develop designs for K^n treatments, K being prime or prime power, with replicates having a rectangular pattern with rows of size K^r and columns of size K^c with $r + c = n$. There are, obviously, many possible configurations. To describe the general case leads to rather complicated notation. For this reason we shall illustrate this type of design by the following example.

Example 18.5 Suppose we have $t = 3^3 = 27$ treatments and replicates with rows of size 3 and columns of size 9. A possible system of row and column confounding would be as follows:

Replicate I:	Rows:	A, B, AB, AB^2
	Columns:	ABC
Replicate II:	Rows:	A, C, AC, AC^2
	Columns:	AB^2C
Replicate III:	Rows:	B, C, BC, BC^2
	Columns:	ABC^2

This leaves each effect or interaction unconfounded in at least one replicate. □

With proper modifications, the analysis proceeds as outlined in Section 18.9 using model (18.32) and the weights

$$w_r = \frac{1}{\sigma_e^2 + K^r\,\sigma_r^2} \qquad w_c = \frac{1}{\sigma_e^2 + K^c\,\sigma_c^2}$$

Estimation of the weights is accomplished easily by following the procedures outlined in Section 11.12 in combination with (18.41) and (18.43). We merely illustrate this for Example 18.5.

Table 18.14 Partial ANOVA for Lattice Rectangle of Example 18.5

Source	d.f.	$E(\text{MS})$
$\boldsymbol{X}_r \mid \mathfrak{I}, \boldsymbol{X}_s, \boldsymbol{X}_c, \boldsymbol{X}_\tau$	24	$\sigma_e^2 + \frac{3}{2}3\sigma_r^2$
$\widetilde{A}_{\text{II,III}}$ vs. $\widehat{A}_{\text{III}}$	2	
$\widetilde{B}_{\text{I,III}}$ vs. $\widehat{B}_{\text{II}}$	2	$\sigma_e^2 + \frac{1}{3}3\sigma_r^2$
$\widetilde{C}_{\text{II,III}}$ vs. $\widehat{C}_{\text{I}}$	2	
$\widetilde{AB}_{\text{I}}$ vs. $\widehat{AB}_{\text{II,III}}$	2	
$\widetilde{AB}^2_{\text{I}}$ vs. $\widehat{AB}^2_{\text{II,III}}$	2	
$\widetilde{AC}_{\text{II}}$ vs. $\widehat{AC}_{\text{I,III}}$	2	$\sigma_e^2 + \frac{2}{3}3\sigma_r^2$
$\widetilde{AC}^2_{\text{II}}$ vs. $\widehat{AC}^2_{\text{I,III}}$	2	
$\widetilde{BC}_{\text{III}}$ vs. $\widehat{BC}_{\text{I,II}}$	2	
$\widetilde{BC}^2_{\text{III}}$ vs. $\widehat{BC}^2_{\text{I,II}}$	2	
$\widetilde{A}_{\text{I}}$ vs. $\widetilde{A}_{\text{II}}$	2	
$\widetilde{B}_{\text{I}}$ vs. $\widetilde{B}_{\text{III}}$	2	$\sigma_e^2 + 3\sigma_r^2$
$\widetilde{C}_{\text{II}}$ vs. $\widetilde{C}_{\text{III}}$	2	
$\boldsymbol{X}_c \mid \mathfrak{I}, \boldsymbol{X}_s, \boldsymbol{X}_r, \boldsymbol{X}_\tau$	6	$\sigma_e^2 + \frac{2}{3}9\sigma_c^2$
$\widetilde{ABC}_{\text{I}}$ vs. $\widehat{ABC}_{\text{II,III}}$	2	
$\widetilde{AB^2C}_{\text{II}}$ vs. $\widehat{AB^2C}_{\text{I,III}}$	2	$\sigma_e^2 + \frac{2}{3}9\sigma_c^2$
$\widetilde{ABC}^2_{\text{III}}$ vs. $\widehat{ABC}^2_{\text{I,II}}$	2	

Example 18.5 (Continued) The partitioning of $\text{SS}(\boldsymbol{X}_r \mid \mathfrak{I}, \boldsymbol{X}_s, \boldsymbol{X}_c, \boldsymbol{X}_\tau)$ and $\text{SS}(\boldsymbol{X}_c \mid \mathfrak{I}, \boldsymbol{X}_s, \boldsymbol{X}_r, \boldsymbol{X}_\tau)$ is indicated in Table 18.14 together with the associated $E(\text{MS})$. With these, w_r and w_c can be estimated in the usual way. □

18.11 RECTANGULAR LATTICES

The class of two-dimensional lattices is applicable when the number of treatments is a perfect square; that is, $t = K^2$. Obviously, this limits the number of cases in which such a lattice design can be used. To remedy this deficiency Harshbarger (1947, 1949, 1951) introduced designs for $t = K(K-1)$ treatments in blocks of size $K-1$, which are referred to as *rectangular lattice designs*. Actually, this is a special case of a more general class of designs for $t = K(K-L)$ treatments in blocks of size $K-L$ $(K > L)$. However, only the case $L = 1$ is really useful from a practical point of view since $(K-1)^2 < K(K-1) < K^2$, thus accommodating intermediate values of t. In contrast to the lattice designs discussed so far there does not now exist a correspondence between the treatments and factorial combinations, a fact that will be reflected in the analysis of these designs.

18.11.1 Simple Rectangular Lattices

We denote the treatments by ordered pairs (x, y) with $x, y = 1, 2, \ldots, K, x \neq y$. We then form two replicates, each with K blocks of size $K - 1$ as follows:

Replicate I: Treatments with the same value for x form the xth block
Replicate II: Treatments with the same value for y for the yth block

Thus each treatment (x, y) appears together in a block with $2(K - 2)$ treatments all of which have one digit in common with (x, y). The remaining $K^2 - 3K + 3$ treatments have either 0, 1, or 2 digits in common with (x, y). Nair (1951) has shown that these groups of treatments define an association scheme for a PBIB(4) design for $K \geq 4$ with the actual resolved PBIB(4) design as given above with $t = K(K - 1), k = K - 1, r = 2, b = 2K$.

More formally, the association scheme and the parameters of the design (see Section 4.3) are as follows: Two treatments (x, y) and (x', y') are said to be

1st associates if $(x = x', y \neq y')$ or $(y = y', x \neq x')$
2nd associates if $x \neq x', x \neq y', y \neq x', y \neq y'$
3rd associates if $(x = y', y \neq x')$ or $(y = x', x \neq y')$
4th associates if $x = y', y = x'$

Thus $n_1 = 2(K - 2), n_2 = (K - 2)(K - 3), n_3 = 2(K - 2), n_4 = 1$ with $\lambda_1 = 1, \lambda_2 = \lambda_3 = \lambda_4 = 0$.

Furthermore

$$\boldsymbol{P}_1 = \begin{bmatrix} 0 & 1 & 0 & 0 & 0 \\ 1 & K-3 & K-3 & 1 & 0 \\ 0 & K-3 & (K-3)(K-4) & K-3 & 0 \\ 0 & 1 & K-3 & K-3 & 1 \\ 0 & 0 & 0 & 1 & 0 \end{bmatrix}$$

$$\boldsymbol{P}_2 = \begin{bmatrix} 0 & 0 & 1 & 0 & 0 \\ 0 & 2 & 2(K-4) & 2 & 0 \\ 1 & 2(K-4) & (K-4)(K-5) & 2(K-4) & 1 \\ 0 & 2 & 2(K-4) & 2 & 0 \\ 0 & 0 & 1 & 0 & 0 \end{bmatrix}$$

$$P_3 = \begin{bmatrix} 0 & 0 & 0 & 1 & 0 \\ 0 & 1 & K-3 & K-3 & 1 \\ 0 & K-3 & (K-3)(K-4) & K-3 & 0 \\ 1 & K-3 & K-3 & 1 & 0 \\ 0 & 1 & 0 & 0 & 0 \end{bmatrix}$$

$$P_4 = \begin{bmatrix} 0 & 0 & 0 & 0 & 1 \\ 0 & 0 & 0 & 2(K-2) & 0 \\ 0 & 0 & (K-2)(K-3) & 0 & 0 \\ 0 & 2(K-2) & 0 & 0 & 0 \\ 1 & 0 & 0 & 0 & 0 \end{bmatrix}$$

For $K = 3$ this design reduces to a PBIB(3) design since then $n_2 = 0$.

The analysis of this design proceeds as outlined in Section 4.5.

18.11.2 Triple Rectangular Lattices

We now denote the treatments by triplets (x, y, z) with $x, y, z = 1, 2, \ldots, K$ and $x \neq y \neq z \neq x$. The triplets are chosen in the following way. We take a Latin square of order K, replacing the Latin letters $A, B, C, \ldots$ by the "Latin" numbers $1, 2, 3, \ldots$, respectively, arranged in such a way that the diagonal contains the numbers $1, 2, \ldots, K$. Then leaving out the diagonal the remaining $K(K-1)$ cells are identified by the row number, x, the column number, y, and the "Latin" number, z. Each such cell corresponds to a treatment (x, y, z), and the treatments, which are allocated to the blocks in the three replicates as follows:

Replicate I:	Treatments with the same x value form the xth block
Replicate II:	Treatments with the same y value form the yth block
Replicate III:	Treatments with the same z value form the zth block

giving a resolved design with parameters $t = K(K-1)$, $k = K-1$, $b = 3K$, $r = 3$.

Nair (1951) has shown that for $K = 3$ and $K = 4$ the resulting design is a PBIB design, but that this is no longer true for $K \geq 5$. For $K = 3$ we have a PBIB(2) design with the following association scheme: Two treatments (x, y, z) and (x', y', z') are said to be 1st associates if $x = x'$ or $y = y'$ or $z = z'$ and 2nd associates otherwise. It then follows that $n_1 = 3$, $n_2 = 2$, $\lambda_1 = 1$, $\lambda_2 = 0$ and

$$P_1 = \begin{pmatrix} 0 & 1 & 0 \\ 1 & 0 & 2 \\ 0 & 2 & 0 \end{pmatrix} \qquad P_2 = \begin{pmatrix} 0 & 0 & 1 \\ 0 & 3 & 0 \\ 1 & 0 & 1 \end{pmatrix}$$

For $K = 4$, the association is as follows: Two treatments (x, y, z) and (x', y', z') are said to be 1st associates if $x = x'$ or $y = y'$ or $z = z'$, 2nd associates if they have three digits alike, and 3rd associates otherwise. This leads to a PBIB(3) design with parameters $n_1 = 6$, $n_2 = 2$, $n_3 = 3$, $\lambda_1 = 1$, $\lambda_2 = \lambda_3 = 0$ and

$$\boldsymbol{P}_1 = \begin{pmatrix} 0 & 1 & 0 & 0 \\ 1 & 2 & 1 & 2 \\ 0 & 1 & 0 & 1 \\ 0 & 2 & 1 & 0 \end{pmatrix} \qquad \boldsymbol{P}_2 = \begin{pmatrix} 0 & 0 & 1 & 0 \\ 0 & 3 & 0 & 3 \\ 1 & 0 & 1 & 0 \\ 0 & 3 & 0 & 0 \end{pmatrix} \qquad \boldsymbol{P}_3 = \begin{pmatrix} 0 & 0 & 0 & 1 \\ 0 & 4 & 2 & 0 \\ 0 & 2 & 0 & 0 \\ 1 & 0 & 0 & 2 \end{pmatrix}$$

The analysis of these triple rectangular lattices follows the procedures given in Section 4.5. For $K \geq 5$ these designs can be analyzed according to the methods developed in Chapter 1 for the general incomplete block design. It should be mentioned, however, that for $K = 5$ treatments are compared with five different variances depending whether they (1) appear together in the same block and have either 1, 2, or 3 digits alike or (2) do not appear in the same block and have 1 or 2 digits alike. It appears then that this design has a structure that is more general than that of a PBIB design (see also Nair, 1951) but that is unknown.

The method used to construct triple rectangular lattices can be generalized to construct rectangular lattices with more than three replicates by using several MOLS (where available) to label the treatments appropriately. For $K - 2$ MOLS (if they exist) we obtain the near balance rectangular lattices with K replicates (Harshbarger, 1951).

18.12 EFFICIENCY FACTORS

It should be clear from our discussion in the preceding sections that the one-restrictional lattice designs are certain types of incomplete block designs, and that the two-restrictional lattic designs are resolvable row–column designs. For both of these types of designs we have discussed in Sections 1.12 and 6.6.7, respectively, how comparisons among competing designs can be made by computing their efficiency factors, and by providing upper bounds for the efficiency factors. These results can, of course, be used here, too.

More specifically, however, Patterson and Williams (1976) give the efficiency factor for a square lattice as

$$E = \frac{(K^2 - 1)(s - 1)}{(K^2 - 1)(s - 1) + s(K - 1)} \tag{18.45}$$

where s is the number of replicates or number of different systems of confounding used. For other one-restrictional lattices (18.45) represents an upper bound.

For a two-restrictional lattice John and Williams (1995) give an upper bound

$$E_{\rho\gamma} = \frac{t - \rho - \gamma + 1}{t - 1} \tag{18.46}$$

where $t =$ number of treatments, $\rho =$ number of rows, and $\gamma =$ number of columns. Alternatives to (18.46) were derived by John and Street (1992, 1993).

CHAPTER 19

Crossover Designs

19.1 INTRODUCTION

We have mentioned earlier (e.g., I.9.1) that a very important aspect of experimental design is the reduction of error in order to improve the inference from experiments. This is particularly true if an experiment involves biological entities, such as animals or humans, since such entities typically exhibit rather larger variability. One way to reduce this natural variability is to use each animal or human, which we shall refer to as subjects, as a block rather than as an experimental unit. Different treatments are then applied successively, that is, in different time periods, to each subject, so that each subject–period combination now represents the experimental unit. This is often referred to as each subject acting as its own control, since now comparisons between treatments can be made within rather than between subjects.

Obviously, for this procedure to be of any value certain conditions have to be fulfilled: (1) a subject reacts to the treatment soon after the treatment has been applied, (2) the treatment effect only lasts for a limited time, (3) after this time the subject is restored to its original state, and (4) the treatment effects are the same in each period. If these conditions are satisfied, we may use some form of block design where the subjects are the blocks and the treatments to be administered to the subject are applied at random. If, however, period effects are suspected, that is, the subjects change systematically over the time of the trial, then we employ some sort of row–column design, with rows representing the periods and columns representing the subjects. This situation may occur, for example, in a dairy cattle feeding trial, where the treatments are applied during the cow's lactation period and where it is known that changes occur during the lactation period regardless of the treatments (Cochran, Autrey, and Cannon, 1941).

It is this latter type of situation that we are concerned with in this chapter. The designs suitable for these situations are called *crossover designs* or *changeover*

Design and Analysis of Experiments. Volume 2: Advanced Experimental Design
By Klaus Hinkelmann and Oscar Kempthorne
ISBN 0-471-55177-5

designs, and we shall discuss some of their features in the following sections. We should mention that these designs are sometimes referred to as *repeated measurements designs*, but we shall reserve this term for designs where repeated measurements are obtained on a subject following a single treatment application (see I.13.7).

Crossover designs were first considered in principle in the context of agricultural experiments. Cochran (1939), for example, alluded to their special features in connection with rotation experiments. Further developments came with applications in animal feeding trials (e.g., Lucas, 1957; Patterson, 1951; Patterson and Lucas, 1959, 1962), biological assay (e.g., Finney, 1956), pharmaceutical and clinical trials (e.g., Senn, 1993), psychology (e.g., Cotton, 1998), and industrial research (e.g., Raghavarao, 1990). For a brief history of this subject see Hedayat and Afsarinejad (1975).

19.2 RESIDUAL EFFECTS

We have pointed out above that one of the advantages of crossover designs is that certain treatment contrasts may be estimated more precisely on a within-subject basis as compared to designs where only between-subject information is available. There are, however, also disadvantages, such as the length of time required for a trial.

Another major disadvantage may arise if the treatments exhibit effects beyond the period in which they are applied. These lingering effects are referred to as *residual effects* or *carryover effects*. If these effects cannot be accounted for, they may bias the estimates of contrasts among treatment effects or, as they also are referred to, *direct effects*. Wash-out periods (with either no treatment or a standard treatment) have been used to eliminate this problem, but that may in some cases be unethical and it also prolongs the duration of the trial even more.

There may be situations in which one is interested to estimate the residual effects, but generally we are interested only in the direct effects. It is therefore important to construct designs that allow us to estimate the direct effects separately from the residual effects.

19.3 THE MODEL

We have argued before (see I.2.2) that the development of the experimental design and the formulation of an appropriate statistical model are intimately connected in that the structures of the treatment design, the error control design, and the sampling and observation design determine essentially the complexity of the statistical model for purposes of analyzing the data. From our discussion, so far it is clear that a model for crossover designs follows that for a row–column design, that is, it contains period, subject, and treatment (direct) effects. In addition, however, we also need to include residual effects, and that requires that we have to make assumptions about the nature of the residual effects. Over how many periods do they extend? Do they interact with the treatments applied in these periods? Do they change over time?

The commonly used assumptions are that the carryover effects last only for one period and that they are constant over time and do not depend on the treatment applied in that period (but see also Section 19.8.6). Thus, if we denote by y_{ij} the observation in period i $(i = 1, 2, \ldots, p)$ on subject j $(j = 1, 2, \ldots, n)$, we write the model as

$$y_{ij} = \mu + \pi_i + \beta_j + \tau_{d(i,j)} + \rho_{d(i-1,j)} + e_{ij} \tag{19.1}$$

where π_i represents the ith period effect, β_j the jth subject effect, $\tau_{d(i,j)}$ the treatment effect, with $d(i, j)$ denoting the treatment applied to subject j in period i using the design d, $\rho_{d(i-1,j)}$ the carryover effect associated with the treatment assigned to subject j in period $i-1$, and e_{ij} the error with mean 0 and variance σ_e^2.

We shall rewrite (19.1) in matrix notation as follows (see Stufken, 1996):

$$\boldsymbol{y} = \mathfrak{I}\mu + \boldsymbol{X}_1\boldsymbol{\pi} + \boldsymbol{X}_2\boldsymbol{\beta} + \boldsymbol{X}_{d3}\boldsymbol{\tau} + \boldsymbol{X}_{d4}\boldsymbol{\rho} + \boldsymbol{e} \tag{19.2}$$

with $\boldsymbol{y}' = (y_{11}, y_{21}, \ldots, y_{pn})$, $\boldsymbol{\pi}' = (\pi_1, \pi_2, \ldots, \pi_p)$, $\boldsymbol{\beta}' = (\beta_1, \beta_2, \ldots, \beta_n)$, $\boldsymbol{\tau}' = (\tau_1, \tau_2, \ldots, \tau_t)$, $\boldsymbol{\rho}' = (\rho_1, \rho_2, \ldots, \rho_t)$,

$$\boldsymbol{X}_1 = \begin{bmatrix} \boldsymbol{I}_p \\ \vdots \\ \boldsymbol{I}_p \end{bmatrix} = \mathfrak{I} \times \boldsymbol{I}_p\,, \quad \boldsymbol{X}_2 = \begin{bmatrix} \mathfrak{I}_p & \boldsymbol{0}_p & \cdots & \boldsymbol{0}_p \\ \boldsymbol{0}_p & \mathfrak{I}_p & \cdots & \boldsymbol{0}_p \\ \vdots & & & \\ \boldsymbol{0}_p & \boldsymbol{0}_p & \cdots & \mathfrak{I}_p \end{bmatrix} = \boldsymbol{I}_n \times \mathfrak{I}_p$$

and the $pn \times t$ design-dependent matrices as

$$\boldsymbol{X}_{d3} = \begin{bmatrix} \boldsymbol{X}_{d31} \\ \boldsymbol{X}_{d32} \\ \vdots \\ \boldsymbol{X}_{d3n} \end{bmatrix} \qquad \boldsymbol{X}_{d4} = \begin{bmatrix} \boldsymbol{X}_{d41} \\ \boldsymbol{X}_{d42} \\ \vdots \\ \boldsymbol{X}_{d4n} \end{bmatrix} \tag{19.3}$$

where the $p \times t$ matrix $\boldsymbol{X}_{d3j}$ denotes the period–treatment incidence matrix for subject j and where $\boldsymbol{X}_{d4j} = \boldsymbol{L}^*\boldsymbol{X}_{d3j}$ denotes the $p \times p$ period–residual effect incidence matrix for subject j, with the $p \times p$ matrix $\boldsymbol{L}^*$ defined as

$$\boldsymbol{L}^* = \begin{bmatrix} \boldsymbol{0}'_{p-1} & 0 \\ \boldsymbol{I}_{p-1} & \boldsymbol{0}_{p-1} \end{bmatrix} \tag{19.4}$$

The form of $\boldsymbol{L}^*$ in (19.4) implies that $\rho_{d(0,j)} = 0$.

19.4 PROPERTIES OF CROSSOVER DESIGNS

We have seen in our earlier discussions that balancedness plays an important and useful role in the construction of designs. It is therefore not surprising—in fact, it is quite intuitive—that "good" crossover designs benefit from certain forms of balancedness.

We shall denote a crossover design d with t treatments, n subjects, and p periods by CO(t, n, p) and the totality of all CO(t, n, p) by $\Omega_{t,n,p}$. In order to describe balancedness more precisely we then give the following definitions, following Cheng and Wu (1980) and Afsarinejad (1990).

Definition 19.1 A design $d \in \Omega_{t,n,p}$ is said to be *uniform on the periods* if each treatment is assigned to λ_1 subjects in each period. □

Definition 19.2 A design $d \in \Omega_{t,n,p}$ is said to be *uniform on the subjects* if each treatment is assigned λ_2 times to each subject. □

Definition 19.3 A design $d \in \Omega_{t,n,p}$ is said to be *uniform* if it is uniform on the periods and uniform on the subjects. □

These definitions imply that

$$n = \lambda_1 t \qquad p = \lambda_2 t \tag{19.5}$$

and hence $n \geq t$, $p \geq t$.

Definition 19.4 A design $d \in \Omega_{t,n,p}$ is said to be *balanced for residual effects* (or *balanced* for short) if each treatment is immediately preceded by each other treatment (but not by itself) λ times. □

Definition 19.5 A design $d \in \Omega_{t,n,p}$ is said to be *strongly balanced for residual effects* (or *strongly balanced* for short) if each treatment is immediately preceded by all the treatments (including itself) λ^* times. □

These definitions imply that

$$n(p-1) = \lambda t(t-1) \qquad n(p-1) = \lambda^* t^2 \tag{19.6}$$

For the important case $p < t$ Afsarinejad (1990) gives the following definitions for balancedness.

Definition 19.6 A design $d \in \Omega_{t,n,p}$ with $p < t$ is said to be *balanced* if it is uniform on periods, balanced for residual effects, and distinct treatments are applied to the subjects. □

Definition 19.7 A design $d \in \Omega_{t,n,p}$ with $p < t$ is said to be *strongly balanced* if it is uniform on periods and strongly balanced for residual effects. □

19.5 CONSTRUCTION OF CROSSOVER DESIGNS

In this section we shall describe some methods of constructing crossover designs.

19.5.1 Balanced Designs for $p = t$

A method of constructing balanced designs for the case $p = t$ was given by Williams (1949). These designs are commonly referred to as *Williams squares.* The method can be described easily by developing the treatment sequence assigned to subject 1 (Bradley, 1958). We distinguish between t even and t odd.

1. *t even*: Let us denote the treatment sequence for subject 1 by $(c_1, c_2, \ldots, c_t)'$. Then the sequence for subject $j+1$ $(j = 1, 2, \ldots, t-1)$ is given by $(c_1 + j, c_2 + j, \ldots, c_t + j)'$ where each element in this sequence is reduced mod t. Now, labeling the treatments $1, 2, \ldots, t$, we have specifically, $(c_1, c_2, \ldots, c_t)' = (1, t, 2, t-1, 3, \ldots, t/2, (t+2)/2)'$. This yields a $\text{CO}(t, t, t)$.
2. *t odd*: Here $(c_1, c_2, \ldots, c_t)' = (1, t, 2, t-1, 3, \ldots, (t+3)/2, (t+1)/2)'$. Developing $(c_1, c_2, \ldots, c_t)$ as in (1) yields t sequences for t subjects. We then take the sequence for subject t, that is, $(c_1 + t - 1, c_2 + t - 1, \ldots, c_t + t - 1)'$ and make the mirror image of it the sequence for subject $t+1$. The sequences for subjects $t+2, t+3, \ldots, 2t$ are then obtained by developing the sequence for subject $t+1$ as described above, always mod t. This yields a $\text{CO}(t, 2t, t)$.

Using arguments based on cyclic groups, one can show that the designs obtained by the methods described in (1) and (2) above are balanced crossover designs with $\lambda = 1$ and $\lambda = 2$, respectively, in (19.6). In addition, the $\text{CO}(t, t, t)$ in (1) is obviously a minimal design since $\lambda_1 = 1$ in (19.5). The same cannot be said for the $\text{CO}(t, 2t, t)$ in (2). In fact, it is known that minimal balanced designs do not exist for $t = 3, 5, 7$ and exist only for some t odd (Dénes and Keedwell, 1974).

19.5.2 Balanced Designs for $p < t$

Even for moderate values of t the designs discussed in Section 19.5 may not always be practicable because the duration of the trial may be too long. Rather than using Latin square designs, as in (1) above, or multiples of Latin square designs, as in (2) above, we need designs that have the form of Youden squares (see I.10.5) or multiples of Youden squares, that is, designs with $p < t$. At the same time it is desirable to consider minimal designs, that is, to keep the number of subjects as small as possible.

We know from (19.5) and (19.6) that for a balanced $\text{CO}(t, n, p)$ we require $n = \lambda_1 t$ and $n = \lambda t(t-1)/(p-1)$, which implies $\lambda_1(p-1) = \lambda(t-1)$. A minimal design is then obtained for $\lambda = 1$. We thus have the following theorem.

Table 19.1 Generating Columns for Minimal Balanced CO(t, n, p)

Column 1	2	3	...	i	...	λ_1
c_1	c_p	c_{2p-1}	...	$c_{(i-1)p-(i-2)}$	...	$c_{(\lambda_1-1)p-(\lambda_1-2)}$
c_2	c_{p+1}	c_{2p}	...	$c_{(i-1)p-(i-3)}$	...	$c_{(\lambda_1-1)p-(\lambda_1-3)}$
$\vdots$	$\vdots$	$\vdots$		$\vdots$		$\vdots$
c_p	c_{2p-1}	c_{3p-2}	...	$c_{ip-(i-1)}$	...	c_t

Theorem 19.1 (Afsarinejad, 1990) The necessary and sufficient conditions for the existence of a minimal balanced CO(t, n, p) with $p < t$ are $n = \lambda_1 t$ and $\lambda_1(p-1) = t - 1$.

We shall describe here two of several general methods of constructing balanced CO(t, n, p).

Method 1

Extending the method of Section 19.5.1, Afsarinejad (1983, 1990) proposed to construct a minimal balanced CO($t, \lambda_1 t, p$) for $p < t$ by developing (mod t) each of the λ_1 columns of Table 19.1 associated with subjects 1, $t+1$, $2t+1, \ldots, (\lambda_1 - 1)t + 1$, where

$$\boldsymbol{c}' = (c_1, c_2, \ldots, c_t) = \begin{cases} (1, t, 2, t-1, 3, \ldots, t/2, (t+2)/2) & \text{for } t \text{ even} \\[1ex] (1, t, 3, t-2, 5, \ldots, (t-3)/2, \\ \quad (t+5)/2, (t+1)/2, (t+5)/2, \\ \quad (t-3)/2, \ldots, t, 1) & \text{for } t = 4\gamma + 1 \\[1ex] (1, t, 3, t-2, 5, \ldots, (t-1)/2, \\ \quad (t+3)/2, (t-1)/2, \ldots, t, 1) & \text{for } t = 4\gamma + 3 \end{cases}$$

We comment briefly on the form and nature of the columns in Table 19.1:

1. The columns consist of interlaced increasing and decreasing sequences of positive integers;
2. For t odd the column consists of three parts, say $\boldsymbol{c}' = (\boldsymbol{c}'_{-1}, c_0, \boldsymbol{c}'_1)$, where $\boldsymbol{c}'_1$ is the mirror image of $\boldsymbol{c}'_{-1}$, and $c_0 = (t+1)/2$ or $(t+3)/2$ for $t = 4\gamma + 1$ or $t = 4\gamma + 3$, respectively;
3. The set $\bigcup_{i=2}^{t}(c_i - c_{i-1})$ contains each nonzero integer mod t. These represent the $p-1$ successive differences between the elements of each of the

λ_1 columns [recall that $\lambda_1(p-1) = t-1$]. After development each column will lead to $(p-1)t$ such differences, so that each ordered pair of distinct treatments will appear exactly once in the design, implying $\lambda = 1$.

We shall illustrate the procedure with the following example.

Example 19.1 Let $t = 7$, $p = 4$, and hence $\lambda_1 = 2$. From Table 19.1 we obtain $(c_1, c_2, \ldots, c_7) = (1, 7, 3, 5, 3, 7, 1)$. Thus the generating columns are $(1, 7, 3, 5)'$ and $(5, 3, 7, 1)'$, and the design is given by

		Subject													
		1	2	3	4	5	6	7	8	9	10	11	12	13	14
Period	1	1	2	3	4	5	6	7	5	6	7	1	2	3	4
	2	7	1	2	3	4	5	6	3	4	5	6	7	1	2
	3	3	4	5	6	7	1	2	7	1	2	3	4	5	6
	4	5	6	7	1	2	3	4	1	2	3	4	5	6	7

□

Method 2

This method is due to Patterson (1952) (see also Patterson and Lucas, 1962) and is based on balanced incomplete block designs. Suppose we have a BIBD $(t, b, k, r; \alpha)$. We then take the treatments in each block as the generating column for a balanced CO(k, k, k) for k even or a CO$(k, 2k, k)$ for k odd as described in Section 19.5.1. Adjoining these b crossover designs leads to the final CO(t, bk, k) with $\lambda = \alpha$ for k even or CO$(t, 2bk, k)$ with $\lambda = 2\alpha$ for k odd.

We illustrate this procedure with the following example.

Example 19.2 Consider the BIBD (5, 5, 4, 4; 3)

Block				
1	2	3	4	5
1	1	1	1	2
2	2	2	3	3
3	3	4	4	4
4	5	5	5	5

Recall that a balanced CO(4, 4, 4) is obtained from the generating column $(1, 4, 2, 3)'$. For a generating column (block) from the BIB design of the form (T_1, T_2, T_3, T_4) we then obtain the CO(4, 4, 4) by using the association $T_1 \equiv 1, T_2 \equiv 4, T_3 \equiv 2, T_4 \equiv 3$. In other words, in the CO(4, 4, 4) given by

1	2	3	4
4	1	2	3
2	3	4	1
3	4	1	2

replace i by the associated T_j successively for each block. Part of the final CO(5, 20, 4) is given below:

		Subject								
		1	2	3	4	⋯	17	18	19	20
Period	1	1	3	4	2	⋯	2	4	5	3
	2	2	1	3	4	⋯	3	2	4	5
	3	3	4	2	1	⋯	4	5	3	2
	4	4	2	1	3	⋯	5	3	2	4

□

19.5.3 Partially Balanced Designs

One possible disadvantage of the designs obtained by method 2 in Section 19.5.2 is that the number of subjects required may be rather large. This is due to the fact that in general the number of blocks in a BIB design is relatively large. We have seen earlier that one "remedy" is to resort to PBIB designs with a smaller number of blocks. The same idea can be used here. Instead of developing the blocks of a BIB design as illustrated in Example 19.2, we may choose to develop in the same manner the blocks of a PBIB(m) design (Patterson, 1952; Patterson and Lucas, 1962). As a consequence, however, comparisons between direct effects from the resulting crossover design are estimated with up to m different variances. Of particular interest here are PBIB(2) designs but also, because of the ease of construction, cyclic PBIB designs (see Section 5.3) as advocated by Davis and Hall (1969). For another class of partially balanced designs see Blaisdell and Raghavarao (1980).

19.5.4 Strongly Balanced Designs for $p = t + 1$

We have pointed out earlier that the main interest in crossover designs is to estimate comparisons among direct effects. There are, however, situations where the residual effects are of interest in themselves as well. To increase the efficiency of estimating residual effect contrasts and to achieve orthogonality between the direct and residual effects (see Section 19.7), we consider strongly balanced designs. The additional important feature of these designs is that each treatment is also preceded by itself λ^* times (see Definition 19.5).

An obvious way to construct a minimal strongly balanced design is to consider the balanced designs CO(t, t, t) and CO($t, 2t, t$) of Section 19.5.1 and add an extra period by simply repeating the treatment assignments from period t. We thus create the so-called extra-period designs (Lucas, 1957) CO($t, t, t + 1$) and CO($t, 2t, t + 1$) with $\lambda^* = 1$ and 2, respectively. Some other extra-period designs based on orthogonal Latin squares were given by Lucas (1957).

19.5.5 Strongly Balanced Designs for $p < t$

Similar to Theorem 19.1 we can state the following theorem.

Theorem 19.2 (Afsarinejad, 1990) The necessary and sufficient conditions for the existence of a minimal strongly balanced $CO(t, n, p)$ with $p < t$ are $n = \lambda_1 t$ and $\lambda_1(p - 1) = t$.

The construction then follows the method described in Section 19.5.2 with the following modifications (Afsarinejad, 1983, 1990):

1. The last element in the λ_1th generating column in Table 19.1 is c_{t+1}.
2. The $\boldsymbol{c}$ vector

$$
\begin{aligned}
&(c_1, c_2, \ldots, c_t, c_{t+1}) \\
&= \begin{cases}
(1, t, 2, t-1, \ldots, t/2, (t+2)/2, (t+2)/2) & \text{for } t \text{ even} \\[2ex]
(1, t, 3, t-2, \ldots, (t-3)/2, (t+5)/2, \\ \quad (t+1)/2, (t+1)/2, (t+5)/2, \\ \quad (t-3)/2, \ldots, t, 1) & \text{for } t = 4\gamma + 1 \\[2ex]
(1, t, 3, t-2, \ldots, (t-1)/2, (t+3)/2, \\ \quad (t+3)/2, (t-1)/2, \ldots, t, 1) & \text{for } t = 4\gamma + 3
\end{cases}
\end{aligned}
$$

3. The set $\bigcup_{i=2}^{t+1}(c_i - c_{i-1})$ contains each number mod t and, hence, each ordered pair of treatments appears exactly once in the design.

We illustrate this procedure with the following example.

Example 19.3 Let $t = 6$, $p = 4$, and hence $\lambda_1 = 2$. Then $(c_1, c_2, \ldots, c_7) = (1, 6, 2, 5, 3, 4, 4)$. The generating columns are (1, 6, 2, 5) and (5, 3, 4, 4) and the final design is

		Subject											
		1	2	3	4	5	6	7	8	9	10	11	12
Period	1	1	2	3	4	5	6	5	6	1	2	3	4
	2	6	1	2	3	4	5	3	4	5	6	1	2
	3	2	3	4	5	6	1	4	5	6	1	2	3
	4	5	6	1	2	3	4	4	5	6	1	2	3

□

19.5.6 Balanced Uniform Designs

Uniform crossover designs (see Definition 19.3) are not only structurally appealing but also an important role in connection with the consideration of optimality (see Section 19.6). It is easy to verify that the CO(t, t, t) and CO$(t, 2t, t)$ given in Section 19.5.1 are uniform crossover designs. From these designs we can obtain, of course, CO(t, c_1t, c_2t) and CO$(t, 2c_1t, c_2t)$, respectively, with c_1, c_2 positive integers.

19.5.7 Strongly Balanced Uniform Designs

One method of construction is provided through the proof of the following theorem.

Theorem 19.3 (Cheng and Wu, 1980) If n is a multiple of t^2 and $p = \lambda_2 t$ with λ_2 even, then there exists a strongly balanced uniform CO(t, n, p).

A simple way to construct such a design (Cheng and Wu, 1980) is to first consider the case $n = t^2$ and $\lambda_2 = 2$. For periods 1 and 2 write down all possible ordered pairs of the t treatments as a $2 \times t^2$ array. To each element in this array add i mod $t (i = 1, 2, \ldots, t - 1)$ to obtain periods $2i + 1$ and $2i + 2$.

As an illustration we consider the following example.

Example 19.4 Take $t = 3$, thus $n = 9$, $p = 6$, leading to the following CO(3, 9, 6) with $\lambda^* = 3$:

		Subject								
		1	2	3	4	5	6	7	8	9
Period	1	1	1	1	2	2	2	3	3	3
	2	1	2	3	1	2	3	1	2	3
	3	2	2	2	3	3	3	1	1	1
	4	2	3	1	2	3	1	2	3	1
	5	3	3	3	1	1	1	2	2	2
	6	3	1	2	3	1	2	3	1	2

It is then obvious how one can obtain in general a CO$(t, ct^2, \lambda_2 t)$ with c a positive integer and λ_2 even. For a different method of construction see Berenblut (1964). □

19.5.8 Designs with Two Treatments

The most widely used crossover designs involve two treatments, A and B say, where in the clinical setting A may be a standard drug or placebo and B may be the test drug. And the simplest design is to consider two sequences of treatment application to a subject: AB or BA. For two subjects the CO(2, 2, 2) is given by

		Subject 1	Subject 2
Period	1	A	B
	2	B	A

It is, of course, obvious by considering the available degrees of freedom that this CO(2, 2, 2) will not allow the estimation of all parameters in model (19.1). Even replicating each sequence r times resulting in a CO(2, $2r$, 2) does not solve this problem. Therefore, designs with additional periods and/or additional sequences will have to be considered.

We shall mention some such possible designs by indicating the number of sequences s and the number of periods p, as well as the sequences themselves in a $p \times s$ array. The final design is then obtained by replicating each sequence r times (we note that equal replication is desirable, but not necessary). The resulting design is then a CO(2, sr, p).

$p = 2$, $s = 4$ (Balaam, 1968):

$$\begin{matrix} A & B & A & B \\ B & A & A & B \end{matrix}$$

$p = 3$, $s = 2$:

$$\text{(i)}\ \begin{matrix} A & B \\ B & A \\ B & A \end{matrix} \qquad \text{(ii)}\ \begin{matrix} A & B \\ B & A \\ A & B \end{matrix} \qquad \text{(iii)}\ \begin{matrix} A & B \\ A & B \\ B & A \end{matrix}$$

Design (i) is, of course, an extra-period design as discussed in Section 19.5.4. Design (ii) is referred to as a *switch-back design.*

$p = 3$, $s = 4$:

Combining various sequences from the designs for $p = 3$, $s = 2$ leads to

$$\text{(i)}\ \begin{matrix} A & B & A & B \\ B & A & B & A \\ B & A & A & B \end{matrix} \qquad \text{(ii)}\ \begin{matrix} A & B & A & B \\ B & A & A & B \\ B & A & B & A \end{matrix}$$

$$\text{(iii)}\ \begin{matrix} A & B & A & B \\ B & A & A & B \\ A & B & B & A \end{matrix}$$

$p = 3,\ s = 6$:

Combine all six sequences from (i), (ii), (iii) for $p = 3,\ s = 2$.

$p = 4,\ s = 2$:

(i)	A	B	(ii)	A	B	(iii)	A	B
	A	B		B	A		B	A
	B	A		A	B		B	A
	B	A		B	A		A	B
(iv)	A	B	(v)	A	B	(vi)	A	B
	B	A		A	B		B	A
	A	B		B	A		B	A
	A	B		A	B		B	A
(vii)	A	B						
	A	B						
	A	B						
	B	A						

Some of these designs may be combined to obtain designs with $s = 4$ or $s = 6$ sequences, for example.

A detailed discussion of the designs given above can be found in Jones and Kenward (2003) and Ratkowsky, Evans, and Alldredge (1993). An extensive discussion of 2-treatment designs is also given by Kershner and Federer (1981).

19.6 OPTIMAL DESIGNS

In the previous sections we have discussed different classes of crossover designs and have given some methods of constructing designs in these classes. Although the designs presented here have certain desirable properties, such as balancedness, for example, and may be quite appropriate from a practical point of view, they may not necessarily be optimal from a statistical point of view. We shall address this question briefly in this section and give a few results.

19.6.1 Information Matrices

In order to write out the information matrices for $\boldsymbol{\tau}$ and $\boldsymbol{\rho}$ [see model (19.2)], we shall follow Cheng and Wu (1980) and Stufken (1996) and use the following notation:

$$
\begin{aligned}
n_{uj} &= \text{number of times that treatment } u \text{ is assigned to subject } j \\
\widetilde{n}_{uj} &= \text{number of times that treatment } u \text{ is assigned to subject } j \text{ in} \\
&\quad\ \text{the first } p-1 \text{ periods}
\end{aligned}
$$

ℓ_{ui} = number of times that treatment u appears in period i

m_{uv} = number of times that treatment u is immediately preceded by treatment v

$r_u = \sum_{j=1}^{n} n_{uj} = \sum_{i=1}^{p} \ell_{ui}$ = number of replications for treatment u

$\widetilde{r}_u = \sum_{j=1}^{n} \widetilde{n}_{uj} = \sum_{i=1}^{p-1} \ell_{ui} = \sum_{v=1}^{t} m_{vu}$ = number of replications for treatment u in the first $p-1$ periods

$s_u = \sum_{i=2}^{p} \ell_{ui} = \sum_{v=1}^{t} m_{uv}$ = number of times that treatment u appears in the last $p-1$ periods

We then have that

$$\sum_u n_{uj} = p \qquad \sum_u \widetilde{n}_{uj} = p - 1 \qquad \sum_u \ell_{ui} = n$$

$$\sum_u r_u = np \qquad \sum_u \widetilde{r}_u = n(p-1)$$

Now we consider (19.2) and assume, without loss of generality, $\mu = 0$ (or, equivalently, define $\boldsymbol{\pi}^* = \mu\mathfrak{I} + \boldsymbol{\pi}$). Then the coefficient matrix for the normal equations for model(19.2) can be written as

$$\begin{bmatrix} X_1'X_1 & X_1'X_2 & X_1'X_{d3} & X_1'X_{d4} \\ X_2'X_1 & X_2'X_2 & X_2'X_{d3} & X_2'X_{d4} \\ X_{d3}'X_1 & X_{d3}'X_2 & X_{d3}'X_{d3} & X_{d3}'X_{d4} \\ X_{d4}'X_1 & X_{d4}'X_2 & X_{d4}'X_{d3} & X_{d4}'X_{d4} \end{bmatrix}$$

$$= \begin{bmatrix} n\boldsymbol{I}_p & \mathfrak{I}_p\mathfrak{I}_n' & \boldsymbol{L}_d' & \widetilde{\boldsymbol{L}}_d' \\ \mathfrak{I}_n\mathfrak{I}_p' & p\boldsymbol{I}_n & \boldsymbol{N}_d' & \widetilde{\boldsymbol{N}}_d' \\ \boldsymbol{L}_d & \boldsymbol{N}_d & \boldsymbol{D}_d & \boldsymbol{M}_d \\ \widetilde{\boldsymbol{L}}_d & \widetilde{\boldsymbol{N}}_d & \boldsymbol{M}_d' & \widetilde{\boldsymbol{D}}_d \end{bmatrix} \tag{19.7}$$

where, dropping the subscript d for convenience and indicating the dimensions instead,

$$\boldsymbol{L}_{t\times p} = (\ell_{ui}) \qquad \widetilde{\boldsymbol{L}}_{t\times p} = \left(\widetilde{\ell}_{ui}\right)$$

with $\widetilde{\ell}_{u1} = 0$, $\widetilde{\ell}_{ui} = \ell_{u,i-1}$ $(i \geq 2)$,

$$N_{t\times n} = (n_{uj}) \qquad \widetilde{N}_{t\times n} = \left(\widetilde{n}_{uj}\right)$$

$$D_{t\times t} = \text{diag}(r_1, r_2, \ldots, r_t) \qquad \widetilde{D}_{t\times t} = \text{diag}\left(\widetilde{r}_1, \widetilde{r}_2, \ldots, \widetilde{r}_t\right) \qquad M_{t\times t} = (m_{uv})$$

We then obtain from (19.7) the information matrix for estimating jointly the direct and residual effects as

$$\begin{aligned} C_{\tau,\rho} &= \begin{pmatrix} D & M \\ M' & \widetilde{D} \end{pmatrix} - \begin{pmatrix} L & N \\ \widetilde{L} & \widetilde{N} \end{pmatrix} \begin{pmatrix} nI_p & \mathfrak{I}_p\mathfrak{I}_n' \\ \mathfrak{I}_n\mathfrak{I}_p' & pI_n \end{pmatrix}^{-} \begin{pmatrix} L' & \widetilde{L}' \\ N' & \widetilde{N}' \end{pmatrix} \\ &\equiv \begin{pmatrix} C_{11} & C_{12} \\ C_{21} & C_{22} \end{pmatrix} \end{aligned} \tag{19.8}$$

where

$$C_{11} = D - n^{-1}LL' - p^{-1}NN' + n^{-1}p^{-1}N\,\mathfrak{I}_n\mathfrak{I}_n'\,N'$$

$$C_{12} = C_{21}' = M - n^{-1}L\widetilde{L}' - p^{-1}N\widetilde{N}' + n^{-1}p^{-1}N\,\mathfrak{I}_n\mathfrak{I}_n'\,\widetilde{N}'$$

$$C_{22} = \widetilde{D} - n^{-1}\widetilde{L}\widetilde{L}' - p^{-1}\widetilde{N}\widetilde{N}' + n^{-1}p^{-1}\widetilde{N}\,\mathfrak{I}_n\mathfrak{I}_n'\,\widetilde{N}'$$

It then follows from (19.8) that the information matrices, that is, the coefficient matrices for the reduced normal equations, for estimating the direct effects (adjusted for residual effects) and for estimating the residual effects (adjusted for direct effects) are given by

$$C_\tau = C_{11} - C_{12}\,C_{22}^{-}\,C_{21} \tag{19.9}$$

and

$$C_\rho = C_{22} - C_{21}\,C_{11}^{-}\,C_{12} \tag{19.10}$$

respectively.

19.6.2 Optimality Results

As we have seen in Section 1.13 optimality considerations are based on the information matrices for competing designs. For some of the classes of designs, which we have discussed in Section 19.5, the matrices (19.9) and (19.10) reduce to quite simple forms so that optimal designs can be characterized if not with respect to $\Omega_{t,n,p}$ then with respect to some suitable subset $\Omega \subset \Omega_{t,n,p}$.

For example, for a uniform CO(t, $\lambda_1 t$, t) Hedayat and Afsarinejad (1978) show that

$$C_\tau = \lambda_1 t\, I - \frac{t}{\lambda_1(t^2 - t - 1)} MM' + \frac{\lambda_1(2-t)}{t^2 - t - 1} \mathfrak{I}_t \mathfrak{I}_t' \tag{19.11}$$

and

$$C_\rho = \lambda_1 \frac{t^2 - t - 1}{t} I - \frac{1}{\lambda_1 t} MM' + \frac{\lambda_1(2-t)}{t^2} \mathfrak{I}_t \mathfrak{I}_t' \tag{19.12}$$

with M such that $M\mathfrak{I}_t\mathfrak{I}_t' = \lambda_1(t-1)\mathfrak{I}_t\mathfrak{I}_t'$. Together with the facts that $C_\tau \mathfrak{I}_t = 0$ and C_τ is completely symmetric enables them to state the following theorem.

Theorem 19.4 A design d^* in $\Omega^u_{t,\lambda_1 t,t}$ is universally optimal for the estimation of direct effects if $M(d^*) = \lambda_1(\mathfrak{I}_t\mathfrak{I}_t' - I)$.

Here $\Omega^u_{t,\lambda_1 t,t}$ denotes the set of all uniform CO(t, $\lambda_1 t$, t) and $M(d^*)$ denotes the M matrix for design d^*. The same result holds with respect to estimating residual effects. Now we know from Sections 19.5.6 and 19.5.1 that for a balanced design d in $\Omega^u_{t,\lambda_1 t,t}$ with $\lambda_1 = 1$ for even t and $\lambda_1 = 2$ for odd t we have $M(d) = (\mathfrak{I}_t\mathfrak{I}_t' - I)$ and $M(d) = 2(\mathfrak{I}_t\mathfrak{I}_t' - I)$, respectively. Thus patching λ_1 (or $\lambda_1/2$) copies of a balanced design d in $\Omega^u_{t,t,t}$ (or $\Omega^u_{t,2t,t}$) for t even (odd) yields a balanced design d in $\Omega^u_{t,\lambda_1 t,t}$ with $M(d) = \lambda_1(\mathfrak{I}_t\mathfrak{I}_t' - I)$. Even more generally, $M(d) = \lambda_1(\mathfrak{I}_t\mathfrak{I}_t' - I)$ if and only if d is balanced (Hedayat and Afsarinejad, 1978). Thus we have the following theorem.

Theorem 19.5 A design d^* in $\Omega^u_{t,\lambda_1 t,t}$ is universally optimal over $\Omega^u_{t,\lambda_1 t,t}$ if it is balanced.

Although balancedness is important, Kunert (1984) argues that for sufficiently large n one can find designs that are better than uniform balanced designs. At the same time he proved the following theorem.

Theorem 19.6 If a uniform balanced design d^* exists in $\Omega_{t,\lambda_1 t,t}$, then

$$\frac{\mathrm{tr}(C_\tau(d^*))}{\sup_{d\in\Omega_{t,n,t}}[\mathrm{tr}(C_\tau(d))]} \geq 1 - (t^2 - t - 1)^{-2} \tag{19.13}$$

where $C_\tau(d)$ is the information matrix for direct effects for design d.

From (19.13) we see that even though a uniform balanced design may not be optimal it is nevertheless highly efficient for estimating direct effects for $t > 2$. More specifically, Kunert (1984) showed the following theorem.

Theorem 19.7 A uniform balanced design d^* with $n = \lambda_1 t$ and $p = t$ is universally optimal for direct effects in $\Omega^u_{t,\lambda_1 t,t}$ if $\lambda_1 = 1$ and $t \geq 3$, or $\lambda_1 = 2$ and $t \geq 6$.

For an extension of this result to $t \geq 3$ and $\lambda_1 \leq (t-1)/2$ see Hedayat and Yang (2003).

The efficiency of uniform balanced designs for $p = t$ is generally smaller for residual effects than it is for direct effects. Here then strongly balanced designs are important.

Theorem 19.8 (Cheng and Wu, 1980) A strongly balanced design d^* in $\Omega_{t,n,p}$ is universally optimal for the estimation of direct and residual effects over $\Omega_{t,n,p}$.

Theorem 19.9 (Cheng and Wu, 1980) Let $n = \lambda_1 t$ and $p = \lambda_2 t + 1$ and let d^* be a strongly balanced design in $\Omega_{t,n,p}$, which is uniform on the periods and uniform on the subjects in the first $p - 1 = \lambda_2 t$ periods. Then d^* is universally optimal for the estimation of direct and residual effects over $\Omega^u_{t,n,p}$.

Such designs include, of course, the extra-period crossover designs discussed in Section 19.5.4.

For the proofs of these and other results on optimal designs we refer the reader to Cheng and Wu (1980), Kunert (1984), and Stufken (1996).

19.7 ANALYSIS OF CROSSOVER DESIGNS

In this section we shall discuss briefly how SAS PROC GLM can be used to analyze data from a crossover design using model (19.1). Since the designs in general are nonorthogonal, we need to use reduced normal equations to obtain estimates of direct and residual effects, as already alluded to in (19.9) and (19.10).

We consider two examples, one a balanced design and the other a strongly balanced design, and use them to point out numerically some differences between them.

Example 19.5 In Table 19.2 we present the data from a Williams square CO(4, 4, 4) (see Section 19.5.1) and the input for analyzing them according to model (19.1). We comment on some features of the input and output in Table 19.2:

1. In the input we identify periods, subjects, treatments, and residuals by P, S, T, C, respectively;
2. Since there is no residual effect in period 1 and since we cannot leave the corresponding place for C blank (as SAS would delete the entire data point), we introduce "level" 0 for that purpose;
3. The general form of estimable functions (see $L2$) and the type III SS for periods show that period effects are partially confounded with residual effects;
4. Treatment and residual effects are nonorthogonal to each other as indicated by the fact that the type I and type III SS for treatments are not identical.

Table 19.2 Williams Square Crossover Design CO (4, 4, 4)

```
options nodate pageno=1;
data Williams;
input P S T C Y @@;
datalines;
1 1 1 0 27.5  2 1 4 1 20.1  3 1 2 4 23.9  4 1 3 2 17.2
1 2 2 0 29.9  2 2 1 2 23.6  3 2 3 1 20.2  4 2 4 3 16.8
1 3 3 0 14.3  2 3 2 3 15.7  3 3 4 2 12.3  4 3 1 4 17.2
1 4 4 0 19.3  2 4 3 4 12.8  3 4 1 3 20.0  4 4 2 1 23.4
;
run;

proc print data=Williams;
title1 'TABLE 19.2';
title2 'WILLIAMS SQUARE CROSSOVER DESIGN';
title3 'CO(4, 4, 4)';
run;

proc glm data=Williams;
class P S T C;
model Y=P S T C/e solution;
title3 'ANALYSIS OF VARIANCE';
run;
```

Obs	P	S	T	C	Y
1	1	1	1	0	27.5
2	2	1	4	1	20.1
3	3	1	2	4	23.9
4	4	1	3	2	17.2
5	1	2	2	0	29.9
6	2	2	1	2	23.6
7	3	2	3	1	20.2
8	4	2	4	3	16.8
9	1	3	3	0	14.3
10	2	3	2	3	15.7
11	3	3	4	2	12.3
12	4	3	1	4	17.2
13	1	4	4	0	19.3
14	2	4	3	4	12.8
15	3	4	1	3	20.0
16	4	4	2	1	23.4

The GLM Procedure

Class Level Information

Class	Levels	Values
P	4	1 2 3 4

Table 19.2 (*Continued*)

S	4	1 2 3 4
T	4	1 2 3 4
C	5	0 1 2 3 4

Number of observations 16

General Form of Estimable Functions

Effect		Coefficients
Intercept		L1
P	1	L2
P	2	L3
P	3	L4
P	4	L1-L2-L3-L4
S	1	L6
S	2	L7
S	3	L8
S	4	L1-L6-L7-L8
T	1	L10
T	2	L11
T	3	L12
T	4	L1-L10-L11-L12
C	0	L2
C	1	L15
C	2	L16
C	3	L17
C	4	L1-L2-L15-L16-L17

Dependent Variable: Y

Source	DF	Sum of Squares	Mean Square	F Value	Pr > F
Model	12	378.8035000	31.5669583	57.26	0.0033
Error	3	1.6540000	0.5513333		
Corrected Total	15	380.4575000			

R-Square	Coeff Var	Root MSE	Y Mean
0.995653	3.781124	0.742518	19.63750

Table 19.2 *(Continued)*

Source	DF	Type I SS	Mean Square	F Value	Pr > F
P	3	53.8875000	17.9625000	32.58	0.0086
S	3	154.5075000	51.5025000	93.41	0.0018
T	3	149.8475000	49.9491667	90.60	0.0019
C	3	20.5610000	6.8536667	12.43	0.0337

Source	DF	Type III SS	Mean Square	F Value	Pr > F
P	2	2.2200000	1.1100000	2.01	0.2790
S	3	111.9660000	37.3220000	67.69	0.0030
T	3	161.8278182	53.9426061	97.84	0.0017
C	3	20.5610000	6.8536667	12.43	0.0337

Parameter		Estimate		Standard Error	t Value	Pr > |t|
Intercept		14.67500000	B	0.76085478	19.29	0.0003
P	1	4.67000000	B	0.66412850	7.03	0.0059
P	2	-0.60000000	B	0.52503968	-1.14	0.3361
P	3	0.45000000	B	0.52503968	0.86	0.4544
P	4	0.00000000	B	.	.	.
S	1	2.86000000	B	0.55066626	5.19	0.0139
S	2	3.59500000	B	0.55066626	6.53	0.0073
S	3	-3.45500000	B	0.55066626	-6.27	0.0082
S	4	0.00000000	B	.	.	.
T	1	5.65000000	B	0.55066626	10.26	0.0020
T	2	6.25500000	B	0.55066626	11.36	0.0015
T	3	-1.28500000	B	0.55066626	-2.33	0.1018
T	4	0.00000000	B	.	.	.
C	0	0.00000000	B	.	.	.
C	1	2.80000000	B	0.66412850	4.22	0.0244
C	2	0.62000000	B	0.66412850	0.93	0.4194
C	3	-1.14000000	B	0.66412850	-1.72	0.1846
C	4	0.00000000	B	.	.	.

NOTE: The X'X matrix has been found to be singular, and a generalized inverse was used to solve the normal equations. Terms whose estimates are followed by the letter 'B' are not uniquely estimable.

5. Since we are considering a balanced design, we have $\text{var}(\widehat{\tau}_u - \widehat{\tau}_{u'})$ the same for each $u, u'(u \neq u')$, with the same holding true for $\text{var}(\widehat{\rho}_u - \widehat{\rho}_{u'})$. A partial verification of this is that $\text{se}(T_u) = \text{se}(\widehat{\tau}_u - \widehat{\tau}_4) = .55$ for $u = 1, 2, 3$, and $\text{se}(C_u) = \text{se}(\widehat{\rho}_u - \widehat{\rho}_4) = .66$ for $u = 1, 2, 3$. □

Example 19.6 In Table 19.3 we present the data for an extra-period design CO(4, 4, 5) (see Section 19.5.4) and the input for analyzing them. Again, we shall comment briefly on some features of the input and output in Table 19.3:

Table 19.3 Extra-Period Crossover Design CO (4, 4, 5)

```
options nodate pageno=1;
data extra;
input P S T C Y @@;
if C=0 then C=4;
datalines;
1 1 1 0 27.5  2 1 4 1 20.1  3 1 2 4 23.9  4 1 3 2 17.2
5 1 3 3 18.5  1 2 2 0 29.9  2 2 1 2 23.6  3 2 3 1 20.2
4 2 4 3 16.8  5 2 4 4 17.1  1 3 3 0 14.3  2 3 2 3 15.7
3 3 4 2 12.3  4 3 1 4 17.2  5 3 1 1 18.0  1 4 4 0 19.3
2 4 3 4 12.8  3 4 1 3 20.0  4 4 2 1 23.4  5 4 2 2 21.4
;
run;

proc print data=extra;
title1 'TABLE 19.3';
title2 'EXTRA-PERIOD CROSSOVER DESIGN';
title3 'CO(4,4,5)';
run;

proc glm data=extra;
class P S T C;
model Y=P S T C/e solution;
lsmeans T/stderr;
title3 'ANALYSIS OF VARIANCE';
run;
```

Obs	P	S	T	C	Y
1	1	1	1	4	27.5
2	2	1	4	1	20.1
3	3	1	2	4	23.9
4	4	1	3	2	17.2
5	5	1	3	3	18.5
6	1	2	2	4	29.9
7	2	2	1	2	23.6
8	3	2	3	1	20.2
9	4	2	4	3	16.8
10	5	2	4	4	17.1
11	1	3	3	4	14.3
12	2	3	2	3	15.7
13	3	3	4	2	12.3
14	4	3	1	4	17.2
15	5	3	1	1	18.0
16	1	4	4	4	19.3
17	2	4	3	4	12.8
18	3	4	1	3	20.0
19	4	4	2	1	23.4
20	5	4	2	2	21.4

Table 19.3 (*Continued*)

```
                    The GLM Procedure

                 Class Level Information

              Class      Levels      Values

              P               5      1 2 3 4 5
              S               4      1 2 3 4
              T               4      1 2 3 4
              C               4      1 2 3 4

               Number of observations  20

          General Form of Estimable Functions

           Effect                Coefficients

           Intercept             L1

           P              1      L2
           P              2      L3
           P              3      L4
           P              4      L5
           P              5      L1-L2-L3-L4-L5

           S              1      L7
           S              2      L8
           S              3      L9
           S              4      L1-L7-L8-L9

           T              1      L11
           T              2      L12
           T              3      L13
           T              4      L1-L11-L12-L13

           C              1      L15
           C              2      L16
           C              3      L17
           C              4      L1-L15-L16-L17

Dependent Variable: Y
```

Source	DF	Sum of Squares	Mean Square	F Value	Pr > F
Model	13	385.1805000	29.6292692	21.77	0.0006
Error	6	8.1675000	1.3612500		
Corrected Total	19	393.3480000			

Table 19.3 *(Continued)*

R-Square	Coeff Var	Root MSE	Y Mean
0.979236	5.995510	1.166726	19.46000

Source	DF	Type I SS	Mean Square	F Value	Pr > F
P	4	56.4080000	14.1020000	10.36	0.0073
S	3	119.2600000	39.7533333	29.20	0.0006
T	3	190.4300000	63.4766667	46.63	0.0001
C	3	19.0825000	6.3608333	4.67	0.0518

Source	DF	Type III SS	Mean Square	F Value	Pr > F
P	4	52.2875000	13.0718750	9.60	0.0089
S	3	167.4140000	55.8046667	41.00	0.0002
T	3	190.4300000	63.4766667	46.63	0.0001
C	3	19.0825000	6.3608333	4.67	0.0518

Parameter		Estimate		Standard Error	t Value	Pr > |t|
Intercept		14.20416667	B	1.02435193	13.87	<.0001
P	1	4.88750000	B	0.96739825	5.05	0.0023
P	2	-0.70000000	B	0.82500000	-0.85	0.4287
P	3	0.35000000	B	0.82500000	0.42	0.6862
P	4	-0.10000000	B	0.82500000	-0.12	0.9075
P	5	0.00000000	B	.	.	.
S	1	3.45000000	B	0.75311852	4.58	0.0038
S	2	3.42500000	B	0.75311852	4.55	0.0039
S	3	-3.70833333	B	0.75311852	-4.92	0.0026
S	4	0.00000000	B	.	.	.
T	1	5.56666667	B	0.75311852	7.39	0.0003
T	2	6.42500000	B	0.75311852	8.53	0.0001
T	3	-0.52500000	B	0.75311852	-0.70	0.5118
T	4	0.00000000	B	.	.	.
C	1	2.67500000	B	0.82500000	3.24	0.0176
C	2	0.87500000	B	0.82500000	1.06	0.3297
C	3	-0.00000000	B	0.82500000	-0.00	1.0000
C	4	0.00000000	B	.	.	.

NOTE: The X'X matrix has been found to be singular, and a generalized inverse was used to solve the normal equations. Terms whose estimates are followed by the letter 'B' are not uniquely estimable.

Table 19.3 (*Continued*)

Least Squares Means

T	Y LSMEAN	Standard Error	Pr > |t|
1	22.3375000	0.5394137	<.0001
2	23.1958333	0.5394137	<.0001
3	16.2458333	0.5394137	<.0001
4	16.7708333	0.5394137	<.0001

1. The input statement 'If C = 0 then C = 4' is included for the sole purpose of obtaining least squares (LS) means (see Ratkowsky, Evans and Alldredge, 1993) since otherwise LS means are "nonestimable" (see also (5) below);
2. With the statement in (1) there is no longer any confounding between period and residual effects;
3. Treatment and residual effects are now orthogonal to each other [independent of the statement in (1)] as indicated by the equality for the type I and type III treatment SS;
4. The balancedness property is again indicated in the standard errors for treatments and residuals;
5. The numerical value for the treatment LS means does depend on the statement in (1) 'If C = 0 then C = u' for $u = 1, 2, 3, 4$, but differences between LS means are independent of the value of u. □

19.8 COMMENTS ON OTHER MODELS

All our discussions up to this point are based on model (19.1) with all effects, except $\boldsymbol{e}$, assumed to be fixed effects. From many points of view—practical and theoretical—this seems to be the most plausible model, but in certain cases good arguments can be made for other models. We shall conclude this chapter by commenting briefly on some of these models without going into all the details.

19.8.1 No Residual Effects

The presence of residual effects not only affect the analysis of the data from a crossover design but also the interpretation of such an analysis. For example, we have mentioned earlier that the 2-treatment, 2-period, 2-sequence (AB, BA) design cannot be analyzed under model (19.1). Subject matter knowledge may suggest that no residual effects exist. Or appropriate wash-out periods can eliminate residual effects, but they may lengthen considerably the duration of the experiment. In any event, the design problem becomes then one of Latin square-type designs (see I.10). Obviously, we no longer need strongly balanced designs.

We should warn, however, that the use of a balanced design in itself does not automatically imply that residual effects can be dropped from model (19.1), as encountered often in the applied literature.

19.8.2 No Period Effects

To overcome the problem mentioned above for the 2-treatment, 2-period, 2-sequence design, it may be quite reasonable to assume that there are no period effects. That may be true for other situations as well where the number of periods is relatively small. In this case the design reduces to a block design (with the subjects representing the blocks) but with restricted randomization so that the balancedness property can be preserved.

19.8.3 Random Subject Effects

In certain situations there may be good arguments that the subject effects are random rather than fixed effects. This is a question that cannot be answered generally but needs to be decided on a case-by-case basis.

If the subject effects are considered to be random effects, then the analysis differs from that presented in Section 19.7 in that now in addition to intrasubject information also intersubject information with respect to direct and residual effects becomes available (this is, of course, akin to the interblock information for block designs as discussed in Section 1.7). Computationally this can be handled by using, for example, SAS PROC MIXED with the appropriate RANDOM statement.

One other consequence of assuming random subject effects is an increase of the standard error for direct and residual least-squares means.

19.8.4 Autocorrelation Error Structure

It is plausible to argue that since we obtain p observations on the same subject the observations are correlated. In the context of model (19.1) this implies that for a given j the e_{ij} are correlated. One way to look at such a correlation structure is to consider the p observations for a subject a short time series and hence the entire set of observations as n time series. The simplest correlation structure in this context is induced by a first-order autoregression with

$$e_{ij} = \rho e_{i-1,j} + \eta_{ij} (i > 1) \qquad e_{1j} = \eta_{1j} \tag{19.14}$$

and $-1 < \rho < 1$. The η_{ij} are assumed to be independently distributed with mean 0 and

$$\text{var}(\eta_{ij}) = \begin{cases} \sigma^2/(1-\rho^2) & \text{for } i = 1 \\ \sigma^2 & \text{for } i > 1 \end{cases}$$

Then the variance–covariance structure for e is given by

$$\text{var}(\boldsymbol{e}) = \sigma^2(\boldsymbol{I}_n \times \boldsymbol{V})$$

where $\boldsymbol{V} = (V_{ii'})$ and $V_{ii'} = \rho^{|i-i'|}/(1-\rho^2)$ for $i, i' = 1, 2, \ldots, p$.

The autocorrelation structure (19.14) leads to an information matrix that is a function of ρ, a generally unknown parameter [for an explicit expression for the joint information matrix for direct and residual effects see Matthews (1987)]. As a consequence it becomes much more difficult to obtain general results for optimality of designs.

Even for $t = 2$ and $p = 3, 4$ the results are very complicated and not very practical (Matthews, 1987). The problem is compounded by the fact that we usually do not know the correct value of ρ in (19.14). A useful strategy then would be to choose a design that has good properties, that is, is efficient, over a reasonable range of ρ rather than choose the optimal design for a particular value of ρ. Along these lines Matthews (1987) has shown that for $t = 2$ and $p = 3$ an equal number of subjects for the sequences ABB, BAA, AAB, BBA yields a highly efficient design for estimating direct effects and ABB, BAA, ABA, BAB for estimating residual effects. For $t = 2$ and $p = 4$ the corresponding result is that the sequences $ABBA$, $BAAB$, $AABB$, $BBAA$ yield a highly efficient design over the full range of ρ for estimating direct and residual effects. As Matthews (1987) pointed out, an interesting property of the design above for $p = 4$ is that if the experiment had to be stopped after three or two periods the resulting design is still highly efficient.

We have seen earlier that balanced crossover designs play an important role. Because their information matrix $\boldsymbol{C}_\tau$ in (19.11) is completely symmetric it follows, of course, that comparisons between two direct effects are estimated with the same variance. Gill (1992) investigated this property for the case of autocorrelated errors (remember that $\boldsymbol{C}_\tau$ is now a function of ρ). He found that for $p \leq t$ and t a prime or prime power the method proposed by Williams (1949) can be used to construct a balanced design with $t(t-1)$ subjects. Let $L_1, L_2, \ldots, L_{t-1}$ be the $t-1$ mutually orthogonal Latin squares of size t in semistandard order; that is, the elements in the first row of L_1, L_2, $\ldots$, L_{t-1} are in natural order. Then adjoining $L_1, L_2, \ldots, L_{t-1}$ yields a balanced CO(t, $t(t-1)$, t).

For further results concerning Williams designs see Kunert (1985).

19.8.5 Second-Order Residual Effects

It may not be very common, but it is entirely possible that the effect of a treatment lingers on for two periods (and perhaps even more). In such a case we are not only dealing with first-order residual effects as we have discussed so far, but also with second-order residual effects. This changes model (19.1) to

$$y_{ij} = \mu + \pi_i + \beta_j + \tau_{d(i,j)} + \rho_{d(i-1,j)} + \gamma_{d(i-2,j)} + e_{ij} \tag{19.15}$$

where $\gamma_{d(i-2,j)}$ indicates the second-order residual effect in period i for the treatment that the design d assigned to subject j in period $i-2$.

Consideration of second-order residual effects obviously further complicates the construction of efficient crossover designs. Williams (1950) proposed a method based on mutually orthogonal Latin squares with the property that each treatment is preceded by each ordered pair of different treatments (see I.10.7.3). For these designs and under model (19.15) the direct effects are nonorthogonal to their first-order and second-order residual effects.

An orthogonal CO(2, 4, 6) was proposed by Quenouille (1953) (see also John and Quenouille, 1977). His idea was extended for $t \geq 2$ by Sharma (1977). His method produces what he refers to as completely balanced designs.

Definition 19.8 A design d in $\Omega_{t,n,p}$ is called a *completely balanced design* if (1) d is uniform, (2) d is strongly balanced for first-order residual effects, (3) each treatment is preceded equally often by every other treatment (including itself) by two periods, and (4) each ordered pair of treatments occurs equally often in the last two periods. □

It is easy to see that a minimal completely balanced design is a CO(t, t^2, $3t$). Sharma (1977) introduced a method of constructing such designs through the use of orthogonal arrays of the form OA[t^2, 3, t, 2; 1]: Such an orthogonal array can be obtained by adjoining t arrays of the form

$$\begin{bmatrix} 1 & 2 & \cdots & t \\ u & u & \cdots & u \\ u & u+1 & \cdots & u+t-1 \end{bmatrix} \qquad (u = 1, 2, \ldots, t) \qquad (19.16)$$

The CO(t, t^2, $3t$) is obtained from developing the full array (19.16) by adding successively $1, 2, \ldots, t-1$ to its three rows, always mod t.

We illustrate this procedure for the probably most useful cases $t = 2$ and $t = 3$ in the following examples.

Example 19.7 For $t = 2$ the array (19.16) (periods 1, 2, 3) and the final CO(2, 4, 6) are obtained as

	Subject 1	2	3	4
Period 1	1	2	1	2
2	1	1	2	2
3	1	2	2	1
4	2	1	2	1
5	2	2	1	1
6	2	1	1	2

or

Subject 1	2	3	4
A	B	A	B
A	A	B	B
A	B	B	A
B	A	B	A
B	B	A	A
B	A	A	B

□

Example 19.8 For $t = 3$ we obtain the following CO(3, 9, 9), with the first three rows constituting array (19.16):

		Subject								
		1	2	3	4	5	6	7	8	9
	1	1	2	3	1	2	3	1	2	3
	2	1	1	1	2	2	2	3	3	3
	3	1	2	3	2	3	1	3	1	2
	4	2	3	1	2	3	1	2	3	1
Period	5	2	2	2	3	3	3	1	1	1
	6	2	3	1	3	1	2	1	2	3
	7	3	1	2	3	1	2	3	1	2
	8	3	3	3	1	1	1	2	2	2
	9	3	1	2	1	2	3	2	3	1

□

19.8.6 Self and Mixed Carryover Effects

One important characteristic of strongly balanced designs is the fact that each treatment is preceded by itself λ_2 times. This means that some observations are a function of the direct and residual effects of the same treatment, whereas other observations are influenced by direct and residual effects from different treatments. Model (19.1) does not distinguish between these two situations in that ρ_u is the same regardless of whether it occurs together with τ_u or $\tau_{u'} (u \neq u')$. This has sometimes been considered as a weakness of the model. One can argue that there might be some form of interaction between direct and residual effects of treatments. This was, indeed, considered by Sen and Mukerjee (1987), but the ensuing model becomes too complicated and not practically useful.

A different approach to this problem, which can be considered as a special form of direct-residual treatment interaction, was first proposed by Afsarinejad and Hedayat (2002) and further considered by Kunert and Stufken (2002). Following their development, we consider two types of carryover effects: *self carryover effect*, which occurs when treatment u is followed by itself, and *mixed carryover effect*, which occurs if treatment u is followed by treatment $u' (u \neq u')$, independent of what u' is. This distinction then leads to the following model [using the same notation as for model (19.1)]:

$$y_{ij} = \mu + \pi_i + \beta_j + \tau_{d(i,j)} + \begin{Bmatrix} \rho_{d(i-1,j)} & \text{if } d(i,j) \neq d(i-1,j) \\ \chi_{d(i-1,j)} & \text{if } d(i,j) = d(i-1,j) \end{Bmatrix} + e_{ij} \tag{19.17}$$

Obviously, $\rho_{d(i-1,j)}$ refers to the mixed and $\chi_{d(i-1,j)}$ to the self carryover effect.

Based on model (19.15) Afsarinejad and Hedayat (2002) provide useful designs for $t \geq 2$ and $p = 2$, and Kunert and Stufken (2002) discuss the optimality of designs for $t \geq 3$ and $p > 2$. We shall report some of their findings, but refer to the original articles for details.

1. $p = 2$ The CO(t, n, 2) consists of two parts, say D_1 and D_2, where in D_1 n_1 subjects receive the same treatment in both periods and in D_2 n_2 subjects receive different treatments in periods 1 and 2, with $n_1 + n_2 = n$. It is obvious that we can estimate differences between the χ_u from D_1, and hence we would require $n_1 \geq t$. Of critical importance then becomes the construction of D_2.

One method to construct a CO(t, n_2, 2) for estimating direct effects is to use each ordered pairs of treatments for $n_2 = t(t-1)$ subjects. This may lead to a rather large design. An alternative method is derived by Afsarinejad and Hedayat (2002) through the use of BIB designs. It can be described as follows: Consider a symmetric BIB design with t blocks of size k. Write the design in the form of a $k \times t$ Youden square, $\boldsymbol{H}$ say. Add to $\boldsymbol{H}$ a new row to obtain a $(k+1) \times t$ array, $\boldsymbol{H}^*$ say, that maintains the Youden square property. Then assign k subjects to each treatment for the first period in D_2. Consider, for example, treatment $u (u = 1, 2, \ldots, t)$. Search for treatment u in the first row of $\boldsymbol{H}^*$, and then assign successively those treatments to the second period that appear in the corresponding column of $\boldsymbol{H}$. The resulting design D_2 is an optimal design in the subclass of CO(t, n_2, 2).

To illustrate this method we consider the following example.

Example 19.9 For $t = 7$ and $k = 3$ we have the symmetric BIB design

$$\boldsymbol{H}: \quad \begin{matrix} 1 & 2 & 3 & 4 & 5 & 6 & 7 \\ 2 & 3 & 4 & 5 & 6 & 7 & 1 \\ 4 & 5 & 6 & 7 & 1 & 2 & 3 \end{matrix}$$

We obtain

$$\boldsymbol{H}^*: \quad \begin{matrix} 3 & 4 & 5 & 6 & 7 & 1 & 2 \\ 1 & 2 & 3 & 4 & 5 & 6 & 7 \\ 2 & 3 & 4 & 5 & 6 & 7 & 1 \\ 4 & 5 & 6 & 7 & 1 & 2 & 3 \end{matrix}$$

Then D_2 is given by the 21×2 array

$$D_2: \quad \begin{matrix} 3 & 3 & 3 & 4 & 4 & 4 & 5 & 5 & 5 & 6 & 6 & 6 & 7 & 7 & 7 & 1 & 1 & 1 & 2 & 2 & 2 \\ 1 & 2 & 4 & 2 & 3 & 5 & 3 & 4 & 6 & 4 & 5 & 7 & 5 & 6 & 1 & 6 & 7 & 2 & 7 & 1 & 3 \end{matrix}$$

and the entire design uses $n_1 + n_2 = 7 + 21 = 28$ subjects. □

2. $p > 2$ The results of Kunert and Stufken (2002) are closely tied to the notions of balanced block designs and generalized Youden designs (see Kiefer, 1958, 1975a, 1975b).

Definition 19.9 A block design with t treatments and blocks of size k is called a *balanced block design* if (1) each treatment is replicated equally often, (2) each pair of treatments occurs together the same number of times, and (3) $|n_{ij} - k/t| < 1$ for all i, j ($i = 1, 2, \ldots, t$; $j = 1, 2, \ldots, b$), where n_{ij} is the number of times treatment i appears in block j (see Section 1.3.1). □

Definition 19.10 A row–column design is called a *generalized Youden design* if it is balanced block design when considering rows as blocks and columns as blocks. □

Investigating the properties of the information matrix $\boldsymbol{C}_\tau$ under model (19.15) with regard to uniform optimality Kunert and Stufken (2002) introduced the following definition.

Definition 19.11 A design d in $\Omega_{t,n,p}$ is called *totally balanced* if the following conditions hold:

1. d is a generalized Youden design;
2. d is a balanced block design in the carryover effects;
3. d is balanced for carryover effects;
4. The number of subjects where both treatment u and u' appear $[p/t] + 1$ times and treatment u' does not appear in the last period is the same for every pair $u \neq u'$. □

To establish universal optimality we then have the following theorem.

Theorem 19.10 (Kunert and Stufken, 2002) For $t \geq 3$ and $3 \leq p \leq 2t$, if a totally balanced design d^* in $\Omega_{t,n,p}$ exists, then d^* is universally optimal for the estimation of direct effects over $\Omega_{t,n,p}$.

No general methods of constructing such designs exist, but they can sometimes be obtained by modifying existing designs considered in earlier sections. We illustrate this with the following example.

Example 19.10 For $t = 3$, $p = 4$, and $n = 6$ the extra-period design

$$d = \begin{bmatrix} 1 & 2 & 3 & 1 & 2 & 3 \\ 3 & 1 & 2 & 2 & 3 & 1 \\ 2 & 3 & 1 & 3 & 1 & 2 \\ 2 & 3 & 1 & 3 & 1 & 2 \end{bmatrix}$$

is universally optimal under model (19.1). Now d is strongly balanced for residual effects, rather than balanced. Hence d is not totally balanced. Changing the arrangement in the last period leads to the following totally balanced design:

$$d^* = \begin{bmatrix} 1 & 2 & 3 & 1 & 2 & 3 \\ 3 & 1 & 2 & 2 & 3 & 1 \\ 2 & 3 & 1 & 3 & 1 & 2 \\ 1 & 2 & 3 & 1 & 2 & 3 \end{bmatrix}$$

Thus d^* is universally optimal under model (19.15). □

Kunert and Stufken (2002) point out the interesting fact that in optimal designs under model (19.15) generally no self carryover effects occur. This shows that optimality may not always be the ultimate criterion because in some applications it may be useful to obtain estimates of self carryover effects, for example, in connection with a single drug therapy for a chronic disease (Afsarinejad and Hedayat, 2002).

19.8.7 Factorial Treatment Structure

So far we have tacitly assumed that we are dealing with t qualitative treatments. It is, however, entirely possible that we could have quantitative treatments. For example, our treatments may be represented by one or more factors at several levels. This fact may be used to construct special designs that achieve certain statistically desirable properties, such as certain types of orthogonality. Such designs were discussed by for example, Berenblut (1967, 1968) and Mason and Hinkelmann (1971).

Here we shall discuss briefly yet another type of treatment structure, namely a factorial treatment structure. We shall limit ourselves to the case of 2^m factorials, although the approach can be extended easily to other factorials. An obvious way of proceeding is to simply equate the 2^m treatment combinations to the t treatments and use the designs we have discussed earlier. This will lead, however, very quickly to a prohibitively large number of periods. To reduce the number of periods we shall make the assumption that some interactions are negligible and then use appropriate systems of confounding.

In order to use this approach it is useful to use in model (19.1) the reparameterization (7.42) and rewrite the model as

$$y_{ij} = \mu + \pi_i + \beta_j + a\big(\boldsymbol{x}_{d(i,j)}\big) + a_r\big(z_{d(i-1,j)}\big) + e_{ij}$$

where $\boldsymbol{x}_{d(i,j)}$ represents the treatment combination $\boldsymbol{x}$ assigned by d to subject j in period i, and $z_{d(i-1,j)}$ the treatment combination z assigned to subject j in period $i-1$. Further, $a(\boldsymbol{x}_{d(i,j)})$ and $a_r(z_{d(i-1,j)})$ represent the true direct and residual effects in terms of main effect and interaction effect components $E^{\boldsymbol{\alpha}}_{\boldsymbol{\alpha}'\boldsymbol{x}} = (A_1^{\alpha_1} A_2^{\alpha_2} \cdots A_m^{\alpha_m})_{\boldsymbol{\alpha}'\boldsymbol{x}}$ and $E^{\boldsymbol{\alpha}}_{r,\boldsymbol{\alpha}'\boldsymbol{x}} = (A_1^{\alpha_1} A_2^{\alpha_2} \cdots A_m^{\alpha_m})_{r,\boldsymbol{\alpha}'\boldsymbol{x}}$, respectively.

The method to construct suitable designs is to first construct the block design confounding selected interactions and their generalized interactions with blocks. These blocks constitute the generating subjects. Each generating subject is then developed into a Williams square as described in Section 19.5.1. Formally we can describe the result as in the following theorem.

Theorem 19.11 (Shing and Hinkelmann, 1992) For a 2^m factorial with $\ell = 2^{q-1}$ effects $E^{\beta_1}, E^{\beta_2}, \ldots, E^{\beta_\ell}$ negligible (this set is made up of q independent interactions and their generalized interactions), where $m \geq q + 1$, there exists a CO(2^m, 2^m, 2^{m-q}) such that all nonnegligible direct and residual effects and interactions are estimable.

To construct such a design we assign 2^{m-q} treatment combinations to each of 2^q subjects in arbitrary order by confounding $E^{\beta_1}, E^{\beta_2}, \ldots, E^{\beta_\ell}$ with subjects. Each of the 2^q initial subjects is then developed into a Williams square. We adjoin the 2^q Williams squares to obtain the final CO(2^m, 2^m, 2^{m-q}).

We illustrate this procedure in the following example.

Example 19.11 Suppose we have a 2^3 factorial with factors A, B, C. Suppose further that we want to use $p = 4 = 2^2$ periods, that is, we have $m = 3$, $q = 1$. Thus we need to confound one interaction with subjects, say ABC. The final design is then given in Table 19.4, with subjects 1 and 5 being the initial subjects, and using the fact that the Williams square, that is, CO(4, 4, 4) is of the form

$$\begin{matrix} 1 & 2 & 3 & 4 \\ 4 & 1 & 2 & 3 \\ 2 & 3 & 4 & 1 \\ 3 & 4 & 1 & 2 \end{matrix}$$

implying that for subject 1 we have the association $(0, 0, 0) \equiv 1$, $(1, 0, 1) \equiv 2$, $(0, 1, 1) \equiv 3$, $(1, 1, 0) \equiv 4$, with a similar association for subject 5.

From the design in Table 19.4 we can estimate A, B, AB, C, AC, BC and A_r, B_r, $(AB)_r$, C_r, $(AC)_r$, $(BC)_r$. We point out, however, that the effects like

Table 19.4 CO(2^3, 8, 4) with ABC Confounded

	Subjects							
Periods	1	2	3	4	5	6	7	8
1	(0,0,0)	(1,0,1)	(0,1,1)	(1,1,0)	(1,1,1)	(0,1,0)	(1,0,0)	(0,0,1)
2	(1,1,0)	(0,0,0)	(1,0,1)	(0,1,1)	(0,0,1)	(1,1,1)	(0,1,0)	(1,0,0)
3	(1,0,1)	(0,1,1)	(1,1,0)	(0,0,0)	(0,1,0)	(1,0,0)	(0,0,1)	(1,1,1)
4	(0,1,1)	(1,1,0)	(0,0,0)	(1,0,1)	(1,0,0)	(0,0,1)	(1,1,1)	(0,1,0)

A_r have no particular meaning, except that they represent linear combinations of residual effects. □

Obviously, for large values of m this method may lead to rather large designs. In that case we may need to use fractional factorials together with a system of confounding. More specifically, if we use a 2^{m-s} fraction and 2^{m-s-q} periods, then we can construct a CO(2^{m-s}, 2^{m-s}, 2^{m-s-q}) by the method described above. However, this must be considered carefully because this may not lead to practically useful designs for all values of m, s, and q.

For another approach to constructing crossover designs with a factorial treatment structure we refer to Fletcher, Lewis, and Matthews (1990). They consider in particular the case of two factors with up to four levels in three or four periods. Their designs are obtained by cyclic generation from one or more suitably chosen initial sequences. A table of such sequences is provided by Fletcher, Lewis, and Matthews (1990).

APPENDIX A

Fields and Galois Fields

The notion of a finite field, in particular a Galois field, plays an important role in the construction of experimental designs (see, e.g., Chapters 3 and 13). We shall give here the elementary notions of fields and Galois fields. For a more thorough discussion of this topic the reader is referred to, for example, Carmichael (1956).

Consider a set of s distinct elements (marks) $\mathfrak{U} = \{u_0, u_1, \ldots, u_{s-1}\}$.

Definition A.1 The set $\mathfrak{U}$ is said to be a *finite field of order s* if its elements satisfy the following conditions:

1. The elements may be combined by addition with the laws

$$u_i + u_j = u_j + u_i$$
$$u_i + (u_j + u_k) = (u_i + u_j) + u_k$$

2. The sum of any two elements is an element in $\mathfrak{U}$.
3. Given two elements u_i and u_k, there is a unique element u_j such that $u_i + u_j = u_k$. Then u_j is said to be determined by subtraction, and we write $u_j = u_k - u_i$.
4. The element having the additive property of zero is u_0 so that

$$u_0 + u_j = u_j \quad \text{or} \quad u_0 = u_j - u_j$$

for all j.

5. The elements may be combined by multiplication with the laws

$$u_i u_j = u_j u_i$$
$$u_i(u_j u_k) = (u_i u_j)u_k$$
$$u_i(u_j + u_k) = u_i u_j + u_i u_k$$

Design and Analysis of Experiments. Volume 2: Advanced Experimental Design
By Klaus Hinkelmann and Oscar Kempthorne
ISBN 0-471-55177-5

Table A.1 Primitive Marks of Finite Fields of Order p

Order of Field (p)	Primitive Mark (u)
3	2
5	2
7	3
11	2
13	2
17	3
19	2
23	5

6. The product of any two elements is an element in $\mathfrak{U}$.

7. From the relations

$$u_0 u_i = u_i u_0 = u_i(u_j - u_j) = u_i u_j - u_i u_j = u_0$$

it follows that u_0 has the multiplicative properties of zero.

8. Given any $u_i (\neq u_0)$ and any u_k, there is a unique u_j such that

$$u_i u_j = u_k$$

Then u_j is said to be determined by division, and u_j is called the quotient of u_k by u_i.

9. The quotient u_i/u_i has the multiplicative properties of unity and is chosen to be u_i. □

The finite field of $s = p$ elements, where p is a prime number, may be represented by $u_0 = 0, u_1 = 1, u_2 = 2, \ldots, u_{p-1} = p - 1$. Addition and multiplication are the ordinary arithmetic operations, except that the resulting number is reduced mod p, that is, the resulting number is replaced by the remainder after dividing by p.

Every element $u \in \mathfrak{U}(u \neq u_0)$ satisfies the equation $u^{p-1} = 1$. If for a given u, $p - 1$ is the smallest power satisfying this equation, then u is called a *primitive mark* of the field. Then the marks $u, u^2, u^3, \ldots, u^{p-1}$ are all distinct and hence represent, in some order, the elements of $\mathfrak{U}$. Examples of primitive marks (roots) for different values of p are given in Table A.1. As an illustration consider the following example.

Example A.1 Consider $p = 7,\ u = 3,\ u^1 = 3,\ u^2 = 3^2 = 2,\ u^3 = 2 \cdot 3 = 6,\ u^4 = 6 \cdot 3 = 4,\ u^5 = 4 \cdot 3 = 5,\ u^6 = 5 \cdot 3 = 1.$ □

The field of order p described above is a special case of a finite field of order $s = p^n$, where p is a prime number. Such a field is referred to as a *Galois field* and is generally denoted by GF(p^n). For $n = 1$ the elements are the residues modulo p as described above. For $n > 1$ the elements are obtained as follows.

Let $P(x)$ be a given polynomial in x of degree n with integral coefficients and let $F(x)$ be any polynomial in x with integral coefficients. Then $F(x)$ may be expressed as

$$F(x) = f(x) + p \cdot q(x) + P(x)\,Q(x)$$

where

$$f(x) = b_0 + b_1 x + b_2 x^2 + \cdots + b_{n-1}\,x^{n-1}$$

and coefficients b_i ($i = 0, 1, \ldots, n-1$) belonging to GF(p), that is, the set $\mathfrak{U} = (0, 1, 2, \ldots, p-1)$. This relationship may be written as

$$F(x) = f(x) \quad \text{mod } p,\, P(x) \tag{A.1}$$

and we say that $f(x)$ is the residue modulis p and $P(x)$. The functions $F(x)$, which satisfy (A.1) when $f(x)$, p, and $P(x)$ are kept fixed, form a class. If p and $P(x)$ are kept fixed, but $f(x)$ is varied, p^n classes may be formed since each coefficient in $f(x)$ may take the p values $0, 1, 2, \ldots, p-1$.

It may be readily verified that the classes defined by the $f(x)$'s make up a field. For example, if $F_i(x)$ belongs to the class corresponding to $f_i(x)$, and $F_j(x)$ corresponding to $f_j(x)$, then $F_i(x) + F_j(x)$ belongs to the class corresponding to $f_i(x) + f_j(x)$. The other operations defined for a field are, likewise, satisfied, any function obtained by ordinary algebraic operations being replaced by its residue modulis p and $P(x)$. In order that division be unique, it is also necessary that p be a prime number and that $P(x)$ cannot be expressed in the form

$$P(x) = P_1(x)\,P_2(x) + p\,P_3(x) \tag{A.2}$$

where $P_i(x)$ are polynomials in x with integral coefficients, and the degrees of $P_1(x)$ and $P_2(x)$ being positive and less than the degree of $P(x)$. Any $P(x)$ that does not admit the representation is thus *irreducible mod p*.

Summarizing the preceding discussion, we have the following definition.

Definition A.2 For p a prime number and $P(x)$ an irreducible polynomial mod p, the residue classes given by (A.1) form a finite field of order p^n. This field is called a *Galois field*, and is denoted by GF(p^n). □

As an illustration we consider the following example.

Example A.2 Let $p = 3$ and consider GF(3^2). Then $P(x)$ is of degree 2, that is, of the form $a_0 + a_1 x + a_2 x^2$ with $a_0, a_1, a_2 \in \{0, 1, 2\}$. For $P(x)$ to be irreducible mod 3 we can exclude $a_0 = 0$. Suppose we take $a_0 = 1$. Then $P(x)$ is of the form $1 + a_1x + a_2x^2$. If $a_1 = 0, a_2 = 2$ we have $P(x) = 1 + 2x^2$, which is 0 for $x = 1$ and can be written as $(1 + x)(1 + 2x) - 3x$, that is, it is of the form (A.2), and hence is not irreducible mod 3. We then consider $a_1 = 0, a_2 = 1$, that is, $P(x) = 1 + x^2$. For $x = 0, 1, 2$, $P(x)$ takes the values 1, 2, 2 mod 3, respectively, and hence is irreducible mod 3. For $P(x) = 1 + x^2$ we then consider the possible $f(x)$'s, which are of the form $b_0 + b_1x$. Hence, the elements of the GF(3^2) are: $u_0 = 0, u_1 = 1, u_1 = 2, u_3 = x, u_4 = 2x,\ u_5 = 1 + x, u_6 = 2 + x, u_7 = 1 + 2x, u_8 = 2 + 2x$. Just as for the GF($p$), there exists as primitive mark u such that the powers u^j $(j = 1, 2, \ldots, p^n - 1)$ represent the nonzero elements of GF(p^n). In the above case, $u = 1 + x$ is such a primitive mark, since

$$u^1 = 1 + x = u_5$$

$$u^2 = 1 + 2x + x^2 = 2x = u_4$$

$$u^3 = 2x(1 + x) = 2x + 2x^2 = 2x + 1 + 2(1 + x^2) = 1 + 2x = u_7$$

$$u^4 = (1 + 2x)(1 + x) = 1 + 3x + 2x^2 = 2 + 2(1 + x^2) = 2 = u_2$$

$$u^5 = 2(1 + x) = 2 + 2x = u_8$$

$$u^6 = (2 + 2x)(1 + x) = 2 + 4x + 2x^2 = 4x = x = u_3$$

$$u^7 = x(1 + x) = x + x^2 = 2 + x + (1 + x^2) = 2 + x = u_6$$

$$u^8 = (2 + x)(1 + x) = 2 + 3x + x^2 = 1 = u_1$$ □

The irreducible polynomials $P(x)$ are not unique. It can be shown, however, that there always exists a particular irreducible $P(x)$ such that $u = x$ is a primitive element. Such a $P(x)$ is called a *primitive polynomial.* A partial list of primitive polynomials is given in Table A.2 [for a more extensive list see Hedayat, Sloane, and Stufken (1999)].

To perform arithmetic operations in a GF it is convenient to first obtain addition and multiplication tables in terms of the powers of the primitive element, x, using the primitive polynomial (see Table A.2). We illustrate this with the following example.

Example A.3 Consider the GF(2^2). The elements are $u_0 = 0, u_1 = 1, u_2 = x, u_3 = 1 + x = x^2$, with $P(x) = 1 + x + x^2$. We then have the following addition table:

	u_0	u_1	u_2	u_3
u_0	u_0	u_1	u_2	u_3
u_1	u_1	u_0	u_3	u_2
u_2	u_2	u_3	u_0	u_1
u_3	u_3	u_2	u_1	u_0

Table A.2 List of Primitive Polynomials for Selected Values of p and n

GF	$P(x)$
2^2	$1 + x + x^2$
2^3	$1 + x + x^3$
2^4	$1 + x + x^4$
2^5	$1 + x^2 + x^5$
3^2	$2 + x + x^2$
3^3	$1 + 2x + x^3$
5^2	$2 + x + x^2$

or

	0	x^0	x^1	x^2
0	0	x^0	x^1	x^2
x^0	x^0	0	x^2	x^1
x^1	x^1	x^2	0	x^0
x^2	x^2	x^1	x^0	0

For example,

$$x^1 + x^2 = x(1 + x) = x^3 = 1 = x^0$$

Similarly, we obtain the following multiplication table:

	u_0	u_1	u_2	u_3
u_0	u_0	u_0	u_0	u_0
u_1	u_0	u_1	u_2	u_3
u_2	u_0	u_2	u_3	u_1
u_3	u_0	u_3	u_1	u_2

or

	0	x^0	x^1	x^2
0	0	0	0	0
x^0	0	x^0	x^1	x^2
x^1	0	x^1	x^2	x^0
x^2	0	x^2	x^0	x^1

For example,

$$x^2 \cdot x^2 = x^4 = x^3 \cdot x = x = x^1$$

With these tables we can then carry out all arithmetic operations in GF(2^2). □

APPENDIX B

Finite Geometries

A finite geometry consists of a finite set of elements, called points, that are subject to certain conditions about points and lines. Such a geometry can be represented in terms of the marks of a Galois field, GF(p^n), where p is a prime. Of primary importance in the context of constructing experimental designs are K-dimensional projective and Euclidean geometries, denoted by PG(K, p^n) and EG(K, p^n), respectively.

PROJECTIVE GEOMETRY PG(K, P^N)

A point of a K-dimensional projective geometry is defined by the ordered set of homogeneous coordinates $(\mu_0, \mu_1, \ldots, \mu_K)$ where $\mu_i \in$ GF(p^n) and at least one $\mu_i \neq u_0 = 0$. The symbol $(\mu\mu_0, \mu\mu_1, \ldots, \mu\mu_K)$ denotes the same point, where $\mu \in$ GF(p^n) and $\mu \neq u_0$. The total number of points in the PG(K, p^n) then is

$$\frac{p^{n(K+1)} - 1}{p^n - 1} = 1 + p^n + p^{2n} + \cdots + p^{Kn} \tag{B.1}$$

since there exist $p^{n(K+1)} - 1$ ordered sets $\neq (0, 0, \ldots, 0)$, and these may be arranged in groups of $p^n - 1$ sets all of whose elements represent the same point.

Now consider two distinct points $(\mu_0, \ldots, \mu_K)$ and $(\nu_0, \ldots, \nu_K)$. Then the line between these points is defined by the points

$$(\mu\mu_0 + \nu\nu_0, \mu\mu_1 + \nu\nu_1, \ldots, \mu\mu_K + \nu\nu_K) \tag{B.2}$$

Design and Analysis of Experiments. Volume 2: Advanced Experimental Design
By Klaus Hinkelmann and Oscar Kempthorne
ISBN 0-471-55177-5

where $\mu, \nu \in \mathrm{GF}(p^n)$, not both 0. The number of possible combinations of μ and ν is $p^{2n} - 1$, so that because of proportionality the line (B.2) contains

$$\frac{p^{2n} - 1}{p^{n-1}} = p^n + 1 \tag{B.3}$$

points.

The system of points and lines given by (B.1) and (B.3) is called a *K-dimensional projective geometry* $\mathrm{PG}(K, p^n)$.

In (B.2) we have defined what we might call a 1-space in the $\mathrm{PG}(K, p^n)$. More generally, we can define an M-space ($M < K$) in $\mathrm{PG}(K, p^n)$ as follows: If the points $(\mu_{i0}, \mu_{i1}, \ldots, \mu_{iK})$ $(i = 0, 1, \ldots, M)$ are any $M + 1$ points not in the same $(M - 1)$-space, then an M-space is defined by the points

$$\left(\sum_{i=0}^{M} \mu_i \mu_{i0}, \sum \mu_i \mu_{i1}, \ldots, \sum \mu_i \mu_{iK} \right) \tag{B.4}$$

where $\mu_0, \mu_1, \ldots, \mu_M \in \mathrm{GF}(p^n)$ and not all equal to 0. Alternatively, (B.4) can be described in terms of the solutions to $K - M$ independent, linear, homogeneous equations

$$a_{i0}\mu_0 + a_{i1}\mu_1 + \cdots + a_{iK}\mu_K = 0 \tag{B.5}$$

$(i = 1, 2, \ldots, K - M)$ with a_{ij} fixed.

In order to find the number of M-spaces in the K-space ($M < K$), we need to find first the number of ways in which the $M + 1$ noncollinear base points can be chosen, such that they do not all lie in any $(M - 1)$-space:

$$\begin{array}{ll} \text{1st point:} & 1 + p^n + \cdots + p^{Kn} \equiv N_{K1} \\ \text{2nd point:} & p^n + \cdots + p^{Kn} \equiv N_{K2} \\ \text{3rd point:} & p^{2n} + \cdots + p^{Kn} \equiv N_{K3} \\ \vdots & \\ (M+1)\text{th point:} & p^{Mn} + \cdots + p^{Kn} \equiv N_{K,M+1} \end{array}$$

and then the ways in which $M + 1$ points of a given $\mathrm{PG}(M, p^n)$ may be selected in a given order so that they do not all lie in any $(M - 1)$-space:

$$\begin{array}{ll} \text{1st point} & 1 + p^n + \cdots + p^{Mn} \equiv N_{M1} \\ \text{2nd point} & p^n + \cdots + p^{Mn} \equiv N_{M2} \\ \vdots & \\ (M+1)\text{th point} & p^{Mn} \equiv N_{M,M+1} \end{array}$$

Hence every PG(K, p^n) contains

$$\psi(M, K, p^n) = \frac{\prod_{i=1}^{M+1} N_{Ki}}{\prod_{i=1}^{M+1} N_{Mi}} \tag{B.6}$$

M-spaces, with N_{Ki} and N_{Mi} as defined above.

Similarly, we can show that every PG(S, p^n) in PG(K, p^n) is contained in

$$\varphi(S, M, K, p^n) = \frac{\left(p^{(S+1)n} + \cdots + p^{Kn}\right) \times \cdots \times \left(p^{Mn} + \cdots + p^{Kn}\right)}{\left(p^{(S+1)n} + \cdots + p^{Mn}\right) \times \cdots \times \left(p^{(M-1)n} + p^{Mn}\right) p^{Mn}} \tag{B.7}$$

PG(M, p^n)'s for $S < M$. In particular, for $S = 0$, every point is contained in $\varphi(0, M, K, p^n)$ PG(M, p^n) in PG(K, p^n), and, for $S = 1$, every line is contained in $\varphi(1, M, K, p^n)$ PG(M, p^n) in PG(K, p^n), or, since with every pair of points also the line containing them is in the PG(M, p^n), it follows that every pair of points is contained in $\varphi(1, M, K, p^n)$ different PG(M, p^n).

EUCLIDEAN GEOMETRY EG(K, P^N)

Consider a PG(K, p^n) and the subset of points (μ_0, μ_1, ..., μ_K) for which $\mu_0 \neq 0$. Take $\mu_0 = 1$, then the points (1, μ_1, ..., μ_K) constitute an EG(K, p^n) consisting of p^{Kn} points. The excluded points (0, μ_1, ..., μ_K) form a PG($K - 1$, p^n) with homogeneous coordinates (μ_1, μ_2, ..., μ_K). Hence, the EG(K, p^n) is obtained by deleting from the PG(K, p^n) a PG($K - 1$, p^n).

Every PG(M, p^n) contained in PG(K, p^n) but not in PG($K - 1$, p^n) becomes an EG(M, p^n) since by deleting PG($K - 1$, p^n) from PG(K, p^n) we also delete PG($M - 1$, p^n) from PG(M, p^n). The number of EG(M, p^n) contained in EG(K, p^n) then is $\psi(M, K, p^n) - \psi(M, K - 1, p^n)$, with ψ defined in (B.6). The number of EG(M, p^n) containing a given EG(S, p^n) with $S < M$ is $\varphi(S, M, K, p^n)$ as defined in (B.7).

APPENDIX C

Orthogonal and Balanced Arrays

ORTHOGONAL ARRAYS

Definition C.1 A $K \times N$ matrix $\boldsymbol{A}$, with entries from a set $\sum$ of $s \geq 2$ elements, is called an *orthogonal array* of strength t, size N, K constraints and s levels if each $t \times N$ submatrix of $\boldsymbol{A}$ contains all possible $t \times 1$ column vectors with the same frequency λ, denoted by OA$[N, K, s, t; \lambda]$, where λ is called the index of $\boldsymbol{A}$ and $N = \lambda s^t$. □

For the case $\sum = \{0, 1, \ldots, s-1\}$ consider the following example.

Example C.1 OA[9, 4, 3, 2; 1]:

$$\boldsymbol{A} = \begin{pmatrix} 0 & 0 & 0 & 1 & 1 & 1 & 2 & 2 & 2 \\ 0 & 1 & 2 & 0 & 1 & 2 & 0 & 1 & 2 \\ 0 & 1 & 2 & 1 & 2 & 0 & 2 & 0 & 1 \\ 0 & 1 & 2 & 2 & 0 & 1 & 1 & 2 & 0 \end{pmatrix}$$ □

There are different methods of constructing orthogonal arrays [for a thorough discussion see Hedayat, Sloane, and Stufken (1999)]. We shall present here a particular method based on projective geometries.

Theorem C.1 If we can find a matrix $\boldsymbol{C} = (c_{ij})$ of K rows and τ columns, whose elements c_{ij} belong to a GF(p^n), and for which every submatrix obtained by taking t rows is of rank t, then we can construct an OA$[p^{n\tau}, K, p^n, t; \lambda]$.

Proof Consider $\tau \times 1$ column vectors $\boldsymbol{\xi}$ whose coordinates belong to GF(p^n). Then there are $p^{\tau n}$ different $\boldsymbol{\xi}_i$. Consider then the matrix $\boldsymbol{A}$ whose $p^{\tau n}$ columns are the vectors $\boldsymbol{C}\,\boldsymbol{\xi}_i$ $(i = 1, 2, \ldots, p^{n\tau})$. $\boldsymbol{A}$ is then an orthogonal array:

Design and Analysis of Experiments. Volume 2: Advanced Experimental Design
By Klaus Hinkelmann and Oscar Kempthorne
ISBN 0-471-55177-5

Suppose A^* is a $t \times p^{\tau n}$ submatrix of A, and C^* is the corresponding $t \times \tau$ submatrix of C. Consider

$$A^* = C^*(\boldsymbol{\xi}_1, \boldsymbol{\xi}_2, \ldots, \boldsymbol{\xi}_s)$$

with $s = p^{\tau n}$, and denote the ith column of A^* by $\boldsymbol{\alpha}_i (i = 1, 2, \ldots, s)$, that is,

$$\boldsymbol{\alpha}_i = C^* \boldsymbol{\xi}_i$$

or

$$\begin{aligned} c_{11}\xi_{i1} + c_{12}\xi_{i2} + \cdots + c_{1\tau}\xi_{i\tau} &= \alpha_{i1} \\ c_{21}\xi_{i1} + c_{22}\xi_{i2} + \cdots + c_{2\tau}\xi_{i\tau} &= \alpha_{i2} \\ &\vdots \\ c_{t1}\xi_{i1} + c_{t2}\xi_{i2} + \cdots + c_{t\tau}\xi_{i\tau} &= \alpha_{it} \end{aligned}$$

Assume without loss of generality that the first t columns of C^* are linearly independent. Then consider all the $\boldsymbol{\xi}_i$ that have the same last $\tau - t$ components. There are p^{tn} different such vectors. Each will give rise to a different $\boldsymbol{\alpha}_i$. But the total number of different $\boldsymbol{\alpha}_i$ is p^{tn}, that is, each possible $\boldsymbol{\alpha}_i$ is obtained exactly once from such a set. Now there are $p^{n(\tau-t)}$ possible sets of vectors for which this holds. Hence each different $\boldsymbol{\alpha}_i$ is obtained exactly $p^{n(\tau-t)}$ times. This implies that in A^* each possible $t \times 1$ column vector occurs $\lambda = p^{n(\tau-t)}$ times and shows that A is an orthogonal array of strength t and index λ. □

The rows of C in Theorem C.1 may be interpreted as the coordinates of a PG($\tau - 1$, p^n) such that no t points lie in a space of dimension $\leq t - 2$. To find a matrix C then means to find K points of a PG($\tau - 1$, p^n) with the above restriction. We illustrate this with the following example.

Example C.2 Consider the case $p^n = 2$. In the PG($\tau - 1$, 2) consider all points that are not in the ($\tau - 2$)-space:

$$x_0 + x_1 + \cdots + x_{\tau-1} = 0$$

There are exactly $2^{\tau-1}$ such points, namely those with an odd number of unity components. No three of these points are on the same line, since in the PG($\tau - 1$, 2) each line has exactly three points, one of which is excluded. Hence we can construct an OA[2^τ, $2^{\tau-1}$, 2, 3; λ].

Consider, for example, the case $\tau = 3$:

$$C = \begin{pmatrix} 1 & 0 & 0 \\ 0 & 1 & 0 \\ 0 & 0 & 1 \\ 1 & 1 & 1 \end{pmatrix}$$

$$(\xi_1, \xi_2, \ldots, \xi_8) = \begin{pmatrix} 0 & 1 & 0 & 1 & 0 & 1 & 0 & 1 \\ 0 & 0 & 1 & 1 & 0 & 0 & 1 & 1 \\ 0 & 0 & 0 & 0 & 1 & 1 & 1 & 1 \end{pmatrix}$$

and hence

$$\boldsymbol{A} = \begin{pmatrix} 0 & 1 & 0 & 1 & 0 & 1 & 0 & 1 \\ 0 & 0 & 1 & 1 & 0 & 0 & 1 & 1 \\ 0 & 0 & 0 & 0 & 1 & 1 & 1 & 1 \\ 0 & 1 & 1 & 0 & 1 & 0 & 0 & 1 \end{pmatrix}$$

with $\lambda = 1$. It is easy to see that $\boldsymbol{A}$ is also an OA$[2^\tau, 2^{\tau-1}, 2, 2; 2]$. □

BALANCED ARRAYS

Balanced arrays, or B-arrays, were introduced by Chakravarti (1956) under the name partially balanced arrays. The following definition is adapted from Srivastava and Chopra (1973).

Definition C.2 A $K \times N$ matrix $\boldsymbol{B}$, with entries from a set $\sum$ of $s \geq 2$ elements, is said to be a *balanced array* of strength t, size N, K constraints, and s levels if for each $t \times N$ submatrix $\boldsymbol{B}_0$ of $\boldsymbol{B}$ and for each $t \times 1$ vector $\boldsymbol{x} = (x_1, x_2, \ldots, x_t)'$ with elements from $\sum$, we have

$$\lambda(\boldsymbol{x}, \boldsymbol{B}_0) = \lambda(P(\boldsymbol{x}), \boldsymbol{B}_0) \tag{C.1}$$

where $\lambda(\boldsymbol{x}, \boldsymbol{B}_0)$ is the number of columns of $\boldsymbol{B}_0$ that are identical with $\boldsymbol{x}$, and $P(\boldsymbol{x})$ is any vector obtained by permuting the elements of $\boldsymbol{x}$.

We denote such an array by BA$[N, K, s, t; \boldsymbol{\lambda}']$, where $\boldsymbol{\lambda}'$ is the vector of the totality of $\lambda(\boldsymbol{x}, \boldsymbol{B}_0)$ as defined by Definition C.1. The term $\boldsymbol{\lambda}'$ is referred to as the index set of the B-array. Note that if all elements of $\boldsymbol{\lambda}'$ are identical, say equal to λ, then the B-array becomes an orthogonal array with index λ. □

For the case $\sum = \{0, 1, 2, \ldots, s-1\}$ we illustrate the notion of a B-array with the following example.

Example C.3 Consider BA[15, 5, 3, 2; 1, 2]

$$\boldsymbol{B} = \begin{pmatrix} 0 & 0 & 0 & 0 & 0 & 1 & 1 & 1 & 1 & 1 & 2 & 2 & 2 & 2 & 2 \\ 0 & 1 & 2 & 1 & 2 & 1 & 2 & 0 & 2 & 0 & 2 & 0 & 1 & 0 & 1 \\ 1 & 0 & 2 & 2 & 1 & 2 & 1 & 0 & 0 & 2 & 0 & 2 & 1 & 1 & 0 \\ 2 & 2 & 0 & 1 & 1 & 0 & 0 & 1 & 2 & 2 & 1 & 1 & 2 & 0 & 0 \\ 1 & 2 & 1 & 0 & 2 & 2 & 0 & 2 & 1 & 0 & 0 & 1 & 0 & 2 & 1 \end{pmatrix}$$

where

$$\lambda(x_1, x_2) = \begin{cases} \lambda_1 = 1 & \text{for } x_1 = x_2 \\ \lambda_2 = 2 & \text{for } x_1 \neq x_2 \end{cases}$$ □

Condition (C.1) can be expressed operationally as follows. Let in $\boldsymbol{x}$ the symbol i occur r_i times $(0 \leq i \leq s-1)$. Then

$$\sum_{i=0}^{s-1} r_i = t$$

and each of the $t!/r_0!r_1! \dots r_{s-1}!$ vectors obtained by permuting $\boldsymbol{x}$ has the same λ value.

For the special case $s = 2$, the index set $\boldsymbol{\lambda}'$ consists of $t+1$ components, λ_0, λ_1, ..., λ_t, corresponding to the number of nonzero components in $\boldsymbol{x}$, which is also referred to as the weight of $\boldsymbol{x}$.

There exist several methods of constructing B-arrays [see, e.g., Chakravarti (1956, 1961), Srivastava and Chopra (1973), Aggarwal and Singh (1981)]. We shall mention here one method that is based on the existence of orthogonal arrays (Chakravarti, 1961).

Suppose, an orthogonal array OA$[N, K, s, t; \lambda]$ can be divided into two disjoint arrays such that one array is a B-array or a degenerate B-array [a degenerate B-array being one that has some but not all $\lambda(x_1, x_2, \dots, x_t)$ equal to zero], with $\lambda(x_1, x_2, \dots, x_t) < \lambda$ for all $t \times 1$ vectors $\boldsymbol{x}$. Then the remaining array is a B-array with λ parameters given by $\lambda^*(x_1, x_2, \dots, x_t) = \lambda - \lambda(x_1, x_2, \dots, x_t)$.

As an illustration consider the following example.

Example C.4 Consider OA[9, 3, 3, 2; 1]

$$\boldsymbol{A} = \begin{pmatrix} 0 & 1 & 2 & 0 & 1 & 2 & 0 & 1 & 2 \\ 0 & 1 & 2 & 2 & 0 & 1 & 1 & 2 & 0 \\ 0 & 1 & 2 & 1 & 2 & 0 & 2 & 0 & 1 \end{pmatrix}$$

Divide $\boldsymbol{A}$ into

$$\boldsymbol{A}_1 = \begin{pmatrix} 0 & 1 & 2 \\ 0 & 1 & 2 \\ 0 & 1 & 2 \end{pmatrix} \quad \text{and} \quad \boldsymbol{A}_2 = \begin{pmatrix} 0 & 1 & 2 & 0 & 1 & 2 \\ 2 & 0 & 1 & 1 & 2 & 0 \\ 1 & 2 & 0 & 2 & 0 & 1 \end{pmatrix}$$

where $\boldsymbol{A}_1$ is a degenerate B-array and $\boldsymbol{A}_2$ then represents a BA[6, 3, 3, 2] with $\lambda_1 = 0$, $\lambda_2 = 1$, where λ_1, λ_2 are as defined in Example C.3. □

In other situations we may be able to obtain a B-array by deleting not only columns but also rows from an orthogonal array (see Chakravarti, 1961).

APPENDIX D

Selected Asymmetrical Balanced Factorial Designs

1. $4 \times 2 \times 2, k = 4$ (Li, 1944)

Replication 1 Block				Replication 2 Block				Replication 3 Block			
1	2	3	4	1	2	3	4	1	2	3	4
000	001	010	011	000	001	010	011	000	001	010	011
101	100	111	110	111	110	101	100	110	111	100	101
210	211	200	201	201	200	211	210	211	210	201	200
311	310	301	300	310	311	300	301	301	300	311	310

$$\lambda(000) = 3 \qquad h(000) = 2$$
$$\lambda(100) = 0 \qquad h(100) = \tfrac{1}{4}$$
$$\lambda(010) = 0 \qquad h(010) = 0$$
$$\lambda(110) = 1 \qquad h(110) = 0$$
$$\lambda(001) = 0 \qquad h(001) = 0$$
$$\lambda(101) = 1 \qquad h(101) = 0$$
$$\lambda(011) = 0 \qquad h(011) = \tfrac{1}{4}$$
$$\lambda(111) = 1 \qquad h(111) = -\tfrac{1}{4}$$

$$E(110) = E(101) = E(111) = \tfrac{2}{3}, \text{ all other } E(z) = 1, E = \tfrac{4}{5}.$$

Design and Analysis of Experiments. Volume 2: Advanced Experimental Design
By Klaus Hinkelmann and Oscar Kempthorne
ISBN 0-471-55177-5

2. $4 \times 2 \times 2, k = 8$ (Li, 1944)

Replication 1 Block		Replication 2 Block		Replication 3 Block	
1	2	1	2	1	2
000	001	000	001	000	001
011	010	011	010	011	010
100	101	101	100	101	100
111	110	110	111	110	111
201	200	201	200	200	201
210	211	210	211	211	210
301	300	300	301	301	300
310	311	311	310	310	311

$$\begin{aligned}
\lambda(000) &= 3 & h(000) &= 2 \\
\lambda(100) &= 1 & h(100) &= \tfrac{1}{4} \\
\lambda(010) &= 0 & h(010) &= \tfrac{1}{2} \\
\lambda(110) &= 2 & h(110) &= -\tfrac{1}{8} \\
\lambda(001) &= 0 & h(001) &= \tfrac{1}{2} \\
\lambda(101) &= 2 & h(101) &= -\tfrac{1}{8} \\
\lambda(011) &= 3 & h(011) &= -\tfrac{1}{4} \\
\lambda(111) &= 1 & h(111) &= \tfrac{1}{8}
\end{aligned}$$

$$E(111) = \tfrac{2}{3}, \text{ all other } E(z) = 1, E = \tfrac{14}{15}.$$

3. $4 \times 4, k = 4$ (Yates, 1937b)

Replication 1 Block				Replication 2 Block				Replication 3 Block			
1	2	3	4	1	2	3	4	1	2	3	4
00	01	02	03	00	01	02	03	00	01	02	03
11	10	13	12	12	13	10	11	13	12	11	10
22	23	20	21	23	22	21	20	21	20	23	22
33	32	31	30	31	30	33	32	32	33	30	31

$$\begin{aligned} \lambda(00) &= 3 & h(00) &= 2 \\ \lambda(10) &= 0 & h(10) &= \tfrac{1}{4} \\ \lambda(01) &= 0 & h(01) &= \tfrac{1}{4} \\ \lambda(11) &= 1 & h(11) &= -\tfrac{1}{4} \end{aligned}$$

$$E(10) = E(01) = 1,\ E(11) = \tfrac{2}{3},\ E = \tfrac{4}{5}.$$

4. $4 \times 3 \times 2$, $k = 12$ (Li, 1944)

Replication 1 Block		Replication 2 Block		Replication 3 Block	
1	2	1	2	1	2
000	001	001	000	001	000
011	010	010	011	011	010
021	020	021	020	020	021
101	100	100	101	100	101
110	111	111	110	110	111
120	121	120	121	121	120
200	201	201	200	201	200
211	210	210	211	211	210
221	220	221	220	220	221
301	300	300	301	300	301
310	311	311	310	310	311
320	321	320	321	321	320

Replication 4 Block		Replication 5 Block		Replication 6 Block	
1	2	1	2	1	2
001	000	000	001	000	001
010	011	011	010	010	011
020	021	020	021	021	020
100	101	101	100	101	100
111	110	110	111	111	110
121	120	121	120	120	121
200	201	201	200	201	200
211	210	210	211	211	210
221	220	221	220	220	221
301	300	300	301	300	301
310	311	311	310	310	311
320	321	320	321	321	320

Replication 7 Block		Replication 8 Block		Replication 9 Block	
1	2	1	2	1	2
000	001	001	000	001	000
011	010	010	011	011	010
021	020	021	020	020	021
100	101	101	100	101	100
111	110	110	111	111	110
121	120	121	120	120	121
201	200	200	201	200	201
210	211	211	210	210	211
220	221	220	221	221	220
301	300	300	301	300	301
310	311	311	310	310	311
320	321	320	321	321	320

$$\begin{aligned}
\lambda(000) &= 9 & h(000) &= \tfrac{23}{3}\\
\lambda(100) &= 3 & h(100) &= \tfrac{1}{3}\\
\lambda(010) &= 3 & h(010) &= \tfrac{1}{3}\\
\lambda(110) &= 5 & h(110) &= -\tfrac{1}{12}\\
\lambda(001) &= 0 & h(001) &= \tfrac{2}{3}\\
\lambda(101) &= 6 & h(101) &= -\tfrac{1}{6}\\
\lambda(011) &= 6 & h(011) &= -\tfrac{1}{6}\\
\lambda(111) &= 4 & h(111) &= -\tfrac{1}{3}
\end{aligned}$$

$$E(101) = \tfrac{26}{27},\ E(111) = \tfrac{23}{27},\ \text{all other } E(z) = 1,\ E = \tfrac{22}{23}.$$

Using replicates 1, 2, 3 only will yield a design with the same relative efficiencies, but the design is not a BFD.

5. $4 \times 4 \times 3$, $k = 12$ (Li, 1944)

Replication 1 Block				Replication 2 Block				Replication 3 Block			
1	2	3	4	1	2	3	4	1	2	3	4
000	010	020	030	000	030	010	020	000	020	030	010
110	100	130	120	130	100	120	110	120	100	110	130
220	230	200	210	210	220	200	230	230	210	200	220

Replication 1 Block				Replication 2 Block				Replication 3 Block			
1	2	3	4	1	2	3	4	1	2	3	4
330	320	310	300	320	310	330	300	310	330	320	300
001	011	021	031	001	031	011	021	001	021	031	011
111	101	131	121	131	101	121	111	121	101	111	131
221	231	201	211	211	221	201	231	231	211	201	221
331	321	311	301	321	311	331	301	311	331	321	301
002	012	022	032	002	032	012	022	002	022	032	012
112	102	132	122	132	102	122	112	122	102	112	132
222	232	202	212	212	222	202	232	232	212	202	222
332	322	312	302	322	312	332	302	312	332	322	302

$$
\begin{aligned}
\lambda(000) &= 3 & h(000) &= 3 \\
\lambda(100) &= 0 & h(100) &= 0 \\
\lambda(010) &= 0 & h(010) &= 0 \\
\lambda(110) &= 1 & h(110) &= 0 \\
\lambda(001) &= 3 & h(001) &= -\tfrac{1}{3} \\
\lambda(101) &= 0 & h(101) &= \tfrac{1}{12} \\
\lambda(011) &= 0 & h(011) &= \tfrac{1}{12} \\
\lambda(111) &= 1 & h(111) &= -\tfrac{1}{12}
\end{aligned}
$$

$$E(110) = \tfrac{2}{3}, \text{ all other } E(z) = 1, E = \tfrac{44}{47}.$$

6. $3 \times 3 \times 2, k = 6$ (Yates, 1937b)

Replication 1 Block			Replication 2 Block			Replication 3 Block			Replication 4 Block		
1	2	3	1	2	3	1	2	3	1	2	3
100	200	000	200	000	100	100	200	000	200	000	100
010	020	120	020	120	010	210	010	110	010	110	210
220	110	210	110	210	220	020	120	220	120	220	020
201	001	101	101	201	001	201	001	101	101	201	001
021	121	011	011	021	121	011	111	211	211	011	111
111	211	221	221	111	211	121	221	021	021	121	221

$$\begin{array}{ll} \lambda(000) = 4 & h(000) = \frac{5}{2} \\ \lambda(100) = 0 & h(100) = \frac{1}{2} \\ \lambda(010) = 0 & h(010) = \frac{1}{2} \\ \lambda(110) = 2 & h(110) = -\frac{1}{6} \\ \lambda(001) = 0 & h(001) = \frac{1}{2} \\ \lambda(101) = 2 & h(101) = -\frac{1}{6} \\ \lambda(011) = 2 & h(011) = -\frac{1}{6} \\ \lambda(111) = 1 & h(111) = -\frac{1}{6} \end{array}$$

$$E(110) = \tfrac{7}{8},\ E(111) = \tfrac{5}{8},\ \text{all other } E(z) = 1,\ E = \tfrac{15}{17}.$$

If only replications 1 and 2 or 3 and 4 are used, then the relative information for $A_1 \times A_2$ is $\frac{3}{4}$ and $A_1 \times A_2 \times A_3$ is $\frac{1}{4}$, but the design is not a BFD.

7. $3 \times 3 \times 2 \times 2, k = 12$ (Li, 1944)

Replication 1 Block			Replication 2 Block		
1	2	3	1	2	3
0100	0000	0001	0101	0001	0000
0111	0011	0010	0110	0010	0011
0201	0101	0200	0200	0100	0201
0210	0110	0211	0211	0111	0210
1001	1100	1000	1000	1101	1001
1010	1111	1011	1011	1110	1010
1200	1201	1101	1201	1200	1100
1211	1210	1110	1210	1211	1111
2000	2001	2100	2001	2000	2101
2011	2010	2111	2010	2011	2110
2101	2200	2201	2100	2201	2200
2110	2211	2210	2111	2210	2211

Replication 3 Block			Replication 4 Block		
1	2	3	1	2	3
0101	0000	0001	0100	0001	0000
0110	0011	0010	0111	0010	0011
0200	0201	0100	0201	0200	0101

Replication 3 Block			Replication 4 Block		
1	2	3	1	2	3
0211	0210	0111	0210	0211	0110
1001	1101	1000	1000	1100	1001
1010	1110	1011	1011	1111	1010
1100	1200	1201	1101	1201	1200
1111	1211	1210	1110	1210	1211
2000	2001	2101	2001	2000	2100
2011	2010	2110	2010	2011	2111
2201	2100	2200	2200	2101	2201
2210	2111	2211	2211	2110	2210

$$\lambda(0000) = 4 \qquad h(0000) = \tfrac{5}{2}$$
$$\lambda(1000) = 0 \qquad h(1000) = \tfrac{1}{2}$$
$$\lambda(0100) = 0 \qquad h(0100) = \tfrac{1}{2}$$
$$\lambda(1100) = 2 \qquad h(1100) = -\tfrac{1}{6}$$
$$\lambda(0010) = 0 \qquad h(0010) = \tfrac{3}{4}$$
$$\lambda(1010) = 2 \qquad h(1010) = -\tfrac{1}{4}$$
$$\lambda(0110) = 2 \qquad h(0110) = -\tfrac{1}{4}$$
$$\lambda(1110) = 1 \qquad h(1110) = \tfrac{1}{12}$$
$$\lambda(0001) = 0 \qquad h(0001) = \tfrac{3}{4}$$
$$\lambda(1001) = 2 \qquad h(1001) = -\tfrac{1}{4}$$
$$\lambda(0101) = 2 \qquad h(0101) = -\tfrac{1}{4}$$
$$\lambda(1101) = 1 \qquad h(1101) = \tfrac{1}{12}$$
$$\lambda(0011) = 4 \qquad h(0011) = -\tfrac{1}{2}$$
$$\lambda(1011) = 0 \qquad h(1011) = \tfrac{1}{6}$$
$$\lambda(0111) = 0 \qquad h(0111) = \tfrac{1}{6}$$
$$\lambda(1111) = 2 \qquad h(1111) = -\tfrac{1}{6}$$

$$E(1100) = \tfrac{7}{8},\ E(1111) = \tfrac{5}{8},\ \text{all other } E(z) = 1.$$

If only replications 1 and 2 or 3 and 4 are used, the relative information for $A_1 \times A_2$ remains $\frac{7}{8}$ and for $A_1 \times A_2 \times A_3 \times A_4$ remains $\frac{5}{8}$, but these designs are not BFDs.

8. $5 \times 2 \times 2, k = 10$ (Shah, 1960)

Block									
1	2	3	4	5	6	7	8	9	10
000	010	010	010	000	010	000	000	010	000
011	001	001	001	011	001	011	011	001	011
100	100	110	110	110	100	110	100	100	110
111	111	101	101	101	111	101	111	111	101
210	200	200	210	210	210	200	210	200	200
201	211	211	201	201	201	211	201	211	211
310	310	300	300	310	300	310	300	310	300
301	301	311	311	301	311	301	311	301	311
410	410	410	400	400	400	400	410	400	410
401	401	401	411	411	411	411	401	411	401

$$
\begin{aligned}
\lambda(000) &= 5 & h(000) &= \tfrac{38}{10} \\
\lambda(100) &= 2 & h(100) &= \tfrac{2}{10} \\
\lambda(010) &= 0 & h(010) &= \tfrac{6}{10} \\
\lambda(110) &= 3 & h(110) &= -\tfrac{1}{10} \\
\lambda(001) &= 0 & h(001) &= \tfrac{6}{10} \\
\lambda(101) &= 3 & h(101) &= -\tfrac{1}{10} \\
\lambda(011) &= 5 & h(011) &= -\tfrac{3}{10} \\
\lambda(111) &= 2 & h(111) &= -\tfrac{2}{10}
\end{aligned}
$$

$E(011) = \frac{24}{25}$, $E(111) = \frac{19}{25}$, all other $E(z) = 1$, $E = \frac{18}{19}$.

APPENDIX E

Exercises

CHAPTER 1

1.1 Consider the following design with $t = 7$ treatments in blocks of size 3 and 4:

	Block 1	2	3	4	5	6	7
Treatments	2	3	3	5	1	7	3
	4	5	6	4	6	2	1
	1	2	4	7	5	6	7
	2			4	1		3

(a) Give the parameters for this design.
(b) Write out the incidence matrix.
(c) Write out the $\boldsymbol{C}$ matrix of (1.9).
(d) Show that all differences $\tau_i - \tau_{i'}$ are estimable.

1.2 Show that $\boldsymbol{C}\mathfrak{I} = \boldsymbol{O}$ for $\boldsymbol{C}$ of (1.9).

1.3 The experiment of Exercise 1.1 was originally planned for blocks of size 3, but it turned out that some of the blocks had actually 4 experimental units. The investigator decided to replicate one of the treatments assigned to those blocks.
(a) Sketch the ANOVA table for this experiment, giving sources of variation and d.f.

Design and Analysis of Experiments. Volume 2: Advanced Experimental Design
By Klaus Hinkelmann and Oscar Kempthorne
ISBN 0-471-55177-5

(b) Discuss how the error d.f. can be divided into two sets and what the advantage of that would be.

1.4 Consider the following incomplete block design with $t = 6$, $b = 6$, and $k = 2$:

	Block 1	2	3	4	5	6
Treatments	1	3	2	4	5	5
	2	1	3	6	6	4

(a) Write out the $\boldsymbol{C}$ matrix.

(b) Show that this is a disconnected design.

(c) Identify a set of estimable functions of the treatment effects.

1.5 Consider the following data obtained from an experiment using an incomplete block design with $t = 9$, $b = 6$, $k = 3$, and 2 subsamples (the numbers in parentheses indicate the treatments):

Block				
	1	53.5 , 54.8 (6)	53.2 , 54.0 (4)	57.7 , 56.4 (7)
	2	53.1 , 52.7 (3)	58.6 , 57.1 (1)	53.9 , 55.1 (2)
	3	57.2 , 56.5 (9)	55.0 , 55.9 (7)	51.5 , 53.2 (8)
	4	53.7 , 52.9 (4)	53.6 , 54.6 (2)	57.9 , 56.8 (9)
	5	54.5 , 53.3 (3)	52.8 , 53.2 (7)	53.3 , 55.0 (5)
	6	48.9 , 47.8 (8)	53.5 , 54.9 (6)	56.7 , 55.4 (1)

(a) Write out a model for this data set.

(b) Obtain the intrablock analysis using SAS.

(c) Obtain the standard errors for all treatment comparisons $\tau_i - \tau_{i'}$.

1.6 Consider an incomplete block design with t treatments, b blocks of size k, r replications per treatment, and m subsamples per experimental unit.

(a) Explain how the derivations in Sections 1.8.2 and 1.8.3 need to be modified for this case.

(b) Using the data from Exercise 1.5 obtain the combined analysis, including treatment comparisons.

(c) Compare the results with those of Exercise 1.5.

CHAPTER 2

2.1 Consider the BIB design

Block		Block		Block	
1	1 4 6 7	7	2 3 6 7	13	1 2 4 9
2	2 6 8 9	8	2 4 5 8	14	1 5 6 9
3	1 3 8 9	9	3 5 7 9	15	1 3 6 8
4	1 2 3 4	10	1 2 5 7	16	4 6 7 8
5	1 5 7 8	11	2 3 5 6	17	3 4 5 8
6	4 5 6 9	12	3 4 7 9	18	2 7 8 9

(a) Give the parameters of this design.
(b) Write out its $\boldsymbol{C}$ matrix.

2.2 **(a)** Describe a scenario where you may want to use the design given in Exercise 2.1 but in such a form that each treatment occurs exactly 4 times in the first 2 positions and 4 times in the last 2 positions over the 18 blocks.
(b) Write out the Latinized form of this design.
(c) Give the SAS statements for analyzing data using the design in (b).

2.3 **(a)** Show that the design given in Exercise 2.1 is a 4-resolvable BIB design.
(b) Give the SAS statements for analyzing data using the design in (a).

2.4 **(a)** Show that the design obtained in Exercise 2.3 can be Latinized.
(b) Give the SAS statements for analyzing data using the design obtained in (i).

2.5 Investigate whether the design given in Exercise 2.1 is locally resistant w.r.t. any of the treatments.

CHAPTER 3

3.1 Give a proof of Theorem 3.2.

3.2 Obtain the complement of the BIB design given in Exercise 2.1 and verify that it is a BIB design.

3.3 **(a)** Using the successive diagonalizing method construct a BIB for $t = 25$ treatment in blocks of size 5.
(b) Obtain the $\boldsymbol{C}$ matrix for this design.
(c) Suppose that the last replicate is missing from the design. Investigate the properties of the residual design w.r.t. estimating treatment differences $\tau_i - \tau_{i'}$.

3.4 Construct the BIB design #25 in Table 3.1.

3.5 Construct the BIB design #18 in Table 3.1.

3.6 Construct the BIB design #22 in Table 3.1.

CHAPTER 4

4.1 Consider the rectangular PBIB design of Example 4.8.

- **(a)** Write out explicitly Eqs. (4.34) and obtain a solution of these equations.
- **(b)** Give explicit expressions for the variances of simple treatment comparisons.
- **(c)** Generate data for this design and, using the g inverse obtained from SAS PROC GLM, verify the results obtained in (b).

4.2 Write out the association scheme for the L_4-PBIB(2) design for $t = 9$ treatments.

4.3
- **(a)** Write out the association scheme for the EGD/(7)-PBIB design with $t = 18$ treatments.
- **(b)** Give the parameters of this design.
- **(c)** Give the $\boldsymbol{P}$ matrices for this design.

CHAPTER 5

5.1 Verify that the design D' of Example 5.1 is a PBIB(2) design and give its parameters, including the $\boldsymbol{P}$ matrices.

5.2 Using the PG(3, 2) construct a GD-PBIB(2) with $t = 2 \cdot 6 = 12$ treatments in 8 blocks of size 6.

5.3 Show that the dual of the BIB design #8 in Table 3.1 is a triangular PBIB(2) design.

5.4 Construct a L_2-PBIB(2) design for $t = 25$ treatments in 20 blocks of size 5.

5.5
- **(a)** Construct a cyclic PBIB design for $t = 7$ treatments in $b = 7$ blocks of size $k = 4$.
- **(b)** Obtain the efficiency factor of this design.
- **(c)** Check whether there exists a more efficient cyclic PBIB design than the design obtained in (a).

5.6 Construct a balanced array BA[15, 6, 3, 2] from OA[18, 7, 3, 2; 2] and give its index set.

5.7 Using the method of balanced arrays construct an EGD/3-PBIB design for $t = 12$ treatments in 12 blocks of size 3.

5.8 Using the method of balanced arrays construct an EGD/7-PBIB design for $t = 12$ treatments in 12 blocks of size 2.

CHAPTER 6

6.1 **(a)** Construct an α design for $t = 15$ treatments in $r = 3$ replicates and blocks of size $k = 5$.

(b) Describe the properties of this design.

(c) Obtain the efficiency factor of this design.

6.2 Construct an α design for $t = 19$ treatments in blocks of size 5 and 4, with $r = 2$ replicates.

6.3 Construct a generalized cyclic incomplete block design for $t = 15$ treatments in blocks of size 5.

6.4 **(a)** Show that the design of Example 6.6 with $r = 3$ replicates is a (0, 1) design.

(b) Obtain the efficiency factor of this design.

6.5 Consider the BTIB design with $t = 4$ test treatments and blocks of size 3 with generator designs

$$\text{GD}_1 = \begin{pmatrix} 0 & 0 & 0 & 0 & 0 & 0 \\ 1 & 1 & 1 & 2 & 2 & 3 \\ 2 & 3 & 4 & 3 & 4 & 4 \end{pmatrix}$$

$$\text{GD}_2 = \begin{pmatrix} 0 & 0 & 1 & 1 & 1 & 1 & 2 & 2 \\ 1 & 3 & 2 & 2 & 3 & 3 & 3 & 4 \\ 2 & 4 & 3 & 4 & 4 & 4 & 3 & 4 \end{pmatrix}$$

$$\text{GD}_3 = \begin{pmatrix} 1 & 1 & 1 & 2 \\ 2 & 2 & 3 & 3 \\ 3 & 4 & 4 & 4 \end{pmatrix}$$

(a) Using the generator designs given above, construct two BTIB designs with $b = 12$ blocks.

(b) Give the parameters of both designs and compare their efficiency factors.

6.6 Consider the GD-PBIB(2) design with 6 treatments and 9 blocks of size 2:

1	3	5	1	6	5	1	3	5
2	4	6	3	2	4	4	6	2

(a) By inspection of the design above give its parameters and the association structure.

(b) Using the PBIB design above, obtain a PBTIB design for $t = 5$ test treatments.

(c) Give the parameters of the design obtained in (b).

6.7 (a) Using the PBIB design of Exercise 6.6, obtain a PTBIB design for $t = 4$ test treatments.

(b) Give the parameters of this design.

6.8 Consider the GD-PBIB(2) design for $t = 6$ treatments in $b = 6$ blocks of size 4:

1	4	2	5	4	1	5	2
2	5	3	6	5	2	6	3
3	6	1	4	6	3	4	1

(a) Investigate the properties of this design as a row–column design.

(b) Obtain the efficiency factor of the design obtained in (a).

CHAPTER 7

7.1 For the 2^4 factorial give explicit expressions for the true response of the treatment combination $\boldsymbol{x}' = (0, 1, 0, 1)$ using (7.42) and (7.49).

7.2 For the 2^5 factorial write out explicitly the contrast that defines the 4-factor interaction $ABCD$.

7.3 For the 2^3 factorial with r replications per treatment and m subsamples per experimental unit write out:

(a) An expression for the interaction AB.

(b) An expression for the sum of squares for AB.

(c) The ANOVA table, including sources of variation, d.f., SS, E(MS).

7.4 Consider the 2^6 factorial with one replication per treatment combination. Assume that all interactions involving three or more factors are negligible.

(a) Write out the ANOVA table including sources of variation, d.f., SS, E(MS).

(b) Assume that there are $m = 2$ subsamples per treatment combination. Write out the ANOVA table, including sources of variation, d.f., SS, E(MS).

(c) For the situation described in (b) give the input statements for SAS PROC GLM and SAS PROC MIXED to perform both the intrablock and combined analyses.

CHAPTER 8

8.1 **(a)** Determine the intrablock subgroup for the 2^5 factorial in blocks of size 8, confounding interactions involving three or more factors.

(b) Write out the remaining blocks.

8.2 Generate the design of Exercise 8.1 using SAS PROC FACTEX.

8.3 Consider the design generated in Exercise 8.1 with $r = 2$ replications.

(a) Assuming that all interactions involving three or more factors are negligible, write out the ANOVA table, including sources of variation, d.f., SS, and E(MS).

(b) Suppose that for each experimental unit we have $m = 2$ subsamples. Modify the ANOVA table in (a) appropriately.

8.4 For the situations described in Exercise 8.3 give the SAS statements for performing the respective analyses, using SAS PROC GLM and SAS PROC MIXED.

CHAPTER 9

9.1 Consider the 2^5 factorial with 3 replicates and in blocks of size 4. Without confounding main effects construct a design that provides as much and as uniform as possible information about 2-factor interactions.

9.2 For the design obtained in Exercise 9.1, assuming that 4- and 5-factor interactions are negligible, give the ANOVA table, including sources of variation, d.f., SS, and E(MS).

9.3 For the design obtained in Exercise 9.1 give expressions for the combined estimators of all 2-factor interactions and the variances of these estimators.

9.4 Discuss some possibilities of arranging a 2^6 factorial in two 8×8 squares, giving the actual designs and the relative information provided for main effects and 2-factor interactions.

CHAPTER 10

10.1 For the 3^4 factorial write out all main effects and interactions together with their 2-d.f. components.

10.2 For the 3^4 factorial write out the parametrization (10.15) for the true response of the treatment combination $\boldsymbol{x}' = (0, 1, 1, 2)$.

10.3 Construct a design for the 3^4 factorial in blocks of size 9 without confounding main effects and 2-factor interactions.

10.4 Consider the 3^4 factorial in blocks of size 3.

(a) Suppose two replicates are available. Using partial confounding, construct a design that obtains at least partial information on all main effects and 2-factor interaction components.

(b) Give the relative information for all main effects and 2-factor interaction components.

(c) Assuming that all 3- and 4-factor interactions are negligible, sketch the ANOVA table for the design obtained in (a), giving sources of variation, d.f., SS, and $E(\text{MS})$.

(d) Give the SAS statements for performing the analysis of an experiment using the design obtained in (a) for PROC GLM (intrablock analysis) and PROC MIXED (combined analysis).

CHAPTER 11

11.1 Consider the 5^2 factorial in 10 blocks of size 5.

(a) Obtain a plan that yields full information on main effects and at least partial information on all interaction components.

(b) Sketch the ANOVA table, including sources of variation, d.f., and SS.

(c) Give the parametrization for $a(1\quad 1\quad 1\quad 1\quad 1) - a(0\quad 0\quad 0\quad 0\quad 0)$ and give the variance of the estimator for this difference in yields.

11.2 As an alternative to the design obtained in Exercise 11.1 consider the design based on the PBIB-L_2 design LS 51 (Clatworthy, 1973) given by

Block						Block					
1	1	2	3	4	5	6	1	6	11	16	21
2	6	7	8	9	10	7	2	7	12	17	22
3	11	12	13	14	15	8	3	8	13	18	23
4	16	17	18	19	20	9	4	9	14	19	24
5	21	22	23	24	25	10	5	10	15	20	25

(a) Set up a correspondence between the treatments 1, 2, …, 25 and the factorial treatments $(x_1, x_2, x_3, x_4, x_5)$.

(b) Explain the major difference between the two designs with respect to the estimation of main effects and interaction effects.

(c) Explain how you would compare the overall efficiency of the two designs.

11.3 Consider the 4^3 factorial in blocks of size 16.

(a) Assuming that 3-factor interactions are negligible, obtain a plan that yields full information on main effects and 2-factor interactions.

(b) Suppose that 8 blocks are available for the experiment. Write out the ANOVA table, including sources of variation, d.f., and SS.

11.4 **(a)** Apply Theorem 11.7 to the case of the 4^n factorial and write out the Fisher plan for the maximum number of factors.

(b) For the plan obtained in (a) identify the resulting system of confounding.

11.5 **(a)** For the 4^3 factorial in blocks of size 8 obtain a system of confounding with orthogonal factorial structure.

(b) Obtain the efficiency factor of the design.

CHAPTER 12

12.1 Consider the $2^2 \times 3^3$ factorial in 18 blocks of size 6.

(a) Assuming that interactions involving three or more factors are negligible, obtain a design using the Kronecker product method.

(b) Explain why the design obtained in (a) is or is not satisfactory, and if it is not satisfactory what you might want to do to fix the problem.

12.2 Consider the $2^2 \times 3^3$ factorial in blocks of size 6.

(a) Using the GC/n method construct a system of confounding.

(b) Obtain the loss of d.f. for the design obtained in (a).

(c) Compare the design obtained in (a) with that obtained in Exercise 12.1.

12.3 **(a)** Apply the method of finite rings to the $2^2 \times 3^3$ factorial in blocks of size 6.

(b) Obtain the variance for the comparison between the treatments (1, 1, 2, 2, 2) and (0, 0, 0, 0, 0).

12.4 Consider the $2^2 \times 4^2$ factorial in blocks of size 8.

(a) Construct a balanced factorial design such that all treatment contrasts are estimable.

(b) For the design of (a) obtain the efficiency factor.

CHAPTER 13

13.1 Consider a 2^{8-2} design.

(a) Give an identity relationship such that all main effects are strongly clear and all 2-factor interactions are either clear or strongly clear.

(b) Obtain the design based on the identity relationship in (a).

(c) Give the resolution of the design.

13.2 Consider the design in Exercise 13.1. Suppose that for the actual experiment blocks of size 16 need to be used.

(a) Identify a suitable system of confounding.

(b) Obtain the actual design.

(c) For the design in (b) sketch the ANOVA table, including sources of variation and d.f., assuming that interactions involving four or more factors are negligible.

13.3 Consider a 3_{IV}^{6-2} factorial based on the identity relationship $I = ABCD = AB^2EF =$ all GI.

(a) Complete the identity relationship.

(b) Write out the alias structure for all main effects and 2-factor interaction components.

(c) Identify all the clear effects.

(d) Obtain the actual design.

13.4 For the design of Exercise 13.3 suppose that blocks of size 9 need to be used.

(a) Obtain a suitable system of confounding and state any assumptions you are making.

(b) For the arrangement in (a) sketch the ANOVA, giving sources of variation and d.f., assuming that all interactions involving three or more factors are negligible.

13.5 Consider the 2_{IV}^{6-2} fraction in 4 blocks.

(a) Using the independent defining words given in Table 13.12, obtain the actual design.

(b) Suppose that the plan in (a) is replicated three times. Sketch the ANOVA, including sources of variation and d.f.

(c) Rather than replicating the plan in (a) three times, discuss other possibilities of carrying out the experiment.

CHAPTER 14

14.1 Construct a saturated orthogonal main effect plan for the 2^7 factorial.

14.2 Construct an OMEP for the 3^4 factorial in 9 runs.

14.3 **(a)** Construct an OMEP for the 2^4 factorial in 9 runs.

(b) Identify the alias structure for this design.

14.4 Using the OMEPs obtained in Exercises 14.1 and 14.2, construct an OMEP for the $2^7 \times 3^4$ factorial.

14.5 **(a)** Construct an OMEP for the $4^2 \times 3 \times 2^2$ factorial.

(b) Suppose the four- and three-level factors are quantitative factors with equidistant levels. Show that the estimators for the linear effects of a four-level factor and the three-level factor are uncorrelated.

14.6 **(a)** Construct an OMEP for the $4 \times 3^2 \times 2^2$ factorial.

(b) Suppose the plan in (a) is replicated twice. Sketch the ANOVA table, including sources of variation and d.f.

14.7 Consider the following main effect plan for the 2^4 factorial in five runs:

Run	A	B	C	D
1	0	0	0	0
2	0	1	1	1
3	1	0	1	1
4	1	1	0	1
5	1	1	1	0

(a) Is this an OMEP?

(b) Show how you would estimate the main effects.

(c) Obtain the efficiency factor of this design.

CHAPTER 15

15.1 Construct a supersaturated design for 14 factors in 8 runs.

15.2 **(a)** Construct a supersaturated design for 20 factors in 12 runs.

(b) Obtain $E(s^2)$ for this design and compare it to the lower bound (15.9).

15.3 **(a)** Construct a three-level supersaturated design for 12 factors in 12 runs.

(b) Obtain the av χ^2 and max χ^2 values of (15.3) and (15.4), respectively.

CHAPTER 16

16.1 **(a)** Using the results of Table 16.1, write out the resolution III.1 design for $n = 5$ factors.

(b) Verify that the conditions of Theorem 16.5 are satisfied.

(c) Show explicitly that, individually, each 2-factor interaction is estimable.

16.2 **(a)** Obtain the resolution V.1 design for $n = 4$ factors.

(b) Verify that the conditions of Theorem 16.4 is satisfied.

(c) Show explicitly that, individually, each 3-factor interaction is estimable.

16.3 Find the search probabilities for the design obtained in Exercise 16.1.

16.4 Fine the search probabilities for the design obtained in Exercise 16.2.

CHAPTER 17

17.1 Consider the case of three design factors A_1, A_2, A_3, each at three levels, and two noise factors B_1, B_2, also each at three levels.

(a) Construct the smallest cross array design such that all main effects are estimable.

(b) Suppose that all five factors are quantitative factors and that only linear $\times$ linear interactions between A_i and B_j ($i = 1, 2, 3$; $j = 1, 2$) are important. Sketch the ANOVA table for the design in (a), including sources of variation and d.f.

17.2 Find an alternative single array design to the design obtained in Exercise 17.1 that uses fewer runs.

17.3 In certain cases when the design and noise factors have two levels it is possible to utilize a $2^{n-\ell}$ fractional factorial to construct a single array design. For example, for three design factors A_1, A_2, A_3 and three noise factors B_1, B_2, B_3 the 2^{6-2} fraction based on

$$I = A_1A_2A_3B_2 = A_1A_2A_3B_1B_3 = B_1B_2B_3$$

yields a suitable design.

(a) Write out the design in terms of control and noise factor level combinations.

(b) Identify the design–noise factor interactions that are clear.

17.4 **(a)** Use the method described in Exercise 17.3 to obtain a single array for four design factors and two noise factors, each at two levels.

(b) Identify all the clear effects.

(c) What additional assumptions will you have to make to estimate all main effects of the control factors and all control $\times$ noise factor interactions?

CHAPTER 18

18.1 **(a)** Construct a balanced square lattice for 25 treatments in blocks of size 5.

(b) Sketch the ANOVA table for this design, including sources of variation and d.f.

(c) Show that this design is a BIB design.

18.2 **(a)** Construct a triple three-dimensional lattice for 64 treatments in blocks of size 4.

(b) Sketch the ANOVA table for this design, including sources of variation and d.f.

(c) Consider the B | T-ANOVA and indicate how you would estimate the weights w and w'.

18.3 Construct a simple square lattice for 81 treatments in blocks of size 9.

18.4 **(a)** Construct a balanced lattice square for 16 treatments.

(b) Sketch the ANOVA table for this design, including sources of variation and d.f.

18.5 Construct a triple rectangular lattice for 30 treatments in blocks of size 5.

CHAPTER 19

19.1 Construct a balanced CO(5, 10, 3).

19.2 Consider the irreducible BIB design for 4 treatments in blocks of size 3. Use this design to construct the CO(4, 24, 3).

19.3 **(a)** Consider the PBIB(2) design in (5.9) and use it as the generator design to construct the CO(6, 12, 3).

(b) Obtain the variances for all simple treatment comparisons.

19.4 Construct the strongly balanced CO(4, 8, 3).

19.5 Suppose we have a 2^4 factorial treatment structure. Assume that the 4-factor interaction is negligible.

(a) Construct the CO(2^4, 2^4, 2^3).

(b) Explain how you would estimate the direct main effects and 2-factor interactions.

References

Abel, R. J. R., A. E. Brouwer, C. J. Colbourn and J. H. Dinitz (1996). Mutually orthogonal latin squares (MOLS). In: *The CRC Handbook of Combinatorial Designs*. C. J. Colbourn and J. H. Dinitz (eds.). Boca Raton, FL: CRC, pp. 111–142.

Abraham, B., H. Chipman and K. Vijayan (1999). Some risks in the construction and analysis of supersaturated designs. *Technometrics*, **41**, 135–141.

Addelman, S. (1961). Irregular fractions of the 2^n factorial experiment. *Technometrics*, **3**, 479–496.

Addelman, S. (1962a). Orthogonal main-effect plans for asymmetrical factorial experiments. *Technometrics*, **4**, 21–46 (Erratum **4**, 440).

Addelman, S. (1962b). Symmetrical and asymmetrical fractional factorial plans. *Technometrics*, **4**, 47–58.

Addelman, S. and O. Kempthorne (1961). Some main-effect plans and orthogonal arrays of strength two. *Ann. Math. Statist.*, **32**, 1167–1176.

Adhikary, B. (1966). Some types of m-associate PBIB association schemes. *Calcutta Statist. Assoc. Bull.*, **15**, 47–74.

Adhikary, B. (1967). A new type of higher associate cyclical association schemes. *Calcutta Statist. Assoc. Bull.*, **16**, 40–44.

Afsarinejad, K. (1983). Balanced repeated measurement designs. *Biometrika*, **70**, 199–204.

Afsarinejad, K. (1990). Repeated measurement designs—A review. *Commun. Statist.—Theory Meth.*, **19**, 3985–4028.

Afsarinejad, K. and A. S. Hedayet (2002). Repeated measurements designs for a model with self and simple mixed carryover effects. *J. Statist. Plann. Inf.*, **106**, 449–459.

Aggarwal, K. R. (1974). Some higher-class PBIB designs and their applications as confounded factorial experiments. *Ann. Inst. Statist. Math.*, **26**, 315–323.

Aggarwal, K. R. and T. P. Singh (1981). Methods of construction of balanced arrays with application to factorial designs. *Calcutta Statist. Assoc. Bull.*, **30**, 89–93.

Atkinson, A. C. and A. N. Donev (1992). *Optimum Experimental Design*. Oxford: Clarendon.

Design and Analysis of Experiments. Volume 2: Advanced Experimental Design
By Klaus Hinkelmann and Oscar Kempthorne
ISBN 0-471-55177-5

Baksalary, J. K. and P. D. Puri (1990). Pairwise-balanced, variance-balanced and resistant incomplete block designs revisited. *Ann. Inst. Statist. Math.*, **42**, 163–171.

Balaam, L. N. (1968). A two-period design with t^2 experimental units. *Biometrics*, **24**, 61–73.

Banerjee, K. S. (1970). Some observations on the White-Hultquist procedure for the construction of confounding plans for mixed factorial designs. *Ann. Math. Statist.*, **41**, 1108–1113.

Banerjee, K. S. (1975). *Weighing Designs*. New York: Dekker.

Baumert, L. D., S. W. Golomb and M. Hall, Jr. (1962). Discovery of an Hadamard matrix of order 92. *Bull. Am. Math. Soc.*, **68**, 237–238.

Bechhofer, R. E. and A. C. Tamhane (1981). Incomplete block designs for comparing treatments with a control: General theory. *Technometrics*, **23**, 45–57 (Corrigendum **24**, 171).

Bechhofer, R. E. and A. C. Tamhane (1983a). Incomplete block designs for comparing treatments with a control (II): Optimal designs for $p = 2(1)6, k = 2$ and $p = 3, k = 3$. *Sankhyā*, Ser. B, **45**, 193–224.

Bechhofer, R. E. and A. C. Tamhane (1983b). Design of experiments for comparing treatments with a control: Tables of optimal allocations of observations. *Technometrics*, **25**, 87–95.

Bechhofer, R. E. and A. C. Tamhane (1985). Tables of admissible and optimal balanced treatment incomplete block (BTIB) designs for comparing treatments with a control. *Selected Tables Math. Statist.*, **8**, 41–139.

Berenblut, I. I. (1964). Change-over design with complete balance for first residual effects. *Biometrics*, **20**, 707–712.

Berenblut, I. I. (1967). A change-over design for testing a treatment factor at four equally spaced levels. *J. Roy. Statist. Soc., B*, **29**, 370–373.

Berenblut, I. I. (1968). Change-over designs balanced for the linear component of first residual effects. *Biometrika*, **55**, 297–303.

Beyer, W. H. (ed.) (1991). *CRC Standard Probability and Statistics Tables and Formulae*. Boca Raton, FL : CRC.

Bhattacharya, C. G. (1998). Goodness of Yates-Rao procedure for recovery of inter-block information. *Sankhyā A*, **60**, 134–144.

Bisgaard, S. (1994). A note on the definition of resolution for blocked 2^{k-p} designs. *Technometrics*, **36**, 308–311.

Blaisdell, E. A. and D. Raghavarao (1980). Partially balanced changeover designs based on m-associate PBIB designs. *J. Roy. Statist. Soc., B*, **42**, 334–338.

Booth, K. H. V. and D. R. Cox (1962). Some systematic supersaturated designs. *Technometrics*, **4**, 489–495.

Bose, R. C. (1939). On the construction of balanced incomplete block designs. *Ann. Eugenics*, **9**, 353–399.

Bose, R. C. (1942). A note on the resolvability of balanced incomplete block designs. *Sankhyā*, **6**, 105–110.

Bose, R. C. (1947a). On a resolvable series of balanced incomplete block designs. *Sankhyā*, **8**, 249–256.

Bose, R. C. (1947b). Mathematical theory of the symmetrical factorial design. *Sankhyā*, **8**, 107–166.

Bose, R. C. and K. A. Bush (1952). Orthogonal arrays of strength two and three. *Ann. Math. Statist.*, **23**, 508–524.

Bose, R. C. and W. S. Connor (1952). Combinatorial properties of group divisible incomplete block designs. *Ann. Math. Statist.*, **23**, 367–383.

Bose, R. C. and K. Kishen (1940). On the problem of confounding in the general symmetrical factorial design. *Sankhyā*, **5**, 21–36.

Bose, R. C. and D. M. Mesner (1959). On linear associative algebras corresponding to association schemes of partially balanced designs. *Ann. Math. Statist.*, **30**, 21–38.

Bose, R. C. and K. R. Nair (1939). Partially balanced incomplete block designs. *Sankhyā*, **4**, 337–372.

Bose, R. C. and T. Shimamoto (1952). Classification and analysis of partially balanced incomplete block designs with two associate classes. *J. Am. Statist. Assoc.*, **47**, 151–184.

Bose, R. C., W. H. Clatworthy and S. S. Shrikhande (1954). Tables of partially balanced incomplete block designs with two associate classes. *North Carolina Agricultural Experiment Station Technical Bulletin* No. 107, Raleigh, NC.

Bose, R. C., S. S. Shrikhande and K. N. Bhattacharya (1953). On the construction of group divisible incomplete block designs. *Ann. Math. Statist.*, **24**, 167–195.

Box, G. E. P. (1988). Signal-to-noise ratios, performance criteria, and transformation. *Technometrics*, **30**, 1–17.

Box, G. E. P. and J. S. Hunter (1961a). The 2^{k-p} fractional factorial designs. Part I. *Technometrics*, **3**, 311–351 (Errata **5**, 417).

Box, G. E. P. and J. S. Hunter (1961b). The 2^{k-p} fractional factorial designs. Part II. *Technometrics*, **3**, 449–458.

Box, G. E. P. and R. D. Meyer (1986a). An analysis of unreplicated fractional factorials. *Technometrics*, **28**, 11–18.

Box, G. E. P. and R. D. Meyer (1986b). Dispersion effects from fractional designs. *Technometrics*, **28**, 19–27.

Box, G. E. P. and K. B. Wilson (1951). On the experimental attainment of optimum conditions. *J. Roy. Statist. Soc., B*, **13**, 1–45.

Bradley, J. V. (1958). Complete counterbalancing of immediate sequential effects in a Latin square design. *J. Am. Statist. Assoc.*, **53**, 525–528.

Bulutoglu, D. A. and C.-S. Cheng (2003). Hidden projection properties of some nonregular fractional factorial designs and their applications. *Ann. Statist.*, **31**, 1012–1026.

Calinski, T. and S. Kageyama (2000). *Block Designs: A Randomization Approach. Volume I: Analysis*. New York: Springer.

Calvin, L. D. (1954). Doubly balanced incomplete block designs for experiments in which the treatment effects are correlated. *Biometrics*, **10**, 61–88.

Carmichael, R. D. (1956). *Introduction to the Theory of Groups of Finite Order*. New York: Dover.

Chakravarti, I. M. (1956). Fractional replication in asymmetrical factorial designs and partially balanced arrays. *Sankhyā*, **17**, 143–164.

Chakravarti, I. M. (1961). On some methods of construction of partially balanced arrays. *Ann. Math. Statist.*, **32**, 1181–1185.

Chang, C.-T. (1989). On the construction of hypercubic design. *Comm. Statist.—Theory Meth.*, **18**, 3365–3371.

Chang, C.-T. and K. Hinkelmann (1987). A new series of EDG-PBIB designs. *J. Statist. Plann. Inf.*, **16**, 1–13.

Chang, L.-C. and W.-R. Liu (1964). Incomplete block designs with square parameters for which $k \leq 10$ and $r \leq 10$. *Sci. Sinica*, **13**, 1493–1495.

Chang, L.-C., C.-W. Liu and W.-R. Liu (1965). Incomplete block designs with triangular parameters for which $k \leq 10$ and $r \leq 10$. *Sci. Sinica*, **14**, 329–338.

Chatterjee, K., L.-Y. Deng and D. K. J. Lin (2001). Two-level search design for main-effect plus two plan. *Metrika*, **54**, 233–245.

Chatterjee, K. and R. Mukerjee (1986). Some search designs for symmetric and asymmetric factorials. *J. Statist. Plann. Inf.*, **13**, 357–363.

Chatterjee, K. and R. Mukerjee (1993). Search designs for estimating main effects and searching several two-factor interactions in general factorials. *J. Statist. Plann. Inf.*, **35**, 131–138.

Chen, H. and C.-S. Cheng (1999). Theory of optimal blocking of 2^{n-m} designs. *Ann. Statist.*, **27**, 1948–1973.

Chen, H. H. and C.-S. Cheng (2004). Aberration, estimation capacity and estimation index. *Statist. Sinica*, **14**, 203–215.

Cheng, C.-S. (1996). Optimal design: Exact theory. In: *Handbook of Statistics*, Volume 13. S. Ghosh and C. R. Rao (eds.). Amsterdam: Elsevier Science, pp. 977–1006.

Cheng, C.-S. (1997). $E(s^2)$-optimal supersaturated designs. *Statist. Sinica*, **7**, 929–939.

Cheng, C.-S., D. M. Steinberg and D. X. Sun (1999). Minimum aberration and maximum estimation capacity. *J. Roy. Statist. Soc.*, *B*, **61**, 85–93.

Cheng, C.-S. and C.-F. Wu (1980). Balanced repeated measurement designs. *Ann. Statist.*, **8**, 1272–1283 [Correction **11**, (1983), 349].

Chopra, D.V. and J. N. Srivastava (1974). Optimal balanced 2^8 fractional factorial designs of resolution V, $37 \leq N \leq 51$. *Sankhyā*, *A*, **36**, 41–52.

Chopra, D.V. and J. N. Srivastava (1975). Optimal balanced 2^7 fractional factorial designs of resolution V, $43 \leq N \leq 48$. *Sankhyā*, *B*, **37**, 429–447.

Clatworthy, W. H. (1955). Partially balanced incomplete block designs with two associate classes and two treatments per block. *J. Res. National Bur. Standards*, 54.

Clatworthy, W. H. (1956). Contributions on partially balanced incomplete block designs with two associate classes. *NBS Appl. Math. Series*, 47.

Clatworthy, W. H. (1967). Some new families of partially balanced designs of the Latin square type and related designs. *Technometrics*, **9**, 229–244.

Clatworthy, W. H. (1973). Tables of two-associate-class partially balanced designs. *NBS Appl. Math. Series*, 63.

Cochran, W. G. (1939). Long-term agricultural experiments (with discussion). *J. Roy. Statist. Soc.*, *B*, **6**, 104–148.

Cochran, W. G., K. M. Autrey and C. Y. Cannon (1941). A double change-over design for dairy cattle feeding experiments. *J. Dairy Sci.*, **24**, 937–951.

Cochran, W. G. and G. M. Cox (1957). *Experimental Designs*, 2nd ed. New York: Wiley.

Cochran, W. G. and G. M. Cox (1992). *Experimental Designs*. New York: Wiley.

Cohen, A. and H. B. Sackrowitz (1989). Exact tests that recover interblock information in balanced incomplete block designs. *J. Am. Statist. Assoc.*, **84**, 556–559.

Connor, W. S. and W. H. Clatworthy (1954). Some theorems for partially balanced designs. *Ann. Math. Statist.*, **25**, 100–112.

Constantine, G. (1983). On the trace efficiency for control of reinforced balanced incomplete block designs. *J. Roy. Statist. Soc., B*, **45**, 31–36.

Corbeil, R. R. and S. R. Searle (1976). Restricted maximum likelihood (REML) estimation of variance components in the mixed model. *Technometrics*, **18**, 31–38.

Cotter, S. C., J. A. John and T. M. F. Smith (1973). Multi-factor experiments in non-orthogonal designs. *J. Roy. Statist. Soc., B*, **35**, 361–367.

Cotton, J. W. (1998). *Analyzing Within-Subjects Experiments*. Mahwah, NJ: Lawrence Earlbaum Associates.

Cox, D. R. (1958). The interpretation of the effects of non-additivity in the Latin square. *Biometrika*, **45**, 69–73.

Cox D. R. and N. Reid (2000). *The Theory of the Design of Experiments*. Boca Raton, FL: Chapman and Hall / CRC.

Daniel, C. (1959). Use of half-normal plots in interpreting factorial two-level experiments. *Technometrics*, **1**, 311–341.

Daniel, C. (1973). One-at-a-time plans. *J. Am. Statist. Assoc.*, **68**, 353–360.

Das, M. N. (1958). On reinforced incomplete block designs. *J. Indian Soc. Agric. Statist.*, **10**, 73–77.

David, H. A. (1963). *The Method of Paired Comparisons*. London: Griffin.

David, H. A. (1965). Enumeration of cyclic paired-comparison designs. *Am. Math. Monthly*, **72**, 241–248.

David, H. A. and F. W. Wolock (1965). Cyclic designs. *Ann. Math. Statist.*, **36**, 1526–1534.

Davies, O. L. (ed.) (1956). *The Design and Analysis of Industrial Experiments*. London: Oliver and Boyd.

Davis, A. W. and W. B. Hall (1969). Cyclic change-over designs. *Biometrika*, **56**, 283–293.

Dean, A. M. and J. A. John (1975). Single replication factorial experiments in generalized cyclic designs: II. Asymmetrical arrangements *J. Roy. Statist. Soc, B*, **37**, 72–76.

Dénes, J. and A. D. Keedwell (1974). *Latin Squares and Their Applications*. London: English Universities Press.

Dey, A. (1985). *Orthogonal Fractional Factorial Design*. New York: Halsted.

Dunnett, C. W. (1955). A multiple comparison procedure for comparing several treatments with a control. *J. Am. Statist. Assoc.*, **50**, 1096–1121.

Dunnett, C. W. (1964). New tables for multiple comparisons with a control. *Biometrics*, **20**, 482–491.

Eccleston, J. A. and C. A. McGilchrist (1986). Algebra of a row-column design. *J. Statist. Plann. Inf.*, **12**, 305–310.

Eccleston, J. A. and K. G. Russell (1975). Connectedness and orthogonality in multi-factor designs. *Biometrika*, **62**, 341–345.

Eccleston, J. A. and K. G. Russell (1977). Adjusted orthogonality in non-orthogonal designs. *Biometrika*, **64**, 339–345.

Ehrenfeld, S. (1953). On the efficiency of experimental design. *Ann. Math. Statist.*, **26**, 247–255.

Engel, J. (1992). Modelling variation in industrial experiments. *Appl. Statist.*, **41**, 579–593.

Fan, J. and R. Li (2001). Variable selection via nonconcave penalized likelihood and its oracle properties. *J. Am. Statist. Assoc.*, **96**, 1348–1360.

Fang, K.-T., D. K. J. Lin and C.-X. Ma (2000). On the construction of multi-level supersaturated designs. *J. Statist. Plann. Inf.*, **86**, 239–252.

Federer, W. T. (1955). *Experimental Design—Theory and Application.* New York: Macmillan.

Fichtner, A. (2000). Auswertung eines Mais-Feldexperimentes zur Untersuchung von Genotypeffekten bei quantitativen Merkmalen. Unpubl. Diplomarbeit: Universität Dortmund.

Finney, D. J. (1945). The fractional replication of factorial experiments. *Ann. Eugenics*, **12**, 291–301 (Correction **15**, 276).

Finney, D. J. (1956). Cross-over designs in bioassay. *Proc. Roy. Soc., B*, **145**, 42–60.

Fisher, R. A. (1940). An examination of the different possible solutions of a problem in incomplete blocks. *Ann. Eugenics*, **10**, 52–75.

Fisher, R. A. (1942). The theory of confounding in factorial experiments in relation to the theory of groups. *Ann. Eugenics*, **11**, 341–353.

Fisher, R. A. (1945). A system of confounding for factors with more than two alternatives, giving completely orthogonal cubes and higher powers. *Ann. Eugenics*, **12**, 283–290.

Fisher, R. A. (1949). *The Design of Experiments*, 5th ed. London: Hafner.

Fisher, R. A. and F. Yates (1957). *Statistical Tables for Biological, Agricultural and Medical Research*, 5th ed. New York: Hafner.

Fletcher, D. J., S. M. Lewis and J. N. S. Matthews (1990). Factorial designs for crossover clinical trials. *Statist. Medicine*, **9**, 1121–1129.

Folks, J. L. and O. Kempthorne (1960). The efficiency of blocking in incomplete block designs. *Biometrika*, **47**, 273–283.

Fries, A. and W. G. Hunter (1980). Minimum aberration 2^{k-p} designs. *Technometrics*, **22**, 601–608.

Fujii, Y. (1976). An upper bound of resolution in symmetrical fractional factorial designs. *Ann. Statist.*, **4**, 662–667.

Ghosh, S. (1981). On some new search designs for 2^m factorial experiments. *J. Statist. Plann. Inf.*, **5**, 381–389.

Ghosh, S. (1982). Robustness of BIBD against the unavailability of data. *J. Statist. Plann. Inf.*, **6**, 29–32.

Ghosh, S. and C. Burns (2001). Two general classes of search designs for factor screening experiments with factors at three levels. *Metrika*, **54**, 1–7.

Ghosh, S. and C. Burns (2002). Comparison of four new classes of search designs. *Austr. New Zealand J. Statist.*, **44**, 357–366.

Ghosh, S. and L. Teschmacher (2002). Comparisons of search designs using search probabilities. *J. Statist. Plann. Inf.*, **104**, 439–458.

Gill, P. S. (1992). Balanced change-over designs for autocorrelated observations. *Austral. J. Statist.*, **34**, 415–420.

Good, I. J. (1958). The interaction algorithm and practical Fourier analysis. *J. Roy. Statist. Soc., B.*, **20**, 361–372 (Addendum **22**, 372–375).

Graybill, F. A. (1969). *Matrices with Applications in Statistics.* Belmont, CA: Wadsworth.

Graybill, F. A. and R. B. Deal (1959). Combining unbiased estimators. *Biometrics*, **15**, 543–550.

Graybill, F. A. and D. L. Weeks (1959). Combining inter-block and intra-block information in balanced incomplete block designs. *Ann. Math. Statist.*, **30**, 799–805.

Grundy, P. M. and M. J. R. Healy (1950). Restricted randomization and quasi-Latin squares. *J. Roy. Statist. Soc., B.*, **12**, 286–291.

Gupta, B. C. (1990). A survey of search designs for 2^m factorial experiments. In: *Probability, Statistics and Design of Experiment*, R. R. Bahadur (ed.). New Delhi: Wiley Eastern. pp. 329–345.

Gupta, S. and R. Mukerjee (1989). *A Calculus for Factorial Arrangements*. Berlin: Springer.

Harshbarger, B. (1947). Rectangular lattices. *Va. Agri. Expt. Stat. Memoir*, **1**, 1–26.

Harshbarger, B. (1949). Triple rectangular lattices. *Biometrics*, **5**, 1–13.

Harshbarger, B. (1951). Near balance rectangular lattices. *Va. J. Sci.*, **2**, 13–27.

Harshbarger, B. and L. L. Davis (1952). Latinized rectangular lattices. *Biometrics*, **8**, 73–84.

Hartley, H. O. and J. N. K. Rao (1967). Maximum-likelihood estimation for the mixed analysis of variance model. *Biometrika*, **54**, 93–108.

Hartley, H. O. and C. A. B. Smith (1948). The construction of Youden squares. *J. Roy. Statist. Soc., B*, **10**, 262–263.

Hedayat, A. and K. Afsarinejad (1975). Repeated measurement designs I. In: *A Survey of Statistical Design and Linear Models*. J. N. Srivastava, (ed.). Amsterdam: North Holland, pp. 229–242.

Hedayat, A. and K. Afsarinejad (1978). Repeated measurement designs II. *Ann. Statist.*, **6**, 619–628.

Hedayat, A. and P. W. M. John (1974). Resistant and susceptible BIB designs. *Ann. Statist.*, **2**, 148–158.

Hedayat, A. S. and D. Majumdar(1984). A-Optimal incomplete block designs for control-test treatment comparisons. *Technometrics*, **26**, 363–370.

Hedayat, A. and E. Seiden (1970). F-square and orthogonal F-square design: A generalization of Latin square and orthogonal Latin square design. *Ann. Math. Statist.*, **41**, 2035–2044.

Hedayat A. S. and W. D. Wallis (1978). Hadamard matrices and their applications. *Ann. Statist.*, **6**, 1184–1238.

Hedayat A. S. and M. Yang (2003). Universal optimality of balanced uniform crossover designs. *Ann. Statist.*, **31**, 978–983.

Hedayat, A. S., M. Jacroux and D. Majumdar (1988). Optimal designs for comparing test treatments with controls (with discussion). *Statist. Sci.*, **4**, 462–491.

Hedayat, A., D. Raghavarao and E. Seiden (1975). Further contributions to the theory of F-squares design. *Ann. Statist.*, **3**, 712–716.

Hedayat, A. S., N. J. A. Sloane and J. Stufken (1999). *Orthogonal Arrays: Theory and Applications*. New York: Springer.

Hemmerle, W. J. and H. O. Hartley (1973). Computing maximum likelihood estimates for the mixed A. O. V. model using the W-transformation. *Technometrics*, **15**, 819–831.

Hinkelmann, K. (1964). Extended group divisible partially balanced incomplete block designs. *Ann. Math. Statist.*, **35**, 681–695.

Hinkelmann, K. (1997). The Kempthorne-parameterization for asymmetrical factorials. *Technical Report 97-7*, Department of Statistics, Virginia Tech, Blacksburg, VA.

Hinkelmann, K. (2001). Remembering Oscar Kempthorne (1919–2000). *Statist. Sci.*, **16**, 169–183.

Hinkelmann, K. and O. Kempthorne (1963). Two classes of group divisible partial diallel crosses. *Biometrika*, **50**, 281–291.

Hinkelmann, K. and O. Kempthorne (1994). *Design and Analysis of Experiments. Volume 1. Introduction to Experimental Design*. New York: Wiley.

Hoblyn, T. N., S. C. Pearce and G. H. Freeman (1954). Some considerations in the design of successive experiments in fruit plantations. *Biometrics*, **10**, 503–515.

Holcomb, D. R., D. C. Montgomery and W. M. Carlyle (2003). Analysis of supersaturated designs. *J. Qual. Tech.*, **35**, 13–27.

Hotelling, H. (1944). Some improvements in weighing and other experimental techniques. *Ann. Math. Statist.*, **15**, 297–306.

Huang, W.-C. L. (1989). Applications of the Chinese remainder theorem to the construction and analysis of confounding systems and randomized fractional replicates for mixed factorial experiments. Unpublished Ph.D. dissertation, Virginia Polytechnic Institute and State University, Blacksburg, VA.

Jacroux, M. (1984). Upper bounds for efficiency factors of block designs. *Sankhyā*, *B*, **46**, 263–274.

Jacroux, M. (1987). On the determination and construction of MV-optimal block designs for comparing test treatments with a standard treatment. *J. Statist. Plann. Inf.*, **15**, 205–225.

Jarrett, R. G. (1983). Definitions and properties for m-concurrence designs. *J. Roy. Statist. Soc.*, *B*., **64**, 1–10.

Jarrett, R. G. and W. B. Hall (1978). Generalized cyclic incomplete block designs. *Biometrika*, **65**, 397–401.

John, J. A. (1966). Cyclic incomplete block designs. *J. Roy. Statist. Soc.*, *B*., **28**, 345–360.

John, J. A. (1969). A relationship between cyclic and PBIB designs. *Sankhyā*, *B*, **31**, 535–540.

John, J. A. (1973). Factorial experiments in cyclic designs. *Ann. Statist*, **6**, 188–194.

John, J. A. (1987). *Cyclic Designs*. London: Chapman and Hall.

John, J. A. and A. M. Dean (1975). Single replicate factorial experiments in generalized cyclic designs: I. Symmetrical arrangements. *J. Roy. Statist. Soc.*, *B*, **37**, 63–71.

John, J. A. and J. A. Eccleston (1986). Row-column α-designs. *Biometrika*, **73**, 301–306.

John, J. A. and M. H. Quenouille (1977). *Experiments: Design and Analysis*, 2nd ed. New York: Macmillan.

John, J. A. and T. M. F. Smith (1972). Two factor experiments in nonorthogonal designs. *J. Roy. Statist. Soc.*, *B*, **34**, 401–409.

John, J. A. and D. J. Street (1992). Bounds for the efficiency factor of row-column designs. *Biometrika*, **79**, 658–661.

John, J. A. and D. J. Street (1993). Bounds for the efficiency factor of row-column designs (amendment). *Biometrika*, **80**, 712–713.

John, J. A. and E. R. Williams (1995). *Cyclic and Computer Generated Designs*, 2nd ed. London: Chapman and Hall.

John, J. A., F. W. Wolock and H. A. David (1972). Cyclic designs. *Appl. Math.* Series 62. National Bureau Standards: Washington, D.C.

John, P. W. M. (1966). An extension of the triangular association scheme to three associate classes. *J. Roy. Statist. Soc., B*, **28**, 361–365.

John, P. W. M. (1976). Robustness of balanced incomplete block designs. *Ann. Statist.*, **4**, 956–959.

Jones, B. and M. G. Kenward (2003). *Design and Analysis of Cross-over Trials*, 2nd ed. Boca Raton, FL: Chapman and Hall/CRC.

Kackar, R. N. (1985). Off-line quality control, parameter design, and the Taguchi method. *J. Qual. Tech.*, **17**, 176–188.

Kageyama, S. (1972). On the reduction of associate classes for certain PBIB designs. *Ann. Math. Statist.*, **43**, 1528–1540.

Kempthorne, O. (1947). A simple approach to confounding and fractional replication in factorial experiments. *Biometrika*, **34**, 255–272.

Kempthorne, O. (1948). The factorial approach to the weighing problem. *Ann. Math. Statist.*, **19**, 238–245.

Kempthorne, O. (1952). *Design and Analysis of Experiments*. New York: Wiley.

Kempthorne, O. (1953). A class of experimental designs using blocks of two plots. *Ann. Math. Statist.*, **24**, 76–84.

Kempthorne, O. (1956). The efficiency factor of an incomplete block design. *Ann. Math. Statist.*, **27**, 846–849.

Kempthorne, O. (1984). Statistical methods and science. In: *W. G. Cochran's Impact on Statistics*, P.S.R.S. Rao and J. Sedransk (eds.). New York: Wiley, pp. 287–308.

Kempthorne, O. and W. T. Federer (1948a). The general theory of prime-power lattice designs. I. Introduction and designs for p^n varieties in blocks of p plots. *Biometrics*, **4**, 54–79.

Kempthorne, O. and W. T. Federer (1948b). The general theory of prime-power lattice designs. II. Designs for p^n varieties in blocks of p^s plots and in squares. *Biometrics*, **4**, 109–121.

Kershner, R. P. and W. T. Federer (1981). Two-treatment crossover designs for estimating a variety of effects. *J. Am. Statist. Assoc.*, **76**, 612–618.

Khare, M. and W. T. Federer (1981). A simple construction procedure for resolvable incomplete block designs for any number of treatments. *Biom. J.*, **23**, 121–132.

Kiefer, J. (1958). On the nonrandomized optimality and randomized nonoptimality of symmetrical designs. *Ann. Math. Statist.*, **29**, 675–699.

Kiefer, J. (1975a). Construction and optimality of generalized Youden designs. In: *A Survey of Statistical Design and Linear Models*, J. N. Srivastava (ed.). Amsterdam: North-Holland, pp. 333–353.

Kiefer, J. (1975b). Balanced block designs and generalized Youden designs. I. Construction (Patchwork). *Ann. Statist.*, **3**, 109–118.

Kishen, K. and J. N. Srivastava (1959). Mathematical theory of confounding in asymmetrical and symmetrical factorial designs. *J. Indian Soc. Agric. Statist.*, **11**, 73–110.

Kramer, C. Y. and R. A. Bradley (1957). Intra-block analysis for factorials in two-associate class group divisible designs. *Ann. Math. Statist.*, **28**, 349–361 (Addenda, **29**, 933–935).

Kshirsagar, A. M. (1966). Balanced factorial designs. *J. Roy. Statist. Soc., B*, **28**, 559–567.

Kunert, J. (1984). Optimality of balanced uniform repeated measurements designs. *Ann. Statist.*, **12**, 1006–1017.

Kunert, J. (1985). Optimal repeated measurements designs for correlated observations and analysis by weighted least squares. *Biometrika*, **72**, 375–389.

Kunert, J. (1997). On the use of the factor-sparsity assumption to get an estimate of the variance in saturated designs. *Technometrics*, **39**, 81–90.

Kunert, J. and J. Stufken (2002). Optimal crossover designs in a model with self and mixed carryover effects. *J. Am. Statist. Assoc.*, **97**, 898–906.

Kurkjian, B. and M. Zelen (1962). A calculus for factorial arrangements. *Ann. Math. Statist.*, **33**, 600–619.

Kurkjian, B. and M. Zelen (1963). Applications of the calculus of factorial arrangements. I. Block and direct product designs. *Biometrika*, **50**, 63–73.

Kusumoto, K. (1965). Hyper cubic designs. *Wakayama Med. Rept.*, **9**, 123–132.

Lenth, R. V. (1989). Quick and easy analysis of unreplicated factorials. *Technometrics*, **31**, 469–473.

Lentner, M. and T. Bishop (1993). *Experimental Design and Analysis*, 2nd ed. Blacksburg, VA: Valley Book.

Leon, R. V., A. C. Shoemaker and R. N. Kackar (1987). Performance measure independent of adjustment: An explanation and extension of Taguchi's signal to noise ratio. *Technometrics*, **29**, 253–265.

Lewis, S. M. (1982). Generators for asymmetrical factorial experiments. *J. Statist. Plann. Inf.*, **6**, 59–64.

Li, J. C. R. (1944). Design and statistical analysis of some confounded factorial experiments. *Iowa Agri. Expt. Stat. Res. Bull.* **333**, 452–492.

Li, R. and D. K. J. Lin (2002). Data analysis in supersaturated designs. *Statist. Prob. Lett.*, **59**, 135–144.

Li, R. and D. K. J. Lin (2003). Analysis methods for supersaturated design: Some comparisons. *J. Data Sci.*, **1**, 249–260.

Lin, D. K. J. (1993). A new class of supersaturated designs. *Technometrics*, **35**, 28–31.

Lin, D. K. J. and N. R. Draper (1992). Projection properties of Plackett and Burman designs. *Technometrics*, **34**, 423–428.

Lin, D. K. J. and N. R. Draper (1993). Generating alias relationships for two-level Plackett and Burman designs. *Comp. Statist. Data Anal.*, **15**, 147–157.

Lin, P. K. (1986). Using the Chinese remainder theorem in constructing confounded designs in mixed factorial experiments. *Comm. Statist.—Theory Methods*, **15**, 1389–1398.

Liu, M. and R. Zhang (2000). Construction of $E(s^2)$ optimal supersaturated designs using cyclic BIBDs. *J. Statist. Plann. Inf.*, **91**, 139–150.

Lorenzen, T. J. and M. A. Wincek (1992). Blocking is simply fractionation. *GM Research Publication* 7709.

Lucas, H. L. (1957). Extra-period Latin-square change-over designs. *J. Dairy Sci.*, **40**, 225–239.

Majumdar, D. (1996). Optimal and efficient treatment-control designs. In: *Handbook of Statistics* 13, S. Ghosh and C. R. Rao (eds.). Amsterdam: Elsevier Science, pp. 1007–1053.

Majumdar, D. and W. Notz (1983). Optimal incomplete block designs for comparing treatments with a control. *Ann. Statist.*, **11**, 258–266.

Margolin, B. H. (1968). Orthogonal main-effect 2^n3^m designs and two-factor aliasing. *Technometrics*, **10**, 559–573.

Margolin, B. H. (1969). Resolution IV fractional factorial designs. *J. Roy. Statist. Soc., B*, **31**, 514–523.

Mason, J. M. and K. Hinkelmann (1971). Change-over designs for testing different treatment factors at several levels. *Biometrics*, **27**, 430–435.

Masuyama, M. (1965). Cyclic generation of triangular PBIB designs. *Rep. Stat. Appl. Res. Un. Japan, Sci. Engrs.*, **12**, 73–81.

Mathew, T., B. K. Sinha and L. Zhou (1993). Some statistical procedures for combining independent tests. *J. Am. Statist. Assoc.*, **88**, 912–919.

Mathon, R. and A. Rosa (1996). 2-(v, k, λ) designs of small order. In: *The CRC Handbook of Combinatorial Designs*, C. J. Colbourn and J. H. Dinitz (eds.). Boca Raton, FL: CRC, pp. 3–41.

Matthews, J. N. S. (1987). Optimal crossover designs for the comparison of two treatments in the presence of carryover effects and autocorrelated errors. *Biometrika*, **74**, 311–320.

Montgomery, D. C. (1991). *Introduction to Statistical Quality Control*, 2nd ed. New York: Wiley.

Montgomery, D. C. (1999). Experimental design for product and process design and development. *Statistician*, **48**, 159–173.

Morgan, J. P. (1996). Nested designs. In: *Handbook of Statistics*, 13, S. Ghosh and C. R. Rao (eds.). Amsterdam: Elsevier Science, pp. 939–976.

Morgan, J. P. and R. A. Bailey (2000). Optimal design with many blocking factors. *Ann. Statist.*, **28**, 553–577.

Muller, E.-R. (1966). Balanced confounding in factorial experiments. *Biometrika*, **53**, 507–524.

Myers, R. H., A. I. Khuri and G. Vining (1992). Response surface alternatives to the Taguchi robust parameter design approach. *Am. Statist.*, **46**, 131–139.

Nair, K. R. (1944). The recovery of inter-block information in incomplete block designs. *Sankhyā*, **6**, 383–390.

Nair, K. R. (1951). Rectangular lattices and partially balanced incomplete block designs. *Biometrics*, **7**, 145–154.

Nair, V. N. (1986). Testing in industrial experiments with ordered categorical data. *Technometrics*, **28**, 283–291.

Neyman, J., K. Iwaszkiewicz and S. Kolodziejczyk (1935). Statistical problems in agricultural experimentation. *Suppl. J. Roy. Statist. Soc.*, **2**, 107–154.

Nguyen, N.-K. (1996). An algorithmic approach to constructing supersaturated designs. *Technometrics*, **38**, 69–73.

Nguyen, N.-K. and E. R. Williams (1993). An algorithm for constructing optimal resolvable row-column designs. *Austral. J. Statist.*, **35**, 363–370.

Ogasawara, M. (1965). A necessary condition for the existence of regular and symmetric PBIB designs. Inst. Stat. Mimeo Series 418: Chapel Hill, NC.

Ogawa, J. (1974). *Statistical Theory of the Analysis of Experimental Designs*. New York: Dekker.

Ohnishi, T. and T. Shirakura (1985). Search designs for 2^n factorial experiments. *J. Statist. Plann. Inf.*, **11**, 241–245.

Paley R. E. A. C. (1933). On orthogonal matrices. *J. Math. Phys.*, **12**, 311–320.

Paterson, L. J. (1983). Circuits and efficiency in incomplete block designs. *Biometrika*, **70**, 215–225.

Patterson, H. D. (1951). Change-over trials. *J. Roy. Statist. Soc., B*, **13**, 256–271.

Patterson, H. D. (1952). The construction of balanced designs for experiments involving sequences of treatments. *Biometrika*, **39**, 32–48.

Patterson, H. D. and H. L. Lucas (1959). Extra-period change-over designs. *Biometrics*, **15**, 116–132.

Patterson, H. D. and H. L. Lucas (1962). Change-over designs. *North Carolina Agr. Exp. Station Tech. Bull.*, 147.

Patterson, H. D. and R. Thompson (1971). Recovery of inter-block information when block sizes are unequal. *Biometrika*, **58**, 545–554.

Patterson, H. D. and E. R. Williams (1976). A new class of resolvable incomplete block designs. *Biometrika*, **63**, 83–92.

Pearce, S. C. (1953). Field experimentation with fruit trees and other perennial plants. *Commonwealth Bur. Horticulture Plantation Crops, Tech. Comm.*, 23.

Pearce, S. C. (1960). Supplemented balance. *Biometrika*, **47**, 263–271 (Corrigenda **48**, 475).

Pearce, S. C. (1963). The use and classification of non-orthogonal designs (with discussion). *J. Roy. Statist. Soc., A*, **126**, 353–377.

Pearce, S. C. (1964). Experimenting with blocks of natural size. *Biometrics*, **20**, 699–706.

Phadke, M. S. (1982). Quality engineering using design of experiments. *ASA Proceedings, Section on Statistical Education*, 11–20. Alexandria, VA: American Statistical Association.

Plackett, R. L. (1946). Some generalizations in the multifactorial design. *Biometrika*, **33**, 328–332.

Plackett, R. L. and J. P. Burman (1946). The design of optimum multifactorial experiments. *Biometrika*, **33**, 305–325.

Prvan, T. and D. J. Street (2002). An annotated bibliography of application papers using certain classes of fractional factorial and related designs. *J. Statist. Plann. Inf.*, **106**, 245–269.

Pukelsheim, F. (1993). *Optimal Design of Experiments*. New York: Wiley.

Quenouille, M. H. (1953). *The Design and Analysis of Experiments*. London: Griffin.

Raghavarao, D. (1960). A generalization of group divisible designs. *Ann. Math. Statist.*, **31**, 756–771.

Raghavarao, D. (1971). *Construction and Combinatorial Problems in Design of Experiments*. New York: Wiley.

Raghavarao, D. (1986). Random balance designs. In: *Encyclopedia of Statistical Science*. S. Kotz, N. L. Johnson and C. B. Read (eds.). New York: Wiley, pp. 505–506.

Raghavarao, D. (1990). Cross-over designs in industry. In: *Statistical Design and Analysis of Industrial Experiments*. S. Ghosh (ed.). New York: Dekker, pp. 517–530.

Raghavarao, D. and K. Chandrasekhararao (1964). Cubic designs. *Ann. Math. Statist.*, **35**, 389–397.

Raghavarao, D. and S. K. Thartare (1967). An inequality for doubly balanced incomplete block designs. *Calcutta Statist. Assoc. Bull.*, **16**, 37–39.

Raghavarao, D. and S. K. Thartare (1970). A new series of doubly balanced designs. *Calcutta Statist. Assoc. Bull.*, **19**, 95–96.

Raktoe, B. L. (1969). Combining elements from distinct finite fields in mixed factorials. *Ann. Math. Statist.*, **40**, 498–504.

Raktoe, B. L. (1970). Generalized combining of elements from finite fields. *Ann. Math. Statist.*, **41**, 1763–1767.

Raktoe, B. L., A. Hedayat and W. T. Federer (1981). *Factorial Designs*. New York: Wiley.

Raktoe, B. L., A. A. Rayner and D. O. Chalton (1978). On construction of confounded mixed factorial and lattice designs. *Austral. J. Statist.*, **20**, 209–218.

Rao, C. R. (1946a). Hypercubes of strength *d* leading to confounded designs in factorial experiments. *Bull. Calcutta Math. Soc.*, **38**, 67–78.

Rao, C. R. (1946b). Difference sets and combinatorial arrangements derivable from finite geometries. *Proc. Nat. Inst. Sci. India*, **12**, 123–135.

Rao, C. R. (1947a). General methods of analysis for incomplete block designs. *J. Am. Statist. Assoc.*, **42**, 541–561.

Rao, C. R. (1947b). Factorial experiments derivable from combinatorial arrangements of arrays. *J. Roy. Statist. Soc., B*, **9**, 128–139.

Rao, C. R. (1950). The theory of fractional replication in factorial experiments. *Sankhyā*, **10**, 81–86.

Rao, V. R. (1958). A note on balanced designs. *Ann. Math. Statist.*, **29**, 290–294.

Rashed, D. H. (1984). Designs for multiple comparisons of control versus test treatments. Unpublished Ph.D. dissertation, Virginia Polytechnic Institute and State University, Blacksburg, VA.

Ratkowsky, D. A., M. A. Evans and J. R. Alldredge (1993). *Cross-over Experiments—Design, Analysis and Application*. New York: Dekker.

Ray-Chaudhuri, D. K. (1965). Some configurations in finite projective spaces and partially balanced incomplete block designs. *Can. J. Math.*, **17**, 114–123.

Riedwyl, H. (1998). Modifying and using Yates' algorithm. *Statist. Papers*, **39**, 41–60.

Roy, J. and K. R. Shah (1962). Recovery of interblock information. *Sankhyā, A*, **24**, 269–280.

Roy, P. M. (1953–1954). Hierarchical group divisible incomplete block designs with *m*-associate classes. *Sci. Culture*, **19**, 210–211.

Russell, K. G., J. A. Eccleston and G. J. Knudsen (1981). Algorithms for the construction of (M, S)-optimal block designs and row-column designs. *J. Statist. Comp. Simul.*, **12**, 93–105.

SAS Institute (1999–2001). *SAS® Proprietary Software Release 8.2*. Cary, NC: SAS Institute.

Satterthwaite, F. E. (1946). An approximate distribution of estimates of variance components. *Biometrics*, **2**, 110–114.

Satterthwaite, F. E. (1959). Random balance experimentation. *Technometrics*, **1**, 111–137.

Seely, J. and D. Birkes (1984). Parametrizations and resolution IV. In: *Experimental Design, Statistical Models, and Genetic Statistics*. K. Hinkelmann (ed.). New York: Dekker, pp. 77–94.

Sen, M. and R. Mukerjee (1987). Optimal repeated measurements designs under interaction. *J. Statist. Plann. Inf.*, **17**, 81–91.

Senn, S. J. (1993). *Cross-over Trials in Clinical Research*. Chichester, U.K.: Wiley.

Seshradi, V. (1963). Combining unbiased estimators. *Biometrics*, **19**, 163–170.

Shah, B. V. (1958). On balancing in factorial experiments. *Ann. Math. Statist.*, **29**, 766–779.

Shah, B. V. (1959). A generalization of partially balanced incomplete block designs. *Ann. Math. Statist.*, **30**, 1041–1050.

Shah, B. V. (1960). On a 5×2^2 factorial experiment. *Biometrics*, **16**, 115–118.

Shah, K. R. (1964). Use of inter-block information to obtain uniformly better estimators. *Ann. Math. Statist.*, **35**, 1064–1078.

Shah, K. R. (1970). On the loss of information in combined inter- and intra-block estimation. *J. Am. Statist. Assoc.*, **65**, 1562–1564.

Shah, K. R. (1977). Analysis of designs with two-way elimination of heterogeneity. *J. Statist. Plann. Inf.*, **1**, 207–216.

Sharma, V. K. (1977). Change-over designs with complete balance for first and second residual effects. *Canadian J. Statist.*, **5**, 121–132.

Shing, C.-C. and K. Hinkelmann (1992). Repeated measurement designs for factorial treatments when high order interactions are negligible. *J. Statist. Plann. Inf.*, **31**, 81–91.

Shirakura, T. (1991). Main effect plus one or two plans for 2^m factorials. *J. Statist. Plann. Inf.*, **27**, 65–74.

Shirakura, T. and T. Ohnishi (1985). Search designs for 2^m factorials derived from balanced arrays of strength $2(\ell + 1)$ and AD-optimal search designs. *J. Statist. Plann. Inf.*, **11**, 247–258.

Shirakura, T., T. Takahashi and J. N. Srivastava (1996). Searching probabilities for nonzero effects in search designs for the noisy case. *Ann. Statist.*, **24**, 2560–2568.

Shrikhande, S. S. (1950). The impossibility of certain symmetrical balanced incomplete block designs. *Ann. Math. Statist.*, **21**, 106–111.

Shrikhande, S. S. (1952). On the dual of some balanced incomplete block designs. *Biometrics*, **8**, 66–72.

Shrikhande, S. S. (1960). Relations between certain incomplete block designs. In: *Contributions to Probability and Statistics*. I. Olkin (ed.). Stanford, CA: Stanford University Press, pp. 388–395.

Shrikhande, S. S. (1965). On a class of partially balanced incomplete block designs. *Ann. Math. Statist.*, **36**, 1807–1814.

Shrikhande, S. S. and D. Raghavarao (1963). A method of constructing incomplete block designs. *Sankhyā, A*, **25**, 399–402.

Sihota, S. S. and K. S. Banerjee (1981). On the algebraic structures in the construction of confounding plans in mixed factorial designs on the lines of White and Hultquist. *J. Am. Statist. Assoc.*, **76**, 996–1001.

Sitter, R. R., J. Chen and M. Feder (1997). Fractional resolution and minimum aberration in blocking factorial designs. *Technometrics*, **39**, 382–390.

Srivastava, J. N. (1965). Optimal balanced 2^m fractional factorial designs. In: *S. N. Roy Memorial Volume*. Chapel Hill, NC: University of North Carolina and Calcutta: Indian Statistical Institute.

Srivastava, J. N. (1975). Designs for searching non-negligible effects. In: *A Survey of Statistical Design and Linear Models*. J. N. Srivastava (ed.). Amsterdam: North-Holland, pp. 507–519.

Srivastava, J. N. (1984). Sensitivity and revealing power: Two fundamental statistical criteria other than optimality arising in discrete experimentation. In: *Experimental Design, Statistical Models, and Genetic Statistics*. K. Hinkelmann (ed.). New York: Dekker, pp. 95–117.

Srivastava, J. N. (1993). Nonadditivity in row-column designs. *J. Comb. Inform. System Sci.*, **18**, 85–96.

Srivastava, J. N. (1996). A critique of some aspects of experimental design. In: *Handbook of Statistics*, 13. S. Ghosh and C. R. Rao (eds.). Amsterdam: Elsevier Science, pp. 309–341.

Srivastava, J. N. and D. A. Anderson (1970). Some basic properties of multidimensional partially balanced designs. *Ann. Math. Statist.*, **41**, 1438–1445.

Srivastava, J. N. and D. V. Chopra (1971). Balanced optimal 2^m fractional factorial designs of resolution V, $m \leq 6$. *Technometrics*, **13**, 257–269.

Srivastava, J. N. and D. V. Chopra (1973). Balanced arrays and orthogonal arrays. In: *A Survey of Combinatorial Theory*. J. N. Srivastava, F. Harary, C. R. Rao, G.-C. Rota, and S. S. Shrikhande (eds.). Amsterdam: North-Holland, pp. 411–428.

Srivastava, J. N. and S. Ghosh (1976). A series of balanced 2^m factorial designs of resolution V which allow search and estimation of one extra unknown effect. *Sankhyā, B*, **38**, 280–289.

Srivastava, J. N. and S. Ghosh (1977). Balanced 2^m factorial designs of resolution V which allow search and estimation of one extra unknown effect, $4 \leq m \leq 8$. *Comm. Statist.—Theor. Meth.*, **6**, 141–166.

Srivastava, J. N. and S. Ghosh (1996). On nonorthogonality and nonoptimality of Addelman's main-effect plans satisfying the condition of proportional frequencies. *Statist. Prob. Lett.*, **26**, 51–60.

Srivastava, J. N. and B. C. Gupta (1979). Main effect plan for 2^m factorials which allow search and estimation of one unknown effect. *J. Statist. Plann. Inf.*, **3**, 259–265.

Srivastava, J. N. and Y. C. Wang (1998). Row-column designs: How and why non-additivity may make them unfit for use. *J. Statist. Plann. Inf.*, **73**, 277–315.

Srivastava, R., V. K. Gupta and A. Dey (1990). Robustness of some designs against missing observations. *Comm. Statist., A*, **19**, 121–126.

Street, A. P. and D. J. Street (1987). *Combinatorics of Experimental Design*. Oxford: Clarendon.

Stufken, J. (1987). A-optimal block designs for comparing test treatments with a control. *Ann. Statist.*, **15**, 1629–1638.

Stufken, J. (1991). On group divisible treatment designs for comparing test treatments with a standard treatment in blocks of size 3. *J. Statist. Plann. Inf.*, **28**, 205–211.

Stufken, J. (1996). Optimal crossover designs. In: *Handbook of Statistics*, Vol. 13. S. Ghosh and C. R. Rao (eds.). Amsterdam: Elsevier Science, pp. 63–90.

Sun, D. X., C. F. J. Wu and Y. Y. Chen (1997). Optimal blocking schemes for 2^n and 2^{n-p} designs. *Technometrics*, **39**, 298–307.

Surendran, P. U. (1968). Association matrices and the Kronecker product of designs. *Ann. Math. Statist.*, **39**, 676–680.

Taguchi, G. (1986). *Introduction to Quality Engineering: Designing Quality into Products and Processes*. Tokyo: Asian Productivity Organization.

Taguchi, G. (1987). *System of Experimental Design: Engineering Methods to Optimize Quality and Minimize Costs*. White Plains, NY: UNIPUB/Kraus Internation.

Taguchi, G. (1993). *Taguchi Methods—Design of Experiments*. Dearborn, MI: American Supplies Institute.

Takeuchi, K. (1962). A table of difference sets generating balanced incomplete block designs. *Rev. Intern. Statist. Inst.*, **30**, 361–366.

Tang, B. and C. F. J. Wu (1997). A method for constructing supersaturated designs and its $E s^2$ optimality. *Canadian J. Statist.*, **25**, 191–201.

Tharthare, S. K. (1963). Right angular designs. *Ann. Math. Statist.*, **34**, 1057–1067.

Tharthare, S. K. (1965). Generalized right angular designs. *Ann. Math. Statist.*, **36**, 1051–1062.

Thompson, W. A. (1956). A note on the balanced incomplete block designs. *Ann. Math. Statist.*, **27**, 842–846.

Thompson, W. A., Jr. (1958). A note on PBIB design matrices. *Ann. Math. Statist.*, **29**, 919–922.

Thompson, W. A. (1962). The problem of negative estimates of variance components. *Ann. Math. Statist.*, **33**, 273–289.

Tjur, T. (1990). A new upper bound for the efficiency factor of a block design. *Austral. J. Statist.*, **32**, 231–237.

van der Waerden, B. L. (1966). *Algebra*. Berlin: Springer.

Vartak, M. N. (1955). On an application of Kronecker product of matrices to statistical designs. *Ann. Math. Statist.*, **26**, 420–438.

Vartak, M. N. (1959). The non-existence of certain PBIB designs. *Ann. Math. Statist.*, **30**, 1051–1062.

Venables, W. N. and J. A. Eccleston (1993). Randomized search strategies for finding optimal or near-optimal block and row-column designs. *Austral J. Statist.*, **35**, 371–382.

Vining, G. G. and R. H. Myers (1990). Combining Taguchi and response surface philosophies: A dual response approach. *J. Quality Tech.*, **22**, 38–45.

Voss, D. T. (1986). On generalizations of the classical method of confounding to asymmetrical factorial experiments. *Comm. Statist., A*, **15**, 1299–1314.

Wald, A. (1943). On the efficient design of statistical investigations. *Ann. Math. Statist.*, **14**, 134–140.

Webb, S. R. (1964). Characterization of non-orthogonal incomplete factorial designs. In: *Appendix B of ARL 65-116, Part I*. Aerospace Research Laboratories: US. Air Force, Wright-Patterson Air Force Base, Dayton, OH.

Webb, S. R. (1968). Saturated sequential factorial designs. *Technometrics*, **10**, 535–550.

Webb, S. R. (1971). Small incomplete factorial experiment designs for two- and three-level factors. *Technometrics*, **13**, 243–256 (Correction **15**, 951).

Westfall, P. H., S. S. Young and D. K. J. Lin (1998). Forward selection error control in the analysis of supersaturated designs. *Statist. Sinica*, **8**, 101–117.

Whitaker, D., E. R. Williams and J. A. John (1997). *CycDesigN: A Package for the Computer Generation of Experimental Design*. Canberra: CSIRO.

White, D. and R. A. Hultquist (1965). Construction of confounding plans for mixed factorial designs. *Ann. Math. Statist.*, **36**, 1256–1271.

Williams, E. J. (1949). Experimental designs balanced for the estimation of residual effects of treatments. *Austral. J. Sci. Res., A*, **2**, 149–168.

Williams, E. J. (1950). Experimental designs balanced for pairs of residual effects. *Austral. J. Sci. Res., A*, **3**, 351–363.

Williams, E. R. and J. A. John (2000). Updating the average efficiency factor in α-designs. *Biometrika*, **87**, 695–699.

Williams, E. R. and M. Talbot (1993). *ALPHA+: Experimental Designs for Variety Trials. Design User Manual.* Canberra: CSIRO and Edinburgh: SASS.

Wolfinger, R. D. and R. D. Tobias (1998). Joint estimation of location, dispersion, and random effects in robust design. *Technometrics*, **40**, 62–71.

Wolock, F. W. (1964). Cyclic designs. Unpublished Ph.D. dissertation, Virginia Polytechnic Institute, Blacksburg, VA.

Worthley, R. and K. S. Banerjee (1974). A general approach to confounding plans in mixed factorial experiments when the number of a factor is any positive integer. *Ann. Statist.*, **2**, 579–585.

Wu, C. F. J. (1993). Construction of supersaturated designs through partially aliased interactions. *Biometrika*, **80**, 661–669.

Wu, C. F. J. and Y. Chen (1992). A graph-aided method for planning two-level experiments when certain interactions are important. *Technometrics*, **34**, 162–175.

Wu, C. F. J. and M. Hamada (2000). *Experiments—Planning, Analysis, and Parameter Design Optimization.* New York: Wiley.

Wu, H. and C. F. J. Wu (2002). Clear two-factor interactions and minimum aberration. *Ann. Statist.*, **30**, 1496–1511.

Yamada, S., Y. T. Ikebe, H. Hashiguchi and N. Niki (1999). Construction of three-level supersaturated design. *J. Statist. Plann. Inf.*, **81**, 183–193.

Yamada, S. and D. K. J. Lin (1999). Three-level supersaturated designs. *Statist. Probab. Lett.*, **45**, 31–39.

Yamada, S. and T. Matsui (2002). Optimality of mixed-level supersaturated design. *J. Statist. Plann. Inf.*, **104**, 459–468.

Yamamoto, S., Y. Fuji and N. Hamada (1965). Composition of some series of association algebras. *J. Sci. Hiroshima Univ., A*, **29**, 181–215.

Yates, F. (1935). Complex experiments. *J. Roy. Statist. Soc., B*, **2**, 181–223.

Yates, F. (1936a). Incomplete randomized blocks. *Ann. Eugenics*, **7**, 121–140.

Yates, F. (1936b). A new method of arranging variety trial involving a large number of varieties. *J. Agri. Sci*, **26**, 424–455.

Yates, F. (1937a). A further note on the arrangement of variety trials: Quasi-Latin squares. *Ann. Eugenics*, **7**, 319–332.

Yates, F. (1937b). The design and analysis of factorial experiments. *Imp. Bur. Soil Sci. Tech. Comm.*, **35**, 1–95.

Yates, F. (1939). The recovery of inter-block information in variety trials arranged in three-dimensional lattices. *Ann. Eugenics*, **9**, 136–156.

Yates, F. (1940a). The recovery of inter-block information in balanced incomplete block designs. *Ann. Eugenics*, **10**, 317–325.

Yates, F. (1940b). Lattice squares. *J. Agric. Sci.*, **30**, 672–687.

Youden, W. J. (1937). Use of incomplete block replications in estimating tobacco-mosiac virus. *Contr. Boyce Thompson Inst.*, **9**, 41–48.

Youden, W. J. (1940). Experimental designs to increase accuracy of greenhouse studies. *Contr. Boyce Thompson Inst.*, **11**, 219–228.

Youden, W. J., O. Kempthorne, J. W. Tukey, G. E. P. Box and J. S. Hunter (1959). Discussion of "Random balance experimentation" by F. E. Satterthwaite and "The applications of random balance designs" by T. A. Budne. *Technometrics*, **1**, 157–184.

Zahn, D. A. (1975). Modifications of revised critical values for the half-normal plot. *Technometrics*, **17**, 189–200.

Zelen, M. (1958). The use of group divisible designs for confounded asymmetrical factorial arrangements. *Ann. Math. Statist.*, **29**, 22–40.

Zhang, R. and D. K. Park (2000). Optimal blocking of two-level fractional factorial designs. *J. Statist. Plann. Inf.*, **91**, 107–121.

Zhou, L. and T. Mathew (1993). Combining independent tests in linear models. *J. Am. Statist. Assoc.*, **80**, 650–655.

Zoellner, J. A. and O. Kempthorne (1954). Incomplete block designs with blocks of two plots. *Iowa State College Agri. Expt. Stat. Res. Bull.*, 418.

Author Index

Design and Analysis of Experiments. Volume 2: Advanced Experimental Design
By Klaus Hinkelmann and Oscar Kempthorne
ISBN 0-471-55177-5

Subject Index

Design and Analysis of Experiments. Volume 2: Advanced Experimental Design
By Klaus Hinkelmann and Oscar Kempthorne
ISBN 0-471-55177-5

WILEY SERIES IN PROBABILITY AND STATISTICS

ESTABLISHED BY WALTER A. SHEWHART AND SAMUEL S. WILKS

Editors: *David J. Balding, Noel A. C. Cressie, Nicholas I. Fisher, Iain M. Johnstone, J. B. Kadane, Geert Molenberghs. Louise M. Ryan, David W. Scott, Adrian F. M. Smith, Jozef L. Teugels*

Editors Emeriti: *Vic Barnett, J. Stuart Hunter, David G. Kendall*

The ***Wiley Series in Probability and Statistics*** is well established and authoritative. It covers many topics of current research interest in both pure and applied statistics and probability theory. Written by leading statisticians and institutions, the titles span both state-of-the-art developments in the field and classical methods.

Reflecting the wide range of current research in statistics, the series encompasses applied, methodological and theoretical statistics, ranging from applications and new techniques made possible by advances in computerized practice to rigorous treatment of theoretical approaches.

This series provides essential and invaluable reading for all statisticians, whether in academia, industry, government, or research.

ABRAHAM and LEDOLTER · Statistical Methods for Forecasting
AGRESTI · Analysis of Ordinal Categorical Data
AGRESTI · An Introduction to Categorical Data Analysis
AGRESTI · Categorical Data Analysis, *Second Edition*
ALTMAN, GILL, and McDONALD · Numerical Issues in Statistical Computing for the Social Scientist
AMARATUNGA and CABRERA · Exploration and Analysis of DNA Microarray and Protein Array Data
ANDĚL · Mathematics of Chance
ANDERSON · An Introduction to Multivariate Statistical Analysis, *Third Edition*
* ANDERSON · The Statistical Analysis of Time Series
ANDERSON, AUQUIER, HAUCK, OAKES, VANDAELE, and WEISBERG · Statistical Methods for Comparative Studies
ANDERSON and LOYNES · The Teaching of Practical Statistics
ARMITAGE and DAVID (editors) · Advances in Biometry
ARNOLD, BALAKRISHNAN, and NAGARAJA · Records
* ARTHANARI and DODGE · Mathematical Programming in Statistics
* BAILEY · The Elements of Stochastic Processes with Applications to the Natural Sciences
BALAKRISHNAN and KOUTRAS · Runs and Scans with Applications
BARNETT · Comparative Statistical Inference, *Third Edition*
BARNETT and LEWIS · Outliers in Statistical Data, *Third Edition*
BARTOSZYNSKI and NIEWIADOMSKA-BUGAJ · Probability and Statistical Inference
BASILEVSKY · Statistical Factor Analysis and Related Methods: Theory and Applications
BASU and RIGDON · Statistical Methods for the Reliability of Repairable Systems
BATES and WATTS · Nonlinear Regression Analysis and Its Applications
BECHHOFER, SANTNER, and GOLDSMAN · Design and Analysis of Experiments for Statistical Selection, Screening, and Multiple Comparisons
BELSLEY · Conditioning Diagnostics: Collinearity and Weak Data in Regression

*Now available in a lower priced paperback edition in the Wiley Classics Library.
†Now available in a lower priced paperback edition in the Wiley–Interscience Paperback Series.

† BELSLEY, KUH, and WELSCH · Regression Diagnostics: Identifying Influential Data and Sources of Collinearity
BENDAT and PIERSOL · Random Data: Analysis and Measurement Procedures, *Third Edition*
BERRY, CHALONER, and GEWEKE · Bayesian Analysis in Statistics and Econometrics: Essays in Honor of Arnold Zellner
BERNARDO and SMITH · Bayesian Theory
BHAT and MILLER · Elements of Applied Stochastic Processes, *Third Edition*
BHATTACHARYA and WAYMIRE · Stochastic Processes with Applications
† BIEMER, GROVES, LYBERG, MATHIOWETZ, and SUDMAN · Measurement Errors in Surveys
BILLINGSLEY · Convergence of Probability Measures, *Second Edition*
BILLINGSLEY · Probability and Measure, *Third Edition*
BIRKES and DODGE · Alternative Methods of Regression
BLISCHKE AND MURTHY (editors) · Case Studies in Reliability and Maintenance
BLISCHKE AND MURTHY · Reliability: Modeling, Prediction, and Optimization
BLOOMFIELD · Fourier Analysis of Time Series: An Introduction, *Second Edition*
BOLLEN · Structural Equations with Latent Variables
BOROVKOV · Ergodicity and Stability of Stochastic Processes
BOULEAU · Numerical Methods for Stochastic Processes
BOX · Bayesian Inference in Statistical Analysis
BOX · R. A. Fisher, the Life of a Scientist
BOX and DRAPER · Empirical Model-Building and Response Surfaces
* BOX and DRAPER · Evolutionary Operation: A Statistical Method for Process Improvement
BOX, HUNTER, and HUNTER · Statistics for Experimenters: An Introduction to Design, Data Analysis, and Model Building
BOX and LUCEÑO · Statistical Control by Monitoring and Feedback Adjustment
BRANDIMARTE · Numerical Methods in Finance: A MATLAB-Based Introduction
BROWN and HOLLANDER · Statistics: A Biomedical Introduction
BRUNNER, DOMHOF, and LANGER · Nonparametric Analysis of Longitudinal Data in Factorial Experiments
BUCKLEW · Large Deviation Techniques in Decision, Simulation, and Estimation
CAIROLI and DALANG · Sequential Stochastic Optimization
CASTILLO, HADI, BALAKRISHNAN, and SARABIA · Extreme Value and Related Models with Applications in Engineering and Science
CHAN · Time Series: Applications to Finance
CHARALAMBIDES · Combinatorial Methods in Discrete Distributions
CHATTERJEE and HADI · Sensitivity Analysis in Linear Regression
CHATTERJEE and PRICE · Regression Analysis by Example, *Third Edition*
CHERNICK · Bootstrap Methods: A Practitioner's Guide
CHERNICK and FRIIS · Introductory Biostatistics for the Health Sciences
CHILÈS and DELFINER · Geostatistics: Modeling Spatial Uncertainty
CHOW and LIU · Design and Analysis of Clinical Trials: Concepts and Methodologies, *Second Edition*
CLARKE and DISNEY · Probability and Random Processes: A First Course with Applications, *Second Edition*
* COCHRAN and COX · Experimental Designs, *Second Edition*
CONGDON · Applied Bayesian Modelling
CONGDON · Bayesian Statistical Modelling
CONOVER · Practical Nonparametric Statistics, *Third Edition*
COOK · Regression Graphics
COOK and WEISBERG · Applied Regression Including Computing and Graphics

*Now available in a lower priced paperback edition in the Wiley Classics Library.
†Now available in a lower priced paperback edition in the Wiley–Interscience Paperback Series.

COOK and WEISBERG · An Introduction to Regression Graphics
CORNELL · Experiments with Mixtures, Designs, Models, and the Analysis of Mixture Data, *Third Edition*
COVER and THOMAS · Elements of Information Theory
COX · A Handbook of Introductory Statistical Methods
* COX · Planning of Experiments
CRESSIE · Statistics for Spatial Data, *Revised Edition*
CSÖRGŐ and HORVÁTH · Limit Theorems in Change Point Analysis
DANIEL · Applications of Statistics to Industrial Experimentation
DANIEL · Biostatistics: A Foundation for Analysis in the Health Sciences, *Eighth Edition*
* DANIEL · Fitting Equations to Data: Computer Analysis of Multifactor Data, *Second Edition*
DASU and JOHNSON · Exploratory Data Mining and Data Cleaning
DAVID and NAGARAJA · Order Statistics, *Third Edition*
* DEGROOT, FIENBERG, and KADANE · Statistics and the Law
DEL CASTILLO · Statistical Process Adjustment for Quality Control
DEMARIS · Regression with Social Data: Modeling Continuous and Limited Response Variables
DEMIDENKO · Mixed Models: Theory and Applications
DENISON, HOLMES, MALLICK and SMITH · Bayesian Methods for Nonlinear Classification and Regression
DETTE and STUDDEN · The Theory of Canonical Moments with Applications in Statistics, Probability, and Analysis
DEY and MUKERJEE · Fractional Factorial Plans
DILLON and GOLDSTEIN · Multivariate Analysis: Methods and Applications
DODGE · Alternative Methods of Regression
* DODGE and ROMIG · Sampling Inspection Tables, *Second Edition*
* DOOB · Stochastic Processes
DOWDY, WEARDEN, and CHILKO · Statistics for Research, *Third Edition*
DRAPER and SMITH · Applied Regression Analysis, *Third Edition*
DRYDEN and MARDIA · Statistical Shape Analysis
DUDEWICZ and MISHRA · Modern Mathematical Statistics
DUNN and CLARK · Basic Statistics: A Primer for the Biomedical Sciences, *Third Edition*
DUPUIS and ELLIS · A Weak Convergence Approach to the Theory of Large Deviations
* ELANDT-JOHNSON and JOHNSON · Survival Models and Data Analysis
ENDERS · Applied Econometric Time Series
ETHIER and KURTZ · Markov Processes: Characterization and Convergence
EVANS, HASTINGS, and PEACOCK · Statistical Distributions, *Third Edition*
FELLER · An Introduction to Probability Theory and Its Applications, Volume I, *Third Edition,* Revised; Volume II, *Second Edition*
FISHER and VAN BELLE · Biostatistics: A Methodology for the Health Sciences
FITZMAURICE, LAIRD, and WARE · Applied Longitudinal Analysis
* FLEISS · The Design and Analysis of Clinical Experiments
FLEISS · Statistical Methods for Rates and Proportions, *Third Edition*
FLEMING and HARRINGTON · Counting Processes and Survival Analysis
FULLER · Introduction to Statistical Time Series, *Second Edition*
FULLER · Measurement Error Models
GALLANT · Nonlinear Statistical Models
GHOSH, MUKHOPADHYAY, and SEN · Sequential Estimation
GIESBRECHT and GUMPERTZ · Planning, Construction, and Statistical Analysis of Comparative Experiments
GIFI · Nonlinear Multivariate Analysis

*Now available in a lower priced paperback edition in the Wiley Classics Library.
†Now available in a lower priced paperback edition in the Wiley–Interscience Paperback Series.

GIVENS and HOETING · Computational Statistics
GLASSERMAN and YAO · Monotone Structure in Discrete-Event Systems
GNANADESIKAN · Methods for Statistical Data Analysis of Multivariate Observations, *Second Edition*
GOLDSTEIN and LEWIS · Assessment: Problems, Development, and Statistical Issues
GREENWOOD and NIKULIN · A Guide to Chi-Squared Testing
GROSS and HARRIS · Fundamentals of Queueing Theory, *Third Edition*
† GROVES · Survey Errors and Survey Costs
* HAHN and SHAPIRO · Statistical Models in Engineering
HAHN and MEEKER · Statistical Intervals: A Guide for Practitioners
HALD · A History of Probability and Statistics and their Applications Before 1750
HALD · A History of Mathematical Statistics from 1750 to 1930
† HAMPEL · Robust Statistics: The Approach Based on Influence Functions
HANNAN and DEISTLER · The Statistical Theory of Linear Systems
HEIBERGER · Computation for the Analysis of Designed Experiments
HEDAYAT and SINHA · Design and Inference in Finite Population Sampling
HELLER · MACSYMA for Statisticians
HINKELMANN and KEMPTHORNE · Design and Analysis of Experiments, Volume 1: Introduction to Experimental Design
HINKELMANN and KEMPTHORNE · Design and Analysis of Experiments, Volume 2: Advanced Experimental Design
HOAGLIN, MOSTELLER, and TUKEY · Exploratory Approach to Analysis of Variance
HOAGLIN, MOSTELLER, and TUKEY · Exploring Data Tables, Trends and Shapes
* HOAGLIN, MOSTELLER, and TUKEY · Understanding Robust and Exploratory Data Analysis
HOCHBERG and TAMHANE · Multiple Comparison Procedures
HOCKING · Methods and Applications of Linear Models: Regression and the Analysis of Variance, *Second Edition*
HOEL · Introduction to Mathematical Statistics, *Fifth Edition*
HOGG and KLUGMAN · Loss Distributions
HOLLANDER and WOLFE · Nonparametric Statistical Methods, *Second Edition*
HOSMER and LEMESHOW · Applied Logistic Regression, *Second Edition*
HOSMER and LEMESHOW · Applied Survival Analysis: Regression Modeling of Time to Event Data
† HUBER · Robust Statistics
HUBERTY · Applied Discriminant Analysis
HUNT and KENNEDY · Financial Derivatives in Theory and Practice
HUSKOVA, BERAN, and DUPAC · Collected Works of Jaroslav Hajek—with Commentary
HUZURBAZAR · Flowgraph Models for Multistate Time-to-Event Data
IMAN and CONOVER · A Modern Approach to Statistics
† JACKSON · A User's Guide to Principle Components
JOHN · Statistical Methods in Engineering and Quality Assurance
JOHNSON · Multivariate Statistical Simulation
JOHNSON and BALAKRISHNAN · Advances in the Theory and Practice of Statistics: A Volume in Honor of Samuel Kotz
JOHNSON and BHATTACHARYYA · Statistics: Principles and Methods, *Fifth Edition*
JOHNSON and KOTZ · Distributions in Statistics
JOHNSON and KOTZ (editors) · Leading Personalities in Statistical Sciences: From the Seventeenth Century to the Present
JOHNSON, KOTZ, and BALAKRISHNAN · Continuous Univariate Distributions, Volume 1, *Second Edition*

*Now available in a lower priced paperback edition in the Wiley Classics Library.
†Now available in a lower priced paperback edition in the Wiley–Interscience Paperback Series.

JOHNSON, KOTZ, and BALAKRISHNAN · Continuous Univariate Distributions, Volume 2, *Second Edition*
JOHNSON, KOTZ, and BALAKRISHNAN · Discrete Multivariate Distributions
JOHNSON, KOTZ, and KEMP · Univariate Discrete Distributions, *Second Edition*
JUDGE, GRIFFITHS, HILL, LÜTKEPOHL, and LEE · The Theory and Practice of Econometrics, *Second Edition*
JUREČKOVÁ and SEN · Robust Statistical Procedures: Aymptotics and Interrelations
JUREK and MASON · Operator-Limit Distributions in Probability Theory
KADANE · Bayesian Methods and Ethics in a Clinical Trial Design
KADANE AND SCHUM · A Probabilistic Analysis of the Sacco and Vanzetti Evidence
KALBFLEISCH and PRENTICE · The Statistical Analysis of Failure Time Data, *Second Edition*
KASS and VOS · Geometrical Foundations of Asymptotic Inference
† KAUFMAN and ROUSSEEUW · Finding Groups in Data: An Introduction to Cluster Analysis
KEDEM and FOKIANOS · Regression Models for Time Series Analysis
KENDALL, BARDEN, CARNE, and LE · Shape and Shape Theory
KHURI · Advanced Calculus with Applications in Statistics, *Second Edition*
KHURI, MATHEW, and SINHA · Statistical Tests for Mixed Linear Models
* KISH · Statistical Design for Research
KLEIBER and KOTZ · Statistical Size Distributions in Economics and Actuarial Sciences
KLUGMAN, PANJER, and WILLMOT · Loss Models: From Data to Decisions, *Second Edition*
KLUGMAN, PANJER, and WILLMOT · Solutions Manual to Accompany Loss Models: From Data to Decisions, *Second Edition*
KOTZ, BALAKRISHNAN, and JOHNSON · Continuous Multivariate Distributions, Volume 1, *Second Edition*
KOTZ and JOHNSON (editors) · Encyclopedia of Statistical Sciences: Volumes 1 to 9 with Index
KOTZ and JOHNSON (editors) · Encyclopedia of Statistical Sciences: Supplement Volume
KOTZ, READ, and BANKS (editors) · Encyclopedia of Statistical Sciences: Update Volume 1
KOTZ, READ, and BANKS (editors) · Encyclopedia of Statistical Sciences: Update Volume 2
KOVALENKO, KUZNETZOV, and PEGG · Mathematical Theory of Reliability of Time-Dependent Systems with Practical Applications
LACHIN · Biostatistical Methods: The Assessment of Relative Risks
LAD · Operational Subjective Statistical Methods: A Mathematical, Philosophical, and Historical Introduction
LAMPERTI · Probability: A Survey of the Mathematical Theory, *Second Edition*
LANGE, RYAN, BILLARD, BRILLINGER, CONQUEST, and GREENHOUSE · Case Studies in Biometry
LARSON · Introduction to Probability Theory and Statistical Inference, *Third Edition*
LAWLESS · Statistical Models and Methods for Lifetime Data, *Second Edition*
LAWSON · Statistical Methods in Spatial Epidemiology
LE · Applied Categorical Data Analysis
LE · Applied Survival Analysis
LEE and WANG · Statistical Methods for Survival Data Analysis, *Third Edition*
LEPAGE and BILLARD · Exploring the Limits of Bootstrap
LEYLAND and GOLDSTEIN (editors) · Multilevel Modelling of Health Statistics
LIAO · Statistical Group Comparison
LINDVALL · Lectures on the Coupling Method

*Now available in a lower priced paperback edition in the Wiley Classics Library.
†Now available in a lower priced paperback edition in the Wiley–Interscience Paperback Series.

LINHART and ZUCCHINI · Model Selection
LITTLE and RUBIN · Statistical Analysis with Missing Data, *Second Edition*
LLOYD · The Statistical Analysis of Categorical Data
MAGNUS and NEUDECKER · Matrix Differential Calculus with Applications in Statistics and Econometrics, *Revised Edition*
MALLER and ZHOU · Survival Analysis with Long Term Survivors
MALLOWS · Design, Data, and Analysis by Some Friends of Cuthbert Daniel
MANN, SCHAFER, and SINGPURWALLA · Methods for Statistical Analysis of Reliability and Life Data
MANTON, WOODBURY, and TOLLEY · Statistical Applications Using Fuzzy Sets
MARCHETTE · Random Graphs for Statistical Pattern Recognition
MARDIA and JUPP · Directional Statistics
MASON, GUNST, and HESS · Statistical Design and Analysis of Experiments with Applications to Engineering and Science, *Second Edition*
McCULLOCH and SEARLE · Generalized, Linear, and Mixed Models
McFADDEN · Management of Data in Clinical Trials
* McLACHLAN · Discriminant Analysis and Statistical Pattern Recognition
McLACHLAN, DO, and AMBROISE · Analyzing Microarray Gene Expression Data
McLACHLAN and KRISHNAN · The EM Algorithm and Extensions
McLACHLAN and PEEL · Finite Mixture Models
McNEIL · Epidemiological Research Methods
MEEKER and ESCOBAR · Statistical Methods for Reliability Data
MEERSCHAERT and SCHEFFLER · Limit Distributions for Sums of Independent Random Vectors: Heavy Tails in Theory and Practice
MICKEY, DUNN, and CLARK · Applied Statistics: Analysis of Variance and Regression, *Third Edition*
* MILLER · Survival Analysis, *Second Edition*
MONTGOMERY, PECK, and VINING · Introduction to Linear Regression Analysis, *Third Edition*
MORGENTHALER and TUKEY · Configural Polysampling: A Route to Practical Robustness
MUIRHEAD · Aspects of Multivariate Statistical Theory
MULLER and STOYAN · Comparison Methods for Stochastic Models and Risks
MURRAY · X-STAT 2.0 Statistical Experimentation, Design Data Analysis, and Nonlinear Optimization
MURTHY, XIE, and JIANG · Weibull Models
MYERS and MONTGOMERY · Response Surface Methodology: Process and Product Optimization Using Designed Experiments, *Second Edition*
MYERS, MONTGOMERY, and VINING · Generalized Linear Models. With Applications in Engineering and the Sciences
† NELSON · Accelerated Testing, Statistical Models, Test Plans, and Data Analyses
† NELSON · Applied Life Data Analysis
NEWMAN · Biostatistical Methods in Epidemiology
OCHI · Applied Probability and Stochastic Processes in Engineering and Physical Sciences
OKABE, BOOTS, SUGIHARA, and CHIU · Spatial Tesselations: Concepts and Applications of Voronoi Diagrams, *Second Edition*
OLIVER and SMITH · Influence Diagrams, Belief Nets and Decision Analysis
PALTA · Quantitative Methods in Population Health: Extensions of Ordinary Regressions
PANKRATZ · Forecasting with Dynamic Regression Models
PANKRATZ · Forecasting with Univariate Box-Jenkins Models: Concepts and Cases
* PARZEN · Modern Probability Theory and Its Applications
PEÑA, TIAO, and TSAY · A Course in Time Series Analysis
PIANTADOSI · Clinical Trials: A Methodologic Perspective

*Now available in a lower priced paperback edition in the Wiley Classics Library.
†Now available in a lower priced paperback edition in the Wiley–Interscience Paperback Series.

PORT · Theoretical Probability for Applications
POURAHMADI · Foundations of Time Series Analysis and Prediction Theory
PRESS · Bayesian Statistics: Principles, Models, and Applications
PRESS · Subjective and Objective Bayesian Statistics, *Second Edition*
PRESS and TANUR · The Subjectivity of Scientists and the Bayesian Approach
PUKELSHEIM · Optimal Experimental Design
PURI, VILAPLANA, and WERTZ · New Perspectives in Theoretical and Applied Statistics
† PUTERMAN · Markov Decision Processes: Discrete Stochastic Dynamic Programming
QIU · Image Processing and Jump Regression Analysis
* RAO · Linear Statistical Inference and Its Applications, *Second Edition*
RAUSAND and HØYLAND · System Reliability Theory: Models, Statistical Methods, and Applications, *Second Edition*
RENCHER · Linear Models in Statistics
RENCHER · Methods of Multivariate Analysis, *Second Edition*
RENCHER · Multivariate Statistical Inference with Applications
* RIPLEY · Spatial Statistics
RIPLEY · Stochastic Simulation
ROBINSON · Practical Strategies for Experimenting
ROHATGI and SALEH · An Introduction to Probability and Statistics, *Second Edition*
ROLSKI, SCHMIDLI, SCHMIDT, and TEUGELS · Stochastic Processes for Insurance and Finance
ROSENBERGER and LACHIN · Randomization in Clinical Trials: Theory and Practice
ROSS · Introduction to Probability and Statistics for Engineers and Scientists
† ROUSSEEUW and LEROY · Robust Regression and Outlier Detection
* RUBIN · Multiple Imputation for Nonresponse in Surveys
RUBINSTEIN · Simulation and the Monte Carlo Method
RUBINSTEIN and MELAMED · Modern Simulation and Modeling
RYAN · Modern Regression Methods
RYAN · Statistical Methods for Quality Improvement, *Second Edition*
SALTELLI, CHAN, and SCOTT (editors) · Sensitivity Analysis
* SCHEFFE · The Analysis of Variance
SCHIMEK · Smoothing and Regression: Approaches, Computation, and Application
SCHOTT · Matrix Analysis for Statistics, *Second Edition*
SCHOUTENS · Levy Processes in Finance: Pricing Financial Derivatives
SCHUSS · Theory and Applications of Stochastic Differential Equations
SCOTT · Multivariate Density Estimation: Theory, Practice, and Visualization
* SEARLE · Linear Models
SEARLE · Linear Models for Unbalanced Data
SEARLE · Matrix Algebra Useful for Statistics
SEARLE, CASELLA, and McCULLOCH · Variance Components
SEARLE and WILLETT · Matrix Algebra for Applied Economics
SEBER and LEE · Linear Regression Analysis, *Second Edition*
† SEBER · Multivariate Observations
† SEBER and WILD · Nonlinear Regression
SENNOTT · Stochastic Dynamic Programming and the Control of Queueing Systems
* SERFLING · Approximation Theorems of Mathematical Statistics
SHAFER and VOVK · Probability and Finance: It's Only a Game!
SILVAPULLE and SEN · Constrained Statistical Inference: Inequality, Order, and Shape Restrictions
SMALL and McLEISH · Hilbert Space Methods in Probability and Statistical Inference
SRIVASTAVA · Methods of Multivariate Statistics
STAPLETON · Linear Statistical Models

*Now available in a lower priced paperback edition in the Wiley Classics Library.
†Now available in a lower priced paperback edition in the Wiley–Interscience Paperback Series.

STAUDTE and SHEATHER · Robust Estimation and Testing
STOYAN, KENDALL, and MECKE · Stochastic Geometry and Its Applications, *Second Edition*
STOYAN and STOYAN · Fractals, Random Shapes and Point Fields: Methods of Geometrical Statistics
STYAN · The Collected Papers of T. W. Anderson: 1943–1985
SUTTON, ABRAMS, JONES, SHELDON, and SONG · Methods for Meta-Analysis in Medical Research
TANAKA · Time Series Analysis: Nonstationary and Noninvertible Distribution Theory
THOMPSON · Empirical Model Building
THOMPSON · Sampling, *Second Edition*
THOMPSON · Simulation: A Modeler's Approach
THOMPSON and SEBER · Adaptive Sampling
THOMPSON, WILLIAMS, and FINDLAY · Models for Investors in Real World Markets
TIAO, BISGAARD, HILL, PEÑA, and STIGLER (editors) · Box on Quality and Discovery: with Design, Control, and Robustness
TIERNEY · LISP-STAT: An Object-Oriented Environment for Statistical Computing and Dynamic Graphics
TSAY · Analysis of Financial Time Series
UPTON and FINGLETON · Spatial Data Analysis by Example, Volume II: Categorical and Directional Data
VAN BELLE · Statistical Rules of Thumb
VAN BELLE, FISHER, HEAGERTY, and LUMLEY · Biostatistics: A Methodology for the Health Sciences, *Second Edition*
VESTRUP · The Theory of Measures and Integration
VIDAKOVIC · Statistical Modeling by Wavelets
VINOD and REAGLE · Preparing for the Worst: Incorporating Downside Risk in Stock Market Investments
WALLER and GOTWAY · Applied Spatial Statistics for Public Health Data
WEERAHANDI · Generalized Inference in Repeated Measures: Exact Methods in MANOVA and Mixed Models
WEISBERG · Applied Linear Regression, *Third Edition*
WELSH · Aspects of Statistical Inference
WESTFALL and YOUNG · Resampling-Based Multiple Testing: Examples and Methods for p-Value Adjustment
WHITTAKER · Graphical Models in Applied Multivariate Statistics
WINKER · Optimization Heuristics in Economics: Applications of Threshold Accepting
WONNACOTT and WONNACOTT · Econometrics, *Second Edition*
WOODING · Planning Pharmaceutical Clinical Trials: Basic Statistical Principles
WOODWORTH · Biostatistics: A Bayesian Introduction
WOOLSON and CLARKE · Statistical Methods for the Analysis of Biomedical Data, *Second Edition*
WU and HAMADA · Experiments: Planning, Analysis, and Parameter Design Optimization
YANG · The Construction Theory of Denumerable Markov Processes
* ZELLNER · An Introduction to Bayesian Inference in Econometrics
ZHOU, OBUCHOWSKI, and McCLISH · Statistical Methods in Diagnostic Medicine

*Now available in a lower priced paperback edition in the Wiley Classics Library.
†Now available in a lower priced paperback edition in the Wiley–Interscience Paperback Series.